THE STEPHEN BECHTEL FUND

IMPRINT IN ECOLOGY AND THE ENVIRONMENT

The Stephen Bechtel Fund has
established this imprint to promote
understanding and conservation of
our natural environment.

The publisher gratefully acknowledges the generous contribution
to this book provided by the Stephen Bechtel Fund.

OCEAN

OCEAN

Reflections on a Century of Exploration

Wolf H. Berger

With contributions by E. N. Shor

UNIVERSITY OF CALIFORNIA PRESS
Berkeley Los Angeles London

University of California Press, one of the most distinguished university presses in the United States, enriches lives around the world by advancing scholarship in the humanities, social sciences, and natural sciences. Its activities are supported by the UC Press Foundation and by philanthropic contributions from individuals and institutions. For more information, visit www.ucpress.edu.

University of California Press
Berkeley and Los Angeles, California

University of California Press, Ltd.
London, England

Library of Congress Cataloging-in-Publication Data

Berger, Wolf H.
Ocean : reflections on a century of exploration / Wolf H. Berger with contributions by E. N. Shor.
p. cm.
Includes bibliographical references and index.
ISBN 978-0-520-24778-9 (case : alk. paper) 1. Underwater exploration—History. 2. Oceans—History. 3. Oceanography—History. I. Shor, E. N. II. Title.
GC65.B47 2008
551.46—dc22 2008055348

Manufactured in the United States of America
16 15 14 13 12 11 10 09
10 9 8 7 6 5 4 3 2 1

The paper used in this publication meets the minimum requirements of ANSI/NISO Z39.48-1992 (R 1997) (*Permanence of Paper*). ∞

TO THE TEACHERS OF OCEAN SCIENCES EVERYWHERE

AND TO THE MEMORY OF
HARALD U. SVERDRUP
AND OF
ROGER R. REVELLE,
WHO TAUGHT ALL OF US

CONTENTS

FOREWORD

One of the major events celebrating the centennial year of the Scripps Institution of Oceanography (2003) was a reunion and symposium commemorating the fiftieth anniversary of Scripps's great MidPac Expedition and its predecessor, Capricorn. I am afraid that my opening remarks may have dismayed the veteran scientists gathered there, for I ventured that their great research cruise brought the age of European geographic exploration, which started with Prince Henry the Navigator 500 years ago, to a definitive end. I borrowed a thought that John Maynard Keynes had applied to Isaac Newton: they were the last generation to look upon the world through the eyes of the ancients.

From our present vantage point in time, when every square meter of earth has been surveyed by satellite and can be "googled" at will, our challenge is no longer to discover what is there, but to understand how the many complex parts of the Earth's systems interact. The challenge is not to push outward but to connect together.

The fact that oceanographers just then had become involved with hydrogen bomb tests already presaged that the world to come was to be very different. A few, especially Roger Revelle, saw where things were going.

We know now what the great pioneers did not: our civilization faces an entirely new situation. Humans have always altered their environment to suit their needs, but their impact reached global proportions only in the last 50 years. New words have since entered public discourse: biodiversity decline, disappearing fish, habitat fragmentation, deforestation, increasing drought, desertification, ozone hole, atmospheric brown cloud, retreating glaciers and mountain snows, melting Arctic ice, sea level rise, and the granddaddy of them all, global warming. The pristine state of nature, if it ever existed, has disappeared, and the world is going to be different tomorrow because of what we humans are doing today.

We also know as never before that Earth science has an unshirkable obligation to help preserve our civilization and the planet that has supported it so lavishly—thus far.

I sat down after my opening remarks wishing that someone would tell that story. I even had the momentary fantasy that I might, but it dissolved in the press of a busy director's mundane affairs. So I greeted with great satisfaction Wolf Berger's proposal to commemorate the centennial by telling the story of the great transition in our science. Wolf and his generation have had the never-to-be-repeated privilege of viewing the world through the eyes of both the ancients and the moderns. For Scripps, it has meant many things, especially the world before and after plate tectonics and climate change. With Elizabeth Shor's assistance, Wolf relives how our great institution earned its place in intellectual history. And he does so with lyrical prose that makes the story both accessible and pleasurable for the general reader.

CHARLES F. KENNEL
Director of Scripps Institution of Oceanography, 1997–2006
Los Angeles and La Jolla
April 2007

Introduction

COPING WITH A SEA OF CHANGE

What ocean scientists do has changed drastically in recent years. Not only are we using tools undreamed of a few decades ago, tools that produce immense data floods of entirely new types of observations, but also now the very objects of study have become moving targets. The various ecosystems of the ocean environment are changing rapidly in unforeseen and practically unpredictable ways. Today's graduate students and their mentors deal with a sea that is quite different from the one we studied even three decades ago.

It is a good time to take stock of what we have learned about the sea, as we enter an age where human impacts on the ocean environment—overfishing, coastal pollution, global warming—begin to exceed all other factors driving long-term change. This new age puts the stamp of "history" on what was, until yesteryear, a more or less familiar part of the planet's workings. The climate system, which governs the ocean's motions and thereby the life cycles of its inhabitants, is changing before our eyes, as the ocean warms and the winds and currents respond. It is in this situation that a historical approach, one that includes change, is especially appropriate. The concept of equilibrium, which pervades traditional studies, remains useful but is really no longer applicable. Perhaps it never was entirely realistic.

To ease the troubled mind, beyond the fleeting images of rapid change are the underlying patterns of behavior of the ocean and its ecosystems, rules and concepts that emerged from the studies in the past century. Of all these studies, those concerned with the history of the ocean are likely to remain highly relevant. Today's changes cannot change history. In fact, the significance of today's changes can only be measured against the range of changes of the past. To define what is "unusual" implies knowledge of history, and the definition of change itself presupposes a comparison with expectations, derived from both experience and theory. Where there is conflict between experience and theory, in assessing future developments, theory must yield, because the understanding of processes is incomplete in all things pertaining to the behavior of the sea.

As a geologist studying the sea, I am comfortable with an emphasis on ocean history in preference to process, whose complexities can be baffling. Before one can get to process, one needs to

FIGURE P.1 Digging up the history of the sea. The drilling vessel *JOIDES Resolution.*

establish what has happened and what is happening, by observation. This is not to deny the crucial importance of process. Deep understanding is umbilically linked to the rules governing process, rules that emerge from experience, from experiment, and from theoretical anticipation.

In the past 50 years, the evolution of concepts that allow us to comprehend physical, chemical, and biological aspects of the workings of the ocean proceeded at an incredible pace. Regarding ocean history, these developments moved so fast and so far in the last several decades that the implications even of fundamental discoveries have not yet been fully integrated into the ocean sciences. A prime example is the narrative of the history of the sea that has emerged from deep-ocean drilling (fig. P.1). There is no question that this narrative is highly relevant to all discussion of the diversity of living marine organisms. Their evolution, in the last 40 million years, took place within the framework of the overall cooling of the planet, the expansion of polar ice (first on Antarctica and later in the Arctic), and the filling of the deep sea with cold water. The fact that the planet has cold poles and a cold ring current around Antarctica is crucial for the productivity of the sea.

It is not really possible, evidently, to summarize what may be called knowledge of the ocean. The last hundred years of scientific developments have been overwhelming in all respects—much more has been learned in that short interval than in all the previous centuries combined. A recently published ocean encyclopedia comprises six volumes with some 3,400 pages of tightly written and well-illustrated text.[1] Yet, on studying these volumes, every expert will undoubtedly decide that his or her own field is less fully represented than it could be. An expansion by a factor of three—to some 10,000 pages—would pose no problem at all, given the amount of important information available.

When I set out (in 1998) to summarize what we have learned about the ocean within the twentieth century, with the help of Scripps his-

FIGURE P.2 Reaching out beyond the breakers: the pier of Scripps Institution of Oceanography in La Jolla, California. Scripps Institution of Oceanography is a part of the University of California, San Diego campus.

torian Elizabeth N. Shor, it soon emerged that there was no point in attempting to pursue the goals appropriate for an encyclopedia or a textbook. Instead, we decided that the book should appeal to the general reader with an interest in the sea, and especially to teachers who communicate the basics of oceanography to pupils who have no background in the Earth sciences. In this context, the aim is not to replace a proper textbook. Instead, the aim is to deepen understanding and appreciation, as in a guided tour of selected exhibits.

The choice of subjects reflects these aims. We picked topics that we judged especially instructive in illustrating the essence of ocean sciences, and within this realm we emphasized discoveries and concepts pertaining to descriptive geological and biological oceanography. In addition, we paid special attention to the history of exploration, notably to developments at our home institution, Scripps Institution of Oceanography (fig. P.2).

In many ways, the developments at Scripps can be taken as representative for all of the ocean sciences. In part, this is so because Scripps is the largest and oldest oceanographic institute in the United States. Mainly, however, it is because trends in research tend to run parallel across the nation and, indeed, across the world. Research from the 1940s to the 1960s at the institution, during a time of major expansion, is well illustrated in the *Scripps Centennial Volume* (1903–2003), which presents the contributions of major figures in oceanography active in La Jolla at that time.[2] Fully one-half of the biographical essays involve geology and geophysics, and most of the rest deal with marine biology and ecology. Our own view of oceanography reflects this same bias. The focus is on central concepts, which generally have more than one parent. Where we refer to biographic backgrounds of individual scientists, it is usually not with the intent to emphasize priority (which calls for more research than we were able to invest), but with the intent to illustrate the process of exploration and discovery.

Each chapter begins with a summary of the contents of the chapter, in an attempt to prepare the reader for what is to follow. Obviously, this type of structure implies some repetition. Also, in order to make the various chapters more or less self-contained, allowing for reading from the back, as it were, some redundancy proved unavoidable. Of course, this is especially true when comparing chapter 2 to other chapters, since chapter 2 represents a kind of executive summary for the entire volume. The perceptive reader will also note some parallelisms between the two chapters on the edge of the sea (chapters 3 and 5) and some duplication concerning chapters 7, 8, 9, and 10 (all of which treat aspects of the productivity of the sea). Likewise one will find some duplication within chapters 11 and 12 (which both have discussions of results from acoustic exploration). Naturally, the two chapters on geologic history (chapters 13 and 14) show some reiteration concerning concepts and methods. The alternative—to send the reader back toward other chapters for checking points arising—was judged more tedious than a modest amount of recapitulation.

The central goal in writing this book is to foster public understanding of the workings of the ocean as a life-support system. It would please me greatly if the book should turn out to be useful in improving the general level of Earth literacy in public education. Despite the rich contributions of ocean sciences to the knowledge of the home planet, there is no evidence that the young today are significantly better educated in respect to Earth knowledge than were our forebears.

For at least two reasons, this situation is unsatisfactory. The first is that all children yearn to learn about their environment, so that there is great incentive to foster Earth literacy. The joyful discoveries that accompany each outing into the mountains or to the seashore are greatly enhanced for one who has learned to observe and to think like a naturalist, all through life. The second is that we, the human species, have become a major geologic agent—in fact, *the* major geologic agent of ongoing change. We need to begin to educate ourselves about the consequences of human impact on the environment, so that we can mitigate the impact where possible, as good stewards of the home planet.

I gratefully acknowledge the assistance received from fellow scientists and staff members at Scripps Institution of Oceanography, my home for many decades. Clearly, this work would not have been possible without the freely rendered expert advice and other help from the La Jolla oceanographic community. Also, I thankfully remember the education received over decades from students and from shipmates, and from colleagues in many other oceanographic institutions, at countless meetings and conferences, and during sabbatical stays in Kiel, Oslo, Jerusalem, Bremen, and Bergen. Of these links to the international community, the ones to Bremen and to Bergen have been especially strong and fruitful in the last two decades.

Special thanks are due to Elizabeth N. Shor, coauthor for chapters 9 and 11 (chapters, in fact, that she initiated). Her assistance and encouragement in the early years of this effort were of crucial importance. Her expertise and abiding interest in the history of Scripps[3] led to the addition of biographic detail in several chapters of the book; for example, material on the role of Boyd Walker in grunion studies (chapter 3), on Francis Shepard and submarine canyons (chapter 5), on Carl Hubbs and the cruise of the Zaca (chapter 7), on the exploits of Jacques Piccard in the *Trieste* (chapter 10), on several aspects of the MidPac Expedition (chapter 12), and on the Mohole project (chapter 14). In the course of working on the book, the focus of the narrative shifted somewhat from Scripps lore toward a general history of ideas and concepts regarding the ocean; however, the many personal vignettes serve as a valuable reminder that such ideas and concepts grow from the struggle and enthusiasm of individuals, working among their peers at a given time and place. Betty's insistence that this is important informs the entire narrative.

Where the word *we* is used in the text, it commonly reflects the advice and collaboration so generously given by Betty in the formative stages of this work. Her counsel and her insistence on strict accuracy in all historical detail were invaluable.[4]

Thanks are due the various copyright holders for giving permission to quote from their texts. We are especially indebted to Princeton University Press in this regard.[5] Also, I made extensive use of the standard text by Sverdrup et al.,[6] with the ready approval of Prentice-Hall. Other heavily used materials are in publications issued by Springer-Verlag[7,8] and by Wiley Interscience[9]; they and coauthors and coeditors of the works in question are herewith gratefully acknowledged. In addition, I thank the staff of the Scripps Archives and of *Explorations* magazine for freely providing material when asked, and the director for permission to use it here.[10] In organizing the material for the purpose of public education, I have greatly benefited from discussions with colleagues at the University of Bremen, and with staff at the Birch Aquarium, the Aquarium of the Pacific (Long Beach), and the Ocean Institute at Dana Point.

Last but not least, I thank my wife, Karen, for great patience and loving support.

NOTES AND REFERENCES

1. J. H. Steele, S. A. Thorpe, K. K. Turekian (eds.), 2001, *Encyclopedia of Ocean Sciences*, 6 vols. Academic Press, San Diego, 3399 pp.

2. R. L. Fisher, E. D. Goldberg, C. S. Cox (eds.), 2003, *Coming of Age: Scripps Institution of Oceanography*. SIO, University of California at San Diego, 213 pp.

3. E. N. Shor, 1978, *Scripps Institution of Oceanography: Probing the Oceans, 1936 to 1976*. Tofua Press, San Diego, 502 pp.

4. Of course, the responsibility for all errors in fact or interpretation remains with the author.

5. For permission to quote rather freely from H. W. Menard, 1986, *The Ocean of Truth*. Princeton University Press, Princeton, N. J., 353 pp. (See chapter 12.) Other permissions were received from University of California Press (regarding Stommel 1965), *Scientific American* (regarding Munk 1955, and Isaacs and Schwartzlose 1975), National Academies Press (regarding Steele 2000), Belser Verlag (regarding Fricke 1972, words by Irenäus Eibl-Eibesfeldt), Columbia University Press (regarding Prothero 1994), *Science* AAAS (regarding Ryther 1969, and McGowan et al. 1998), Woods Hole Oceanographic Institution (regarding Alvin photo of a black smoker, modified), Hauff Urweltmuseum (regarding image of an ichtyosaur fossil), Munich Reinsurance Company (regarding image of hurricane tracks in and around the Caribbean), the Geological Society of America (regarding images in Hedgpeth 1957), and the American Society for Limnology and Oceanography (regarding a figure in Reid 1962). Permissions and encouragement also were received from several individuals, regarding quotes or figures: Walter Munk, Warren Wooster, Joseph Reid, Peter Franks, Fred Spiess, Eugen Seibold, and Gerold Wefer. Thanks to all.

6. H. U. Sverdrup, M. W. Johnson, R. H. Fleming, 1942, *The Oceans: Their Physics, Chemistry, and General Biology*. Prentice-Hall, Englewood Cliffs, N.J., 1087 pp.

7. E. Seibold, W. H. Berger, 1996, *The Sea Floor: An Introduction to Marine Geology*, 3rd ed. Springer Verlag, Berlin, 356 pp.

8. G. Wefer, W. H. Berger, G. Siedler, D. J. Webb (eds.), 1996, *The South Atlantic: Present and Past Circulation*. Springer Verlag, Berlin, 644 pp.

9. W. H. Berger, V. S. Smetacek, G. Wefer (eds.), 1989, *Productivity of the Ocean: Present and Past*. Wiley-Interscience, Chichester, U.K., 471 pp.

10. SIO Archives: Deborah Day; SIO Publications Office: Chuck Colgan; Director of SIO (1997–2006): Charles F. Kennel.

ONE

Discovering the Ocean

OF FISH AND SHIPS AND PEOPLE

Lobster, scallop, and tuna are among the more expensive items on the seafood menu, and for good reasons. We like to eat these things, and there are many of us, and not so many of them any more. In fact, with regard to fish suitable for fine dining, there are now roughly 10 times fewer in the sea than only a few decades ago.[1] Many other animals of the sea once or recently heavily exploited are similarly diminished, including, for example, sea turtles and large whales. Jellyfish, however, remain in sufficient abundance (fig. 1.1). Their nutritional value, in relation to weight, is low. Nevertheless, they are already finding their way into seafood restaurants, thus confirming a long-term trend away from catching highly prized predators such as tuna and swordfish, toward netting plankton-eaters and invertebrates.[2]

For millennia, the relationship between people and the sea has been determined largely by human fondness for seafood. Fishermen had the most intimate knowledge of winds and currents and of the changes that come with the seasons. Also, of course, they knew where to go to find fish and crabs and oysters. Their prey was found in view of the land. Later on, with bigger vessels, fishing moved out into the open ocean away from visual contact with the coast. Fishermen discovered new riches: enormous aggregations of cod, schools of herring. In high northern latitudes, whales have long been part of marine hunting cultures. Not so long ago, in the nineteenth century, whaling was an important business in many coastal communities throughout the world. In New England, shore whaling off Nantucket and Long Island made a start around 1690. The business grew into a worldwide enterprise in the early 1800s, centered at New Bedford, where more than 400 whalers were registered at the Custom House in 1857.[3] Yankee whalers knew the sea and the habits of whales.[4] They were the sages of the sea well before oceanography emerged as a branch of the Earth sciences. Their knowledge was

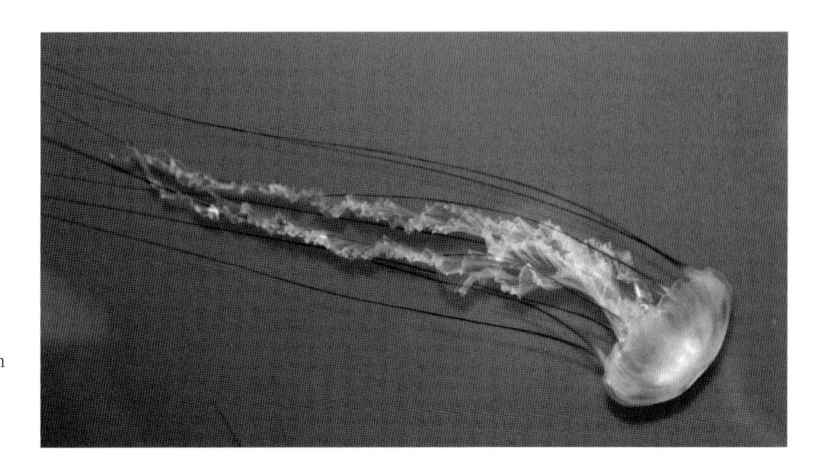

FIGURE 1.1. Pelagic jellyfish include some of the largest and most ancient forms of animal life in the sea.

closely tied to purpose, a link that has largely persisted into modern ocean sciences. The main lesson of the story of whale hunting, often retold, is that the sea's resources are not inexhaustible, and that overexploitation will not engender restraint but will stop only after collapse of the resource or from outside intervention (in this case, the discovery of petroleum).

As we have learned more about the ocean, motivated by the needs of fisheries, navies, and shipping, and also by curiosity, a new planet has entered our awareness, one where the ocean is the dominating feature of conditions on Earth. Winds from the sea bring the rain that determines where plants and animals on land shall thrive. Their patterns of distribution, in turn, determine the life-style of humans dependent on agriculture.

Awareness that our planet has an enormous ocean, with island continents, starts with the discovery of the World Ocean around AD 1500. Knowledge that the deep ocean is cold and that its salt is the same everywhere are achievements of the nineteenth century. Since then, the science of the sea has expanded rapidly. The general pattern of this expansion, as seen in the large oceanographic institutes around the world, is nicely reflected in the history of Scripps Institution of Oceanography ("Scripps" for short), one of the largest and oldest of the genre. As it happens, Scripps is just over 100 years old: it was founded in 1903.

A BOUNTIFUL SEA?

The lesson learned from whaling—that the ocean's resources are large but limited—has recently emerged again with respect to many other prey items that once yielded millions of tons of food, such as herring and cod (fig. 1.2).

The collapse of these fisheries in the North Atlantic was not entirely unexpected. Warnings were sounded back in the early twentieth century.[5] But optimism prevailed. As recently as 1967, the American fishery scientist W. M. Chapman averred that the total harvest of fish from the ocean each year could be expanded to some 2 billion tons, given the right technology. At the same time, Scripps oceanographer John D. Strickland preferred a much lower number, admitting a maximum of 600 million tons.[6] Both guesses were much too high, exceeding reality by more than a factor of 20 and of 6, respectively. In the decades since, it has become clear that the total catch has stagnated around 100 million tons, despite ever increasing efforts. At the catch of 100 million tons, already, the sea is being fished out. Large fish are becoming rare, and the food web in the sea is changed accordingly.[7]

To become an honorary citizen of Newfoundland (a distinction bestowed by the natives during happy hours to many a visitor over a suitable number of drinks) one must pass several tests, one of which consists of kissing the lips of a dead codfish.[8] There are good

FIGURE 1.2. For centuries, herring and cod have supported major fisheries of the North Atlantic.

reasons for demanding that any citizen of Newfoundland, real or honorary, should demonstrate allegiance to this grim-faced denizen of the sea. The settling of the coast of Newfoundland by Europeans, and indeed of all the shores north of Cape Cod, owes a great debt to the rich cod fisheries of the Grand Banks off Newfoundland, and to those along much of the coast of Atlantic Canada and New England.[9]

The Atlantic cod *(Gadus morhua)*, until very recently the King of the Grand Banks, is a member of the gadoid fishes, an extremely important group of predators obtaining much of their food at the bottom of the sea, in fertile regions. The cod is commonly caught at weights up to 25 pounds or so but is able to grow to a size of 6 feet and 200 pounds.[10] Other gadoids are the haddock, the pout, the poor cod, the coalfish or saithe, the pollock (the British "Pollack"), the whiting, and the silver cod. The various species of the genus *Gadus* differ in color and markings and appearance of the lateral line, and the presence or absence of a barbel (the appendage on the lower jaw that is the hallmark of the cod).

Fishing for cod off New England was part of the negotiations between the new nation of the United States of America and the King of England, following the revolution. At the insistence of John Adams, the United States was granted fishing rights on the Grand Banks by England and permission to land in Nova Scotia and Labrador to salt-dry the catch. Georges Bank is largely in international waters (even today, notwithstanding the expanded economic zone). It is an extension of the continental shelf, out to some 200 miles off New England, covered with debris piled there by enormous glaciers that once moved out from the Hudson Bay region. The shelf ends at an underwater precipice where cold Arctic waters flowing southward meet the warmer water of the Gulf Stream. The meeting of cold and warm eddies provokes much turbulence and mixing, aided by strong currents interacting with the shallow bottom, and this turbulence moves nutrient-rich deep waters into the sunlit zone at the surface. The rich supply of nutrients and the strong sunlight in spring and early summer start the food chain that supports the fish and crustaceans that cod feed on.[11]

The Atlantic cod ranges throughout the North Atlantic, from the North Sea and around the British Isles and the Bay of Biscay to Iceland and up to northern Norway and into the Barents Sea. An important problem for fishery biologists has been to define the degree to which the stocks in different regions are separate populations. There was a time, in the early 1970s, when Atlantic cod was the second most important species, by weight, in the world's fish catch. Only the Peruvian anchovy yielded a higher tonnage.[12] By 1980, the catch had fallen from more than 3 million tons to just over 2 million, and the cod moved from second to sixth place. The cod's less-exploited cousin, the Alaska pollock, moved from third to first, the Peruvian anchovy fishery having collapsed in 1973. In 1992, Peruvian anchovy (by then recovered) and Alaskan pollock shared first and second place, while cod dropped off the list of the top 10 species. The fisheries of Atlantic Canada and Newfoundland suffered accordingly.[13] Equally important, the entire ecology supporting the bounty of the sea in the North Atlantic has been affected.[14]

A fishery can decline, in principle, for a number of reasons, including changing preferences of the fishermen, overfishing, unfavorable environmental change, or a combination of these. The evidence suggests that the rise of very large industrial fishing vessels with enormous nets and advanced facilities for finding and processing fish is the main reason for the demise of the cod fisheries since the 1970s. Serious fish removal efforts started in 1951, with the 2,600-ton *Fairtry* from Britain. The technology of the *Fairtry* incorporated insights gained during the Antarctic whale fishery, which by then had little future left.[15] For example, an enormous chute at the heck of the ship led up from the water line to the deck, to facilitate the hauling in of large nets, in a manner analogous to hauling up freshly killed whales. Likewise, processing at sea, a technology long used to advantage in harvesting whale products, was now adapted for fish. Other nations soon followed in building large long-distance floating fish factories, equipped with modern acoustic detectors developed during World War II. By the 1970s close to a thousand factory ships were operating at sea, nearly half of them with a homeport in the Soviet Union, with the next largest fleets being from Japan and Spain. Thus, the time-honored activity of harvesting fish was replaced with systematic fish removal on an enormous scale using sophisticated industrial methods.

When discussing the collapse of fisheries, the influence of climate change cannot be neglected. Historical records suggest that climate change had profound effects on the abundance of cod through the centuries. For example, off the Faroe Islands the cod fishery failed in 1625 and 1629, and the cod disappeared entirely for many years after 1675, presumably because of cooling associated with the harshest period within the Little Ice Age. Catches of cod were rather low for the entire time between 1600 and 1850, during the reign of this climatic period, which is known for having brought expanded sea ice around Iceland and abundant violent storms that discouraged sea voyages in the northern North Atlantic.[16] The climate has been warming since, especially in the North Atlantic realm.[17] Such warming has effects on the ecology of the North Atlantic, including recovery of fish stocks.[18] However, both the sense and the magnitude of such effects on the various stocks are quite uncertain. Climate change introduces a wild card into the betting game called "fisheries management."

The strain from the new methods of fish mining on the fisheries in the northwestern Atlantic soon had political consequences. In 1977 Canada proclaimed exclusive use of its coastal waters to 200 nautical miles offshore, following similar action by Iceland. Since then a 200-mile zone of exclusive economic use has been claimed by other nations bordering the sea, greatly expanding a kind of privatization of the ocean, with the coastal nations as owners. While there are positive aspects to taking ownership in terms of managing resources, the effect on ocean sciences has not been beneficial, on the whole. Neither has the potential advantage for rational management played out in a desirable manner. After annexing the coastal ocean, Canada promptly expanded its own factory fishing fleet. The reasons included misconceptions about the ocean's productivity, along with misguided economic considerations. (The prominent Canadian fisheries scientist Daniel Pauly refers to such calculations as the "march of folly."[19]

The second most important food fish of the North Atlantic realm traditionally has been the Atlantic herring; its peak catch (in 1966) exceeded that of the cod (in 1968) by 5 percent, at a tonnage of more than 4 million. The Atlantic herring *(Clupea harengus)* is a plankton-eating fish traveling in schools (like sardine and anchovy). It has a close relative in the Pacific, the Pacific herring *(C. pallasi)*. Another important member of the genus is the sprat (or brisling). Herringlike fishes, including anchovy, sardines, and pilchard, strain the water for small organisms, swimming through clouds of plankton. The fishes in this group (clupeids) use specialized comblike tools in the gill region (called gill rakers) to trap the food before it can exit with the water streaming from the oral cavity through the gills. Their food includes microscopic algae and small zooplankton. Thus, they feed at a lower level in the food chain than do the fishes in the cod family, which as adults prey on other fish. Feeding low on the food chain (that is, on small plankton) is the clue to the enormous abundances of herringlike fishes, and their correspondingly large representation in the global catch.[20] Herring became economically important when ways were found to preserve the fish at sea, using salt.

The herring has long been an important food item in northern Europe; in the Middle Ages, the wealth and power of the Hanseatic League of cities in northern Germany (with outposts in other countries including Norway) was closely tied to the herring trade. All through the nineteenth century and during much of the twentieth, the herring continued to be of great importance in the economies of the fishing nations of northern and western Europe. In their classic book on marine biology, first published in 1928, the British zoologists F. S. Russell and C. M. Yonge wrote:

> No fish in the sea are caught in such great numbers as the herring. One boat may catch over 100,000 fish a day and the total catch on such a day for Yarmouth would be 30,000,000. The great fishery gives employment to an army of workers on land, chief among which are the Scottish fisher girls. . . . They work all day cleaning and gutting the herring harvest with razor-sharp knives wielded in their dexterous hands. . . . The value of herring landed in England alone in the year 1924 was about four and a half million pounds.[21] The total value of the herring fisheries of all the nations of Northern and Western Europe was nearly ten million pounds, or a little under a quarter of the value of the whole sea fisheries.[22]

Herring occurs in great aggregations, which greatly facilitates its capture by purse-seining.[23] Thus, the evolution of schooling, which emerged in response to predation (much as the flocking in birds), is proving to be ruinous with the appearance of a predator training his sights on the entire school.[24] In recent decades the herring fishery in the North Atlantic has lost its preeminent position as one of the great fisheries in the world. In the early 1990s it was still among the top species in the total yield, but with distinctly lower tonnage than during the peak years. It has since deteriorated further.

Effective management action is economically painful. It is easier to move elsewhere after collapse of the resource—as long as an elsewhere is available. When "elsewhere" offers no solution, management issues become serious. In the case at hand, the uncertainties regarding the effects from year-to-year changes in food supply and in the mortality of juveniles led to endless and inconclusive discussions about the advisability of restricting fishing efforts. The fisheries biologist R. Bailey and the marine ecologist J. Steele[25] commented:

> Rather than considering the restriction of fishing effort, much of the scientific discussion at this time hinged on the need for controls on industrial fisheries for juvenile herring and on fishing for herring on the spawning grounds. . . . Evidence on changes in recruitment [of juvenile fish to the stock of adults] was equivocal. There was some indication of a decrease in year class strength for the North Sea as a whole, but it was uncertain whether this was simply part of the natural variability in recruitment or related to the egg production of the stock.[26]

Thus the problem: small stocks can produce sufficient eggs to make sufficient juveniles to allow the stock to replenish, provided the survival of juveniles is good. So, why restrict fishing on the stock? Nothing can be done about the survival of juveniles. In the case of predatory fish such as cod, recruitment may actually benefit from removal of adults, because the adults eat the juveniles. Such arguments are difficult to refute, especially when dynamics of stocks and of interactions between stocks are poorly understood.

Not surprisingly, therefore, the assessments concerning sustainable fisheries vary considerably, even among expert scientists. As a result, advice to regulatory agencies regarding the behavior of fish stocks is bound to be somewhat tentative. The outcome is that fisheries science cannot release the regulators from responsibility for their decisions by giving them hard facts with which to pound the table when pressed by special interests (that is, by the fish removal industry). Thus, the regulators quickly discover that doing little or nothing that would impede whatever exploitation is going on commends itself as the least painful strategy for action. Too often, this strategy results in setting "limits" near or even above the actual catch that is economically feasible.

There are many lessons to be learned from the collapse of cod and herring fisheries in the North Atlantic. Mainly, the failure to regulate access to common resources results in unsustainable removal. In moving from local fisheries to industrial clear-fishing, harvesting turns into looting. As a result of untrammeled exploitation, the resource vanishes, to the detriment of those most in need of it. Since gain and loss accrue to different players, there is no inherent control mechanism on the process of overexploitation. It seems futile, therefore, to put much hope on voluntary restraint.

A NEW PLANET

Every generation of environmental scientists—practitioners of meteorology, oceanography, geochemistry, ecology—gets to study a new planet, not only in concept but in reality as well.

The way we humans view our home planet has been changing from century to century, and lately from decade to decade. In the early Middle Ages, the planet was a disk in people's minds, and if one sailed to the edge of it, one was liable to fall off. Monsters occupied the depths of the sea, and they could emerge without warning to wreck a ship and swallow the people. On the whole, the sea was a dangerous place, well beyond human control and understanding. A semblance of control could be achieved by the outstanding seamanship of a competent captain, and by the crew doing nothing that might invite bad luck.

With the arrival first of steamships and then of motor vessels, the balance shifted, and the ocean was no longer dangerous. Perceptions changed fundamentally. Unlike for sailing ships, there is no problem in moving against even strong winds and currents when employing a diesel engine packing the power of thousands of horses. Also, a 100-ton monster whale poses no danger to a vessel that is more than 10 times heavier. On the contrary, in the nineteenth century, steam vessels began to endanger the survival of creatures in the sea that had lived there for millions of years.

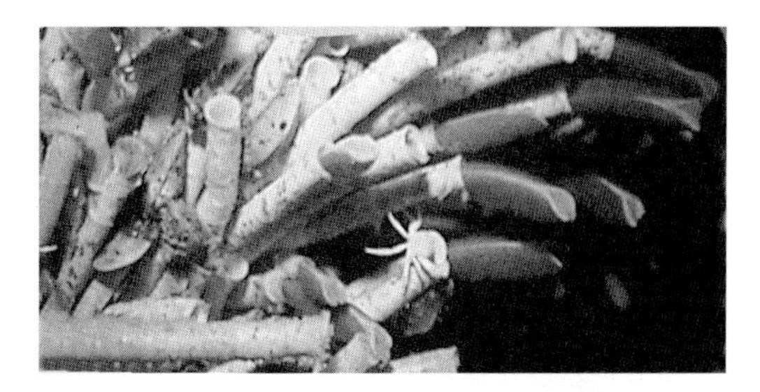

FIGURE 1.3. Gutless tube worms processing sulfurous emissions from a hot vent, with the help of symbiotic bacteria living within their bodies. Also a small crab.

The new habit, started in the eighteenth century, of systematically describing everything in the sea and labeling each organism with a latinized name, soon resulted in a growing list of species populating the sea. A growing list of species, naturally, changes the ocean in concept: each species discovered adds a new dimension to an already complex system. But extinction, whether global or regional, objectively changes the system, and ecological extinction is now commonplace.[27] What is being studied today never existed before, and the same will be true a few decades from now.

In addition, there is much about our ocean planet that has existed for millions of years, is of fundamental interest, and yet has escaped discovery until very recently. Once discovered, such previously hidden features become part of a new conception of the planet, as well.

The presence of strange and wonderful organisms along hot vents in certain places of the deep-sea floor is a case in point. It is a discovery that is well known and rightly celebrated (fig. 1.3). The discovery was made from a small fragile-looking submarine diving at great depth, and this has bestowed to it a flavor of romantic adventure. Diving in *Alvin*, the scientific minisub operated by Woods Hole Oceanographic Institution, in places where hot basalt makes new seafloor, scientists found hot springs and a host of animals surrounding them. The hot vent fauna includes giant clams, gutless tubeworms, eyeless shrimp, and bacteria processing

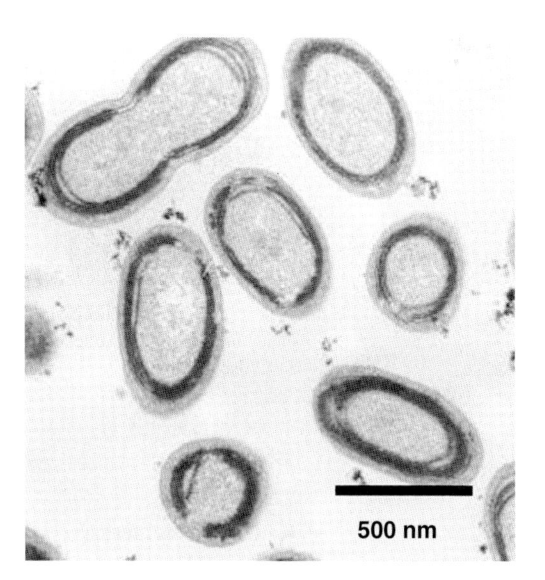

FIGURE 1.4. The smallest and the most abundant: cyanobacteria of the genus *Prochlorococcus*.

sulfur. A new type of ecology was born on that day in 1979, when these strange gardens were first seen. These communities represent an ecosystem that is built not on organisms deriving their energy from the Sun, but on bacterial activity associated with the chemical interactions between hot basalt and seawater.[28]

It is perhaps not so surprising that we should find something new and exciting in places where no one has looked before. But equally noteworthy and surely more marvelous are the instances where fundamental discoveries are made in places where everyone looked intently for nearly a century, while missing vital ingredients of the system. A striking illustration of such a major gap in knowledge, now filled, is the former obscurity of micron-sized photosynthesizing bacteria in the sea.[29] Millions are present in every liter of seawater (fig. 1.4). But the prominence of *Prochlorococcus* as the single most abundant light-consuming creature of the ocean has been discovered only quite recently. Along with the discovery that iron is a limiting nutrient whose rarity impedes production of diatoms and other light-using microbes over large areas of the sea,[30] such new insights demand rethinking of much that was taken as established textbook wisdom.

Lately, much of what we are learning about the ecosystems of the sea is tied to human impact.[31] Large-scale experiments are being performed, unwittingly, through the introduction of invasive species into estuaries and semi-enclosed seas,[32] and through overfishing and coastal pollution. In the meantime, physicists, geologists, and chemists are expanding our horizons using sophisticated instruments developed after World War II and more recently, including satellites, drilling vessels, and automatic analyzers. Modern computing makes it possible to process the enormous amounts of data that are generated with automated devices, and also to simulate complex physical and biological systems and their interactions.

These various developments certainly warrant introducing the notion of a "new planet." But the notion arises especially forcefully in the context of the "great geophysical experiment"[33] that is being performed on the home planet. It is the large-scale release of greenhouse gases to the atmosphere, which physically changes the planet from one decade to the next. As the permafrost melts around the Arctic and the sea ice retreats, a new and unfamiliar planet emerges, for people and for polar bears.[34] In the warmer regions of the global ocean, as well, many changes are afoot that have been ascribed to various types of human impact. One of these is the deterioration of coral reefs in the Caribbean and also elsewhere. To recognize such changes and develop the appropriate policy is a major challenge that calls not just for more science but also for whole-scale mobilization of public awareness.[35]

There was a time, in the 1950s and 1960s, when being a student at Scripps Institution of Oceanography included the opportunity to collect dinner from just outside the breakers, including lobster, abalone, and big territorial fish in the kelp forests. Today, finding any of these within the hour would be quite unlikely, and taking them would be against the law. Things have changed—today's students discover a different ocean. Compared with the ocean of half a century ago, many large vertebrate and invertebrate species are ecologically extinct, that is, they no longer play a role in the ecosystems of the sea. Scripps ecologist Paul Dayton refers to "ghost species," whose erstwhile existence provides explanations for the ecology of a residual community of organisms, in the kelp forests offshore of Scripps and elsewhere along the coast of California.[36] The entire ocean, as it were, is now full of such ghosts, from the once whale-rich seas around the Antarctic and Spitsbergen to the traditional fishing grounds off Newfoundland, and to the Caribbean Sea that once teemed with turtles. To understand the functioning of present ecosystems, we must take into account the response of each system to the removal of key players within it.

Sharks and turtles are well on the way to the realm of ghosts in many regions of former abundance. Both sharks and marine turtles have been part of the ocean's scenery and ecology for hundreds of millions of years—sharks since before the Devonian (see the appendix 4 for a geologic time scale). Marine turtles are offspring from terrestrial reptiles that evolved during the Devonian.[37] The turtles have seen the great marine saurians come and go—ichthyosaurs, plesiosaurs, and mosasaurs. They survived where others failed, and accommodated themselves to the rise of the marine mammals in the last 40 million years. But turtles and sharks are now witnessing the arrival of a new planet with new rules for survival set by the reigning superpredator: humans. The rules are changing faster than ever before. Traditional rates of adaptation cannot cope with such rapid change.

Scripps oceanographer Jeremy Jackson, who was diving on Caribbean reefs some 40 years ago, recalls the exquisite richness of these ecosystems. Even the best-preserved reefs, according to Jackson, have been greatly impacted by human activities including overfishing and pollution. Most strikingly, the large species of branching coral that dominated shallow reefs for at least half a million years have declined dramatically since the 1980s.[38] Thus, there is now no way to explore the natural ecosystem functioning of Caribbean reefs. What can be discovered is the additional human impact on an already impacted system that is rapidly changing.

The overexploitation of the oceans has its origin in the doctrine of the "freedom of the seas," which emerged with the discovery of the World Ocean and its use for trade and fishing.[39] This freedom has been considerably curtailed since, especially by the 200-mile extensions of jurisdiction by coastal nations in the 1970s and 1980s. Yet most of the open ocean remains in the state of an unregulated commons. The late Garrett Hardin, University of California economist and biologist, argued that the exploitation of an unregulated commons by entirely rational parties inevitably leads to collapse of the resource.[40]

Hardin's prediction is well supported by the recent collapse of cod and herring fisheries. An additional dynamic, beyond the "tragedy of the commons" paradigm of Hardin, is the "march of folly" process emphasized by Daniel Pauly of the University of British Columbia. Pauly suggests that ruinous policies pursued by government agencies ensured a subsidy-driven destruction of successive fisheries, with considerable impact on all marine ecosystems.

Marine biologists are not the only ocean scientists who are discovering a new ocean every decade and, indeed, every year. All oceanographers do. The ocean itself—its surface temperature, its currents, the winds driving its waves—is changing all the time. Much of this change is natural, but a substantial portion, not known in detail, is now man-made.[41] Humans have become a major player among the forces of nature on the face of the planet. This fact was not an issue in the first half of the twentieth century but started to emerge in the 1950s and 1960s, mainly among geochemists.[42]

Exciting discoveries have been made over the last several decades regarding various types of oscillations within the climate of the sea. These affect the global distribution patterns of warm and cold water at the surface of the sea, and the delivery of nutrients to surface waters from waters below. The El Niño phenomenon, the Pacific Decadal Oscillation, and the North

Atlantic Oscillation are prime examples of substantial variations in the workings of wind and weather and currents, in various parts of the ocean, and have implications for droughts and floods on adjacent continents.[43]

Just as we discover these patterns, they are already being modified in poorly understood ways by the effects of global warming. In fact, it is conceptually difficult to separate natural variation from human-induced climate change. This difficulty will not go away with additional research. The reason is that the range of natural variation—poorly known in the first place—cannot emerge from studying recent climate variation, because natural variation is no longer the only player in climate change. The implication for marine biology is that the ocean's variability generates ecologic variability, which masks the human impact. Under these conditions, the benefits of management are difficult to assess.

Whenever the benefits of management actions are hard to recognize, there is no ready defense against those who favor business as usual, claiming that the economic costs of regulating the commons are excessive. This is why Hardin's theoretical treatment of the problem is so important. On a global scale, similarly, lack of proof of benefit of restraint may be anticipated in the context of the climate change conundrum. That is why computation of possible and probable changes in the behavior of ocean and atmosphere for different scenarios has become a crucial tool in climate research. The skills needed to produce relevant models have been growing exponentially, based on a rapid expansion of computing power, of raw observational data, and of the understanding of the physics of climate. Unfortunately, the knowledge needed to appreciate the nature of the results of computations is not readily imparted or acquired.[44]

Despite all the expansion of the knowledge bases in recent years, when measured against the complexities of today's rapid global change, our understanding of planetary physics and biology is still woefully inadequate, and this is true also for the economics and the politics of the environment. Our lack of knowledge is especially vexing in regard to the ongoing change in the ocean, the most massive part of the climate system, and the largest habitat for life, hosting the most ancient life-forms. Nevertheless, we must attempt to gain a sense of the nature of the sea and its role in the ongoing changes. If we ignore the ocean, we cannot be good stewards of the Earth.

DISCOVERY OF THE WORLD OCEAN

The Greek tradition, which is the cradle of modern science, held a view of the world dominated by enormous landmasses (Eurasia and Africa) surrounding a smallish sea, the appropriately named Mediterranean. This view changed dramatically 500 years ago with the discovery of the Atlantic, Indian, and Pacific oceans, discoveries that made islands out of continents.[45]

In the western tradition, the single most important step in discovering the World Ocean was the daring expedition into the unknown by three modest-size vessels, the *Santa Maria*, the *Niña*, and the *Pinta*, which set out from Spain to discover a westward passage to India, 500 years ago. When the Genoese explorer and adventurer Christopher Columbus[46] made landfall in the Bahamas on that fateful October day in 1492, he fell on his knees to kiss the ground and give thanks to God. He had good reason. For quite some days many of his men had been ready to throw him overboard, to end the horrifying voyage into the uncharted void. In the nick of time, the cannon had gone off on the *Pinta*, announcing the sighting of land. "What will we get to see?" Columbus wrote into his logbook. "Marble bridges? Golden-roofed temples? Spice gardens? People like us, or some strange race of giants? Did we reach an island or Japan itself?"

Spice gardens, of course, would have been better than even pure gold. What they found were harmless natives, "mighty forests," a clear brook, and "enormous unknown fruits"—but no signs of wealth. Columbus named the island that saved him from mutiny San Salvador, made contact with the friendly "Indians," claimed their land for the Spanish Crown, and promptly started looking for Japan.

One might argue whether Columbus had in fact discovered America on that day. There is, however, little question that he and his disgruntled crew manning the *Santa Maria*, the *Niña*, and the *Pinta* had discovered the Atlantic, crossing it where it is widest, and where the trade winds are most favorable for the task. As far as discovering America, we might, along with many historians, give priority to Leif Ericsson, who made landfalls on Baffin Island (in AD 1001), on Labrador, and on Newfoundland (Vinland). However, Leif's exploits did not result in maps or useful reports for later explorers. His voyage traversed the Atlantic where it is narrowest.

What Columbus did by crossing the Atlantic Ocean at its widest, not once but repeatedly, was to bring that body of water to the attention of the western world. Entirely new possibilities for trade routes opened, and other explorers soon followed Columbus's lead. One of these was Amerigo Vespucci (1451–1512), a Florentine merchant who sailed in the service of Spain and of Portugal. He explored the South American coast and realized that this was not Asia, but a "new world." By 1507 his name was used to refer to the New World. Its inhabitants, however, remained "Indians," and the islands first seen by Columbus the "West Indies"—that is, the Indies found by sailing west.

Columbus thought he had crossed the World Ocean, having reached islands he deemed close to India and Japan. But the awesome truth became known a few years later (in 1520) when Ferdinand Magellan (ca.1480–1521) rounded Patagonia through the straits that now carry his name, and crossed the ocean that Balboa had seen from the shores of Panama—an ocean more immense by far than the Atlantic. A new hemisphere—one covered by water—had to be added to the globe. Mediterranean-centered geography was finally scrapped. Within three decades, Columbus's landfall in the Bahamas, Vasco da Gama's voyage into the Indian Ocean around the tip of Africa,[47] and especially Magellan's crossing of the Pacific had transformed the world.[48] "Planet Earth," it turned out, really is "Planet Ocean." Today, this insight is commonplace. As our satellites view the Earth from a position above the Pacific, we see almost nothing but water.

Right into the eighteenth century, the Mediterranean view of the world, that land encircles the sea, informed excursions into the unknown. Thus, cartographers extended Ptolemy's hypothetical southern continent all the way across the southern Pacific.[49] This enormous Terra Australis finally evaporated when the great navigator and explorer Captain James Cook (1728–1779) tried to find it. During three major expeditions (1768–1771, 1772–1775, and 1776–1780), he established the boundaries of the Pacific, mapped the coast of Australia, and put a great number of islands on the map, including Hawaii and New Caledonia. A series of southward forays took him and his crew into the furious and icy storms of the regions around Antarctica, and even beyond the polar circle. Within one decade of intense exploration, Cook sailed the Terra Australis off the world map and put the Southern Ocean in its place. (He did leave room for Antarctica.)

In the process of discovering the Southern Ocean, Cook noted mammal life of a density never seen before. Soon after, large-scale slaughter by sealers and whalers would considerably reduce that abundance. Such is the fate of explorers; they set in motion events far beyond their vision and intent.

After charting the western coast of North America, James Cook had, essentially, completed the world map. He had, moreover, done so in a style that sets him sharply apart from most previous explorers, seekers of riches who hazarded their ships and crews and had no regard whatsoever for the people they came in contact with. Cook's foremost concern was the safety and health of his men, and he came in peace wherever he made landfall. His sad awareness of the changes he was bringing to native peoples of the islands he discovered and revisited ("we introduce among them wants and perhaps diseases which they never before knew") illustrates his compassionate humanity.[50]

Nevertheless, Cook served king and country. His voyages set the stage for the remarkable global expansion of British sea power, which culminated in the British Empire spanning the globe. The careful and comprehensive mapping of seaways and shores on a global scale was a high priority from the very beginning. Having fewer ships and cannons than the Spanish, the Portuguese, the Dutch, and even the French, England trusted to superior seamanship, and better science. Precision chronometry (to determine longitude) and scurvy-fighting foodstuffs (lemons, sauerkraut) have been part of the naval toolkit ever since Cook. The chronometry, combined with astronomical tables, provided for precise positioning. The tables were the responsibility of the Astronomer Royal in Greenwich, near London. He has long been relieved of that duty: the officer on the bridge now uses satellite positioning and does not need his advice.[51]

Only the polar regions remained poorly known into the nineteenth century. The many efforts of earlier explorers to find a northeast or a northwest passage from the Atlantic to the Pacific (to avoid Spanish and Portuguese galleons interdicting trade) were unsuccessful. Eventually it became clear that to realize any passage in the north, one had to sail into the Arctic Ocean, along a narrow ice-free zone in summer. Most arctic summers were short in the period between 1600 and 1850 AD, and winters severe. The period is referred to as the Little Ice Age and is the coldest on record for the last 8,000 years.

The last great ocean basin to be discovered was the Arctic Ocean. In June of 1883, Nansen left Norway in the *Fram*, built with the support of the Norwegian parliament and many individuals, including the King of Norway.[52] The vessel was built to withstand the crushing pressure from pack ice, her hull specially strengthened and shaped. Nansen had studied the ice drift and concluded that it moved from off Siberia across the pole to between Greenland and Spitsbergen (today's Fram Strait). He sailed the *Fram* to eastern Siberia and let her be frozen in (by 27 September). She stayed that way for the next 35 months. Slowly she drifted toward the pole. However, when it became clear she would miss the pole by some considerable distance, Nansen and his companion F. Hjalmar Johansen set out to reach the pole by sled. They had to turn back 268 miles from the pole. They wintered in a stone hut on the coast of Franz Josef Land, in the Barents Sea. A British exploring expedition picked them up and returned them to Norway. The *Fram* arrived late in summer in 1896, one week after Nansen. The drift of the *Fram* established the Arctic Ocean as a deep-sea basin centered on the North Pole. With this feat, the map of the World Ocean was complete.

EARLY OCEANOGRAPHY AND THE *CHALLENGER* EXPEDITION

The history of oceanography is conveniently divided into "early" and "modern" by the British *Challenger* Expedition (1872–1876). The distinction was appropriate till into the 1960s. Today's postmodern period is characterized by massive data collection using satellites and automated observing platforms (some floating, others swimming), by precision measurement of highly diluted trace substances in the sea, by the application of molecular methods to marine biology, and by ubiquitous and massive computing power. The reason to look at the results of the *Challenger* Expedition today is not so much for the data it still provides, but for the fact that this famous world-circling voyage established many of the basic concepts about the ocean: its overall depth, its generally frigid temperature below a thin warm surface layer, its almost invariant salinity, the abundance of organisms in surface waters and at depth, and the nature of the bottom sediments.

It is interesting to speculate why the largest expedition of the nineteenth century, the U.S. Exploring Expedition (1838–1842) is not credited with laying the foundations for modern oceanography. The Exploring Expedition involved six ships and 346 men, sent to the Pacific by the U.S. Navy. Unfortunately, the commander of the expedition, Lieutenant Charles Wilkes, was not suited as a leader of a collaborative scientific venture, a fact of which he was unaware.[53] More importantly, the tasks of the "Ex. Ex.," as it was commonly referred to, were focused on charting islands and the shores of Antarctica (Wilkes Land), rather than on exploring the deep ocean. Thus, by consensus, the founding feat of modern oceanography is the circumnavigation by the British vessel HMS *Challenger*.

Of course, ocean sciences existed well before that, largely for the benefit of navigation. Benjamin Franklin (1706–1790) published the first map of the Gulf Stream (in 1769), in an effort to help speed ships between England and America. Commander Matthew Fontaine Maury (1806–1873) expanded this task to a global scale. He was the first genuine armchair oceanographer engaged in data processing. Information from the logbook entries from merchant and whaling ships and from navy vessels served him to make charts of winds and currents. These charts delighted the navigators of the time, and his book *Physical Geography of the Sea*, first published in 1855, sold well for decades, in many editions. Maury's legacy lives on in the maps of currents found in every ocean textbook. Such maps offer a pattern of prevalent conditions; whether one finds a current running where it is shown, on any given day, is another matter.

In the nineteenth century, marine biological research was greatly advanced by the British naturalist Edward Forbes (1815–1854), who established that bottom-living organisms tend to prefer their own depth zones. Hence, as one descends deeper into the water along a slope, one would collect different types of animals. Noting that the number of animals in dredges decreased rapidly with depth, and extrapolating to depths with little or no information, Forbes proposed an "azoic zone" in the deepest parts of the ocean, where, he thought, high pressure and low temperature were hostile to life. One of the tasks of the *Challenger* Expedition was to test this idea. Forbes's depth zone scheme proved invaluable. But his notion about a lifeless abyss turned out to be wrong. Ironically, he is more often cited for his one wrong idea than for his many good ones, a fate he shares with the distinguished French naturalist Jean-Baptiste Lamarck (1744–1829), who proposed evolution before Darwin did but failed in defining the mechanism.

Another pre-*Challenger* scientist whose work remains very influential is Christian Gottfried Ehrenberg (1795–1876), professor of medicine in Berlin. He observed that many of the marine limestone layers on land are composed of the remains of microscopic organisms. To find such organisms in the present ocean, he filtered seawater through fine gauze. He established that most of the life in the ocean is of microscopic size—a fact that now dominates our understanding of the productivity of the sea. His work apparently greatly influenced John Murray (1841–1914), the young naturalist of the *Challenger* Expedition, who subsequently pursued plankton studies with outstanding success.

In the years before the *Challenger* set out on her circumnavigation, pressing new questions had arisen after the publication of Charles Darwin's book *The Origin of Species* (1859). Debate now focused on the evolution of life, which gave meaning to taxonomy and biogeography in ways not appreciated earlier. Darwin's theory of natural selection established biology as a field ranking intellectually with the established sciences of physics and chemistry and beyond, as it impacted the search for meaning in human existence.

The *Challenger* Expedition was the first sent out to satisfy curiosity about the ocean, rather than pursuing economic or military needs.[54] The expedition was carefully planned and prepared by highly motivated scientists and was funded for postcruise studies, for sample handling, and for publication. It was to establish a firm basis for accepting or rejecting many of the tentative ideas put forward previously in the early days of ocean study. Its sampling protocol provided a template for subsequent major expeditions. One important focus was a systematic survey of deep-living organisms in the sea. Would there be an azoic zone as proposed by Forbes? Would there be "living fossils," that is, creatures hitherto only known from their remains in ancient rocks? Darwin's influence was palpable.

The HMS *Challenger* was a substantial vessel, a three-masted sailing corvette with steam engines, a 2,300-ton displacement, and a length of 200 feet; it had been built in 1858. The ship, with a complement of about 240 men, was at sea for 41 months, sailing almost 69,000 nautical miles—the equivalent of three times around the world. During these more than three years, its crew was measuring temperatures, taking water samples and bottom samples, and building up an enormous collection of the organisms of the open ocean. After the *Challenger* returned, scientists were occupied for years working up the samples and publishing the results, in 50 thick volumes. These were edited by John Murray and published from 1885 to 1895. Many of them are still regularly used and referred to in the scientific literature.

The work on the *Challenger* was hard and tedious. There were no cranes, as on today's vessels, to do the lifting of equipment. Every depth determination necessitated letting out a hemp rope and bringing it back by a man-powered winch. Lowering a dredge to capture animals or rocks from the seafloor took hours. The work came to be called "drudging" by the crew, though the dredge was retrieved by a donkey engine. But even the toiling sailors took interest in the samples coming on deck, it is reported, at least in the early phase of the expedition.

At every temperature station the great difference between surface and bottom water was confirmed. Measurements showed that chilly temperatures dominate the water column: only the uppermost layer of the ocean is warm (fig. 1.5). William Benjamin Carpenter (1813–1885), one of the organizers of the expedition, had suggested that the entire ocean might prove to be filled with frigid waters from polar areas. He was right. (It is said that the first application of the new insight about a cold ocean was the cooling of wine bottles by suspending them below the ship on sufficiently long lines. Discoveries have applications!)

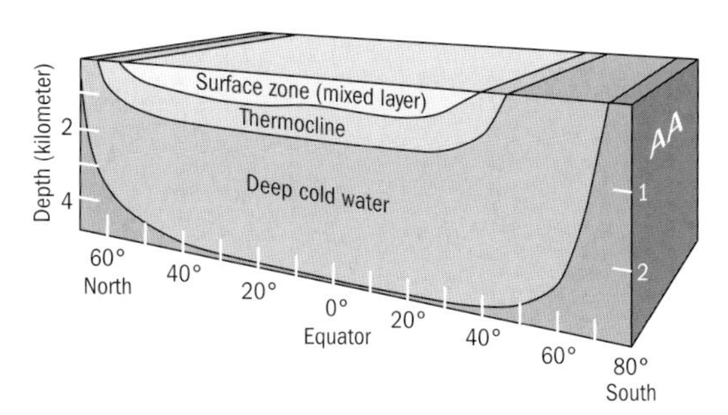

FIGURE 1.5. The ocean is filled with cold water. A thin warm layer covers it, except in high latitudes. Cold and warm water are separated by a transition zone, the "thermocline."

Patterns of seafloor topography started to emerge from the depth soundings. Away from the continents and islands, the ocean was consistently very deep, around 4,000 meters, with a common range between 3,000 meters and 5,500 meters. Interestingly, relatively shallow depths seemed to be typical for the central Atlantic. Eventually, when plotting all soundings and considering a slight difference in temperature of abyssal waters in the eastern and western Atlantic, it was realized that the central Atlantic region rises above the general depth. A deep trough exists on either side of the rise, so that there are two basins in the Atlantic, rather than one. The first step toward the discovery of the Mid-Atlantic Ridge had been taken.

The sounding lead at the end of the wire was constructed so as to retain some of the bottom mud it struck. The majority of samples consisted of the familiar vanilla-colored ooze, found in the North Atlantic by earlier expeditions. The ooze is full of the tiny shells of single-celled animals called foraminifers (fig. 1.6). Their remains were well known from the Cretaceous chalk deposits on the shores of southern England. It was thought by scientists of the time that these organisms had populated the seafloor and left their shells in the bottom sediment upon death. Could the deep-sea floor be teeming with life? Well, in fact the teeming is of a modest sort: there is not much food available in the vast deserts of the abyss. John Murray solved the puzzle. He showed, using a net for sampling surface waters, that the shells came

FIGURE 1.6. Foraminiferal shells extracted from deep-sea sediments. The size is that of sand grains. The shells within the sediment are a means by which the ocean remembers its history.

from the foraminifers in the surface waters of the ocean, where they live together with the rest of the plankton. Murray explained that the shells of the foraminifers fall to the seafloor after death of the animal within.[55] Many decades later, the shells became the chief means by which to reconstruct the history of the ocean.

Soundings from very great depths brought samples not of light gray foraminifer ooze, but of red clay, an extremely fine soil deposit with a dark reddish brown color. Where did that come from? Wind-blown dust? Fine suspended matter brought by rivers? Weathering of volcanic ashes? The scientists had a great time guessing and arguing. They could not see the particles making up the red clay, which are too small for study by a standard microscope. X-ray analysis (which allows study of clay minerals) was to come 60 years later. In fact, it turns out that all of the sources considered by the scientists at the time do contribute to red clay in different proportions, depending upon the region studied. The reddish color is owing to the presence of iron oxide, testimony to the abundance of oxygen in a cold ocean.

Within the very first months of the *Challenger* Expedition, the dredge brought up organic structures that biologist Rudolf von Willemoes-Suhm (1847–1875) (who died later in the voyage) identified as tube-building worms. The depth was almost 3,000 fathoms—well over 5,000 meters. Living creatures from the abyss! After some more such hauls Forbes's concept of an azoic zone was finished. Right there, before the very eyes of the crew (who wondered, one suspects, why the scientists became so excited about some worms) the azoic zone attained the unenviable status of one of the curious notions in the junkyard of science (fig. 1.7).

The scientists of the *Challenger* also discovered one of the deepest places in the Atlantic: the Puerto Rico Trench, and one of the deepest places in the Pacific: the Challenger Deep in the Mariana Trench, near Guam. (Serendipity works, on occasion, even at sea.) In each case, thermometers fractured at the end of the line, as pressures kept mounting to enormous values. What could possibly be the origin of these great chasms? The time for this question had not yet come. Not in their most speculative moods did the thought occur to these explorers that here the seafloor might move downward and disappear into the interior of the Earth.

The *Challenger* Expedition still ranks as the most successful venture of its kind. Acclaim was great for the returning scientists, albeit not precisely unanimous. In this case, however, any "taxpayer's concern" was unjustified. In connection with working up the *Challenger* materials, John Murray came across some phosphatic samples from Christmas Island (in the Line Islands of the central Pacific). Subsequently, the Crown annexed the phosphate-bearing island, and Murray obtained a concession to work the deposits. The taxes from this venture made up for the expedition costs. Also, the income allowed Murray to finance much additional oceanographic research on his own. (Murray's achievement remains inspiring but unfortunately unique.)

The rich harvest in terms of marine biology showed that the organisms everywhere in the sea are thoroughly modern. The scientists studying the material gathered by the *Challenger* could not know that the deep ocean used to be warm through much of geologic time and turned cold only beginning 40 million years ago. Thus, there is every reason that the deep-sea fauna should be geologically young. But these discoveries came a century later.

FIGURE 1.7. Creatures (sea cucumber, octopod) brought up by the dredge on the *Challenger*, showing there is life at great depth.

POST-*CHALLENGER* EXPEDITIONS

The glorious example of the *Challenger* Expedition soon invited replication and follow-up. Among the most important of these were Nansen's *Fram* Expedition (1893–1896), referred to earlier, and a number of expeditions directed by Albert I, Prince of Monaco, focused on marine biology. Many followed in the first half of the twentieth century, for example, the German *Meteor* Expedition (1925–1927), a series of British expeditions to Antarctica (*Discovery* I and II, 1924–1934), the Swedish *Albatross* Expedition (1947–1948), and the Danish *Galathea* Expedition (1950–1952). Scripps joined the spirit of the time by launching the MidPac Expedition to the Mid-Pacific Mountains (1950) and the Capricorn Expedition (1952–1953) to the western Pacific.

Through the *Fram* Expedition, Nansen discovered that the Arctic is a deep, wide ocean rather than a shallow sea studded with islands. Also, the two-layer structure of this ocean, with a brackish layer on top of normal seawater filling the basin, was firmly established. The expeditions of Albert I, near the end of the nineteenth century, documented the rich diversity of sea creatures within the North Atlantic, with some emphasis on the poorly known fauna below the sunlit zone.

The *Meteor* Expedition (1925–1927) was the first thoroughly modern oceanographic expedition, laid out in a grid pattern. It systematically surveyed the Atlantic Ocean in three dimensions, collecting information on physical and chemical properties, plankton and bottom life, seafloor morphology, and sedimentation. The German government financed this expedition at a time of severe economic depression. Possibly, the supersalesmanship of the physical chemist Fritz Haber played a role. (Haber was famous for showing how to make money from air, by generating nitrogen fertilizer.) He proposed that there is a lot of gold in the ocean, which was exactly what the treasury needed during the aftermath of World War I.[56]

For 27 months the *Meteor* crisscrossed the Atlantic between latitude 15° N and the ice limit to the south (64° S), logging a distance of 125,000 kilometers, about the same as the HMS *Challenger* did. A newly developed method, quasi-continuous echo-sounding, showed that the Mid-Atlantic Rise discovered by the *Challenger* is a rugged mountain system, easily exceeding the Alps in size (see chapter 11).

The measurements of temperature and salinity at depth (by Nansen Bottle) showed a complicated layered structure of the water. Closest to the bottom there is extremely cold water, coming from the Antarctic; then there is a layer of rather more saline water, not quite so cold, which comes from sources around Greenland. Other layers, from less-frigid source regions, follow upward, and finally there is a top layer of warm tropical or subtropical waters.

Measurements of the most important dissolved gases—oxygen and carbon dioxide—showed that the ocean breathes like a living organism. Oxygen enters the system at high latitudes, with sinking cold water. It is then used up by decay and respiration at depth, resulting in production of carbon dioxide. In the process, nitrate and phosphate are released to the water, from the decaying organic matter. The discoveries regarding the marine carbon cycle and the associated nutrient cycles contributed to the understanding of the productivity patterns of the sea. The central subtropical regions were recognized as deserts bearing mainly minute organisms ill suited to feed fish. Coastal oceans, in contrast, showed high nutrient contents in surface waters, supporting diatoms, the "grass of the sea." Where such "grass" is abundant, grazers can thrive and feed the larger predators including seabirds and mammals.

At the time of the *Meteor* Expedition, the British started a series of cruises into Antarctic waters (1924–1934), with *Discovery* I and *Discovery* II, to investigate physical oceanography and marine biology, largely in the context of whaling, which had become an important industry. The expeditions established the high seasonal productivity of the waters of the Antarctic and worked out the food chain of the giant baleen whales (which is based on diatoms and krill), as well as patterns of migration. These expeditions also determined the structure of the circumpolar ocean, which turned out to be highly complex. Stripped to the essentials, we may think of the Southern Ocean as a ring of water, poorly stratified, running from west to east around the Antarctic and mixing vertically right to the bottom. This ring serves as a mixing station for the waters of the World Ocean and, thereby, governs the rate of deep circulation. Different types of water enter it, and a more or less uniform mixture leaves it, at various places and depths. This mixing process keeps the deepwater masses of the several ocean basins at close to average values with regard to temperature and salinity.

The Swedish *Albatross* Expedition (1947–1948) and the Danish *Galathea* Expedition (1950–1952) were the last significant voyages of the traditional globe-encircling type. The *Albatross* logged 70,000 kilometers in 15 months. Its chief contribution was a large number of long sediment cores from the tropical zones of all ocean basins, including the Red Sea and the Mediterranean. The analysis of these cores showed clearly that the last 500,000 years were characterized by a regular succession of ice ages separated by warm periods. Our own time, for the last 10,000 years, is in one of those warm intervals.

The chief contribution of the *Galathea* Expedition was to provide the first survey toward establishing the global distribution of plankton productivity. The biologist E. Steemann Nielsen did the work, using a novel method he had invented, based on radiocarbon. He collected, in a glass container, a sample of the microscopic algae in the plankton. He then added a solution of radioactive bicarbonate. As the organisms grew, while exposed to light, they would take up some of the radioactive carbon and become radioactive themselves. By measuring their radioactivity, Nielsen could determine their rate of growth. He found that this rate varies greatly between different regions of the ocean, with upwelling areas having the highest rates, and warm tropical waters having the lowest, excepting the regions of equatorial upwelling.

Large expeditions provided for quantum steps in the progress of oceanographic science in the first half of the twentieth century—but they are not the whole story. Many institutions throughout the world carried out regional researches that provided valuable information about currents, tides, seasonal changes in productivity, and the distribution and natural history of marine organisms. In the 1930s, there were two sizeable ocean-research centers in the United States, Scripps Institution of Oceanography, in La Jolla (San Diego, California), and Woods Hole Oceanographic Institute, in Massachusetts. Woods Hole, established in 1930, owned the 142-foot ketch *Atlantis,* which explored the Gulf Stream, among other things, discovering the nature of this largest of currents in the Atlantic.

Scripps had a smaller vessel, the 64-foot fishing vessel *Scripps.* Unfortunately, it burned at the dock in November 1936, shortly after the meteorologist and oceanographer Harald U. Sverdrup arrived from Norway to lead the institution. Within a year, Scripps was mobile again, thanks to a generous gift by Robert P. Scripps, who donated a 100-foot auxiliary schooner, which was renamed *E. W. Scripps* in honor of his father.[57] While the *Atlantic* studied the Gulf Stream, the Caribbean, and the Gulf of Mexico, the *E. W. Scripps* worked the California Current, the California Borderland, and the Gulf of California.[58]

Some of the ventures of the *E. W. Scripps* may serve as an example of the type of regional work in progress at the time in many parts of the world. The first long-distance expedition mounted at Scripps was to the Gulf of California, for two months in 1939. The primary objective was to study the currents and water structure of the Gulf and the relationships with the adjoining Pacific Ocean. Water samples and bottom samples were taken at more than 50 stations. A grid of 2,500 soundings showed unexpected ridges, basins, and troughs in the southern part of the Gulf. Surface samples demonstrated that the well-known intermittent red color of this region's water (which invited the label of "Vermilion Sea") is due to certain types of plankton. The waters were rich in organisms as a result of upwelling.

Sediments retrieved from below upwelling areas smelled of hydrogen sulfide. It turned out that the high supply of organic matter to the seafloor overwhelms the oxygen supply, stripping the bottom water of oxygen and making it anaerobic. At that point, bacterial decay proceeds by using oxygen from sulfate to burn the carbon of the organic matter. One result of tearing the sulfate apart is the production of poisonous hydrogen sulfide. The finely laminated shale in marine deposits of hundreds of millions of years of geologic history now had a simple explanation: accumulation in anaerobic conditions, with hydrogen sulfide keeping out burrowing organisms. The mother rocks for petroleum typically consist of such shale.

Scripps geologist Roger R. Revelle was especially interested in these processes and obtained funds from the Geological Society of America to carry out additional work in the area, in a cruise lasting 78 days in the late fall of 1940, on the same ship. The geologist Francis P. Shepard joined him on that cruise, continuing his studies on submarine canyons (which he had begun along the California coast), and bringing along his graduate students Robert S. Dietz and Kenneth O. Emery. The proceedings for this cruise were published after the war (in 1950).[59]

The work by individual institutions—as opposed to national expeditions—became ever more important as a result of World War II. The emphasis shifted from large-scale exploration to systematic surveying, from mapping of distributions to studying processes. Oceanographers were sought out for their expertise, and oceanography became a profession. When German submarines threatened the support lines of the Allies across the Atlantic, the government was forced to develop means to combat this threat. The physics of the ocean suddenly became a prime topic of study, notably the physics of sound transmission.

Waves, and especially the breaking of waves along shores, proved to be important in planning the landing of assault troops. The success of the invasions of Sicily and Normandy by the Allies owed much to the wave forecasts by professional oceanographers back in California. Right after the war, the atomic tests on Bikini Atoll were monitored (at some risk) by a large contingent of oceanographers. Effects of explosions on waves and on the regional ecology, and the dispersion of radioactive materials were the subjects of study.

Following this period of military engagement, the U.S. Navy remembered how useful the ocean sciences had been and remained in partnership with the oceanographic institutions, supporting their work. Several new institutions were established. The war also had brought a quantum jump in instrumentation, especially electronics. Thus, in the 1950s, sophisticated survey equipment came into use, such as seismic profiler systems that send sound to below the seafloor and record the echoes, and towed magnetometers that detect the slightest change in the strength of the magnetic field. The measurements made with such instruments led to a major revolution in the Earth sciences in the 1960s. Deep-sea drilling and satellite oceanography provide additional examples of the continuing advances in technology that now shape the course of the ocean sciences.

Concern about the economics of fisheries and, more recently, about pollution, as well as a strong and growing interest in weather and climate prediction, have contributed to the rapid expansion of oceanographic facilities in the last several decades. These concerns motivate many of the large interinstitutional and international projects that are at the hub of oceanographic activities today. In fact, the problems arising in the context of global warming require integration not just across institutes and nations, but across disciplines, notably climatology, biogeochemistry, atmospheric chemistry, bacteriology, all aspects of ecology, and geologic history. Oceanography, along with the other Earth sciences, is converging toward Earth system science. It may seem like a long voyage, but it took only a little over a century, from the spotty measurements of the *Challenger* to the satellite surveys of today.

SCRIPPS: EVOLUTION OF A MARINE RESEARCH CENTER

The history of Scripps Institution of Oceanography nicely reflects the general major trends in oceanographic research in the past 100 years. Research proceeded from marine biological work concerned with the study of nearshore organisms, to worldwide operations investigating the geophysics of the seafloor and the life-support systems of the planet.

Scripps Institution of Oceanography is located some 10 miles north of the Navy harbor of San Diego, where its research vessels are berthed when in port. The campus enjoys a pleasant setting just north of the palm-studded seashore village of La Jolla. Scripps vies with Woods Hole Oceanographic Institution for the status of most distinguished oceanic research center in the world. Its research, like that of its sister institute on the East Coast, is directed mainly toward the deep sea and global problems, without, however, neglecting the coastal ocean and the seashore. Scripps is part of the University of California at San Diego, one of the top research universities in the nation, founded in the early 1960s.

Researchers and students at Scripps (more than a thousand scientists) investigate a broad sweep of problems in the ocean sciences and ocean engineering, as well as in Earth sciences and atmospheric sciences. Scripps's world-renowned library holds an enormous store of knowledge about the ocean and the planet in general. Much of the more recently acquired knowledge is too vast to be printed: it is stored electronically for downloading into the computers of specialists using the data.

The institution started as a marine biology station for summer fieldwork, with the modest goal of making a biological survey of the coast of California. More ambitious goals soon involved the California Current, the Gulf of California, and finally the entire Pacific basin, the Indian Ocean, and the World Ocean. Today, ocean studies are but one part of global studies on Earth systems, which include the seas, the air, the land, and the great ice fields at the poles. These days, Scripps scientists (and their colleagues at similar institutions) are members of national and international projects involving deep-sea circulation, ocean-air interaction, satellite-based measurements of gravity and productivity, as well as all aspects of the carbon cycle on Earth.

More than a century ago, in 1903, Berkeley's zoology professor William Emerson Ritter (1856–1944) established the Marine Biological Association of San Diego, with the help of newspaper magnate E. W. Scripps and his sister Ellen Browning Scripps, and physicians Fred and Charlotte Baker, as well as other prominent citizens of San Diego.[60] Ritter, eager to start exploring the sea within the constraints of available resources, put forward a vision to make a biological survey of the Pacific Ocean adjacent to the coast of California.

The investigation of the coastal zone and its organisms, of course, had been the first step in ocean exploration for a number of similar stations on the shores of Europe. E. W. Scripps found cheap land for the necessary facilities north of La Jolla, about an hour's walk from the village. By 1910 the marine terrace north of La Jolla bore a two-story concrete laboratory, and a graded road provided access. Ellen Browning Scripps (1836–1932) promised a generous endowment.[61] When the University of California accepted the property in 1912, the regents acknowledged the primary donors to the laboratory with the name "Scripps Institution for Biological Research.[62]

By then, in the San Diego region, Ritter and his colleagues had collected 862 species of marine organisms and described 328 of them as new. Aware of the vastness of the ocean and the modest means of his fledgling institution, Ritter wrote wistfully about the infinite complexity of cause and law in the sea. Ritter was a great admirer of Darwin and diligently pursued the implications of adaptation through evolution, emphasizing the importance of habitat and environment in attempting to understand organisms.[63] His specific research interests centered on tunicates and acorn worms. Ritter took evolution to be the great unifying theory for all manifestations of life, including human behavior. It is perhaps not by chance that Ritter's favorite objects of study, the acorn worms, are among the most primitive of humankind's own phylum, the chordates.[64]

The distinguished paleontologist Thomas Wayland Vaughan (1870–1952) succeeded Ritter as director of Scripps, in 1924. Vaughan pushed to have Scripps take a major role in the scientific exploration of the Pacific Ocean, partly by urging scientists to hitch rides on ships of federal agencies. In 1925, Vaughan purchased a 64-foot fishing boat and named it *Scripps*. She was a small vessel to face so large an ocean but was very useful for coastal work.[65]

In 1936, Harald Ulrik Sverdrup (1888–1957), who was to become the most prominent ocean scientist of his generation, became the third director of Scripps. He carried the institution through the war years and beyond until 1948. Sverdrup's unique background in meteorology and oceanography enabled him to tackle large-scale problems, and in several disciplines. His many years of experience in the Arctic had taught him how to make measurements in the most difficult of circumstances. As a Norwegian, naturally, he was very familiar with fishing and fisheries. His tenure at Scripps set the institution on a new course of integration of physics and biology, and on thinking big. Sverdrup had a low opinion of the research vessel acquired by Vaughan; he wanted a boat that could venture farther out to sea. As noted above, he soon obtained the 104-foot *E. W. Scripps* and thus could lead exploration well offshore.[66]

When, in 1938, the U.S. Bureau of Fisheries proposed a project to explore the spawning activity of the sardine, Sverdrup readily accepted the challenge. The Pacific sardine is a smallish herringlike fish less than 12 inches long, with a glittering dark green to blue back and an iridescent silvery belly. It travels in large schools not far from shore. The sardine fishery had expanded considerably in previous decades and was producing fishmeal and fertilizer at a profitable rate. In the 1936/37 fishing season 726,000 tons of sardines were hauled into California harbors. It was the most productive season on record, about one-quarter of the U.S. tonnage of fish landings. Some fisheries scientists were concerned that overfishing might soon drive down sardine abundances.[67] This was the situation that called for research on sardine reproduction patterns, and systematic studies of the California Current system were initiated accordingly.

From 1938 to mid-1941, Scripps scientists monitored the Current, repeatedly occupying a grid of 40 stations in a regular fashion. A full integration of physical, chemical, and biological studies was now accomplished—the very task that Sverdrup had set himself on arriving at Scripps. Sverdrup soon longed to go back to Norway, as did his wife Gudrun. However, the invasion of Norway by German troops in 1940 closed this option, and the Sverdrups stayed on for the remainder of the war, and three more years after that.

Sverdrup's masterful opus, *The Oceans*, coauthored with the biologist Martin W. Johnson and the chemist Richard H. Fleming, was completed during the war.[68] It was started in 1938, and it was the first book to offer a comprehensive view of Pacific oceanography. It aided the U.S. Navy in the war effort and was restricted in its distribution, initially. The hefty volume (more than 1,000 pages) defined the field of oceanography for the following decades, till about 1970, before the arrival of satellites, deep-ocean drilling, data processing, and ecosystem modeling. It is still useful, after more than half a century. When oceanography became a popular subject of instruction at the college level, in the 1970s, many courses (and textbooks written for these courses) owed their contents and their organization to that book. Most oceanographers keep a copy of it handy.[69]

Thus, the transition from marine biology and fisheries science to oceanography was accomplished. After the war, the navy took a strong interest in all aspects of oceanography, giving massive support to the expansion of all sorts of ocean research, with special emphasis in those areas related to its mission. In addition, the survey of the California Current resumed, under the auspices of the state, initiating the longest-running detailed documentation of the history of any such large region in the world, with most interesting results.

Led by Roger Randall Revelle (1909–1991), Scripps's fifth director (from 1950 to 1964), Scripps participated fully in the navy-funded postwar expansion, with several ocean-going vessels, contributing significantly to the new global view of the ocean emerging in the postwar years. Marine geology, the stepchild in prewar oceanography, benefited enormously and ended up creating a major revolution in its mother science, as a result of intense global exploration of the seafloor in the 1950s and 1960s. The years that followed are marked by the growth of climate sciences, and of all fields connected to human impact on the ocean's life-support systems—physical, chemical, and biological.

NOTES AND REFERENCES

1. P. Ward, R. A. Myers, 2005, *Shifts in open-ocean fish communities coinciding with the commencement of commercial fishing.* Ecology 86, 835–847; B. Worm, M. Sandow, A. Oschlies, H. K. Lotze, R. A. Myers, 2005, *Global pattern of predator diversity in the open oceans.* Science 309, 1365–1369.

2. The trend has a name: fishing down marine food webs. D. Pauly, V. Christensen, J. Dalsgaard, R. Froese, F. Torres, 1998, *Fishing down marine food webs.* Science 279 (5352), 860–663.

3. *Encyclopaedia Britannica*, 1974, *Macropaedia* 19, Whaling.

4. J. R. Spears, 1908, *The Story of the New England Whalers.* Macmillan, New York, 418 pp.; C. W. Ashley, 1942, *The Yankee Whaler.* Halcyon House, Garden City, N.Y., 156 pp.; J. T. Jenkins, 1921, *A History of the Whale Fisheries.* Whitherby, London, 336 pp.; E. K. Chatterton, 1926, *Whales and Whaling.* William Farguer Payson, New York, 248 pp.; P. Budker, 1959, *Whales and Whaling.* Macmillan, New York, 182 pp. For a lively account of the whale hunt in the nineteenth century, see H. Melville, 1851, *Moby-Dick.* Harper and Brothers, New York; and N. Philbrick, 2000, *In the Heart of the Sea: The Tragedy of the Whaleship Essex.* Penguin Putnam, New York, 302 pp. An early example of the modern scientific approach to the natural history of whales is provided by K. S. Norris (ed.), 1966, *Whales, Dolphins, and Porpoises.* University of California Press, Berkeley, 789 pp. A scholarly and poetic description of life in the sea from the point of view of a sperm whale is presented by V. B. Scheffer, 1969, *The Year of the Whale.* Charles Scribner's Sons, New York.

5. E. S. Russell, 1942, *The Overfishing Problem.* Cambridge University Press, London, 130 pp.

6. Quoted in V. B. Scheffer, *The Year of the Whale,* p. 125.

7. D. Pauly, J. Maclean, 2003, *In a Perfect Ocean: The State of Fisheries and Ecosystems in the North Atlantic Ocean.* Island Press, Washington, D.C., 175 pp.

8. Another is to pronounce *Newfoundland* correctly, with the stress on the last syllable.

9. At the very end of the fifteenth century, John Cabot (born Giovanni Caboto, of Venice) found the Grand Banks, sailing from England to Newfoundland (or thereabouts). After Cabot, the English no longer had to fish off Iceland and argue with the powerful Hanseatic League, a trust of cities centered in northern Germany that had the annoying

habit of laying claim to good fishing grounds everywhere in the eastern North Atlantic. Cabot discovered the fishing grounds for England, but apparently Basque fishermen had been working the area, and this is how Cabot learned about fishing with weighted baskets (*fide* Fagan, 2000, p. 77). In any case, the Basques had been selling salt cod for years, but they chose not to tell anyone where they fished (Kurlansky, 1997, p. 29). Brian Fagan, 2000, *The Little Ice Age.* Basic Books, New York, 246 pp.; Mark Kurlansky, 1997, *Cod, A Biography of the Fish That Changed the World.* Penguin Books, New York, 294 pp.

10. The first effect of heavy fishing pressure is to decrease the typical size of adult fish. For example, concerning the smallish Atlantic cod from kelp forests in the coastal Gulf of Maine, we know from the leftover cod bones in Indian middens that the typical length in pre-European times was around 3 feet or more. In contrast, for the last 50 years or so a 1-foot-long fish is a perfectly respectable catch. J. B. C. Jackson, M. X. Kirby, W. H. Berger, K. A. Bjorndal, L. W. Botsford, B. J. Bourque, R. Bradbury, R. Cooke, J. Erlandson, J. A. Estes, T. P. Hughes, S. Kidwell, C. B. Lange, H. S. Lenihan, J. M. Pandolfi, C. H. Peterson, R. S. Steneck, M. J. Tegner, R. R. Warner, 2001, *Historical overfishing and the recent collapse of coastal ecosystems.* Science 293, 629–638.

11. The word *fish* is used both for singular and plural in common English. The context prescribes when to use *fishes.* Fishermen catch fish, while scientists study fishes.

12. Food and Agriculture Organization statistics, summarized in P. Weber, 1994, *Fish, Jobs, and the Marine Environment.* Worldwatch Institute, Washington, D.C., 76 pp.

13. M. Kurlansky, 1997, *Cod.* Penguin Books, New York, 294 pp.

14. D. Pauly, J. MacLean, *In a Perfect Ocean.*

15. C. S. Woodard, 2000, *Ocean's End: Travels through Endangered Seas.* Basic Books, New York, 300 pp.

16. B. Fagan, *The Little Ice Age.*

17. P. D. Jones, K. R. Briffa, T. P. Barnett, S. F. B. Tett, 1998, *High-resolution palaeoclimatic records for the last millennium: Interpretation, integration and comparison with general circulation model control-run temperatures.* The Holocene 8, 455–471.

18. N. C. Stenseth, A. Mysterud, G. Ottersen, J. W. Hurrell, K.-S. Chan, M. Lima, 2002, *Ecological effects of climate fluctuations.* Science 297, 1292–1296. K. T. Frank, R. I. Perry, K. F. Drinkwater, 1990, *The predicted response of northwest Atlantic invertebrate and fish stocks to CO_2-induced climate change.* Transactions of the American Fisheries Society 119, 353–365.

19. D. Pauly, 2003, presentation at the Scripps Symposium for Marine Biodiversity, November 2003.

20. It is the same on land: rabbit and deer feed low on the food chain and are correspondingly more abundant than foxes, wolves, and mountain lions.

21. Assuming Keynesian inflation, this is between $100 and $200 million in today's money.

22. F. S. Russell, C. M. Yonge, 1936, *The Seas: Our Knowledge of Life in the Sea and How It Is Gained.* Frederick Warne, London, 379 pp. F. S. Russell was a leading figure in marine biology in the first half of the twentieth century, and the director of the Plymouth Laboratory of the Marine Biological Association, the world-renowned center for sea research in Scotland. C. M. Yonge was a distinguished British marine zoologist and ecologist.

23. Purse-seining is a fishing method whereby the vessel sets a long and deep ribbon of net around a school of fish, encircling the prey. When the circle is complete, the bottom is pulled shut by shortening the string at the lower limit of the net, as when closing a purse.

24. Fish aggregate in schools, presumably to avoid or confuse natural predators. They have no defense against purse-seining because this type of attack is not part of their evolutionary history.

25. John Steele was director of Woods Hole Oceanographic Institution from 1977 to 1989.

26. R. Bailey, J. Steele, North Sea Herring Fluctuations: in M. H. Glantz, L. E. Feingold (eds.), 1990, *Climate Variability, Climate Change and Fisheries Project.* National Center for Atmospheric Research, Boulder, Colo., pp. 213–220.

27. *Ecologic extinction* describes the loss of an important role of a given species in structuring regional ecologic systems. For example, many of the great baleen whales are ecologically extinct. So are groupers in the kelp offshore of Scripps. In contrast, the great auk (a flightless bird of the northern North Atlantic) is terminally extinct, from overexploitation. Of course, it is ecologically extinct as well.

28. The initial discovery of the hot vents was by geologists and geochemists wondering about the interaction of seawater with hot basalt. They wanted to know why seawater has the chemistry it does. Biologists soon took over with a more spectacular story. See C. L. van Dover, 2000, *The Ecology of Deep-Sea Hydrothermal Vents.* Princeton University Press, Princeton, N.J., 424 pp.

29. T. Platt, W. K. W. Li (eds.), 1986, *Photosynthetic picoplankton.* Canadian Bulletin of Fisheries and Aquatic Sciences 214, 1–583; S. W. Chisholm, R. J. Olson, E. R. Zettler, J. Waterbury, R. Goericke, N. Welschmeyer, 1988, *A novel free-living prochlorophyte*

abundant in the oceanic euphotic zone. Nature 334, 340–343. Much of the carbon fixation in the sea is by the newly discovered chlorophyll-bearing prokaryote *Prochlorococcus:* K. Suzuki, N. Handa, H. Kiyosawa, J. Ishizaka, 1995, *Distribution of the prochlorophyte Prochlorococcus in the central Pacific Ocean as measured by HPLC.* Limnology and Oceanography 40 (5), 983–989.

30. J.H. Martin, S.E. Fitzwater, 1988, *Iron deficiency limits phytoplankton growth in north-east Pacific subarctic.* Nature 331, 341.

31. R.M. May (ed.), 1984, *Exploitation of Marine Communities.* Dahlem Konferenzen, Springer-Verlag, Berlin, 366 pp.

32. A striking example of a large-scale experiment of this sort is the introduction of the ctenophore *Mnemiopsis leidyi* into the Black Sea, from ballast water of cargo ships. The common name for ctenophores is gooseberry jellyfish. They are small and feed on minute plankton. The invasive predator, lacking predators of its own, became extremely abundant. By eating the food for fish larvae, and the fish eggs and larvae too, it destroyed important fisheries in the region. The appearance of another ctenophore, a species of *Beroë* that preys on smaller ctenophores, moved the system toward recovery in the mid-1990s. J. Travis, 1993, *Invader threatens Black, Azov seas.* Science 262, 1366–1367; A. E. Kideys, 2002, *Fall and rise of the Black Sea ecosystem.* Science 297, 1482–1483.

33. The phrase "great geophysical experiment" is ascribed to Roger Randall Revelle (1909–1991), director of Scripps from 1951 to 1964.

34. R.A. Kerr, 2002, *A warmer Arctic means change for all.* Science 297, 1490–1492. The same issue of *Science* has several articles concerning climate change in polar regions. On NASA's Web site *Visible Earth,* there are images relating to the changing Arctic sea ice cover (*30 years of Arctic warming;* last updated 26 Feb 2005). While temperature changes are less drastic in low latitudes, warming here may result in widespread coral bleaching events. C. Wilkinson, 2001, *Status of coral reefs of the world: 2000.* Web site for Global Coral Reef Monitoring Network, Australian Institute of Marine Science.

35. Proposals for policy implications and stewardship are made in Pew Oceans Commission, 2003, *America's Living Oceans: Charting a Course for Sea Change.* Pew Oceans Commission, Arlington, Va. (summary and full report on CD-Rom).

36. P.K. Dayton, M.J. Tegner, P.B. Edwards, K.I. Riser, 1998, *Sliding baselines, ghosts, and reduced expectations in kelp forest communities.* Ecological Applications 8 (2), 309–322.

37. The Devonian period lasted from 410 to 360 million years ago. See the geologic time scale in appendix 4.

38. J.B.C. Jackson et al., *Historical overfishing.*

39. M. Orbach, 2003, presentation at the Scripps Symposium for Marine Biodiversity, November 2003.

40. G. Hardin, 1968, *The tragedy of the commons.* Science 162, 1243–1248.

41. The well-known difficulties in separating natural background variations in climate from human impact based on the release of greenhouse gases from automobiles and power plants has given rise to intense political discussions (sometimes dressed up as scientific debates by special interests, and in the media).

42. *The Encyclopedia of Oceanography,* edited by R.W. Fairbridge (Reinhold, New York, 1966), has no entry for "pollution" but makes reference to human impact in the essay entitled *Carbon dioxide cycle in the sea and atmosphere,* by Taro Takahashi. Takahashi cites works by G.N. Plass (1956) and R. Revelle and H. Suess (1957) and refers to studies by C.D. Keeling, among others. He writes (p. 171): "A change in the CO_2 concentration in the atmosphere could appreciably affect the thermal budget of the surface of the Earth, and might cause a long-term change in the weather and climate due to the greenhouse effect." By the 1970s, Scripps geochemist E.D. Goldberg was treating human input to biogeochemical cycles as one more factor in the cycling of elements on the planet. E.D. Goldberg, 1975, *Marine pollution,* in J.P. Riley, G. Skirrow (eds.), *Chemical Oceanography,* 2nd ed., vol. 3. Academic Press, London, pp. 39–89.

43. J.W. Hurrell, Y. Kushnir, G. Ottersen, M. Visbeck (eds.), 2003, *The North Atlantic Oscillation: Climatic significance and environmental impact.* Geophysical Monograph 134, 1–279; H.F. Diaz, V. Markgraf, 1992, *El Niño: Historical and Paleoclimatic Aspects of the Southern Oscillation.* Cambridge University Press, Cambridge, U.K., 476 pp.

44. The lack of such knowledge is painfully evident in many editorials criticizing the results of climate computations, even in some prominent newspapers.

45. There are indications that Chinese seafarers reached the Americas sometime in the early fifteenth century, well before the period of global European discovery associated with the names of Columbus, Magellan, and da Gama. Gavin Menzies, 2002, *1421, the Year China Discovered America.* HarperCollins, New York, 552 pp.; D.J. Boorstin, 1983, *The Discoverers.* Random House, New York, 745 pp. Clearly, if this is true, Chinese rulers were aware of the World Ocean long before Magellan crossed the Pacific.

46. Cristoforo Colombo (1451–1506), leader of the expedition to the West Indies, discovered the trade

wind route across the Atlantic, about 20° north of the equator.

47. Vasco da Gama (ca. 1460–1524), seafarer, trader, and buccaneer, opened the way from Portugal to India, for the spice trade, sailing around the Cape of Africa, later called the Cape of Good Hope.

48. Ferdinand Magellan (1480?–1521), supported by King Carlos I of Spain, discovered a passage from the Atlantic to the Pacific, through Tierra del Fuego, at 52° S. He sailed across the Pacific to the Philippines, where he was slain by natives.

49. Claudius Ptolemaius (second century AD), Greek astronomer and geographer, described the motions of the Sun, Moon, and planets and created a global map reflecting the limited knowledge of his time.

50. James Cook (1728–1779), captain in the British navy, explored the boundaries of the Pacific and charted the islands within it, including Hawaii (where he was killed by natives). He was one of the great explorers of all of history.

51. The present distinguished holder of the position of Astronomer Royal, Sir Martin Rees thinks deeply about the life history of the universe, past and future. M. Rees, 2001, *Our Cosmic Habitat.* Princeton University Press, Princeton, N.J., 205 pp.

52. Fridtjof Nansen (1861–1930) established that the Arctic Sea is a deep ocean basin. He discovered that ice drifts to the right of the wind, an observation that was crucial in the development of theories linking currents to wind.

53. N. Philbrick, 2003, *Sea of Glory: America's Voyage of Discovery—The U.S. Exploring Expedition, 1838–1842.* Viking, New York, 452 pp.

54. Edmond Halley's 1698 *Paramore* Expedition, sponsored by the Royal Society to take observations of magnetic declination, is sometimes accorded the honor of the first sea journey undertaken for a purely scientific object. It is well to remember that ships were steered by compass, and thus navigation depended on knowledge of declination. In consequence, research on magnetism had immediate use in a military and commercial context.

55. In actuality, most of the shells apparently derive from individuals that disappeared into their offspring, by dividing into hundreds of new cells or gametes and leaving an empty house.

56. There is indeed gold in seawater—there is a bit of everything dissolved in it—but the concentration is very low, and hence it would be very difficult to extract.

57. E. N. Shor, 1978, *Scripps Institution of Oceanography: Probing the Oceans, 1936 to 1976.* Tofua Press, San Diego, 502 pp.

58. There are many outstanding oceanographic institutions with a long history. A survey is in E. M. Borgese (ed.), 1992, *Ocean Frontiers: Explorations by Oceanographers on Five Continents.* Abrams, New York, 288 pp. Scripps Institution of Oceanography is represented with an article by its former director, the late Roger Revelle.

59. C. A. Anderson, J. W. Durham, F. P. Shepard, M. L. Natland, R. R. Revelle, 1950, *1940* E. W. Scripps *Cruise to the Gulf of California.* Memoir 43, Geological Society of America, New York, 398 pp.

60. Edward Willis Scripps (1854–1926), owner of many so-called penny newspapers in the United States , designed to bring news to the general public, was also a yachtsman who sought relaxation at sea. He befriended marine biologist William E. Ritter and helped plan the founding of the Marine Biological Association of San Diego (the forerunner of Scripps Institution of Oceanography) in the early 1900s.

61. Ellen Browning Scripps (1836–1932), a major benefactor in the San Diego area and beyond (hospital, college, science, and other projects), she played a leading role as a sponsor of the newly founded Marine Biological Association of San Diego. E. W. Scripps was her younger half-brother; she had written a column for his newspaper chain for some years before she retired to La Jolla.

62. The early years of the laboratory and its transformation into an ocean-going institution are summarized in H. Raitt, B. Moulton, 1967, *Scripps Institution of Oceanography: First Fifty Years.* Ward Ritchie Press, Los Angeles, 217 pp.

63. D. Day, 2002, *Scripps benefactions: The role of the Scripps family in the founding of the Scripps Institution of Oceanography,* in K. R. Benson, P. F. Rehbock (eds.), *Oceanographic History: The Pacific and Beyond.* University of Washington Press, Seattle, pp. 2–6.

64. Ritter was director from 1903 to 1923. Ritter's holistic approach to biology did not readily provide for testable hypotheses. At least one historian claimed that it did not produce a clear research focus—neither for "regular" marine biology, nor for oceanography. K. R. Benson, 2002, *Marine biology or oceanography: Early American developments in marine science on the West Coast,* in K. R. Benson, P. F. Rehbock (eds.), *Oceanographic History: The Pacific and Beyond.* University of Washington Press, Seattle, p. 301.

65. T. W. Vaughan, the second director of SIO (1924–1936), was a paleontologist and a recognized authority on corals and coral reefs.

66. H. U. Sverdrup, the third director of SIO (1936–1948), was the lead author on the book that defined the scope of oceanography for the three decades that followed.

67. W. L. Scofield, F. N. Clark of the California Department of Fish and Game; *fide* J. Radovich, 1982, *The collapse of the California sardine fishery. What have we learned?* California Cooperative Oceanic Fisheries Investigations Reports 23, 56–78.

68. H. U. Sverdrup, M. W. Johnson, R. H. Fleming, 1942, *The Oceans, Their Physics, Chemistry, and General Biology*. Prentice-Hall, Englewood Cliffs, N.J., 1087 pp.

69. No survey data are available, but a cursory inspection of bookshelves of colleagues at Scripps suggests that this is true at least for gray-haired oceanographers.

TWO

A Portrait of the Ocean Planet

ELEMENTS OF OCEAN LITERACY

Earth literacy has at least one big benefit: a sense of delight at living on this very special planet. Also, clearly, Earth literacy is useful when trying to follow discussions about human impacts on the planet and what to do about it.[1] Knowledge of the ocean is crucial to Earth literacy. Ocean literacy starts with the knowledge that the ocean covers 70 percent of the planet's surface and is typically 2.5 miles (4 kilometers) deep, with the bulk of the water being very cold, and in the dark. It includes knowing why fishing is good in some parts of the sea and bad in others.

The ocean contains over 99 percent of life's habitat space. On a high taxonomic level (class to phylum), animal diversity in the sea greatly exceeds that on land. In other words, there are many more fundamentally different animals in the sea than on land. As the main reservoir for water, the ocean is at the center of all water-governed processes. Water is the lifeblood of the planet; it is the main limiting factor for the growth of land plants and all the organisms dependent on them. Water vapor in the air powers the climate processes that govern our environment. Frozen water in high latitudes and in the high mountains helps maintain the temperature gradients responsible for the wind patterns and for the geographic patterns of living things. Our home planet is an oasis for life in space because of the abundance of water.

The ocean basins are made of basaltic rock that is geologically young, while the continents are made of granitic rocks that include extremely ancient types.

A SKETCH OF MAJOR FEATURES

The ocean planet, our home, is the largest of the four inner planets of the solar system. It was born some 4.6 billion years ago, coalescing from rocks and dust that were traveling around the Sun in a disk. It is not known when or how Earth acquired its moon. What is known is that the Moon keeps the spin axis of the mother

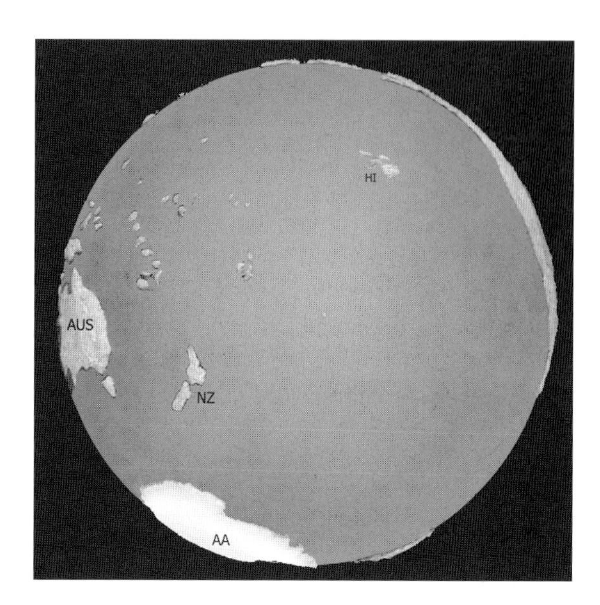

FIGURE 2.1. The enormous size of the Pacific and the fact that planet Earth is really planet Ocean emerges when centering the view from space on the islands of Hawaii in the Pacific.

planet reasonably close to vertical to the orbital plane, which helps constrain the range of seasonal variations. Water has been present for more than 3.5 billion years. Earth was able to retain it because of its strong gravity, and because its distance from the central star is such that the water could make an ocean layer as it emerged from the slowly churning mantle or arrived from space in icy comets. Much closer to the Sun, and the water would have been boiled away. Much farther and it would all be ice, covered with volcanic debris.

Basalt, the rock that makes the ocean basins and underlies the seafloor, is the main rock type of the planet; all of the oceanic islands in the Pacific are made of it, through volcanism. In Hawaii different types of basaltic rocks can be readily inspected, on any mountain path and all along the beaches. On the Big Island, we can see fresh black lava, glistening in the sun, and we can smell the sulfurous gases emanating from vents not far from the visitor center in Hawaii Volcanoes National Park. The message is clear: the Earth is hot and alive deep inside. And here, in the middle of the enormous Pacific Ocean (fig. 2.1), a hot basaltic message arrives at the surface creating the Hawaiian Islands.

New basaltic seafloor is being made all the time, through seafloor spreading (fig. 2.2, upper panel). The longest mountain range on Earth, the Mid-Ocean Ridge, is made of geologically young basalt.[2] It marks the place where new and hot material comes up from the mantle and makes new ocean crust.

Seafloor spreading determines the bathymetry of the ocean floor. The Mid-Ocean Ridge (with the Mid-Atlantic Ridge and the East Pacific Rise as the most prominent sections) spans Earth like the broken seams on a bursting baseball. New seafloor is generated here by basalt derived from rising mantle material. As it moves away from its line of origin, the seafloor cools and sinks to greater depth.

If the mantle were to make new seafloor through geologic time, and do nothing else, the Earth would have to expand to accommodate the growing ocean basins.[3] However, this is not the case. An equal amount of seafloor is destroyed, mainly in the subduction zones rimming the Pacific basin (fig. 2.2, lower panel). Here, the seafloor dives back down into the mantle, rumbling with earthquakes large and small as it does so, and generating a deep trench at the line of collision with the opposing island chain (e.g., Japan) or continental margin (e.g., Chile).

Importantly, the collisions result in making new continental rock, within an enormous mixing machine driven by volcanic processes.

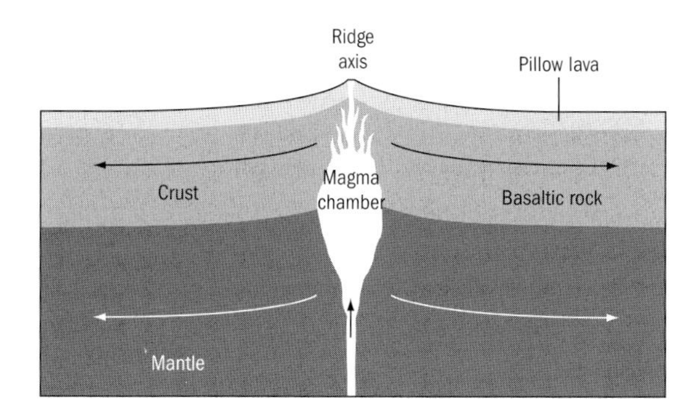

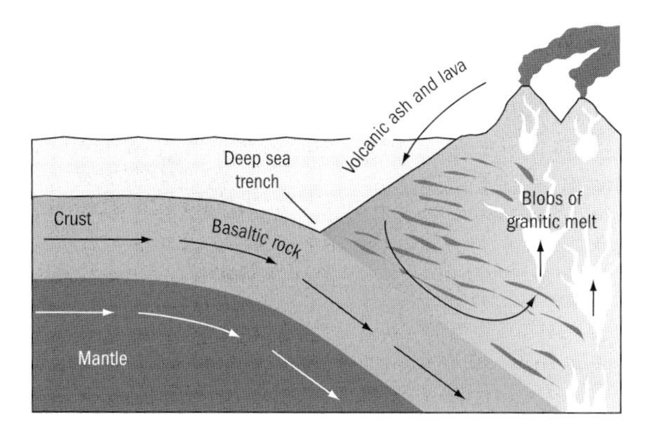

FIGURE 2.2. Seafloor is being made and destroyed all the time: at the Mid-Ocean Ridge, there is seafloor spreading *(upper graph)*, and in the zones marked by deep-sea trenches, there is subduction and associated mountain building *(lower graph)*.

Prominent results of such processes are seen in the high mountain chains near the western edges of North and South America. The mountains in the south, the Andes, have given their name to the type of material that grows new continent: andesitic rock. Its chemistry is intermediate between basalt and granite. Much of the granite ultimately derives from boiling off the more soluble elements in the basaltic oceanic crust, and sending the solutions into the mountain roots before the oceanic crust dives back into the mantle. The relatively more soluble materials are silicon oxide and alkalic elements (sodium, potassium, calcium), and these are consequently enriched in andesite (and in granite) relative to basalt.

The coast of California is a product of such collision between seafloor and continent. At its shores we can see both volcanic rocks and granitic rocks pounded by the waves of the Pacific (fig. 2.3). These rocks were produced when there was a trench offshore and subduction was going on. Things have quieted down for some time; the collision zone is now confined to Oregon and Washington. The granitic mountain roots that formed landward of the ancient collision zone have had time to rise, buoyed by their lighter weight relative to basaltic material. They kept rising until increasing erosion checked their skyward progress. They are the High Sierras, and snow now covers their most lofty peaks, which is why we call them Sierra Nevada, the Snowy Mountain Range. The snow covers rocks that originated many miles down in the heat of the collision of seafloor and continent.

With this very brief background on the effects of mantle convection at the surface of the planet, we now understand why water and land are segregated so well, that is, why we do not have an ocean planet with nothing but islands. Seafloor spreading produces ocean basins, and seafloor subduction keeps the continents piling up in a few places. What we are looking at when contemplating the global patterns of ocean basins and continents is the scale of the mantle convection, which is imprinted on the planet's surface. That scale is huge, as huge as the Pacific itself.[4]

FIGURE 2.3. The cliffs of Big Sur at the shores of central California expose a mélange of very different types of rocks, both black oceanic basalt and light-colored continental material. The mixture (made in the subduction zone) is shot through with the products of hydrothermal alteration (from reaction of the rocks with hot water). This is how continents are built.

The composition of seawater is another long-standing puzzle that has found its solution thanks to the new understanding based on seafloor spreading and seafloor subduction worked out in the 1960s and 1970s. New seafloor is hot. It cools while interacting with seawater. The interaction between basalt and seawater helps determine the composition of the salt dissolved in the seawater. How this composition is maintained through geologic time was one of the unsolved problems in marine geochemistry even into the 1970s. Only with the discovery of the hot vents in the eastern Pacific did scientists realize that basalt-seawater interaction must be considered a dominant factor in keeping the chemistry of seawater near its present condition.

SOME LESSONS FROM THE SEASHORE

Clearly, "ocean literacy" for the general public cannot encompass more than a rather small part of the body of knowledge that makes up the program of modern ocean science.[5] We must choose. Prominence must be given to items that are readily accessible to direct observation and that notably affect human health and activities. Thus, the natural history of the coastal ocean may best serve as the door to ocean literacy. The coastal ocean is in plain view and accessible for direct experience and observation, and coastal processes affect people everywhere. We could hardly do better than to start our quest for ocean literacy by going to the edge of the sea to walk along a beach, in discourse with the waves that break in the surf zone, and with the shore life all around us.[6]

When walking along the edge of the sea, most people will note these two things: the arrival of waves, and the activities of shorebirds looking for food. Waves are part of the physical setting of the ecologic theater; the birds are actors in the ecologic play.[7]

Waves are generated in storms and transmit some of the energy of these storms over large distances. They do so quietly till they break. At that point, they noisily dissipate the energy, setting up a complicated pattern of nearshore circulation (with dangerous consequences for inexperienced swimmers). The waves are bigger

FIGURE 2.4. Waves breaking at the shore in Hawaii.

the longer strong winds blew in their area of origin, and the longer the distance over which winds blew. When waves reach the coast, they bend, being slowed in shallow water. Wherever a wave crest gets close to the shore its speed diminishes, and thus a wave makes a stately turn like a marching band, so the crest becomes more nearly parallel to the beach. On approaching the beach, the lower part of a wave slows, while the upper part (which is still in deeper water) keeps going. The wave breaks. Where it does so, its height is about two-thirds of the depth of the water (fig. 2.4).

The waves wash up debris, including living and dead organisms that may serve as food for the shore birds patrolling the edge of the beach and also for the many organisms hidden within the sand, including worms and crustaceans.

Many of the birds we see along the shore migrate over long distances. This is true, for example, for the marbled godwit *(Limosa fedoa)*, which is a common visitor to the beaches of California in fall and winter. Its long bill allows it to retrieve sand crabs, worms and mollusks hiding within the sand and mud of beaches and mudflats. The godwit is among the most common and largest of the shore birds in San Diego County; they number in the thousands. Nonbreeding (presumably immature) individuals are seen in summer as well but are not very abundant. The godwits commute between the Great Plains (including southern Canada) and the shores of California, feeding along the ocean and breeding in the grasslands, on the ground.[8] A hundred years ago, before they were hunted to near extinction there, they were abundant on the East Coast as well.[9] From its migration habit, it seems reasonable to assume that the godwit (along with others in the sandpiper family) started out as a prairie bird and discovered the seashore in the course of evolution, building on its experience of feeding in wetlands. The godwits, then, are dependent on several quite different environments—a prairie with grasshoppers, wetlands for intermediate stops along the migration route, and a productive coastal ocean. That multiple dependence makes them vulnerable to disturbance in any one of their haunts.

A much greater commitment to the sea has been made by the pelicans and cormorants, which roost along the shore and dive for fish (fig. 2.5). Their sea-loving ancestry goes back a long time—tens of millions of years. They depend on the availability of fish, and safe spaces for breeding. Pelicans and cormorants are a fairly common sight along the West Coast. In contrast, sightings of pelagic seabirds are rare; albatrosses and petrels stay well offshore.

Birds are appealing: they are the action heroes of the seashore. We appreciate their behavior, which includes the familiar activities of foraging, socializing, quarreling, courting, and providing for offspring. Also, as vertebrates they are distant cousins of ours, having descended from the same air-breathing fishes that

FIGURE 2.5. Common sights along the shores of California: brown pelicans *(upper)* and cormorants *(lower)*.

lived a few hundred million years ago. From the perspective of Scripps ichthyologist Richard Rosenblatt, we are not all that far apart. Rosenblatt points out (with a wink, over a cup of coffee) that the basic diversity of fishes is great enough so that amphibians, reptiles, birds, and mammals could all qualify as air-breathing fishes, without undue strain on the general taxonomy.[10]

Flying like a bird has long been a dream of humans. And some of us do fly. The hang-glider enthusiasts make use of the sea breeze, which is set up by the different rate in heating of ocean and land. Water warms but slowly in the sun, while land heats up quite rapidly. By noon, the air rises over the land to the east along the mountains. A landward breeze ensues, which fills the coastal lowlands with marine air. If this marine air is cool enough to resist convection it just sits there, moist and foggy, and lets the landward wind pass above it. Because of its relatively small volume it readily fills up with exhaust, which produces smog. Whenever the offshore waters are cold, therefore, the chances for fog and smog increase.

As we continue our walk along the ocean, we are likely to find tangles of kelp, uprooted by large waves from the seafloor just beyond the breakers (fig. 2.6). Perhaps we should not use the word *uprooted* because these large plants have no roots. They are giant algae, and what looks like roots is called holdfasts. They take up nutrients through their leaflike fronds and grow in the light as photosynthesizing primary producers. To keep the fronds in the surface layer while being attached to the bottom, they have gas floats. It is the floats that spell their doom when big waves come in and bring the floats into the circular motion that defines waves. If the holdfast happens to be attached to a smallish rock, or to a ledge with a crumbling crust ready to detach from the mother rock, the plant

FIGURE 2.6. Gas float of elkhorn kelp along with small floats of giant kelp on La Jolla beach.

leaves the bottom and drifts ashore. For the naturalist beach walker it is a bonanza. The fronds and holdfasts yield brittle stars, isopods, worms, and bryozoans for inspection.

Of special interest are the rocks exposed to wave action: they are the habitat for a great many creatures, especially mussels and barnacles. Small crabs and various types of mollusks move around between the firmly attached forms, wherever there is room. There is a clear zoning of organisms, related to exposure of their habitat to waves and air. The rock habitat is a favorite subject for ecologic studies, because of the diversity of environmental conditions in a narrow band between land and sea, at a place that is very accessible.

The organisms we are inspecting live in the intertidal zone, that is, the band along the shore that is flooded and exposed in a quasi-diurnal rhythm. It is not a daily rhythm, because the Earth's rotation is but one element in making the tides. The other is the Moon. The Moon runs eastward in the sky, parallel to Earth's rotation, but it takes about 29 days to complete the circle. So, each day it has changed its position in the sky by the equivalent of one twenty-ninth of 24 hours, and the Earth's rotation has to make up the difference to get to the previous configuration, which takes about 50 minutes. Because there are two tides per day, the typical cycle is then 12 hours and 25 minutes.

The reason why there are two tides rather than one each day bears explaining. If the direct gravitational pull of Moon and Sun were the only factors to consider, we would expect one cycle per day. Imagine a water-covered Earth. The Moon would raise a bulge of water on the surface, and the Earth would rotate under this bulge, which would consequently seem to move westward. Reality is more complicated, of course, because the tidal wave cannot move freely but is constrained to move in circles within the various ocean basins. But here we only look at the principle. Why should there be a bulge in Earth's ocean at the side opposite to the direction of the Moon? The answer is that we are not dealing only with gravitational pull, but also with the centrifugal forces that keep Earth and its moon separated. It is the balance between the two forces that matters. On the side facing the Moon, we have gravity somewhat exceeding the centrifugal forces. On the side opposite the Moon, centrifugal forces are stronger. Thus, a bulge results here too. The overall balance is achieved only on average, but not at the points closest to the Moon and farthest from it. The imbalance in forces results in making the wave that is the tide.

The tidal range is large whenever Moon and Sun are in line (new or full Moon), because then the gravitational and centrifugal forces line up and sum up to maximum strength.

In San Diego County, as is true for much of California, the edge of the sea commonly is a narrow beach next to steep cliffs. In contrast, much of the Atlantic seashore has wide beaches and a broad belt of lagoons and marshes. Dunes form where beach sand is piled up by winds. In the west, the sea cuts terraces into a rising coast, terraces that accommodate the beach and, after uplift, bear coastal vegetation, roads, and settlements. In the east, the land slowly subsides and receives the sediments brought by rivers and trapped in lagoons, marshes, and estuaries. The fact that the western shore is rising, while the eastern shore is sinking, readily translates into fundamentally different coastal landscapes (fig. 2.7).

To appreciate the fact that uplift is taking place along the coast of California, all we have to do is look at the exposed rocks in the landscape, which were made deep in the crust, or represent ancient marine sediments. To understand the sinking of the East Coast and its consequences, we have to take cores in the sediments that contain the history of changes in marshlands and lagoons. The uppermost sediments tell a history of a large rise in sea level, about 100 meters in the last 15,000 years, caused by the melting of the great northern ice sheets of the last glacial period. We see the result: the drowning of the eastern seashore in a great marine invasion, making estuaries and marshes and lagoons.[11]

While discussing the general geology of the coasts we should mention that California has

FIGURE 2.7. West Coast versus East Coast shorescape: land rising, land sinking. (Northern California vs. Long Island, N.Y.)

offshore petroleum and the East Coast does not. Walking along the beach in Santa Barbara County, we might actually step on a lump of tar, and looking out to sea, we may note a drilling rig in the mist. Why this difference in petroleum riches? The reason is that the coastal ocean off California is highly productive and has been so for the last 20 million years. High production in coastal waters implies high delivery of organic carbon to the seafloor offshore. Burial of organic carbon within the sediments accumulating on the seafloor in offshore basins started the process that resulted in making petroleum. The tar on the beach is most likely not from a drilling platform or from a ship, but originated from natural seeps. Native coastal people, for thousands of years, used such tar to make their boats watertight.

WINDS AND CURRENTS

The world's ocean basins are filled with very cold water, except for a rather thin warm layer on top, in tropical and temperate latitudes. This

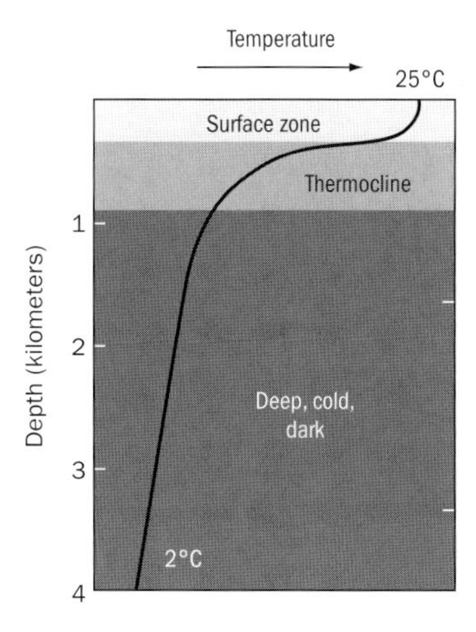

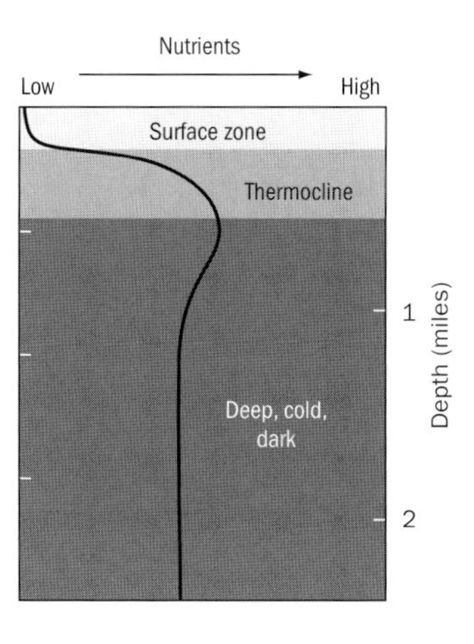

FIGURE 2.8. Vertical structure of the ocean, in low to temperate latitudes, and the associated pattern of nutrient abundance.

two-layer structure, with the "thermocline" the zone of separation, governs the grand patterns of water motion, water chemistry, and life in the sea. From the point of view of life habitat, the thermocline is the second-most important boundary zone on the planet, after the shoreline, which separates land and sea.[12] There are several reasons why this boundary plays such an important role in the natural history of the sea. The most conspicuous one is the fact that nutrients are trapped below it (fig. 2.8).

The thin layer of warm surface waters overlying the thermocline has temperatures between 15 and 29 °C (60 to 85 °F). The water below the thermocline, the bulk of all water masses, has temperatures of less than 5 °C (41 °F). The thermocline typically lies between depths of 100 and 1,000 meters. By impeding vertical mixing, the thermocline hinders the flow of nutrients from deep waters to the surface layer, thus setting the productivity of the ocean at a low level. In areas where the thermocline is shallow, or is disturbed by eddies, nutrients can come up and fertilize the sunlit zone, where algae can grow. Such areas occur at the margins of the ocean basins and along the equator.

What are the forces producing this overall temperature distribution? What is the origin of the thermocline? In principle, the cold water that fills the ocean basins originates in high latitudes, at the surface, which is the only place where the temperature of the water can be changed significantly, through heating or cooling. Cold water is heavy and sinks and fills the basins. The Sun heats the surface of the sea in low latitudes, and the wind stirs the warm water layer and thickens it. Where cold and warm water meet, a few hundred meters down, we have the thermocline, the zone of pronounced change in temperature.[13] As part of the redistribution of heat on the planet, warm water moves into high latitudes to replace the cold water moving down to fill the basins. In this fashion, an overall deep circulation is set up, the deep-sea circulation, commonly referred to as thermohaline circulation.[14]

In a very elementary way, deep-sea circulation can be represented as a continual filling of an ocean basin with cold water from polar regions (fig. 2.9). Complications arise from the fact that cold sources are few and feed a network of basins that exchange water with each other, and that wind stress drives much of the circulation of upper water layers (including feeding the thermocline directly from the surface). Exchange between basins is commonly represented in textbooks by a global loop conveyor,[15] which leaves out the most important aspect of the deep circulation: the mixing of deep and shallow waters in the Southern Ocean. The deep ring current around Antarctica is the central mixing station for all of the ocean's water masses. The Great Ring Current dominates the patterns of exchange between the basins, while

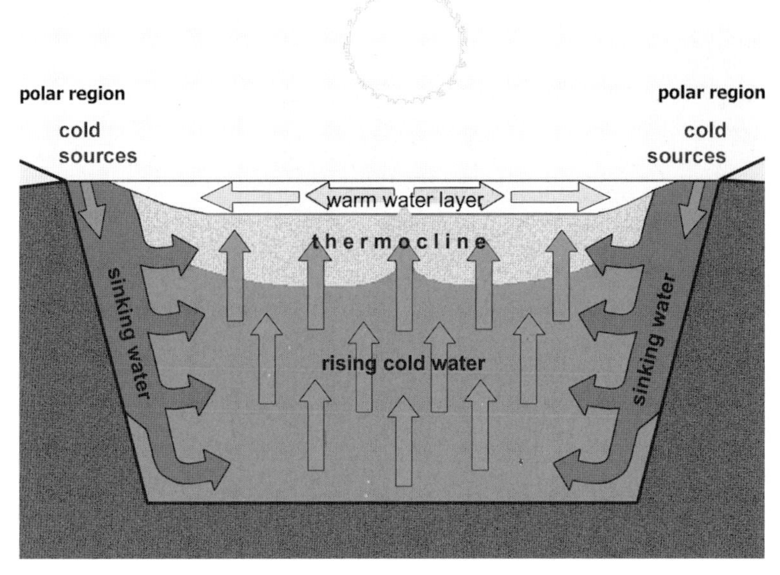

FIGURE 2.9. Why deep water is cold. Elementary representation of the nature of deep circulation.

the Great Loop Current is best understood as the main modification to a basic arrangement, whereby the major basins are appendices to the southern Ring (fig. 2.10).

The sense of the exchange between the major basins is immediately obvious from the position of the source regions of deep water: they are all at the cold ends of the Atlantic! Thus, the deep Atlantic is filled with water that was at the surface not so long ago, and exchanged gases with the air. So, it has a good fill of oxygen. In contrast, the deep waters of the North Pacific are farthest away from the sources. Much of the oxygen has been used up along the way; the water is "old." When oxygen is used up, nutrients are set free from the organic matter that is being oxidized. Therefore, the deep North Pacific has the highest nutrient contents in the ocean, while the deep North Atlantic has the lowest.

But why does the deep cold water rise at all? Why does it not stay there, made during the coldest of winters and reluctant to make room for later additions of cold water? The answer is "deep mixing."

As deep currents move over uneven seafloor, along the continents and along the Mid-Ocean Ridge, they generate turbulence, which results in the mixing of the coldest waters at the bottom with less-cold waters higher up in the water column. This provides an opportunity for new cold water to invade the vertical sequence of deepwater layers at the appropriate density level, which keeps the circulation going. We have already met the chief mixing machine in the sea: the Great Ring Current. The faster it runs, the more it mixes, and the faster the deep water can be replaced by new dense water from the cold sources. But, in addition, mixing at depth owes much to tidal motions in the deep sea, generated by the Moon.[16] The Moon helps in stirring the ocean, which in turn stimulates deep circulation. The rate of global upward motion of the deep water is of the order of 1 centimeter per day.[17] This upward movement sets a baseline for the productivity of the ocean, being responsible for recycling the nutrients trapped in deep waters back into the sunlit zone, where they can be used. From this simple chain of argument, we can see that both the Great Ring Current and the Moon have a role in fostering the productivity of the sea.

But before we turn to the fascinating topic of productivity, we must consider the surface

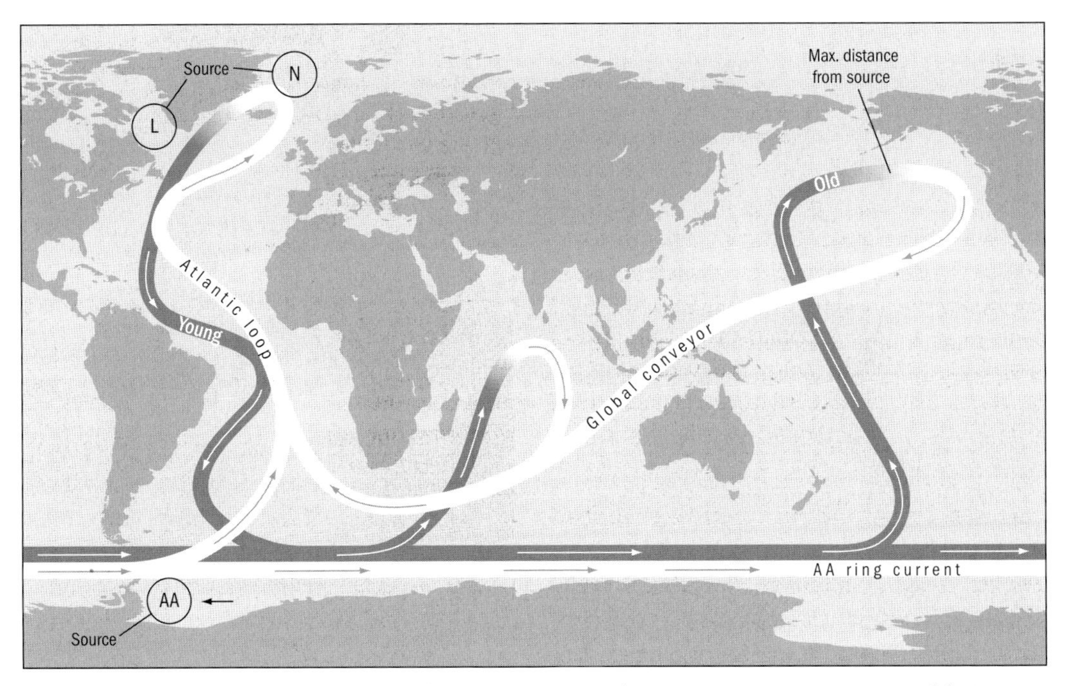

FIGURE 2.10. How the basins exchange water. The main connection is the Great Ring Current moving around the Antarctic Continent (AA). The deepwater sources are at the cold ends of the Atlantic basin. The loop current (Global Conveyor) determines how much warm upper water gets into the North Atlantic, providing a heat subsidy. L, Labrador Sea; N, Nordic Sea; AA, Antarctic Weddell Sea; Young, oxygen-rich deep water; Old, oxygen-poor deep water. The AA Ring Current is mixed to great depth.

currents involving the water layer *above* the thermocline, that is, the currents that are driven by the wind. The fact that currents follow the wind, on the whole, is well appreciated. It was pointed out by Lieutenant Maury more than a century ago.[18]

The surface of the sea, of course, is the part of the ocean that has been studied the longest and what we are most familiar with, from swimming, fishing, or sailing. The most striking feature of the sea surface is its ever-changing topography, dominated by waves large and small. Waves are generated by wind (except for tsunami waves, which result from earthquakes). The stronger and longer the wind blows, and the larger the area over which it is active, the more energy is imparted to the waves. Close to a storm, we find short and choppy waves, as well as long and powerful ones, with everything in between. Far away from the storm area we find mainly the long waves, the so-called swell. This shift to the long waves is somewhat analogous to the contrast between the sharp and hard crackling sound made by nearby lightning and the rumbling of distant thunder.

The biggest waves are raised by the fierce winds around Antarctica, and the swell from these waves pervades the central Pacific. This was well known by the Polynesian seafarers, who learned to steer by the southern swell, and to interpret its disturbance by islands in its path.[19] Among all the planetary winds the ones of the Southern Ocean stand out as the strongest; they blow from the west and they make the Great Ring Current run eastward, with the largest transport of all currents (fig. 2.11). Equally important in explaining the overall patterns of surface currents are the trade winds, which blow from east to west and toward the equator in all basins (as shown in the graph, fig. 2.12).[20]

Winds work on the roughness of the sea surface, the waves, to impart to the surface waters the energy that drives currents. The winds

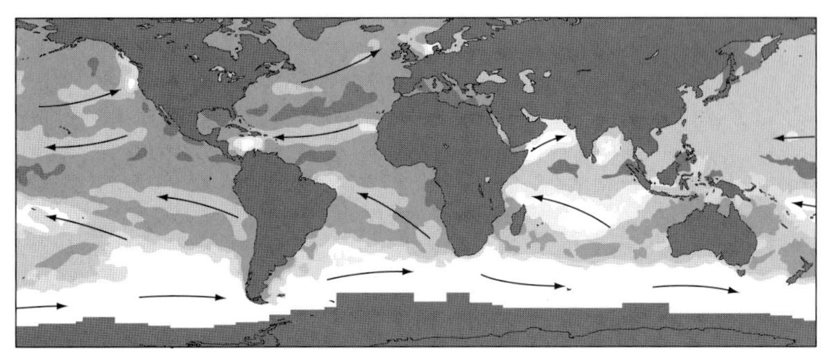

FIGURE 2.11. The global wind field as seen from space. Areas with strong winds are shown white. Note that the winds are stronger in the Southern Hemisphere than in the Northern Hemisphere, overall.

themselves result from differential heating of Earth's surface. It is warm at the equator because here the Sun is high in the sky, and it is cold in high latitudes because the Sun is low on the horizon, so that the same amount of sunlight is spread over a larger surface. The winds arise from this unequal distribution of heat. In turn, winds and their currents redistribute heat, decreasing the contrasts in temperature. The zone of maximum wind strength, around Antarctica, owes to the fact that Antarctica is covered by ice, making the continent into a stable cold center on the planet.

The way in which winds move heat is by taking up water vapor from the sea surface in warm regions, and moving it to cool regions, where the vapor condenses. The energy contained in vapor and released in making rain is called latent heat. The latent heat within the air is extremely important, but its presence is not sensed readily because it is not reflected in the temperature of the air. Warm air has more sensible heat than cold. But sensible heat is much less important in atmospheric heat transport than is latent heat. It is the rain in Iceland that keeps the island from freezing over, not so much the warmth of the marine air.

The vapor content of the air is highest in the tropical regions, and lowest in polar areas. Thus, any winds that blow across latitudes must transport heat, or heat deficit.[21] In addition, water vapor is higher over the ocean than over land, on average, for obvious reasons. Thus, any exchange of air between continents and oceans will move heat to the continents, and heat deficit to the ocean surface. The unequal heating of land and sea sets up coastal winds and monsoon winds, which act to bring the land closer to the temperature of the surrounding ocean. The hot summers and cold winters in the deep interior of large continents reflect the distance to the sea and the lack of exchange of air with the ocean. The water cycle is part of this heat exchange machine. Drought in the interior of large continents is another hallmark of the distance from the sea.

To complete the overview of winds on the planet, we must mention the westerlies, that is, zonal winds that blow poleward of the trades. The trades blow from the east, the westerlies from the west (fig. 2.12). The currents driven by these winds cannot run around the globe but must close to form large central gyres within the various ocean basins.[22] The central gyres dominate the surface circulation patterns of all oceans, except in the Arctic and Antarctic regions. Together they represent the major ocean currents in low to temperate latitudes (fig. 2.13).

The Gulf Stream is the most prominent of the gyre-linked currents; it represents the western section of the central gyre of the North Atlantic. The Gulf Stream is second in size only to the circumpolar current of the Southern Ocean (that is, the Great Ring Current). The Gulf Stream has velocities up to 4 knots (jogging speed)[23] and transports some

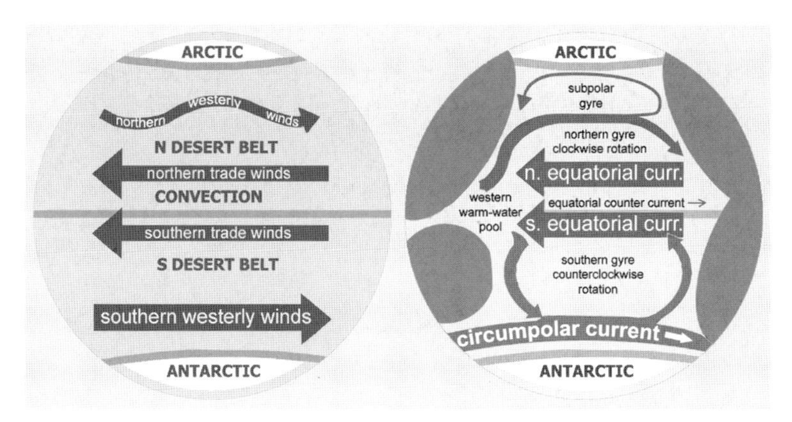

FIGURE 2.12. Zonal winds of planet Earth *(left)* and the currents they generate *(right)*.

100 million tons of water per second—more than 100 times the outflow of all of the world's rivers combined.

The wind fields are not exactly symmetrical between the Northern Hemisphere and the Southern Hemisphere, because the temperature gradients from equator to poles are different for the two hemispheres. The southern icy regions are larger than the northern ones, and the southern westerly winds face much less obstruction from landmasses than the northern ones. The overall result of the unequal distribution of landmasses and wind strengths on the planet is that the convection zone in the tropics, where air rises, is shifted distinctly north of the equator. The rising air pulls in some of the replacement from south of the equator, and this implies that the Northern Hemisphere as a whole receives a heat subsidy from the southern one. Because of this general heat piracy of the north, the Southern Hemisphere is colder than the northern one, at any given latitude. The fact that the hurricane lane in the Atlantic (from off Africa toward the Caribbean) is well north of the equator is the most obvious consequence of this heat transfer from south to north.

In each hemisphere, the west-wind and trade-wind belts are separated by the desert zones of the planet, which are centered near 28° N and 24° S, respectively. On land, the desert zone has low rainfall and little vegetation. The most striking example of vast dry regions is in the Sahara belt in northern Africa, a region that extends eastward into Saudi Arabia and Iran. In the ocean, the desert zone is represented by the five great gyres, warm-water lenses that rotate in response to the driving zonal winds. These gyres are starved of nutrients; they are the deserts in the sea. The best-known example is the Sargasso Sea in the western North Atlantic (fig. 2.13, "S").

The high-production areas of the ocean are outside the gyres, mainly in the coastal oceans, where eddies pump nutrients upward from the thermocline into sunlit waters. Here, boundary currents rub against the continents, making the swirls and eddies that do the work. Also, nutrient-rich waters rise along the equator, where surface waters part (due to Earth's rotation) to let deeper water come up, and in the high latitudes where winter winds mix the water deeply, bringing up nutrients. As we consider these processes of nutrient fertilization, we enter the topics of productivity and life-support systems in the sea.

LIFE IN THE SEA

The ocean is the great central life habitat on the planet. Taking up 70 percent of the surface and with a thickness of 4,000 meters, it exceeds the terrestrial habitat by a factor of 200 or so.[24] Growth of green organisms at the base of the food chain is restricted to the sunlit zone near the surface. The rest of the ocean's habitat, more than 95 percent of it, is poorly lit and not suited for photosynthesis. Thus, in most of the ocean

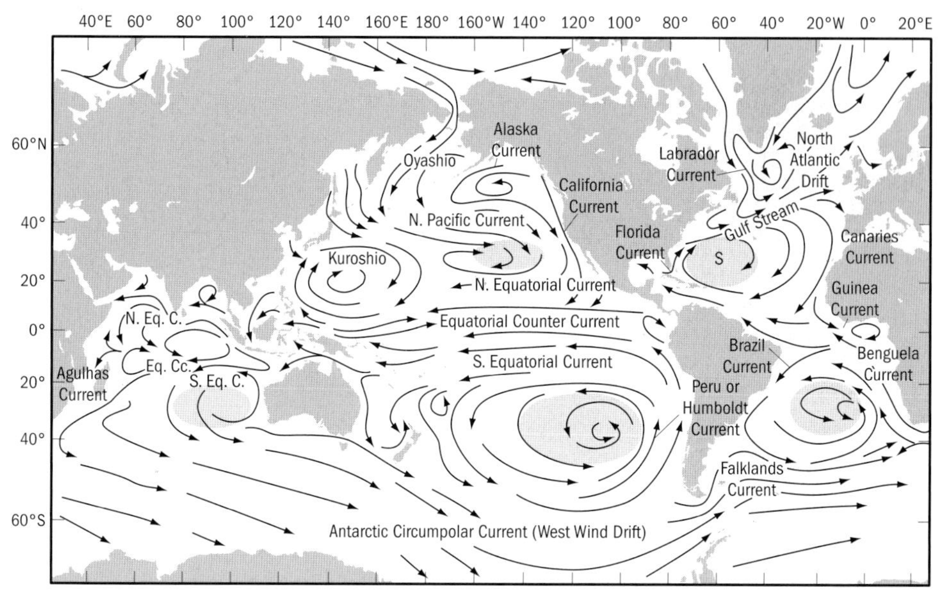

FIGURE 2.13. Surface currents of the World Ocean. Note the prominence of the central gyres (darkened for identification). S, Sargasso Sea (the center of the subtropical gyre in the North Atlantic).

habitat, organisms live off the scraps falling down from the surface, or else off the animals living on scraps, in a food chain that starts with scavengers. The deserts of the sea—the central gyres—deliver but little organic material to deep-living forms, and thus the great expanses below the gyres are only thinly populated. But the regions of high production in coastal seas and estuaries, and along the equator and in high latitudes, produce a rich rain of food for animals living at depth and on the seafloor. It is here that we see high abundances of organisms both in surface waters and in the waters below, and especially on the bottom.

When talking about life in the sea, it is good to know something about the major groups of organisms involved.[25] Many of these are familiar from lakes and rivers—water bodies on loan from the ocean, as it were—but many are not. In general, the diversity of animal life in the sea is greater than on land, except when considering diversity at the species level. In other words, there are many more fundamentally different types of animals in the sea—a testimony to the great geologic age of the habitat. But there are more marginally diversified organisms on land, as a result of the greater isolation of populations, and the prolific development of insects.[26]

The animals that immediately come to mind as typically marine are the fishes (fig. 2.14).[27] The most abundant types of fishes in the sea belong to a group called ray-finned.[28] They make up by far the bulk of fish landings, and they include such familiar kinds as herring, mackerel, cod, tuna, eel, seahorse, salmon, mudskipper, and halibut. The ray-fins, whose fins are made of spines and skin and no flesh, belong to the class of bony fishes. The other two classes are the cartilaginous fishes (with sharks and rays) and the jawless fishes (the group including the hagfishes).[29]

The class of bony fishes includes forms with more ancient aspects, for example, the lobe-finned fishes (mostly extinct, with the living genus *Latimeria*) and the lungfishes. Sometime in the Devonian one species of fish of the lungfish line became the ancestor of primitive amphibians, and hence of the terrestrial vertebrates, including reptiles, birds, and mammals.

The ray-fins, on the whole, depend on a swim bladder for buoyancy control. Their repro-

FIGURE 2.14. Common fishes in the sea. *Upper:* ray-finned bony fishes (lionfish, mudskipper, flying fish, flatfish, salmon, rattail, tuna, deepwater fishes), showing a wide range of adaptations (poison points, tree climbing, gliding in air, probing mud, pursuit hunting, ambush). *Lower:* cartilaginous fishes (leopard shark, ray, pelagic shark).

duction is linked to high fecundity, that is, thousands or even millions of eggs. The great bulk of living ray-fins belong to the teleosts; that is, they have undergone marked modification from the original ray-fin stock within the last 150 million years.[30]

The next-largest group of fishes, but with much fewer species than the ray-fins, consists of the cartilaginous fishes (fig. 2.14, lower panel). Sharks, rays, and skates differ from the bony fishes in a number of important aspects, reflecting a distance to common ancestry comparable to that between lungfishes and mammals. Ecologically, an important aspect of the sharks and their relatives is low fecundity. Pelagic sharks bear live young, which grow within the mother's body to a size at which they can swim away and fend for themselves (rather than entering the plankton as fish eggs). Thus, the rate of reproduction depends greatly on the number of adults present in a population. Nearshore representatives of cartilaginous fishes lay large eggs within a horny encasement, for protection. Fertilization is internal.

Some sharks are fear-inspiring predators. The white shark, reputed to grow to 36 feet in length (11 meters) is the largest predator among modern fishes.[31] The largest fish living today is the whale shark, which grows to a length of well over 40 feet, with an estimated weight of around 13 tons. It is a peaceful plankton-eater.[32] Skates and rays feed on bottom animals (crustaceans, mollusks, worms), as is immediately evident from their habit of bottom contact and the position of their mouth. They find their prey in shallow water.

Some of the air-breathing offspring of ancient lungfishes that once conquered the terrestrial environment and diversified into reptiles, mammals, and birds have reinvaded the sea. We know them as sea turtles, dolphins, and penguins. Sea turtles are much more ancient invaders than the toothed whales called dolphins; they were already at sea in the Mesozoic more than a hundred million years ago, while the whale ancestors went marine in the early Tertiary, some 50 million years ago. Marine reptiles other than turtles are among crocodiles, snakes, and lizards.

The marine turtles and the penguins come on land to lay their eggs; otherwise they are fully adapted to the sea, spending their life there. The dolphins bear their young at sea, as do all whales, toothed or not. Whales, penguins, and turtles are expert divers. We have been learning much about the abilities of mammals and birds to dive in recent years.[33] Sperm whales are the champion divers, finding food at enormous depths.

The largest whales are those that have sieve plates (baleen) hanging from the roof of the mouth. They use the baleen for filtering the water, after swallowing great gulps of it, with the purpose of retaining krill and other zooplankton within the mouth. The largest of the baleen whales is the blue whale, the biggest and heaviest animal that ever lived on our planet. The best-known baleen whale is the humpback whale, whose males sing during courtship, a fact discovered in the 1960s, and one that drew enormous public attention to whales and their plight from overhunting.[34] Protective measures followed.

The baleen whales evolved within the last 30 million years or so from toothed ancestors to take advantage of exploding plankton populations that accompanied the overall cooling of the planet and the development of the Great Ring Current. Seals and sea lions are a recent addition to the roster of marine mammals; they still breed on land. The polar bear exemplifies a terrestrial mammal at the edge of becoming a marine one. (Its peculiar status is reflected in its name: *Ursus maritimus*.) Polar bears hunt seals, ambushing them when they come up for air in holes in the sea ice. As sea ice shrinks in the Arctic Sea, they will have to adjust their method of hunting or go hungry.

Fishes and air-breathing vertebrates are but one of the large groups of animals in the sea, albeit the most conspicuous one. The invertebrates most closely related to this group are the ascidian tunicates and salps, whose larvae resemble tadpoles in general structure. Of the two, the tunicates—sea squirts—are the more familiar kind. They are quite abundant in places on the seafloor, in shallow waters, where they filter the water for small food particles. Salps are part of the plankton; they too filter the water for small edible stuff.

Many organisms are filter feeders among the animals attached to the bottom. Filtering the passing water for food is the way of life for sponges, sea fans (gorgonians, related to corals), sea lilies (echinoderms), barnacles (arthropods), and mussels (mollusks). Within each of the major marine groups (phyla; singular, *phylum*) that these filter-feeding organisms are classified with (cnidarians, echinoderms, arthropods, mollusks), there are a host of other related forms with similar building plans and similar development, but with entirely different lifestyles. Jellyfishes (related to sea fans) are deathtraps for zooplankton that get tangled in the trailing tentacles studded with poisonous stinging cells. Sea cucumbers (related to sea lilies) eat mud. Copepods (related to barnacles) gather small plankton organisms, while crabs (in the same phylum) forage on the seafloor. Snails and chitons (relatives of mussels) scrape microbes off rocks. Some snails drill a hole into the shells of clams to get meat. Some hunt in the open sea for small zooplankton in the sunlit zone. The brainiest among the invertebrates, the cephalopods, are hunters as well: octopods eat crabs; squids pursue smaller fishes.

It may seem daunting to keep track of this incredibly diverse universe of living organisms, and it is, given that there are tens of thousands of species in each major group doing thousands of different things. Thus, for the generalist (such as oceanographers) it will not do to become expert in more than a few well-chosen smallish groups, otherwise there is overload. From the point of view of marine ecology, the focus is on how these various organisms make a living and how they interact with each other and, thereby, help set the environment of survival and evolution for each other.

However, some knowledge of taxonomy of marine organisms, on the level of "fish" or "bird" or "mammal," surely is appropriate when pursuing the goal of ocean literacy. Indeed common language guides us well in this task. *Lobster, crab, shrimp, barnacle* are familiar terms, as are *mussel, snail, clam, squid, sea urchin, sea star, jellyfish,* and *coral.* Most people will readily identify these animals; even if they cannot say which phylum they are a part of. Fortunately, such assignment is not particularly difficult. There are only a very few phyla whose representatives are abundant and occur both on land and in the sea, and there are another few that are almost equally prolific in quality and quantity but are mainly or entirely restricted to the marine realm. The big phyla are the chordates (fishes, reptiles, mammals, birds), the arthropods (copepods, crabs, shrimp, barnacles), and the mollusks (snails, clams, squid, octopods), with representatives on land and in the sea, and the echinoderms (sea urchins, sea cucumbers, sea stars, brittle stars) and cnidarians (jellyfish, siphonophores, soft and hard coral), abundant in the ocean.

The single most prolific phylum on the planet, in terms of variety of animals, life-styles, and abundance, is the phylum of the arthropods (fig. 2.15). For species diversity, none other can match it because it contains the insects, the most successful group of animals in the terrestrial environment. Beetle species alone fill books, as do butterflies. Mosquitoes and flies are ubiquitous and some are dangerous. Social insects (ants, bees) offer important insights on regulation and communication in automated cooperatives. In the sea, smallish arthropods called copepods (millimeter size, typically) are near the base of the food chain and provide food for a host of miniature predators in many different phyla, including fish larvae, crab larvae, arrow worms, gooseberry jellyfish, polychaete worms, and larvae of medusae, mollusks, and echinoderms. The most familiar arthropods of the sea are lobsters and crabs, popular items on the seafood menu. Baleen whales gorge themselves on krill during the growing season in the Southern Ocean, a type of shrimp 2 inches long (euphausid shrimp). Krill also feed penguins, and the so-called crab-eater seal, the most abundant of all seals (whose name should be krill-eater seal, obviously). Copepods have been the object of study of plankton ecologists since early in the twentieth century—what they eat, how they find it, how they partake in vertical migration. In the last few decades we have realized that if one wishes to understand life in the Antarctic, one has to study the life history of the diatom-eating krill.

The third of the centrally important phyla is that of the mollusks (fig. 2.15). Again, it is familiar from wet environments on land: snails (including air-breathing ones), slugs, and freshwater mussels are common sights. But this phylum has its greatest diversity in the sea. With its shell-forming representatives among snails and bivalves and especially the shelled cephalopods (ammonites), it has long dominated studies on stratigraphy and evolution in the geologic record. While ammonites and belemnites (cephalopods) became extinct at the end of the Cretaceous, 65 million years ago, their extant cousins, the squid, dominate midwater ecology today. Their tentacles, their stealth and speed, their large eyes, and their brain (remarkably large for invertebrates) make them formidable predators. In fact, the largest of deep-sea predators are giant squids, whose only enemy, presumably, is the giant sperm whale.

The largest and most conspicuous of the purely marine phyla is that of the echinoderms (fig. 2.16). As is true for the mollusks, the

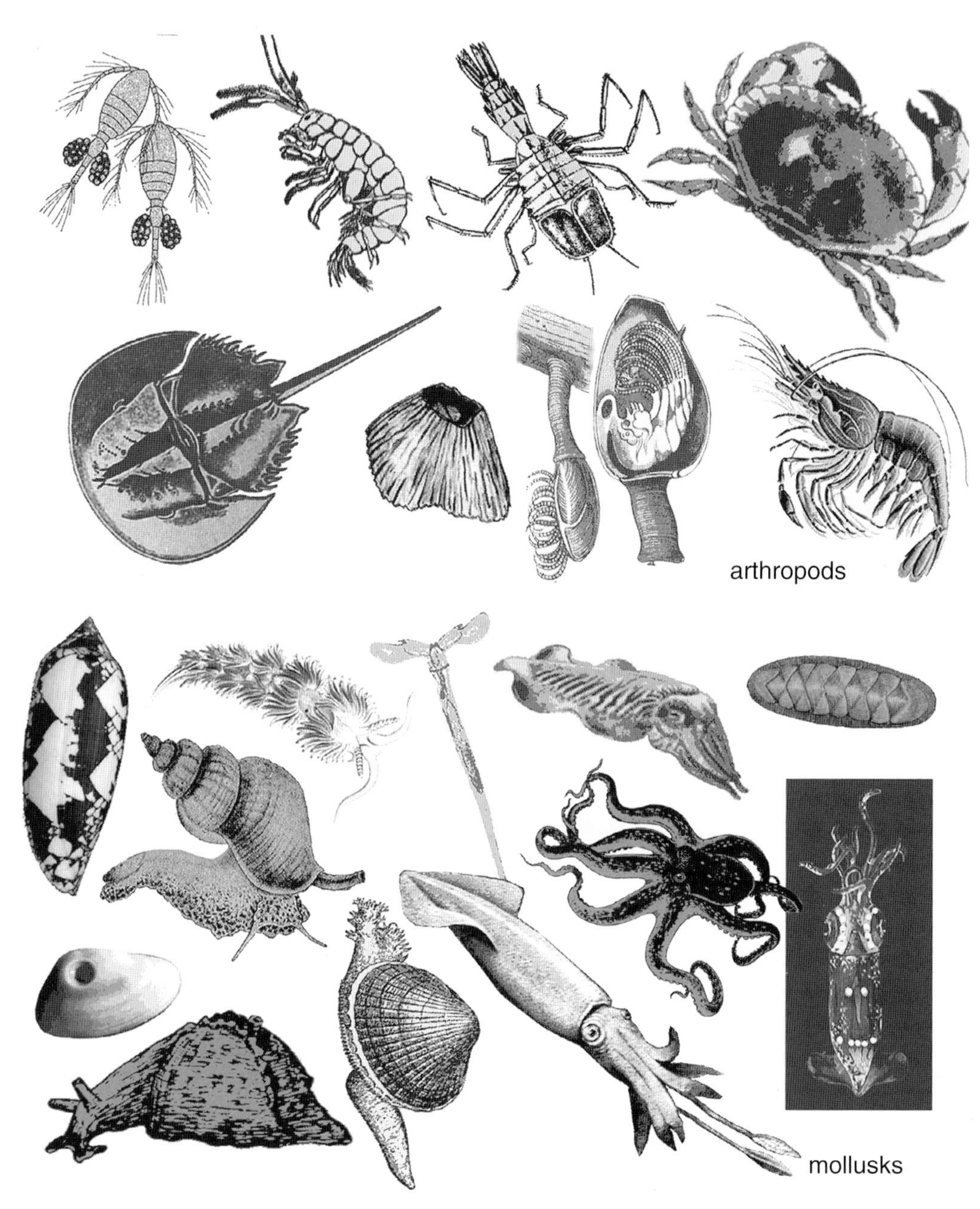

FIGURE 2.15. Common marine representatives of the two giant phyla, arthropoda and mollusca. *Upper:* two copepods, two amphipods, crab, horseshoe crab, barnacles, shrimp. *Lower:* snail shell, nudibranch, pteropod, cuttlefish, chiton, snail, octopus, deep-sea squid, clam shell (drilled), sea hare, cockle, squid.

phylum's geologic pedigree is impressive and reaches far back into the early Phanerozoic, with many forms that are now extinct. Brainless creatures, they know little and learn nothing but yet are programmed well for survival by hunting (sea stars), filtering (brittle stars, sea lilies), and deposit feeding and grazing of algae (sea urchins). They live on the seafloor, although their larvae are quite abundant in the plankton of the coastal oceans.

The most ancient type of predators in the present sea are the jellyfish, medusae that

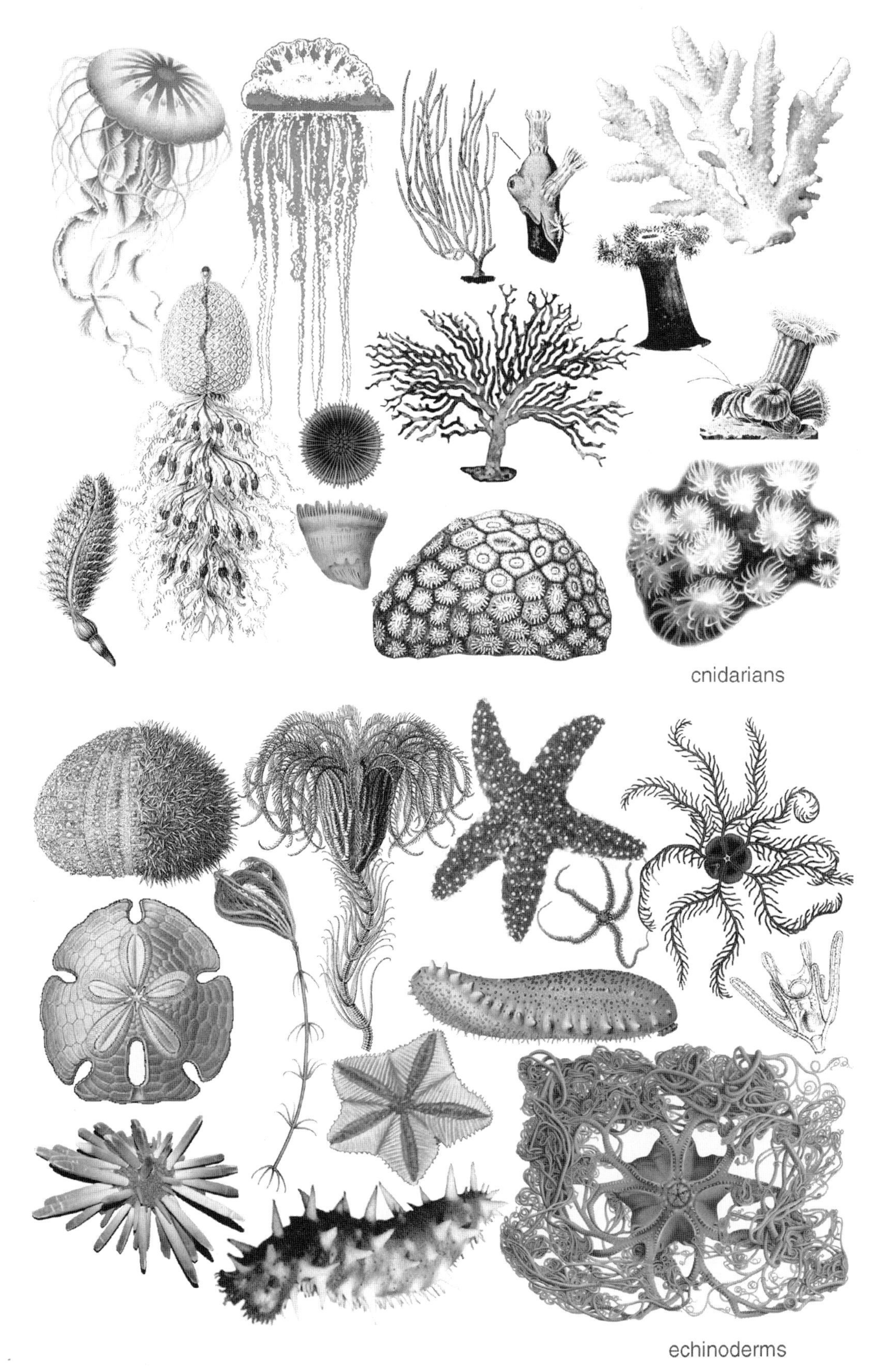

FIGURE 2.16. Animals with stinging cells (cnidaria) and sea stars and kin (echinodermata). *Upper:* jellyfish, siphonophore, gorgonian, sea anemones (one on hermit crab), branched stone coral, sea pen, siphonophore, deep-sea solitary coral, gorgonian and massive stone coral, anemone colony. *Lower:* sea urchin, sea lily, sea star, brittle stars (two, and larva), sea urchin, sea lily, sea star, sea cucumber, sea urchin (pencil spines), sea cucumber (spiny), basket star.

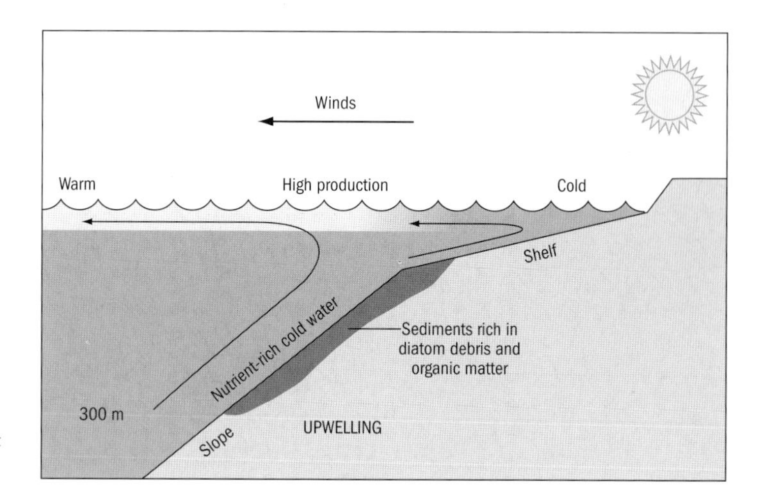

FIGURE 2.17. Upwelling of cold water brings nutrients to the sunlit zone.

propel themselves through the water by pulsed contraction of their dome-shaped bell, sampling the environment with their dragging tentacles (fig. 2.16). The siphonophores, likewise with tentacles and stinging cells, have a similar feeding strategy. Both belong to the phylum of the cnidaria, animals with stinging cells, along with the various types of corals. One of the more familiar types of siphonophores is the notorious Portuguese man-o-war, abundant in the Sargasso Sea. It has a float that doubles as a sail, dragging endlessly long tentacles through the water, and entangling and killing small animals that come in contact with them. As the larger of the predatory fishes are being removed from the sea by modern fishing methods, jellyfishes and siphonophores presumably will gain in importance as the remaining large predators.

The category commonly referred to as *coral* (upside-down medusae, as it were) includes a host of different types of creatures built upon the simple principle of a stomach encased in a double-walled tube with one opening, surrounded by tentacles. Many forms are colonial, with each polyp a clone of the next, and arranged on a hard substrate. The stone corals among these (for which the substrate is calcium carbonate) build the largest biological edifices on the planet. The biggest one of these is the Great Barrier Reef off eastern Australia. The reefs actually are cooperative projects between the stony corals and their symbionts—minute algae within the coral bodies that feed their host in return for protection and supply of nutrients. Coral, in life, are not white but take on the color of their algae: olive green, brown, yellow, red. Reefs are oases in the ocean desert. Exactly how they operate, and how they maintain the enormous diversity of organisms that make up reef ecosystems, is the subject of much ongoing research (see chapter 4).

From the viewpoint of marine ecology, jellyfishes are low-energy predators, while birds and mammals have high energy requirements. Drifting with the currents while slowly dragging a web of tentacles costs next to nothing. But searching for fishes and diving to catch them is quite costly in terms of effort spent. Such a life-style is supported only in regions of high productivity. When the coastal ocean became highly productive, owing to planetary cooling, a high-energy life-style became an attractive option. Cooling started about 40 million years ago; it intensified 20 million years ago and again 10 million years ago. Cooling produced the thermocline (that is, a lid on nutrient-rich water) and also the winds that provide for upwelling in privileged regions (fig. 2.17). The effect was to encourage terrestrial birds and mammals living along the coast to make their living at sea in such regions.

Ultimately, the opportunity for a high-energy life-style arises because of a short food chain in the upwelling regions of the sea (fig. 2.18). In such a chain, organic matter is efficiently transferred from primary producers to zooplankton

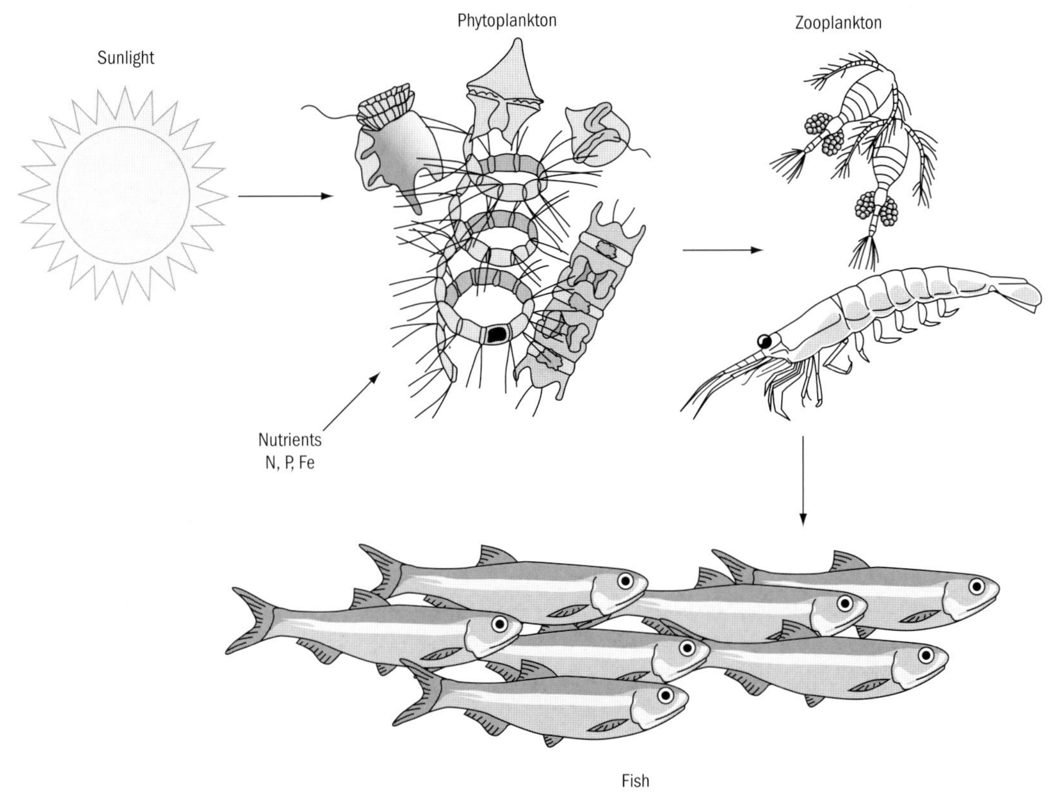

FIGURE 2.18. The short food chain of upwelling regions, which supports herringlike fish, which in turn feed mammals and birds. Arrows indicate the flow of energy and matter. Phytoplankton: dinoflagellates (top three) and diatoms (bottom). Zooplankton: copepods (two), krill.

to fishes and to the top predators such as birds, seals, and dolphins. This type of short food chain is the basis for support of highly active predators. The upwelling of cold water, by bringing nutrients to the sunlit zone, provides the environment that allows the phytoplankton (diatoms and dinoflagellates) to flourish, which in turn feeds a rich soup of zooplankton. Herringlike fishes strain the mixture of phytoplankton and zooplankton, filtering the food from the water, and in turn they become the prey of seabirds. Upwelling is why we have *piedras blancas* along California, the white rocks covered by the droppings of seabirds.

CHEMISTRY AND FERTILITY OF THE SEA

To try to understand the patterns of productivity of the ocean, we have to study the dissolved substances in seawater—not so much the major ones, which are responsible for the salinity (see appendix 2), but the minor ones, which serve as nutrients: phosphate, nitrate, silicate. The issue is, which of these are limiting to growth and which are in plentiful supply. Significantly, all these nutrients are greatly depleted within much of the surface layer of the sea and are concentrated below it, in the thermocline layer and below. This pattern means that the surface layer, which is the only one that has sunlight for powering photosynthesis, lacks nutrients, while the bulk of the lightless ocean has nutrients in abundance. In consequence, much of the overall productivity of the sea relies on the recycling of nutrients within the sunlit layer, with chlorophyll-bearing microbes fixing carbon and nutrients, and other microbes recycling the matter produced. It is only in the last several decades that the importance of this internal microbial loop has been fully realized.[35]

There is another type of loop, in which the microbial primary producers (phytoplankton) are a hundred times larger than the photosynthesizing bacteria of the microbial loop. In this loop, well recognized since the middle of the twentieth century, there is transfer of organic matter from the sunlit zone, where all production takes place, to the dark and cold depths of the deep ocean, where the habitat is huge but food is scarce. In essence, then, there are two distinct modes of biological carbon cycling: one for which production is by photosynthesizing bacteria traditionally called blue-green algae (cyanobacteria), and the other for which diatoms and dinoflagellates are the main producers. The first mode, the closed loop, is ubiquitous. It dominates in the deserts of the sea, since it can manage with extremely low nutrient contents. The second, the open loop that includes export, dominates in the fertile regions, with diatoms abundant in upwelling areas. It is the diatom-dominated open loop that supports the high-energy life-style of mammals and birds by providing for fish. We see seals and pelicans along the shores of California because upwelling along the coast brings nutrients into the system, including the silicate that supports the growth of diatoms.

The open loop, the cycle that includes upwelling of nutrients, produces plenty of organic matter for transport to deeper waters and to the seafloor. In places, so much organic carbon arrives at the bottom that its decay uses up all available free oxygen in the water in contact with the sediments.[36] Specialized bacteria then strip oxygen from sulfate, producing hydrogen sulfide and other foul-smelling gases. Methane gas is produced, as well. Such sediments have the potential of becoming sources for hydrocarbons, when subjected to heating. For this reason, continental margins with a long history of upwelling productivity are good places to search for petroleum.

How do we identify nutrients? Many organisms make shells or skeletons of carbonate. Thus, clearly, calcium and carbonate are nutrients. In fact, organisms need pretty much some amount of every one of the major elements in seawater, which are present as negatively and positively charged ions (in order of abundance: chloride, sodium, sulfate, magnesium, calcium, potassium, bicarbonate). Yet, we do not usually reckon these as "nutrients." The reason is, they are not in short supply, and thus their presence is taken for granted.[37]

Seawater, in essence, is a solution of plain table salt, with some Epsom salt mixed in (magnesium sulfate), making it bitter. Ultimately, the salt comes from volcanic emissions, from river influx, and from the reaction of seawater with the basaltic rocks of the floor of the ocean. The abundance of chlorine gives away the volcanic connection. The order of abundance reflects the solubility of the substances involved, rather than the rates of supply from the sources. This simple concept led to the calculation of seawater composition from the first principles of physical chemistry, in the 1960s.[38]

The ratio between the major ions is constant throughout the oceans. But the nutrient salts, which are present but in traces, vary greatly, as mentioned. Thus, their abundances do not follow the rules of chemical equilibrium. The reason why nutrients are depleted in warm surface waters and enriched at depth is extremely simple. In the sunlit zone, phytoplankton remove the nutrients from the water and incorporate them into their body substance. Phytoplankton are eaten, and a portion of what is eaten sinks into deep waters below the sunlit zone. Bacteria work on the settling organic matter and set the nutrients free. Thus, deep waters become nutrient rich. In fact, the easiest way to recognize limiting nutrients (short of doing the experiments with organisms) is to measure concentrations in shallow and deep water. If they are the same, the substance is not a limiting nutrient. If the substance is rare in surface waters and many times more abundant at depth, it is most likely a limiting nutrient for at least some organisms.

Typical nutrient profiles, then, show the nutrient depletion in surface waters, and a rapid rise of nutrient abundance to a maximum in

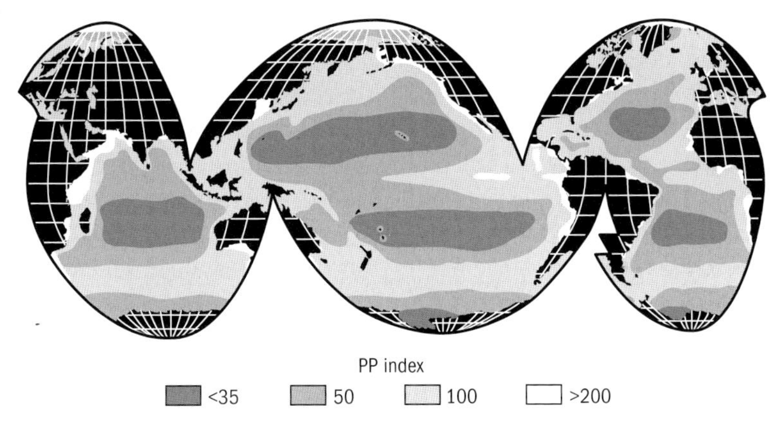

FIGURE 2.19. Global patterns of productivity, generalized. Dark regions: the deserts of the sea with low phytoplankton concentrations. White regions along continental margins: high production areas. Index numbers approximate grams of carbon fixed per square meter and year.

subsurface waters, within the thermocline (fig. 2.8). The nutrient maximum is associated with an oxygen minimum. The reason for this counterpoint abundance is that decay of organic matter takes up oxygen at the same time. Thus, quite generally, we find a negative correlation between nutrient content and oxygen in subsurface waters.[39]

Photosynthesis makes organic matter from carbon dioxide and water and gives off oxygen in the process. During decay the process is reversed: organic matter is oxidized and carbon dioxide and water are generated in the process. There is no net gain of oxygen from this cycle. Thus, the popular notion that photosynthesis in the ocean is responsible for keeping up oxygen values in the atmosphere is incorrect.[40] It is the sedimentation of reduced carbon that is important, on geologic time scales.

The global patterns of productivity are readily seen from space; they are mapped by recording the amount of green color in the light reflected from the surface of the sea. Coastal regions, high-latitude areas, and the narrow belts along the equator stand out as regions of high fertility (fig. 2.19). Coastal regions benefit from mixing and upwelling. Along the equator, surface waters move north and south away from the equator, owing to the combination of easterly winds and the rotation of the Earth. The divergence makes room for water from below. The high-latitude regions are fertile because winter storms mix the water to great depth, bringing up nutrients, and the surface waters receive much sunlight in the following spring and summer. The seasonal warming and provision of sunlight for initially nutrient-rich waters produces the famous "spring bloom," which starts the chain that feeds fishes and seabirds and whales. In winter, with the Sun low on the horizon, there is very little production.

The overall productivity of the sea may be defined as the rate at which carbon is taken up by photosynthetic organisms. Unfortunately, this rate is difficult to measure, and the results depend on the type of measurement made. Also, the ecologic significance of production rates depends on what is being produced and how long it remains in the system. The photosynthetic bacteria contribute little or nothing directly to the food supply of larger animals, or to fisheries. Because of the confusing mixture of different types of production cycles in the sea, published values on rates of carbon fixation are largely without precise meaning. Such rates are not readily comparable to carbon fixation rates in grasslands or forests. Thus, the exercise of contrasting the productivity of the sea and the land has no merit, unless we compare the two for some purpose, for example with regard to producing protein for human consumption. (In this, the ocean's potential is 3 to 10 times less than that of the land.)

Comparing the productivity of ocean and land is not a matter of matching numbers. It is a matter of appreciating the fundamental differences in how energy moves through the food webs. There are but few areas in the ocean that are the equivalent of fields of cabbage and corn. The highly productive kelp forests come closest. They are restricted in their distribution to shallow areas with nutrient-rich waters. In addition, the path from primary producers in the sea to fish is much longer and, therefore, much less efficient than the path from grass to cattle. On land most production is by vascular plants (grasses, herbs, trees). But in the sea, as mentioned, the organisms responsible for the primary production of the ocean are microscopic algae, mainly diatoms and dinoflagellates, in addition to the minute nannoplankton and submicroscopic bacteria. Diatoms have a glassy, porous shell made of opal, a substance related to the mineral quartz, but with lots of water in a disordered mineral structure. Dinoflagellates have an organic cell wall, commonly armored with platelets of cellulose. Nannoplankton cells have calcitic platelets called coccoliths. Producers among the bacterioplankton are too small to study under an ordinary microscope.

The primary producers are eaten by small grazers within the zooplankton, which comprise both single-celled and many-celled organisms of many different types. The small zooplankton is eaten by larger zooplankton, and both feed fish. Typically, the number of steps in the transfer of organic carbon from primary producer to edible fish is near four. At each transfer, only about 15 percent of the food value is used for growth. The result is that of 1,000 grams of carbon fixed in primary production, less than 1 gram ends up as harvestable fish. In contrast, for the short chain of grass to cattle, the yield is a hundred times greater. This simple fact was overlooked, a few decades ago, when hopeful commentators touted the sea as a great source of food for a burgeoning human population. Only if we agree to eat copepod soup or krill salad is there a potential for large expansion of the harvest. This, of course, is the strategy used by the baleen whales.

Because the primary producers of the sunlit layer ultimately deliver the sustenance for everything else, fish are abundant where primary productivity is high. Fishing, therefore, is commonly most successful close to land. This is one important reason for the 200-mile economic zone of coastal states. (The other is the potential for petroleum.) Coastal waters are rich in the larvae of benthic organisms such as worms, mussels, crabs, sea stars, and many types of fishes. Most sessile organisms on the seafloor disperse in their larval stage. In this, they resemble plants that entrust their seeds to the wind. The larvae are able to capture very small food particles and are, in turn, eaten by filter-feeding organisms. Many coastal organisms produce millions of eggs per female each year. Thus, the chance for survival to reproduction for each individual larva is minuscule.

The largest life habitat on Earth is in the poorly lit and entirely dark and cold realm of the waters of the deep sea. By definition, there can be no primary producers here, only scavengers and predators. A modest portion (10 percent or less) of the organic matter produced in surface waters finds its way to below the sunlit zone. From there ever smaller portions get down to great depths, delivering food to abyssal plankton (colored red or black) and to benthic organisms, mostly small worms and protozoans. Many of the fish that live at great depth look unlike their relatives in surface waters. Some have huge mouths, most do not swim very efficiently, and many carry bioluminescent organs. Between the abyssal depths that are the abode of these mini-monsters, and the productive sunlit zone, there is an intermediate gray zone. Many of the inhabitants of this intermediate zone come up at night to feed, in the protection of darkness. Squids are prominent among the predators engaged in such daily migration. They can be caught at night, attracted by lanterns. Others live too far down for daily migration to the surface. Their habits are poorly known.

OCEAN AND CLIMATE THROUGH TIME

The Earth has icy poles and an ocean with an average temperature just above freezing. These facts dominate climate dynamics and life on the planet today. In geologic history, this cold condition has arrived rather recently, starting in earnest some 40 million years ago with ice buildup on Antarctica. For the last 3 million years the Arctic realm, as well, has acquired permanent large ice sheets. Since then the climate has changed periodically, on scales of tens of thousands of years, moving between conditions when only Greenland had ice to conditions when Canada and Scandinavia were deeply covered with ice. The origins of life forms in the sea today are linked to both the overall long-term cooling of the planet since 40 million years ago, and to the great ice age fluctuations of the last 3 million years.

Human activities are changing the climate, as a result of general warming from the release of greenhouse gases. Given the range of natural variation, the level of human impact is difficult to assess. Note that this is not the same as saying it is unimportant.[41] Naturally, one would like to know what the future might hold on a warming planet. The climate history of the geologic past, as read from deep-sea sediments recovered by drilling, suggests that a warmer ocean has less-intense upwelling and less oxygen at depth. Thus, one would expect a shift from ecosystems with high-energy predators (warm-blooded vertebrates, fast-moving fishes) toward systems with predators having low energy requirements (cold-blooded animals, slow-moving fishes, jellyfishes).

At present, whatever water is not in the ocean (3 percent) is mostly locked up in the solid seas of ice on land, dominated by the great Antarctic ice sheet (2 percent of all the water). As a result, the sea level is some 70 meters (200 feet) lower than it would be if the planet were warm in high latitudes. In the last million years, whenever ice expanded in the north and covered Canada, sea level fell another hundred meters. It last rose by that amount between about 18,000 and 8,000 years ago, as Canada and Scandinavia became ice free. The overall average rate of rise was 1 meter per century during this time span. Sea level will rise in the near future, as ice melts in Greenland, on a warming planet.[42] Looking back in time for a warm period similar to ours, some 400,000 years ago, we can obtain information on the natural range of rates of sea level change when starting from a level like that of the present. The range inferred (from the oxygen isotope record in deep-sea sediments) includes a rise of around 1 meter a century, lasting several centuries.

Cold polar regions have sea ice as well as ice on land. While not much water is locked up in the sea ice, it too is an important cog in the climate machine, because of its great extent and seasonal variability. Its presence helps stabilize the position of the polar front, which is linked to the strength of zonal winds. The fierce winter storms in the North Atlantic are a result of moving the polar front southward, till Arctic air meets tropical winds. These storms provide for deep mixing, which brings nutrients into surface waters, thus preparing them for the spring bloom that starts the food chain leading to the North Atlantic fisheries. When sea ice melts in summer, this helps stratify the uppermost water layer, which promotes growth of diatoms and other photosynthesizing organisms.

Our planet has a sea centered on the North Pole, a continent centered on the South Pole, and much more land in the north than in the south. This fundamental asymmetry in geography has consequences for climate, of course, both for the present and for the course of the ice age fluctuations. First of all, the thickness of land ice on Antarctica is measured in kilometers; that of the sea ice in the Arctic in meters. Naturally, the polar ice in the north is more vulnerable to removal upon warming than the polar ice to the south (and also more ready to expand upon cooling). The geographic differences between north and south make the far north less stable than the far south, and rather more responsive to changing conditions. For this reason, the ice age fluctuations are driven by mechanisms that are active mainly in high northern latitudes. Also, with regard to the ongoing warming of the planet, the far north is bound to show the largest changes, as it does indeed (as seen, for example, in the recent collapse of houses built on permafrost, in Alaska).

At this juncture, it is useful to clarify the issue of what research regarding global warming is about. It is largely about the problem of how to decrease uncertainty in projecting present trends into the future. It is not about the fact of greenhouse warming itself. Our planet is a good place to live *because* of greenhouse gas in the atmosphere. Without it, the ocean would be frozen solid except where the water is in contact with active volcanism. Thus, by the most elementary logic, an addition of greenhouse gas warms the planet. There is no controversy whatever about this expectation.

The physical background is exceedingly simple and readily summarized, as follows. The Sun sends energy as sunlight. A portion of this light (30 percent) is reflected back to space. The rest (70 percent) warms the planet. The planet's surface radiates infrared (heat radiation) toward space, which allows it to stay at the same temperature despite the continuous stream of energy received from the Sun. Greenhouse gases (mainly water molecules and carbon dioxide molecules) intercept some of the outgoing infrared radiation. This heats the atmosphere, which in turn radiates infrared in all directions. When the atmosphere is warm enough so that the portion radiated to space equals the incoming sunlight, balance is restored.

The basic physics, then, tells us that increasing the greenhouse gas content of the atmosphere *must* increase global temperature. The question is by how much. Credible estimates put the overall effect between 2 and 3 °C (4 to 6 °F) for a doubling of carbon dioxide, with an uncertainty of a factor of two.[43] A doubling of carbon dioxide over natural background is expected before the end of the century, at present rates of emission. The present amount over background is more than one-third of the background value. The warming of the air then increases the water vapor, whose concentration depends on the temperature. The measured warming of the Northern Hemisphere over the last century has been slower than expected, because the ocean takes up some of the excess heat. Also, an increased amount of dust and other particles in the atmosphere has shaded the ground, and the increased water vapor changes the cloud cover. Thus, despite the simplicity of the principle of greenhouse warming, the outcome of the process as a whole is very complicated. When simulating climate change in computer models one has to make a number of assumptions about poorly understood details, which unsurprisingly yields results of only tentative validity. When faced with this conundrum, some may be tempted to rely on guessing in preference to computing. Such guessing is of course no more reliable than the computations it is meant to replace.

When using long-term history as a guide to expectations for the future, we run up against a serious problem: for centuries before the nineteenth, we do not have measurements of temperatures and wind speeds, summer rain and winter snow. All these things have to be reconstructed from proxies, that is, indirect indicators. By far the bulk of proxies are the remains of living organisms. We assign to the organisms preferences in terms of environmental conditions, and then study their changing abundances, their growth rates, and changes in their composition. From these observations we deduce what the environment was like when and where they grew. Not surprisingly, the results of such reconstruction are open to question and discussion: the records derived from trees and corals are not as reliable as those from instruments.

As far as climate history on a geologic time scale, we live in a time of enormous climate fluctuations. This becomes evident when looking at the fjords of Arctic and Antarctic coastlines, or visiting high mountains anywhere on the planet. Unfailingly, the landscapes we see in

FIGURE 2.20. Coastal landscapes in high latitudes: fjords carved by glacial ice. *Upper:* Ushuaia, Patagonia, Argentina; *Lower:* Bergen, Norway.

high latitudes or at high elevations bear the legacies of the ice age. For example, fjords are drowned ice-carved valleys (fig. 2.20).

Reconstruction of the maximum extent of glacial ice fields in the Northern Hemisphere, as they were some 20,000 years ago, is of great interest in the attempt to understand the overall rules of climate change. The most striking fact is the difference in latitude of the ice margins in North America and northern Europe. The Canadian ice shield covered the Great Lakes, with its southward rim reaching the fortieth parallel at maximum extent. The Scandinavian ice shield expanded across the British Isles and northern Europe, but only to the fiftieth parallel. Heat from the western tropical Atlantic was being delivered to northern Europe by the Gulf Stream system and associated winds, then as now (fig. 2.21).

Thus, the east-west asymmetry was maintained: polar bears ranged farther south along the American margin of the North Atlantic than along the European margin. Also, in the subtropical belt, the Caribbean stayed reasonably warm (many corals survived the first glacial cold spells in the early Quaternary and thrived since), and upwelling increased off North Africa, making it colder there and augmenting productivity.

But we cannot assume that the cross-equatorial heat transport from south to north—the North Atlantic heat piracy—was as active during glacial periods as it is in the present interglacial. The entire system was more symmetrical during glacial times, with equal amounts of ice in the two hemispheres. The effect was to push the winds back toward the equator, hindering the transfer of warm winds from south to north.[44] Applied to the future, these clues suggest that the asymmetry will be increased through warming; that is, additional heat will be collected in the North Atlantic from cross-equatorial heat transport as the climate zones move northward. Monsoon effects (from unequal seasonal heating and cooling of land and sea) should increase. On the whole, tropical storms will gain energy, but zonal winds (which drive upwelling) should become weaker in the Northern Hemisphere.[45]

The ocean keeps meticulous records about climate change in its sediments on the seafloor, at least in those regions where fossils are preserved. Traditionally, the most useful fossils for climate reconstruction have been the remains of a group of organisms known as foraminifers. These single-celled amoebalike creatures build chambered shells made of calcium carbonate. They have representatives both in the plankton and on the seafloor. In the plankton, they record changes in temperature and productivity, both through changes in relative abundance of different species, and in the chemical composition of their shells. On the seafloor, they record largely changes in the rate of food supply, and in the availability of oxygen. With these clues, much detective work can be done in reconstructing surface currents, upwelling intensity, and deep circulation.

Before 1968, the retrieval of information about climate change recorded in seafloor sediments

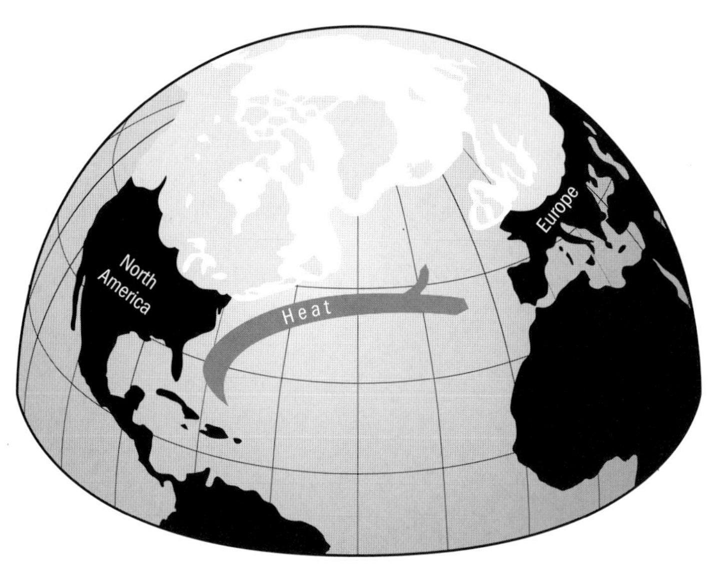

FIGURE 2.21. Maximum ice extent on land and in the sea on the northern hemisphere during glacial periods. Note the east-west asymmetry of ice margins across the North Atlantic.

relied on steel tubes that were pushed into the sediment and pulled up full of mud. In 1968 one of the great enterprises in the Earth sciences began: drilling into the deep-sea floor. The methods of drilling were perfected over the years, and eventually many detailed records of the last several million years were retrieved. As we go back in time, the record is less well documented, but the history of the ocean for the time of the evolution of modern organisms in the sea (the last 40 million years) is now quite well known.

Geologists will argue about what might have been the single most important thing that was learned as a result of the drilling. One insight that is very important in the present context, with focus on climate, is that the present sea level is unusually high. We live in a warm period. Only a few percent of geologic history over the last million years had times warmer than now. At those times, as far as we can tell, sea level apparently was higher by several meters. The message from the record is this: it is difficult to make a climate that is warmer than today's, but it is possible, and when it happens it comes with a sea level distinctly higher than today's.

One rather unexpected benefit of sampling in detail a record tens of millions of years long is the discovery that history offers examples of different kinds of catastrophes of varying severity. The single most spectacular example of catastrophe is the transition from the Cretaceous period (of the Mesozoic era) to the Tertiary period (of the Cenozoic era). It is abrupt, and it saw the extinction of most of the tropical plankton, of ammonites, and of the grand reptiles that used to dominate the Earth and its oceans. The list of doom includes the dinosaurs—except for some that had evolved into birds. The great extinction was the result of collision with a visitor from space, a big chunk of rock several miles across. This discovery revolutionized thinking about the role of catastrophe in the evolution of life on Earth.[46] Mass extinction cleared the planetary stage for the radiation and predominance of mammals, including the primates and their offspring *Homo sapiens*.

We are the sons and daughters of catastrophe.

NOTES AND REFERENCES

1. Literacy: *Literate* and *literacy* refer to the ability to read, hence learning. By analogy, *Earth literacy* implies command over a basic vocabulary of the Earth sciences and knowledge of the major concepts expressed in that vocabulary. Illiteracy is readily recognized; it expresses itself, for example, in an inability to

explain why there are seasons. The concept of having the major elements of ocean literacy up front in this book arose in connection with a workshop on the topic sponsored by CORE and convened by Jerry Schubel, Director of the Aquarium of the Pacific in Long Beach, on 15 June 2005.

2. The Mid-Ocean Ridge is most prominently displayed in the Atlantic, where it bisects the ocean basin with geometric precision.

3. A number of geologists, in fact, proposed an expanding Earth to accommodate new seafloor, in the early twentieth century and up to 1960. A summary of developments in plate tectonics in the second half of the twentieth century, by the people involved, is in N. Oreskes (ed.) with H. Le Grand, 2001, *Plate Tectonics: Seventeen Original Essays by the Scientists Who Made Earth History*. Westview, Boulder, Colo., 424 pp. For background on all things concerning physical geology and plate tectonics, see the latest edition of the textbook by Press and Siever. F. Press, R. Siever, 1994, *Understanding Earth*. W.H. Freeman, New York, 593 pp.

4. When the German geologist Alfred Wegener first published his famous book *Origin of Continents and Oceans*, in 1915, he did not know about convection and thus could not address this aspect of the origin of the face of the planet. What he talked about instead is that once there was one large supercontinent, which he called Pangaea, and one large superocean, Panthalassa, and the supercontinent broke up into big pieces that made today's continents. The ocean basins are the real estate between the continents, covered by water.

5. E.M. Borgese (ed.), *Ocean Frontiers. Explorations by Oceanographers on Five Continents*. Abrams, New York, 288 pp.

6. When working in teacher education at SIO, through University of California Extension, I first take our visitors to the beach. It gets their attention and provokes questions.

7. Paraphrasing a statement of the ecologist G. E. Hutchinson, who considered that environment sets the stage for the play of evolution.

8. P. Unitt, 2004, *San Diego County Bird Atlas*. Proceedings of the San Diego Society of Natural History, No. 39, Ibis Publishing Company, San Diego, 645 pp.

9. J. K. Terres, 1995, *The Audubon Society Encyclopedia of North American Birds*. Wings Books, Avenel, N.J., 1109 pp.

10. Having studied fishes for many decades, Richard Rosenblatt tends to see the world from a somewhat deeper perspective than the rest of us.

11. Worldwide, sea level rose by about 120 meters during a transition lasting some 12,000 years, for a typical rate near 1 meter per century. The rise flooded the shelves and drowned the river valleys, making broad estuaries such as Chesapeake Bay. The amount of sea level rise varies between regions because of local uplift or sinking of the land. That the history of sea level rise is preserved in nearshore sediments of the East Coast was recognized early in the twentieth century, for example by Alfred Redfield, one of the earliest WHOI staff members (see A.G. Gaines, 1992, *The Woods Hole Oceanographic Institution, Massachusetts*, in E. M. Borgese (ed.), *Ocean Frontiers*, pp. 55–93.) On the West Coast, Scripps geologist Francis Shepard and his student and colleague K. O. Emery studied the history and effects of sea level change. F. P. Shepard, K. O. Emery, 1941, *Submarine topography off the California Coast*. Geological Society of America, Spec. Pap. No. 31; Emery, K. O., 1958, *Shallow submerged marine terraces of southern California*. Bulletin of the Geological Society of America 69, 39–60.

12. The word *thermocline* is derived from the Greek words for *heat* and *gradient*. The thermocline is the single most important general feature of the ocean. The fact that it is missing from the dictionary in the dominant word-processing program tells us something about the state of Earth literacy in our culture. (However, *Webster's New Collegiate Dictionary* does have an entry for the subject, so there is hope.)

13. The origin of the thermocline is not clear in all details; a balance of downward mixing of warm water heated by the Sun and upward motion of cold water coming from polar regions is important, as well as lateral addition of water from moderately high latitudes through wind action. This dual nature of the thermocline was recognized in the 1950s and 1960s. P. Welander, 1959, *An advective model of the ocean thermocline*. Tellus 11, 309–311; A. Robinson, H. Stommel, 1959, *The oceanic thermocline and the associated thermohaline circulation*. Tellus 11, 295–308; W. Munk, 1966, *Abyssal recipes*. Deep-Sea Research 13, 207–230. More complicated models followed these early pioneering efforts.

14. The word *thermohaline* is derived from the Greek words for *heat* and *salt*. Temperature and salinity determine the density of seawater. Salt is involved because the warm waters moving poleward have increased their salt content during evaporation in the desert zone (the subtropics) and because salt is added to cold waters when sea ice forms, a process that rejects salt, thus increasing the density of waters below the ice.

15. The deep conveyor concept was used by Wallace Broecker of Lamont to explain certain abrupt climate changes seen in the North Atlantic at the end of the last glacial. W. S. Broecker, G. H. Denton, 1989, *The role of ocean-atmosphere reorganizations in*

glacial cycles. Geochimica et Cosmochimica Acta 53, 2465–2501. He subsequently elaborated on the implications of a global conveyor for the future behavior of the ocean as global warming proceeds. W. S. Broecker, 1997, *Thermohaline circulation, the Achilles heel of our climate system: Will man-made CO_2 upset the current balance?* Science 278, 1582–1588. The concept builds on earlier work by Henry Stommel (WHOI), Joseph Reid (Scripps), and Arnold Gordon (Lamont), among others.

16. B. Sjöberg, A. Stigebrand, 1992, *Computations of the geographical distribution of the energy flux to mixing processes via internal tides and the associated vertical circulation in the ocean.* Deep-Sea Research 39, 269–291; W. H. Munk, C. Wunsch, 1998, *Abyssal recipes II: Energetics of tidal and wind mixing.* Deep-Sea Research 45, 1977–2010.

17. The rate of general upward motion is readily guessed once the age of deep waters is known from the radiocarbon content, which was determined around 1960, at Scripps (Hans Suess) and Lamont (Wallace Broecker). The typical age is close to 1,000 years; that is, it takes 1,000 years to renew the deep-water masses of the ocean. The deep ocean layer has a thickness of 3 kilometers (from the bottom of the permanent thermocline at 1,000 meters to the average depth of 4,000 meters). The water must rise 3,000 meters in 1,000 years; that is, 3 meters a year or about 1 centimeter per day, on average. Within the thermocline layer the renewal of water masses is faster, by a factor of three or so, and the rate of ascent is correspondingly greater. Also, the rise within the upper layer is geographically concentrated in areas of upwelling, in coastal waters and along the equator.

18. Matthew Fontaine Maury (1806–1873), superintendent of the U.S. Navy's Depot of Charts and Instruments, created a series of charts of winds and currents that proved very useful to sea captains. His book, *The Physical Geography of the Sea,* first published in 1855 (Harper and Brothers, New York) may be considered the first textbook in physical oceanography.

19. D. Lewis, 1994, *We, the Navigators.* University of Hawaii Press, Honolulu, 464 pp.

20. In sailor's talk, *blowing trade* meant blowing steady. Winds are named for the direction they come from, which determines how a sailing vessel is to be steered. The currents are labeled with where they go, which determines the error (the offset) in the intended course of the vessel.

21. Heat deficit is carried when the air is dry. Such air will extract heat from the sea surface when taking up water.

22. The central gyres are the largest features of the surface of the sea, of the same scale as continents and ocean basins. The word *gyre,* used by poets in the seventeenth century, means "ring" or "vortex" and implies motion. The word rhymes with *fire.* The cognate modern English word, of Greek derivation, is *to gyrate,* to revolve around a point or axis. F. C. Mish et al. (eds.), 1991, *Webster's Ninth New Collegiate Dictionary.* Merriam-Webster, Springfield, Mass. Scripps oceanographer Walter Munk introduced the word as shorthand for the large rotating circulatory systems that dominate ocean basins. *Fide* H. Stommel, 1966, *The Gulf Stream: A Physical and Dynamical Description,* 2nd ed. University of California Press, Berkeley, 248 pp.

23. A knot is 1 nautical mile per hour, or half a meter per second.

24. For the purpose of this comparison, the height of the terrestrial habitat is taken as less than 40 meters, on average. Few trees are higher, and few birds and insects fly well above the trees.

25. The systematic arrangement of organisms, according to similarities in anatomies, was once central to biological discussion. Today's biochemists classify by genetic endowment (DNA sequences). Much of the existing system of classification of organisms is based on morphology and anatomy, rather than descent. It does not always correctly reflect ancestry, therefore, especially at the family and genus levels.

26. Comparisons between the diversity of land and ocean are useless without a definition of taxonomic level considered. Animal life in the sea is twice as old as animal life on land, as far as the fossil record, and many organisms were not able to leave the sea (including, for example, jellyfishes and echinoderms).

27. A beautifully illustrated book on fishes is that by N. B. Marshall, 1965, *Life of Fishes,* Weidenfeld and Nicolson, London, 402 pp. For fishes in and around North America, see J. O. La Gorce (ed.), 1939, *The Book of Fishes.* National Geographic Society, Washington, D.C., 367 pp. It has many delightful paintings by the artist Hashime Murayama, as well as excellent photographs. For a readable introduction to plankton and fisheries, consult the two well-illustrated volumes by A. C. Hardy, 1965, *The Open Sea: Its Natural History.* Houghton Mifflin, Boston, 2 vols., 335 and 322 pp. Marine botany is summarized on a somewhat more technical level in E. Y. Dawson, 1966, *Marine Botany.* Holt Rinehart Winston, New York, 371 pp.

28. The scientific term is actinopterygii, which is Greek for "spiny finned."

29. In scientific terminology, the three classes are agnatha, chondrichthyes, and osteichthyes, which mean "jawless," "cartilaginous fishes," and "bony fishes" in Greek.

30. *Teleost* denotes "far" and "bony," based on Greek, signifying most developed of the bony fishes.

31. The maximum size reported in A. S. Romer, 1959, *The Vertebrate Story*. 4th ed., University of Chicago Press, Chicago, 437 pp.

32. A. S. Romer, *The Vertebrate Story*, p. 53.

33. Scripps biologist Gerald Kooyman has helped pioneer this field of research. G. Kooyman, 1999, *Diving animals*, in J. N. Miller (ed.), *The Lung at Depth*. Marcel Decker, New York, pp. 587–620.

34. Roger S. Payne was the leading pioneer in these developments. R. S. Payne, 1996, *Among Whales*. Random House, New York, 431 pp.

35. Pioneer work on the topic of the microbial loop has been done in the group of Scripps biologist Farooq Azam. F. Azam, 1998, *Microbial control of oceanic carbon flux: The plot thickens*. Science 280, 694–696.

36. As is the case, for example, within the basin off Santa Barbara, California.

37. The concept of "limiting nutrient" goes back to experiments by the German chemist Justus Liebig (1803–1873), who showed that plants do not grow if deprived of any one of its essential nutrients.

38. By the Swedish chemist Lars Gunnar Sillén. L. G. Sillén, 1967, *The ocean as a chemical system*. Science 156, 1189–1197.

39. The amount of oxygen used in generating the observed abundance of nutrients is the apparent oxygen utilization (AOU).

40. What is correct is that a warm ocean has much less oxygen at depth than a cold one, because of the temperature-dependent solubility of oxygen. There will be always plenty of oxygen in the air, but the same cannot be said for the future of the deep sea.

41. Examples concerning the confusion of *difficult to assess* and *unimportant* are not hard to find. Editorials and opinion pieces concerning global warming in the newspaper media are a good place to look for this type of muddle.

42. With the term *near future* I mean the next few centuries, but starting in the present one. The rate of sea level rise could be greater than 1 meter per century, by up to a factor of two, according to experience based on the last deglaciation of Canada, 12,000 years ago. Greenland has enough ice to raise sea level by about 7 meters, and this ice is available for melting on the time scale of several centuries. Potential Antarctic contributions are difficult to assess but are unlikely to be negligible. The possibility of transient buildup of ice from increased snowfall in this century must be considered.

43. Consensus on such estimates is difficult to come by. A massive effort in this direction is by the Intergovernmental Panel on Climate Change, which has released a number of reports in the 1990s and since. J. T. Houghton, G. J. Jenkins, J. J. Ephraums (eds.), 1990, *Climate Change, The IPCC Scientific Assessment*. Cambridge University Press, Cambridge, 365 pp. The latest assessments (2007) are on the Web.

44. Arguments concerning the importance of global north-south asymmetry in climate evolution, by a distinguished meteorologist, are in H. Flohn, 1985, *Das Problem der Klimaänderungen in Vergangenheit und Zukunft*. Wissenschaftliche Buchgesellschaft, Darmstadt, 228 pp.

45. Tropical storms feed on the power contained in water vapor in warm marine air. Energy for maintaining this reservoir is in warm surface waters.

46. The hypothesis of mass extinction by asteroid impact was put forward by the physicist Luis Alvarez (1911–1988), his son Walter, and their associates, based on chemical studies on a section across the Cretaceous-Tertiary boundary in pelagic limestone beds near Gubbio, Italy. L. W. Alvarez, W. Alvarez, F. Asaro, H. Michel, 1980, *Extraterrestrial cause for the Cretaceous-Tertiary extinctions*. Science 208, 1095–1108. Deep-ocean drilling supported the hypothesis and provided much detail on the sequence of events.

THREE

Life at the Edge of a Fertile Sea

THE BIRTHPLACE OF MARINE SCIENCE

Exploration of the sea starts at the shore, whether for food or for science.

Seabirds are in evidence along practically all coasts and invite much observation and marveling about different life histories of these feathered vertebrates. Some breed locally and raise their young in full view (fig. 3.1). Others stop over on long migrations, using marine wetlands to rest and to feed. Yet other, permanent residents, add drama to the scene (fig. 3.2). But beyond the many types of birds that enliven the scenery, an incredible diversity of creatures can be found along almost any rocky, sandy, or muddy shore. Anthozoans, arthropods, mollusks, vertebrates, and polychaete worms tend to be conspicuous, as are echinoderms. Of course, the species differ from one place to another. One would not expect to find the very same organisms on the shores of Alaska and of Hawaii. However, the similarities are in many ways more striking than the differences. Thus, the exploration of shore life in any one region opens the door to the understanding of life in many such regions.

The environmental challenges for nearshore organisms depend on the nature of the seasons, with the major contrast between subpolar regions and the tropics, as well as on the nature of the shore itself, that is, whether it is flat and muddy, or rugged and rocky. Geology determines that nature: the flat coastal landscape is typical for general sinking, the rugged one for general uplift. Sinking shores have drowned river mouths and large estuaries. Rising shores have narrow beaches and cliffs and terraces, and small lagoons at the end of cliff-bounded river valleys.

As a pattern of exploration, the study of coastal ecology has proceeded from taking stock of what is there, to marveling about adaptations of the different species, to studying interactions between species, and finally to assessing the impact of human activities. Humans are now a

FIGURE 3.1. Life is about food and reproduction. Gull chick begging for food, La Jolla Cove.

dominant factor in determining the nature of life along the shores of the world. Such activities include intense fishing, introduction of nutrients and mud, and disturbance of habitat by the sheer numbers of people finding recreation along the edge of the sea. To preserve some of the rich heritage of the seashore, marine reserves have been set up, where fishing is not allowed or is strictly regulated. Questions about how large and restricted such reserves must be to be effective in regard to conservation goals are at the center of modern ecological studies along the edge of the sea.

WHERE PEOPLE MEET THE SEA

The edge of the sea—from the nearshore wetlands and shorebird nesting sites, to the surf zone—is the major ecologic boundary on the planet. It has always attracted people, because of opportunities for fishing and collecting in the tidal zone, and because of the rich environment for discovery and adventure that it offers to the bold.

The opportunities for fishing are on the decrease, but there is still plenty to discover. The exploration of life along the edge, by walking in the tidal zone or snorkeling armed with a facemask, is as exciting as ever to the young. Experiences will differ greatly depending on geography—a dive into a kelp forest is unlike a dive into coral gardens, though equally inspiring. For the experienced, there is a feeling that "things aren't as they used to be." The big predators—the giant groupers of the kelp, the large sharks in the reefs—are missing. There are no lobsters or abalone in the kelp, and in some reefs in the Caribbean much of the coral has been replaced by green algae gently waving to and fro in the surging surf. The stories of timeless drama that intrigued ecologists for the better part of the past century are gone; the tale now being told is one of increasing impact of human activities and of climate change.

The emphasis in marine ecology has changed along with the changing scenery of life at the edge of the sea, within the last two decades. Scripps's reef expert Jeremy Jackson, an experienced diver, puts it this way: "Every marine ecosystem I have studied during my 30-year career is unrecognizably different from the way it used to be. I want to know why."[1] Ecologist Paul Dayton, who has studied the kelp forests off Scripps's shores for decades, has expressed similar thoughts. He has found that human impact, especially overfishing, dominates the recent history of kelp forests. The removal of large predators, both in kelp forests and in coral reefs, results in a shift of the food web, such that grazers can increase and damage the base of the food web structure.[2] Park rangers in the Rocky Mountains have known about this type of food web dynamics for more than half a century: once mountain lions and wolves are removed, elk populations can increase rapidly and damage the forest and the entire watershed.

Half of us humans, it has been said, live at the coast, and the other half come to visit.[3] Thus, naturally, much of what people know about the sea derives from experience at the shore. The same is true for the beginnings of ocean sciences; the initial work was done along the coast, by wading in the tidal zone and from small boats, and from small fishing vessels. Bigger, motor-driven ships provided platforms to venture farther out, for the study of coastal seas and beyond. In the 1950s, a new world was opened through diving in shallow waters, thanks to the development of self-contained underwater breathing apparatus (scuba) by pioneers such as Jacques Cousteau

FIGURE 3.2. A brown pelican coming in for a share of a big fish caught by a sea lion below (out of view). Galapagos Islands, 2007. Brown pelicans are abundant along the shores of California.

and Hans Hass.[4] Subsequently, the popular image of what oceanography is all about was strongly shaped by reports featuring photographs of curious fish in coral gardens, sharks with rows of sharp teeth, and frenzied squid in the process of mating. In the popular literature, the business of oceanography has been strongly linked to diving in shallow water ever since. The physical oceanographer Robert Stewart, who works with satellite data on a global scale, is mildly amused by this. "People should know," he says, "that many oceanographers doing serious ocean science don't go to sea, and only a very few go diving, except perhaps for fun."[5]

Satellite data, which invaded all aspects of oceanography in the 1980s, hold the undisputable evidence for why life at the edge of the sea is so rich: The most fertile parts of the ocean are right next to the shore. When plotting the pattern of chlorophyll (the strength of green in the reflected light from the sea surface), the continental edges are invariably nicely outlined (fig. 3.3). The green ocean extends out to some distance, typically a hundred miles or so, but in places the blue ocean comes much closer. Off the West Coast, for example, offshore blue waters move toward the coast in the Southern California Bight, as part of a large eddy system.

In warm waters, wherever productivity is low, coral reefs flourish. They need blue water—they are ecosystems that recycle nutrients internally. Florida's coral reefs, at the southernmost tip of the peninsula, benefit from the proximity of the warm nutrient-depleted

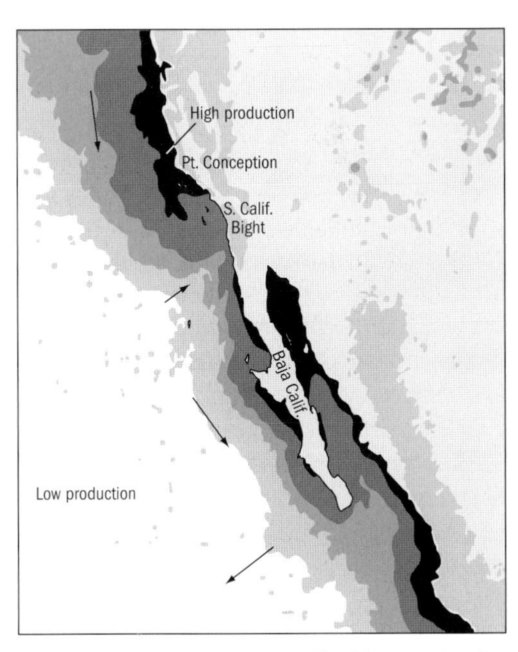

FIGURE 3.3. Chlorophyll patterns off California and in the Gulf of California. Dark regions denote green waters where production is high. Arrows show general sense of current flow.

waters of the open ocean. Kelp forests, in contrast, are restricted to high-production environments with a continuous nutrient supply; they produce organic matter at a furious pace. Much of this organic matter is lost to the coastal ecosystem in the debris that travels to the outer shelf and from there to the continental slope, where it helps feed a rich community of benthic organisms. Thus, abundances of benthic organisms are rather closely linked to the green surface patterns seen in the satellite images, as well.

We have to ask why the nearshore waters are so unusually productive, and why the coastal zone is relatively narrow. The answer is that all motions of the sea against the immobile coast produce swirling and mixing, and mixing brings water to the surface from below the warm surface layer. The zone where friction occurs between the offshore current and the seafloor, and where eddies are made, is narrow because the current touches the floor only in shallow waters. In the process, nutrients are brought up, which fertilize the surface water, and in the sunlight algae large and small then grow. In places a thick warm surface layer prevents cold waters from rising. In such cases, the water that is mixed upward into the sunlit zone is not particularly rich in nutrients, and as a result the fertilization effect is modest. Under such conditions corals can grow along the shore, as off Florida.

The mixing processes involve currents that run along the coast, and also internal waves. These are waves that move along the main boundary between warm and cold water, the thermocline. The waves can be seen from space, by their effects on surface temperatures. They are enormous, and they occur throughout the coastal ocean, stimulated by interactions between ocean currents and topographic obstacles. On breaking against the seafloor in offshore waters, they contribute to mixing. More directly, the currents along the coast produce eddies, swirls big and small, just like a river does along its banks. The eddies stir the water and mix thermocline water into the surface water, much like swirls in a cup mix coffee and

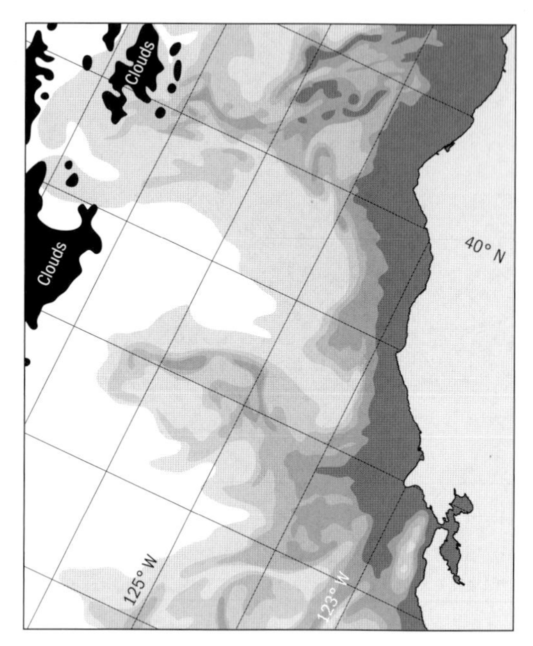

FIGURE 3.4. Mixing in coastal waters: eddies and filaments in the coastal waters off California, seen in thermal infrared from space (dark is cold and denotes rising water along the coast).

milk. The swirls of the coastal ocean can be seen from space, because the mixing lowers the temperature, which lowers the infrared radiation from the surface of the sea. In addition, we can see how the swirls feed filaments of cool water that penetrate far out into the coastal ocean, making it fertile (fig. 3.4).

The mixing mechanisms described are not the whole story. The wide shelves on both sides of the North Atlantic have very fertile waters and a rich nearshore fauna, including that of highly productive estuaries and wetlands. Here some of nutrients are delivered from land, others come in from the sea, brought by a pattern of estuarine circulation, which sets in when freshwater is added to the nearshore environment. In this circulation pattern, the freshwater runs out toward the open ocean but takes some of the more salty water with it, from below the fresh layer. The missing saltwater then is replaced from offshore, tapping into nutrient-rich waters (fig. 3.5). Once nutrients have entered the estuarine system, they have difficulty leaving it, because the outward-running

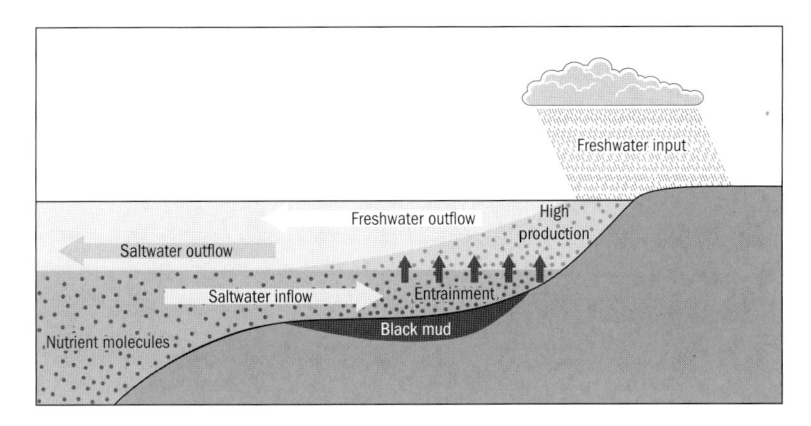

FIGURE 3.5. In estuarine circulation, the entrainment of saltwater into the outward flowing surface layer brings deeper water rich in nutrients inshore. Nutrients are trapped in the inner end of the estuary.

surface water is depleted of nutrients by the algae growing in the sunlight, and the nutrients return to the deeper layer with sinking organic matter. Thus, estuarine systems are nutrient traps, and this makes them unusually productive. Familiar examples of such systems are Chesapeake Bay on the East Coast, and the Baltic Sea bordering Germany and Poland. The principle applies to every river mouth that opens into the sea.

Regarding the nature of coastal ecosystems, then, the most important differences seen are due to warm versus cold, fertile versus blue water, and muddy versus rocky. The greatest conceivable contrast, by these criteria, is that between coral reefs (warm, blue water, homemade rocks) versus mud flats in high latitudes (cold, large seasonal contrast, green water, no rocks). In warm regions with muddy flat shores and a modest tidal range, seams of mangrove thickets are common. Armed with these simple concepts, we can readily develop expectations concerning what a coastal ecosystem might look like, from noting the geology of the coastal landscape, the temperature of offshore waters (cold waters have nutrients, warm waters do not), and the seasonal variations of the region. We can then also ask why something that we expect to see is in fact not there—a type of observation whose significance is quite commonly underestimated. Most advances in science, by far, come from checking on expectations rather than from stumbling on something interesting.

In a story emphasizing Californian oceanography, we might usefully explore along the shores of the coastal waters of the California Current, which are cold and fertile waters. Clearly, coral reefs warrant their own chapter for discussion of their very special conditions in a warm desert environment. As it happens, high-production ecosystems are right next to Scripps, and their study has occupied a great number of ecologists over the years.

A FISH SPAWNING ON THE BEACH AND OTHER LUNATIC ANIMALS

The shores of La Jolla do not lack for drama. A high-production coastal ocean unfailingly has seabirds populating the seashore, which makes for plenty of action as the animals gather food and feed their young. At the shores of La Jolla we have pelicans, cormorants, seagulls, sandpipers, sanderlings, and the occasional egret. Of these, the first two make their living by fishing. The spectacular dives of pelicans, straight down from a height of 30 feet, with wings flattened against the body, are a common sight at La Jolla. The cormorants, commonly, fish farther out to sea, and they come and go in unending commuter traffic between roosts in the cliffs and kelp offshore, rapidly and energetically flapping their black stiff wings in somewhat labored flight.

There is drama in the water as well, of course, but it is not readily observed. A sea star pulling apart a mussel to digest it, while a dramatic act in

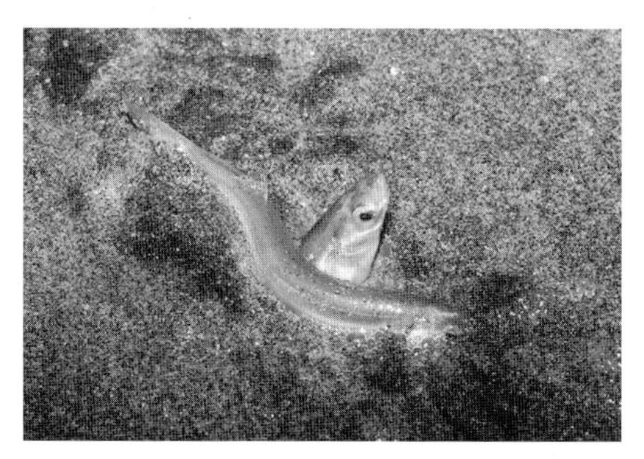

FIGURE 3.6. Spawning on the beach. A female grunion has burrowed into the sand; a male grunion releases sperm to fertilize the eggs being laid.

a sense (at least for the participants), is decidedly unexciting to watch. There is little to see, and the process takes hours. However, there are some spectacular sights involving fishes when the time is right. Along the beaches of southern California, few shows are more thrilling than a grunion run during springtime (fig. 3.6). It happens at night shortly after a full Moon or new Moon. On a sandy beach, at the peak of the tide, the ground is littered with hundreds of silvery fish, about 6 to 8 inches long, wriggling about in a frenzy, some arriving with a new wave, others heading back to the water. They are engaged in the business of reproduction and oblivious to the human observers or hunters. The females burrow into the sand, tail first, to lay the eggs. One or more males curl about the bodies of the females, releasing the milt containing fertilizing sperm. Their job done, the fish slither toward the waves, frantically beating their tail in the wash to gain speed.

At the cost of exposing themselves to gulls, raccoons, foxes, and people, the grunion parents have given their offspring a good start. They did so at night, presumably to minimize the exposure. The eggs, buried in the sand at the edge of high tide, can develop much more safely than if released into the coastal waters as plankton.

The spectacular habits of the grunion greatly intrigued the naturalist and ichthyologist Carl L. Hubbs (1894–1979) when he first saw the drama in action. Even when he was still a student at Stanford he had written a paper on the spawning of the grunion *(Leuresthes tenuis)*. After joining Scripps (in 1944) he decided to study the matter in depth. That is, more precisely, he decided to get a graduate student to do a thesis on the subject. The student he chose was Boyd Walker.[6] Walker realized he needed help in coping with the numbers involved. He engaged a large group of volunteers. Soon, on appropriate grunion nights during the spring of 1947, more than a hundred enthusiasts gathered on the sandy beaches along southern California, moving in on the spawning fish. Unlike others in those nights, these hunters were not dashing and splashing about to catch grunion for frying or pickling. They were here to count the fish, determine their sex, and clip their fins for later identification. By enlisting helpers for three successive years in this fashion, Walker gathered the data for his study on the biology of this enigmatic fish that comes on land to breed.

Walker found that the grunion spawned from late February to mid-August, most heavily from March to June. The runs use beaches from north of Point Conception in central California to Punta Abreojos in Baja California. The cycles of spawning are tied to the phases of the Moon, and the same fish reappear on successive runs. Each female produces between 1,000 and 3,000 eggs at any one event. The eggs in the sand hatch about 2 weeks after spawning, but only when their hiding place is agitated by waves. This ensures that the hatchlings emerge when the spring tide that follows the initial spawning churns the sand and carries the brood away to the sea. If the high tide fails to reach the eggs, the larvae wait for another cycle (or even two) before hatching. After that they perish if left in the sand.

The grunion is a shining example of Moon-dominated reproductive cycles along the fringes of the ocean. The fishes come onto the beach to spawn on the night after the peak tide and on the following two or three nights. This makes sense. If they were to spawn before the highest tide, the eggs would have no time to develop before being washed out to sea. Eggs deposited at the highest tide might just stay in place, high and dry, with subsequent tides falling short of releasing them.

Somehow, the grunion keeps track of the best time to spawn, using an internal clock set by the moonlight and possibly other clues.

Spawning runs similar to those of the grunion, by the horseshoe crab, may be observed on the East Coast on nights of the new and full Moon in May and June. *Limulus polyphemus* is not a crab, despite its name, but a relative of scorpions, and more closely related to spiders than to lobsters. Considered a living fossil, this hard-shelled foot-long and hoof-shaped creature with a pointed tail is a relict from the era of the dinosaurs. Look-alikes are readily recognized, for example, in the finely bedded Jurassic limestone of Solnhofen in northern Bavaria. For most of the year, the horseshoe crabs feed on small bottom-living organisms in deep water offshore. In the spring, they head for shallow water, crawling along the bottom, in pairs or in chains. Finally, when the tide is right, they come ashore by the thousands to spawn. The females deposit eggs in a shallow hole in the sand. The males release sperm at the site, to fertilize the eggs. Within 2 weeks those eggs that were not eaten by birds and other foragers are ready to hatch at the next high tide.

The feats of timing of both grunion and horseshoe crab are astonishing. But these beach spawners are by no means alone in knowing their tide tables. The mussel beds along the West Coast, from Alaska to Mexico, harbor a predaceous worm, *Nereis vexillosa*, with powerful chitinous jaws for grasping prey. The worm is an annelid (that is, a cousin of the earthworm) that grows to a foot long. Scripps biologist Martin W. Johnson looked into its reproductive habits, in the early 1940s. When ready to reproduce, this worm moves out to sea, to seed the water with sperm and eggs during nights of new Moon in summer.

In the Samoan and Fijian islands of the South Pacific, another annelid—the palolo worm—similarly relies upon the Moon to time its reproduction. However, it sends only the reproductive portion of the body to swarm, while the segment with head and vital organs stays put within its coral shelter. At the lunar signal, the headless bodies spin toward the surface in vast numbers. Eggs burst out from the female segments, while sperm spews from the male ones. The islanders are ready, in their canoes, scooping up the pieces for a feast of worm roe. They know the precise timing of the event: they too watch the Moon. How many millions of years did it take the worm to get the Moon into its genes? How many centuries did it take for people to figure out the connection? How would one find out?

Lunar reproductive cycles are ubiquitous. We find them in coral reefs and in mass spawnings of unrelated species, and we find them in pelagic animals, even in the open ocean. Fishermen quickly take advantage of such regular events. In the Caribbean, for example, Nassau groupers *(Epinephelus striatus)*, fish that reach 3 feet in length and weigh about 15 pounds, congregate to spawn in certain reef areas during the full Moon in the winter months. The water becomes cloudy white with eggs and sperm. Whenever such spawning areas are discovered by fishermen, populations are quickly reduced to a small fraction of the former size. Regulations are now in place in the Bahamas and elsewhere to allow the populations to recover and persist.[7]

Let's go back to the grunion. How did it adapt so well to the tides? How could such a complicated breeding behavior have evolved? The professional beach-combing biologist Edward F. Ricketts, a keen observer of nearshore life, suggested that when faced with Moon-directed reproduction, we are "confronted with a mystery beside which any of Sherlock Holmes' problems seem pale and insipid."[8] Actually, Holmes's sidekick, Dr. Watson, might have come up with an answer, perhaps along the following lines.

First of all, it seems apparent that the tidal spawning of this little fish is but a special case of Moon-synchronized reproduction. Fishes releasing eggs and sperm into the surroundings need to make sure that plenty of members of the same species are about. The more concentrated the spawners, and the less dispersed the eggs and sperm, the more likely it is that full fertilization will occur. Also, with so many eggs in the water, predators will soon be sated. Thus, at the beginning of this evolutionary narrative, the fish in any one population that responded more precisely to a lunar or tidal cycle and congregated in a more restricted region, such as the surf zone, had the higher reproductive rates. Milt and eggs were most concentrated for those fish that spawned right at the edge of the surf, next to the beach. The tide was high during spawning, because the only distinctive lunar signals are full and new Moon: anything in between does not mark a special event. During high tide, some of the eggs washed onto the beach. Here they could disappear into the rocks (in a gravel beach), or (if lucky) were buried in the sand by successive waves. Within the beach sediment, they were relatively safe from being eaten. Again, with some luck, the next high tide was able to free the eggs, with the fully developed larvae.

From this scenario, we can easily spin a tale of further development. Females began to release their eggs ever farther up the beach, and the males followed them. Those females that made an effort to release the eggs within the sediment (rather than just on top of it) could move from gravel beaches (where eggs drop down by themselves) to sandy beaches (where they do not). The females that did this right had more offspring than did their sisters.

Paleobiologists fondly call this type of reconstruction by narration a "just-so story," in honor of that great storyteller Rudyard Kipling. Is our just-so story true? We do not know. Whenever we talk about a string of events that happened a long time ago, we are just guessing. A good guess predicts something not yet observed. In the case at hand, our tale suggests that related species, or at least some fish within the same habitat, might be expected to spawn at the edge of the surf zone, minding the Moon, and preferring gravel beaches.

Moon-directed reproduction has yet other advantages beyond getting lots of eggs and sperm together. By overloading the environment with eggs, losses from predation are greatly reduced. Whatever creatures eat eggs will be well fed on the occasion but will have to do without such feasting for the rest of the lunar cycle. To seize this particular advantage it is beneficial to synchronize release of eggs and sperm not just with members of the same species, but also with other animals in the vicinity. Hence, one finds mass spawning of different species, as has been observed in coral reefs of the South Pacific. For adults that have to come out of hiding to find a mate, a similar case can be made that all should come out of hiding at once, so that predators will take a lesser toll.

So, by Watsonian logic based on plausibility, we can make sense of reproductive lunacy in the ocean. Human reproduction, of course, also is linked to the Moon. It reminds us that we share many things with other organisms on this planet, including the Moon.

WHY ANIMALS ARE WELL ADAPTED TO THEIR HABITAT

The conceptual tools that allowed our hypothetical Dr. Watson to take up Rickett's challenge and come up with a plausible answer regarding the grunion's behavior are straight from *The Origin of Species*, published in 1859. (As it happens, this also was the year of the birth of Sir Arthur Conan Doyle, the creator of Sherlock Holmes and Dr. Watson.) *The Origin of Species* is the crowning work of the great naturalist Charles Darwin, and the founding feat of biology as a science.

The central idea of Darwin's book is that species are adapted to their environment because they change as the environment changes. This adjustment to external change is by natural selection acting on the competing members of the same generation. The members of any one species vary in their capacity to leave offspring. The ones with the most offspring (the ones that cope best) will gradually replace the others. Darwin's theory explains the fine-tuning of organisms to their environment.[9]

Darwin was not the first to introduce the idea that life on Earth evolves. The French naturalist Jean Baptiste Lamarck insisted early in the nineteenth century that all classification reflects, or should reflect, ancestry.[10] But the crucial step on the path to evolutionary theory was to present overwhelming evidence that organisms do indeed change, and to find an acceptable mechanism for making new species. Darwin was struck by the fact that the level of success of different species seems to differ. Some animals are very abundant, presumably a measure of their success. Other animals are quite rare, and perhaps not quite so successful. Something is checking the abundance of organisms, each of which, through exponential growth, could easily overpopulate the world if given the resources to do so. Darwin realized that the forces balancing the potential for exponential population growth must act slightly differently on the various members of a given population. The ones that are slightly more able to cope will prevail in the competition for the survival of offspring. Thus, a changing environment (or migration into a new environment) will subtly shape the composition of a species till it is quite different from the ancestral type.

The major steps in going from classification to evolution are familiar to every naturalist; they constitute part of our great intellectual heritage of the last several centuries. For astronomy, there is the fact that Earth circles the Sun, for physics, that one type of energy can change into another. For biology, there is evolution. Nothing we see makes much sense without it.

The Swedish botanist Carolus Linnaeus established the modern system for naming and classifying plants and animals. He published it in his *Systema Naturae*, in 1735.[11] He defined species and introduced the convention whereby each species receives a genus and species name (as in *Mytilus edulis*, the edible mussel). He also grouped genera into higher categories. His scheme formed the basis for systematic hierarchies, and we now have the following sequence:

Kingdom ≫ Phylum ≫ Class ≫ Order ≫
Family ≫ Genus ≫ Species

The class level contains familiar common names such as coral, sea star, barnacle, bird, and shark, and some less familiar ones such as crustacean, gastropod, and cephalopod. Order and family levels divide these large categories further (e.g., slugs and snails, squid and scuttle fish) till we get to genus, and finally species. The most familiar example of Linnaean classification is our own systematic position, as humans: Eukaryotes (subkingdom animals), chordates (subphylum vertebrates), mammals, primates, hominids, *Homo, H. sapiens.* The French zoologist Lamarck offered an explanation as to why such groupings exist at all. In his book *Philosophie Zoologique* (published in 1809) he proposed that all organisms arose by evolution and are related to each other by common ancestry. The more similar the organisms, the more closely related they are. Half a century later, the British naturalists Darwin and Wallace (in 1858) proposed a mechanism for evolution.[12] Darwin (1859) affirmed that classification reflects common ancestry and is the result of evolution through geologic history. The mechanism driving evolution is adaptation to changing conditions through natural selection, whereby the best adapted have the most offspring.

The history of life is the history of extinction of some species and the generation of new species from ancestral ones. All available evidence supports these two concepts. Furthermore, there is no alternative explanation that could clarify why some organisms whose remains we find in the geologic record are no longer extant, and why present forms have antecedents, as fossils, that look similar but gradually less so as we go back in time. Common ancestry also explains the many striking similarities between the larvae of different groups of organisms in different orders but in the same class.

It is a matter of celebration, in a book extolling the science of the sea, that Darwin's epochal insights go back to a world-encircling sea voyage. The voyage of the *Beagle* is forever joined to the revolution in thought that Darwin wrought. Darwin's monumental contributions to evolution and ecology have become central to understanding life on Earth. His great conceptual leap was to introduce mechanistic principles (unthinking automatic processes) to the complex problems dealing with the distribution of species and with their origins. In doing so, he implicitly rejected miraculous divine intervention. This corollary was not lost on the clerics of the time, and some objected vociferously. Physicists, astronomers, and geologists had already long declined to call on miraculous intervention to explain poorly understood phenomena. Their insistence on a strictly scientific approach prevailed. It has been more difficult to convince everyone in the public arena that the same rules apply in biology.[13]

The basic idea underlying Darwin's mechanism driving evolution is the observation that there are limits to expansion of a population, limits that any one population has to cope with and that are constantly challenging its extent.

The edge of the sea is a cauldron of evolution where sea life invades the land, and where land life invades the sea. The diversity of seashore habitats and the different opportunities for making a living arising from this diversity sustain a great variety of species. Each species has to solve the problem of obtaining the basic requisites of survival: a suitable place to live, enough to eat, protection from predators, opportunity for reproduction. The strategies adopted to achieve this end range from the seemingly simple habits of mud-eating worms to the complex behaviors of territorial fishes and sea gulls. In some environments, coping with the physical environment is the main challenge, while in others interactions between organisms dominate the ecologic scene. Crudely generalizing: at the edge of the sea, biological interactions are predominant in warm waters, while physical factors become very important in cold waters, where conditions change markedly through the seasons.

The types of organisms and the communities that form at the edge of the sea are as varied as the types of seacoasts and their climates. How can we bring order to the relationships between environment and life forms, and to the relationships between the organisms themselves, without drowning in descriptive information? It is a difficult task. The scientists who study this confusing scene, who discover why organisms live in one place and not another, and observe what they do to each other, are ecologists, and their field of study is ecology.[14]

In recent years the term *ecology* has acquired a broader meaning beyond the study of the ways of life of organisms and their communities. *Ecology-minded* may actually refer to someone recycling paper and beer cans. We have come to realize that resources are limited: energy and raw materials, clean air and clean water, food free from harmful chemicals—nearly everything that is desirable is in short supply. Ecologists have always known this. In fact, their studies revolve largely around this very business of resources being limited. Limited resources are the reason why individual species cannot expand their population forever, as emphasized by the young Darwin in his journal. Limits on required resources govern the number of mussels on the rocks, the number of crabs scurrying about, the algae, the sponges, the corals, the sea urchins, the fishes, and the sea gulls. Each and every species has its own set of limits, because each has its own set of requirements and survival skills.

Each shore type, with its various habitats, presents special challenges to the organisms that would live there. However, there are some generic problems that are typical for life at the shore, regardless of habitat type. The challenge, for any organism, is to grow to adulthood and leave offspring—as many as possible. That is the goal. The programs for achieving this differ, depending on conditions.

MEETING THE CHALLENGES OF A ROCKY SEASHORE

In southern California, the typical shorescape is a narrow rocky beach in front of cliffs, and the same is true for the scenery seen from the Scripps pier (fig. 3.7). The nearshore organisms are adapted to strong wave action and a high tidal range. There is a marked zonation of animals, with respect to the elevation above the low end of the tidal range.

Why is this so? First and foremost, there is the danger of drying out. The conspicuous zonation of intertidal organisms along rocky

FIGURE 3.7. Beach and cliffs north of La Jolla, a favorite site for studying the rocky intertidal and other shore life.

shores largely reflects the greater or lesser ability of the creatures to do without water. The higher up the organisms live, the greater is their exposure to air and sun, and the better they have to be at dealing with the potential loss of water. Organisms fight desiccation by withdrawing into a shell and closing up tight (acorn barnacles, bay mussels, periwinkles), or by pressing their shell against the rock on which they sit (limpets, chitons), or by carrying a supply of moisture to bathe the gills by which they breathe (rock crab). The small acorn barnacle (fig. 3.8), which lives in the uppermost part of the intertidal zone, closes up so tightly that it will not be harmed if, in experiment, it is briefly submerged in lethal preservative fluid. Most of the hardy creatures living in the spray zone in fact would be perfectly happy to live farther down, where there is more water and more food. However, there are also more competitors there, and more predators.[15]

Another problem is the danger of becoming overheated. Every organism, of course, is adapted to the temperature range of its specific habitat. The creatures in polar areas like it cool, those in the tropics like it warm. The challenge for many nearshore organisms comes from a very large range of temperature, or from sudden large change. Like drying out, this hazard is most acute in the rocky intertidal, where animals are fully exposed to the Sun. To avoid heat problems, the larvae of the species clinging to rock are particular about where to settle. As every real estate agent knows, location is everything. The larvae like it where there is some current, even backwash from spray. Also, they look for cracks. Both preferences steer them away from dry, hot places.

FIGURE 3.8. Shelled animals on rock in the intertidal zone: mussels, gooseneck barnacles, limpet. Mussels and limpet carry the small acorn barnacle *Chthamalus fissus*.

FIGURE 3.9. Two kinds of sea anemones: colonial and solitary.

A common sight on rocks in the middle tidal range is the colonial sea anemone *(Anthopleura elegantissima)* (fig 3.9, upper). When the tide is out, these animals close shop, and their shell-covered sides make a roof keeping the sun out and slowing desiccation. In the same colony, animals will still be open and ready to catch food if they are submerged. Thus, it is the exposure to the air (rather than the sun) that makes the difference in their behavior. The closely related solitary green sea anemone *(A. xanthogrammica)*, prefers a slightly deeper and shadier place, being unable to close up in the fashion of its shell-collecting cousins (fig. 3.9, lower). When exposed, its oral area fills up with sand, which is washed off when the waves return. The wet sand keeps the animal moist.

As many a body surfer has found out when being slammed onto the ground by a large breaker, wave shock is an important element of the environment on the Pacific coast. The vast expanse of unbroken sea surface over which the westerly winds can blow (the fetch) can make for large waves and for pounding surf. To be able to live where breakers crash against rocky shores, an organism needs special adaptations. On the rocks, clinging animals use various devices for attachment: sticky strings (byssus hairs) for mussels, a form of cement for acorn barnacles, suction feet for the sea stars, a foot like a suction disc for limpets and chitons. (Limpets are snails with a cuplike shell; chitons are in a class by themselves, somewhat snail-like but with a segmented shell.) The rock oyster simply cements its entire shell to the rock.

The opposite of the risk of drying out is the danger of osmotic drowning. Chemically, each organism is a salty fluid enclosed by a semipermeable membrane bag. If the concentration of dissolved salt within the bag is higher than the concentration outside it, water diffuses into it, and this inward diffusion of water leads to internal drowning. For this reason, a marine organism in low-salinity water has to accept an increased (and hence diluted) fluid content, or get rid of excess water by excreting it, or lose salt, or use a combination of these strategies. Presumably, the reason that sea stars, sea urchins, and sea cucumbers avoid brackish water and freshwater is their inability to cope with these requirements.

For many nearshore organisms, osmotic stress becomes limiting in the attempt to settle in estuaries, especially in the upper reaches. The marine creatures invade the estuary at the bottom, within the tongue of heavy saline water coming in from the open sea (fig. 3.5). One by one, the invaders drop out and fail to move farther up the estuary. The species dropping out first are called stenohaline, because they are tied to a narrow (Greek: *stenos*) range of salt *(halos)* content. Species that occur over a broad *(eury)* range of salinity are called euryhaline. An organism can be both stenohaline and euryhaline at different stages of development. For example, experiments show that the larvae of the California mussel cannot tolerate brackish seawater in which the adults can survive quite well.

Salinity tolerances of marine organisms change with changing environmental conditions: animals that are too hot or too cold for comfort, or are starved, have lower tolerances. This relationship applies to all forms of stress, not just salinity, and to all organisms, including people. (Noise and air pollution lower resistance to stress.)

The fundamental ecological classification of organisms is a "who eats whom," exemplified by the basic food chain starting with primary producers and moving through herbivores to carnivores. This kind of scheme works well on land, where we have the series plants-rabbits-foxes, or plants-caterpillars-birds. In the ocean, assignment of trophic levels (the rungs on the ingestion ladder) can be quite difficult. Many greenish (or brown) protists bearing chlorophyll—that is, producers—also capture and consume prey; in other words, they are also predators. It is as though, in the food chain on land, most plants were trapping insects. Filtering organisms such as mussels feed on a mixture of plants (microscopic chlorophyll-bearing protists, actually) and animals (protozoans and all sorts of larvae including their own), making them both grazers and predators. Sardines strain the water through their gill rakers, eating diatoms—which makes them grazers—and also small zooplankton—which makes them second-order predators. Thus, in marine ecology, we deal with food webs rather than food chains. Trophic levels within this web are statistical concepts, defined in terms of number of feeding steps since photosynthesis.

There are several ways to escape being eaten. Making oneself invisible is one, by hiding in a burrow or crevice, or by adopting protective coloration. Running away fast is another, used by shore crabs, for example. Another defense ploy is to produce a solid shell that is difficult to handle because of knobs and spines. Making oneself difficult to dislodge is yet another way to resist becoming dinner. Octopus mothers care for their eggs, defending and cleaning them. The small coastal squid *Loligo* makes its eggs distasteful: when fed to a sea anemone they were rejected unharmed (16). The defensive ink ejected by an octopus put in an aquarium with a moray eel was observed to disable the eel's sense of smell.[17] All such strategies cannot completely discourage predation, but they can slow the process down, and thereby direct predators toward other food, more easily gathered. Thus, defense is a competition between potential prey, the goal being to come in last in the race to oblivion.

Success in life is tied to an organism's abundance. According to Darwin, as we have seen, rarity may be the first step to extinction. By this criterion, few seashore organisms on the West Coast are more successful than the California mussel, covering acres and acres of rocky shore exposed between the tides, from Alaska to Mexico. Clearly, the mussel is able to extract the resources it needs from its environment in a very efficient manner.

The mussel-bed assemblage is in fact the most conspicuous shore-life community along the West Coast, commonly a densely packed carpet of the California mussel *(Mytilus californianus)* and the gooseneck barnacle *(Pollicipes polymerus)* (fig. 3.10). Other remarkable members are the colorful common sea star *(Pisaster ochreus),* which can be purple, brown, orange, or yellow, and various types of snails, sea anemones, tube-building worms, encrusting barnacles, and crabs. The community looks similar from Alaska to Mexico, although different species or varieties are involved. The dense clustering of mussels and gooseneck barnacles is noteworthy. Presumably it is a preference developed in defense against the main predator, the common sea star. The sea star opens mussels by applying hydraulic pressure through its suction feet on both valves. When the mussel finally opens, unable to cope with the sustained pressure, the sea star digests the prey within the shell, extruding its stomach for the purpose. Densely packed mussels are difficult to get at, so sea stars commonly work on the specimens at the lower boundary of a bed. As an additional benefit for the predator, the mussels here tend to be relatively large, and exposure during low tide is minimized.[18]

Life on the rocks is just one part of the life cycle. A future mussel must find a suitable place to live while drifting about as a larval explorer,

FIGURE 3.10. The mussel-bed assemblage of the Californian rocky shores: California mussel and gooseneck barnacles. The sea star *Pisaster* preys on the mussels.

waiting to be carried ashore to a solid object. These larvae are part of the temporary plankton, called meroplankton. Coastal waters are rich in organisms filtering the water for food; hence, almost all mussel larvae end up in someone's stomach. The adult mussels compensate for this loss by producing prodigious numbers of offspring. The larvae are good at hunting for the proper habitat. They are sensitive to light, to currents, to pressure changes, to salinity, to the roughness and bacterial cover of the substrate, and to the presence of other animals, especially adult California mussels. If they like what they feel, they settle. If not, they drift on. After settling, metamorphosis sets in. The little mussels grow byssal hairs, which are tough, hornlike threads extruded by a gland in the foot, to attach themselves firmly to the rock. Most probably the snails crawling about will soon eat the newcomers. The few that survive can grow to a length of around 6 inches and stay around for more than a decade. The history of growth is recorded in the chemistry of their shell and can be recovered by measuring the abundance of the stable isotopes of oxygen and carbon through time.[19] The same method also allows one to determine at what time of the year mussels were collected for food by aboriginal people, by analyzing the shells in the middens they left behind.[20]

A grown mussel will filter up to 100 liters (25 gallons) of seawater a day, extracting diatoms, dinoflagellates, and the meroplanktonic larvae of nearshore animals, including those of mussels. In a sense, the mussels send food gatherers offshore: their own larvae.

The California mussel is eminently edible (as is its well-named cousin, *Mytilus edulis*, the bay mussel, fig. 3.11). Mussels were a primary food item for the earliest Californians, at least back to 7,000 years BP. Shell middens along the shores of southern California are typically 90 percent mussel shells, according to Scripps biologist Carl L. Hubbs, who inspected them in the 1950s and 1960s. (Much of the remainder consists of fragments of the giant owl limpet *[Lottia gigantea]*—a tough chew—and of abalone shells.) Unfortunately for humans (and presumably fortunately for mussels), mussels can be poisonous during times when certain dinoflagellates are abundant in their food, especially in northern California. Blooms of such noxious microscopic algae occur almost solely in the summer months; their origin is not clear but apparently has to do with the establishment of a thin, warm surface layer over cold water. In any case, ingestion of harmful algae by the mussels makes these inedible, and this is why mussels and bivalves in general are shunned in the summer by the wise. Some early explorers, unfamiliar with regional conditions, lost their lives to paralytic poisoning when eating mussels out of season.[21]

The fact that mussels can become poisonous reflects the efficiency with which these animals extract whatever is suspended in the ocean. In fact, their body tissues provide a sampling of what is in the sea, including materials introduced through pollution. In the 1970s, Scripps chemist Edward D. Goldberg recognized the value of using the mussels and other bivalves as collectors of foreign substances in the sea and, hence, as monitors of water quality. He and his associates analyzed mussels and oysters collected on the West, East, and Gulf coasts for trace metals, chlorinated hydrocarbons, petroleum hydrocarbons, and radionuclides. Because of the minute concentrations of these substances, especially painstaking precautions must be taken to avoid any contamination. Many interesting and useful results emerged from this investigation,

FIGURE 3.11. Dense aggregation of the bay mussel, *Mytilus edulis*.

dubbed "Mussel Watch," which documented the changing concentrations and the pathways of a number of industrial pollutants, and their dispersal in the coastal ocean.[22]

The mussel-watch approach subsequently proved very fruitful within a large national program organized by the U. S. National Oceanographic and Atmospheric Administration (NOAA). Of special interest is the fate of halogenated hydrocarbons such as DDT and PCB.[23] These compounds are biologically very active and interfere with reproduction, larval growth, and immune responses of organisms exposed. The marine biologist Rachel Carson first drew attention to the fact that such poisons are concentrated up the food chain.[24] This is the reason they are found even in the remotest parts of the sea, for example in organisms living at abyssal depths in the Atlantic, as well as in Antarctic penguins and Arctic seals. The best-known example both of the bioconcentration effect and the interference with reproduction is the failure of pelicans along California to lay viable eggs as a result of poisoning with DDT. A thinning of the eggs was especially obvious, leading to breakage. (Other predatory birds also were affected.) As a hazard to human health, DDT was largely banned in developed countries in the early 1970s, but its use continues elsewhere, where alternative means to prevail against insects (for example, malaria-carrying mosquitoes) may be difficult to obtain.

The mussel-watch program allowed the monitoring of the effects of the phasing out of chlorinated hydrocarbons, with respect to the various regions along the coastline of the United States. These pollutants tend to linger in the environment, even after active release has ceased.

BEGINNINGS OF A MARINE BIOLOGICAL LABORATORY

Imagine arriving at an unfamiliar seashore, never before studied. As a marine biologist, what should you do? Find out what is living there, of course, and record the abundance, distribution, and habits of the species found. This is precisely what E. W. Ritter, professor of zoology in Berkeley, had in mind when he moved his summer field operations to San Diego in 1903. The questions arising when doing marine biology naturally are embedded in the intellectual climate of the time. Inventory, taxonomic systematics, and the question of evolutionary relationships provided focus for biological studies. Ritter was part of this environment, but he was also ahead of his times.

Ritter's intent was to establish a year-round research program on marine biology. Unlike the summer stations elsewhere, it would not include undergraduate students or temporary outside investigators (or, at most, just a few). He knew that "a large amount of work in systematic zoology" lay ahead—something that called for dedicated experts. Because of his great interest in habitats and in animal-habitat interaction, he also wanted physical, chemical, and hydrographic researchers. It was this breadth of view—quite unusual in a period focused on classification—that made his program oceanographic rather than solely biological. Ritter was an early ecologist. He believed in studying the organisms within the system that sustained them. This fundamental credo laid the foundations for the strategy of research at Scripps and gave it the distinction of being the first oceanographic institution in the United States.

Scripps was not, of course, the first marine biological laboratory in North America. There already were a few of those at the end of the nineteenth century. Several summer marine stations had been established on the East Coast, following the example of Louis Agassiz's summer laboratory program for biologists and geologists at Penikese Island, off the coast of Massachusetts, in 1873.[25] In 1881 the Woman's Education Association, through the Boston Society of Natural History, established a summer Sea-side Laboratory at Annisquam near Gloucester, Massachusetts. Men were welcome, too. In 1886 the program was moved to Woods Hole, Massachusetts, to be near the fisheries laboratory of the U. S. Fish Commission.[26] The program metamorphosed into the Marine Biological

Laboratory and thus started, in 1888, a distinguished record of research that continues to the present, side by side with one of the great oceanographic institutions of the world.

Other precedents existed in Europe and in Japan. An outstanding example for a full-fledged marine biology laboratory was the Zoological Station in Naples, founded by the German zoologist Anton Dohrn in 1872 as an international research institute where scientists would go to study marine creatures. In 1887, the Imperial University of Tokyo opened a Marine Biological Station at Misaki. In 1888, a research station (the Marine Biological Association) was established in Plymouth, England, for the study of fisheries and general marine biology. Thus, the time was right for taking up the challenge in southern California, as well, where the Pacific coast offered so much uncharted territory for this type of work.[27]

The proper home for a marine biologist is a laboratory next to the sea, with facilities for collecting and keeping marine organisms. (These days, of course, access to a chemical lab and to computing facilities is equally important, but we are talking here about the first decade of the past century.)

So, Ritter first set out to find a site to build a laboratory. There was a piece of land, 170 acres of "barren brown hills, canyons and arroyos" north of La Jolla, a suburb of San Diego, that seemed pretty useless to everyone else but that would do fine for Ritter's purpose. The property had a relatively narrow terrace next to the sea, just high enough to be safe from the pounding surf below. It seemed ideal for building the research station that Ritter had in mind. And Ritter had more than a great idea: he had a friend who was ready to help in bringing it about. E.W. Scripps, wealthy owner of many newspapers, was eager to bring a world of learning to San Diego.[28] The two thought big. They talked about research buildings, cottages, a botanical and zoological garden, a museum of natural history, and even a college. Ritter saw the pristine sandy beach at the edge of the land, and those rich tide pools along the rocky shores just five minutes away. He knew that this was the ideal place to start "a biological survey of the waters of the Pacific adjacent to the coast of Southern California."

Ritter chose well. The Scripps campus is centered between the two heads of a great submarine canyon, which would provide immediate access to deep water to divers half a century later. Gray whales can be seen and counted from the top of the station during their migrations to and from Baja California. The area has some of the best beaches for grunion spawning, and rich kelp beds offshore. The rocky shore nearby, while limited in extent, has great diversity of organisms. Interesting lagoons and tidal flats are not far away. Many decades later, two marine preserves were to be set aside, to protect these various assets.

The city of San Diego allowed the Marine Biological Association of San Diego to acquire the land, as the sole bidder, for $1,000. It proved a good investment. By 1910 there was a laboratory, and cottages were built at the station in 1913 and 1915 for some of the staff members. A truck belonging to the station brought groceries and mail from the town, 10 minutes away by dirt road.[29] The station had a research vessel, the 85-foot-long *Alexander Agassiz*, built in 1907. It was used to collect biological specimens and water samples in the coastal waters offshore.

As mentioned, in walking distance there were the tide pools and rocks with a rich sampling of the great abundance and diversity of life that characterizes Californian shores (fig. 3.12). As is typical for all nearshore environments, the master phyla of the sea are well represented: sea anemones (cnidarians), mussels, snails, nudibranchs, chitons, octopus (mollusks), barnacles, crabs, isopods, flies (arthropods), and fishes and birds (chordates). We do not know whether Ritter, for the benefit of newcomers and students, posted a simple key to the most common creatures found along the shores near Scripps. Of course, no publications were available to help in this, but his experience with shore biology in Pacific Grove to the north would have helped greatly: the shore life there is

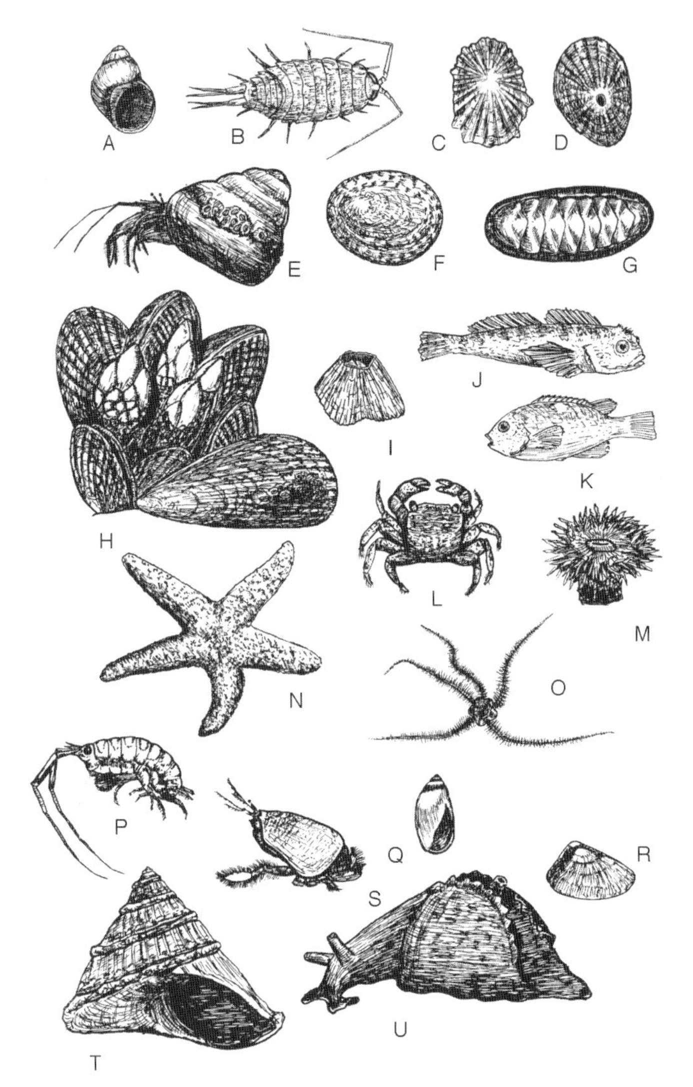

FIGURE 3.12. An amateur's guide to the animals of the seashore near Scripps illustrates the diversity of marine life in the vicinity of the Institution. Shown are: A, periwinkle (gastropod); B, "rock louse" (isopod); C, rough limpet (gastropod); D, volcano limpet (gastropod); E, hermit crab in black turban shell (decapod, gastropod); F, owl limpet (gastropod); G, chiton or "sea cradle" (amphineuran); H, California mussel (pelecypod) and gooseneck barnacle (cirriped); I, acorn barnacle (cirriped); J, tide pool sculpin (bony fish); K, opal eye (bony fish); L, rock crab or striped shore crab (decapod); M, sea anemone, solitary and aggregate (anthozoans); N, common sea star (asteroid); O, brittle star (ophiuroid); P, beach hopper (amphipod); Q, olive snail (gastropod); R, bean clam (pelecypod); S, sand crab (decapod); T, wavy top shell (gastropod); U, sea hare (gastropod).

not much different from that off Scripps. One drawback in attempting to provide instructional materials: desktop mechanical duplication of printed matter was not yet an option.

We have already met many of the organisms populating the rocky shores and discussed the factors, physical and biological, responsible for their zonation within the intertidal. Of these factors, physical ones (such as length of exposure to air) are the ones most readily determined, and their influence was discovered first. The more subtle biological factors remained for later exploration.

Many interesting questions arise just from observing the behavior and interactions of the various animals. For example, when exactly do the snails, limpets, and chitons that are lodged in cracks and depressions move about to feed? How do the hermit crabs *(Pagurus)* acquire their shells when it is time to move to a new house? Who eats rock oysters? Why do shorebirds congregate in small feeding areas on the lower beach? And so on, building a question on every common observation. Usually, the answers include constraints by the physical environment (desiccation, in the intertidal zone), external opportunities (for example, empty shells washed into a heap by waves in some tidal pool), sporadic predation (sea stars, birds), competition by conspecifics and others (holding on to space), and concentration of food items (for example, sand crabs aggregating in areas of increased suspension). That is, both obvious and subtle physical processes and biological interactions are at work testing the adaptations—the fine-tuning of capabilities—of every organism in its environment, every day. After finding out just what lives in a location, the way organisms cope with their challenges was the object of study in early ecology.

A perennial object of study is the relationship between organisms. Here the question is just how closely these animals are related. It is a question that has maintained its relevancy ever since Darwin.

The answer, in almost all cases, has already been given by means of traditional systematics. If the two animals in question are placed in the same genus, they are closely related. If in the same phylum, they are related, but distantly. But classification is only a reflection of our understanding, not a precise mirror of evolutionary history. As our understanding improves, classification must change. Initially, classification was based on morphology (what things look like). Then ontogeny was added (how animals change during growth from egg to adult), including larval stages, if any. But we now have information from the genetic code itself, looking into the chemistry of the instructions for building and running an organism, which are contained in every one of its cells. Not surprisingly, perhaps, some organisms that look alike are not so closely related, and others that seem quite dissimilar are in fact close cousins, judging by their genetic code.

An emphasis on morphology, in the early days of marine biology, led to the production of beautiful illustrations showing great respect for detail and, in fact, a love for the art behind the science (fig. 3.13). Perhaps no example of such appreciation is more remarkable than that displayed in the book by the German biologist Ernst Haeckel entitled *Art Forms of Nature*.[30] In this collection, Haeckel illustrated many of the

FIGURE 3.13. Mollusk morphology as art, at the beginnings of marine biology (E. Haeckel).

forms that he placed into his tree of life, based on existing classifications and his own studies of anatomy and development. His insistence that ontogeny is a major indicator of relationships proved useful and fruitful, although he has been faulted for overstating his case.[31]

Morphology alone is not a completely reliable guide to relatedness. The barnacles, shaped like miniature volcanoes, would seem to have little in common with their arthropod cousins, the crabs, isopods, amphipods, and insects. However, the larval stage of the barnacle, having the typical nauplius morphology, gives away its origin within this group.[32]

There was still the matter of a suitable guidebook that could help in identifying organisms found along the shore. In the 1920s, Myrtle Elizabeth Johnson, a schoolteacher in California (and a former student of Ritter), decided that a regional seashore book was needed. She and a fellow teacher, Harry James Snook, began spending their summers at Scripps compiling material from researchers at the institution and from others elsewhere along the West Coast.

The results of this work, the *Seashore Animals of the Pacific Coast,* were published in 1927, subsidized by Ellen Browning Scripps. Arranged in the customary systematic fashion by animal phyla, the book had good descriptions of coastal animals (except fishes) of the Pacific coastal states, with habits as well as habitats and ranges, and was illustrated with many line drawings and photographs, a few of which were in color. S. F. Light, professor of zoology at the University of California at Berkeley, published a more technical manual for identification of intertidal invertebrates, in 1941.[33]

But the best of guidebooks was yet to come. In the 1930s, the professional collector Edward Ricketts, who lived in Monterey, brought a new approach to the documentation of seashore life. Ricketts set out to capture the essence of life at the edge of the sea in his book *Between Pacific Tides,* richly illustrated by coauthor Jack Calvin (fig. 3.14). Ricketts broke tradition by arranging the text by habitat and, thus, emphasized the different types of environment in determining the diversity of shore life and the many challenges to survival. Everyone studying the intertidal zone along the West Coast has a dog-eared copy of this delightful work.[34] Published in 1939 by Stanford University Press, it went through many editions and printings (despite the objection by early reviewers that its arrangement was awkward). Later editions were revised and expanded by the marine biologist Joel W. Hedgpeth.

FIGURE 3.14. Example of an illustration in the famous guidebook by Ricketts and Calvin: a green abalone (then common, now quite rare). Calvin was the photographer.

We have already referred to the low temperature and the high productivity of the waters along the shores of the West Coast. The relationship between low temperature and high productivity emerged around 1915, when Scripps scientists George Francis McEwen (1882–1972) and Ellis LeRoy Michael (1881–1920) announced their finding that in spring and early summer along the coast of California, cool water rises to the surface from depths of a few hundred feet, bringing nutrients. They built on earlier studies by McEwen on upwelling.[35] To make the connection between physics and biology, they designed a contraption combining plankton net, water sampler, and thermometer and obtained samples from specific depths in the coastal waters, exploring the possibility of depth zonation in the plankton.

The collaboration between McEwen, a physicist and mathematician, and Michael, a biologist, and also Erik Gustaf Moberg (1891–1963), a chemist, proved most fruitful. This meeting of the minds vindicated Ritter's philosophy of mixing the disciplines at the institution. McEwen had learned about the work of the German oceanographer Hermann Thorade, who had found a pattern of low temperatures near the coast, in measurements made by German ships.[36] Thorade related these findings to the work of the Swedish oceanographer V. W. Ekman, who had established mathematical formulae describing the sideways drift of moving surface water that would result in coastal upwelling (a process now called Ekman pumping).[37] The crucial ingredient in Ekman's formulation is the effect of Earth's rotation, through the Coriolis force, which off California results in an offshore component in the southward motion of surface waters.[38]

The Scripps scientists, eventually, agreed that upwelling was related to prevailing seasonal

FIGURE 3.15. Martin Johnson *(left)* and assistant, getting ready to put the plankton net into offshore waters to obtain samples of meroplankton and other zooplankton, in the 1950s.

winds and that it causes summer fogs by cooling moisture-laden coastal air masses. Low clouds then hug the coast, owing to stable stratification in the lowermost air layer. Most importantly, they decided that upwelling is responsible for the tremendous productivity of the coastal waters, which supports the abundant marine life along the seashore. Physical circulation was thus tied intimately to nearshore marine ecology.

No one did more to elucidate the dynamics of this ecology, as it relates to the nearshore organisms and the dispersal of their larvae within the eddying current offshore, than Martin Johnson, the biologist of the trio of authors of the *Oceans* text (fig. 3.15). He established the intimate connection of the nearshore ecology with the physics and biology of the offshore waters. The larvae are as seeds in the wind (fig. 3.16); the current offshore serves to disperse them but also harbors abundant predators eating them. The distribution of larvae of nearshore organisms provided the first delineation of the seaward extent of the coastal ocean.[39]

LIFE IN SAND AND MUD: A SECRET EXISTENCE

Sand beaches and mud flats are much more common on the eastern and southern coasts of North America than on the west coast. The chief reason is the aforementioned sinking of the coastal plains in the east and the south, which provides vast areas for depositing sand and mud brought by rivers, in part reworked by waves. As

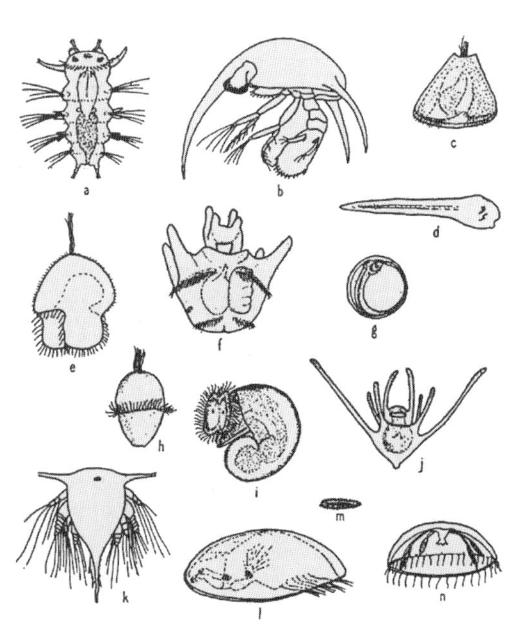

FIGURE 3.16. Larvae of the meroplankton: a, annelid; b, sand crab; c, bryozoan; d, tunicate; e, nemertean worm; f, sea urchin; g, fish egg; h, scale worm; i, snail; j, brittle star; k and l, barnacle; m, cnidarian; n, medusa of hydroid. (As given by M. Johnson.)

a result, coastal plains, lagoons, marshes, broad estuaries, and wide beaches dominate the eastern shorescape. In contrast, at the western shores, the land is rising, and wave action produces cliffs bordered by narrow beaches. Lagoons and mudflats are found in the lower reaches of the rivers, where they run into the sea.

The tidal flats and lagoons of California, in essence, originate from the filling up of bays with sediments brought in by rivers in flood. The bays, in turn, owe their existence mainly to the cutting of broad canyons into earlier lagoons during the last lowering of sea level, when water was tied up in the enormous ice caps on Canada and Scandinavia. Thus, the California coast has but few wetlands and beaches, but there is enough of every type of sandy and muddy habitat to provide marine biologists with rich opportunities for study.

Sandy beaches tend to look bare to the uninitiated visitor, as do mud flats. But soon one notices that shorebirds know better: sandpipers, godwits, turnstones, curlews, and plovers assiduously probe the ground for mole crabs, beach bloodworms, bean clams, and olive snails, some using extraordinarily long bills to get at their prey at the edge of the water. A favorite target, the common egg-shaped sand crab or mole crab *(Emerita)*, about an inch long, stands on end within the sand, with only its stalked eyes and antennae protruding, as it filters the outgoing waves for edible particles. The turbulence around the exposed antennae causes small V-shaped depressions in the sand, as the backwash runs over them, which defies the attempt of the small crabs to remain completely hidden. When removed from the sand, they quickly burrow back into it, using their spadelike legs. Clusters of the crabs leave a rough patch on the seaward edge at the water line. Presumably these are gatherings produced involuntarily by wave action and nearshore currents.

Other animals are more difficult to spot; they burrow well into the sediment (as, for example, the bloodworm, a polychaete), or, if too small to move sediment, they live within the tiny interstitial spaces between sand grains. While the animals clinging to the rocks have evolved devices for attachment, the animals in sand and mud have evolved mechanisms for forcing their way through the sediment. Two prominent mechanisms are "swimming" through the sand, using specialized appendages (many arthropods have these), and a kind of peristalsis that might be called "anchoring-and-pulling," which is common in worms and mollusks.

Burrowing organisms living at the very edge of the sea, in the narrow zone where the sand is wet but intermittently exposed, have a problem to solve. As the tide rises and falls, this zone moves landward and seaward. To stay in the preferred zone, the animals have to be ready to move with the tides. They have to have a sense of direction (knowing where the sea is), an internal clock, which is reset by tidal action, and a program for knowing when to burrow and when to come out. A simple experiment with isopods that live in beach sand (pill bugs in the garden are familiar relatives) nicely illustrates the presence of an internal clock synchronized with the tides. One can fill a container with sand and seawater and add a number of the isopods collected from the beach. Even when away from the ocean, punctually at high tide they will show their peak swimming activity.

What precisely sets the internal clock of these inconspicuous denizens of the beach? Is it the Moon? Is it some "subtle geophysical influence"? In the 1960s, Scripps biologist James T. Enright investigated the matter. He took the beach isopod *Excirolana chiltoni* to the laboratory to test its responses to various stimuli. The most effective way to change the internal clock of the animal turned out to be the mechanical swirling and stirring of the sand. So, the organism apparently senses the tides through their effects on the waves, which are stronger when the tide-deepened water lets them arrive at the beach with greater energy.[40]

Mudflats occur at the river mouths and in bays and estuaries. They may look barren compared with the densely covered rocky surfaces, but in fact microscopic algae and bacteria are growing in profusion on the wet surface. When stirred into the water by the tides, the algae and

FIGURE 3.17. Solutions to the problem of combining hiding and feeding in a mud flat.

bacteria make a rich broth that, in turn, sustains armies of filter-feeding animals (called suspension feeders). Food is found on the surface and in the water, but safety increases with depth in the sediment. This poses an optimization problem for the animals desiring both. How can a burrowing organism stay deep and yet get plenty of food?

One way to deal with food-and-safety optimization is to make an open burrow through which water is pumped, bringing the food (fig. 3.17). This is the strategy adopted by ghost shrimps and lugworms, for example. Some of the worms ("inn-keepers") host other species as unbidden guests (free-loaders called commensals by ecologists). These take advantage of the safety of the burrow and the stream of water bringing food. Another solution to the safety-versus-food problem is to keep the main body deep but send long thin siphons to the surface where the food is, the strategy of many clams. The contrasting needs for deep burial and shallow feeding gave rise to some fairly odd forms, such as the long-long-necked gaper, a clam. An extreme case is the geoduck *(Panopea generosa)*,[41] which weighs up to 12 pounds. The geoduck buries itself deeply within the mud, commonly 3 feet below the surface, sending its long siphons up to the surface. It is famous for being difficult to capture. Presumably, like other clams, it escapes downward when disturbed, by extending its strong wedge-shaped foot into the mud below and then inflating it to make an anchor for pulling the rest of the body down.

Some animals, such as certain snails, hunt for prey while moving through the mud. Much dead animal and plant matter is dumped on mudflats by the tides, and consequently scavenging becomes an important way of life in this environment. Deposit feeding, of course, is typical for the soft-bottom habitat and is entirely restricted to it. The diversity of animals in the mud flat arises from these various feeding strategies—filtering, hunting, scavenging, deposit feeding—combined with the defense strategies—shell formation, burrowing, escape reactions. Here, as everywhere, adaptations are driven by the need to eat while avoiding being eaten.

The stirring of the mud by burrowing organisms is called bioturbation. This process controls the diversity of organisms on muddy substrates, that is, who can and who cannot make a living there. Also, by irrigating sediments down to a foot below the surface, bioturbation brings oxygen into the mud, which in turn is used by bacteria to process organic matter, bringing carbon and nutrients back into circulation. The resulting recycling, played out in all the wetlands and marshes of the world, is of fundamental importance to the productivity of the ocean and the chemistry of the atmosphere.

The churning of the sediment makes for a soft ground, which is unsuitable for organisms that need a hard substrate. Thus, deposit feeders and suspension feeders have different needs and do not get along very well. However, the intertidal exclusion principle is not strictly enforced. Quite commonly, there are mounds sticking up above the thin layer of extremely muddy water, right above the sediment surface. Here suspension feeders can settle. Also, storm waves or tidal currents can expose the shells of burrowing bivalves and thereby provide pavements. Sometimes patches of mussel beds or oyster beds can develop from a base of just a few exhumed shells. These can coalesce into large beds and continue to build up into modest-size mussel or oyster "reefs." Oyster reefs are large-scale water conditioners; they filter enormous amounts of water, clearing it of algae and detritus.

And thereby hangs a story—the story of Chesapeake Bay.

Chesapeake Bay, bordered by Maryland and Virginia, is (or perhaps we should say used to be) the paragon of an estuary, supporting an incredible diversity of shore life, including freshwater organisms and fully marine organisms, and everything in between. Historical records show that gray whales, dolphins, manatees, river otters, sea turtles, alligators, giant sturgeons, sharks, and rays all were once abundant within the Bay. But they are no longer seen there.[42] The quality of the water has deteriorated, beginning in the nineteenth century and greatly accelerating in the twentieth. The obvious manifestation of the deterioration is that the water is not clear, but murky—green and brown from a proliferation of phytoplankton and an abundance of organic detritus in the water.

To understand the precarious situation in Chesapeake Bay, we have to go back to the principle governing estuarine fertility, which is estuarine circulation (fig. 3.5). This circulation, marked by the inflow of saline water at the bottom and outflow of mixed freshwater and saltwater at the top, represents a type of upwelling, with the additional aspect that nutrients coming from the river from land and nutrients coming with the seawater from the ocean are trapped within the estuarine system till they build to high concentrations. A rich growth of phytoplankton ensues, feeding dense aggregations of copepods and other zooplankton. The organic-rich detritus accumulating on the bottom from the growth of plankton in the sunlit zone poses a problem: it decays and takes up oxygen. In short, the more nutrients there are and the more plankton blooms, the less oxygen remains for creatures living in the salt portion of the estuary, in the deeper water.

So who needs oxygen? The marine fishes do that enter the estuary from the sea, within the salty layer at the bottom. These are, for example, fishes coming to spawn in the inner parts of the system, where freshwater meets saltwater, such as several species of herring (shad, alewife, blueback) and the striped bass *(Morone saxatilis)*. Others spawn at sea, but their larvae migrate into the upper estuary to escape into the hiding places provided there and feed on the rich plankton diet (menhaden, spot, croaker, weakfish). Whether the juveniles are from parents spawning in brackish water or at sea, they require oxygen-rich waters in the various coves of the inner estuary.

The history of environmental deterioration in Chesapeake Bay (which is similar to that of other temperate estuaries) suggests that the removal of key species from the ecosystem plays an important role. First, an increase in nutrient supply predisposed the Chesapeake system for plankton blooms and oxygen stress. Even before 1800 there was an increase in sedimentation and burial of organic carbon in the bottom deposits of the bay, indicating soil erosion and increased supply of nutrients from the hinterland, presumably largely due to deforestation through expansion of agriculture by European colonists. Sea grasses and diatoms living on the seafloor (and using sunlight) declined, while phytoplankton (and associated detritus in the water) increased at that time. At some time later, nutrient supply again increased dramatically, presumably from the use of artificial fertilizer, which eventually reached the estuary through river-borne runoff. Lack of oxygen became widespread in the 1930s, as the flux of organic matter to the bottom exceeded the ability of the recycling system to oxidize the material. Bottom-living organisms died off during periods of severe oxygen shortage and were unable to recolonize the sediment surface.

What exactly happened in the 1930s? Was nutrient supply the crucial factor of collapse? Perhaps not. Up into the 1920s, masses of oysters were lining the bay's floor over vast areas, clearing the entire water column every few days or so. They held off the feared "green soup" phenomenon (called eutrophication by lake ecologists) and thus helped keep oxygen levels high. Alas, late in the nineteenth century, mechanical harvesting of oysters by dredging started to reduce their numbers considerably. Soon the familiar harvest collapse began, with the oyster yield shrinking to a few percent of peak values by the early twentieth century. After the demise of the oyster fishery, the troubles began in earnest: the water became green soup, and the bottom collected stinking black mud. Outbreaks of oyster parasites followed in the 1950s: when short of oxygen, oysters are stressed and susceptible to infection.

Presumably, reestablishing the oyster reefs would be a first step to restoring a healthy ecosystem in the bay. The alternative of unchecked deterioration may result in an overabundance of jellyfishes, such as observed in many stressed estuarine ecosystems. The low oxygen requirements of their polyp stages on the seafloor, and the absence of predators (which are warned off by the lack of oxygen), would seem to favor high reproductive rates for these ancient and efficient predators. The overall effect of overstressing estuarine ecosystems is to bring the environment back into a Precambrian state, when life was simple, and bacteria and jellyfishes ruled the world.

LIFE IN THE KELP FOREST: COPING WITH BOOM AND BUST CYCLES

At Scripps, the kelp forest offshore from the institution has long served as a laboratory to study nearshore ecosystem changes, both from natural causes (sporadic warming of the offshore waters and influx of tropical waters from the south) and from human impact, especially removal of lobsters and large fishes. Diving ecologists Paul Dayton and Mia Tegner carried out a large number of such studies over many years, in part with graduate students. Their chief conclusion is that fishing out the top predators made the kelp plants vulnerable to attack from grazers, that is, sea urchins.[43]

Kelp forests are quite common along the shores of San Diego and elsewhere along the California coast, just beyond the breakers. They are among the most productive ecosystems of the world. Invisible from the beach, they make their presence known after storms, when great masses of tangled mats and mounds of kelp litter the shoreline, decaying afterward for many weeks, surrounded by enormous swarms of kelp flies.[44] Kelp beds are easily accessible to divers, and a diving visit to a kelp forest is an unforgettable experience. The Sun's rays, broken into flickering beams of light by the rough water surface, filter through the great canopies of the "trees" of this forest, primarily the giant kelp *(Macrocystis pyrifera)*. On the seafloor, specks of light dance across carpets of colorful algae dotted with mollusk shells, while fishes dart in and out of the shadows to inspect the intruder.[45]

The species making up the kelp forest are macroalgae—organisms in a primitive nonflowering plant phylum. These plants have no roots, stems, or leaves, only holdfasts, stipes, and fronds. The holdfast anchors the plant to the seafloor, commonly on a rock outcrop or a boulder. The fast-growing stipe moves the fronds upward toward the light.

The fronds are the sites of photosynthesis; they take up nutrients and carbon dioxide from the water, using these substances and light to grow more fronds. Tall kelp species shade out the others and become dominant. Off the shores of southern California, *M. pyrifera* is the dominant form (fig. 3.18). It towers high above the rest, using the moored-buoy principle to stand erect, with many small floats keeping the mooring lines taut (the height is measured in tens of feet). Each plant has multiple stipes, from which hollow gas-filled bulbs branch off, each with a single bladelike frond. A giant kelp plant lives only about five years, and individual fronds die and fall in a few months. Reproductive strategies in kelp plants vary and commonly involve alternations between sexual and asexual generations, which differ greatly in size and appearance. The less abundant elkhorn kelp (*Pelagophycus porra*) has a single main stipe, kept taut by a single large gas-filled float on the top, from which branch out the various leads to the enormous fronds, which hang down into the water like so many banners.

Not all, or even most, species of kelp have floats or stipes. Closer inshore, where wave action becomes very important, some seaweed species have a stiff stipe standing up a short distance above the seafloor, with fronds waving to and fro in the surf. Others stay close to the bottom, avoiding excessive strain from wave attack. Yet other algae stay small, producing a kind of miniforest composed of different types of red algae, some with calcareous skeletons. Encrusting forms also are abundant, completing the roster of nearshore macroalgae.

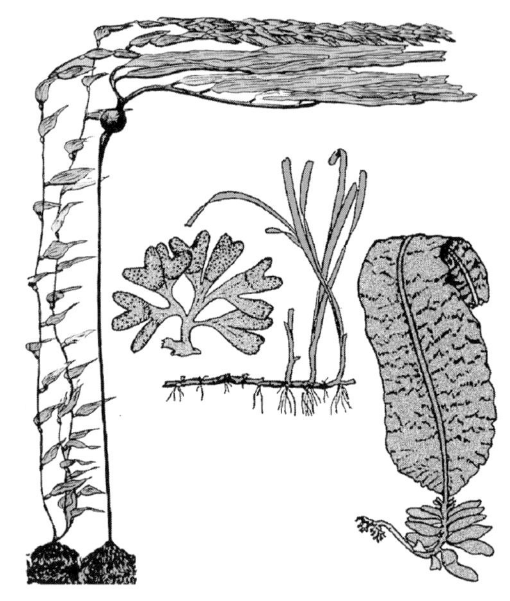

FIGURE 3.18. Plants of the kelp forest and rocky surf zone: giant kelp, elkhorn kelp, the marine algae *Fucus* and *Alaria*, and the sea grass *Zostera*. Not to scale.

Some years are better for growing kelp than others. This was soon realized during commercial harvesting. Thus, during the 1910s the California Department of Fish and Game and in the 1950s the kelp-harvesting company Kelco supported research programs at Scripps to determine the reasons for the variability. The diving ecologist Conrad Limbaugh (who initiated the diving program at Scripps) was among the researchers making detailed observations in the kelp forest. He was among the first to use underwater photography in documenting the kelp habitat.[46]

The early scuba divers marveled at the enormous specimens of sea bass, the hundred-pound groupers, and the countless spiny lobsters and abalone that lived in and around the kelp forests. In fact, many of these creatures were so abundant they helped feed staff and students well into the 1960s. Divers never went hungry.

In the mid-1950s, Scripps biologist Wheeler North (1922–2002) initiated a systematic study (supported by the Department of Fish and Game) to find out why the kelp beds off

San Diego appeared to be declining. That study, over several years, concluded that harvesting the kelp did not interfere with fishing (as claimed by some) but actually improved the catch.[47] Presumably, the effect results from increasing the availability of sunlight, which stimulates much growth on the shallow seafloor. Also, kelp cutting reduced the amount of loose kelp on the beach, thus helping to reduce the abundance of kelp flies. The decline in coverage with living kelp was found to be parallel to an increasing number of sea urchins, which graze kelp at the holdfast level. Encouraged by the kelp industry, commercial divers began spreading quicklime in areas where sea urchins were especially common. This left open the question why sea urchins increased in the first place. Were their natural predators being removed? Lobsters eat urchins!

In the late 1960s, Japan Airlines proposed to the Sea Grant program at Scripps that the red urchins *(Strongylocentrotus franciscanus)* be harvested for their roe to be sold in Japan. A substantial fishery grew up from that suggestion, which reduced the grazing on kelp. The fishery declined after some time, when the red urchin became much less abundant. Wheeler North also noted the effects of an El Niño year in 1957/58, the first that was seriously studied by Scripps scientists. The increase in temperature, and the associated decrease in upwelling, which decreased the nutrient supply, clearly was a likely cause of the decline of the kelp.

Thus, by 1960 or so, much had been learned about the processes controlling the productivity of the kelp forest and the diversity of the organisms within it. Yet, its ecology was still poorly known. How fast do the patches of forest reestablish themselves, after destruction by storms? Do patches of kelp persist for long, with a given composition of species—or do they come and go within a few years? Such questions called for long-term observation by divers and for field experiments.

The various growth rates of kelp plants and inhabitants, and their interactions, were documented through many years of careful observation by Scripps biologists Paul Dayton and Mia Tegner, in a number of locations. Off Point Loma in San Diego, they found a complicated interplay of a number of factors governing the health of kelp forests. The main factor is the supply of nutrients from cool upwelling water. In a healthy patch, more than half of the living mass of the forest of giant kelp is within the uppermost 3 feet of the water column. (This makes harvesting easy.) Here nutrients are taken up by the fronds, in the sunlight, during photosynthesis. The forest declines when ocean temperatures increase and the supply of nutrients to the uppermost water layer is reduced. Also, kelp suffers from storms and from grazing by amphipods and sea urchins. Temperature, nutrient supply, and storms are related to climate; they follow large-scale patterns that involve much of the North Pacific. Complications in attempting to unravel the workings of the ecosystem arise not only from climate change, but also from the fact that fishing has changed and is constantly changing the structure of the system.[48]

Within the kelp forest, competition for light is important. Individual plants are readily shaded out by neighbors, especially larger neighbors. The strategy of growing large, to catch nutrients and light, carries the cost of becoming vulnerable to wave attack. Thus, the kelp forest is restricted to the zone that is shallow enough to provide sunlight to new recruits, on the seafloor, and deep enough to avoid the intense wave action typical for the shore. On a wave-cut terrace near a depth of 30 feet, such a zone may be quite wide and long. In such cases, the interior of the forest will tend to receive water low in nutrients, these having been taken up by upstream neighbors. In compensation, the interior plants will be less bothered by sea urchins and by encrusting moss animals, their larvae having settled nearer the edges of the forest.

The influence of human activities has become increasingly important in understanding the state of affairs within the kelp forests. No longer do scuba divers find giant sea bass, groupers, kelp bass, or sheephead. Also, abalone and lobsters are rarely seen. Mia Tegner considered overfishing (commercial, not sport fishing) the greatest threat to the kelp forest. In the early 1990s the use of gill nets was banned within 3 miles of the southern California coast. A few years later Tegner persuaded the legislature to forbid the taking of any abalone south of San Francisco for a decade.[49]

North of Point Conception, sea otters *(Enhydra lutris)* are the chief predators of sea urchins in the kelp beds. They are the keystone species protecting the kelp from collapse through grazing. Sea otters are voracious consumers of *both* sea urchins and shellfish—including abalone. Their appetite for abalone has detracted somewhat from their merits as kelp protectors.[50]

Obviously, cessation of upwelling leads to a demise of kelp forests. A well-documented instance of virtual collapse of kelp forests occurred during the El Niño of 1957–1959, as mentioned. This event brought warm tropical waters into the bight of southern California, with creatures normally found only off Baja California, well south of the border. Interestingly, as the kelp disappeared, sea urchin populations vastly increased. Presumably, the urchins were feasting on the dying kelp, and their populations were expanding much faster than those of their predators (which were being fished out). It took several years before conditions returned to something like normal—a concept that remains elusive in any case.

The normal state of affairs, always poorly defined in systems dominated by climatic fluctuations, has become quite impossible to assess. To begin with, through the years of increased human activity along the coast, there has been large-scale removal of important kelp forest inhabitants, including the once-common abalones, lobsters, and large fishes. The ramifications of these various changes in the ecosystem are not known and are now impossible to assess. There is yet another reason why we shall be frustrated in the attempt to establish what is normal. There are indications that the entire system comprising the California Current and its nearshore habitats has changed in fundamental ways during the last 30 years or so. Long-term records from the Marine Life Research Program at Scripps indicate that the California Current has become warmer and less fertile. To what extent this development is due to global warming, and to what extent it represents natural variations in a system prone to change, is an open question. Most likely, answers on this point will remain controversial.

NOTES AND REFERENCES

1. Quoted in the Scripps publication *Explorations,* 2001, 8 (2), 3.

2. M. J. Tegner, P. K. Dayton, 1999, *Ecosystem effects of fishing.* Trends in Ecology and Evolution 14, 261–262.

3. Pew Oceans Commission, 2003, *America's Living Oceans: Charting a Course for Sea Change.* Summary Report. Pew Oceans Commission, Arlington, Va, 166 pp.

4. J. Y. Cousteau, F. Dumas, 1953, *The Silent World.* Harper and Row. Reprinted, 1954, Pocket Books, New York, 225 pp.; H. Hass, 1958, *We Come from the Sea.* Doubleday, Garden City, New York, 288 pp.

5. Pers. comm., 2005. Scripps alumnus Stewart is a pioneer in the application of satellite-derived information to the large-scale physics of the ocean. R. H. Stewart, 1985, *Methods of Satellite Oceanography.* Scripps Studies in Earth and Ocean Sciences. University of California Press, Berkeley, 360 pp. He is engaged in public education, as well. R. H. Stewart, 2004, *What Every Student Ought to Know about the Oceans—the Oceanographers Perspective.* Presentation at the National Marine Educators Association Annual Meeting, Saint Petersburg, Fla., July 2004.

6. Boyd Wallace Walker (1917–2001) obtained his Ph.D. in 1949 at University of California at Los Angeles. His thesis was entitled *Periodicity of spawning by the grunion,* Leuresthes tenuis, *an atherine fish.* He went on to a distinguished career as a biology professor at UCLA.

7. C. Woodard, 2004, *Changing tides—targeted fishing bans protect Nassau grouper.* Nature Conservancy 54 (2), 19.

8. Edward F. Ricketts (1897–1948) made a living collecting marine animals and supplying specimens to laboratories and schools, from Pacific Grove and Monterey, California. The book he wrote on life along the California seashore, with photographer Jack Calvin, *Between Pacific Tides,* became an all-time classic of intertidal biology. Ricketts was a close friend of the writer John Steinbeck. They coauthored the book *Sea of Cortez,* first published in 1941 (by Viking Penguin).

9. Charles Darwin (1809–1882) started his career as the naturalist on the HMS *Beagle,* which circum-

navigated the globe in the years between 1831 and 1836. He published his experiences in 1839 in *A Naturalist's Voyage on the Beagle,* which was well received by his peers and the public. His book on evolution, *The Origin of Species,* published in 1859, revolutionized the biological sciences, including taxonomy, paleontology, and ecology. Darwin realized his ideas would stir controversy and delayed publication for that reason. Eventually, his book had to be published in some haste because of an article by the young naturalist Alfred Russell Wallace, describing evolution as the outcome of natural selection, which Wallace sent to Darwin for comment. Wallace graciously accepted the priority of the older man, who had already worked on the problem for about two decades. To some, Darwin's book is the most important scientific opus ever published. Others have found much to criticize. (One prominent biologist, a hundred years later, has said that Darwin wrote an interesting book but left out the origin of species. One prominent geologist has recently pointed out that Darwin missed the importance of mass extinction in driving evolution.) Darwin's revolutionary ideas were vigorously defended by his friend Thomas Huxley in Britain, and by the biologist Ernst Haeckel in Germany, both well-spoken advocates.

10. Jean Baptiste Pierre Chevalier de Lamarck (1744–1829), Professor of Zoology of Insects, Worms, and Microscopic Animals at the Muséum National d'Histoire Naturelle in Paris, saw taxonomic classification as reflecting ancestry. His ideas were rejected by his younger contemporary, the famous naturalist Georges Cuvier, whose opinion prevailed. This prevented the idea of evolution from taking hold in France (or elsewhere in Europe). Lamarck is now mainly cited for the erroneous notion that acquired traits can be inherited. The prime textbook example is the long neck of the giraffe: In Lamarckian evolution the parents stretch their necks to get at the leaves, and pass this experience on to offspring. In Darwinian evolution, the giraffes with longer necks have more offspring.

11. Linné (1707–1778) latinized his name as Linnaeus when he wrote for publication. Also, he wrote in Latin, which was then the lingua franca. His first interest was plants, but he also named many animal species. Botanists still describe new species in Latin, following the example of the Master.

12. Alfred Russell Wallace (1823–1913) discovered the striking difference in fauna and flora of Asian and Australian affinities, in Indonesian islands. As a young man, he sent a letter to Darwin proposing natural selection as a mechanism of evolution. Later, he wrote perceptively on the processes governing the biogeography of organisms in his book *Island Life* (1880, 1892) emphasizing the effects of changes in climate through geologic time.

13. The problem keeps emerging in the news as a topic for discussion by some school boards. Biologists no longer struggle with the concept of evolution; it is their chief tool in explaining the patterns of life on the planet. Removing evolution from biology would destroy our ability to understand the process of creation. To teach some sort of "alternative theory" would be like offering phlogiston theory as an alternative to thermodynamics in chemistry.

14. The word *ecology* derives from the Greek word for "house" or "home": *oikos.* The word *economy* has the same root. Both deal with needs and how they are satisfied in a world of limited resources.

15. *Chthamalus fissus* readily grows at a lower level in the tidal zone, where there is much more food, less heat, and less desiccation. In fact, many do on the shells of mussels. But they cannot make it on the rock lower down because of being crowded out and because of being eaten. Their ability to withstand more heating and drying than their competitors and predators makes the space farther up a safer place to live. So that is where the "smart" larvae settle: high enough to escape severe competition and predation but not so high that they will starve or die from exposure. J. H. Connell, 1961. *The influence of interspecific competition and other factors on the distribution of the barnacle* Chthamalus stellatus. Ecology 42, 710–723. P. Dayton, 1971. *Competition, disturbance, and community organization: the provision and subsequent utilization of space in a rocky intertidal community.* Ecological Monographs 41, 351–389.

16. G. E. MacGinitie, N. MacGinitie, 1968, *Natural History of Marine Animals,* McGraw-Hill, New York, 533 pp.

17. Ibid., p. 396.

18. Why should the same (or very similar) animal associations exist over such a wide range of latitudes? Ricketts, who posed this question, also provided an answer: Ocean currents bring warm water north, to Washington, British Columbia, and Alaska, and cold water south, to Alta and Baja California. Coastal upwelling along California, in a sense, brings the high-latitude conditions right into the subtropics. The upwelling water is cold. It is cold because it sank below the surface layer somewhere near 50° N and did not see the Sun after that. Thus, in a sense, the mussels and their associates all live in the same type of water, from Alaska to Mexico. E. F. Ricketts, J. Calvin, J. W. Hedgpeth, 1968, *Between Pacific Tides,* 4th ed. Stanford University Press, Stanford, Calif., 614 pp.

19. J. S. Killingley, W. H. Berger, 1979, *Stable isotopes in a mollusk shell: Detection of upwelling events.*

Science 205, 186–188. The method was introduced in the 1950s by a group of chemists and paleontologists assembled by the nuclear chemist Harold Urey, in Chicago. Samuel Epstein and Heinz Lowenstam were prominent members. See G. Wefer, W. H. Berger, 1991, *Isotope paleontology: growth and composition of extant calcareous species.* Marine Geology 100, 207–248.

20. J. S. Killingley, 1981, *Seasonality of mollusk collecting determined from O-18 profiles of midden shells.* American Antiquity 46, 152–158.

21. Jack Calvin, researching Russian records from Sitka, Alaska, found numerous references to fatal mussel poisoning of early Russian explorers. Ricketts et al., *Between Pacific Tides,* p. 187.

22. S. Fowler, 1990, *Critical review of selected heavy metal and chlorinated hydrocarbon concentrations in the marine environment.* Marine Environmental Research 29, 1–64.

23. E. D. Goldberg, 1991, *Halogenated hydrocarbons, past, present and near-future problems.* Science of the Total Environment, 100, 17–28.

24. Rachel Carson, 1962, *Silent Spring.* Houghton Mifflin, Boston, Mass.

25. Agassiz died at the end of that year and the program lasted only one year more.

26. The U.S. Fish Commission was established by Congress in 1871. Spencer Fullerton Baird (1823–1887) was its first director.

27. In California, biologists from Stanford University held a summer biological program in 1892, at Pacific Grove, which became Hopkins Marine Laboratory. That same summer William Emerson Ritter, together with colleagues and students from the University of California, set up a wood-and-canvas tent at Pacific Grove. Ritter's group went on looking for a more congenial (and less crowded) place and by 1903 had settled in San Diego, where, with strong endorsement by the city's businessmen, the Marine Biological Association of San Diego was founded on 26 September of that year. Scripps Institution of Oceanography dates its origin from that event and thus celebrated its centennial in 2003.

28. The memory of Edward Willis Scripps (1854–1926) and the entire Scripps family is honored in the name of San Diego's Scripps Institution of Oceanography. The story of the building of his newspaper empire (and much else) is told in V. H. Trimble, 1992, *The Astonishing Mr. Scripps The Turbulent Life of America's Penny Press Lord.* Iowa State University Press, Ames, 547 pp. For a brief review of William E. Ritter's interactions with E. W. Scripps and other friends and benefactors see the centennial issue of Scripps Institution of Oceanography magazine *Explorations,* vol. 10, No. 1, pp. 4–19.

29. The first building of 1910 is now a historic monument and still in use after restoration in the late 1970s.

30. Ernst Haeckel (1834–1919) was among the leading zoologists of his time. He is still cited for his taxonomic studies on radiolarians, shelled plankton that makes microscopic fossils. The art work is in E. Haeckel, 1904, *Kunstformen der Natur.* Leipzig. Reprinted, 2006, Prestel Verlag, München. The work has some of the best drawings of marine organisms available anywhere.

31. For example, by the paleontologist S. J. Gould. S. J. Gould, 1977, *Ontogeny and Phylogeny.* Harvard University Press, Cambridge, Mass., 501 pp.

32. Charles Darwin was much intrigued with barnacles and first described the common little gray acorn barnacle, *Chthamalus.*

33. M. E. Johnson, H. J. Snook, 1927, *Seashore Animals of the Pacific Coast.* Macmillan, New York, 659 pp.; S. F. Light, 1941, *Laboratory and Field Text in Invertebrate Zoology.* Associated Students of the University of California, Berkeley. Updated and currently available on demand as R. I. Smith, J. T. Carlton (eds.), 1975, *Light's Manual: Intertidal Invertebrates of the Central California Coast,* 3rd ed. University of California Press, Berkeley, 721 pp. Fourth printing, with corrections, 1989.

34. Edward Flanders Robb Ricketts (1897–1948)—the inspiration for "Doc" in John Steinbeck's *Cannery Row*—was a born naturalist. He had extraordinary gifts of observation and he wrote lively prose.

35. G. F. McEwen, 1912, *The distribution of ocean temperatures along the west coast of North America deduced from Ekman's theory of the upwelling of cold water from the adjacent ocean depths.* Internationale Revue der gesamten Hydrobiologie und Hydrographie 5, 243–286.

36. H. Thorade, 1909, *Über die Kalifornische Meeresströmung.* Annalen der Hydrographie und Maritimen Meteorologie 37, 17–34, 63–76.

37. The name of Vagn Walfrid Ekman (1874–1954) is recalled in *Ekman pumping* and *Ekman drift.* V. W. Ekman, 1905, *On the influence of Earth's rotation on ocean currents.* Ark. f. Mat. Astr. och Fysik. Arkiv for Matematik Astronomi och Fysik 2 (11), 1–52.

38. The deflecting force is named after the French engineer and mathematician Gaspard Gustave de Coriolis (1792–1843). He described the apparent deviation of a body moving within a rotating frame, from its inertial path.

39. Martin Wiggo Johnson (1893–1984) was a distinguished expert on the nearshore fauna. He joined the staff in 1934. He took on the difficult task of determining the taxonomy of the zooplankton, which includes larval stages of many of the coastal marine animals (the meroplankton). Johnson painstakingly

worked out the life history of local species. He was widely appreciated for his extensive knowledge on the whole spectrum of local invertebrates. Of course, as coauthor of the famed text *The Oceans*, he also had broad knowledge of life in the sea in general.

40. J. T. Enright, 1965, *Entrainment of a tidal rhythm*. Science 147, 864–867.

41. Pronounced "goyduck," apparently a word taken from a northeastern native American language.

42. The disturbing story of Chesapeake Bay is told in J. B. C. Jackson et al., 2001, Science, 239, 629 ff., with emphasis on overfishing and the effects of oyster removal.

43. M. J. Tegner, P. K. Dayton, 1999, *Ecosystem effects of fishing*. Trends in Ecology and Evolution 14, 261–262.

44. Because tourists (and residents, too) prefer clean beaches, washed-up kelp is commonly removed and trucked elsewhere for dumping.

45. The La Jolla Underwater Park, just south of Scripps and adjoining La Jolla, was set aside (in 1972) for those who wish to enter this charming realm.

46. Conrad Limbaugh (1925–1960) is best known, perhaps, for his work on cleaning symbiosis. C. Limbaugh, 1961, *Cleaning symbiosis*. Scientific American 205 (2/August), 42–49. Through patient observation, he realized that big fishes come to "cleaning stations" where specialized "cleaner" shrimps and fishes remove infected tissue from sore spots, as well as parasites.

47. W. D. North, C. L. Hubbs (eds.), 1968, *Utilization of kelp-bed resources in southern California*. Fisheries Bulletin 139, California Department of Fish and Game.

48. P. K. Dayton, M. J. Tegner, P. B. Edwards, K. L. Riser, 1999, *Temporal and spatial scales of kelp demography: The role of oceanographic climate*. Ecological Monographs 69 (2), 219–250; M. J. Tegner, P. K. Dayton, *Ecosystem effects of fishing*.

49. Tegner's vigorous efforts in behalf of conservation led to occasional irate phone calls to the Scripps administration by frustrated would-be fishermen. Mia Jean Tegner (1947–2001) received her Ph.D. from Scripps in 1974 and continued to work at the institution. Much of her research involved scuba diving. She died in a diving accident.

50. For centuries, Native Americans hunted the otters for food and for their magnificent fur. The dense, glossy golden brown fur attracted the attention of Russian hunters, who discovered the marine mammals in what is now Alaska, in the mid-1700s. They began a hunt that led to near extinction from there to as far south as Baja California. By 1911, when the otter population dropped to 2,000 animals from an estimated 300,000 worldwide (mostly in the fertile northern waters), conservationists began calling for regulations, and the animals were protected from slaughter. Otters have not reappeared in southern California and Baja California (except perhaps for an occasional stray) but have steadily increased from central California northward, to about 100,000 individuals. By 1999 the numbers were again decreasing in central California. Marine biologists were scrambling to determine the reasons. Was it net-entanglement? Starvation from a scarcity of prey? Disease? To what extent have humans been a factor? Chances are that such questions will be increasingly difficult to answer, as the ecological systems within the coastal zone change simultaneously in response to human uses of the coastal environments, and in response to climatic change. An overall decrease in upwelling (as suggested for the last 30 years or so) would have many ramifications, including a drop in sea otter populations. If upwelling is on the rise again (as suggested by some), the otters might well come back. But then, global warming is for real, and the type of upwelling seen till the 1970s may never return in the next several centuries. See C. B. Lange, S. K. Burke, W. H. Berger, 1990, *Biological production off southern California is linked to climatic change*. Climatic Change 16, 319–329; D. Roemmich, J. McGowan, 1995, *Climatic warming and decline of zooplankton in the California Current*. Science 267, 1324–1326; K. D. Hyrenback, R. V. Veit, 2003, *Ocean warming and seabird communities of the southern California Current System (1987–98)*: Response at multiple temporal scales. *Deep-Sea Research II 50, 2537–2565*. D. B. Field, T. R. Baumgartner, C. D. Charles, V. Ferreira-Bartrina, M. D. Ohman, 2006, *Planktonic foraminifera of the California Current reflect 20th-century warming*. Science 311, 63–66.

FOUR

Of Coral Reefs and Atolls

STONE GARDENS OF TROPICAL SEAS

Reef-forming stony corals potentially can grow everywhere in the tropical and subtropical realm of the planet—about one-third of its surface—where the water is warm throughout the year (greater than 20 °C).

For stony corals to grow, the seafloor has to be shallow and the ground firm. Also, the water has to be clear, so that plenty of sunlight reaches the seafloor. The reason is that the coral animals bear inside their bodies, within the outermost cells, photosynthesizing microscopic algae (symbiotic dinoflagellates called zooxanthellae), which need light. This association between animals and green microbes (which gives most corals olive green and olive brown colors) is one of the most impressive symbiotic arrangements on the planet.[1] It is one reason why many corals tend to look like plants (fig. 4.1). The symbiosis allows corals to flourish in the nutrient-starved warm-water deserts of the sea, by setting up ecosystems where animals (corals) capture nutrients through predation and recycle them with minimum loss to captive photosynthesizing organisms (dinoflagellates). The coral-dinoflagellate symbiosis makes the reefs and, thus, provides the basis for a proliferation of animals and plants that live within these stony gardens.[2] Reefs are living oases in a vast desert.

Tropical reefs are striking centers of marine diversity, across all types of reef organisms, led off by a great diversity in "coral" itself (fig. 4.2) and extending to fishes, snails, and other mollusks, crustaceans, echinoderms, bryozoans, foraminiferans, and many types of coralline algae.[3]

Reefs, as ecosystems, are in serious trouble in many parts of the world, wherever there is strong impact from human activities. Because of the requirement for shallow water, most reefs are next to land and are readily accessible to commercial exploitation, mainly fishing, but also collecting of corals and mollusks for the shell market. The removal of large fishes benefits the stony

FIGURE 4.1. Many types of corals are shaped much like plants on land.

FIGURE 4.2. Different types of organisms called "coral." The gem-quality red coral named *Corallium rubrum* (by Linné), a reef-building stone coral (called *Madrepora* by Haeckel), a stony hydrozoan known as "fire coral" because of its painful sting upon touching it.

coral's competitors for space: algae, sponges, and soft coral. With fewer predators checking their expansion, they tend to take over. Runoff from agriculture, soil erosion, and sewage bring nutrients into the system, and this removes the competitive advantage of corals in a nutrient-starved environment. Algae, sponges, and tunicates take advantage. With a dwindling cover of coral, a reef system becomes degraded and finally "collapses," that is, it changes irrevocably into a new and much less diverse system.[4]

Another, more general threat to the health of coral reefs is coral bleaching, that is, the loss of photosynthesizing symbionts, which accompanies unusually high water temperatures. Bleaching events apparently have become more abundant and more widespread with the progressive global warming that characterizes climate change for the last quarter century. The process is not well understood. Invasive species, brought from far away in bilge waters, are a problem in some regions.

Stony corals are the masons of the sea; they build enormous cities that offer food, shelter, substrate, and diversity of habitat. They have done so for many millions of years, and the

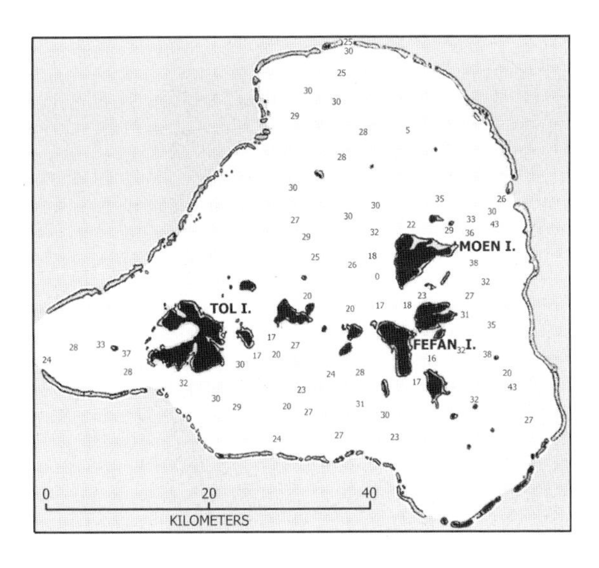

FIGURE 4.3. Extension of girth of the remains of a sinking island by reef growth, as drawn by R. A. Daly (1936). Numbers (maximum, 43) are depths in fathoms.

remains of ancient coral cities are now found in mountain belts wherever the shallow deposits of warm seas partook in the collision of great plates on the surface of Earth, to make new land. Thus, a hike in the Alps may well include a picnic on a reef structure made in the Mesozoic!

Some continents have large accumulations of reef rubble off their tropical shores. The biggest of these accumulations is off northeastern Australia and forms the foundations for the Great Barrier Reef. Tropical islands extend their girth by acquiring a thick mantle of coral rock, bearing active fringing reefs and barrier reefs in the sunlit zone at the top of the volcanic structure (fig. 4.3). Some of them have turned into atolls, quasi-circular barrier reefs rimming lagoons in the middle of the sea, edifices built on sunken volcanoes.

WHY STUDY REEFS?

The word *reef* describes an obstruction lurking below the sea surface, whether made of rocks or sand or organisms.[5] In this sense, reefs are everywhere along the shores of the world's oceans. But in warm, clear, and shallow waters the reefs offshore are biological edifices: stony coral gardens built on their own rubble, thriving cities many thousands of years old.

Biologists and ecologists study coral reefs for the sheer abundance of different kinds of organisms and their life cycles, and for the many intricate interactions between the many reef dwellers. Much of the allure of reefs is the option of freely diving within the stony forests of the sea, experiencing the scene much like a big fish on patrol. Reef ecosystems, like forests, build their own environment, greatly modifying the initial surroundings once they dominate the scene. In the case of forests, tall plants dominate above ground and fungi pervade the soil below, in a symbiosis that traps water and nutrients and extracts mineral matter from the ground to build large physical structures made of wood. In the case of coral reefs, stony plant-like structures dominate the landscape of the shallow seafloor, in a symbiosis that recycles nutrients and builds large physical structures made of calcium carbonate.

Actually, forests and reefs are not merely structurally similar in many respects; they are both intimately linked to the global carbon cycle. The buildup and decay of wood and leaves directly affect the carbon dioxide content of the atmosphere, on many time scales, from annual variation to ice age fluctuation.[6] Likewise, the buildup and destruction of reefs in a world with fluctuating sea level affects the chemistry of the ocean and hence the carbon dioxide content of the atmosphere, which is in constant exchange with the sea.[7] The resulting variations of the

FIGURE 4.4. The astonishing diversity of coral gardens. Palmate, branching, and massive corals in the tropical Pacific, close to the sea surface, and readily accessible for inspection (and use), therefore. Swimmer for scale.

greenhouse gas carbon dioxide contribute to the fluctuations in this gas that accompany large changes in the overall temperature of the planet, as seen in ice cores from Antarctica.[8]

Historically, the chief reason to study coral reefs—which started with mapping their distribution—is to avoid running ships into them. Another is that fishing is excellent along reefs and banks (or used to be). Early exploitation included diving for pearls and for sea cucumbers in the reefs of the western Pacific. The sea cucumbers (holothurians, a group within the echinoderm phylum, which includes starfish and sea urchins) were gathered in the northern parts of the Great Barrier Reef and sold to Asian markets.[9] More recently, many reefs have been severely depleted of large fishes, including sharks, and this removal of key species has had dire consequences in places. Quite generally, many or most of the studies of modern reefs are now focused on the effects of human impact and the response of reefs to stress.[10]

The diversity of stony corals, and the poorly known factors governing their patterns of distribution are intriguing topics to both biologists and geologists (figs. 4.4 and 4.5). In the western tropical Pacific and Indonesian realm, we have more than 400 species; in Hawaii only around 10, and in the Caribbean typically 50.[11] Why is the diversity more than five times lower in the Caribbean than in the western Pacific? What special adaptations do the species in Hawaii possess that allow them to cope with temperatures at the edge of the limit? How are the species on Bermuda reseeded from the Caribbean realm? What controls the growth rates of species in the center and at the edges of coral reef distribution? These and many questions about the interaction between species are problems that have stimulated thought and investigation for many decades.[12]

For geologists, there are more prosaic reasons also for studying reefs. Most of the petroleum in the Middle East, for example, is pumped from the caves and hollows of ancient reefs of Cretaceous age. In Mexico, too, and elsewhere, ancient reefs long buried have been prime targets for oil exploration for many decades. Besides, reefs are among the most interesting repositories for fossils from which to read geologic history.

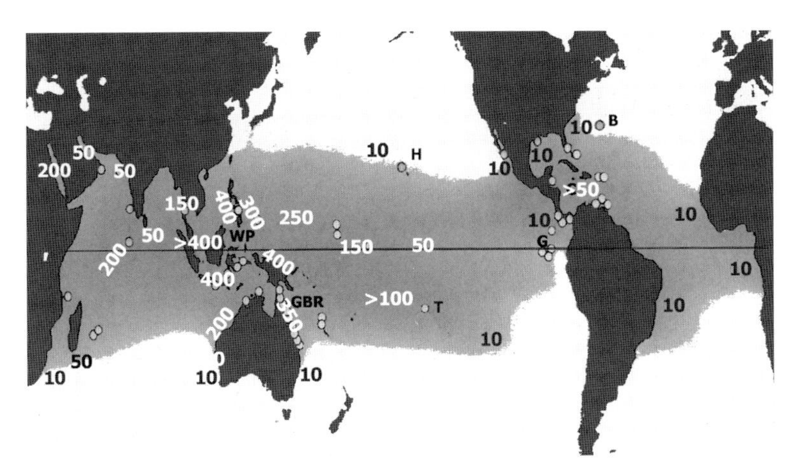

FIGURE 4.5. Distribution limit (dark gray) and diversity of coral reefs (number of coral species). WP, warm pool of the western Pacific; GBR, Great Barrier Reef; H, Hawaii; T, Tahiti; G, Galapagos Islands; C, Caribbean (>50); B, Bermuda.

For a long time, the most enigmatic reefs of the ocean were the atoll islands dotting the vast expanses of the tropical Pacific. Discussions about their origin have continued ever since Charles Darwin proposed his famous theory of atoll formation, in his book on coral reefs published in 1842.[13] A century after Darwin's study, atoll research received a big boost from unexpected quarters: surveys in connection with nuclear weapons tests. The U.S. government used Enewetak and Bikini for a series of such tests in the 1940s and 1950s.[14] Radioactive contamination of the environment, of fish, and of people has been the subject of considerable study (and controversy) since.

The most readily understood reason to study coral reefs in our time is the fact that they may constitute an early warning system on global warming. There is little doubt that warming of tropical waters stresses coral reefs in Tahiti and in the Caribbean, leading to the spectacular coral bleaching observed in the last few decades.[15] One urgent task is to identify the meaning of such bleachings: Are they recurring phenomena, or are they typical for our own time only? What precisely happens during bleachings? Does bleaching do lasting damage?

It now seems that such questions have to be asked on a global scale. Reefs are in danger.[16] This message has only emerged in the 1990s.[17] But time to confront the issue is already running out.

REEFS UNDER SIEGE

Over large areas where reefs are close to highly populated coasts (as in Indonesia and the Philippines, for example), many of them are in poor shape; only about 5 percent or so are considered in excellent condition. Other similarly highly populated regions have similar problems; the reefs are overused, overfished, overstressed. As the fishes disappear, increased efforts are expended to make up for the economic shortfall by intensifying the exploitation of reefs. Along some coasts it is the muddy runoff, from agriculture and deforestation, which damages the reefs. Along yet others it is the nutrient supply from sewage outfalls.[18]

Fishes have been taken from reefs for food, and corals have been collected for jewelry since ancient times, as have mollusk shells. With few exceptions, such fishing and collecting apparently had little or no discernible effect on the abundance of species below the intertidal zone. There are two reasons for this. First of all, it was difficult to hunt and collect on a reef without the equipment now available for the purpose. Second, there was no large worldwide market for fishes and shells, and local demand was quickly satisfied.

Within the last 50 years all that has changed profoundly. Sophisticated technology now supports intense fishing and collecting. The fish harvest can be transported over long distances, using modern refrigeration methods. Shells and coral can be turned into cash in the souvenir market. Live specimens of reef organisms are used to stock seawater aquaria thousands of miles away.

As part of the International Year of the Reef, proclaimed for 1997 for the purpose of increasing the awareness of the public of these threatened underwater habitats, amateur divers made a spot check of 300 different reefs along the coasts of 30 countries throughout the world. They estimated the amount of ground covered by coral, and they looked for reef fishes, lobsters, octopuses, and any other prominent reef organisms. The results are alarming. Reefs with one-third coral cover are commonplace—in fact, they are among the ones considered reasonably well off. A healthy reef, based on older reports and the study of protected reefs, is supposed to have well over one-half of the ground covered by stony corals. Ominously, large predators have become very rare. They have fallen prey to the relentlessly growing demand for specialty foods, expressed in high market prices. After removal of groupers, sea bass, sea perch, and other similar top predators, sharks are now sharing the same fate of ecologic extinction. How did we get there? What can be done?

Intensive fishing on coral reefs began well before the twentieth century, according to Scripps ecologist Jeremy Jackson.[19] Already by about 1800, he says, many large vertebrates (especially turtles) had been decimated in the northern and central Caribbean, and by the early nineteenth century subsistence fishermen were effectively decimating reef fish populations. Jackson's concern is that we shall never know how an Atlantic reef community functioned before it was altered by human impact, before scientific study began. The analogy he uses is powerfully telling: "Studying grazing and predation on reefs today is like trying to understand the ecology of the Serengeti [grasslands in Africa] by studying the termites and the locusts while ignoring the elephants and wildebeests." Large vertebrates—fishes, turtles, and manatees—are crucial for reef maintenance according to Jackson. Unlike on land, where large wild animals were replaced with domesticated cattle and sheep, there are no coral reef livestock to take the place of missing large vertebrates once they are removed.

The main problem, Jackson emphasizes, is to fail to realize that reef resources cannot sustain intense exploitation for long. He identifies two "dangerous stupidities": the first being the "placebo of sustainable use for everyone" and the second the "fallacy of a 'pristine' coral reef." There is no public perception of the magnitude of the loss concerning the reefs, he says, because there is so little memory of what things were like, and therefore what things could be like.[20] Geologic studies by Jackson and by his associate J. M. Pandolfi indicate that for tens of thousands of years *Acropora palmata,* the elkhorn coral, was the dominant cover of the shallow seafloor in the Caribbean, in the tidal zone and just below it. It is still dominant in a few places but has been disappearing from many reefs since about 1980.[21]

The ruined reefs of Jamaica, witness to the dynamics of collapse, amply illustrate the fate of overused and abused coral reefs. Their story exemplifies the unrelenting progression toward destruction wrought by the combined actions of humans and of the forces of nature with the power of Greek tragedy. The basic driving force of ruin is the rapid increase of people and the corresponding increase in demand for food. The Jamaican population went from 300,000 at the end of the eighteenth century to more than 2.6 million at the end of the twentieth. At the same time, all fisheries became motorized, and thus highly mobile.

In the early eighteenth century, green turtle meat from the nearby Cayman Islands was a major meat source in Jamaica. By the end of the eighteenth century, the turtle fishery had gone out of business with the collapse of their target populations. Subsequently, fishing pressures greatly increased, resulting in the removal of large fishes from reefs—snappers, groupers, sea perch, sea bass, sharks. These used to eat the smaller fishes, many of which feed on creatures that attempt to grow on bare rock, including newly settled coral larvae. Other big fishes, though smaller than the big predators, grazed on the algae that compete with corals for space. Hawksbill turtles eat sponges, which also compete with corals for space. On the whole, the big vertebrates created an environment in which corals could thrive. Removing the large vertebrates thus preconditioned the reef community for collapse.[22]

For some time after removal of the large vertebrates, an uneasy equilibrium was maintained by a great abundance of the long-spined black sea urchin, a sea urchin armed with impressive needles tipped with poison *(Diadema antillarum)*. This animal feeds on seaweeds, preventing them from overgrowing and smothering corals. In a sense, the urchin was the coral's best and only remaining friend. With little pressure from predation from specialized fishes feeding on urchins (such as porgies and toadfish), and with greatly reduced competition for algal food, the urchins did very well throughout the 1950s into the early 1980s. They kept the coral fields weeded, like good gardeners.

Then, suddenly and without warning, the sea urchins died off in droves from a species-specific disease that swept throughout the Caribbean in the years 1982 to 1984. In Jamaica some 99 percent of the population succumbed, leaving the reefs with a greatly reduced density of urchins. As a result, reefs suffered greatly. The worst off were the reefs around Jamaica; they are in ruins. They now bear less than 5 percent of coral cover and are overgrown to more than 90 percent by fleshy macroalgae.[23]

Coral expert Terence Hughes—a tireless observer who has monitored the reefs from Australia to the Caribbean—notes that before the *Diadema* die-off, fleshy large algae typically covered less than 5 percent of the ground in waters shallower than 25 meters, except in the intertidal range, where damsel fishes have their garden patches. (The damselfishes destroy coral to make room for edible algae.) But after the sea urchins had experienced mass mortality from the infection, algal mats took over the space usually reserved for corals. In addition, mortality of corals was greatly increased from bleaching events in 1987, 1989, and 1990, according to Hughes. He thinks that the decline since 1990 will not be reversed for many decades, even if conditions become favorable for some reason.[24]

In summary, no fish, no turtles, no sea urchins—no coral. The algae take over.

There are yet more sources of stress for coral. Not only were most of the friends of the coral removed by fishing, leaving them dependent on one single, lowly sea urchin for protection, but also the quality of the water has changed. Deforestation, increased nutrient supply from sewage, and oil pollution are to blame. The removal of mangrove forests is another source of stress.[25] Deforestation in the hinterland causes muddy runoff, which makes corals work harder to stay clean. Nutrient supply favors algal growth. Oil pollution interferes with the chemical signals that guide many interactions in the reef, including reproduction. The end result is widespread takeover by algae.

The increased stress from human activities may be to blame, at least in part, for an apparently increased propensity for corals to sustain damage from bacterial or fungal infections. Black-band disease, first reported from Belize in 1973, kills both brain corals and star corals, removing the soft tissue from the skeleton. White-band disease, first reported in 1982 from Saint Croix, destroys elkhorn and staghorn coral over wide areas of the Caribbean and in Florida.[26] The appearance of the pathogens was a surprise, as was the spread of the epidemic. Yellow-band disease, patchy necrosis, white pox, white plague, and rapid wasting disease are other equally disturbing coral diseases involving the removal of soft tissue and leaving a white, bare skeleton for colonization by algae.

To the long list of human insults to reefs, some of which go back more than a century, must be added another now, one of unknown but potentially dismal proportions: global warming.

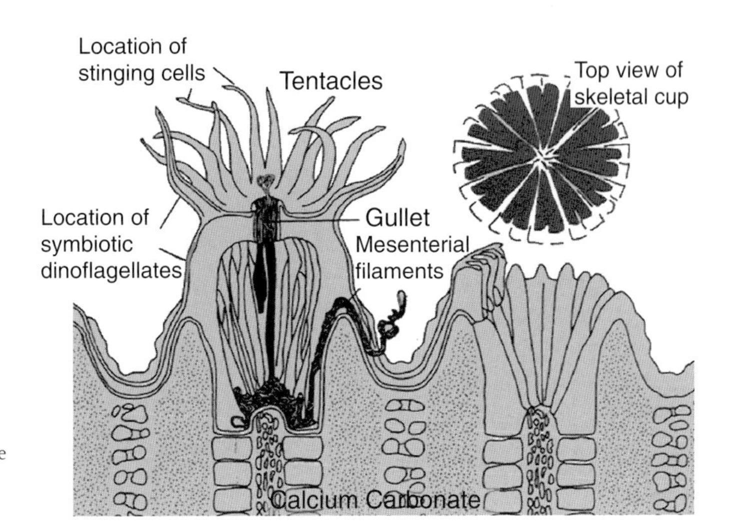

FIGURE 4.6. Coral polyps sitting in their self-made stony cups. They host dinoflagellates (microscopic algae) for photosynthesis and use stinging cells (cells with nematocysts) on their tentacles for catching prey. The filaments are used in catching prey and digesting it outside of the stomach cavity.

Corals grow best in warm waters near the upper limit of their temperature range. When unusual warming occurs, corals lose their microscopic algal symbionts and turn white. This is the notorious coral bleaching, which has made the headlines more than once in the last two decades. The process of coral bleaching is poorly understood. Obviously, it involves a disturbance of the symbiotic relationship between coral animals and their symbiotic algae, but current knowledge is insufficient to tackle the question in some depth.[27]

An even more difficult question concerns what will happen as the chemistry of surface waters is changed from the take-up of excess carbon dioxide in the atmosphere. The carbon dioxide content, which is rising in response to our use of carbon-based energy (about 1 ton of carbon each year per person in the world, on average), is now one-third higher than background. It would be still higher but for the fact that carbon dioxide enters seawater and forms carbonic acid there. So far, the rise in acidity is modest. But continuing use of carbon energy will likely produce changes in chemistry that could interfere with the building of coral skeletons and mollusk shells.[28]

LIFE IN THE REEF

Coral reefs achieve their exalted status as high-diversity ecosystems for several reasons: 1) The reefs are environments where biological activity changes the submarine landscape in fundamental ways, such that the diversity of habitat is greatly increased; 2) the background physical environment is more or less predictable, allowing specialists to flourish (rather than opportunists); and 3) communities are geologically ancient, implying that much time has been available to evolve all sorts of interactions.

But the chief element of the reef environment—the overarching characteristic from which all else follows—is the ability of coral colonies to combine photosynthesis with catching prey using the same large solid surfaces, with minimum expenditure of energy (fig. 4.6). The feat requires the presence of hard ground, which the corals produce themselves. The lead organisms of coral reefs, the stony corals, build the underpinnings for a system of solar collectors, hosting cells that capture solar energy, and they maintain the cells by delivering the nutrients they need. They get the nutrients by acting as a huge trap for plankton in the sea, at night. In this fashion, otherwise unproductive regions of the oceans are turned into oases packed with life.

Biologists have identified some 5,000 species of reef-building coral, so far, and counting.[29] The wild and beautiful varieties of shapes and sizes of their skeletons represent different solutions to the problems faced by corals: a need for sunlight, resistance to mechanical attack, and defense against predators such as coral-eating fishes, sea stars, nudibranchs, and other assorted grazers of the coralline gardens. We marvel at the resulting diversity of forms. In the attempt to cope with this diversity, popular names invoke unlikely associations, as in "brain coral," "staghorn coral," "elkhorn coral," "mushroom coral," "organ pipe," "button coral," "bubble coral," "jeweled toes," and even "dead man's fingers."

The classical problems of coral reefs, which have attracted the labors and thoughts of biologists and geologists alike for a century before the focus shifted to human impact, have to do with diversity and ecology of reef organisms, and with the age and origin of the fantastic underwater structures these organisms built through geologic time and are building still. What exactly are corals—plants or animals? Why make carbonate structures? How many different kinds of organisms are at home in coral reefs? How do they interact? How fast do reefs grow? How old are the reef structures? What is the origin of carbonate platforms and atolls in the middle of the ocean?

Answers to these questions have emerged in the twentieth century, thanks to diving, intensive sampling and observation, new methods of dating, and deep drilling.

Nothing has advanced public awareness of the ocean sciences more than the rise of scuba diving and underwater photography over half a century ago. Facemask and snorkel opened a new world for everyone willing to get wet. It is impossible to forget the first time one drifts over a coral reef, looking down through a facemask and seeing the coral structures close up, like a hawk gliding above the canopy of a forest. The experience is exhilarating, comparable to finding oneself in the high mountains for the first time, having grown up in the plains. Descriptions of reef scenery in the popular literature are apt to be rather flowery (as befits the anthozoan subject), evoking images from the art of the impressionists, and the poetry and music of a romantic era. Words like *serene*, *peaceful*, and *wondrous* suggest themselves.

Reality is otherwise. There is nothing peaceful or miraculous about the reef environment. Here, as elsewhere in the biosphere of the planet, the organisms face a simple task: get sustenance and shelter and reproduce, or go extinct. In the reef community, which evolved through millions of years, every organism becomes another's resource or nemesis, or both. The reef is a dynamic web of partnerships in production and destruction, hostile takeovers and standoffs, extortion and exploitation, trust and treachery, comparable to the complexities of human societies in anarchy. Life in the reef is not serene. Life is tough.

In the reef, stinging cells rule. Invented eons ago, they are the main weapon for hunting and for defense of a multitude of organisms classified as cnidaria, that is, animals bearing nematocysts, the scientific label for stinging cells. Many or most of them have two stages in their life cycles: the polyp form and the medusa form. For corals the polyp form is the dominant manifestation. Colonies grow by budding. Stony corals make a hard substrate from carbonate; soft corals make a flexible skeleton, which may have carbonate or consist entirely of organic substance. For jellyfishes the medusa is the dominant manifestation.[30] All of the animals armed with stinging cells live by waiting for small prey, which they immobilize and trap using these cells.

Some animals, lacking stinging cells and yet finding them useful as a means of defense, have developed close association with cnidarians. Thus, some hermit crabs insist on carrying around sea anemones on the back of their adopted snail shells. A number of fishes swim among the tentacles of jellyfishes such as the Portuguese man-o-war. Certain nudibranch snails feed on cnidarian hosts, obtaining unfired cysts that they employ for their own protection.[31] Colorful small fishes (clown fish) of the genus *Amphiprion* seek protection within the field of waving tentacles of sea anemones. The sheltered fishes in turn attack potential anemone eaters.[32]

Not every space in the reef environment is equally suitable for growth. From a regular stone coral's point of view, the most desirable space is well lit, protected from the worst pounding, but with sufficient water movement to bring nutrients and food particles, and offering the stability that comes with a firm substrate. Mud and shifting sand are poor places for anchoring and growing coral. Good space is in short supply, and the competition for such space is a slow but deadly war. The coral polyps extrude long filaments that secrete digestive enzymes. This system of external digestion, which is routinely used on prey too large to swallow, comes in handy when defending space or claiming new space for growth. Whoever has the more efficient enzyme gets to digest the competitor, by penetrating it with digestive filaments. Upward growth and branching reduce contact. Of course, reef corals compete for space not only with each other, but also with attached algae, sponges, and soft corals (that is, those anthozoans that do not secrete a skeleton). Soft corals are more resistant to exposure to air and to sediment than are their stony cousins. Hence, they compete well in the less favorable marginal areas.

Algae-eating fishes give the edge to coral in the alga-coral war for space. In addition, they maintain an increased diversity within the coral, by keeping down the faster-growing and taller take-over types.

Claiming real estate quickly whenever it becomes available is of the essence. Spreading by budding is the answer. Sexual reproduction (commonly coordinated among many different species on a lunar cycle) makes sure that larval seeds reach other areas for colonization. But budding makes for rapid growth locally. Many plants on land have similar strategies for expanding their territory.

The scene of competition for space between corals and their competitors is shifting all the time, naturally. The stony corals are favored where fertility is low and where the water is clear. The competing algae, sponges, and soft corals tend to take over when nutrient or sediment load increase. As mentioned, human activities tend to provide for shifts away from coral in the direction of algae and sponges.

The triumvirate of factors that governs reef life consists of stinging cells, budding, and the association with symbiotic algae, all three being vital. Corals are not alone in having discovered the advantages of hosting minute algae and have them capture the Sun's energy, under one's skin. Many invertebrate phyla have representatives benefiting from such symbiosis, from the amoebalike single-celled foraminiferans and radiolarians to sponges and flatworms and the large and peaceful giant "killer clam" *(Tridacna)*.

With the insight that three things matter—catching prey, photosynthesis, and the potential for rapid growth—we can readily address another question: why do corals have the shapes they do, and why do other organisms, unrelated, have similar shapes?

Colonial corals, hard and soft, are shaped the way they are because they thrive on having surface areas, for catching prey and for exposing their algal symbionts. Actually, either one of these reasons is sufficient to explain branching, leafing, fingering, and the other adaptations to increase surface area. This is evident from the shape of coralline algae (which do not capture prey) and that of deepwater corals (which do not bear symbionts). In addition, there is the matter of the beating of the surf, in places. Clearly, no matter what the imperative for increasing surface, delicate fingers and branches will be readily broken off—unless the skeleton is flexible. Thus, where the surf is heavy, even occasionally, massive coral and soft coral gain an advantage in holding the real estate. However, the more delicate branching hard corals also can gain from occasional storms: they regenerate colonies from fragments, which permits invasion of already settled areas.

Life in the reef does rely very much on the presence of the various coral organisms, but it also immensely exceeds the scope of the lifestyles that are directly related to stinging cells, algal symbiosis, and colony formation. With all major phyla richly represented, even a cursory description of reef creatures and their life histories and interactions would readily fill volumes.[33] With thousands of species interacting with another, a plethora of patterned interactions readily emerges. Any one topic—say, camouflage and signaling in fishes—would deliver sufficient material for decades of research. Choosing a few topics out of this treasure chest of options is highly subjective, naturally. One way of making a choice is to focus on major threats to reef health.

Corals not only have competitors but also predators. Snails, sea stars, and fishes eat coral polyps. The tips of branching corals are more accessible than the smooth surface of encrusting corals. Thus, coral-eating fishes give the advantage to the latter, in the competition for space.

Perhaps the best known among the predators of corals is the crown-of-thorns starfish, *Acanthaster planci,* which in the 1970s created havoc in the coral reefs on the Great Barrier Reef, on Guam, and elsewhere. Fearing damage to tourism from the loss of live coral, local governments started intensive destruction programs. Collecting the spiny creatures is a tedious task (they grow to a diameter of 2 feet, and their spines are poisonous), so injection with formalin was used to try and stop their progress. Regardless of this effort, huge areas of living coral reef were destroyed by the sea star. Fortunately, over time, the coral reef communities do reestablish themselves, given that the solid reef buildings are still there, with bare rock for larvae ready to settle. However, stressed reefs appear to take longer for recovery from crown-of-thorns outbreaks, as observed in the Great Barrier Reef.[34] Perhaps, in analogy to fire ecology (where stressed forests are the more vulnerable ones), stressed reefs represent inviting targets to begin with.

The *Acanthaster* plague is a reminder that even in the most mature ecosystems there is no such thing as equilibrium. The path of a single *Acanthaster,* where all coral organisms have been removed, is a bare patch of rock where succession can set in, that is, a more or less regular sequence of different organisms, from algae and sponges to coral.[35] The paths from many such sea stars make a big place of bare rock; the paths of tens of thousands convert a living coral reef to a dead one. As long as there are harbors of refuge for the coral elsewhere, and as long as conditions remain favorable for reef growth, the dead reef will eventually be resettled by the larvae from the refuge. Naturally, removing predators of the starfish, such as the giant Triton conch, a favorite of the souvenir trade, greatly increases the starfish's potential for wreaking havoc. But cause and effect, for any specific event, are commonly difficult to establish.

Corals build reefs, and reefs provide great physical diversity of habitat. Part of the building is done, paradoxically, by destruction of coral skeletons. The finished product of reef formation consists of both a solid framework and of debris filling the interstices. The parrotfish is a major factor in producing debris; it takes bites out of the coral itself, skeleton and all. Parrotfish have an exceptionally strong jaw and a hard beaklike mouth to do this. Also, boring algae and boring sponges, as well as other organisms, are busy destroying what reef corals build. Thus, reef growth and reef decay are intimately intertwined, and this greatly increases the types of environment within the reefal landscape and, hence, the opportunities for diversification. Biological interaction comes in as a positive feedback on such habitat-driven diversification, producing the astounding diversity observed.

Species interactions drive diversity. Predator-prey relationships are ubiquitous—animals have to eat. Also, there is a rich horror scene of parasitism involving isopods and copepods and worms. Practically all animals have parasites, that is, organisms modified to live on or within their host, using its physiological machinery for their own advantage. More pleasant to contemplate, and no less pervasive, are symbioses favorable to both parties. Among these, there is an intriguing interaction called cleaning symbiosis. This type of symbiosis is familiar from the association between the Nile crocodile and the Nile bird, already known to Herodotus some 24 centuries ago. While the crocodile rests on shore, it holds its jaws agape, allowing the bird to clamber about the teeth and to pick parasites and food remains.

In tropical reefs, many kinds of large fishes assume the role of customer seeking to be cleaned, and they are relieved of parasites by

FIGURE 4.7. Cleaning shrimp working on a customer (moray eel).

certain types of small fishes and shrimp, the "cleaners" (fig. 4.7). These creatures and their activities became known through the observations—in different parts of the sea—by two pioneer diving biologists, the Austrian ecologist Irenäus Eibl-Eibesfeldt and the Scripps diving biologist Conrad Limbaugh. They found that cleaners have "stations," much like a barbershop, where the customers come to be freed of parasites and diseased skin. Long lines sometimes form at such stations. Limbaugh found that as many as 300 customers were serviced in six hours. A large number of species are specialized as cleaners.

Irenäus Eibl-Eibesfeldt, a pioneer in the study of animal behavior, had his first encounter with cleaning symbiosis during a diving expedition with Hans Hass, the German diving pioneer. In the spring of 1953, at the Los Roques Islands off the coast of Venezuela, he entered the coral gardens there, equipped with a light oxygen apparatus. Moving down below the canopies of elkhorn coral, he reached the reef edge, "a wall of massive coral blocks falling off steeply into deep water." Here, about 15 meters below sea level, he rested. Below him, he recalled, "The reef disappeared in the deep blue of the abyss. Fish swarms flooded past me, materializing out of the blue and vanishing back into it. A shark patrolled in front of me, moving back and forth, expressionless, eyeing me curiously. A green parrotfish with yellow spots on its gill covers was eating a staghorn coral as though it were a carrot".[36]

While thus sitting and observing, he noticed a big spiny perch making its appearance, "propelled by the lazy movements of its floppy pectoral fins, an evil-looking fellow with menacing jaws." The big fish approached the observer, stopping directly in front of him above a coral mound. It opened its mouth wide. Next, two small blue fishes swam toward the predator and began to pick around on its body. To Eibl-Eibesfeldt's utter amazement, one of the little fish made directly for the open jaws of the motionless perch and entered right between them! He expected the perch to swallow the little fellow immediately—but nothing of the sort happened. The predator remained still while the little blue fish swam about in the oral cavity picking up something here and there. After a while the big fish moved its jaws slightly, whereupon the small one swam outside. The perch then shook its body, and took off.

While still recovering from his surprise, Eibl-Eibesfeldt observed the arrival of a second perch, and the entire sequence of interaction with the slim blue fishlets was repeated. He then realized that the little fishes are cleaners that free the big ones from parasites and infections. A third big fish then appeared, and a fourth. "I had landed," he wrote, "in the barber shop of the reef."

Both cleaners and customers use signals to indicate when they are ready for business. The cleaners possess a distinct color pattern that advertises their profession. In addition, they may execute certain characteristic motions—a kind of dance—presumably to solicit peaceful behavior from their formidable customers. ("Remember me? I'm the good guy.") Customers include the occasional shark and even the enormous manta ray. The presence of cleaners on a reef is an important attraction for large fishes. To test this idea, Limbaugh removed cleaners from two isolated patch reefs in the Bahamas. Within a few weeks most of the larger fishes were no longer in the area, and the remaining ones looked unhealthy, bearing sores and frayed fins.[37]

The cleaning symbiosis is based on trust—the cleaner cleans, the customer holds still. But where there is trust there is fraud. Early in their studies, the diving biologists noted that a customer would sometimes dart away after being touched by a cleaner. In fact, it turned out that the supposed cleaner was an imitator, a look-alike that used mimicry to get close to the big fish and tear out a piece of meat with sharp teeth!

The types of observations made by Eibl-Eibesfeldt and by Limbaugh would be difficult to make today, especially so in reefs that are readily accessible. The big fishes that used to come to the coral garden barbershop are no more.

RATES OF GROWTH

Quite generally, the most rapidly growing coral species are also the dominant forms. This accounts for the great abundance of *Acropora* and related forms.

Questions about the rates of coral growth, and coral reef buildup, arose early in the nineteenth century. At the time of Louis Agassiz, the question of rate was tied to that of evolution: how old are the raised reefs of Florida, and do we see changes in the species involved? Also, Darwin's hypothesis of the origin of atolls involved rates: the balance between the sinking of the volcanic islands and the upward growth of the reefs rimming it. Subsequently the realization that sea level rose by tens of meters within a few thousand years, from the melting of northern ice caps, brought up the problem of whether reefs would have been able to keep up and stay in sunlit waters. In recent times, the discovery of the unexpected youth of the Great Barrier Reef—less than a million years—shone a spotlight on rates of reef growth. And finally, as mentioned, the fact that the growth and decay of carbonate reefs affect the carbon dioxide content of the atmosphere stimulates interest in the topic.

To answer questions about rates of carbonate production and reef buildup, we must focus on the carbonate-secreting organisms within the reef community. Several widely different kinds of organisms contribute to the buildup of a coral reef: the corals themselves (stony corals and the related gorgonians), calcareous hydrozoans, calcareous algae, foraminifers, bryozoans, mollusks, echinoderms, and a few others. Large coherent masses forming the structure of a reef are mainly the work of stony corals. The reef rubble, however, commonly makes up the bulk of the reefal carbonate masses that constitute a reef.

Before diving allowed for direct observation, rates were deduced mainly from growth on dated structures (such as wrecks and abandoned anchors), and later by experiments on specimens grown in seawater aquariums.

The nineteenth century American zoologist James Dwight Dana, in his book on corals and coral islands, presented a number of observations gleaned from the literature (as Darwin had earlier).[38] Some of these are quite specific, as when he reports that a head coral grew to a diameter of 12 inches in 12 years, at Key West in Florida, with an upward growth of half an inch in a year. A much faster rate was observed in a branching coral growing on a wreck of a ship that sank 64 years earlier. The observer, quoted by Dana, was surprised to find a thick forest of *Acropora* (Madrepora) on a ship that had not sunk all that long ago. Indeed, we now know that the genus *Acropora* contains some of the fastest-growing species among the stony corals (fig. 4.8).

Interestingly, the predominance of *Acropora* among corals is a quite recent phenomenon, geologically speaking, dating from sometime in the middle of the Quaternary (1 million years or so).[39] Reefs made by fast-growing coral are a result of the ice ages with their great fluctuations of sea level, through a range exceeding 100 meters. In earlier times, when the sea level was less mercurial, there was less of a premium on rapid growth, perhaps. Presumably, the ability to grow very fast comes with some cost, such as susceptibility to destruction by storm waves.

Next to Dana, the Swiss-American naturalist Louis Agassiz contributed important observations regarding the growth of coral reefs, especially in

FIGURE 4.8. Champion of fast growth: the acroporid staghorn coral.

his *Report on the Florida Reefs,* transmitted to the superintendent of the Coast Survey, in 1851.[40] Agassiz's main goal in estimating the rate of reef growth had nothing to do with reefs but everything to do with the most fundamental topic in all of biology and the Earth sciences: the evolution of life. He aimed to show that much of Florida is made of coral reef, that the age of the platform covering the peninsula is very great, and that there is no evidence that the organisms that made the platform are of a kind different from the organisms growing along the shores right now.[41] To his own satisfaction, he succeeded in this task, which allowed him to reject the "mutability" of species, in his typically powerful prose. As on another occasion (when inventing the Great Ice Age) his admirable eloquence did not save him from being quite misled. In the case at hand, he misjudged the time scale on which evolution operates by a factor of ten.

Much information on growth rates has been collected since Darwin, Dana, and Agassiz considered the matter.[42] At Scripps, T. Wayland Vaughan, the second director, contributed to the topic.[43] A readily applied method to learn about the growth rate of corals is to section a coral (that is, cut it with a rock saw) and study its growth lines. The method is analogous to counting tree-rings to measure the growth rate of trees. As in trees, the growth lines of coral contain valuable information on climatic change over the last several hundred years.[44]

With the advent of free diving came many opportunities to measure the growth rates of various organisms directly on the seafloor where they live. One ingenious way to do this is to stain the organism (or, rather, its carbonate skeleton) on a given date and then return later to see how much unstained material has been deposited since. The investigator simply covers the organism (for example, a calcareous alga) with a plastic tent and injects an appropriate stain into the enclosed space. The stain in the skeleton marks the time of the experiment. The tent is then removed. After a few months, the experimenter returns, the organism is harvested, and the added new growth determined. The rate of growth is then known.[45]

There is yet another way to measure rates. It relies on the determination of the seasonal progression of carbonate secretion, by reconstructing the temperature fluctuations recorded within a shell, by chemical means.[46] The basic method is the brainchild of the nuclear chemist Harold Urey.[47] In essence, it consists in measuring the relative abundance of two types of oxygen atoms (called isotopes, because they occupy the same place in the periodic table). Within the calcite crystals of skeletons and shells, the ratio of the oxygen isotopes changes as a function of the temperature at which the carbonate is precipitated. Because temperature is tied to seasons, the seasons can be recognized in coral skeletons, mollusk shells, and even in rather small organisms, across the lines of growth, by the isotopic pattern. In this fashion can the determination of isotopes using modern mass spectrometry contribute to solving century-old questions.[48]

Coral records are now available from a large number of locations,[49] and many interesting insights are emerging, especially in regard to the occurrence of El Niño events.[50] But the rules of interpretation are still developing. What the corals are recording, in any given location and circumstance, is not always clear because they respond to many different factors.[51] One problem, not just with coral but with all biological climate recorders, is biased reporting.[52] The

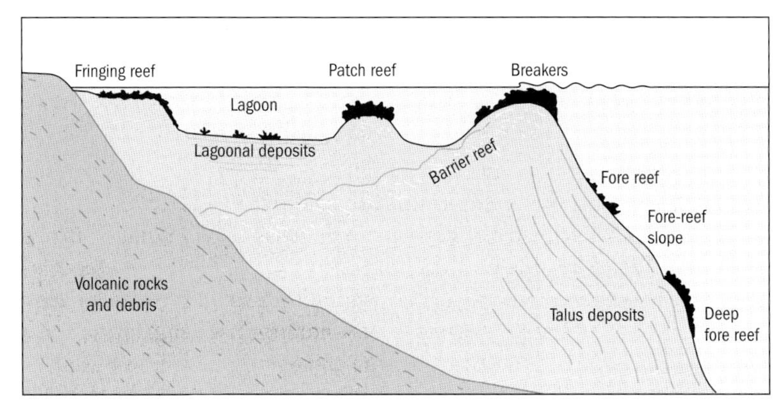

FIGURE 4.9. Main environments of a typical reef structure. The width varies between hundreds of meters and several kilometers. Highest growth rates are in the barrier reef section, which delivers material both to the lagoon and to the fore-reef talus. At the edge of the reef the corals have access to plankton of the open sea.

recording organism keeps adequate notes whenever it is growing well. Calcifying organisms stop building skeleton when conditions are averse, just like trees in cold climates stop making wood in winter. It is a bit like having diary entries for the good days, and little or nothing for the bad ones.

What, then, are the typical rates of growth of the various contributors to reef mass? It depends on the contributor and on the location along the reef structure. The typical growth rate of corals building the outer reef is between a centimeter and an inch per year—a modest-size colony will be several decades old, a large, massive coral head several hundred years. In terms of upward growth, for a dense cover of coral, the potential rate for reef accumulation in shallow water (less than 10 meters deep, say) is a few meters per thousand years. The potential for upward growth in different portions of a reef varies, but on the whole the structure simply grows up to sea level, where growth stops. Whatever additional material is produced is added to the rampart of the fore-reef slope (fig. 4.9).

The various shapes of coral species are typical for their site of growth, with respect to the reef edge. On the rampart in front of the edge and just below surf action, sheltered from the direct onslaught of breakers, large palmate corals can exist. On the edge itself, only massive corals survive. Farther inward, on the platform, bushy but robust forms do well, especially in cracks and ravines. In the lagoon, finally, branching corals are quite safe (except during violent storms). All these forms, naturally, grow at different rates.

The lagoon environment typically has calcareous algae, which can grow quite rapidly. Also, lagoons may host fast-growing coral such as *Porites*. Debris is delivered to the lagoon from the reef flat on top of the barrier, and plankton delivers additional sediment, mainly shells from minute calcareous algae. A potential fill rate of up to a few meters per thousand years seems a reasonable estimate, although much sediment is lost, in places, during tidal flushing of channels guiding water into and out of the lagoon.

By constantly exporting large amounts of debris, the reef makes room for more growth, which provides more debris, expanding the periphery. Thus, when contemplating the dynamics of expanding reefs, we have to keep an eye on both the fast-growing producers and builders, and the complementary activities of the demolition crews. It is as though our reef cities were organized to provide for maximum production of rubble, ruins, and rubbish on which to build new suburbs.

QUESTIONS ABOUT ORIGINS

The first question about origins regarding coral reefs and their inhabitants concerned the nature

FIGURE 4.10. Coral reefs are plankton traps that depend on vertical migration. At night, plankton animals are in surface waters and some drift into reefs.

of the carbonate-secreting anthozoans proper—are they plants or animals? They turned out to be animals hosting algal symbionts, thus being both, in essence. In fact, corals are plankton traps that also photosynthesize. Next, the origin of atolls engendered much discussion, with Darwin proposing sinking of the base and upward growth of coral, and later workers emphasizing sea level changes during the ice ages as factors governing atoll morphology. A host of other questions arise: How and why does the nature of reef building change through time? How and why are Caribbean reefs different from Indo-Pacific ones? Why have rapidly growing coral species evolved in the last million years? How did the Great Barrier Reef get started and grow so rapidly in the last half million years? Let us examine the first two of these questions—the nature of coral and the origin of atolls—and then briefly comment on diversity patterns. The enigma of the Great Barrier Reef is discussed in the section that follows.

The original "origins" question about the nature of anthozoan corals has long been answered: they are animals.[53] In fact, as reef organisms, they are part of an enormous plankton trap scheme that takes advantage of the daily migration of zooplankton. During the day, much of the small zooplankton—copepods mainly—are below the sunlit zone, to avoid predation. They come up during the night to feed on the minute algae within the uppermost layer. Early in the morning it is time to descend. Of course, those among the zooplankton that happened to drift into a reef at night will descend right into the tentacles of anthozoans (fig. 4.10). The corals form a carpet of open mouths and mucus strands on the shallow seafloor, trapping anything that touches the bottom.[54]

We can now assess the origin of the fascinating and tumultuous activity that characterizes the fore-reef and has inspired so many artists. The animals of the spectacular fore-reef gardens, like all others on the reef, largely depend for their sustenance on photosynthesis by dinoflagellates, within the sunlit zone, and on the trapping of migrating zooplankton by anthozoans. They basically live off the organic

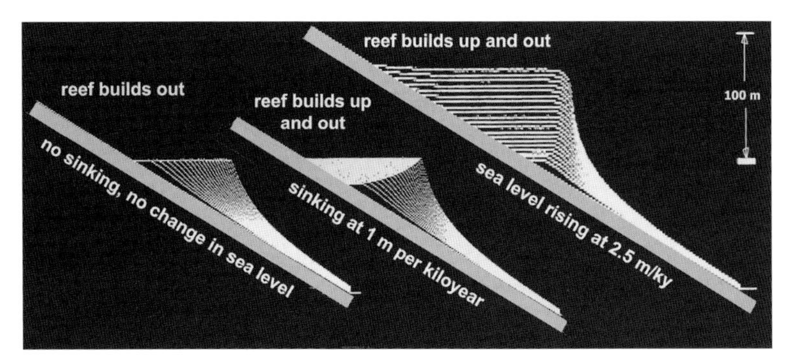

FIGURE 4.11. Computer-aided thought experiments on growing a fringing reef, first with sea level constant, then with the base sinking slowly (1 meter in 1,000 years), and finally with sea level rising (2.5 meters in 1,000 years). Time interval of calculation: 1,000 y (= 1 kyr). Maximum rate of upward growth assumed: 4 meters per 1,000 years.

stuff that comes down from the highly productive barrier reef in the surf zone above. Those corals that live in the dark, of course, do not have photosynthesizing symbionts.[55]

Many of the questions about origin are closely linked to the manner in which reefs are constructed. Like Jerusalem and Rome, coral reefs are built on their own rubble. Thus, there are two aspects to the study of reef construction: the processes of reef building tied to the living reef, and the nature of the dead reef on which this activity takes place. The analogy with cities holds nicely as far as the rate of mound growth. A buildup of a foot per century is nothing unusual for an ancient city. A coral reef can grow at about the same rate under favorable conditions. Because coral reefs grew for hundreds of thousands of years and cities only for thousands, corals made undersea mountains while cities made mounds.

The archetypical form of a coral reef is the fringing reef. We can make a simple thought experiment regarding the growth of such a reef. If we were to visit an island margin at intervals of 1,000 years, we would see a ledge near sea level building outward, first on top of other massive coral, then on top of earlier rubble, held together by bryozoans, encrusting algae, and microscopic precipitations of unicellular algae. Gradually, the beach would widen, and the ridge behind the beach (bearing storm deposits) would grow seaward, slowing in its advance as the source of debris moves farther offshore.

After 10,000 years, we would stand on a small intertidal platform several hundred feet wide, bounded seaward by a steep ramp of coral and shell debris. The entire mass of carbonate would be wedge-shaped in profile, like a city dump at the edge of a mesa (fig. 4.11). Growth would have slowed through time, as more material was lost over the edge on the steepening slope off the reef.

Fringing reefs just like the one we built in imagination are in fact seen surrounding young volcanoes. How do we get from there to a barrier reef, with a lagoon separating the outer reef from the shore? Clearly, the processes that we considered so far—building up to sea level and redistributing and cementing rubble—will simply keep growing a larger and larger platform offshore, perhaps with a broadening storm deposit on top, but no lagoon. A sinking seafloor or a rising sea level will not change that outcome, unless we introduce some additional factor. Charles Darwin thought he had the answer: if an island is sinking, and reef growth is most vigorous at the seaward edge of the reef, then a depression could form between shore and outer reef. Furthermore, if the island sinks till all its land is submerged, the reef ring with its storm ramparts would form an atoll, and a lagoon would now mark the site of the former island. This is Darwin's hypothesis of atoll formation, published in

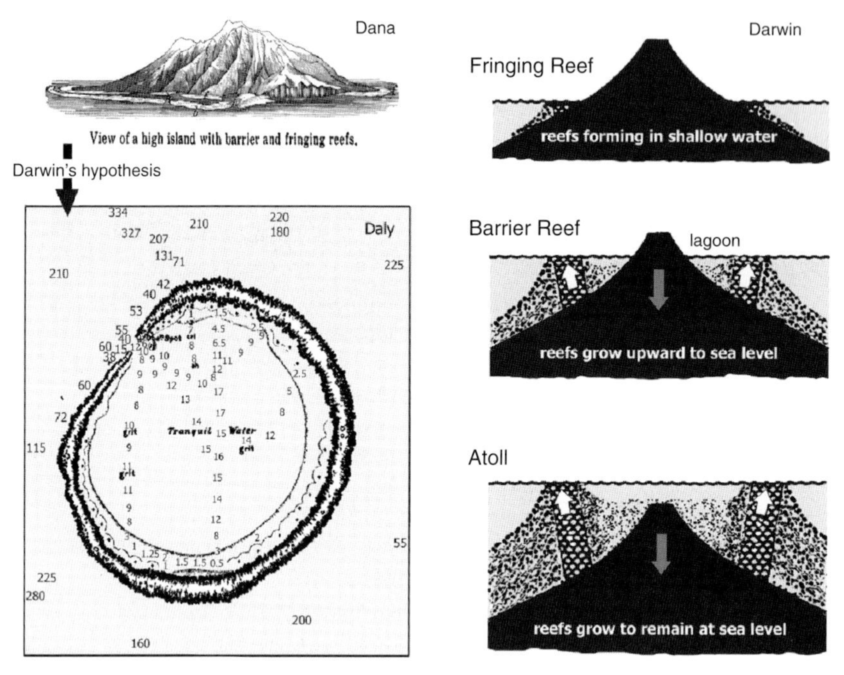

FIGURE 4.12. *Right:* Darwin conceptually put an arrow between fringing reef and atoll, postulating the sinking of islands. *Left:* drawings by Dana and Daly.

1842 (fig. 4.12). Darwin's first ardent supporter on the subject of atolls was James Dwight Dana, who read about the hypothesis during the U.S. Exploring Expedition and found that it "threw a flood of light over the subject."[56]

What Dana missed is that Darwin had no valid explanation for the presence of lagoons. Darwin seemed to think that the vigorous growth at the outer rim of the reef contains the answer. If so, this would imply that the rate of upward growth has a problem keeping up with the rate of sinking, except at the edge of the reef. However, islands are sinking extremely slowly, when compared with coral growth rates. Thus, upward growth is not a problem, at least not in today's reefs. They can easily build up at the required rate, with plenty of material left over to fill the lagoons, as well.

The lagoons, then, need a cause, and this cause was identified, in principle, by Reginald A. Daly, geologist at Harvard, in his book on the ice age published in 1936.[57] He suggested that the lowered sea level, whenever the northern ice caps built up, exposed the reefs to weathering and led to erosion behind the solid reef rim, digging out the floor for the future lagoons, which then formed when sea level rose again. He also noted that the new coral would preferentially grow on preexisting knolls and knobs, something that beginning geologists now learn on their first field trip into ancient reefs. A substantial portion of the erosion (perhaps most of it) is by dissolution of the exposed reef (when sea level is down) by rainwater and runoff.[58]

Returning to the overall diversity patterns, with centers of diversity in the western equatorial Pacific and in the western tropical Atlantic (fig. 4.5), we first note that this is where the warm water piles up, driven by trade winds. The lesser diversity in the Atlantic is readily rationalized: during glacial intervals, setting in around 3 million years ago and growing more severe till about 700,000 years ago, temperature-sensitive species succumbed to unprecedented cooling and became extinct. With the closing of the Panama Isthmus, any option of reseeding from the Pacific was closed.

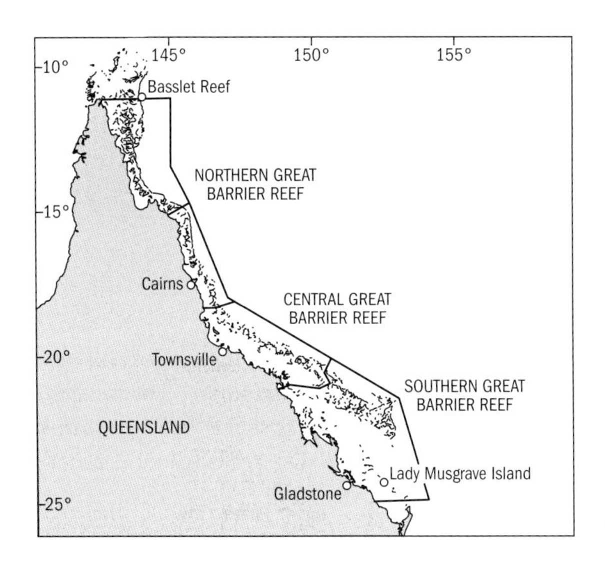

FIGURE 4.13. Geographic position and extent of the Great Barrier Reef.

The problem with this simple explanation is that modern research suggests that Caribbean corals are not particularly closely related to Pacific ones. It now appears that a large contingent of Atlantic and Pacific corals diverged more than 30 million years ago.[59] If this is so, the notion that regional species loss in the Caribbean could be made up by restocking from the Pacific, if only the Isthmus were still open, becomes rather unattractive. Instead, a long-term trajectory of evolution and diversification, informed by the smaller area available in the Caribbean as compared to the Indo-Pacific realm, must be contemplated. Clearly, before this question about the contrast in diversity between the western equatorial Pacific and the Caribbean can be answered, one must firmly establish the degree to which the organisms are related, that is, when their last common ancestor lived.[60]

An important factor in the origin of diversity patterns within each of the major ocean basins is the dispersal of larvae.[61] Looking closely, we can determine that isolated places at the edges of distribution, such as the Hawaiian Islands, develop their own endemic species. In contrast, Bermuda in the North Atlantic has no endemic species but is merely impoverished. Here, colonization is by supply of larvae through the Gulf Stream, from the Caribbean. Off Brazil, there is no warm current from the Caribbean for restocking, so that separate evolution took place. About half of the 20 or so species of coral present are endemic, the rest constitute a small selection of hardy species from the Caribbean realm.[62]

ON THE ORIGIN OF THE GREAT BARRIER REEF

The Great Barrier Reef (fig. 4.13) is the largest edifice built by organisms on our planet. It was discovered (for the western world) by Captain James Cook, in 1770, with painful consequences for his ship.

The great reef is still being discovered: by diving ecologists, from the air, from space, and recently by drilling into it. It turned out to be incredibly young: its bulk was built within the last million years or so. This strongly suggests that it owes its existence to factors having to do with the ice age climate fluctuations. An expansion of the western warm-water pool, together with the fluctuation of sea level through a large depth range, probably provided the necessary environment for rapid reef buildup.

At the beginning of the science of the Great Barrier Reef there is Captain Cook and the 370-ton Whitby cat *Endeavour.* It was late in May, in 1770. The water was calm, glistening in the moonlight, as the vessel, sails reefed, slowly made its way northward along the eastern coast of Australia, pushed gently ahead by a southeasterly breeze. She had left Botany Bay on 6 May, provisioned with fresh water, meat, and vegetables, and carrying a rich assortment of plants new to science. For the last three weeks, James Cook and his crew had sailed within the inner passage of the Great Barrier Reef, a maze of channels between hidden submarine peaks and banks rising abruptly to just below the surface. It was a dangerous way to travel; any moment some unseen obstacle could rip into the bottom of the sturdy vessel. Against all odds, by setting out boats to sound ahead and by exhorting his crew to intense vigilance, Cook had managed to sail hundreds of miles past the Tropic of Capricorn deep into the great reef, without incident.[63]

That night Cook's luck ran out. The date was 11 June 1770. Turning east of north late on that day, to get away from islands emerging ahead, the *Endeavour* gingerly entered deeper water between 6 and 9 p.m., moving from 14 to 21 fathoms depth. As always, Cook planned to anchor for the night, but the deeper water promised safety; with more than 100 feet of water below the flat bottom of his vessel he thought the risk was small. The breeze was just right, the night clear and moonlit; it could not hurt to keep going a little longer to find good anchorage. For the next two hours Cook's guess proved correct. The depth remained satisfactory, the sounding showing 17 fathoms a few minutes before 11 p.m. But then, before a new sounding could be completed, Cook's worst fears materialized. Without warning, a terrible scraping sound enveloped the shuddering vessel, and then a horrible shock stopped the ship dead, sending everything tumbling that was not tied down. The *Endeavour* sat fast on a reef. What was worse, she had run aground during high tide; the falling level of the sea would maroon her for at least 12 hours, exposed to wave attack in the most vulnerable position imaginable.

Throwing overboard whatever they could do without—cannons, iron and stone ballast, barrels—the crew lightened the vessel as much as possible. In the meantime, the ship was taking water, its hull damaged. All available hands worked furiously at the pumps. To the great consternation of all, the following tide was insufficient to lift the vessel off its perch. Cook calmly gave his orders to the despondent crew, doing everything to minimize damage and to prepare for plugging holes. The weather stayed calm; the next high tide, 23.5 hours after running aground, finally floated the vessel off its perch. Anchors secured, drawing much water, the *Endeavour* made for shore—a shore unfortunately inhabited by hostile natives. She found Cook Harbour on the Endeavour River (near today's Cooktown) and was beached for repairs. Kangaroo steak and turtle soup consoled the exhausted party. Cook's sangfroid (and mild weather and a good portion of luck) had narrowly avoided disaster.

Cook had truly discovered the Great Barrier Reef of Australia. Many seamen would do so after him, and many with less good fortune. After this incident and another narrow escape from the treacherous reef, Cook had seen enough of it. He never returned to the place in his subsequent voyages, preferring to revisit Tahiti and New Zealand.

The Great Barrier Reef at the margin of northeaster Australia has become, in the public consciousness, the archetype of coral reef. It is a system of reef structures that includes examples of all types of morphology resulting from the building activities of stony corals and their associates. The Great Barrier Reef, then, subsumes a collection of thousands of different reefs whose "barrier" character emerges only as a pattern on a very large scale (fig. 4.14). As mentioned, there are more or less navigable waters behind the barrier. The dimensions of the Great Barrier Reef are truly gigantic: some 1,200 miles long and up to 100 miles wide in places, it covers an area near 80,000 square miles, from south of the Tropic of Capricorn to the coast of New Guinea. It makes up almost one-fourth of the total area occupied by living reefs (ca. 360,000 square miles).

The Reef has long been the site of extensive biological exploration, first by the British and subsequently by Australian scientists. The Great Barrier Reef Expedition of the Royal Society

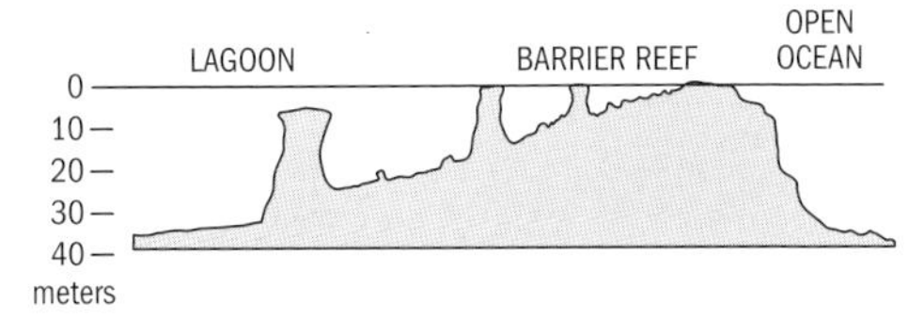

FIGURE 4.14. A cross section of the Great Barrier Reef drawn by R. A. Daly, in 1936.

(1928–1929) contributed an enormous amount of knowledge about the creatures in the reef and their habits. The popular book summarizing results of the expedition, by its leader Sir C. Maurice Yonge, is a delight to read.[64] Systematic investigation by Australian researchers began around 1970, with the founding or expansion of several marine facilities. An important aspect of this development is a government mandate for protection of the Reef. One welcome result of this mandate, to the study of reef ecology, is the necessity to document ongoing change. This approach replaces snapshot inventory (as is the typical result of expeditions) with monitoring and a consequent appreciation for dynamics, such as the role of disturbance in maintaining reef diversity.[65]

The Great Barrier Reef is a complex of coral cities proclaiming, in their architecture, the presence of life on Earth and its powers to change the geography of the planet. What do we know about the origin of this structure? About its age, the time it first started to grow? About the reason why it extends farther south than any other group of reefs? Why is it a barrier complex rather than an enormous carbonate platform? Until very recently, we knew very little that is relevant to these questions, mainly because we had no idea that the Great Barrier Reef is actually very young. This discovery is a result of drilling into the Reef by the Ocean Drilling Program, in an expedition led by the geologists Judith McKenzie and Peter Davis.[66]

What they found is that the Reef expanded southward only within the last million years or so, so that the present configuration—the bulk of the Great Barrier—is a product of the ice ages. Why should this be so?

The current state of understanding is reflected in the summary provided in the Ocean Drilling Program Long Range Plan issued by the Ocean Drilling Program:

> The initiation of the Great Barrier Reef temporally coincides with a major change in climate response to orbital forcing. The advent [of reef building] is correlated with a regional shift in oxygen isotope data indicating a 5 °C rise in surface-water temperature. The current working hypothesis is that this temperature rise was caused by an intensification of the equatorial Trade Winds which drive warm tropic waters into the Western Pacific.[67]

The crucial point is that warm water invaded the region off eastern Australia, from the equatorial zone. At present, the presence of an extensive warm-water pool governs the environment of the entire tropical western Pacific. This pool gathers warm surface waters brought westward by trade winds. The question is whether a strengthening of the trade winds could have caused the observed expansion of the warm-water pool to the eastern margin of Australia.

An important point to consider when addressing this question is that the expansion seems to have proceeded *without regard to whether conditions were glacial or interglacial*. This is appreciated when comparing the oxygen-isotope data cited in the quote above with similar data from close to the equator, from an earlier drilling expedition.[68] The difference between the two records keeps growing steadily through time. If indeed the trade winds were responsible, this would imply that they did not change much during any one of the 100,000-year ice age cycles. This is in conflict with other evidence.

Thus, it is unlikely that a change in trade winds is the main or sole reason for the expansion of the warm-water pool and the great stimulation in reef growth that this expansion engendered.

We have to ask why there is a warm pool in the first place. One needs a place to pile up the pool. The great western Pacific bight can hold on to the warm water because the outlet toward the west, that is, toward the Indian Ocean, is

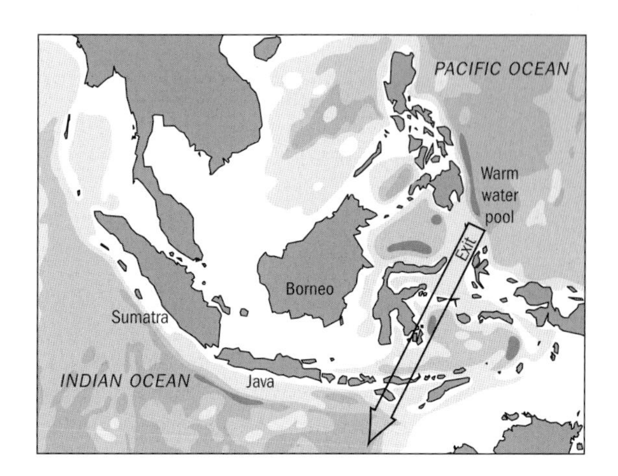

FIGURE 4.15. Map of the Indonesian passage, where warm water moves from the western Pacific warm pool to the Indian Ocean. All land here is rimmed by coral reefs.

obstructed (fig. 4.15). If, in the last million years, this obstruction became more effective, the warm pool could expand in response.

How might this have happened? It seems reasonable to call on variation in sea level to increase reef rubble production, which would help clog the exit of warm water from the Pacific to the Indian Ocean. Large, long-term cycles in sea level variation started 900,000 years ago.[69] These large cycles greatly increased the output of reef materials by providing as much opportunity for rubble formation as for reef buildup. Using our earlier city mound analogy, we can now envisage a regular alternation between periods of building and periods of wrecking. Such a city with varying fortunes, as we know from archeology, grows the higher mound. The outlets for the warm-pool water into the Indian Ocean all are flanked by coral reefs. It is these outlets that will clog, from the increased efficiency of the reefal rubble factories. Restricted outlets will then favor expansion of the warm-water pool, which ultimately helps grow the Great Barrier Reef. In this hypothesis, then, the cause for the Great Barrier Reef must be sought in the overall shallowing of the Indonesian archipelago.

The considerable expansion of the genus *Acropora* within the last million years in the tropical Pacific and elsewhere fits well into this scenario. Rapid sea level variation favors rapidly growing coral, especially during the rapid rises that accompany the wasting of the northern ice caps. *Acropora* is the rapid-growth specialist among the stone corals. By becoming dominant in the coral community, it must have greatly increased the potential for the production of coral rubble.

The clogging-the-exit hypothesis of the origin of the Great Barrier Reef calls on increased amplitudes of sea level variation in the latter half of the Pleistocene as the driving force. Reef expansion in Indonesia increasingly restricted the outflow of warm water from the western Pacific, which resulted in warm-pool expansion, which in turn stimulated growth along the northeastern margin of Australia. If the hypothesis holds up, the corals off Australia owe their existence to the reef-building activities of their cousins in Indonesia.

On a planetary scale, the expansion of the western Pacific warm pool must have greatly influenced climate everywhere, mainly by increasing the potential for interannual variations, such as El Niño oscillations. Furthermore, the expansion of *Acropora* would have increased the amplitudes of ice age variations in carbon dioxide.[70] If these proposed connections hold up to scrutiny, the evolution of *Acropora* and other fast-growing coral emerges as a major feedback mechanism on climate change.

SUNKEN REEFS OF THE MID-PACIFIC

Not everyone signed on to Darwin's explanation of atolls as enthusiastically as did Dana. John Murray, naturalist of the *Challenger* Expedition, took strong exception to Darwin's theory of the origin of atolls by sinking of volcanoes and upward building of coral reefs. Atolls are widespread. What evidence was there that the seafloor is sinking everywhere? None whatever. Darwin (in Murray's view) had simply made up a plausible story, a speculation—no independent evidence for the hypothesis was available.

What Murray could not know (nor could Darwin) is that the evidence for the sinking seafloor actually does exist over large areas of the Pacific. The evidence rests with sunken flat-topped islands, crowned by dead coral that once grew in shallow water. It took another hundred years after the publication of Darwin's book on coral reefs and atolls until these sunken islands would be discovered on echo soundings, by the geologist and U.S. Navy Reserve Officer Harry H. Hess (1909–1969). After another few years, a Stanford graduate student, Edwin L. Hamilton (1914–1998), who had returned to school from Navy duties, went out to verify the nature of the flat-topped seamounts during the Scripps MidPac Expedition (1950), by dredging their crests.[71]

Hess, running the echo sounder of the supply ship he commanded during World War II, had identified some 20 seamounts at depths from 1 to 2 kilometers between Hawaii and the Mariana Islands, all flat on top. He named them *guyots* (pronounced "ghee-yohs").[72] Hess published his findings in 1946, adding 140 similar examples from sounding records at the U.S. Navy Hydrographic Office. He interpreted guyots as drowned islands and claimed that they had once been at the surface of the ocean, planed off by erosion and rimmed by reef flats.[73]

Hess's findings—provided the interpretation was correct—nicely fit with Darwin's idea of widespread subsidence in the central Pacific.[74] With Hess's sunken islands, the enigma of widespread seafloor subsidence, first argued by Darwin, now had been widened to regions outside the atoll belt. Clearly, such widespread sinking of the seafloor was something geologists ought to understand—it involved a substantial portion of the planet's surface. The obvious task was to determine the nature and the age of the cover of Hess's seamounts.

In the first years after the war, when oceanographers were getting ready to tackle big problems in a big way, the enigma of the guyots was among the most intriguing questions in deep-sea geology. The answer (or rather, a first step toward a comprehensive answer) would come from the results of the MidPac Expedition, which put to sea in July 1950, from San Diego. Two ships were involved: the *Horizon*, a 143-foot modified tugboat, seaworthy but definitely less than comfortable, and a navy vessel (with a code for a name), 220 feet long. The geologists on board, on the whole, thought the Pacific was very old—there was no reason to think otherwise. Regarding the guyots, one of the participants, Robert S. Dietz (1914–1995) of the U.S. Navy Electronics Lab at San Diego speculated that they were Cretaceous islands. H. William Menard (1920–1986), in his autobiographical book *The Ocean of Truth*, recalled:

> [Robert Dietz] reasoned that most of the mountain building around the margins of the Pacific was Cretaceous and that if something happened at the edges it probably happened in the middle, too. This was early Dietz but a good vintage. The reasoning was wild, somewhat plausible, and only incidentally wrong, and the conclusion was correct. He was thirty-six.[75]

The mid-Pacific mountains, between Hawaii and Guam, were reached at the end of August, and sampling of guyots from the *Horizon* began on 1 September 1950. On the first attempt, the cable parted and the dredge was lost. The following attempts brought up disappointingly small samples, chips, manganese, basalt, and limestone barren of fossils. Finally, the luck turned, and the dredge brought up exciting finds:

> The ninth day of dredging, on the thirty-second try, Ed Hamilton was suddenly confident that he had a doctoral thesis. The dredge was full of Middle Cretaceous reef corals. . . . Ed's professors at Stanford later told him he could have submitted his results on the back of an envelope and gotten a degree.[76]

Hamilton reported the results from the dredgings in 1956.[77] Hess had been right about the guyots; they were indeed drowned islands. But rather than being geologically ancient, these features had been islands only 100 million years ago. Rather than sinking at an imperceptible few meters per million years, they sank at a rate 10 times faster. The question what drowned the reefs remained, however.[78]

For the geologists on the expedition, the geologic youth of the sunken islands became a central issue of great bewilderment. For the next 15 years, after many more dredgings, nothing older than the reef rocks recovered during the MidPac Expedition was brought up. How old was this Pacific basin anyway? Perhaps much of the water now in the sea had emerged from the mantle during the last 100 million years.[79] Seafloor spreading would eventually provide an answer, but it was still more than a decade away. Dietz would be one of the several geologists proposing seafloor spreading (and the one to name the concept).

In terms of ocean history (rather than seafloor history) the intriguing thing is that the reefs drowned at all, given the slow sinking and their subtropical position. The dying and drowning of the reefs, near 100 million years ago, is still an unsolved problem. Most likely some untoward event, resulting in widespread stress on reef organisms, is to blame. One candidate is the emergence of large plumes of hot magma, eating their way through the oceanic crust and resulting in enormous outflows of basalt.[80] There are indications that the outpouring led to global warming (presumably from the release of carbon dioxide and from flooding the continents as sea level rose) and resulted in widespread oxygen deficiency within the sea, below surface waters.[81] Occasional incursions of foul water from the thermocline into the sunlit zone, owing to deep mixing by storms, could have stressed reefs repeatedly until they succumbed, over a short geologic period.

Drilling into several guyots in the Mid-Pacific Mountains, in the 1990s, resulted in detailed information on the age and nature of reefs and other sediments but did not resolve the fundamental problem of why the reefs ceased to keep up with the sinking of their base, given the modest rate of subsidence. Of course, the time unfavorable for reef growth needed to be only long enough to let the mounts subside to below 30 meters or so—after that, a return of favorable conditions would not matter, since rapid growth depends on photosynthesizing symbionts. The time interval needed for this amount of sinking without growth is between 1 and 2 million years, depending on the age of the seafloor. (Sinking rates diminish with age of the seafloor.) The time span required to kill reefs by foul-water stress can be accommodated within the duration of spells of lack of oxygen observed in the record.

How trustworthy are such guesses regarding the causes of the demise of the mid-Cretaceous reefs?

In the modern ocean, as we have seen, coral reefs are killed by heat (by bleaching), by cold winters (as shown by the limits of distribution), by excess nutrient supply (sewage kills reefs), and by excess siltation (no reefs off muddy rivers). For deadly biological agents, we have the example of the crown-of-thorns *(Acanthaster planci)* outbreaks in the western Pacific and on the Great Barrier Reef (episodic since the 1960s) and the microbial attacks on coral in the Caribbean. Excess heat and excess nutrient supply, by sporadic upwelling of oxygen-poor waters in an oxygen-starved sea, would both seem good candidates for factors driving reef collapse.

Whatever the reason or reasons for the demise of the reefs on the sunken islands of the Mid-Pacific mountains, Hamilton's dredge samples showed that Darwin's conclusion that large areas of the Pacific must be sinking was correct. Independently, Darwin's theory had obtained striking support from drilling on Enewetak Atoll, in 1952. The drill encountered basalt after penetrating 1,300 meters of reef materials.[82] Yet, Darwin's theory had to be modified somewhat: The variations in sea level that accompany the buildup and decay of large northern ice sheets have to be considered, as urged by the alpine geologist Albrecht Penck (1858–1945) and the American geologist Reginald Aldworth Daly (1871–1957), early in the twentieth century. During times of low sea level, the central lagoon area is drained of seawater, and the rain dissolves carbonate and erodes the center of the carbonate platform. The eroded center can then make a lagoon as sea level rises, which results in the typical atoll configuration.

Atolls are the legacy of both seafloor sinking and ice age sea level fluctuation.

NOTES AND REFERENCES

1. The word *anthozoans*, the name of the animal class of corals, means "flower animals." Originally coined because the tentacles of polyps look somewhat like flower petals, the designation turns out to be a good one: flower animals participate in photosynthesis thanks to their algal symbionts.

2. Technically, *symbiosis* means "living together," which can be of advantage to both, or detrimental to one of the organisms involved. The word is used here in the older common meaning of a mutually beneficial arrangement.

3. Naturalists of the nineteenth century distinguished madrepora, stony structures with regularly spaced identical polyps anchored within their cups, from millepora, stony structures with abundant pinholes bearing very small polyps of different size and function. The madrepora are stony corals in the modern sense, while the millepora belong to the "fire corals," so called because their sting penetrates the skin on touch. Fire corals are more closely related to jellyfishes than to stony corals. Many forms closely related to the stony corals do not make a solid carbonate skeleton. They are referred to as "soft corals." Best known among these are the sea anemones, the sea pens, the sea fans, and the horny branching corals such as the precious "black coral."

4. J. B. C. Jackson, M. X. Kirby, W. H. Berger, K. A. Bjorndal, L. W. Botsford, B. J. Bourque, R. Bradbury, R. Cooke, J. Erlandson, J. A. Estes, T. P. Hughes, S. Kidwell, C. B. Lange, H. S. Lenihan, J. M. Pandolfi, C. H. Peterson, R. S. Steneck, M. J. Tegner, R. R. Warner, 2001, *Historical overfishing and the recent collapse of coastal ecosystems.* Science 293, 629–638.

5. The Vikings called it *rif*, a ledge of rocks. *Rif* is related to the word "rib." This use illustrates how body parts serve to describe geography, along with "headland," "river mouth," "neck," and "ridge," the latter having lost its original meaning of "backbone" in English.

6. R. B. Bacastow, C. D. Keeling, T. P. Whorf, 1985, *Seasonal amplitude increase in atmospheric CO_2 concentration at Mauna Loa, Hawaii, 1959–1982*, Journal of Geophysical Research 90, 10540; N. J. Shackleton, 1977, *Carbon-13 in Uvigerina: Tropical rainforest history and the equatorial Pacific carbonate dissolution cycles*, in N. R. Andersen, A. Malahoff (eds.), *The Fate of Fossil Fuel CO_2 in the Oceans*, Plenum Press, New York, pp. 401–427.

7. W. H. Berger, 1982, *Increase of carbon dioxide in the atmosphere during deglaciation: The coral reef hypothesis.* Naturwissenschaften 69, 87–88.

8. J. R. Petit, J. Jouzel, D. Raynaud, N. I. Barkov, J.-M. Barnola, I. Basile, M. Bender, J. Chappelllaz, M. Davis, G. Delaygue, M. Delmotte, V. M. Kotlyakov, M. Legrand, V. Y. Lipenkov, C. Lorius, L. Pépin, C. Ritz, E. Saltzman, M. Stievenard, 1999, *Climate and atmospheric history of the past 420,000 years from the Vostok ice core, Antarctica.* Nature 399, 429–436.

9. The sea cucumbers were sometimes referred to as *bicho do mar*, Portuguese for sea-slug, or *bêche-de-mer* in French.

10. R. A. Kinzie III, R. W. Buddemeier, 1996, *Reefs happen.* Global Change Biology 2, 479–494; R. W. Buddemeier, D. G. Fautin, 2002, *Large-scale dynamics: The state of the science, the state of the reef, and the research issues.* Coral Reefs 21, 1–8; O. R. Wilkinson, 1996, *Global change and coral reefs: Impacts on reefs, economies and human cultures.* Global Change Biology 2, 547–558.

11. N. E. Chadwick-Furman, 1996, *Reef coral diversity and global change.* Global Change Biology 2, 559–568.

12. D. R. Bellwood, T. P. Hughes, 2001, *Regional-scale assembly rules and biodiversity of coral reefs.* Science 292, 1532–1534; C. M. Roberts and 11 others, 2002, *Marine biodiversity hotspots and conservation priorities for tropical reefs.* Science 295, 1280–1284. For general background regarding discussions on biodiversity, see R. M. May, 1973, *Stability and Complexity in Model Ecosystems.* Princeton University Press, Princeton, N. J.; K. S. McCann, 2000, *The diversity-stability debate.* Nature 405, 228–233.

13. Charles Darwin, 1842, *The Structure and Distribution of Coral Reefs.* Smith Elder, London. 3rd ed. 1889, 344 pp.

14. K. O. Emery, J. I. Tracey, Jr., H. S. Ladd, 1954, *Geology of Bikini and nearby atolls.* U.S. Geological Survey Professional Paper 260-A, 1–265.

15. Coral bleaching is closely tied to the El Niño/Southern Oscillation phenomenon. It contributes importantly to all the other stressors. An overview of reef stress is in D. Bryant, L. Burke, J. McManus, M. Spalding, 1998, *Reefs at Risk: A Map-Based Indicator of Threats to the World's Coral Reefs.* World Resources Institute, Washington, D.C. For scientific background see P. W. Glynn, 1996, *Coral reef bleaching: Facts, hypotheses and implications.* Global Change Biology 2, 495–509.

16. D. R. Bellwood, T. P. Hughes, C. Folke, M. Nyström, 2004, *Confronting the coral reef crisis.* Nature 429, 827–833.

17. J. B. Jackson quoted in the *New York Times*, 26 April 2005, p. D2.

18. T. P. Hughes, 1994, *Catastrophes, phase shifts, and large-scale degradation of a Caribbean coral reef.* Science 265, 1547–1551.

19. J. B. C. Jackson, 1997, *Reefs since Columbus.* Coral Reefs 16, Supplement, S23–S32.

20. This fateful error in public perception, of course, is familiar to everyone engaged in any aspect of conservation and land use, where a resetting of the baseline, after each major change, is commonplace. The term *shifting baseline*, coined by the fishery biologist Daniel Pauly, has recently gained acceptance to describe the general process. D. Pauly, 1995, *Anecdotes and the shifting baseline syndrome of fisheries.* Trends in Ecology and Evolution 10, 430.

21. J. M. Pandolfi, 2002, *Coral community dynamics at multiple scales.* Coral Reefs 21, 13–23.

22. The concept of preconditioning ecosystems for collapse is outlined in J. B. C. Jackson et al., *Historical overfishing.*

23. T. P. Hughes, *Catastrophes*, p. 1547.

24. According to Hughes and to Jackson, the sea urchins have been very slow in coming back. Individuals are large and well fed and have well-developed reproductive organs. However, for eggs and sperm to meet upon ejection into the water, it is best if the urchins occur in clumps and clusters. With so few urchins left, clumps do not readily form, so that the replenishment of the populations is much slower than what one might expect when considering the number of eggs each female urchin produces.

25. P. J. Mumby and 11 others, 2004, *Mangroves enhance the biomass of coral reef fish communities in the Caribbean.* Nature 427, 533–536.

26. W. F. Precht, 2002, *Endangered acroporid corals of the Caribbean.* Coral Reefs 21, 41–42.

27. B. E. Brown, J. C. Ogden, 1993, *Coral bleaching.* Scientific American 268, 64–70; T. J. Goreau, R. I. Hayes, 1994, *Coral bleaching and ocean "hot spots."* Ambio 23, 176–180; P. W. Glynn, *Coral reef bleaching.* At Scripps, coral bleaching is being studied by marine biologist Nancy Knowlton and her graduate students. She finds that the story is greatly complicated by there being many different species of symbiotic algae involved, which react differently to stress (pers. comm., 2006).

28. F. Marubini, C. Ferrier-Pages, J.-P. Cuif, 2003, *Suppression of skeletal growth in scleractinian corals by decreasing ambient carbonate-ion concentration: A cross-family comparison.* Proceedings of the Royal Society of London B 270, 179–184.

29. Modern methods of DNA analysis are revealing that many look-alike forms are in fact different species, and the list is expanding rapidly (Nancy Knowlton, SIO, pers. comm., 2002).

30. For background see any general zoology text. For a survey of Caribbean reef organisms and ecology see the excellent Peterson Field Guide by E. H. Kaplan, 1982, *Coral Reefs, Caribbean and Florida*, Houghton Mifflin, Boston, 289 pp.

31. P. Billeter, 1982, *Intimate relationships on the reef*, in E. H. Kaplan (ed.), *Coral Reefs*, pp. 242–250.

32. The symbiosis between the colorful little clown fish and its coral host is now common knowledge thanks to Hollywood's movie *Nemo*, which describes the adventures of a hypothetical roving individual. In actuality, anemone fishes are stay-at-homes. The males are territorial and defend their adopted living castle against potential rivals.

33. Books that treat reef organisms include A. Longhurst, D. Pauly, 1987, *Ecology of Tropical Oceans.* Academic Press, San Diego, 407 pp.; Z. Dubinski (ed.), 1990, *Coral Reefs.* Elsevier, Amsterdam, 550 pp.; J. Stafford-Deitsch, 1991, *Reef: A Safari through the Coral World.* Sierra Club Books, San Francisco, 200 pp.; C. Birkeland (ed.), 1997, *Life and Death of Coral Reefs.* Chapman and Hall, New York, 536 pp. For fishes in the reefs see P. F. Sale, 1991, *The Ecology of Fishes on Coral Reefs.* Academic Press, San Diego, 754 pp.; N. V. C. Polunin, C. M. Roberts (eds.), 1996, *Reef Fisheries*, Chapman and Hall, London, 477 pp.; P. F. Sale (ed.), 2003, *Coral Reef Fishes.* Academic Press, San Diego, 549 pp.

34. R. M. Seymour, R. H. Bradbury, 1999, *Lengthening reef recovery times from crown-of-thorns outbreaks signal systemic degradation of the Great Barrier Reef.* Marine Ecology Progress Series 176, 1–10.

35. *Succession* denotes the consecutive appearance of different assemblages of organisms in the same place, for example, starting with a freshly exposed rock surface. *Climax* is the label applied to the community existing after succession has run its course.

36. Irenäus Eibl-Eibesfeldt's account of his unique adventure is in H. W. Fricke, 1972, *Korallenmeer*, Belser Verlag, Stuttgart 224 pp. Quotes are translations.

37. C. Limbaugh, H. Pederson, F. A. Chace Jr., 1961, *Shrimps that clean fishes.* Bulletin of Marine Science of the Gulf and Caribbean 11 (2), 237–257; C. Limbaugh, 1961, *Cleaning symbiosis.* Scientific American 205 (2), 42–49.

38. James Dwight Dana (1813–1895) was naturalist on the Great U.S. Exploring Expedition, 1838–1842. He authored *Corals and Coral Islands* (1872, New York) and *Manual of Geology* (1862, Philadelphia), among other works.

39. The incompleteness of the fossil record prevents assigning a more precise date for the origin of *Acropora* dominance, *fide* Jeremy Jackson (pers. comm., 2002).

40. Jean Louis Rodolphe Agassiz (1807–1873), a prominent zoologist and geologist of his time, came

to the U.S. from Switzerland, accepting a professorship at Harvard in 1847. Among geologists, Agassiz is known as the scientist who promulgated the Great Ice Age, which replaced the Great Flood as the purported chief agent producing the vast deposits formerly called diluvial and now recognized as glacial.

41. One of the more interesting results of the study by Agassiz was the realization that much of Florida is indeed underlain by ancient reefs.

42. Early work on rates of growth is summarized by the British biologist C. M. Yonge, in a report presenting results of the Great Barrier Reef Expedition, 1928–1929. C. M. Yonge, 1940, *The biology of reef-building corals.* Scientific Report 1 (13), 353–391. Typically, growth rate values for fast-growing coral vary around 1 centimeter per year, by a factor of two. Of course, it does not follow (as Yonge seems to imply) that a reef will be built at such a rate: to make a useful estimate for overall reef growth one would have to know the overall production rate of carbonate rubble, per square meter per year.

43. Vaughan (1870–1952) made extensive studies on coral growth rates in the 1910s. He wrote the definitive review of American corals at the end of the decade T. W. Vaughan, 1919, *Fossil corals from Central America, Cuba, and Porto Rico, with an account of the American Tertiary, Pleistocene, and Recent coral reefs.* U.S. National Museum Bulletin 103, 189–524. Thomas G. Thompson, Vaughan's biographer, calls him "the unquestioned authority on the Mesozoic, Cenozoic, and Recent corals of the United States, eastern Mexico, the West Indies, and Panama." T. G. Thompson, 1958, *Thomas Wayland Vaughan, 1870–1952.* National Academy of Sciences Biographical Memoirs 32, p. 402.

44. J. Pätzold, T. Bickert, B. Flemming, H. Grobe, G. Wefer, 1999, *Holozänes Klima des Nordatlantiks rekonstruiert aus massiven Korallen von Bermuda.* Natur und Museum 129, 165–177.

45. For example, G. Wefer, 1980, *Carbonate production by algae Halimeda, Pencillus, and Pedina.* Nature 285, 323–324.

46. G. Wefer, W. H. Berger, 1991, *Isotope paleontology: Growth and composition of extant calcareous species.* Marine Geology 100, 207–248.

47. Harold Clayton Urey (1893–1981) won the Nobel Prize in Chemistry (1934) for the discovery of deuterium. After his retirement from the University of Chicago, he joined the chemistry department at Revelle College, University of California at San Diego, where his researches focused on the history of the solar system and the origin of the Moon. The principle of the "paleothermometer" proposed by Urey is as follows: When calcium carbonate precipitates from seawater, to make a shell, it incorporates oxygen, according to its formula that prescribes one atom of calcium, one atom of carbon, and three atoms of oxygen for every molecule of carbonate. As far as oxygen, there is a choice between two types of atoms (called isotopes): the regular one with 8 protons and 8 neutrons in the nucleus (called oxygen-16), and a rare one with 8 protons and 10 neutrons (called oxygen-18). (There is a third one, very rare, which is ignored.) The heavier isotope (oxygen-18) becomes fixed into the crystal structure a little more readily than the lighter one (oxygen-16); thus, the skeletal carbonate is enriched with the heavy isotope, relative to the seawater from which it precipitates. At higher temperatures, however, the behavior of the two isotopes becomes more similar (they both get more agitated) so that the enrichment effect is diminished. By measuring changes in the ratio of the two isotopes, then, within a shell, one can tell how temperature changed through time. The method is sensitive to about 1 °F (0.5 °C).

48. The pioneer work on oxygen isotopes was done by Sam Epstein and Heinz Lowenstam, who worked with Urey. S. Epstein, H. A. Lowenstam, 1953, *Temperature-shell-growth relations of recent and interglacial Pleistocene shoalwater biota from Bermuda.* Journal of Geology 61, 424–438.

49. R. B. Dunbar, J. E. Cole (organizers), 1993, *Coral Record of Ocean-Atmosphere Variability; Report from the Workshop on Coral Paleoclimate Reconstruction.* NOAA Climate and Global Change Program Special Report 10, 37 pp. (This report has a long list of references for paleoclimatic studies of coral records.)

50. J. E. Cole, R. G. Fairbanks, 1990, *The Southern Oscillation recorded in the oxygen isotopes of corals from Tawara Atoll.* Paleoceanography 5, 669–683.

51. D. J. Barnes, J. M. Lough, 1996, *Coral skeletons: Storage and recovery of environmental information.* Global Change Biology 2, 569–582.

52. G. Wefer, W. H. Berger, *Isotope paleontology.*

53. A discussion of thoughts on the topic of coral animal versus coral plant is given in T. F. Goreau, N. I. Goreau, C. M. Yonge, 1971, *Reef corals: Autotrophs or heterotroph.* Biological Bulletin 141, 247–260; see also C. Maurice Yonge, 1980, *The Royal Society and the study of coral reefs,* in M. Sears, D. Merriman (eds.), *Oceanography: The Past.* Springer, New York, pp. 438–447.

54. C. Wild, M. Huettel, A. Klueter, S. G. Kremb, M. Y. M. Rasheed, B. B. Jørgensen, 2004, *Coral mucus functions as an energy carrier and particle trap in the reef ecosystem.* Nature 428, 66–70.

55. J. P. Dawson, 2002, *Biogeography of azooxanthellate corals in the Caribbean and surrounding areas.* Coral Reefs 21, 27–40.

56. J. D. Dana, 1872, *Corals and Coral Islands.* Dodd and Mead, New York, 398 pp.

57. R.A. Daly, 1936, *The Changing World of the Ice Age.* Yale University Press, New Haven, Conn., 271 pp.

58. E.G. Purdy, E.L. Winterer, 2001, *Origin of atoll lagoons.* GSA Bulletin 113, 837–854; E.G. Purdy, E.L. Winterer, 2006, *Contradicting barrier reef relationships for Darwin's evolution of reef types.* International Journal of Earth Sciences 95, 143–167.

59. H. Fukami, A.F. Budd, G. Paulay, A. Solé-Cava, C.A. Chen, K. Iwao, N. Knowlton, 2004, *Conventional taxonomy obscures deep divergence between Pacific and Atlantic corals.* Nature 247, 832–835.

60. Current research is turning to DNA patterns to establish relationships between species. The new information is what puts conventional taxonomy in question; see H. Fukami, A.F. Budd, G. Paulay, A. Solé-Cava, C.A. Chen, K. Iwao, N. Knowlton, *Conventional taxonomy.*

61. C. Mora, P.M. Chittaro, P.F. Sale, J.P. Kritzer, S.A. Ludsin, 2003, *Patterns and processes in reef fish diversity.* Nature 421, 933–936.

62. H. Schuhmacher, 1976, *Korallenriffe.* BLV Verlagsgesellschaft, München, 275 pp. Also see J.H. Steele, S.A. Thorpe, K.K. Turekian (eds.), 2001, *Encyclopedia of Ocean Sciences.* Academic Press, San Diego, 3399 pp. In particular, see the entries *Coral reefs* (J.W. McManus), *Coral reef and other tropical fisheries* (V. Christensen and D. Pauly), and *Coral reef fishes* (M.A. Hixon). Numbers of described species tend to increase with time.

63. Captain James Cook (1728–1779) mapped the Pacific Ocean's boundaries and islands. This was his first voyage (1768–1771).

64. C.M. Yonge, 1930, *A Year on the Great Barrier Reef: The Story of Corals and of the Greatest of their Creations.* Putnam, London, 246 pp.

65. For discussion of this transition from inventory to monitoring, see P. Mather, 2002, *From steady state to stochastic systems: The revolution in coral reef biology,* in K.R. Benson, P.F. Rehbock (eds.), *Oceanographic History, The Pacific and Beyond.* University of Washington Press, Seattle, pp. 458–467. For a recent summary of ongoing research, see T.P. Hughes, A.H. Baird, E.A. Dinsdale, N.A. Moltschaniwskyj, M.S. Pratchett, J.E. Tanner, B.L. Willis, 1999, *Patterns of recruitment and abundance of corals along the Great Barrier Reef.* Nature 397, 59–63.

66. J.A. McKenzie, P.J. Davies, A.A. Palmer-Julson, and shipboard party, 1993, *Northeast Australian Margin.* Proceedings of the Ocean Drilling Program, 133. Ocean Drilling Program, Texas A&M University, College Station, 903 pp. Also see International Consortium for Great Barrier Reef Drilling, 2001, *New constraints on the origin of the Australian Great Barrier Reef: Results from an international project of deep coring.* Geology 29, 483–486.

67. JOIDES Planning Committee, 1996, *Understanding our Dynamic Earth through Ocean Drilling.* Joint Oceanographic Institutions, Inc., Washington, D.C., p. 27. The reference to a major change in climate response to orbital forcing presumably is to the Mid-Pleistocene Climate Shift, 900,000 years ago, which marks the onset of long-period cycles (80,000 to 120,000 years).

68. W.H. Berger, G. Wefer, 2003, *On the dynamics of the ice ages: Stage-11 paradox, Mid-Brunhes climate shift, and 100-ky cycle.* AGU Geophysical Monograph 137, 41–59.

69. W.H. Berger, G. Wefer, 1992, *Klimageschichte aus Tiefseesedimenten: Neues vom Ontong-Java- Plateau (Westpazifik).* Naturwissenschaften 79, 541–550.

70. The growth and decay of reefs influences the carbon dioxide content of the atmosphere. B.N. Opdyke, J.C.G. Walker, 1992, *Return of the coral reef hypothesis: Basin to shelf partitioning of $CaCO_3$ and its effect on atmospheric CO_2.* Geology 20, 733–736. Whether this effect is important enough to make a difference in climate fluctuations is not clear at this time.

71. Graduate student Hamilton was part of a team of unusually prominent geologists: Roger R. Revelle, Robert Dietz, H. William Menard, and K.O. Emery. More details about this expedition appear in chapter 12.

72. When visiting Hess after the war, one would enter Guyot Hall, at Princeton, a building named after the Swiss geologist who founded the geology department there.

73. H.H. Hess, 1946, *Drowned ancient islands of the Pacific Basin.* American Journal of Science 244, 772–791.

74. Discovery of the youth of the seafloor was still in the future. Hess thought the guyots were geologically ancient, and sinking therefore had been slow. This naturally raised the question why the reef organisms had not kept up with the sinking to produce atolls.

75. H.W. Menard, 1986, *The Ocean of Truth.* Princeton University Press, Princeton, N.J., 353 pp.

76. *Ibid.*, p. 54

77. E.L. Hamilton, 1956, *Sunken islands of the Mid-Pacific Mountains.* Geological Society of America Memoir 64, 1–97.

78. Compared with the potential rate of reef growth, 2 inches per thousand years is not impressive. Even had the sinking been faster by another factor of 10 (or the ancient reef builders less efficient by a factor of 10), the organisms should have been able to keep up, making an atoll rather than a drowned island. The question why they did not is still open.

79. As suggested by R.R. Revelle, 1955, *On the history of the oceans.* Journal of Marine Research 14,

446–461. It was deemed unlikely by H. W. Menard, 1964, *Marine Geology of the Pacific*. McGraw-Hill, New York, 271 pp.

80. Several basaltic plateaus apparently originated in the middle Cretaceous, around the time in question. The largest of these is the Texas-size Ontong Java Plateau, east of New Guinea.

81. The widespread black shale deposits in the middle Cretaceous are witness to oxygen deficiency. Oxygen isotopes show unusually high temperatures in the sea. Both observations have been tied to the outpouring of basalt, as summarized in R. L. Larson, 1991, *Geological consequences of superplumes*. Geology 19, 963–966.

82. H. S. Ladd, E. Ingerson, R. C. Townsend, M. Russell, H. K. Stephenson, 1953, *Drilling on Eniwetok Atoll, Marshall Islands*. Bulletin of the American Association of Petroleum Geolologists 37, 2257–2280.

FIVE

The Zen of the Beach

MUSINGS ON A RIVER OF SAND

Few things are more pleasant than walking along the beach barefoot, feeling the sand between the toes.

Millions enjoy these simple delights every year. They come and spread their blankets on the sand and watch the children build sand castles, fated to be washed away by the waves of the rising tide. They watch the shorebirds hunting for worms and crabs hidden within the sand.

But what is sand? What is the nature of the grains that feel gritty between toes and fingers? What stories do they have to tell?

The answer depends on the location, of course. The beaches of southern California generally have mineral grains derived from the weathering of rocks in nearby mountains to the east, mixed with similar grains from the sandy deposits that make up the cliffs rising landward of the beaches. The mountains are largely made of igneous and metamorphic rocks, that is, the minerals they deliver upon weathering were made deep inside the Earth. Material locally derived from cliff erosion is commonly marine sediment, with ground-up shell mixed in. Some beaches hardly have any sand but consist of pebbles. Off Scripps, the beach consists of a layer of well-sorted sand, several feet thick, with pebbles at the base of the sand layer. The beach rests on a terrace cut into the land by waves. The terrace grows as the cliffs keep retreating. In turn, this retreat threatens houses built on the edge of cliffs.

Among the mineral grains making up the beach sand, quartz is the most conspicuous. Quartz is typical for many beaches fed from terrestrial sources, for the simple reason that quartz is resistant to abrasion and chemical destruction and outlasts most other types of grains. Sand, then, tells stories about its origin and about its travel to the site where we find it. But the sand is usually not just sitting there—it is in transit. Waves wash the sand, move it seaward and landward according to season, and move it along the shore, as well. The general direction of travel is south, because winter waves from the north are the most effective in moving the sand along the shore. Eventually, the sand comes up against an obstacle, such as a promontory. It then has no place to go but down into the deep sea, commonly within a canyon carved into shelf and slope. Exactly how this vanishing act is accomplished was a complete mystery for a long time; some elements of the process still are obscure despite much study. The layers formed at the final resting place of the sand—in a basin offshore or at the foot of the continental slope—tell a story of giant muddy floods invading a usually quiet environment. To survive such events, animals living on the bottom subject to episodic flooding have to be able to escape upward through the layer of mud left by a flood. Those that fail to escape make fossils.

Beaches on the West Coast, if present at all, tend to be rather narrow, and the beaches of San Diego are no exception. As a thin band of bright sand, they separate the vast ocean to the west from the former wetlands and elevated terraces to the east, now largely developed along a wide coastal strip. No longer a lonely marine biology station as it was in its early years at the beginning of the twentieth century, Scripps is now situated at the northern rim of a thriving coastal metropolis.

PEOPLE AND THE COAST

More than half—perhaps two-thirds—of the people in the world live within a few tens of miles from the shore.[1] The coastal regions of California have always been attractive to people. Ancient shell middens along the shores of California testify that humans lived here for thousands of years, including on the shores of San Diego.[2] A coastal location has many advantages, even if seafood is no longer the main motivator. In southern California, the mild climate along the coast, with cool summers and mild winters, is especially appealing. Farther inland, the seasonal contrast increases markedly. Most of the largest cities of the world are on the coast and have large harbors to facilitate trade; those away from the ocean commonly are connected through lakes or rivers—Chicago and New Orleans, for example.

What sets southern California apart from other sunny shores is the relatively low temperature of the water offshore. The subtropical location and the clear skies, of course, result in a strong warming of surface waters. But these warm surface waters tend to drift offshore, as a result of prevailing winds. Colder waters rise from below to make up the deficit. The outcome is that water along the coast is quite cool, as is readily seen when recording infrared radiation by satellite. In summer, when air rises over the heated land, starting in the middle of the day, marine air moves inland and brings cool relief. Conversely, winter nights get less cold along the coast than farther inland, because of the heat stored in the water offshore.

One of the greatest assets of a coastal location with green offshore waters—generally not fully appreciated—is the ability of the ocean to take up enormous amounts of waste without sustaining lasting damage. Of course, there is a limit to the capacity of the sea for processing waste in any one location. Where broad shelves or bays prevent ready dispersion of waste and its disappearance into deep waters, or where reefs thrive in clear offshore waters, there is potentially much trouble from waste effluents, with negative effects from pollution overload. In many populated places, waste disposal along the coast can conflict with business interests linked to seashore tourism.

For California, tourism is a substantial part of the economy, and its beaches have value measured in dollars, therefore.[3] Thus, for good reasons coastal communities show a strong interest in keeping nearshore waters clean and in keeping the sand on the beach. A keen interest in cliff erosion is similarly motivated by practical consideration.

Overbuilding the coastal zone, with its attendant problems of coastal erosion and filling of wetlands, generates familiar problems. Such problems occupy planners, developers, and

FIGURE 5.1. Storm waves readily move across narrow beaches and erode structures along the shore, as here in northern San Diego County.

homeowners in many parts of the coast in southern California and elsewhere. People who build (or buy) in flood plains and on unstable cliffs usually bear the consequences, although the public interest is involved in many instances also. For example, a common strategy against the dangers of flooding is to build on elevated fill in the flood plain or to insist that a river be channeled, which impacts the general environment. Likewise, to stem cliff erosion, sea walls are built, which impacts the public beach. Sea walls can protect the foot of a cliff but not its upper reaches, unless the entire cliff is modified at great expense.

Cliff failure is a perennial source of anxiety for homeowners who enjoy watching the surf, and for government agencies responsible for maintaining roads next to the sea (fig. 5.1). Failure is quite unpredictable. A reasonable guess is that if it has recently happened in any given place, it is likely to happen there again, or next to it, given the right conditions, including storm action and earthquakes. Rainfall, which can result in landslides, is an important factor. Overall, the rate of retreat of the cliffs may be quite low, perhaps on the order of a foot per century. But while 50 years can pass without anything moving, a 20-foot-wide chunk of land can slide down the face of the cliff in an afternoon.

The El Niño event of 1997/98 provided an opportunity for a detailed study of the response of coastal erosion along California to the increased rain and storm activity. Its arrival was predicted well ahead of the event, and thus a multiagency program could be set in place to monitor coastal erosion. Participating agencies included NASA, NOAA, and the USGS.[4] Airplanes carrying laser altimeters surveyed 1,200 kilometers of the western coast in October 1997 and in April 1998, that is, before and after the winter rains. The most extreme change noted was a 13-meter cliff retreat at a point in Pacifica in coastal central California. Twelve houses had to be condemned due to sliding. Other impacts consisted in the denudation of sandy beaches, removal of sand from some places and pile-up of sand in others, and abrupt collapse of some cliffs quite similar to others nearby that showed no change.[5]

River floods laden with mud and rushing into the sea are an integral part of the geology of our beaches. The bigger the floods, the more sand they carry.[6] The terms *hundred-year flood* and *fifty-year flood* have a comforting ring to them, suggesting that the big ones are going to be seen only once in a lifetime, at most. Quite so—it is indeed possible that many of us shall never see a truly great flood. On the other hand, the next rainy period might just be the one to provoke such an event. In any case, episodic flooding moves a lot more mud (and sand) than would a steady flow. River flow in southern California is decidedly episodic and varies with climate change.[7]

How do we estimate the size of a hundred-year flood, when the records go back only a hundred years? We cannot, actually. We can make crude guesses from the patterns of variation in rainfall. The average rainfall along the coastal strip of San Diego County (in which Scripps Institution resides) is a little less than 10 inches a year (for the past 50 years, that is). Since records began in 1875 at coastal downtown San Diego, the measured annual rainfall has varied from 3.46 inches in 1960/61 to 25.97 inches in 1883/84. The second-highest year, 1940/41, had 24.74 inches. Thus, low extremes, on the scale of a century, apparently are near 2.5 times less, and high extremes are near 2.5 times greater than a typical "average" value. From such information, statisticians can make estimates about how often to expect 5 times or 10 times the average rainfall.[8]

For the hundred-year flood, however, the annual rainfall in San Diego may not be all that relevant. What counts is how much precipitation there is in the mountains, over a short period of time. The rainfall in the mountain areas of San Diego County fluctuates greatly. Occasionally, a single storm system can dump almost a year's worth of water in a given area. The resulting flood is correspondingly impressive, racing down the normally dry canyons straight toward the sea. Many of the small coastal cities which have sprung up within the last few decades presumably are ill-prepared for the hundred-year flood, which by definition is outside their experience.

To control the floods, and to store water for the dry season, we have dams on the river courses. In the meantime, wetlands and lagoons close to the sea are being filled with sediment from surrounding developments and their runoff, and also deliberately, to prepare space for hotels and golf courses. Marshes that were once filled with migrating ducks and herons are now parking lots or shopping malls. What we see today in terms of natural wetlands, on the whole, is a pitiful remnant of what there was within human memory.

Many new challenges are appearing for the intelligent application of what we have learned about coastal processes over the past 60 years. One of these challenges involves changes in sea level. Of course, small changes in sea level are exceedingly difficult to measure: sea level is a moving target, affected by tides, air pressure, the strength of offshore currents, and changes in coastline. Historically, the interest in knowing the level where water meets land was for establishing and maintaining harbors. The depth of the water determined which ships could dock and which could not. A tide gauge was installed at the dock to record the ups and downs of the water surface. The sea level at a particular point is the mean position of the water surface over the circa-19-year Moon cycle, relative to a nearby point on land. Worldwide compilations of sea level measurements of this sort have attained significance in discussions of global warming.

Sea level is projected to rise by some 10 inches along the coast of California within the present century, from projecting the rise over the last several decades. This estimate may be safely considered a minimum; twice that value is more likely as warming expands the water in the sea, and even 5 feet or so is not out of the question, as will be explained in chapter 15, on climate change. The impact of storm surges will be intensified by a rise, however modest, even if the storms stay at the same level of strength.[9] In the Sacramento–San Joaquin Delta, 1,100 miles of levees protect farmlands, towns, and highways from flooding, on land claimed from marshes.[10] According to a study by the Union of Concerned Scientists and the Ecological Society of America, a 1-foot rise in sea level would transform a 100-year high tide into one that occurs every decade, greatly increasing the likelihood of levee failure. The requirement, of course, would be a buildup of the levees. However, not all challenges will be met quite so easily. Of special interest to people in southern California, who depend on a supply of water from the north, is the conclusion of the California Energy Commission that, because of climate warming, 700,000 acre-feet of freshwater may be required to offset saltwater intrusion into areas protected by the levee system.[11]

The problem here touched on is that low-lying land, relative to sea level, is susceptible to having its groundwater replaced by salty seawater. Worldwide, the rise of sea level is about 8 inches per century at present (2005), and the impact is being noticed in many regions, especially where the land is sinking in delta-type regions, such as in Egypt and Bangladesh.

The contemplation of the "river of sand," to which we turn next, will show the intimate interconnectedness of a host of natural processes, from mountain uplift to river flooding, from distant storms to breakers on the beach, from submerged canyons to the finely

FIGURE 5.2. The beaches of southern California have pebbles and sand largely derived from crystalline rocks. Sand from the Scripps beach is rich in quartz grains.

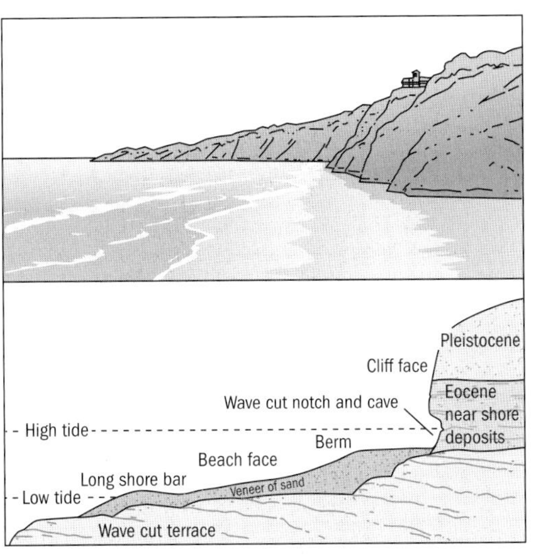

FIGURE 5.3. What the beach looks like to a tourist *(upper)* and to a geologist *(lower)*. The profile implies the cutting of a terrace by the waves.

layered rocks in the cliffs that recede before the attack of the waves. Within the last hundred years we, the human species, have become an important, even dominant, part of this system. Our understanding of the natural cycles, gained through such great efforts within the last hundred years, is no longer sufficient to allow us to cope with the problems arising. To manage the environment, we have to learn to manage our own impact on natural systems. Managing people may turn out to be much more difficult than coping with the forces of nature.

SAND STORIES

The beach is more than a playground; it is the arena of live geologic performance. Every handful of sand, every mineral grain, contains an intriguing story of origins deep in the Earth, of uplift and erosion, of transport by rivers and waves (fig. 5.2). The Zen of the beach, for a geologist, is to listen to the sand grains tell their story. Where do they come from? Where do they go? Where do the waves come from that move them about? Are the grains on the beach the same as in the cliffs, and how fast do the cliffs retreat as they feed the beaches?

The study of questions such as these have intrigued many of the geologists who took up residence at Scripps, on the staff or as guests. After all, this is our front yard. Besides, nowhere are the processes that shape the coastal landscape acting with greater vigor than here, where the ground trembles in the grip of mountain-building forces that raise the land high above the sea, and where powerful waves sent by fierce storms at sea assault the land, tirelessly gnawing at the base of its ramparts. It is geology in action, and there is nothing very subtle about it. One can *hear* the waves doing their thing in the clatter of the pebbles and the swishing of the sand. Just sitting there, watching the surf and the surf riders, one is surrounded by the ancient battle sounds of land and sea.

The first scientists at Scripps who studied the nearby beaches established that the beach is a thin veneer of sand sitting on top of a wave-cut terrace (fig. 5.3). They also soon noted that the sand moves inshore and offshore in a seasonal pattern. Beaches are nice and wide in late summer—the sand is soft and deep. But in winter, in many places, the beaches may be narrow, and much of the beach is pebbly (fig. 5.4). The sand veneer thins in winter. Winter waves from storms to the north, with their great power and turbulence, suspend the sand, denying it a place in its normal habitat and forcing it to move outside the wave-swept zone where it settles in the quieter waters outside the breakers. Summer

FIGURE 5.4. What the beach looks like after storm waves remove the sand. *Upper:* enrichment of hard-to-move pebbles. (Note protective board in front of view window.) *Lower:* exposed layers of rock of the marine terrace normally bearing a beach.

waves, more gentle, take up the sand again from the offshore bars and deposit it shoreward, where the environment is now quiet enough to let the sand stay. If the winter storms happen to be unusually strong and numerous, much of the beach ends up offshore, and it takes a while to bring it back.

During the last three decades or so, the beaches have tended to shrink. Strong storms, more abundant in recent years than earlier, have taken the sand out, but it has not come back as readily as before, and it is not being replenished by material delivered by the rivers entering the sea. "Why are the beaches disappearing?" has become a rather common question; almost everyone living in southern California has heard it. Before one can attempt an answer, one must learn where the sand comes from, and where it goes.

In a nutshell, sand is delivered in the mud brought to the sea by the rivers, especially during floods, and it is moved along the shore by wave action, especially during storms. The overall transport is from north to south. Surprisingly—at first glance—this is true for both the western and the eastern side of the North American continent. The reason is that the waves produced by nearby winter storms, which rage in the north, are more powerful than summer storms, on average, and also more powerful than the waves arriving from distant winter storms from the southern Pacific. The northern winter storms send the stronger waves, which move beach sand southward. Eventually the "river of sand" that makes up the beach comes to a place where it is deflected offshore, into a submarine canyon. It then moves along the canyon axis into the deep basin where the canyon ends. Here the sand finds its final resting place far from shore.

The main elements of this intriguing story were worked out over many years at Scripps Institution of Oceanography, between the 1940s and 1970s. It is a story that touches on all the Earth processes that shape our landscapes in the coastal zones of North America and elsewhere.[12]

The concept of a river of sand arose from the studies of Douglas Inman, who became a graduate student at Scripps in 1947 and began exploring the interaction between waves and beaches.[13] Inman first clearly defined the beaches of southern California as ribbons of sand extending from their river sources to their terminations at the head of a canyon. The river of sand that is the beach is fed from rivers of mud that bring the weathering products from the mountains to the sea. The most important thing about these muddy rivers is that they are highly seasonal, and that they turn into mighty torrents from time to time, during periods of unusually strong rains. It is during such torrential episodes that the rivers move debris down to the sea in a serious manner. The rest of the time nothing much is happening. The sand we are walking on is largely the product of river floods, sifted and sorted during its path southward.

Much of our beach is quartz sand. But this is not so for many other beaches.

In San Diego, the sand consists of bits and pieces of mineral released from the weathering of granitic rocks, in nearby mountains. We find quartz, feldspar, hornblende, and mica, for example. On Pacific islands, we might find bits of basaltic rock (black sand) or grains of olivine released from basalt (green sand) or all kinds of shell fragments, ranging from mollusks to sea urchins to foraminifers (white sand). The very white sand on some Florida beaches (insofar as it is natural) is made of quartz grains, durable survivors from repeated erosion and transport, originally starting in the Appalachian Mountains. In southern Florida, on the Atlantic side, we might find sand made of reef debris, instead. If quartz is present, its abundance will normally increase with time, because it is very resistant to both weathering and abrasion. Feldspar is more easily damaged by chemical attack. Dark minerals have iron, which tends to rust out. Quartz is the survivor in a kind of elimination game played on the minerals. Only granitic rocks or their derivatives can deliver quartz—there is none in basalt. Thus, quartz sands are abundant in the beaches of continents, but not of oceanic islands.

Such is the beach sand along much of southern California. It is a sample of the minerals formed deep within the Earth, when the mountains to the east were young, when roots were built from hot magma rising from far below the crust. These roots are now exposed as domes of granite covered with chaparral or pines. The granite was uplifted and unroofed, after millions of years deep within the Earth, to meet its fate of weathering and erosion, attacked by acids in rain and soil, broken up by freezing water and the mighty hydraulic power of roots.

WAVES AND THE MOVING SAND

Waves are responsible for driving along the river of sand. But the interplay between storm waves, breakers, and nearshore currents is a complicated symphony of watery motion. Just what are the forces driving the river of sand, and how fast does it move?

There is no question, of course, that the waves move the sand. They do so by picking it up from the bottom, suspending it, transporting it over some distance, and letting it settle out again, especially in the breaker zone. Outside of the breakers there is much less transport. Still, the waves produce a to-and-fro motion of water along the bottom, which moves sand right over the floor, resulting in ripple marks parallel to the beach. On the beach face, uprushing water moves thin sheets of sand upslope right before our eyes, and backwash takes sand back toward the sea, aided by gravity. This to-and-fro motion leaves fine beach laminations, wherein heavy and light minerals are nicely separated, the entire beach face acting like a giant gold pan.[14]

Waves, then, are the driving force for the river of sand, and to understand beach processes we have to study waves—where they come from, how they change through the seasons, how they get here, how they break, and how a portion of their energy ends up moving beach material.

Waves are children of the wind. In the crashing of the breakers along the beach we hear the fury of storms far out to sea, where the waves were born. The length and height of a set of waves is related to the intensity and duration of the winds that made it: the stronger and longer the wind blows, the higher and longer the waves. Another vital component is the fetch, that is, the distance over which the wind travels. Under a continuous 30-knot wind (strong but not so unusual), 90 percent of the waves are between 3.5 and 18 feet high; their average length is about 250 feet, and their average period about 10 seconds. Such waves are not uncommon in winter; they can be observed breaking at the end of Scripps Pier on occasion (fig. 5.5). In a 60-knot wind (not recommended for sailing), 90 percent of the waves are from 14 to 72 feet high, and an occasional one may reach well over a hundred feet; their wavelengths are many hundreds of feet, with typical periods about 20 seconds.[15]

FIGURE 5.5. Waves that break at the end of the Scripps Pier are about 10 feet high. They are generated by winter storms offshore, to the north. They readily move sand.

The greatest storms are generated around the Antarctic, in the Southern Ocean. Waves from this source can be traced across the entire Pacific Ocean, and they are present throughout much of the year.[16] The fierce cyclones of the "roaring forties" and "screaming fifties" can produce waves that are more than half a mile between crests. Off the coast of South Africa the combination of waves, currents, and the continental shelf results in enormous "rogue" waves, so steep in height and trough that they have broken large ships in half. Such rogue waves are not restricted to the sea around South Africa, of course. One freak wave was measured as 112 feet high, in the Pacific, by a U.S. Navy tanker.[17]

Waves move out from their source area and travel across the ocean until they hit an obstruction, usually a beach, where they break. As every surf-enthusiast knows, the breaking waves have quite different characteristics in different seasons and during different kinds of weather. Along the Pacific coast, powerful winter surf is generated by storms in the north. The summer surf is regular and stately (except during tropical storms); much of this summer wave energy originates in the far-away storm centers of the Southern Ocean. These swells typically have intervals of 14 to 20 seconds.

Tracking the origin of waves arriving at the beaches of California is a game invented by Scripps oceanographer Walter H. Munk.[18] Together with Scripps engineer Frank E. Snodgrass, who designed sensitive pressure meters able to record wave height above the bottom, Munk endeavored (in 1963) to map the path and speed of wave trains across the ocean, by setting up stations from Alaska to New Zealand, including one on the floating instrument platform FLIP, which was deployed northeast of Hawaii.[19]

Munk found that waves took about two weeks to get from their point of origin in the Indian Ocean sector of the Southern Ocean to the northernmost station in Alaska.

Long waves travel faster than short ones and they last longer. Thus, at distant beaches, long swell is what remains from the mixture produced by a storm. When a deep-water wave

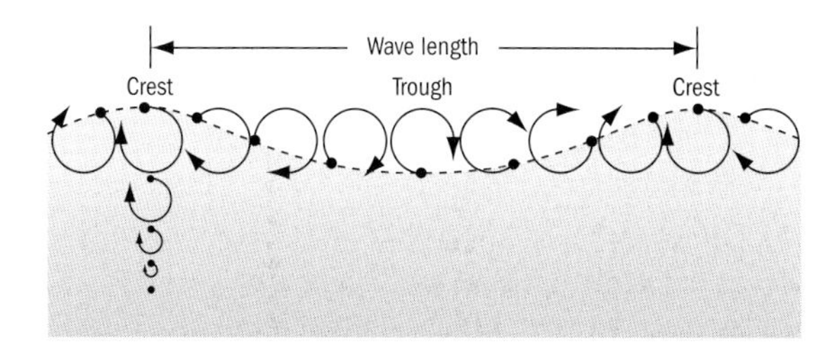

FIGURE 5.6. Waves in the open ocean move energy, with water particles traveling in circles.

reaches the shore, it feels bottom, which slows it down. Since the wave energy is concentrated over a smaller area as a wave shortens due to slowing, the wave height must increase. Eventually, the slowing near the bottom and the continued forward rush at the crest result in the breaking of the wave.

When a wave approaches the beach at an angle, the part closest to the beach slows down first, while the part farthest from it is still in deep water and runs at its normal speed. Thus, the wave is bent and becomes nearly parallel with the contours of the seafloor. This is why waves always break almost parallel to the beach. The remaining angular component of the wave helps to set up a longshore (or littoral) current. For the strong winter waves that are generated in the northern North Pacific, the current moves southward and takes with it some of the sand suspended by turbulence in the zone between beach and breakers.

Waves in the open ocean move energy, not water (fig. 5.6). However, when a wave finally breaks along the shore, it throws water toward the beach, and the water piles up and seeks to return to the sea. The return flow is usually concentrated in a narrow current, called a rip current (fig. 5.7). One can see such currents moving outward through the surf, at places along the beach that are a little deeper than elsewhere (and have low wave action, therefore). Rip currents are fairly easy to spot, when viewing the beach from high up on a cliff, and also when swimming in the surf zone. Scripps oceanographers studying beach dynamics soon got caught up in determining the cause of the rip currents, especially since they pose a serious hazard for inexperienced swimmers. A mythology of surf-zone dynamics had already grown around this hazard—which some called rip tide, invoking a menacing undertow. Scripps geologist Francis P. Shepard, who for years enjoyed a daily swim in the ocean, pointed out that the phenomenon had no relation to tides and urged the term *rip current.*[20] He noted that a rip current could carry a swimmer offshore at speeds as high as 2 miles an hour.[21]

Rip currents may be dangerous, but not to an experienced surf rider. He lets the rip current take him out beyond the surf and uses the big waves to come back in. Swimmers caught in the rip current should swim to one side of it, as quickly as possible, to avoid being carried offshore.

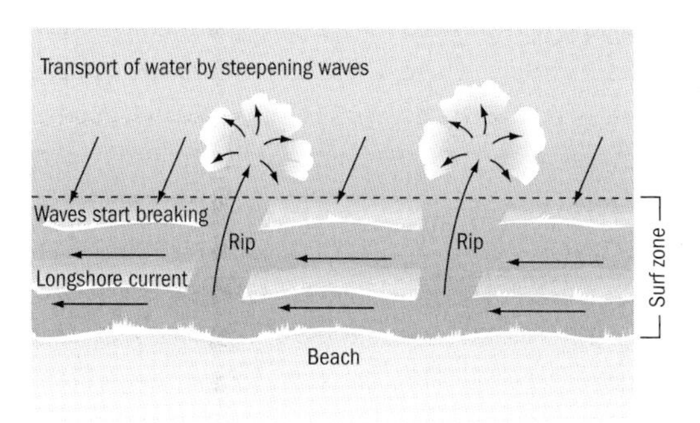

FIGURE 5.7. Rip currents take water piled up on the beach back out to sea. Waves break at a water depth 1/3 greater than wave height because the water particles in the crest outrun the wave (being in deeper water than the rest). Water is carried into the surf zone, where the influx sets up a longshore current that feeds rip currents. Breakers are highest on submerged ridges and are low over canyons because of wave refraction: waves travel more slowly over the shallower areas and therefore bend toward such areas.

Naturally, exploring the dynamics of the surf zone is good fun besides being good science, and there always was much interest in participating in this endeavor. Everyone among the Scripps geologists in the early days contributed to the best of his abilities, but none could measure up to Roger R. Revelle. In the 1930s, when equipment budgets were extremely lean, he would find himself enlisted by the students Robert Dietz and Kenneth Emery as their wave-measuring staff. They found him to be relatively stable and a convenient 2 meters in height.[22]

The war brought a demand for knowledge—and funding. The navy needed to know how to deal with the surf environment during amphibious assaults; forecasting waves became a very high priority. Research during the war and in the years immediately after the war, by Harald U. Sverdrup, Walter H. Munk, John D. Isaacs, and Willard Bascom, among others, greatly advanced the state of the art. Knowledge about surf and beach dynamics became increasingly important in coastal engineering, the protection of harbors, marinas, and beach property, and the preservation of the beaches themselves.

One of the group's projects was again rip currents, which they concluded were caused by the interaction of incident surface waves with longshore edge waves—a series of standing waves set up by the surf, like a monstrous organ pipe stimulated by gusts of wind. Such interaction forms nearshore circulation cells and modifies the beach topography, generating the puzzling series of evenly spaced spits seen along many straight beaches that are ornamented with a sequence of regular crescents. After the war, Douglas L. Inman and his colleagues and students carried out many quantitative studies on these topics, capturing the action of waves and sand where the sea meets the shore.

One of the researchers working with Inman is his former student Reinhard E. Flick, an expert on beach processes. He is frequently asked about the lack of sand on the beach. Where did it go? Damming the rivers for flood control and for water storage is not the only factor reducing sand supply, he says. "Massive coastal development from the 1930s onward created large quantities of sand, much of which ended up on the beach. The extent of that development has practically ended: the available coastal strip is built out, or protected. Thus, the wide beaches we have considered as 'normal' are not so wide any more."[23] Flick notes that dredged sand dumped in water 30 feet deep off Del Mar (a few miles north of Scripps) moved toward shore within a few months, suggesting that offshore emplacement of dredge spoils can help replenish beaches. The politics involved in using such spoils to bolster the sand supply to beaches is rather complicated, however, for various reasons.

Beach sand invariably disappears into deep sites of deposition far offshore. Thus, it needs to be replenished continuously. Bringing in sand from places that have plenty of it can help. But it is expensive. Another approach, widely used, is to try to keep whatever sand is there in place with jetties, groins, or breakwaters. Of course, such structures also are quite expensive. In addition, jetties and groins cut off the flow of sand downriver, which normally leads to increased erosion there, which can lead to serious complaints. Some of the complaints end up in court: preventing sand from moving downstream to protect your own beach is somewhat analogous to pumping water from a small river to irrigate your own fields.

Southern Californians are not the only ones worrying about the disappearance of the beaches. For a number of reasons (including the use of beach materials for building purposes) beaches are in trouble in other regions as well. Detailed studies of beach dynamics were made in the 1990s on the East Coast, along the Outer Banks of North Carolina. Headed by Robert T. Guza, former Scripps student and successor of Inman as director of the Center for Coastal Studies, the project included researchers from Scripps and from 15 other agencies, with support from the Office of Naval Research, the U.S. Army Corps of Engineers, the U.S. Geological Survey, and the National Science Foundation.[24] The pioneers of nearshore dynamics would probably be greatly surprised at the rich assortment of instruments

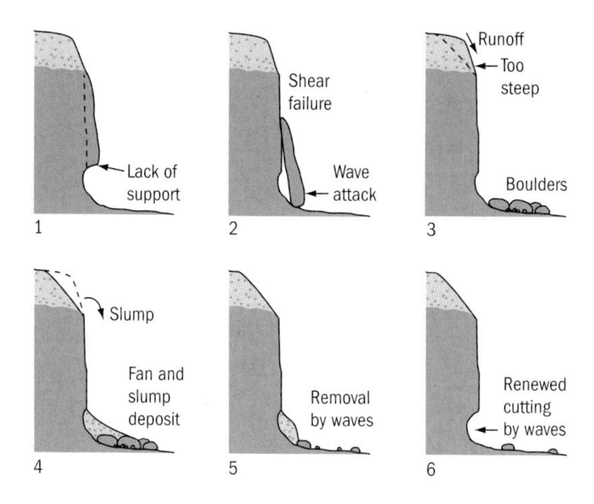

FIGURE 5.8. Erosion at the foot of the cliffs keeps them steep.

now being used to study waves and sand transport: sonic altimeters, pressure sensors, current meters, and several types of amphibious vehicles. Only modern computing methods can handle the torrents of data that are obtained in such investigations, and can also provide the means to model the complex and turbulent interactions of waves, currents, and the beach.

We have come a long way from the personal wave staff.

THE GREAT WALL

When one approaches the shores of southern California from the sea, the land almost everywhere emerges over the horizon as an impressive wall some 200 feet high, typically capped by an ancient terrace or a mesa. The terrace defines a coastal zone that abuts against the mountains to the east. The mountains, in turn, are a link in the sheer endless chain of rugged highlands looking out over the Pacific, from Alaska to Baja California. In some places the wall is not quite so high, and a terrace at intermediate heights, say, 50 feet or so, provides a broad ledge that can bear a road or buildings or both. In other places, lagoons and marshes have breached the wall, where intermittently flowing rivers enter the sea. The forces controlling the origin and evolution of the wall are uplift (related to earthquakes), and erosion by wave action, runoff, and slumping (fig. 5.8).

This type of seashore is unlike anything on the East Coast or the Gulf Coast. On the West Coast, the continent presents itself as a fortress checking the advance of the ocean. It shares but little ground with the sea—a few marshes, some lagoons, narrow strips of pebbly or sandy beach. On the East Coast, in contrast, there is no sharp line separating land and sea. Instead, there is a broad zone of beaches and marshlands, and abundant shallow-water bodies, including broad estuaries admitting seawater far inland, at the bottom of slow-moving rivers flowing throughout the year.

The main reason for this fundamental difference between the coastal landscapes of California and the eastern seaboard is exceedingly simple: the continent rises at its western edge, making mountains. In contrast, at its eastern and southern boundaries the continent is sinking, providing room for deposition of erosion products from the interior over broad expanses of coastal plains. At the western coast the inexorable uplift brings former beaches, marshes and lagoons out of the ocean to greater elevation, taking the terraces well above sea level. As the waves cut into the rising land, they constantly undermine the older raised sediments, and keep the cliffs steep. At the foot of the retreating cliffs, beach sand accumulates on the freshly cut terrace. Because this terrace is young, it provides for but limited accommodation: the beach is a narrow ribbon of sand and not very deep, typically only 5 feet or so, often less.

What is the nature of the cliffs? Are they stable? Or do they retreat and contribute to the beach sand at their base?

The material making up the cliffs certainly is of the right kind to provide beach materials. We can easily verify that the cliffs consist of raised marine sediments, since they are readily accessible from the beach. North of the Scripps Pier are marine deposits of Eocene age, as determined by the study of fossils (microscopic plankton remains, among others). The Eocene strata—mudstones and sandstones—are between 40 and 50 million years old. Ancestral whales already cruised offshore then, but they had four limbs and looked less elegant and efficient than our dolphins, or even our gray whales. Perhaps they went ashore to breed, like modern elephant seals do.

The cliffs well south of the Scripps Pier, just north of La Jolla, consist of Cretaceous sediments, deposited at considerable depth, perhaps in a basin like the San Diego Trough offshore. The sediments consist of alternating thin layers of silty sandstones and shales. Spacing of the layers is nearly perfect, and there is little or no disturbance from burrowing within layers. The astute observer will note that many of the sandy beds are graded, that is, grains are coarser at the base of each individual layer than at the top. This is a clue to the mode of deposition. Geologists envisage dilute mud slurries racing down the continental slope, within a canyon, and emerging into a basin where the mud slurry dumps its sediment, making thin sandy layers across the floor of the entire basin, coarse grains falling out first. Studies in ancient marine sediments in southern California were important in the development of this concept—referred to as turbidity currents—which revolutionized the science of sedimentation in the 1950s and 1960s.

The Cretaceous strata are some 70 million years old. No whales were blowing along the coast when they were deposited. Indeed, there were no large mammals around at all, although some rodentlike creatures may have foraged along the shore, constantly in fear of predatory reptiles and birds. If anything was blowing like a whale it was one of the large swimming reptilian predators—a plesiosaur or a mosasaur. At times a pod of ichthyosaurs might have passed offshore, jumping dolphinlike, chasing fish. Instead of the occasional dead seal, as today, enormous dead ammonites up to several feet in diameter would wash up on the beach to the east, from time to time. Flying reptiles (pterosaurs) would gather to claim their share of the carrion. The climate was tropical.

But we digress. The stories in the cliffs are not the stories of the present beach.

How do the cliffs stay steep? By breaking up the fallen rubble at the foot of a cliff, waves deny protection to the cliff and prepare the way for renewed attack. The charge is led by storm waves using pebbles for tools. They cut into the cliff base, producing a wave-cut notch. When the notch is deep enough, the overhanging part of the cliff fails and falls to the beach. Subsequent wave attack wears down the fallen rocks. A wave-cut terrace results from the retreat of the cliff.

Thus proceeds the battle. Only the most bookish (or digitally preoccupied) geologist could possibly withstand the lure of occasionally walking along the shore and watching the everlasting contest between land and sea.

Francis P. Shepard was an enthusiastic observer of geologic drama. In fact, his ebullience was legendary, and his untiring energetic pursuit of all kinds of geologic phenomena and processes earned him the nickname "Frantic Fran" from some of his more sedate colleagues. Shepard's zest for exploration was highly infectious: several of his students made major contributions over a broad range of geologic research.[25] However, his exuberance could occasionally be dangerous. His famous photo of 1 April 1946 is a case in point. It shows a lone man standing next to a palm tree facing an incoming tsunami wave some 30 feet high, in Hilo on the island of Oahu. Shepard was so enthralled with photographing this enormous quake-generated sea wave bearing down on him that he almost failed to run to safer ground in time.[26]

Shepard had a strong interest in all processes shaping coastal landscapes and loved to take photos along the shore, to document changes through time, such as the rate of cliff erosion. The process of cliff erosion proved to be highly sporadic in space and time. During the early decades of Shepard's career, changes were modest. But from the 1970s on, a series of intense storms resulted in cliff failure in many places, with substantial sections crumpling to the beach. In large part this was due to episodes of heavy rainfall, which resulted in sediment blocks sliding seaward on top of wet clayey layers. Wave attack also did its part. Many places, however, essentially showed no effect at all. Thus, attempts to determine the rates of cliff erosion by small-scale experiments (such as driving a nail into the cliff, to see whether it would stick out or fall out with time) were shown to be entirely inadequate. Instead, photographic documentation and comparison of maps turned out to be the best way to answer the question.[27]

For whimsy in geologic measurement it is hard to best Kenneth O. Emery (1914–1998), Shepard's student before World War II, and later a renowned marine geologist. Emery invented a way to make a certain type of vandalism useful in determining the wearing away of the cliff face by wind and wave erosion. He sought out the cliff carvings by people who, with a knife, commemorated their visit to the beach with name and date. (The date is necessary for this, of course.) He then plotted the depths of carvings against the year of record and deduced a typical rate of abrasion from the gradual decrease in the depth of the letters. The approach assumes that knives work equally well through the decades, and that people tend to spend roughly the same energy on this activity, regardless of the year. The study was duly published,[28] perhaps not so much for the results—which are in fact of very modest value in the context of cliff erosion—but for the entertaining notion that unsuspecting tourists can be recruited as geologic assistants. Forty years later, he returned to the matter in a more serious vein, using historical data to estimate cliff retreat.[29]

A good way to obtain an idea about the rate of cliff retreat, of course, is to interview longtime residents with houses near the cliff edge. In 1975 I talked to an elderly lady in a house with an ocean view, in Encinitas (about 15 miles north of Scripps), to verify that an entire row of houses had disappeared, as implied by an older map. Mrs. M. commented that she became aware of a cliff erosion problem in the 1940s, when parts of the plumbing broke off from the bottom of her house. "When I saw the beach through the hole in the toilet, I knew it was time to move," she said. She found an out-of-state buyer eager to be as close to the beach as possible. With the proceeds of the sale she promptly bought the house eastward across the street. Two years later she regained the ocean view that she had previously enjoyed. She calculated that her newly purchased house would last to the end of her days. It was a reasonable guess, for a 30-foot setback, and a life expectancy of less than 100 years.

Given that the cliffs do retreat, what can we say about the contribution of cliff erosion to the river of sand? It is difficult to estimate this contribution. Spot measurements, either in time or in space, cannot describe the long-term retreat of the sea cliffs, which depends largely on undercutting at the base, slumping, and landslides. Seemingly identical cliff sections can retreat at different rates, at least during the interval for which observations are available. In the long run, of course, the rate of retreat will be similar all along the coast, maintaining the overall trend of the coastline.

In any case, adding up the supply of sand from estimated rates of cliff erosion shows that this process is not negligible, but quite insufficient for maintaining the beach. We do need the rivers to keep beaches supplied.

THE RIVERS, THE MOUNTAINS, AND SEA LEVEL CHANGE

The coastal landscape of California reflects general uplift (fig. 5.9) and the effects of fluctuating sea level from the buildup and decay of large ice masses in the recent geologic past. The contri-

FIGURE 5.9. Uplift is at the heart of the Californian coastal landscape. A wave-cut terrace a few feet above the sea level suggests a recent episode of uplift north of Santa Barbara. The layers of sediment, originally horizontal, were tilted by mountain-building forces.

bution of rivers to the beach varies greatly. River flow is highly variable, and practically all sediment that reaches the sea from the mountains is transported during floods.

The coastal zone of southern California enjoys a semiarid Mediterranean-type climate, with much sunshine and modest amounts of winter rain during most of the years. During the summer, a high-pressure zone offshore of California blocks entry for moist winds from the Pacific. During winter this high-pressure zone weakens and shifts to the south, opening northern and central California to storms generated in the subarctic Pacific. Nevertheless, rainfall stays comparatively low in southern California, thanks to its position at the northern boundary of the desert belt comprising the Southwest and northern Mexico. Normally, that is. However, in irregular intervals of somewhere between two and seven years, southern California (and adjacent regions) experience the effects of a substantial warming of the surface ocean over large parts of the eastern tropical Pacific. This condition is referred to as El Niño, and it makes for drastic changes in the rules for rainfall distribution.

El Niño (Spanish for "the child") is a name used by Peruvian fishermen to describe the warming of the ocean off their shores, usually setting in close to Christmas ("the Christ Child"), when it occurs.[30] The warming shuts down the supply of cold nutrient-rich waters upwelling off Peru from below the surface-water layer, and this results in a crash of fish populations. During Niño conditions, the California Current warms off our shores, as well. Warm moisture-laden air then reaches southern California from tropical and subtropical regions in the Pacific. Enormous downpours can ensue, such as occurred in the years 1883/84, 1940/41, 1979/80, and 1997/98.

In 1916, the San Diego River, normally barely more than a wet place in the center of the valley, flooded to a height of 6 feet, in response to massive downpours.[31] The impact of a rainfall such as experienced in 1916 today would be enormous—the population is vastly increased, and the river's floodplains, where cattle once grazed, are solid in hotels and shopping malls. The river's bed is now a concrete channel; nevertheless, every storm that brings more than 2 inches of rain interferes with the daily traffic and parking.

Floods are important in sediment transport, because of their great power for moving soil and rocks, due to their volume and velocity, but also due to the fact that they are laden with mud, which makes the water heavier than normal, so it can carry objects (gravel, sand, trucks, etc.) more easily. With but little exaggeration it can be said that practically all transport takes place during flooding, and none during the regular course of seasonal flow.

Floods are unwelcome in areas that are settled. They drown cattle and other livestock, uproot trees, destroy bridges and railway dams, wash out roads, take down buildings, and generally wreak havoc on existing conditions. They also erode riverbanks (or cliffs along the valley) and can even move the course of a river, cutting a new channel. The preferred instrument of flood control is building one or several dams along the course of the river. As a rule, in southern California, if a river valley has water it also has at least one dam. During normal years, dams increase the availability of freshwater, a resource that is in high demand and short supply. During years of heavy rainfall, the dams serve to control downstream damage from flooding.

The dams, of course, make lakes, and the lakes trap sediment that would otherwise get to the seashore. Whatever amount passes the dams (not much) tends to stay in the river valley below because there is no flood to move it. The result is a decrease of sand supply at the beaches.

Except for the dams, the present geologic and climatic conditions are ideal for maximum supply of sediment to the sea. The combination of strong seasonality—freezing winters, hot summers—in the mountains, with occasional forest fires and brush fires, and with occasional torrential rains, makes for a high rate of weathering and erosion and the delivery of plenty of material for transport. The high relief resulting from mountain-building uplift imparts great power to the rivers during flooding stages. The lower valleys of the rivers are already well filled with sediment, so that there is little space to entrap materials being brought down. The sediment has no choice but to move out into the sea, producing fan deposits off the river mouths. The waves then wash these deposits, take the fine particles directly out to greater depths, and deliver much of the sand to the beach. This process starts the river of sand on the terrace in front of the cliffs.

The source of the material transported by the river floods is in the mountains to the east. These are young mountains, rising rapidly, and erosion has exposed the crystalline interior of the Earth, rocks that were formed several miles down in the infernal foundry of mountain building. The igneous rocks (a word related to "fire," as in "ignition") contain, in their minerals, the clues to the temperature and pressure of their fiery birth. Upon exposure, the rocks are weathered and eroded, and their bits and pieces end up as mud and sand in the river valleys and in the sea. Most of this debris moves into the basins off southern California, making deposits several kilometers thick, typically. A corresponding amount is missing from the roof of the mountains.

At present, then, geological and climatic conditions are favorable for making beaches. The mountains are a rich source of material. The rivers bring it to the sea when flooding. Sea level is fairly stable, and waves have had time for several thousand years to deepen the terrace on which the sand can rest.[32]

Circumstances were not always so favorable for making a beach. Only a few thousand years ago, much more of the sediment brought by the floods was trapped in estuaries cut during low stages of sea level stand, in glacial time. These estuaries have since filled up with mud and beach deposits, making lagoons and marshes. As a consequence, the beaches must have been starved throughout the early Holocene.

The most recent low stand of sea level was about 20,000 years ago, during the last major glaciation. The fact that large ice caps were still present this recently (yesterday, geologically) was determined using radiocarbon dating.[33] During this last ice age, as during half a dozen previous ones typically lasting 50,000 years or so, almost all of Canada and much of northern Europe were covered with thick ice sheets.

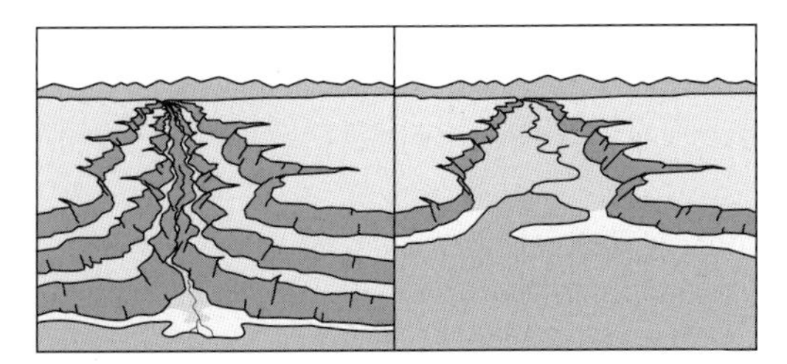

FIGURE 5.10. When sea level stood low, rivers cut deeply into the canyons between the mesas. When sea level rose at the end of the last ice age, bays formed, trapping sediment. They filled up to make wetlands.

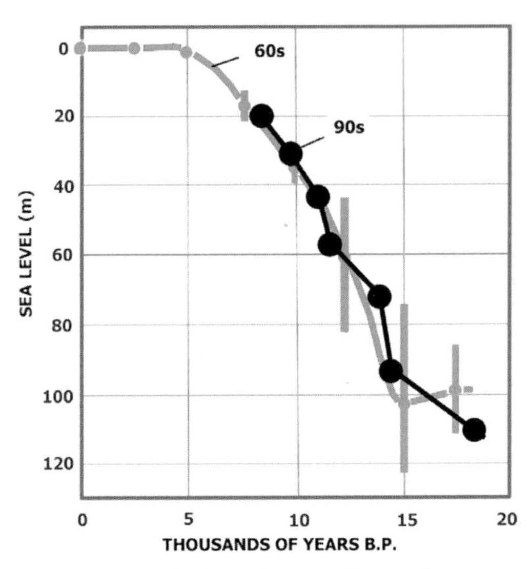

FIGURE 5.11. Sea level rose between 16,000 and 7,000 years ago. Note the large range bars for the reconstruction of the 1960s. In the 1990s it was confirmed that there were two major steps of deglaciation.

The water to make the great ice caps was taken from the ocean, so during the maximum build-up of the ice the sea level was about 400 feet lower than now.

With the sea level so low, the river valleys deepened greatly (fig. 5.10). Much sediment was poured into the ocean, filling the deep offshore basins. Sea level started rising again some 16,000 years ago. Francis P. Shepard and Joseph R. Curray[34] made a concerted effort to reconstruct the rise of sea level during the last 16,000 years. They used radiocarbon dates for carbon-containing matter (seashells, peat, tree stumps) originally deposited close to sea level, and plotted the ages against present depth, with good success.

The dating procedure is straightforward. One needs to determine the amount of radioactive carbon (radiocarbon, with the atomic weight 14), relative to normal carbon (with the weight 12), within the specimen to be dated. While an organism lives, it takes up radiocarbon from the atmosphere. When it dies, it ceases doing so. Through radioactive decay, one-half of the radiocarbon then disappears every 6,000 years. Thus, for objects 12,000 years old, only one-fourth of the radiation intensity is found compared with a recently buried shell or tree trunk. From such measurements in the 1960s and 1970s, Shepard and Curray concluded that the sea rose to its present level in the period between 16,000 and 8,000 years ago, and that the rise was uneven.[35] Many details have been added since (fig. 5.11). It was found, by dating coral deposits, that the rise contains at least two distinct steps. This confirmed earlier suggestions that the big ice sheets did not melt gradually but wasted catastrophically.[36]

Some geologists proposed that sea level actually rose as much as 10 feet above the present level, during the past 5,000 years. Shepard and Curray set out to test this idea. During the Carmarsel Expedition into the tropical Pacific (in 1967) the two of them, often accompanied by colleagues, managed to visit 35 Micronesian islands east of the Mariana Trench. They searched for raised coral reefs that would indicate elevated stands of sea level. In several places they did find chunks of coral several feet higher than present sea level. But after close inspection they came away convinced that these had been dislodged and flung there by violent storms. The question of whether sea level was somewhat higher than today, some 5,000 years ago, is in fact still unresolved.

In southern California, as mentioned, the rise of sea level at the end of the last glacial period resulted in profound changes of the coastal landscape. The ocean invaded the valleys between the mesas, forming large and small bights, which by and by filled up with mud and sand brought by the rivers during floods, and also with beach materials brought into the bays by tidal currents and wave action. When the rise ended about 7,000 years ago, wetlands started to form at the inner part of the bays, wetlands that were rich in marine life such as shrimps, crabs, and clams. These offered excellent foraging for the aboriginal people who had settled the region in the meantime. These people left their shell refuse heaps (middens) for us to study. Over the next several thousand years, the wetlands expanded and eventually reached an equilibrium level, where floods remove as much material as accumulates during quiet times.

For thousands of years, the wetlands provided breeding grounds for many fishes, and forage for indigenous and migrating birds. Destruction of wetlands by human activities began about a century ago in California and has accelerated since. Today, less than 10 percent of the wetlands remain in their original state.

The recent rise in sea level, by some 400 feet since 16,000 years ago, of course, did not impact just the coastal landscape of California, but all coastal zones everywhere. Nearshore environments throughout the world bear the stamp of this rise. The East Coast has extensive wetlands, with marshes and lagoons and drowned estuaries such as Chesapeake Bay, all no more than a few thousand years old. The East's wide beaches are framed by ridges or dunes that are equally young. The broad expanses of sand and mud flats of the Gulf Coast, fed by high sediment supply from the Mississippi drainage system, likewise are of recent origin—they could not have formed when the sea level was low. The same is true for the swamps of Louisiana; that is, wetlands with forests (such as bald cypress) adapted to having their roots permanently covered with water. These trees grew in a narrower belt far off today's shore during times of low sea level.

CANYONS UNDER THE SEA

The ultimate sink for the sediments sent by the mountains are the basins offshore. But only a modest portion of the material makes its way in an unorganized way across the shelf, toward these deep resting places. The bulk of sediments use a rapid transit system, running down the canyons that connect the deep basins to the shore (fig. 5.12).

Submarine canyons are intriguing features of the Earth's surface; they occur on practically all the slopes off the continents, and they are invariably involved in funneling the products of erosion into the deep ocean. Scripps Institution of Oceanography, by sheer luck, is located halfway between the submerged heads of the two branches of La Jolla Canyon, an impressive sea valley running into the San Diego Trough to the south, offshore from the city of San Diego. For two decades, the La Jolla canyon was one of the best-studied underwater valleys in the world and stood at the center of a lively geologic debate.

By the middle 1930s enough echo-sounding recordings had been taken (mostly by the U.S. Coast and Geodetic Survey) to show the outline of the seafloor off the coast of the United States. Among the features that appeared on the East Coast, and a few on the West Coast, were many deep offshore "canyons"—a word put in quotes by the famed (and combative) Harvard geologist Reginald Aldworth Daly.[37] In 1936 Daly protested the theory put forward by some geologists, to the effect that the submarine canyons are river valleys cut at a time of greatly lowered sea level.[38] True, rivers had to run across the shelf when sea level was lowered during glacial times, and thus could cut canyons down to the shelf edge, some 120 meters (400 feet) below the present level. But such cutting was difficult to envisage at greater depths. The crux of the problem is that many of the canyons extend well below the shelf edge, to more than 1,000 meters below current sea level. Daly and other geologists rejected the possibility of a sea level drop of hundreds of meters. There was no place for the water to go. Ice age ice masses were not large enough to contain the mass of water implied in the drastic sea level drops proposed.[39]

Daly suggested that sediment-laden water—mud slurries, in essence—rushed down on the seafloor across the continental slopes driven by

FIGURE 5.12. Geologic setting of the beach (after D. G. Moore): waves make beach sand from river mud, and the sand travels south till it finds a canyon that funnels it down into a deep-sea fan.

their excess weight, and that these density currents, or turbidity currents as they were called later on, carved the canyons.

No one had ever seen one of these currents, or indeed any evidence for their existence (other than the canyons they were supposed to explain). These catastrophic mud slurries were children of geologic imagination.[40] This is not to say that the muddy currents were pure fantasy. Mud slurries running downhill had been observed and described after rainfalls in the desert, moving at high speed and carrying enormous loads of material. Basically, Daly was speculating about how such muddy floods would behave under water. As it turned out, the reasoning was sound. Like in the desert, the muddy floods build up enormous fans, which dominate the landscape of the continental slope.

Shepard came to work on the canyons off Scripps in 1937, as a visiting investigator from the University of Illinois. He took a rowboat and ran a sounding line up and down within the valley and its branches, assisted by his graduate students K. O. Emery and R. S. Dietz. Location was determined by triangulation, using markers onshore. It took some tenacity to get these angles right, on a rocking boat. A surprising narrowness of the canyons soon emerged—their walls seemed essentially vertical in places.

The war abruptly changed this idyllic work pattern. The U.S. Navy developed an intense interest in the topography of the seafloor and the conditions in the deep ocean in general. Shepard went on leave from the University of Illinois to join the University of California Division of War Research in San Diego and to aid the Navy in its quest for information. Later he recalled this period as follows:

> Research carried out under naval auspices differed somewhat in character from ordinary scientific studies. Haste made it necessary to arrive at conclusions without being too particular about elaborate proofs. The large size of the military budgets, however, made it possible to obtain information which would have been beyond the grasp of ordinary scientific work such as was conducted before the war.[41]

The days of rowboat and sounding line were over.

Shepard was soon involved in the detailed charting of continental shelves and upper-slope morphology and of sediment cover off the West

Coast. He also undertook a library search of records all around the northern Pacific coast, including the Asiatic coast from Kamchatka on south and on to the Bay of Bengal, in the Indian Ocean.[42] All through this work he rarely lost sight of the first great problem that had fascinated him early on, that is, the origin of submarine canyons. Eventually, he provided a comprehensive summary of the world's submarine canyons and sea valleys in a book with his former student Robert F. Dill.[43] He worked on the subject until the end of his long career, making observations on the motion of tides and internal waves within the canyons.

The Navy's sponsorship of submarine geography in the 1940s resulted in rapid expansion of knowledge about the regions off California and also off New England, where scientists from Woods Hole Oceanographic Institution, especially Henry C. Stetson and his associates, made detailed investigations. Both the margins beyond the West Coast and those beyond the East Coast have canyons, suggesting that the causes are not related to mountain building, but rather to erosion. A number of important insights emerged from the detailed studies of canyons and especially the heads of the two canyons north and south of Scripps. Detailed multiple surveys of the canyon heads showed that they filled with sand and emptied intermittently. In situ observations demonstrated that the walls of the uppermost canyon are nearly vertical, even overhanging in places.

Direct observations were a challenging feat in the years before the self-contained underwater breathing apparatus (scuba) gave divers freedom from helmet and tether. Shepard wrote about this experience:

> One of the latest innovations in submarine geology has been the use of diving suits and helmets to study the bottom at first hand. This practice was made use of by K. O. Emery in the coral-reef studies of the Bikini atom-bomb expedition. It is also being employed at Scripps Institution. The first eyewitness description of a submarine canyon was recently made by Frank Haymaker, of the staff at Scripps Institution. His description by telephone of the precipitous rock-walled canyon which is located directly off the Scripps Institution aroused great envy among his listeners who were unable to go down because of their lack of training in deep-diving methods.[44]

Since the 1960s, it is a matter of getting into a wetsuit and putting on an air tank with a regulator, and a facemask, to go get a direct look. When one is swimming over the rippled sandy floor just outside the breakers and parallel to shore, half a mile to the south of Scripps, nothing indicates the proximity of the deep canyon. Suddenly the ripples disappear and one is soaring over the dark depths of the canyon, like a bird taking off from the cliff edge. It is an exciting, almost eerie, experience. Entering the chasm, one glides down along a vertical wall made of sedimentary rocks, riddled with small holes from which the watchful eyes of small fishes, like goblins, follow one's every movement.

Diving with scuba began at Scripps in 1948. Anxious to see the canyon with his own eyes, Shepard tried diving himself, but ear problems made him give it up. His students soon began helping him study "his" canyon—mapping it, observing fishes and mollusks eroding the walls, bringing carefully selected samples, photographing everything that might help solve the puzzle of origin. Shepard finally entered Scripps Canyon, in the diving saucer *Soucoupe* of Captain Jacques Cousteau, in 1964. During two dives to 800 feet on a "soul-satisfying" day, he gleefully observed the rocky vertical cliffs, tributary canyons, sandy ledges, and overhangs, as well as watched bat rays, sea anemones, gorgonians, and shrimps. "I felt I learned more about canyons that day than in 30 years of remote control study from the surface," he wrote later.[45]

Looking at the canyon walls, directly or indirectly, did not automatically resolve the question as to the origin of La Jolla Canyon, or any other canyon. Did muddy slurries—turbidity currents—race down the canyons and help carve the walls? Shepard's students and associates decided to put the idea to the test, by experiment. The plan was to set off a turbidity current using explosives in the canyon head. "We stuck a number of 2-inch pipes deep into the sediment and filled them with TNT," said Douglas Inman, who worked with Shepard then, continuing, "Then we moved our ships offshore and lowered the gear into the canyon, to observe the current coming through the canyon. The explosives went off more or less as planned, but nothing whatever happened within the canyon. The sediment settled, having lost the gas generated by all the kelp within it, and we now had a firm plug in the upper canyon."[46]

On another occasion, the group installed a current meter in the upper canyon, mounted to a steel bar anchored within the opposing canyon walls. Inman recalled:

> The canyon at this place was only 8 feet wide, and there was no problem in installing the bar; the length could be adjusted. The current meter had rotating blades, to measure current speed, and a blade like a wind vane to turn into the current. It faithfully recorded both speed and direction of any current in the canyon. We assumed, naturally, that the current would go either shoreward or seaward, parallel to the canyon. However, the measurements showed the current moving at an angle to the direction of the canyon. We were quite puzzled by this. There was nothing wrong with the current meter, so something had to be going on in the canyon that we did not understand. So, we had a look. We put on our diving gear and stayed around to watch the meter. Pretty soon a small predatory fish, a blenny, took up residence on the flat part of the current meter, next to the vane. It waited for prey to swim by underneath, so it could pounce on it. While sitting on its perch, it moved the vane to one side, which gave the faulty reading on the current direction.

The bar and the current meter never did record the occurrence of a turbidity current. The equipment eventually disappeared, presumably removed by a debris flow taking it down the canyon.

ABYSSAL CATASTROPHE

Density currents moving sediment, which soon came to be called turbidity currents, had first been observed in the 1930s, when Lake Mead, behind Hoover Dam, was starting to fill up with the deposits introduced by the Colorado River. Muddy water was seen to flow downhill, hugging the floor of the lake, and dropping thin layers of sediment in the deepest part of the lake, behind the dam. These early observations remained the only observations of such currents for many decades. Yet the existence of such currents, and their importance in moving sediments into the deep sea, has become a cornerstone of deep-sea sediment studies at least since the early 1960s.

The Dutch marine geologist Philip Henry Kuenen[47] began laboratory experiments on such flows in 1937, to test Daly's concept of muddy currents rushing down submarine canyons. The experiments were quite straightforward: he and his students released dilute clay-suspensions into a body of water and observed the progress of the suspension. Much later, Kuenen wrote:

> These early experiments cannot be claimed to have done more than establish a physical possibility and to have guided attention and speculation in the right direction. Marine geology and the study of flyschlike formations have thus been aided in gathering the pertinent evidence which shows that the mechanism of turbidity currents plays an important part in Nature.[48]

Concerning the nature of turbidity currents, some of the most important contributions to the discussion came from the marine geologist Bruce Heezen.[49] Heezen, like Shepard, believed above all in making plenty of observations. Heezen and his colleague Charles Hollister carefully sampled the materials within submarine canyons and just below the canyons, where they issue on the lower slope and into the adjacent basins. They found a mixture of shallow-water and deep-sea components, which they interpreted (correctly) to mean that shallow-water materials had been funneled into the deep sea by way of the canyons. Similar sediments, in principle, are in the San Diego Trough, where kelp and beach sand are found mixed with plankton remains falling down from the sunlit zone above, and with mud brought in during rainstorms. The proper explanation of such mixtures rendered absurd a strange concept that had arisen when beach-derived deposits were recovered from the deep basin of the Gulf of Mexico. Geologists describing the samples had seriously proposed (in the 1950s) that the entire basin floor was at sea level not so long ago, and had since quickly sunk to its present depth. The concept was soon recognized as erroneous (and derided as "elevator tectonics" by those familiar with it).

Heezen was keen on testing the question whether turbidity currents actually rushed down the continental slope, regardless of whether they were responsible for the origin of submarine canyons. While working on his master's degree (received in 1952), he became greatly intrigued with the fact that many submarine cable breaks had been observed downslope from the Grand Banks, following the Grand Banks earthquake of 1929. He interpreted the regular sequence in these failures as a result of a powerful bottom-hugging turbidity current. In promulgating this scenario, published in 1952,[50] Heezen succeeded in raising the discussion of marine sedimentation to a new level of drama and excitement. Here was evidence that sedimentation had a violent, even catastrophic component. All those layers in the deep sea did not just record a quiet uniform environment. Much of the sediment in the abyssal plain off the East Coast, according to this interpretation, was the result of distinct depositional events involving fearfully large and powerful currents, similar in principle to the floods rushing through desert washes after a big rainstorm in the nearby mountains. Each major event was in fact a catastrophe wiping out the benthic organisms on the seafloor over vast regions, by covering them with mud.

Large, rare events, thus the implication of the new hypothesis, are very important in producing the record of geologic history. The emerging concept of sporadic turbidity flow was a serious challenge to the then prevailing paradigm, which stressed gradual change in the framework of "uniformitarianism," the doctrine that observable everyday processes shape the surface of the planet. This doctrine had held sway ever since the eloquent British naturalist Charles Lyell announced and publicized it with forceful prose, in the 1830s.[51] Lyell's doctrine was designed to fight off the invocation of miracles to explain geologic phenomena, early in the nineteenth century. In this he succeeded, but at the price of excluding highly unusual events from consideration when reconstructing geologic history.

As a result of Heezen's work, a troubling question arose: could it be that most layers seen in sedimentary rocks are the legacy of unique catastrophic events?

Recall that Kuenen mentioned that his experiments have implications for the origin of flyschlike formations. What was he talking about? The word *flysch* is conspicuous by its absence from many or most modern geology textbooks; its place has been taken by the terms *graded bedding* and *graded layers*. In common English, we do not call the sediments involved *flysch*, but instead speak of formations liable to make landslides, a somewhat lengthy translation of the original *flysch*.[52]

Thick sequences of fine-grained, finely bedded sedimentary rock that make slides when soaked (gliding on wet shale layers within the sequence) are quite common along parts of the California coast. Such rock sequences ultimately stem from the flow of muddy waters rushing down into the troughs rimming the active margins of a continent. In such a flow, as the mud settles out, the coarser grains drop first, then the fines, producing the graded layers. Thus, in principle, the sediment layers at the bottom of San Diego Trough are flysch, as are the Cretaceous rocks outcropping along the Cove of La Jolla, and the rocks in the hills of Ventura north of Los Angeles.

The origin of the graded layers remained unknown for the first hundred years of geologic study of the Alps, where flysch formations are abundant. The endlessly repetitive sequences, which appear at quite different ages from the late Cretaceous to the early Tertiary, suggested some process or conditions special to the Alps, but no one could figure out what these might be. That flysch is deposited in deep-sea troughs along the rising Alps, at least in part, was admitted in the 1950s in the leading German-language geology textbook of the time.[53] An earlier tradition had it that no such troughs existed, but the sharp-eyed (and sharp-tongued) geologist Gustav Steinmann insisted that deep-sea deposits existed within these sequences; he saw opposing views as resulting from a lack of acuity, and said so.[54] But there is no mention in the 1959 standard textbook of the action of turbidity currents with reference to these sequences.

To be blunt about it, not only was the origin of the mountains themselves unknown, in the 1950s, but alpine geologists had no idea of how to explain the layering of most of the sediment sequences they were looking at. Alpine geology was being taught at the universities, of course, notwithstanding the lack of understanding of basic phenomena. There was plenty to talk about: the types of rocks, their ages, their fossil content, and their distributions. But the question of origins was as yet inaccessible: the processes involved had not yet been studied in today's ocean.

The most telling clue to the origin of flyschlike deposits was the realization that the graded sandy beds contain shallow-water fossils displaced over long distances, while the fine-grained layers separating the graded beds contain in situ fossils grown in place, or supplied from the plankton in overlying waters. Perhaps no one did more to solve this particular conundrum of the origin of the mixed-up fossil assemblages than the Californian geologist and paleontologist Manley Natland.[55] During the late 1920s and early 1930s, he organized one-man oceanographic expeditions to solve the problem. He had already studied the Pliocene flyschlike sediments in the Ventura Basin, for his thesis, and had found that the sandy layers contained benthic foraminifers entirely different from those in the fine-grained layers. Next he went out to the Santa Barbara Channel, in a rowboat, to take grab samples from the seafloor in order to see where these foraminifers (or closely related forms) live today. It was not long before he realized that one set of foraminifers lives in shallow waters, even near the beach, and the other set on the floor of the basin, several hundred meters down.

Aha! The coarse layers represent pulsed input from the shallow-water environment; the fine layers represent the "normal" conditions.

University of California geologist Edward L. Winterer, then working on his Ph.D.,[56] put the puzzle together in a comprehensive study of flyschlike deposits of the Ventura Basin, building on Natland's insights. He was aware of Kuenen's backyard experiments on muddy flow, showing that turbidity currents could well exist, as concerns the physics. In the late 1940s all three scientists had a chance to discuss these matters, the field evidence, and the evidence from experiments: Kuenen came to Ventura County to visit the outcrops in Ventura, led by Natland and Winterer. Here one could plainly see the evidence that turbidites exist on a large scale, outside of Lake Mead, of Lac Léman, and of Kuenen's laboratory. The puzzle, in essence, was solved, and henceforth geologists looked at sedimentary layers with a new kind of awareness.[57]

By the 1960s, a consensus had developed. Graded layers (such as the flysch of the Alps) represent sequences of deposits generated by sporadic turbidity currents that in turn stem from debris flows picking up increasing amounts of water as they move down the continental slope. Remaining questions concerned details—and there were many of these. Just how does the debris flow get started, just how does it get converted into a slurry and a turbidity current, just how fast does the current flow within a canyon, and just how does it behave when it leaves the canyon and enters the basin? Furthermore, just how long does it take till the process of making a graded layer is complete, how much erosion is there of earlier layers in the basin before deposition of the new one sets in, and why are there little ripples at the top of the layer? A flood, as it were, of articles appeared on these various topics.

Among the several questions concerning detail, the speed of the hypothetical current is the most interesting. Turbidity currents are rare events, as mentioned. Thus, it is unlikely that one can measure their speed directly, nor is it advisable to get close to such a current for direct observation. The oceanographer Paul Komar, then a graduate student at Scripps, had a splendid idea on how to attack the problem of the speed of turbidity currents. Komar proposed to use the morphology of the lower part of Monterey Canyon off central California as a source of information on the speed of canyon-filling currents. The canyon starts out as a huge ravine cut into bedrock, on the scale of the Grand Canyon of the Colorado. But in its lower portion, it runs in a channel of its own making, framed by levees.[58] One side is higher than the other, and thereby hangs the tale. It suggests that the surface of the muddy water, when filling the channel, is tilted, since the levees are built from the mud in such flows. The asymmetry can be explained by calling on both centrifugal forces and the Coriolis force (an effect of the Earth's rotation) to act on the muddy fluid rushing down the bent channel. But the equations can be set up properly only if the density and the velocity of the current are known. By guessing the muddiness of the water, Komar was able to come up with a range of velocities, which nicely matched the earlier estimates of Heezen, based on the cable breaks after the Grand Banks earthquake.

His fellow graduate students were impressed when he showed them, in a seminar, how to solve the problem.[59]

NOTES AND REFERENCES

1. A striking illustration of this fact is seen in satellite images of the night side of Earth: the lights of cities outline the coast in many parts of the world.

2. J. M. Erlandson, 1994, *Early Hunter-gatherers of the California Coast.* Plenum Press, New York, 336 pp.; P. M. Masters, D. Gallego, 1997, *Environmental change and coastal adaptations in San Diego County during the Middle Holocene,* in J. M. Erlandson, M. A. Glassow (eds.), *Archaeology of the California Coast during the Middle Holocene.* Institute of Archaeology, University of California, Los Angeles, pp. 11–21.

3. In California, tourism brings billions of dollars into the economy. The importance of a pleasant beach in attracting tourists has led to beach replenishment projects in many coastal communities of the country, in spite of the expense involved.

4. The National Aeronautics and Space Administration, the National Oceanographic and Atmospheric Administration, and the U.S. Geological Survey.

5. A. H. Sallenger, W. Krabill, J. Brock, R. Swift, M. Jansen, S. Manizade, B. Richmond, M. Hampton, D. Eslinger, 1999, *Airborne laser study quantifies El Niño–induced coastal change.* EOS 80, 90, 92–93.

6. The regulation of river flooding by dams has interfered with such supply, through the decades, and therefore, the beaches tend to be starved. At least that is the conventional wisdom. Reinhard Flick at SIO, an expert on coastal engineering, thinks that the history of housing development and associated earth movements must be considered also when discussing the changing amounts of sand on the beach. Bulldozers generate sand supply. R. Flick, pers. comm., 2005.

7. D. L. Inman, S. A. Jenkins, 1999, *Climate change and the episodicity of sediment flow of small California rivers.* Journal of Geology 107, 251–270.

8. Educated guesses are made, for example, by assuming a "normal distribution" on a logarithmic scale. The mathematics of the so-called normal distribution was defined by Karl Friedrich Gauss (1777–1855), who also pioneered the study of Earth's magnetism. Many of nature's more complex phenomena (where many factors are involved) follow a log-normal distribution, at least approximately.

9. Chances are that storms will get stronger as the climate gets warmer. Storms derive their energy from water vapor in marine air. The vapor content increases as the sea surface warms.

10. The term commonly used is *reclaimed,* as though land had somehow been lost to marshes at an earlier period. The history of this usage may go back to the shores of northwestern Europe, where the sea indeed invades the coastal lowlands, which are then reclaimed by building dams and by encouraging sediment accumulation using salt-tolerant plants.

11. C. B. Field, G. C. Daily, F. W. Davis, S. Gaines, P. A. Matson, J. Melack, N. L. Miller, 1999, *Confronting Climate Change in California.* Union of Concerned Scientists, Cambridge, Mass., and the Ecological Society of America, Washington DC, 62 pp., p. 50.

12. Much of this work, up to 1962, is summarized in a book by Francis P. Shepard (1897–1985), professor of geology at SIO from 1942. F. P. Shepard (with chapters by D. L. Inman and E. D. Goldberg), 1963, *Submarine Geology,* 2nd ed. Harper and Row, New York, 557 pp.

13. Douglas L. Inman obtained his Ph.D. at SIO in 1953 and promptly joined the faculty. He founded the

Center for Coastal Studies at Scripps, in 1980. Earlier, he designed the Hydraulics Laboratory, a facility that enables students and researchers to study the interaction between wave action and sediment transport experimentally, among other things.

14. In a few places, gold actually has been found concentrated in this fashion, as off Nome, Alaska. The process of making placer deposits is briefly summarized in E. Seibold, W. H. Berger, 1996, *The Sea Floor, An Introduction to Marine Geology*, 3rd ed. Springer Verlag, Heidelberg, 356 pp.

15. For a brief summary of waves, as relevant to beach processes, see D. L. Inman, 1963, *Ocean waves and associated currents,* chapter 3 in F. P. Shepard, *Submarine Geology*, 2nd ed.

16. The Polynesians used the long southern waves as an aid in navigation between islands: David Lewis, 1972, *We, the Navigators*. University Press of Hawaii, Honolulu, 347 pp.

17. The 478-foot-long vessel survived. S. Junger, 1997, *The Perfect Storm*. W. W. Norton, New York, p. 156. The *Queen Mary* once had its wheelhouse flooded by a single wave, 90 feet above the surface. The same thing happened to the 12,000-ton drilling vessel *JOIDES Resolution,* off southeast Greenland, with a wheelhouse some 40 feet above the water line. A wave broke the window.

18. The Austrian-born physical oceanographer Walter H. Munk was a student of Harald U. Sverdrup in the 1940s, at Scripps, and became a faculty member after he received his Ph.D. in 1947. He founded the University of California at San Diego branch of the Institute of Geophysics and Planetary Physics, established in 1960, and was its first director. Munk made many fundamental contributions to understanding waves, currents, the nature of the thermocline, and the driving forces of deep circulation. Most recently, he has been engaged in the acoustic exploration of the deep ocean.

19. FLIP is a very long cylindrical buoy with laboratories at the upper end. When towed, it is horizontal and looks a bit like a ship, albeit strangely tube-shaped. When flipped, it is vertical, the bottom part flooded with seawater, while the top provides a stable platform relative to the sea surface. Scripps scientists Fred N. Spiess and Fred Fisher designed the platform (see chapter 11).

20. F. P. Shepard, 1948, *Submarine Geology*. Harper and Bros., New York, 348 pp.

21. Francis Parker Shepard (1897–1985) grew up in New England. He received a degree in geology from Harvard in 1919 and a Ph.D. from the University of Chicago in 1922. He then taught at the University of Illinois. On a sabbatical in 1933 he visited Scripps Institution and became intrigued with its offshore canyon. He returned in 1937 as a visiting investigator, along with his graduate students Kenneth O. Emery and Robert S. Dietz. Eventually, he joined the Scripps faculty, after World War II. He was indefatigable as a sea-going oceanographer and made many important contributions to the understanding of the morphology of continental margins and their sediments, together with his many students. His book *Submarine Geology* went through several editions, before the emphasis of marine geology changed toward seafloor spreading and its ramifications.

22. R. S. Dietz, K. O. Emery, 1976, *Early days of marine geology*. Oceanus 19 (4), 19–22.

23. Quoted in C. Colgan, 1998, *Sands of time running out*. Explorations 5 (1), 10–17. *Explorations* (1995–2007) is a publication of the SIO office for public relations.

24. C. Colgan, 1995, *Sand wars*. Explorations 1 (3), 9–15.

25. Some of Shepard's students became prominent geologists: K. O. Emery, Robert Dietz, Douglas Inman, and Joseph Curray, among others.

26. The photo was a standard exhibit in his marine geology class, which he taught until the middle 1960s. When his students asked what happened to the person in the picture, the answer was that he disappeared, as far as he knew.

27. F. P. Shepard, U.S. Grant IV, 1947, *Wave erosion along the southern California coast*. Bulletin of the Geological Society of America 58, 919–926; F. P. Shepard, H. R. Wanless, 1971, *Our changing coastlines*. McGraw-Hill, New York, 579 pp.

28. K. O. Emery, 1941, *Rate of surface retreat of sea cliffs based on dated inscriptions*. Science 93, 617–618.

29. K. O. Emery, G. G. Kuhn, 1980, *Erosion of rock shores at La Jolla, California*. Marine Geology 37, 197–208.

30. The correct transcription of *El Niño* into the English alphabet is *El Ninyo* which preserves the phonetics (as in canyon). The commonly seen newspaper version *El Nino* is incorrect.

31. The winter rains of 1915 were late, and the reservoirs were much depleted. The city council decided to hire a rainmaker, to wit, one Charles M. Hatfield, who claimed to know the trade. Hatfield soon went to work, early in January. By the end of the month, in a succession of downpours, San Diego had received more than 7 inches of rain near the coast. Inland, one recording station is reported to have received 44 inches within a month. Nearby, a dam broke on the Otay River, and 50 people were drowned. Dams held on the San Diego River, but the river (normally close to dry) expanded to become a mile wide near its mouth and crested at more than 6 feet. The city refused to pay Hatfield—severe flooding was

not what the council had in mind when making the contract.

32. Presumably, a part of the terrace now at sea level was cut at earlier occasions, when sea level happened to intersect the rising coast at this particular level. W. H. Berger, 2004, *Terraces on a rising coast*, in R. L. Wernli, C. F. Kennel (conveners), *Oceans 2003*, Proceedings (on CD). Holland Enterprises, Escondido, Calif., 1209–1212.

33. The Austrian-born chemist Hans E. Suess (1909–1993) made important contributions to this field, even before he joined Scripps in 1955. He discovered, by dating logs buried in terminal moraines, that the last ice age ended only about 11,000 years ago. This was of the greatest interest to everyone concerned with prehistoric man, with the ice ages, and with climate history in general. It also was crucial information for sea level studies.

34. Joseph Ross Curray received his Ph.D. at SIO in 1959 and stayed on, eventually joining the Scripps faculty. Career interests included dispersion of sediments on and across shelves, the rise of sea level during the transition from glacial to postglacial time, and the history of the Bengal Fan, which is fed by the Ganges and Brahmaputra rivers. At present he is working on a synthesis of Bengal Fan studies.

35. F. P. Shepard, J. R. Curray, 1967, *Carbon-14 determination of sea level changes in stable areas.* Progress in Oceanography 4, 283–291.

36. R. G. Fairbanks, 1989, *A 17,000-year long glacio-eustatic sea level record: Influence of glacial melting rates on the Younger Dryas event and deep-ocean circulation.* Nature 342, 637–643; E. Bard, B. Hamelin, R. G. Fairbanks, A. Zindler, 1990. *Calibration of the* ^{14}C *timescale over the past 30,000 years using mass spectrometric U-Th ages from Barbados corals.* Nature 345, 405–410.

37. R. A. Daly (1871–1957) is perhaps best known for his theory of the origin of atolls, which involves fluctuating sea level, as outlined in the previous chapter. The hypothesis of turbidity currents was another of his prominent contributions to marine geology.

38. R. A. Daly, 1936, *Origin of submarine canyons.* American Journal of Science 31 (186), 401–420.

39. Shepard was partial to the river-cutting scenario at one time in the 1940s. His views apparently were greatly influenced by contemplating canyons around islands in the Mediterranean. In fact, sea level had been lower by more than a mile in this sea, only 6 million years ago, which supports his reasoning. Unfortunately, he renounced his hypothesis before the drying out of the Mediterranean Sea was discovered, during Leg 13 of the Deep-Sea Drilling Project, in 1970. There is a lesson in this episode: explicitly renouncing one's own hypothesis is usually unnecessary; if wrong, it will not last anyway. Disavowal merely makes certain that one is wrong, either then or later.

40. The imagination of the Swiss hydrologist François Alphonse Forel (1841–1912), to be precise, who published the idea in 1887, based on observing a channel at the bottom of Lac Léman.

41. F. P. Shepard, *Submarine Geology*, p. 7.

42. Much of this was later worked into the second edition of his book, *Submarine Geology*, published in 1963.

43. F. P. Shepard, R. F. Dill, 1966, *Submarine Canyons and Other Sea Valleys.* Rand McNally, Chicago, 381 pp.

44. F. P. Shepard, *Submarine Geology*, p. 8. Frank Haymaker was a former navy diver (with helmet and tether), with an interest in zoology. Working for Shepard, whenever Haymaker found interesting rocks during a dive, he was provided a sledgehammer and a chisel and lowered from a rowboat, so that he could break off the rocks and send them up in a bag. Haymaker also took underwater photographs of the canyon, using a camera developed by Maurice Ewing, at Lamont.

45. Francis P. Shepard, 1980, *Autobiography.* Typescript in Archives of Scripps Institution of Oceanography, 200 pp, p. 153. He thoroughly enjoyed several other diving-saucer opportunities in canyons in the next several years, as deep as 4,150 feet.

46. D. L. Inman, pers. comm., 2003.

47. Philip Henry Kuenen (1902–1976) was the dominant figure in the discussion centered on turbidity currents and their deposits, the turbidites. He was born in Leiden, the Netherlands, and taught geology in Groningen from 1934 till his retirement in 1972.

48. Ph. H. Kuenen, 1965, Experiments in connection with turbidity currents and clay suspensions, in W. F. Whittard, R. Bradshaw (eds.), *Submarine Geology and Geophysics.* Butterworths, London, p. 48. The reference to flyschlike formations is to sequences of thinly bedded sediments in the Alps and the Appenine mountains.

49. Bruce Charles Heezen (1924–1977), marine geologist at Columbia University, was a pioneer in the study of seafloor morphology and Quaternary sedimentation. He died from a heart attack while diving in a Navy research submarine off the coast of Iceland.

50. B. C. Heezen, M. Ewing, 1952, *Turbidity currents and submarine slumps, and the Grand Banks earthquake.* American Journal of Science 250, 849–873.

51. C. Lyell, 1830–1833, *Principles of Geology*, 3 vols. Lyell (1797–1875) was trained as a barrister and knew how to argue well. Also, he was very familiar with the geologic literature of the time, and with the people working in the field. He held a position at King's College, England, from 1830.

52. *Flysch* is a term given by Swiss farmers to certain well-bedded deposits consisting of sequences of hundreds of thin, uniform graded beds of muddy sandstones alternating with fine-grained shales and mudstones, in the northwestern Alps. The name is related to *fliessen*, which is German for "to flow." It describes what is important to the farmers, which is that these sediment packets tend to become unstable when soaked during the spring melt and tend to move downhill, sliding on lubricated surfaces provided by the boundaries between shale and sandstone. Similar sliding is readily observed in many parts of California, in the coastal ranges, after times of heavy rain.

53. R. Brinkmann, 1959, *Abriss der Geologie*, vol. 2, *Historische Geologie*, 8th ed. Ferdinand Enke, Stuttgart, 360 pp.

54. G. Steinmann, 1925, *Gibt es fossile Tiefsee-ablagerungen von erdgeschichtlicher Bedeutung?* Geologische Rundschau. In English: *Are there ancient deep-sea deposits of geologic significance?* International Journal of the Earth Sciences 91, 18–42, 2002 (transl. by W. H. Berger).

55. Manley L. Natland (1906–1991) worked at the Research Center of Atlantic Richfield Oil Company in Los Angeles. He discovered depth zones for benthic foraminifers off California, showing that the nature of assemblages changes with habitat, as well as with time. M. L. Natland, 1933, *The temperature and depth-distribution of some recent and fossil Foraminifera in the southern California region.* Bulletin of the Scripps Institution of Oceanography, Tech. Ser. 3 (10), 225–230.

56. Edward L. Winterer received his Ph.D. from UCLA in 1948 with a thesis on the sedimentary sequences in the Ventura Basin. He joined the geology faculty at Scripps, in 1965, and eventually became a world expert on the marine geology of the Pacific and ancient deep-sea sediments in general. He used the new knowledge gathered through deep-ocean drilling to help elucidate the geology of the Alps.

57. M. L. Natland, Ph. H. Kuenen, 1951, *Sedimentary history of the Ventura Basin, California, and the action of turbidity currents* in J. L. Hough (ed.), *Turbidity Currents and the Transportation of Coarse Sediments to Deep Water. A Symposium.* Spec. Pub. 2. Society of Economic Paleontologists and Mineralogists, Tulsa, Okla., pp. 76–107; Ph. H. Kuenen, 1959, *Turbidity currents a major factor in flysch deposition.* Ecologae Geolologae Helvetica 51 (3), 1009–1021.

58. Levees are river banks raised above the surrounding landscape. They are natural dikes, made during floods by the dumping of sediments on top of the river banks. The Mississippi and the Yellow rivers have levees in their lower courses. If levees break, the river invades the surrounding flood plains. To prevent this from happening, many river levees are artificially enhanced.

59. Others were impressed also. Paul D. Komar subsequently was invited to join the faculty at Oregon State University, in Corvallis, where he pursued a distinguished career in studying coastal processes. His early work is summarized in his book *Beach Processes and Sedimentation* (1976, Prentice Hall, Englewood Cliffs, N.J., 429 pp.).

SIX

Unraveling the Gulf Stream Puzzle

ON A WARM CURRENT RUNNING NORTH

A mighty ocean current runs north and east offshore from the East Coast, between the Florida Keys and Cape Hatteras (fig. 6.1).

It moves warm water northward, from the Gulf of Mexico into the northern North Atlantic, on the way entraining warm water from the Sargasso Sea. It is one of the two most powerful warm currents on the planet.[1] The heat it transports feeds the storms of the northern Atlantic and helps maintain low air pressure in the vicinity of Iceland, stabilizing the Iceland Low. In turn, the anticlockwise winds running about the Iceland Low direct the heat-bearing marine air and associated rainstorms into northwestern Europe, from the British Isles and the Netherlands to Denmark and Norway.

The origin of this great weather-making current was a mystery well into the twentieth century. In the 1940s, it was realized that the current represents the western boundary of the spinning wheel of warm water that is centered on the Sargasso Sea, in the desert belt of the North Atlantic. This wheel is made to spin by winds, but not any winds over the Gulf Stream itself. As long as there are trade winds and westerlies, the great wheel will spin, and the Gulf Stream will move warm water from low to temperate latitudes.

THE WAY NORTH

The city of Bergen in western Norway, the second largest of its country and once the nation's royal capital and its main port, has a number of outstanding assets. As a port, it is ice free all year, despite its latitude, which is just slightly poleward of 60° N. The equivalent position on Greenland is Julianehåb, surrounded by glaciers terminating in the ocean. In Canada, the 60° line runs through the center of Hudson Bay, which is full of sea ice until early summer. In Alaska, Glacier Bay is at this latitude, an enormous fjord system where tourists on cruise ships watch glacier tongues feed icebergs to the

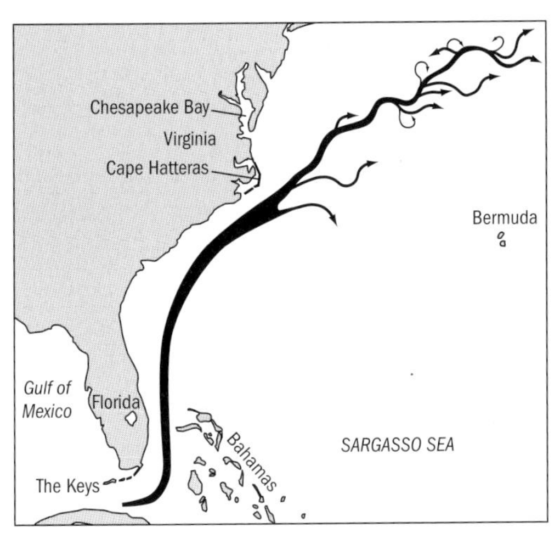

FIGURE 6.1. Path of the Gulf Stream. It starts in the Gulf of Mexico (GM), as the Florida Current south of the Florida Keys, and leaves the coast east of Cape Hatteras. On leaving the East Coast, the Gulf Stream becomes indistinct, making filaments and eddies.

sea. In the Southern Hemisphere, the latitude of 60° is positively uninviting—it runs through the desolate and windswept islands north of the Antarctic Peninsula, where sea ice is common.

The point is that the citizens of Bergen benefit from the greatest warm-water anomaly on the planet (fig. 6.2). The sea off their shores is some 5 °C (9 °F) warmer than is typical for the latitude. Springtime in Bergen arrives in May as a glorious spectacle of greening and flowering, with daylight breaking early and extending far into the late evening. Front yards are blazing with lilacs, hedge roses, poppies, and ubiquitous rhododendron and azalea, while dandelions grace niches and cracks in sidewalks and church walls. At the fish market, pansies are for sale next to gutted mackerel and live crabs.

Bergen's festive spring—celebrated with pageants of parades and musical events—contrasts with May in Hudson Bay, at the same latitude, a bleak scene with but little green, where sea ice still rules the seascape.

The warm water bathing Bergen's shores, with a temperature near 10 °C (50 °F), arrives from the south with the Norwegian Current, and this water has access deep into the countryside through the fjords, which act like so many pipes of a central heating system. The warm Norwegian Current, in turn, is a branch from the North Atlantic Current, which receives heat from the Gulf Stream and carries some of it toward northwestern Europe. All of northwestern Europe depends on this heat supply, much of which was gathered in the subtropical Atlantic and in the sunny Caribbean. The westerly winds of the North Atlantic drive the currents bringing heat, and they themselves bring even more heat—heat locked into the vapor they bear. These are the same winds, of course, that powered the sailing vessels making their way back from the New World to the Old before steam and diesel engines took over this task.

The delivery of subtropical heat to high northern latitudes by the winds and currents toward the shores of Norway generates a great temperature anomaly (fig. 6.3). As mentioned, it is the largest such anomaly on the planet, and Bergen happens to be at its center. Offshore of Norwegian shores, in the channel between the coast proper and a series of offshore islands, the water stays ice free all along Norway and into the Barents Sea. For centuries, fishermen and traders have used this sheltered lane as the road to the north. Without the warm water from the south, there would be no North-Way.

What keeps the Nordic warm-water pool in place year after year? Will the anomaly persist as our planet warms in response to the increase of greenhouse gases in the atmosphere? Or will it weaken as a result of the climate change now under way? Will ice begin to advance in Norway even as the rest of the world is warming?

It is quite reasonable to ask such questions. Anomalies, presumably, are among the more sensitive items within the climate system. When the system changes, anomalies are expected to respond. So far, within the last two decades or so, the response of the Nordic heat attractor has been to become stronger—more rain, and hence more heat, has been arriving in Bergen.[2]

Scandinavians, naturally, have long been interested in the weather—its behavior is crucial for the bounty provided by harvest and fishing at these latitudes. It is this vital interest that led

FIGURE 6.2. The colorful Bergen harbor at the city's fish market is open year-round, thanks to heat import in northern Norway from a warm current offshore and from moist southwestern winds.

them to study the dynamics of the atmosphere and of the ocean. One central aspect of this fact is a wholesale transfer of concepts from meteorology to oceanography—the physics involved in driving winds is applicable, with some adjustments, to the motions in the sea. The Scandinavian influence, of course, is readily apparent in the monumental text by the Norwegian Harald U. Sverdrup and his associates, the text that defined modern oceanography. Sverdrup started work on it in the late 1930s, and it was published a few years later.[3] The book became the template for education in oceanography and for many introductory textbooks that followed.

The most widely known among the ocean-going Scandinavian scientists is Fridtjof Nansen, the Norwegian Arctic explorer. His greatest feat was drifting across the Arctic in the ice-bound *Fram*, in an attempt to reach the North Pole. The *Fram* set out in 1893 and permitted itself to be locked by sea ice off eastern Siberia. All during the drift, the scientists on board were busy making geophysical observations. The ship arrived in Norway three years later. The expedition provided important new information, including the role of ice in shaping the margins of the Arctic Sea, the deflection of ice drift by the Earth's rotation, the production of deep water,[4] and the properties of water at depth (using the "Nansen bottle"[5]). But the most important fundamental contribution was the discovery that the Arctic is as deep as other basins, and not a shallow sea.

Sverdrup, like Nansen, went through the harsh school of year-round Arctic exploration, braving the solitude of polar winter in the icy wastes of a forbidding and unknown sea. In fact, Sverdrup set something of an endurance record for professional oceanographers, on Roald Amundsen's seven-year *Maud* Expedition (1918–1925). He brought back a rich harvest of observations, which constituted much of the body of geophysical knowledge of the Arctic environment for some time.

One of the all-time heroes of meteorology and oceanography is the Swedish physicist Vagn Walfrid Ekman, whose name appears in *Ekman transport, Ekman pumping, Ekman layer,* and the intriguing *Ekman spiral,* all of which are

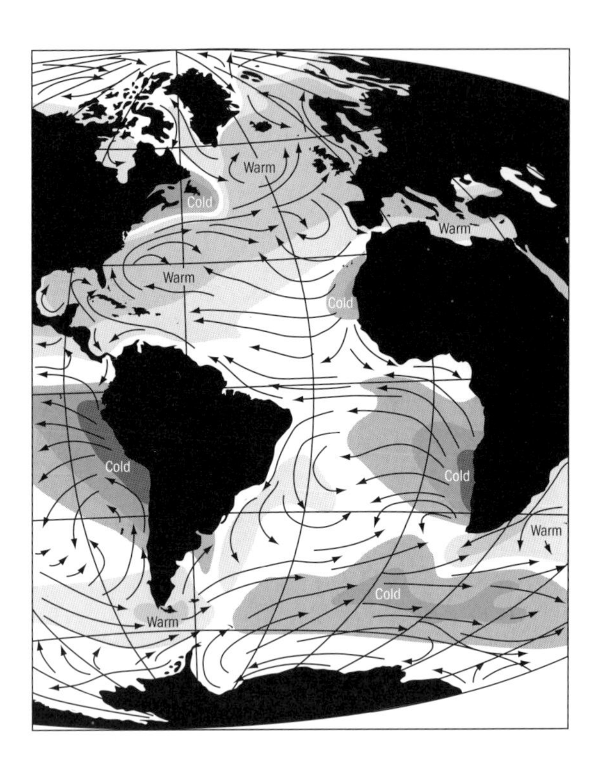

FIGURE 6.3. Great temperature anomalies: much warmer than latitudinal averages in the North Atlantic, much colder off Peru and Namibia. (After G. Dietrich: anomalies shaded.)

important concepts in physical oceanography, and closely related. Prompted by Nansen's observation that polar ice drifts to the right of the wind, Ekman showed that such a deflection is to be expected from the effects of Earth's rotation (Ekman drift) and can be calculated precisely by making certain assumptions about the effects of mixing of water in the upper wind-driven layer.[6] In doing this, he built on the earlier theoretical work of his mentor Vilhelm Bjerknes, a Norwegian physicist working on the motions of fluids, who brought the concepts of this specialty to bear on problems in meteorology and oceanography.[7]

Nansen had surmised, correctly, that floating ice drifts to the right of the direction of the wind because of the Coriolis force. This force must be invoked when describing motions in a rotating framework. In the Arctic, the sea surface is close to right angles to the axis of rotation of the Earth, and the rotational effect on moving bodies is stronger than elsewhere, therefore. (The effect vanishes at the equator.) The distance that any body on the planet's surface travels during a day, due to the Earth's rotation, rapidly increases away from the pole. Thus, a body of water moving away from the pole cannot keep up with its surroundings and falls behind. This is seen as a deflection to the west, that is, to the right of the motion. A body of water moving toward the pole outruns its surroundings, which shows as a deflection to the east. Again, this is to the right of the motion.[8]

Ekman used the Coriolis principle and friction between (idealized) water layers at different depths to set up the equations that would allow calculating the effect observed by Nansen. He first noted that a steady wind will set the water at the surface in motion[9] and that this motion will next result in a Coriolis force.[10] The impulse imparted to the water at the surface of the sea by the wind is then passed on to the water below. This is done by exchange of blobs of water between adjacent levels, a process called turbulent diffusion. In Ekman's equation, this process appears as friction. The friction transfers forces downward, which can now act on deeper layers. At each deeper level, the direction of the force transmitted from the layer above combines with the Coriolis force at the receiving level to produce additional deflection to the right. In addition, the velocity decreases with depth. The result is a spiral with decreasing amplitude downward—the famous Ekman spiral that appears in all textbooks.

The Ekman spiral is not seen in the real ocean, normally. But it nicely describes what ought to be seen if the ocean were simple and disturbances from other processes could be neglected. Thus, Ekman theory remains useful as an ingredient in the mathematical description of many of the large-scale features of the ocean's circulation.

The result of Ekman's calculations was that the uppermost water layer moves at 45° to the right of the wind, after equilibrium is established.[11] The overall transport in the upper part of the water column (adding up all the motions described by Ekman's spiral down to where the direction is opposite to the wind) is to the right of the wind forcing, on the Northern Hemisphere.

Ekman's wind drift theory[12] was just the beginning of the attempt to describe the motions of the sea as a logical result of wind action. As it turned out, Ekman drift is needed to build up the enormous subtropical warm-water lenses called central gyres, which dominate the circulation pattern of much of the ocean. In turn, these lenses each support a 1- to 2-meter-high shieldlike mountain of low-density water, which sets up pressure fields that, together with the Coriolis force, drive the major persistent currents in low and temperate latitudes. Equations describing such pressure-driven currents (called geostrophic currents) were introduced by Sandström and Helland-Hansen (1903) and put to use by Helland-Hansen and Nansen (1909), Wüst (1924), and Sverdrup (1947), among others.[13] The various concepts introduced by these scientists eventually led to a mathematical understanding of the overall patterns of circulation sketched at the beginning of the twentieth century (fig. 6.4).

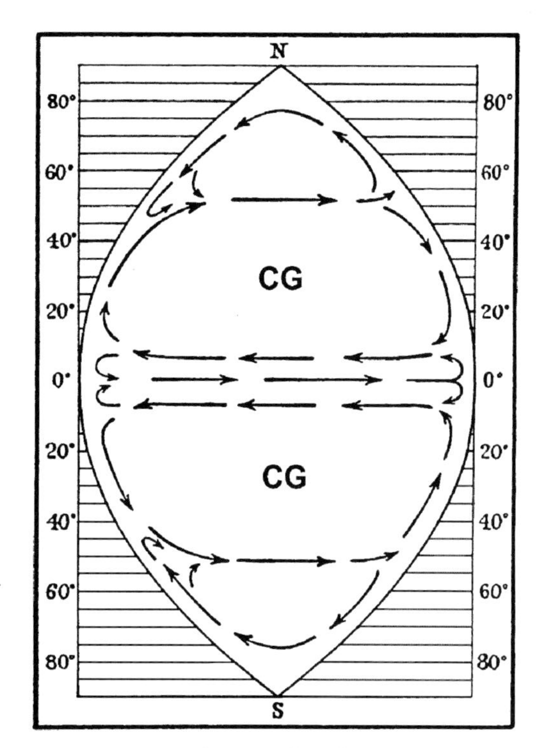

FIGURE 6.4. General pattern of circulation within an idealized ocean basin, as sketched by O. Krümmel in 1907. CG, position of central gyres (here added).

The great central gyres have a warm current in the west, and a cold current in the east, along with eastern-boundary upwelling. Thus, in explaining central gyres, principles embedded in Ekman's equations help explain both the mighty Gulf Stream and the upwelling off California, topics of special interest to oceanographers at Woods Hole and in La Jolla, respectively.[14]

Sverdrup's move to Scripps in 1936 constituted a major "technology transfer" of the most up-to-date concepts and ideas about the dynamics of ocean and atmosphere, largely developed in Scandinavia. It was indeed fortunate for the subsequent development of oceanography at Scripps, the West Coast, and the entire nation, that he accepted the invitation to come to La Jolla. His unique background allowed him to lead Scripps into a new direction, with a focus on ocean currents and their role in biology.

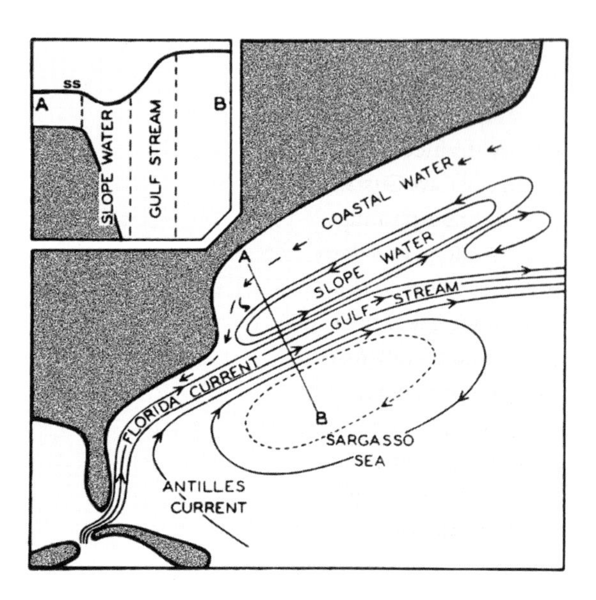

FIGURE 6.5. The Gulf Stream System as sketched by H. Sverdrup in 1942. The Stream is seen to form the edge of a gyrating Sargasso Sea, with warm water piling up to the right (*inset:* ss, sea surface). Cold water moves south along much of the East Coast.

THE GULF STREAM: A RIVER IN THE SEA?

The Gulf Stream is the archetype of ocean currents—none have been more intensely studied in the past century. Its name derives from the fact that its source waters exit from the Gulf of Mexico, through the Florida Straits, in the Florida Current.[15]

The Florida Current turns abruptly north upon leaving the Straits, flowing along the coast of Florida and toward Cape Hatteras (fig. 6.5). As it does so, it gains in strength by taking up water from the open Atlantic. It moves farther out to sea past Cape Hatteras, on a northeastward course. On approaching the banks south of Newfoundland, it turns east and feathers out into filaments and eddies. A portion of the eastward-running warm waters, mixed with other waters in the North Atlantic, eventually moves northeast and ends up in the Norwegian Sea. Most of the Gulf Stream waters do not partake in that northward course, but contribute instead to the southward and southeastward flow as part of the great central gyre that dominates the North Atlantic Basin. The westward return flow is largely through the North Equatorial Current, driven by the trade wind, which gathers surface waters in the tropics well north of the equator, and piles them into the Caribbean Sea. From there, surface waters rush out through the seaway between Yucatan and Cuba, to feed the Florida Current.

The Gulf Stream has long held special fascination for both European and American seafaring folk and also, of course, for oceanographers. The reasons that sailors should care are obvious. The Spanish captains following in the wake of Columbus soon found out that the currents off Florida can be faster than a sailing vessel—no use to try to stem them.[16] Oceanographers have grappled, above all, with questions concerning the nature of the Stream—its sources, the origin of its energy, and its varied manifestations along its extended path. And they have wondered about the implications of the northward movement of warm water for the heat budget of the planet, and especially for the heat budget of northwestern Europe, where so many of them had their homes.

Before 1930, Scandinavian and German oceanographers dominated the discussion about the nature of the Gulf Stream. But in the 1930s, Woods Hole adopted the Gulf Stream as its natural laboratory and carried out extensive surveys, initially led by the yachtsman and oceanographer C. O'D. Iselin, commanding the *Atlantis*.[17] Iselin was soon ably assisted by

Frederick ("Fritz") Fuglister and by Valentine ("Val") Worthington, people who loved going to sea and making observations.[18] Ray Montgomery, trained as a meteorologist under the tutelage of Carl-Gustaf Rossby, provided the theoretical underpinnings.[19] Montgomery had an eager student in Henry Melson Stommel, an explorer at heart, with a knack for mathematics.[20] With Stommel, Woods Hole acquired a natural leader who rapidly rose to national and international prominence.

Stommel made understanding the Gulf Stream a central goal of his career. In his book *The Gulf Stream*, published in 1965, he disparages the traditional view that the great current is a convenient source of heat for Europe:

> There is scarcely any more firmly rooted idea in the mind of the layman than the notion that the Gulf Stream keeps the European climate warm. So long as it was believed that the Gulf Stream was a river of warm water, this idea did make sense. It is no longer possible to be so certain of the direct climatologic influence of the Gulf Stream, for it now seems that it is not so much the Stream itself that is important, as the position and temperature of the large mass of water on its right-hand flank.[21]

Stommel here emphasizes that the nature of the Gulf Stream is tied up with the great rotating warm-water lens in the central North Atlantic. It is not a river in the sea but the rim of an immense wheel. It is the nature of this gyre, he advises, that we should focus on when talking about properties of the Stream or about getting heat into Europe.[22] The message he gave was twofold. First, all you Europeans are too much concerned about what the Gulf Stream can do for you. Second, the Woods Hole team will show you how the Gulf Stream really works.

What happened? How had the Gulf Stream lost its proper mission of bringing heat to Europe? Was it now morphing into a large-scale experiment off the American East Coast for the benefit of the scientists at Woods Hole? Stommel's reference to the "mind of the layman" is a nice touch, given that we are talking about 100 years of European literature.

The new team at Woods Hole insisted on a new credo: testing hypotheses by making well-planned observations. As Stommel explained: "Too much of the theory of oceanography has depended upon purely hypothetical processes. Many of the hypotheses suggested have a peculiar dreamlike quality, and it behooves us to submit them to especial scrutiny and to test them by observation."[23] One cannot help but think of Ekman's spiral as an example of the "dreamlike quality" of hypotheses proposed. Geostrophic flow (of the type traditionally charted to show average current flow) also belongs into that category: what is calculated can rarely be observed.

What Stommel said made sense, of course. Besides, the hypothesis-testing approach was an extremely practical path to follow for the new Woods Hole team of physical oceanographers. Focusing on the changing behavior of the Gulf Stream, they took full advantage of the location of the new institution and could look forward to a rich harvest of interesting results.

The brash new approach of the Woods Hole oceanographers included a disregard of a traditional (European) preoccupation with academic credentials.[24] Eventually, their efforts would result in a new type of oceanography, emphasizing eddy motions at scales of a hundred miles over the idealized stately averages mapped on current charts. As Scripps oceanographer Walter Munk has pointed out when reviewing progress in physical oceanography in the second half of the twentieth century, this new emphasis led to a revolution in the understanding of motion in the sea.[25]

It is now realized that by far the bulk of the kinetic energy in the ocean is contained not within currents but within eddies. Such eddies accompany the Gulf Stream in abundance (fig. 6.6). To study these types of motions, we need to go to a much smaller scale than that reflected in the geography of the large-scale currents mapped in *The Oceans*. Synoptic maps derived from satellite remote sensing have opened up entirely new opportunities in this regard.[26] As a result of the greatly increased density of sam-

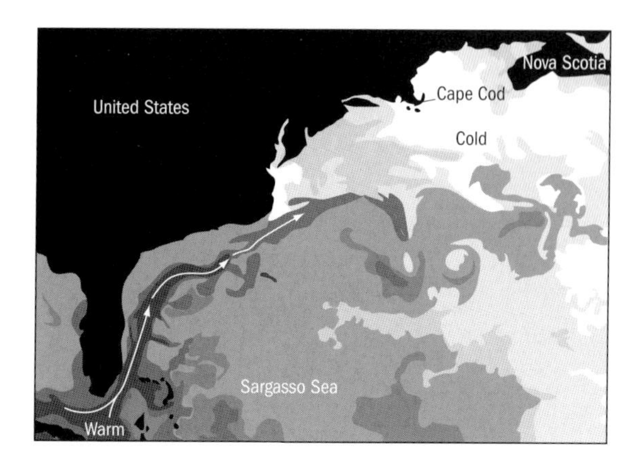

FIGURE 6.6. Gulf Stream eddies south of Cape Cod, seen in temperature distributions, from space. Warm eddies are dark.

pling and observation, within the last two decades or so, there is an overwhelming flood of data. Fortunately, there has been an equivalent rise in the ability to process data, by ever-greater computing power.

The question of just how much the Florida Current (the one between Florida and Cuba) contributes to the Gulf Stream has puzzled many oceanographers since the Scottish naturalist James Croll first raised the issue, in 1875.[27] The matter is taken up in *The Oceans*, as follows:

> If these figures [the estimates for the transport within the Gulf Stream, off Chesapeake Bay and off Woods Hole] are correct, they indicate that between 38 and 57 million m³/sec of Sargasso Sea water and deep water have been added to the Florida–Gulf Stream after the Antilles Current, carrying 12 million m³/sec, joined the flow of 26 million m³/sec through the Straits of Florida.[28]

Sverdrup summarized the patterns of flow in well-designed graphs that became standard textbook items (fig. 6.7).

How did Sverdrup arrive at these various estimates? Both the estimates he put forward, as well as those that he quoted, were based on the type of calculation referred to earlier: a balance between pressure forces and Coriolis force.[29] The concept is to calculate the velocity that provides the balance between the two forces. The pressure force or hydrostatic head (reflected in the topography of the sea surface) can be estimated from the distribution of the density of water masses, which is given by its temperature and salinity. The Coriolis parameter (which

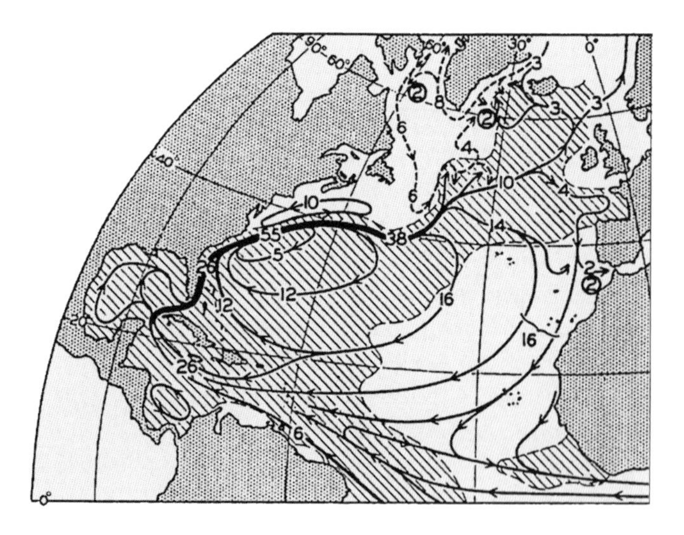

FIGURE 6.7. Volume transport in the Gulf Stream System according to Sverdrup (1942), in millions of cubic meters per second. Note the asymmetry of the central wheel of currents in the subtropical Atlantic.

yields the Coriolis force when multiplied with velocity) is a simple function of the latitude; it captures the whirling associated with the rotation of Earth, for each location considered. The velocity required to achieve the "geostrophic balance" of the opposing forces emerges as the ratio between the pressure force and the Coriolis parameter. Once the motions of the water are mapped in this fashion, of course, mass transport readily emerges. It is important to realize that this mapping explains nothing about the origin of the motions; instead, it describes the average field of motion that is appropriate for the observed distribution of temperature and salinity, given that the Earth rotates.[30]

Sverdrup's vision for the nature of the Gulf Stream anticipated, in its major aspects, the concept of the Gulf Stream as part of a great central wheel of motion in the subtropical Atlantic, with a narrow and intensified flow in the west. In this view, the warm current that brings heat to northwestern Europe is more or less incidental to the basic pattern, rather than the most notable feature of North Atlantic circulation (fig. 6.8).

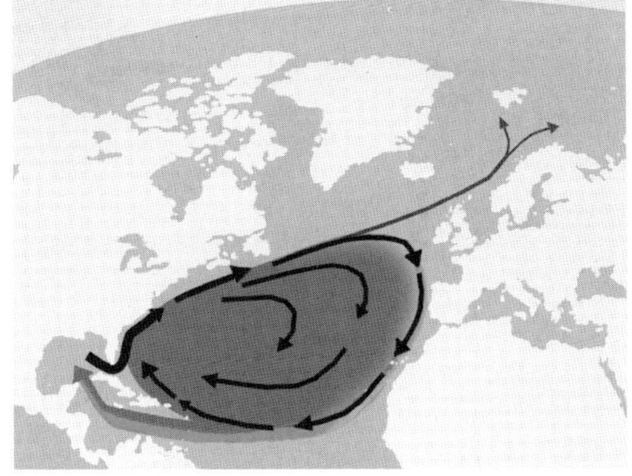

FIGURE 6.8. Two concepts of the Gulf Stream. *Upper:* Gulf Stream as part of the heat supply system for northern Europe (after a contemporary museum exhibit, in Europe). *Lower:* Gulf Stream as part of the great central gyre, with incidental outlier to the north.

It is one thing to calculate average currents, based on the distribution of densities of the water, and another to show that these calculations have some resemblance to the real world. As Sverdrup suspected, reality is much more complicated, and such reality emerges once the motions are sampled in greater detail in space and time. Average motion tends to be entirely masked by short-term fluctuation in velocity at any one location. Eddy motion dominates the field, not steady flow.[31] Two bottles thrown overboard almost simultaneously anywhere in the Gulf Stream will drift apart, follow different paths, and find themselves on different shores after a rather erratic voyage characterized by much eddying on different spatial scales.

Nevertheless, the mighty motions identified by the geostrophic calculations are not just figments of the imagination. The migrations of the common European eel are witness to their existence (fig. 6.9). Many of the larval stages of eels originating in the breeding grounds of the deep Sargasso Sea and destined for European rivers will eventually reach their destination, regardless of all the eddies—and they will do so on a more or less predictable time scale.[32] Statistically, then, and averaged over a sufficient length of time, the currents are real and have been real for more than a million years. And warm water does flow from the Gulf Stream eastward toward Ireland and northward into the Norwegian Sea.[33] And Bergen has flowers, and no glaciers.

The typical velocity of the Florida Current, the fastest portion of the Gulf Stream System, is near 100 centimeters per second (2 knots or 3.6 kilometers per hour). The Gulf Stream off the East Coast is not simply an extension of the Florida Current, of course. Yet, certain aspects of it do conform to a jet stream introduced off

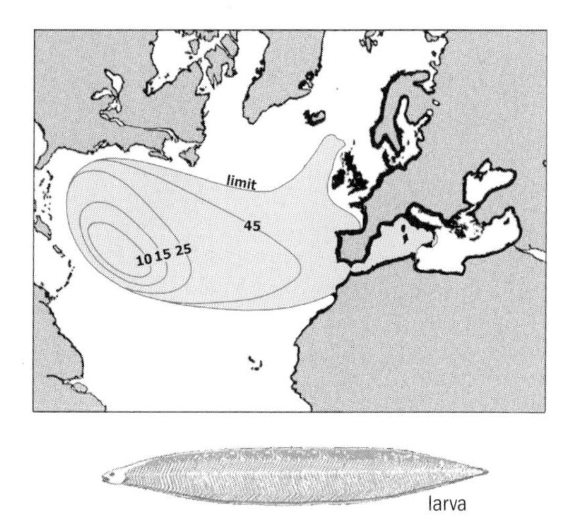

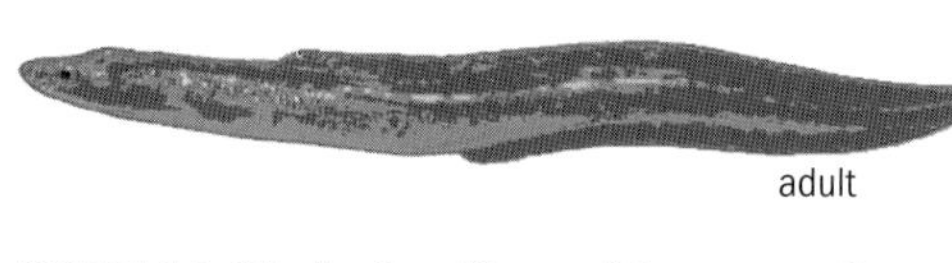

FIGURE 6.9. Distribution of larvae of the common European eel in the central subtropical gyre of the North Atlantic, based on the studies of Johannes Schmidt. Numbers show size of larvae in millimeters. Heavy line along coast shows where elvers invade streams.

Florida and then left to interact with surrounding water, like the jet from a hose at the surface of a swimming pool.[34] As mentioned, the transport off Cape Hatteras is more than twice of that through the Florida Straits. Where does the extra water come from? Sverdrup answers this question as follows:

> After emerging from the Straits of Florida, the current is soon joined by the Antilles Current which, according to Wüst (1924), carries about 12 million m^3/sec.[35]
>
> According to Iselin (1936) the current increases steadily in volume by absorption of Sargasso Sea water, and as it leaves the Blake Plateau both the depth and the volume suddenly increase as a result of the joining in of water of a temperature considerably below 8° [C] which comes from the southwestern Sargasso Sea. . . . A satisfactory theory of the Gulf Stream must . . . account for the increase in volume transport in the direction of flow and the fact that this increase takes place without evidence of strong inflow from the southeast.[36]

Sverdrup's tentative representation of Gulf Stream dynamics well reflects the difficulties encountered in the attempt to understand the nature of this puzzling phenomenon. The attempt to balance the transport within the Gulf Stream off Cape Hatteras by the inflow of tributaries farther to the south well illustrates the underlying concept of a "river in the sea" that had dominated thinking ever since Alexander von Humboldt and Matthew F. Maury considered the matter, early in the nineteenth century.[37] With Sverdrup, we find both vestiges of this thinking and a realization that a new approach is called for.[38]

The solution to the conundrum emerged in Woods Hole, in 1948, a few years after the publication of *The Oceans*. Stommel (1948) showed that the Gulf Stream has to be understood as part and parcel of the entire basinwide circulation. The question about where the Gulf Stream gets its water was misleading to some extent. The important question is why the Gulf Stream is so narrow and reaches such high speed, when the wind field is entirely inadequate to provide an answer. In Stommel's words: "The Gulf Stream is not a river of hot water flowing through the ocean, but a narrow ribbon of high-velocity water acting as a boundary that prevents the warm water on the Sargasso Sea (right-hand) side from overflowing the colder, denser waters on the inshore (left-hand) side."[39]

In this sense, the Gulf Stream is a jet along a front. It does not matter, for this picture, where the water comes from—although it does matter in other contexts, such as the transport of heat, of course, and the distribution of organisms. One implication of Stommel's solution is that one could close the Florida Straits, and the Gulf Stream would hardly notice.

The concept of the Gulf Stream as a boundary phenomenon, a dynamic wall between warm and cold water, is already hinted at in the work of C. O'D. Iselin (1936).[40] Iselin refers to "the incorrect popular belief that the current resembles a warm river flowing through colder seas" and continues, "The fact is, that by far the greatest part of the northeasterly moving water is relatively cold, while even the southeastern side of the current (except at the surface) is not quite as warm as the water at corresponding depths in the Sargasso Sea."[41]

THE GREAT MERRY-GO-ROUND

If the Gulf Stream is not a "river in the sea," what exactly is it? Why is it relatively narrow and fast off the East Coast but has poorly defined source regions (other than the Florida Current)? Why does it break up as soon as it turns eastward?

The short answer is that the Gulf Stream is narrow and fast where it runs across latitudes. The cause for this phenomenon is "western intensification," a process that is active in all major ocean basins. The Gulf Stream is but one example (albeit the most striking one) of a western boundary current of a subtropical (central) gyre.

Stommel's new way of looking at the Gulf Stream was introduced in 1948, in an article entitled "The Westward Intensification of Wind-Driven Ocean Currents." Two years later, Walter Munk generalized the new concept in his paper "On the Wind-Driven Ocean Circulation."[42] Ever since, the underlying scheme is commonly referred to as the Stommel-Munk theory of circulation. The central pillar of the theory is Stommel's discovery that westward intensification of the overall circular current structure in the subtropical realm results from the fact that the Coriolis parameter varies with latitude. A parcel of water moving north off Florida finds itself increasingly in an anticlockwise rotating environment. The reason is that the Earth's rotation makes up an increasing portion of the movement of water layers as the pole is approached. The parcel of water moving north, then, will increasingly exhibit a clockwise rotation relative to the environment. The Gulf Stream as a whole is part of a clockwise rotation (from 8:30 to 11:30 on the basinwide clock face); thus, the motion of water within it benefits from this input of rotary momentum offered by the crossing of latitudes. It runs fast, and it runs just fast enough so that the resulting friction forces (which increase with velocity) become large enough to provide a balance for the overall wind stress, which would keep spinning up the great Sargasso-centered water wheel if unopposed.

Analogous reasoning explains the sluggish and ill-defined southward flow within the great gyre, in its eastern section. The water moving southward within the gyre carries excess anticlockwise rotation into its new environment. This opposes the overall clockwise flow (from 1 to 5 on the basinwide clock face). The result is that the gyre is strongly asymmetrical, the northward flow being represented by a jet in the west (the Gulf Stream), and the southward flow by a generalized broad-scale motion that hardly qualifies as a current. At the center of the gyre, in the Sargasso Sea, the warm water piles up as a consequence of the Coriolis force, which keeps deflecting surface waters toward the right of its intended motion. The gyre center is shifted strongly to the west, relative to the center of the basin. Here the warm-water layer is several hundred meters thick—that is, much thicker than elsewhere in the ocean.

Stommel credits Ray Montgomery for getting him started on investigating the marked east-west asymmetry of the surface circulation in the North Atlantic and elsewhere.[43] He began the work in 1946. The first task was to simplify the ocean to a degree where calculations would not be prohibitively tedious (computers were huge, inefficient, and mostly unavailable). So he built (in his mind and on paper) a flat rectangular ocean representing the North Atlantic basin, with the trade winds dominating the equatorial half of the basin and the prevailing westerlies the poleward half. To the wind stress, which sets water in motion, he added friction, which slows the water, and he made it proportional to velocity. He introduced the condition that the basin is closed (making water move in a closed ring), and added the Coriolis force from the rotation of Earth. Voilà. The western current was narrow and fast in the resulting calculated circulation around the box.

Building on these principles, Munk extended the new approach to the entire ocean, in an attempt to represent the circulation as a necessary consequence of the observed wind field. Such a tour de force had not been attempted previously, and it turned out to be hugely successful. According to the prominent physical oceanographer Albert Defant: "There is no doubt that the Stommel-Munk theory of ocean circulation explains the large-scale geographic picture of the horizontal ocean currents in all oceans as a direct effect of the permanent wind system over these oceans."[44] Munk's formulation differed from that of Stommel mainly in the introduction of a more sophisticated term for friction, emphasizing the lateral dissipation of eddy energy. He used actual wind data to show that the large-scale patterns of clockwise and anticlockwise "gyres" (Munk's term) seen in the ocean are the result of zonal winds.[45]

Munk's calculation not only recreated the Gulf Stream as the western boundary current of a great gyre turned by zonal winds, but also showed that subpolar gyres should develop north of the subtropical ones, turning in the other sense. Such gyres are indeed seen in the subarctic realm (fig. 6.4). In addition, Munk's scheme explained the Equatorial Counter Current, as a balancing flow resulting from the pile-up of water in the western equatorial region, by the trade winds. The pattern calculated is compatible with the pattern observed.

Thus, to a first approximation, all the major currents in the world can be assigned to the flow patterns expected from Munk's model. There are some complications, of course. One is that the calculated transport tends to be off for various reasons. For the Gulf Stream, it is too low, presumably because the Gulf Stream also transports water that comes across the equator, and ends up making deep water in the north. Also, at the eastern boundaries, observations show stronger currents than expected. This has to do with an eastward intensification of the wind field, which results from a buildup of high atmospheric pressure over the cold side of the subtropical belt within each ocean basin. The cold side is on the east, because there surface waters move from high to low latitudes and because the Coriolis-driven Ekman drift moves surface water offshore, admitting the rise of cold water from the thermocline layer, along the shores.

But the important thing is that with the Stommel-Munk approach to the general circulation the problem of westward intensification was essentially solved, and thereby also the main aspect of the Gulf Stream conundrum. Experimental confirmation came from an ingenious laboratory model built by Woods Hole oceanographer von Arx.[46] He used a rotating pan to simulate an ocean basin, which he made bowl-shaped to compensate for unwanted centrifugal forces. A still layer of water on the whirling pan simulates the ocean, with the North Pole at the center. Air hoses are positioned to blow winds over the surface, and the motion of the water is traced out by colored filaments emanating from dissolving permanganate crystals. Munk was pleased with the results of the experiment:

> Von Arx's model faithfully reproduces the gyres of the North Atlantic and the North Pacific, including the intense western boundary currents. The model is especially interesting because topography and winds can be varied to show possible circulations of the oceans in the past, when conditions were different; for instance, one can investigate how the Gulf Stream might have behaved at a time when there was a separation between North and South America in the place of the present Isthmus of Panama.[47]

Analogous experiments are now being performed by powerful computers, and there is no problem in varying geography and winds to study the possible effects of changes on the ocean's circulation. As might be expected, when a gap is open between North and South America (as existed before 5 million years ago), much warm water moves out of the Caribbean into the Pacific, warming the shores of Middle America. At present, the warmth of the upper Gulf Stream owes much to the contribution from the Florida Current, which in turn depends on receiving warm water from the Caribbean. Thus, the closing of the isthmus has greatly increased northward warm-water transport. In fact, when the isthmus started to close (roughly 5 million years ago), the climate of the North Atlantic became unusually warm for the next 2 million years,

FIGURE 6.10. Common paths of tropical storms according to the insurance company Munich Re (1998). Light shades of gray indicate high frequency of strong storms. Tropical storms feed on vapor from a warm sea surface. The Gulf Stream provides a warm-water path. *Inset:* Hurricane as seen from space.

first reversing initial ice buildup in Greenland, and then preventing such buildup till about 3 million years ago, when global cooling trends reasserted themselves.[48]

Through the closure of the Panama Strait, then, the Caribbean Sea became an enormous trap for warm water, collected by the easterly trade winds blowing across the tropical Atlantic, and piled up behind the Antilles. This warm pool fuels the hurricanes that wreak havoc in the Caribbean realm and in the Gulf of Mexico and, by supplying heat to the Gulf Stream, helps fuel storms that move up along the southern East Coast.[49] For obvious reasons, the paths, strength, and frequency of tropical storms are of vital interest to the insurance industry (fig. 6.10).

On the positive side, the warm pool in the Caribbean provides the ideal habitat for coral reefs. The piling up of warm water in this basin is greatly aided by the fact that the convergence zone for warm and moist tropical air is well north of the equator. During glacial periods, that was not necessarily so, and coral growth could then come under stress, with loss of the more sensitive species (at least early on, when northern ice first started to grow). Once the more sensitive species were gone, presumably the system stabilized at lower diversity, being maintained by the resilient forms.

So, after all that has been learned, does the Gulf Stream warm Europe or not? It does indeed, but not just by moving tropical waters north within the North Atlantic basin and sending some of it off toward Scotland and Norway. It warms the entire northern North Atlantic because it serves as a conduit of warm water diverted from south of the equator into the temperate latitudes of the northern basin. This heat transfer is a central issue when contemplating the circulation of the Atlantic.[50] There is a reason why the Caribbean has hurricanes, and why there are none in the tropical South Atlantic. Heat piles up in the North at the expense of the South, and heat produces storms. Some of the heat pirated by the North Atlantic comes from as far as the Indian Ocean, around the Cape of Good Hope (fig. 6.11).

WINDS GOOD AND BAD

As we have seen, the great patterns of circulation on the surface of the ocean result from the

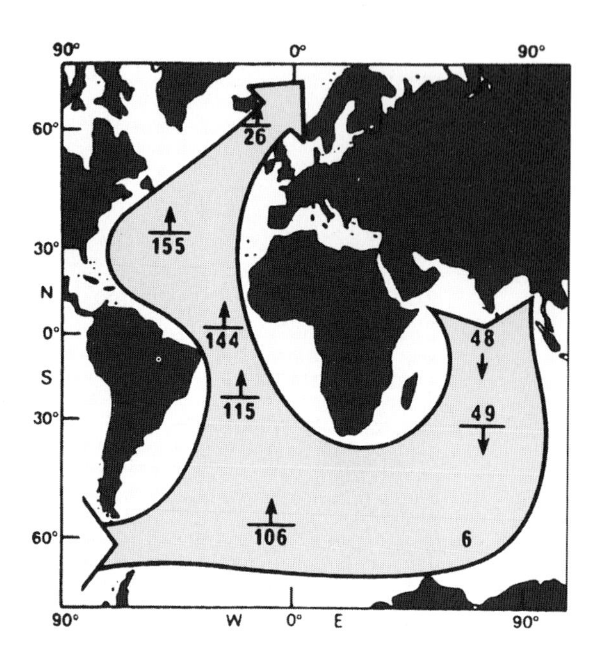

FIGURE 6.11. Heat piracy of the North Atlantic, as envisaged by Henry Stommel (in 1980). Shown is the heat transport across latitudes in deca-trillion watts (10^{13} W).

great patterns of the atmospheric circulation, that is, the winds. This much was intuitively obvious to the sailors and explorers of earlier centuries, and very much to Lieutenant M. F. Maury, who mapped winds and currents. But the question remained why some of the currents, notably the Gulf Stream, apparently do not depend on the wind. Maury thought, rightly, that the diurnal rotation of the Earth somehow was involved in this conundrum.

The enigma was solved only by appropriating the mathematical methods developed in hydrodynamics and meteorology, methods that capture the response of water and air to pressure gradients on a rotating planet. A new kind of physical oceanography thus emerged, as exemplified by the discoveries of Walfrid Ekman, Bjørn Helland-Hansen, Carl-Gustaf Rossby, Harald Sverdrup, Henry Stommel, and Walter Munk. With Munk's demonstration that the wind field can explain practically all the major features of the surface circulation, the central problem of the origin of surface currents was resolved.

At the same time, the manner in which the fickle winds impart their energy to the slow-moving ocean became a question of some importance. The problem has been attacked since the nineteenth century, by direct observation (comparing wind speed with current speed), by experiment (blowing air over a surface in a pond), and theoretically (calculating the stress exerted by moving air on a water surface). The processes involved are rather complex, involving feedbacks between the rippled water surface and a turbulent air layer above this surface.

On a planetary scale, the interaction between ocean and air stabilizes the wind fields through heat exchange. When winds blow over the sea surface, they load up with water by evaporation, cooling the sea and carrying away the corresponding energy as latent heat in the shape of water vapor. Upon condensation of the vapor into clouds, the latent heat becomes available to raise the temperature of the air. Warm air rises, and this attracts more vapor-laden air from the surroundings. The resulting winds bring warm surface waters to the regions of high precipitation, which tends to make the pattern permanent. It is not by chance that Bergen, the city most benefiting from the great warm anomaly of the northern North Atlantic, is also known for long-enduring rain and drizzle.[51]

Warming the air through condensation of water vapor produces convection as warm air rises. In turn, such convection is part of a global

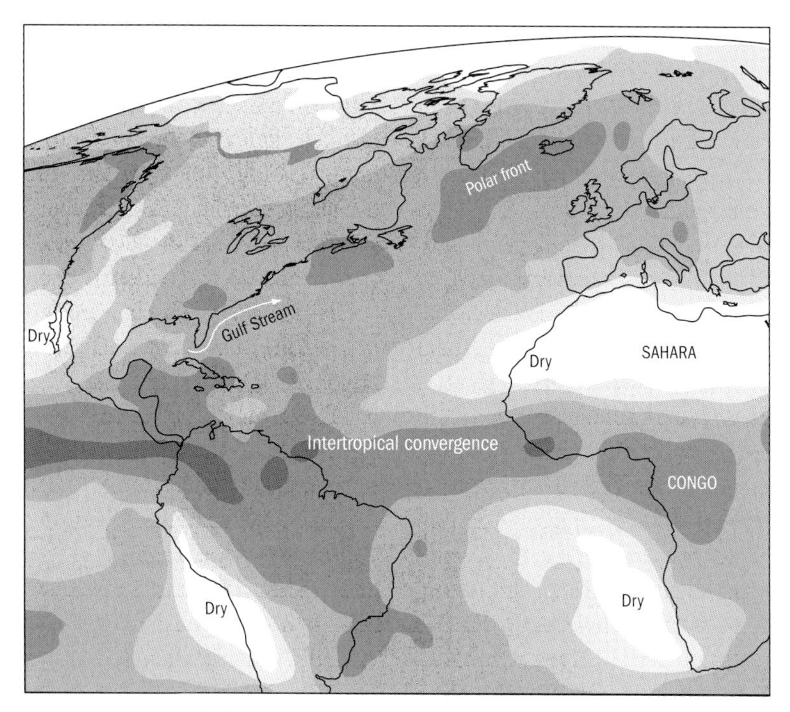

FIGURE 6.12. High (dark) and low (white) precipitation regions in and around the North Atlantic (after D. R. Legates). Note that anomalously warm regions in the sea (see fig. 6.03) are wet, and cold regions are dry (D).

atmospheric circulation and helps to drive the winds. Thus, the motions of atmosphere and ocean are intimately coupled. For this reason, it is exceedingly difficult to decide whether the ocean or the atmosphere is more important in redistributing heat on the planet. The experts are still arguing about this, with estimates of the ocean's contribution ranging from 50 percent to 20 percent![52]

As mentioned, the North Atlantic is engaged in heat piracy, importing heat brought with warm water diverted north at the bow-shaped coast of Brazil, water that moves into the Caribbean. But this is not the only way heat comes north. The atmospheric circulation is involved also and is indeed crucial in this piracy. Warm moisture-laden winds cross the equator from south to north, especially during summer, to dump their load of water some 10 to 15° N in the Intertropical Convergence Zone (fig. 6.12). Nor is the North Atlantic alone in stealing heat from the south. The entire Northern Hemisphere does so. Thus, there is a profound asymmetry between the Northern and the Southern hemispheres, with cherries in Japan and wine in France, but with glaciers coming almost to sea level in New Zealand, at the same latitude. The reason is that the heat equator is displaced well to the north of the geographic equator. In turn, this displacement is a result of the distribution of land and sea on the planet.

Land surfaces are concentrated in the Northern Hemisphere. In summer, they heat up quickly, sending warm air aloft and pulling air masses in at low elevations, in part from the south. In the Southern Hemisphere, one of the large landmasses sits precisely on the South Pole, hosting an enormous ice cap. The effect of this ice shield, surrounded by fierce westerly winds and a correspondingly strong circumpolar current, is to shift the zonal wind systems equatorward.[53] The pull of the monsoon in the north, and the push by the strong zonal winds in the south, collaborate to move the heat equator northward.

The fact that the heat equator is displaced northward of the geographic equator implies that the trade winds (the tropical drive of the great gyres) are similarly displaced, with the southern trades blowing close to the equator and even across it. Together, these winds pile warm water into the western tropical regions of Atlantic and Pacific. As it happens, the recipient regions—the Caribbean Sea in the Atlantic and the warm-water pool east of Indonesia and around New Guinea—also favor an accumulation of these waters north of the equator, by the geographic situation of the corresponding warm-water traps.

The trade winds are the most conspicuous part of the wind system on the planet, and the most steady ("the wind blows trade," that is, on track). In 1492, Christopher Columbus greatly profited from their reliability; they carried his three small sailing vessels across the Atlantic at its widest, from the Canary Islands to the Bahamas, in 36 days. Centuries later, the Norwegian seafarer and amateur archeologist Thor Heyerdahl showed (in 1970) that the same trade winds are capable of blowing a sailing vessel built of reed (made to resemble Egyptian craft) from Morocco to the Caribbean in less than two months.[54]

The trade winds are the most conspicuous expression of the low-latitude atmospheric circulation, which starts with rising air in the tropics. This rise is driven by the energy received from the Sun, which is virtually overhead all through the year at the equator. The updraft of air implies convergence; that is, the air rising from above the sea surface is replaced by winds blowing from higher latitudes. In turn, these winds originated in a belt of falling air, the subtropical zone of high pressure, clear skies, and deserts. At high altitudes, the rising air in the tropics moves poleward to replenish the falling air in the subtropics. In this fashion, we get a closed circulation, a "cell." This general mechanism for making trade winds was first postulated by the famous astronomer and atmospheric physicist Edmund Halley (1656–1742), in 1686.

Halley had the basic mechanism right but could not explain why the trade winds blow from an easterly direction. This puzzle was solved a generation later by the English meteorologist George Hadley (1685–1768), and this is why we talk about a *Hadley cell* rather than a *Halley cell*, which would be equally appropriate. Hadley realized that wind particles moving toward the equator would come from a region of lower eastward velocity and enter a region of higher eastward velocity. The ocean surface, as it were, would run away under the wind, toward the east. Thus, the wind would have a westward motion, as observed. Hadley's explanation clearly invoked what we would now call the Coriolis effect. He published his theory in a famous paper, "Concerning the Cause of the General Trade Winds," in 1735. This was exactly 100 years *before* Coriolis produced the equations describing motions in a rotating coordinate system.

Hadley had the principles right, and in some ways his treatment of the trade-wind problem was the founding act for the science of meteorology. However, the picture painted by Hadley also had a major flaw: he missed the positive feedback on the air circulation that arises from the fact that latent heat is brought toward the equator, by the converging trade winds. It is not just the (somewhat) greater solar heating at the equator that drives the cell. It is mainly this gathering of heat from the sea surface well north and south of the equator, into the Intertropical Convergence Zone, that makes for the vigorous convection. And because of the easterly component of the trades, this convection is most vigorous in the western ocean regions. Rapidly rising air and tropical rainfall at the western end of the tropical zone in the two major ocean basins, then, cannot be explained from regional effects, but only as part of a larger system, just as is true for the Gulf Stream.

Hadley's picture was reasonably realistic as concerns the trades and the doldrums. It had to be modified considerably, however, to account for the other global wind belts (fig. 6.13). The most important air currents, next to the trades, are the westerlies, with their complicated pattern of high and low pressure centers following each other, and their dependence on the changing position of the westerly jet stream along the polar front. The "Bergen School" around Vilhelm

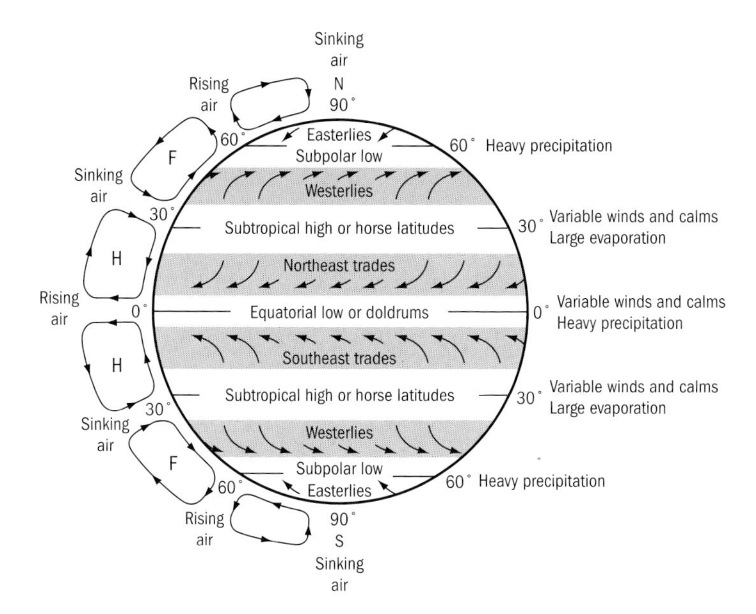

FIGURE 6.13. A traditional representation of the global wind system, emphasizing symmetry about the equator: Hadley cells (H), Ferrel cells (F), trades, and westerlies. (After Fleming, 1957.) Reality is more complex.

Bjerknes and his son Jacob, and the meteorologist-oceanographer Carl-Gustaf Rossby, worked out the elements of this system from the 1920s to the 1940s, thus founding modern meteorology.

To understand the westerlies—which together with the trade winds drive the great central gyres—we have to consider the so-called Ferrel cell. This cell connects the sinking air in the arid zone to the upward motion of warm air riding up the cold air masses along the polar front. The American meteorologist William Ferrel (1817–1891) showed how the tendency of winds to move in circles on a rotating planet gives rise to cyclones that pull the air in from the warmer regions, toward the polar front, thus driving the west winds. The mechanism is somewhat analogous to making the trades (in that latent heat drives the convection that sucks in the air to make winds). But in this case heat is moved poleward, and consequently the motion of the wind is eastward, owing to the Coriolis force.[55]

As Ferrel realized, air that is pulled in toward a low pressure center (where the air rises, cools, and loses water) moves along a spiral path, a cyclone. The westerlies can be understood as being powered by low-pressure centers forming along the polar front. Two of these are both large and quasi-stationary: the one known as the Iceland Low, centered on Iceland in the North Atlantic, and the Aleutian Low centered on the Aleutian Islands in the North Pacific. The westerlies carry along smaller cyclonic eddies—storms—that are largely spawned by these centers. The overall pattern shows winds moving westward around the North Pole, in a wave pattern, with cyclonic motion (that is, counterclockwise motion) around the Icelandic and Aleutian low-pressure centers. The eddies moving with the westerlies bring rain (cyclones) or sunshine (anticyclones) and are marked as the familiar moving L and H patterns on weather maps. The L centers show where warm air has penetrated poleward into colder air masses. The moist air being sucked in brings heat. Thus, rain makes the climate mild (as in western Norway). The H centers show where tongues of colder air are moving toward low latitude, bringing clear skies.

Trade winds and westerlies are the zonal winds that dominate the wind field of the troposphere, that is, the well-mixed layer of the lower atmosphere that we look down on when traveling long distance in a jet airplane. Another major wind pattern is the one produced by monsoon effects. Monsoon winds feed on temperature differences between large land surfaces and the adjacent ocean. Early during summer, when the Sun's position moves high into the sky and both ocean and land receive much sunshine, the land heats up much faster than the sea. One reason is the great heat capacity of water: it takes five times more energy to warm a given mass of water by one degree than an equal mass of rock.

Another reason for the unequal heating is less obvious and yet more important. The surface of the ocean, as far as diurnal heating is concerned, is much thicker than that of the land. The sunlight readily penetrates the ocean surface, and light is absorbed within a layer more than 10 meters thick, while the thickness of the soil layer that is warmed by the Sun (by downward conduction of surface heat) is more like 10 *centimeters*. The land surface, then, is effectively more than 100 times thinner than the sea surface. Depending on the wetness of the soil, the land surface is between 100 and 500 times less resistant to temperature change than the sea surface, given a certain solar radiation input. Now it is clear why diurnal temperature ranges in the deserts of California can readily go over 30 °C, while they stay small along the coast.

Monsoon winds are giant sea and land breezes. Summer is analogous to daytime: the land heats up quickly as noon is approached, air warms and rises, and winds bring moist air from the sea, making clouds. Winter corresponds to nighttime. Summer monsoons bring moisture and rainfall, while winter monsoons tend to bring drought. The strongest monsoons are in the tropics and are associated with large landmasses. The seasonal winds within the Indian monsoon system extend over East Africa, Arabia, Pakistan, India, and Bangladesh, and the entire Arabian Sea. A large number of people are affected by the precipitation patterns associated with these winds.[56]

The wind fields are changing with global warming, but the patterns of change are as yet obscure and a subject of much discussion among the experts. From first principles, one can say this much: The westerlies live off the temperature gradient between subtropics and polar regions. In the Northern Hemisphere, as this temperature gradient weakens from warming in the Arctic, we should expect that the westerlies will weaken along their present path and the monsoon centers will move farther north. At the same time, the greater availability of water vapor in a warmer atmosphere should favor the development of stronger storms. On the whole, then, warming could well result in a less zonal (and hence less predictable) wind field in the north, with higher extremes.

VARIABILITY IN SPACE AND TIME

As mentioned, somewhere south of Newfoundland there is an enormous eddy factory, where much of the kinetic energy of the Gulf Stream is fed into meanders, eddies, and vortices of all shapes and sizes, marking the transition to the North Atlantic Current (fig. 6.6). The extent of this conversion of flow type from jetlike to eddy dominated was realized in the years after World War II, especially through the mapping of quasi-instantaneous temperature fields obtained from bathythermographs. The first multiple-ship survey taking advantage of this technology was the Operation CABOT.[57]

With some notable exceptions resulting from detailed surveys in limited regions,[58] the importance of eddy motions and variability of currents was not generally appreciated when discussing currents in the sea, before Operation Cabot. Munk calls this shift in perception the mesoscale evolution in physical oceanography. He writes:

> We now know that more than 99 percent of the kinetic energy of ocean currents is associated with variable currents, the so-called mesoscale of roughly 100 km and 100 days. Incredible as it may seem, for one hundred years this dominant component of ocean circulation had slipped through the coarse grid of traditional sampling. Our concept of ocean currents has changed from something like 10 ± 1 cm/s to 1 ± 10 cm/s.[59]

Munk credits F. C. Fuglister and the British physical oceanographer John C. Swallow, in their work on the Gulf Stream, with demonstrating

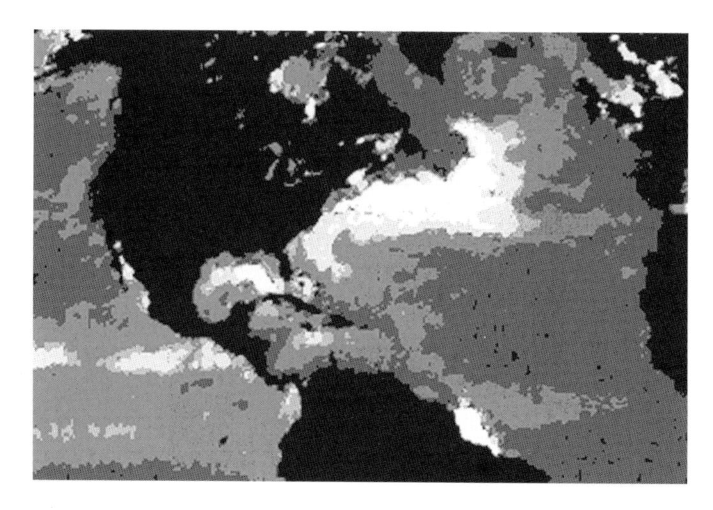

FIGURE 6.14. Variability of the sea surface topography, North Atlantic, from TOPEX satellite data (1992–1998). White: most variable.

that variability in space and time was important and intense near boundary currents.[60] He continues: "We now think of this mesoscale variability as the ocean weather and the underlying circulation as the ocean climate. . . . Climate came first, weather later (rather the opposite of what happened in meteorology."[61]

Greatly increased computing power now allows the construction of circulation models that are able to resolve the mesoscale variability. Munk recalls the "excitement of seeing B. Holland's first spontaneously unsteady wind-driven circulation model."[62] Mesoscale and small-scale processes are represented in increasingly realistic ways, as computing evolves. An equally important technical development has been the gathering of information by satellite surveys. No longer are the subtle changes of sea-surface topography calculated from the density distributions found at scattered stations in the ocean. The surface topography is now observed directly by high-resolution radar—both the overall pattern and the pattern of variability emerge in striking clarity from such surveys.

It turns out, from such surveys, that one of the most variable regions (judging from variability of sea surface topography) is the transition area between the Gulf Stream and the North Atlantic Current (fig. 6.14), precisely where Fuglister found the eddy-generating machine during Operation CABOT. Similarly variable regions (that is, eddy-generating places) occur in the analogous position at the end of the Kuroshio off Japan, at the end of the Brazil Current off Argentina, and in the turbulent region of the Agulhas Retroflection, south of the Cape of Good Hope. The eddy fields typically have a variability of up to a foot of sea level height difference. Stable areas (the eastern portions of the great central gyres) stay within 5 cm or less of the mean.[63]

Variability occurs on all sorts of time scales, from lunar tides to eddies, from seasons to decadal climate oscillations, and from centuries to thousands of years. The Gulf Stream, a creature of the wind, must respond to all of these, although its size and inertia swallow the shorter fluctuations and average their effects. Varying supply of heat to the Norwegian Sea, as a result of varying strength of ocean currents, was of special interest to the geophysicists in Norway, of course. Long before anyone else, they investigated the connection between varying sea-surface temperatures and atmospheric response.[64]

One time scale that is of special interest to humans is that of decades. Changes in the typical weather on this scale affect our economies and our well-being. Prolonged drought, a series of wet summers, a series of harsh winters bringing nasty storms or icy cold, a shortened or lengthened growing season—these are things that have considerable impact. We now have a distinct warming in middle to high northern latitudes for the last 25 years or so. We would like to know to what degree this warming might reflect natural variability, or if it is mainly man-made, and how this change might itself influence variability.

To address such questions we must turn to the long-term variations in climate or, more precisely, to interdecadal fluctuations. (*Long-term* means different things to meteorologists and geologists.)

In the North Atlantic, the leading cause for fluctuations on a decadal scale is the North Atlantic Oscillation, which is also called Arctic Oscillation, because its effects reach deep into northern Siberia. The oscillation consists of a more or less regular (but mainly aperiodic) variation in the difference in air pressure at sea level off Iberia and in Iceland. When the difference is large, the Iceland Low is strong and so is the Azores High. This is called the positive phase. When the difference is small, both the Iceland Low and the Azores High are weak. This is called the negative phase.[65]

We can think of the positive phase as keeping Norway warm, certainly a positive effect for the citizens of Bergen, including the meteorologists and oceanographers there. A strong Iceland Low pulls in moisture-laden air from the south, warmed by the Gulf Stream system, and ready to dump rain (and heat) in western Norway. The corresponding winds, roughly following the lines of equal pressure, are southwesterly east of Iceland and drive warm water into the Norwegian Sea and up toward the islands surrounding Spitsbergen (Svalbard archipelago). Westerly winds are generally strong north of the central gyre, and the Gulf Stream runs strong also. A strong Azores High helps drive strong trade winds, which help turn the gyre (and empower the Gulf Stream). Upwelling of cold water off Iberia and northwestern Africa delivers nutrients to the sea surface; algae grow and feed fishes, and fishing is good.

Whenever the Iceland Low weakens, and the Azores High, winters become cold in western Norway, because the warm air and the warm current are no longer driven northward with vigor. The belt of maximum rain moves south and benefits Iberia. Trade winds, westerlies, and the Gulf Stream all weaken. The cold Labrador Current also weakens in this phase, so we might expect fewer icebergs off Newfoundland. For northwestern Europe, on the whole, dry cold winters are in store. Upwelling along Iberia and off northwestern Africa weakens and local fisheries suffer. Thus the negative phase of the North Atlantic Oscillation.

The exploration of the North Atlantic Oscillation goes back only some two decades,[66] but it has roots that reach back into the 1930s. The oscillation is part of a complicated pattern of "teleconnections" of air-pressure variations over the entire Northern Hemisphere, a pattern that was systematically explored by the American meteorologist Jerome Namias, first at the U.S. Weather Bureau, and subsequently (from 1972) at Scripps Institution of Oceanography.[67] The changing range of contrast of air pressures between the subarctic and the subtropics mapped by Namias (1981) is nowhere greater than in the Atlantic.[68] The fundamental reason is the access of warm water (from the Gulf Stream system) to very high latitudes, where the input of heat can stabilize a very strong low-pressure region (the Iceland Low). Not only does the Gulf Stream send heat to northwestern Europe—it becomes part of an important oscillator because it does so unreliably. The Iceland Low, the centerpiece of the Nordic heat pump that produces the biggest warm anomaly in the Northern Hemisphere, depends on this heat for food. Its efficiency varies through the decades, and this produces the North Atlantic Oscillation in long-term weather patterns in western Europe and (to a lesser degree) all over the Northern Hemisphere.

But why does the Iceland Low engage in a seesaw with the Azores High? To guess at this, we have to try to understand how an oscillator works. It represents a back-and-forth between two quasi-stable states, which differ significantly from each other. Each of these, when established, tends to persist. However, in addition, each quasi-stable state accumulates the seeds of its own destruction, within the appropriate period. It is not necessary to assume fundamental changes in the way the climate system works each time we observe a mode switch. The oscillation is itself an expression of how the climate system works.[69]

The central problem is to find the seeds of destruction for each state, along with the timescale involved. It is fair to say that this search for the causes for climate oscillations is still in its beginnings, and that it is unlikely that simple answers will soon emerge. The reason is that we are dealing here with complicated interactions between atmosphere and ocean, involving (from top down) stratospheric jets and the polar vortex, tropospheric fronts and storm activity, rainfall and snow cover, sea-ice formation and transport, transport of freshwater and heat in surface currents, and changing locations and intensity of deepwater formation. With so many processes intertwined, simple explanations are bound to be wrong or seriously incomplete and can at best only serve as an illustration of the type of feedback that should be considered.

To illustrate, let us assume that a decrease in deep convection in the Nordic Sea (between Greenland, Iceland, and Norway) would weaken the "pulling" of warm waters from the south and, hence, cut down on the supply of warm moist winds to Iceland, thus starving the Iceland Low. In this scenario, to get from a positive to a negative phase, we need to put low-salinity water on top of the Nordic Sea, which provides for stable stratification. Can a strong Icelandic Low, running the North Atlantic Oscillation in positive phase, result in an increased influx of freshwater, after some years? Perhaps. A strong Norwegian Current pushes water deep into the Arctic Sea, which will respond by delivering low-salinity water laden with sea ice to the coastal ocean off Greenland, moving south and then west, mingling with low salinity water brought south by the Labrador Current. After some time, the entire region north of the North Atlantic Current (within the subpolar gyre) will acquire low salinity in surface waters. At that point, the high-salinity water delivered by the Gulf Stream and driven eastward by westerly winds will be diluted by low-salinity waters entering the system from the north. As the warm water now arrives with a lesser salinity in the Nordic Sea, it has some trouble sinking, and thus the pull off Norway weakens and stays weak until the low-salinity anomaly in the subpolar gyre is dissipated.

In support of such a scenario, we might note two things. One is that the time scale of such a lagged negative feedback (or "seeds of destruction") is roughly of the right order. Bringing waters from the Arctic Sea to the region off Newfoundland should take several years, especially since the delivery of low-salinity water needs to go on long enough to build up a sizable anomaly. Dissipation of the salinity anomaly should take another few years. Thus, a 5- to 10-year cycle (as observed) would be a plausible consequence of such a series of events. The other piece of evidence is the observation of the Great Salinity Anomaly, an extensive freshening of the upper 500- to 800-meter layer of the far northern North Atlantic Ocean, which moved from its origins north of Iceland in the 1960s around the subpolar gyre and back into the Greenland Sea in the 1980s.[70] The key factor in generating the Great Salinity Anomaly apparently was the fact that the Iceland Low was unusually strong in the 1960s.[71]

Regardless of our level of understanding of how the climate system works in the northern North Atlantic (or anywhere else), the Great Geophysical Experiment is proceeding: we humans are warming the planet. One of the possible consequences, emphasized by Wallace S. Broecker of Lamont[72] is that convection in the Nordic Sea will be negatively affected, leading into a permanent shift toward a negative phase of the North Atlantic Oscillation. So far, the northern warming has been associated with the opposite effect: the overall NAO has become unusually positive in the last 25 years.[73] The rainfall in Bergen has increased. Bring your umbrella and leave your Arctic gear at home when visiting this wonderful place.

CURRENTS OF THE ABYSS

The transport of warm water into the northern North Atlantic (whether as "Gulf Stream extension" or some other means) and the return of much of this water as dense, cold water at depth is a crucial ingredient in the attempt to explain why northern Europe is so warm. In principle, warm water moves north, gives off heat and sinks as cold water, and returns south, making room for more surface water to move north. The basic process has been understood for some time. It was elaborated on by Fridtjof Nansen, in the early twentieth century, among others. In detail, things are complex because deepwater production may shift location, may occur in pulses, and may depend not just on cooling but on assistance from the production of brines during the formation of sea ice.

The first step in sorting out the deep circulation of the ocean is to ask, simply: Why is the bulk of the ocean at temperatures close to freezing? And why are deep waters, on the whole, slightly fresher than waters at the surface? The answer is that the deep oceans are constantly being refilled from high latitudes, with very cold surface waters. These are slightly less saline than average, because on the whole, freshwater transport by the atmosphere is from warm (evaporation) to cold regions (precipitation). The overall sense of the deep-water circulation prescribes sinking in high latitudes and a corresponding rise everywhere else in the ocean. The rate of this rise depends on the mixing of deepwater layers with overlying water. The mixing decreases density and provides opportunity, thereby, for new dense water from high latitudes to make a new layer of dense water at the bottom.

The underlying principle was recognized by Benjamin Thompson (later Count Rumford), an eighteenth-century scientist of renown. (Also, Thompson had an interesting political career.[74]) He thought of the deep circulation as a kind of global convection pattern: water cooled at high latitudes must sink and flow equatorward at depth under warmer surface waters, which must therefore drift poleward. Rumford's hypothesis was confirmed when it was realized that the deep ocean is very cold everywhere, a fact that became indisputable with the data from the *Challenger* Expedition.[75]

Ultimately, the deep cold ocean is a result of the cooling of Earth's polar regions, which started 40 million years ago. This cooling trend (a result of continents moving into high latitudes and the rise of mountains) eventually produced an icing over of the poles, with an ice cap in the south and sea ice in the north. The boundary zone between the warm surface waters of the sea and the cold waters at depth approached the sunlit zone sometime in the early to middle Miocene, around 20 million years ago, judging from a spurt in plankton evolution. If so, the rate of supply of cold waters to the deep ocean increased over millions of years, reaching a critical level at that time.

What keeps the boundary between cold and warm water in place? This question is difficult to answer. In principle, the boundary zone (that is, the thermocline) represents a dynamic equilibrium between the downward flux of heat from the Sun's radiation, and the upward flux of cold water, which is forced by the continuous supply of cold water at depth. The concept of polar cold-water sources trying to fill the entire ocean, only prevented to do so at the top, where solar warming and wind-mixing take over, is useful for visualizing the general pattern.[76] However, the actual deep circulation is rather more complicated than this simple concept would suggest, with thermocline waters being brought in through large-scale horizontal motion, as well.[77]

The supply of cold abyssal waters depends heavily on just two major source areas: one in the Weddell Sea off Antarctica, the other off Norway and Greenland. The cold water sinking in the northern North Atlantic spreads southward, riding over the even colder water from the Weddell Sea, and eventually feeds into the Circumpolar Current, mixing with other water masses therein (fig. 6.15).

Not all of the densest water is made in high latitudes, however. The other place in the Atlantic where a considerable amount of unusually dense water is made is the Mediterranean Sea. Here strong evaporation combines with seasonal cooling to make surface waters dense, so they can sink and fill the basin.[78] The water of the deep Mediterranean is relatively warm (at 13 °C) but very salty (at >38 per mil, or more than 3.8 percent salt). It is denser than the water outside of the Mediterranean and flows out over the sill at the Straits of Gibraltar. The outflow rapidly runs toward the bottom of the Atlantic but is unable to reach it because of mixing with the surrounding waters. This mixture contributes to the intermediate and upper deep waters at around 1,500-meter depths in the North Atlantic. Scripps oceanographer Joseph Reid [79] has traced the path of the Mediterranean influx and suggested that it contributes to the making of North Atlantic Deep Water, by delivering a high-salinity component into the mix that sinks in the Norwegian Sea.[80]

In summary, three major deepwater sources fill the deep Atlantic: Antarctic Bottom Water, North Atlantic Deep Water, and Mediterranean Water. In addition, there is Antarctic Intermediate Water (to about 20° N) and Subarctic Intermediate Water, at depths above 1,500 meters and underlying the warm-water gyres. The stratifica-

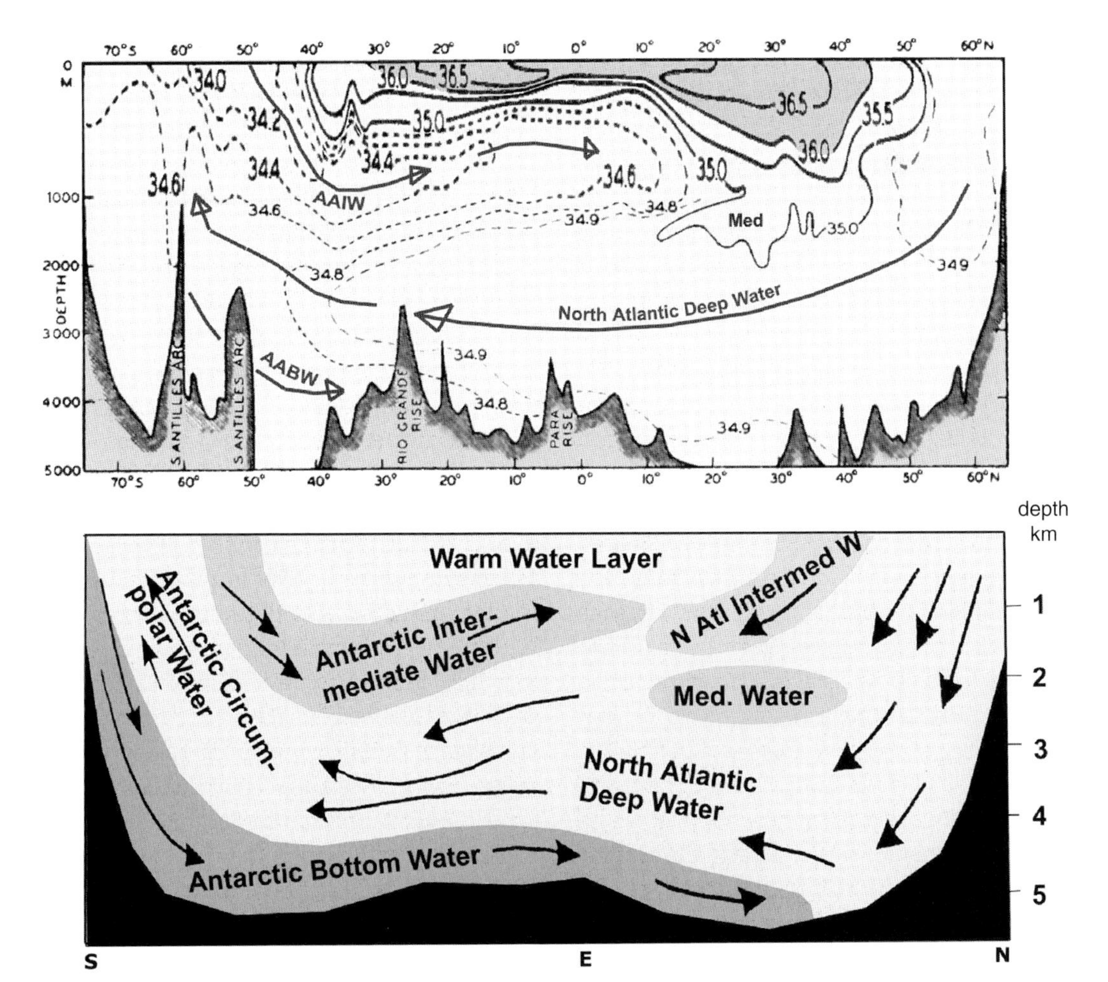

FIGURE 6.15. Stratification of the deep Atlantic, reflecting the various deepwater sources. *Upper:* Stratification as seen in salinity, according to Georg Wüst, as given by Sverdrup et al. (1942). Arrows (here added) denote flow direction according to "core method" of Wüst. Labels (here added): AABW, Antarctic Bottom Water, AAIW, Antarctic Intermediate Water. High salinity (>36 per mil): dark shade. Note asymmetric distribution of highly saline surface waters. Note change of scale at 1,000 meters (factor of two, here redrawn). *Lower:* modern textbook-type representation of water layers in the Atlantic.

tion of water layers and their motions is quite complicated and has by no means been sorted out, with modern estimates for northward and southward transport still varying considerably.[81]

The first oceanographer to make a systematic effort to map the major layers of the Atlantic and their motions was the German oceanographer Georg Wüst,[82] who worked up the unique data set of the *Meteor* Expedition (1925–1927).[83] To obtain a sense of the direction of motion of the various water bodies at depth, Wüst looked for extremes in the properties of the water (temperature, salinity, oxygen) and determined the direction of flow by noting in which direction the extreme in question decays the least. The most consistently satisfying pattern emerged from salinity distributions. By finding the maximum salinity at any one level in the deep water of the Atlantic, the southward path of the North Atlantic Deep Water can be discovered. By finding the minimum salinity down to 2,000 meters, the northward path of the Antarctic Intermediate Water is revealed.

Wüst's profiles soon became iconic illustrations, especially after having been used by Sverdrup in *The Oceans*. Besides providing a picture of the architecture of the deep sea, Wüst made an important discovery concerning the dynamics of deep flow. When he examined in detail the patterns emerging from his simple method (dubbed the core-layer method) for the different depth levels in the Atlantic, it became clear that flow at all depths is westward intensified. Among these findings is one (published in 1936) showing that there is a strong southward flow below the Gulf Stream, at 2,000 meters (corresponding to the upper North Atlantic Deep Water). Defant gave additional documentation in 1941.[84] Stommel (in 1955) subsequently calculated that such a flow should be expected, but that it should hug the bottom, as a "deep western boundary current."[85] Soon after, Swallow and Worthington (in 1957) made the observations confirming that the predicted undercurrent exists, by following a neutrally buoyant float (devised by Swallow) that moves with the water at preset depths.[86]

The striking observations with the Swallow float did much to discourage the notion that the deep ocean has no strong currents. It may be taken as the beginning of the many direct observations of the deep-sea habitat, which in the following decades established that the abyss is the site of dynamic and intriguing processes.[87]

Following up on the new insights regarding the importance of deep western boundary currents, Stommel (fig. 6.16), and Stommel and Arons (1960) proposed a general pattern of deep-sea circulation whereby the deep waters in the Pacific and the Indian Ocean are provided by the Circumpolar Current, which is fed, in turn, by the deep-water sources in the northern North Atlantic, in the Weddell Sea, and in the throughflow at Drake Passage.[88] The proposed Deep Western Boundary Currents, on the whole, have been confirmed since by direct observation.[89] The postulated slow poleward flow in the interior has been difficult to evaluate, because eddy motions tend to dominate the current flow.[90]

The concept that the deep-mixing Circumpolar Current (which may be considered a window to deep waters) receives North Atlantic Deep Water and distributes it (after mixing) to the Indian Ocean and to the Pacific was familiar to Sverdrup,[91] who cites earlier work by Wüst and Deacon. A salinity maximum near 4,000-meter depths, starting in the Atlantic and persisting into the Pacific, was used to trace this deep-water motion, by Joseph L. Reid and associates.[92] They emphasized that the filling of the Pacific Basin is not from the very bottom (that is, with Antarctic Bottom Water), but from the shallower Circumpolar Water, representing "a mixture of Antarctic waters with the warm, saline, oxygen-rich, and nutrient-poor deep waters from the North Atlantic."[93]

Thus, the mixing of deep waters in the Antarctic circumpolar system is an essential element in driving deep circulation. To the extent that the Gulf Stream is linked to deep circulation, it is also linked into the processes around Antarctica! In a general way, the vertical circulation in the Atlantic (with shallow warm water entering the North Atlantic across the equator,

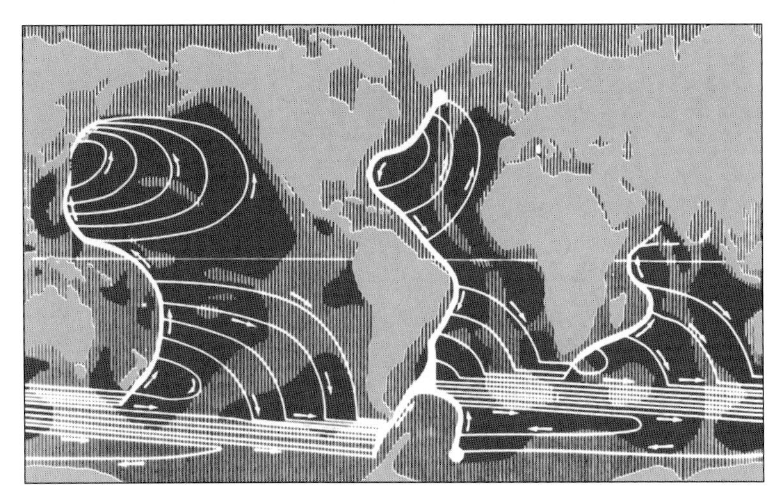

FIGURE 6.16. Stommel's solution for deep circulation (1958), based on fixed deepwater sources and including effects from the Earth's rotation. The water sinking at the two equal sources (off Greenland and in the Weddell Sea) is assumed to rise evenly over the entire ocean. Note the crucial importance of the circum-Antarctic current, and the opposite directions of flow along the western slopes of Atlantic and Pacific basins.

and cold water leaving that basin at depth) is yet one other manifestation of intense mixing in the Antarctic ring current, which lets deep cold water rise to the surface, thus making room for new supply from the North Atlantic (fig. 6.17).

A crucially important piece of information about deep circulation is how fast it renews the deep water-masses. The answer may be expressed in terms of the average time a parcel of water will spend away from contact with the atmosphere. The discovery of this time scale was intimately linked to the measurement of radioactive carbon within the ocean. Radioactive carbon is produced in the atmosphere from the interaction of cosmic rays with nitrogen. The radiocarbon in the air enters surface waters, along with regular carbon. When surface waters sink, they are disconnected from the atmosphere and thus leave behind the source of radiocarbon. At that point, the submerged waters lose radiocarbon at the rate of one-half of the atoms present every 6,000 years, which is a little more than 1 percent each century. As concerns the age of deep water, therefore, the time when it left the surface can be determined from the amount of radiocarbon that is missing. Translating this into the rate of deep-sea circulation is not entirely straightforward for several reasons, one of which is that surface water itself has an age of several hundred years, and when it sinks to make deep water, it entrains subsurface waters as well.

The first measurements of radiocarbon were made in Chicago, in the laboratory of the chemist Willard Libby in 1949.[94] It soon became evident that Libby had provided a powerful new method of dating, one by which oceanographers could determine the sense and rate of flow of deep water. At the invitation of Roger Revelle, the Austrian-American chemist Hans Suess came to La Jolla from the U.S. Geological Survey in 1957 to establish a radiocarbon laboratory and take advantage of the new possibilities.[95] Measurements on the radiocarbon content of deep waters in the Pacific and Indian Ocean established, by about 1960, that the difference in age in deep water-masses has to be counted in centuries and that the Pacific basin has waters that are very much older than those of the Atlantic, in agreement with expectations from oxygen distributions.[96]

The stage was thus set to map out the path of the global circulation based directly on measurement rather than on theoretical considerations. Radiocarbon data for the Atlantic Ocean were gathered at Lamont through the efforts of Wallace S. Broecker, then a rising young star in oceanog-

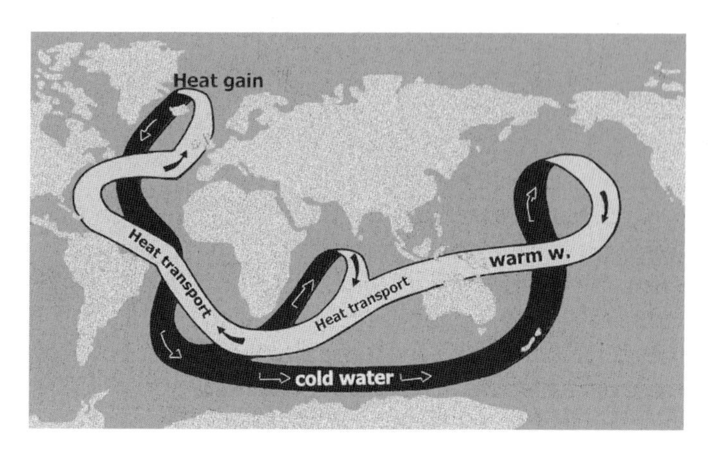

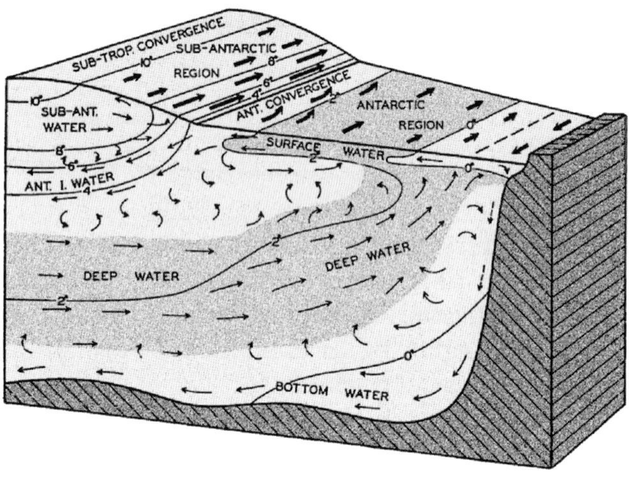

FIGURE 6.17. The Atlantic portion of the global deepwater circulation and its link to the Antarctic circumpolar circulation. *Upper:* schematic representation of the North Atlantic heat piracy (shallow water in, deep water out) within the general "conveyor" scheme of W. S. Broecker. *Lower:* the rise of deep water around the Antarctic continent, as drawn by Sverdrup et al. (1942). The rising of the deep water acts as a long-distance pump on the vertical circulation, and hence for production of North Atlantic Deep Water.

raphy and geochemistry. In a summary paper prepared for the First International Oceanographic Congress (in 1959), Broecker presented the new insights emerging from comparison of radiocarbon distributions in Atlantic and Indo-Pacific.[97] He introduced a box model to describe large-scale oceanic mixing, whereby the flow of water is characterized as an exchange between adjacent boxes.[98] The entire ocean is represented by seven reservoirs, each filled with well-mixed waters. The biggest reservoir is the deep Indo-Pacific, the second biggest the deep Atlantic. There are three surface-water reservoirs: Indo-Pacific, South Atlantic, and North Atlantic. Exchange between deep and surface waters is through the Antarctic and Arctic reservoirs, which have no surface-water lid and provide access of deep water to the atmosphere.[99]

Broecker gave the mean residence time of water molecules in each of the reservoirs as follows: surface waters, 10 to 25 years; Arctic reservoir (source area for North Atlantic Deep Water), 45 years; Antarctic reservoir (Circumpolar Ocean), 100 years; North Atlantic Deep Water, 600 years; Indo-Pacific deep waters, 1,300 years. He concluded as follows[100]: "Regardless of the simplicity of the model and lack of coverage in the Pacific and Indian Oceans it is clear the deep oceans are being renewed with a time scale of the order of magnitude of 1000 years." It was a major discovery.

Armed with these insights, we can now make a simple check on what we should expect in terms of deep-water production. The volume of deep water in the World Ocean may be taken as three-fourths of the total ocean without adjacent reservoirs, which is very close to 1,000 million cubic kilometers.[101] Thus, for a renewal time of 1,000 years, we get 1 million cubic kilometers per year. The year has 31.5 million seconds, so this flux, in millions of cubic meters per second, comes to almost 32 sverdrup.[102] Giving the northern and southern deep-water sources one-half of the task yields 16 sverdrup for the production at the northern and southern source each, a reasonable estimate given what is known about the deep circulation. In regard to the Gulf Stream, the corollary of this rough estimate of deepwater production is that between 10 and 20 percent of the water carried in the Stream off Cape Hatteras is involved in deep circulation.

A systematic study of the deep-sea radiocarbon distribution during the Geochemical Ocean Sections Study in the 1970s (GEOSECS) confirmed and refined the patterns realized earlier.[103] Following GEOSECS, detailed surveys of radiocarbon distributions allowed ever-greater refinement of motions within ocean basins; notably the data from the Transient Tracers in the Ocean program, in the early 1980s, and from the World Ocean Circulation Experiment in the 1990s, a multinational effort.[104]

The fundamental assumption of the pioneering box-model approach, that the ocean may be represented by a few large well-mixed reservoirs, does not hold up well when compared with the results from these surveys. Instead, and not surprisingly, things are more complicated than box models and loop currents would imply. As concerns the nature of the Gulf Stream, to understand it we must go well beyond the central gyre concept (the Stream as the western part of the great wheel) and include all of the deep circulation (the Stream as a part of the vertical circulation). Thus, it is clear that to predict the response of the Gulf Stream to global warming, we should have to predict not only the changes in the North Atlantic wind field (which drives the central gyre), but also the changes in deep circulation throughout the World Ocean. It is a daunting task, and it is still beyond reach at this time.

NOTES AND REFERENCES

1. The other great warm current is its dynamic sibling, the Kuroshio, running northward from the Philippines to Japan.

2. According to rainfall data from Bergen (2002), courtesy Atle Nesje, Bjerknes Center for Climate Research.

3. H. U. Sverdrup, M. W. Johnson, R. H. Fleming, 1942, *The Oceans—Their Physics, Chemistry, and General Biology*. Prentice-Hall, Englewood Cliffs, N.J., 1087 pp. Analyzing the citation patterns in this book, in the chapters on physical oceanography, one notes a strong dominance of Scandinavian and German work in the field before World War II. Citation patterns in a leading postwar text (A. Defant, 1961, *Physical Oceanography*, vol. 1. Pergamon Press, New York, 729 pp.) suggest that important American contributions to physical oceanography began in the late 1940s, with papers by Henry Stommel (Woods Hole) and Walter Munk (La Jolla).

4. F. Nansen, 1912, *Das Bodenwasser und die Abkühlung des Meeres*. Internationale Revue der gesamten Hydrobiologie und Hydrographie 5 (1), 1–42.

5. The Nansen bottle became oceanography's most widely used instrument all through the twentieth century. It is a combined water sampler and thermometer that proved superior to the earlier designs it was based on. In a typical Nansen cast, several (or many) bottles are attached to a steel wire with a weight at its end. Each bottle is "tripped" by a "messenger" when all bottles are at their proper depth. The messenger is a metal weight riding down the wire when released at the surface. It hits a release on the first bottle, which encloses a water sample and turns upside down to reverse the thermometer, fixing the temperature measurement. The pivoting of the bottle releases another messenger, which travels to the next bottle down the wire. When all bottles have been tripped, the line is retrieved and the bottles are detached by hand as they come in reach above the surface.

6. F. Nansen, 1902, *The oceanography of the North Polar Basin*, in *The Norwegian North Polar Expedition 1893–1896*, vol. 3. Jacob Dybwad, (ed.) Kristiania (Oslo), p. 369; V. W. Ekman, 1902, *Om jordrotationes inverkan paa vindströmmar i hafvet*. Nytt Magasin för Naturvidenskaperne 40 (1) 37–63, Kristiania [Oslo]. [*On the effect of the Earth's rotation on wind-driven currents in the sea.*]

7. Vilhelm Friman Koren Bjerknes (1862–1951) is a central figure in the history of meteorology. Together with his son Jacob A. B. Bjerknes (1897–1975) and a number of other young meteorologists (including Halvor Solberg and Tor Bergeron) he founded, after World War I, what became known as the Bergen school of meteorology. This group

proposed new concepts concerning the origin and fate of cyclonic storms (which are common around Iceland and have a tendency to arrive in Norway after relatively short travel). Eventually the group published a manifesto on geophysical fluid mechanics that became the standard reference in the field. V. Bjerknes, J. Bjerknes, H. Solberg, T. Bergeron, 1933, *Physikalische Hydrodynamik, mit Anwendung auf die dynamische Meteorologie.* Julius Springer, Berlin, 797 pp.

8. The Coriolis force is named for the French engineer Gaspard Gustave de Coriolis (1792–1843). The deflecting force is given by the product of velocity and the Coriolis parameter, which describes the rotation of a flat body on the surface of the Earth that is due to Earth's daily spin. The parameter is at maximum at the pole, where the surface is at a right angle to the rotational axis of the planet and is zero at the equator, where the surface is parallel to the axis.

9. This had been worked out earlier to something like one-eighth the speed of the wind for the more common conditions.

10. Because the Coriolis force depends directly on velocity.

11. For science historians it should be of some interest that the Nansen drift—something recorded in the field by an observer with unusual stamina and acuity—became an "Ekman drift" through the addition of mathematics, within a single day of calculations (as narrated in Bjerknes et al., *Physikalische Hydrodynamik*). As the German poet and statesman Johann Wolfgang Goethe has said on occasion, mathematicians can turn a known phenomenon into something entirely new and different.

12. Ekman, V. W., 1905, *On the influence of the Earth's rotation on ocean-currents.* Arkiv för Matematik Astronomi och Fysik, 2 (11), 1–53.

13. J. Sandström and B. Helland-Hansen, 1903, *Über die Berechnung von Meereströmungen.* Reports of the Norwegian Fishery and Marine Investigations 2 (4), 43 pp.; B. Helland-Hansen, F. Nansen, 1909, *The Norwegian Sea: Its physical oceanography based upon the Norwegian researches 1900–1904.* Reports of the Norwegian Fishery and Marine Investigations 2 (2), 390 pp.; G. Wüst, 1924, *Florida und Antillenstrom: Eine hydrodynamische Untersuchung.* Veröffentlichunden des Institutes für Meereskunde, Universitat Berlin, Neue Folge A, 12, 48 pp.; H. U. Sverdrup, 1947, *Wind-driven currents in a baroclinic ocean; with application to the equatorial currents of the eastern Pacific.* Proceedings of the National Academy of Sciences of the U.S.A. 33, 318–326. On the East Coast, geostrophy was used to construct current patterns in the Gulf of Maine, by the oceanographer H. Bigelow. H. B. Bigelow, 1927, *Physical oceanography of the Gulf of Maine.* Bulletin of the U.S. Bureau of Fisheries 40 (2), 511–1027. Henry Bryant Bigelow (1879–1967) was the first director of the Woods Hole Oceanographic Institution. His background was in fisheries biology. For his forays into physical oceanography, Bigelow had help from Edward H. Smith of the International Ice Patrol (and later director of WHOI), who had studied with Helland-Hansen. See R. C. Beardley, W. C. Boicourt, 1981, *On estuarine and continental-shelf circulation in the Middle Atlantic Bight,* in B. A. Warren, C. Wunsch (eds.), *Evolution of Physical Oceanography.* MIT, Boston, pp. 198–233.

14. This is not to say that the oceanographers in La Jolla are not interested in the Gulf Stream or the North Atlantic in general. On the contrary, see J. L. Reid, 1994, *On the total geostrophic circulation of the North Atlantic Ocean: Flow patterns, tracers, and transports.* Progress in Oceanography 33, 1–92; M. K. Flatau, L. Talley, P. P. Niiler, 2003, *The North Atlantic Oscillation, surface current velocities, and SST changes in the subpolar North Atlantic.* Journal of Climate 16, 2355–2369.

15. In 1936 the Woods Hole oceanographer C. O'D. Iselin proposed to apply the name *Florida Current* to the Gulf Stream off Florida up to Cape Hatteras and restrict the name *Gulf Stream* to its complicated northern portion, beyond Cape Hatteras. This is the portion that is farthest away from the Gulf but closest to Woods Hole. For a succinct review of nomenclature, geography, and physics of the Gulf Stream, see P. Cornillon, 1992, *Gulf Stream,* in W. A. Nierenberg (ed.), *Encyclopedia of Earth System Science,* Academic Press, San Diego, pp. 465–480.

16. The recognition of the importance of the Gulf Stream to navigation is one of the many achievements of the polymath and statesman Benjamin Franklin (1706–1790). Franklin published the first map of the Gulf Stream (in 1769), in an effort to speed ships between England and America.

17. Columbus O'Donnell Iselin (1904–1971) was employed at Woods Hole Oceanographic Institution from its beginning in 1930, as general assistant to director Henry Bryant Bigelow. He designed the first ship of WHOI, the *Atlantis,* a 142-foot auxiliary ketch built in Copenhagen. She sailed on 2 July 1931, under Iselin's command. Iselin loved to spend time at sea. In 1940 he became director of WHOI when Bigelow retired, guiding the institution for the next 10 years. The study of currents, and especially of the Gulf Stream, was the focus of his research. C. O'D. Iselin, 1936, *A study of the circulation of the western North Atlantic.* Papers in Physical Oceanography and Meteorology 4 (4), 101 pp.; C. O'D. Iselin, 1940, *Preliminary report on long-period variations in the transport of the Gulf Stream system.* Papers in Physical Oceanography and Meteorology 8 (1), 40 pp.

18. Frederick C. Fuglister (1909–1987) oceanographer and artist, was a member of the powerful post-

war Gulf Stream team at Woods Hole, which included L. Valentine Worthington (1920–1995) and Henry Stommel (1920–1992). Fuglister discovered the fraying of the Gulf Stream at its northern end south of Newfoundland and a strong seasonality in flow. F. C. Fuglister, 1951, *Multiple currents in the Gulf Stream.* Tellus 3, 230–233; F. C. Fuglister, 1951, *Annual variations in current speeds in the Gulf Stream System.* Journal of Marine Research 10, 119–127. With Worthington he published observations on the meandering of the Gulf Stream. F. C. Fuglister, L. V. Worthington, 1951, *Some results of a multiple ship survey of the Gulf Stream.* Tellus 3, 1–14. Worthington, in addition to studying the Gulf Stream, made important contributions to the knowledge of deep circulation, including deepwater production in the North Atlantic. A summary of this work is in L. V. Worthington, 1976, *On the North Atlantic Circulation.* The Johns Hopkins University Oceanographic Studies 6, 110 pp. Neither Fuglister nor Worthington had formal training in oceanography.

19. Raymond B. Montgomery (1910–1988) was on the maiden voyage of *Atlantis* in 1931; he worked at Woods Hole from 1940 to 1960. He used tide gauge records to estimate Gulf Stream flow from the slope of the sea surface, applying principles introduced by Sandström, in 1903. R. B. Montgomery, 1938, *Fluctuations in monthly sea level on eastern U.S. coast as related to dynamics of western North Atlantic Ocean.* Journal of Marine Research 1, 165–185. His best-known publication was published in 1958: *Water characteristics of Atlantic Ocean and of World Ocean.* Deep-Sea Research 5, 134–148.

20. Henry Melson Stommel (1920–1992) worked at Woods Hole from 1942 until his death. Stommel had an undergraduate degree from Yale in astronomy. His explanation of the nature of the Gulf Stream as westward intensification of the grand North Atlantic gyre was a major milestone in the history of oceanography and brought him worldwide recognition. H. Stommel, 1948, *The westward intensification of wind-driven ocean currents.* Transactions of the American Geophysical Union 29, 202–206. Another work of central importance is H. Stommel, 1958, *The abyssal circulation.* Deep-Sea Research 5, 80–82; and another is his book *The Gulf Stream: A Physical and Dynamical Description,* first published in 1958.

21. H. Stommel, 1966, *The Gulf Stream,* 2nd ed. University of California Press, Berkeley, 248 pp.

22. The problem of heat transport, of course, includes the winds as well. Much of the heat carried by winds is in fact transferred from the sea to the air when winds pick up moisture from warm water, using the water's heat to power evaporation. Thus, the fact that warm water is available off the northeast coast is of great interest when contemplating the heat content of westerly winds.

23. H. Stommel, 1965, *The Gulf Stream,* p. 178.

24. Neither Stommel nor Fuglister nor Worthington had a Ph.D.

25. W. Munk, 2000, *Achievements in physical oceanography,* in J. Steele, convenor, 50 *Years of Ocean Discovery, National Science Foundation 1950–2000.* National Academy Press, Washington, D.C., pp. 44–50. The revolution was in some ways presaged by the study of variability in ocean currents in the Norwegian Sea, by Helland-Hansen and Nansen, in 1909. B. Helland-Hansen, F. Nansen, *The Norwegian Sea;* B. Helland-Hansen, F. Nansen, 1920, *Temperature variations in the North Atlantic Ocean and in the atmosphere, introductory studies on the causes of climatological variations.* Smithsonian Miscellaneous Collection 70 (4), 408 pp.

26. I. S. Robinson, 1985, *Satellite Oceanography, an introduction for oceanographers and remote sensing scientists.* Ellis Horwood, Chichester, 455 pp.; R. H. Stewart, 1985, *Methods of Satellite Oceanography.* University of California Press, Berkeley, 360 pp.

27. J. Croll, 1875, *Climate and Time in Their Geological Relations. A Theory of Secular Changes of the Earth's Climate.* Daldy, Isbister, and Co., London, 577 pp.

28. H. U. Sverdrup et al., *The Oceans,* p. 676. Also see Sverdrup's chart, ibid., fig. 187, here reproduced.

29. The equations of motion that allowed Sverdrup to calculate the approximate transport of water along the flow lines shown in the chart are derived in some detail in *The Oceans,* in the chapter on dynamics of ocean currents. In principle, they are set up as a balance of forces, including wind stress, buoyancy force, pressure gradient force, Coriolis force, vertical eddy friction, horizontal eddy friction, and friction at the bottom.

30. The calculation of ocean currents was labor intensive before computers were generally available. Mechanical desk calculators, now mainly found in museums, were used into the 1960s, before programmable electronic calculators took their place. Numbers were entered by hand. Nowadays, measuring instruments store results electronically, and the data are downloaded into computers.

31. As mentioned, this was evident in the detailed work of Helland-Hansen and Nansen, in the Norwegian Sea. But it could not be demonstrated elsewhere because of lack of sampling density.

32. The discovery of the life history of eels, their breeding grounds in the Sargasso and the migration of larvae to rivers in Europe, was a major feat of marine biology with enormous implications for geology, physical oceanography, and evolution. The pioneer performing the feat was the Danish biologist J. Schmidt

(e.g., 1922, *The breeding places of the eel.* Philosophical Transactions of the Royal Society B 211, 179–208).

33. M. Rhein, 2000, *Drifters reveal deep circulation.* Nature 407, 30.

34. This idea was discussed by the Swedish-American oceanographer Carl-Gustaf Rossby (1898–1957), in the 1930s. It may be another one of the "dreamlike" hypotheses referred to by Stommel.

35. G. Wüst, *Florida und Antillenstrom.*

36. H. U. Sverdrup et al., *The Oceans,* pp. 675 and 677.

37. Both Alexander von Humboldt (1769–1859) and Matthew Fontaine Maury (1806–1873) ascribed to the Gulf Stream distant sources in the South Atlantic, in addition to proximal ones. Humboldt wrote as follows (*Kosmos,* vol. 1. J. G. Cotta, Stuttgart, 1845, p. 327, transl. from German):

> The Atlantic Gulf Stream, whose first beginning and initiation is to be found south of the Cape of Good Hope, and which in its great revolution issues from the Sea of the Antilles and the Mexican Bight through the Bahama Strait, moving from south-south-west to north-north-east, gradually leaving the shores of the United States and, deflected eastward at the Bank of Newfoundland, often throws tropical seeds *(Mimosa scandens, Guilandina bondue, Dolichos urens)* upon the shores of Ireland, the Hebrides and Norway. Its northeastern-most extension contributes beneficially to the lesser cold of the seawater and the climate of the northernmost cape of Scandinavia.

In his book *Physical Geography of the Sea,* Lieutenant Maury considered the Gulf Stream "one of the most marvelous things in the ocean" (p. 39). He emphasized evidence that the Gulf Stream draws from many distant sources, citing the path of drift bottles (p. 29). Also, he was convinced of the great importance of the daily rotation of Earth in determining the nature of the Gulf Stream. M. F. Maury, 1857, *The Physical Geography of the Sea.* Harper and Brothers, New York, 360 pp.

38. Regarding the contribution of the Antilles Current to the Gulf Stream, Woods Hole oceanographer Bruce Warren comments as follows (1992, p. 158): "[Sverdrup] must take some blame for prolonging the myth of the Antilles Current, because, while he cited Wüst's transport figure for an 'Antilles Current,' he omitted to mention Wüst's equal and opposite 'countercurrents' on either side of it." B. Warren, 1992, *Physical oceanography in the oceans.* Oceanography 5, 157–159.

39. H. Stommel, *The Gulf Stream,* p. 21.

40. C. O'D. Iselin, 1936, *A study of the circulation of the western North Atlantic.* Papers in Physical Oceanography and Meteorology 4 (4), 101 pp.

41. C. O'D. Iselin, *A study of the circulation,* p. 12.

42. H. Stommel, 1948, *The westward intensification of wind-driven ocean currents.* Transactions of the American Geophysical Union 29, 202–206; W. H. Munk, 1950, *On the wind-driven ocean circulation.* Journal of Meteorology 7, 79–93.

43. H. Stommel, *The Gulf Stream,* p. 87.

44. A. Defant, 1961, *Physical Oceanography,* vol. 1. Pergamon Press, New York, 729 pp. A further generalization of the Stommel-Munk theory is in W. H. Munk, G. F. Carrier, 1950, *The wind-driven circulation in ocean basins of various shapes.* Tellus 2, 158–167.

45. Away from the western boundary currents, the calculated currents resembled those obtained from a solution to the wind-stress problem proposed by Sverdrup (1947), whose equation neglects the friction term. H. U. Sverdrup, 1947, *Wind-driven currents in a baroclinic ocean; with application to the equatorial currents of the Eastern Pacific.* Proceedings of the National Academy of Sciences., Washington, D.C. 33, 318–326.

46. William S. von Arx (1916–1999) is best known from his excellent textbook *Introduction to Physical Oceanography* (Addison-Wesley, Boston, 1962), which went through numerous printings.

47. W. H. Munk, 1955, *The circulation of the oceans.* Scientific American 193 (3), 96–104.

48. The opposite concept, that the closing of the Panama Isthmus helped produce ice buildup in the far north, is quite commonly invoked when explaining the onset of the northern ice ages. The idea, based on a presumed shortage of water vapor for making snow in the north, seems quite implausible. Heat melts ice, and rain does not make snow, unless it is cold enough.

49. A "typical" hurricane, according to the entry in *Encyclopedia of Climate and Weather* (S. H. Schneider, ed., Oxford University Press, 1996) originates from a tropical disturbance off West Africa and travels westward at around 20 mph along 10° N, until, after a few days, it becomes a tropical depression. Several hundred miles before reaching the Lesser Antilles Islands, at about 15° N, it grows into a tropical storm (and receives a name). The storm is now heading toward the northern islands in the chain, at 15 mph. Soon after that it turns toward the northwest, acquiring an eye and the familiar symmetrical shape. It is now one week old, from the initial disturbance. It keeps on turning northward, staying well off Puerto Rico and the Bahamas, while attaining speeds of 120 mph, in its maximum development. As it moves north and east, it weakens as it encounters colder waters, and eventually peters out in a disturbance moving east. (T. P. Eichler, *Hurricanes,* p. 409). In the Atlantic, there are about a dozen tropical storms each season,

half of which reach hurricane strength. In 2005, there were 27 named tropical storms and hurricanes in the North Atlantic, an all-time record.

That the warm waters of the Gulf Stream provide a path for tropical storms has been appreciated for more than a century: "The most furious gales of wind sweep along with it;" wrote Maury in 1857 (*The Physical Geography of the Sea*, p. 58). Also, Maury was aware of the general path of hurricanes and of their origination in the trade wind belt.

50. W. H. Berger, G. Wefer, 1996, *Central themes of South Atlantic circulation*, in G. Wefer, W. H. Berger, G. Siedler, D. J. Webb (eds.), *The South Atlantic: Present and Past Circulation*. Springer-Verlag, Berlin, pp. 1–11.

51. In Oslo, Bergen's successful rival for the status of capital of Norway, and a city with a more continental climate, one might hear the following joke: A visitor to Bergen after experiencing seemingly interminable drizzle, asks a local youngster in exasperation: "Does it rain here always?" The kid looks up somewhat puzzled and replies: "I don't know, I am only eight years old." In fairness, the Sun does shine in Bergen also.

52. K. E. Trenberth, J. M. Caron, 2001, *Estimates of meridional atmosphere and ocean heat transports*. Journal of Climate 14, 3433–3443.

53. The implications of the heat asymmetry are discussed in some detail in H. Flohn, 1985, *Das Problem der Klimaänderungen in Vergangenheit und Zukunft*. Wissenschaftliche Buchgesellschaft, Darmstadt, 228 pp.

54. Heyerdahl's reed-based vessel sailed 3,270 miles in 57 days, from Morocco to Barbados. T. Heyerdahl, 1979, *Early Man and the Ocean*. Doubleday, Garden City, N.Y., 438 pp.

55. A westerly wind moves eastward. An easterly current also moves eastward. For a sailor, it is important where the wind comes *from* and where the current takes the ship *to*.

56. The zonal and monsoonal winds are in a kind of competition. Clearly, a strong summer monsoon blowing from the southwest in the North Atlantic or in the northern Indian Ocean will interfere with the easterly trade winds. If we ask how the wind field will change as a result of global warming, this interference between the two wind systems must be considered.

57. F. C. Fuglister, L. V. Worthington, 1951, *Some results of a multiple ship survey in the Gulf Stream*. Tellus 3, 1–14.

58. Norwegian Sea: Helland-Hansen, Nansen, *The Norwegian Sea*; California Current: Sverdrup et al., *The Oceans*.

59. W. Munk, *Achievements in physical oceanography*. Munk defines *revolution* as meaning that "an oceanographer totally familiar with the topic at the beginning of the period, but with no further learning experience, would flunk a freshman exam at the end of the period." (ibid., p. 45, footnote 5). This is true, of course, for any active scientific field, depending on the length of the period considered. From experience in geology, a revolution is underway when the students realize that their professor is seriously out of step with current developments in the field he is teaching.

60. J. C. Swallow, 1955, *A neutral buoyancy float for measuring deep currents*. Deep-Sea Research 3, 74–81.

61. W. Munk, *Achievements in physical oceanography*, p. 46.

62. William "Bill" Holland obtained his Ph.D. at SIO, in 1965, and then went to Boulder to do modeling at the National Center for Atmospheric Research, one of the leading Earth system computing centers in the world.

63. R. E. Cheney, 2001, *Satellite altimetry*, in J. H. Steele, S. A. Thorpe, K. K. Turekian (eds.), *Encyclopedia of Ocean Sciences*. Academic Press, San Diego, pp. 2504–2510.

64. B. Helland-Hansen, B., F. Nansen, *Temperature variations*.

65. J. W. Hurrell, 2001, *North Atlantic Oscillation (NAO)*, in J. H. Steele, S. A. Thorpe, K. K. Turekian (eds.), *Encyclopedia of Ocean Sciences*. Academic Press, San Diego, pp. 1904–1911.

66. J. C. Rogers, 1984, *The association between the North Atlantic Oscillation and the Southern Oscillation in the Northern Hemisphere*. Monthly Weather Review 112, 1999–2015; J. Hurrell, 1995, *Decadal trends in the North Atlantic Oscillation: Regional temperatures and precipitation*. Science 269, 676–679.

67. Jerome Namias (1910–1997) joined Scripps in 1972 after his retirement from NOAA, at the invitation of Director Nierenberg. Before that year, extended visits produced much research on the statistics of weather patterns in the North Pacific. He had a special interest in the question to what degree such patterns can improve weather predictions for North America, by teleconnections, that is, statistical correlations. Namias was a student of Carl-Gustaf Rossby (in the 1930s, at MIT), and he credits Rossby with stimulating his interest in hemispherewide patterns. J. Namias, 1975, *Preface*, in *Short Period Climatic Variations—Collected Works of J. Namias, 1934 through 1975*, vol. 1. University of California, San Diego.

68. J. Namias, 1981, *Teleconnections of 700 mb Height Anomalies for the Northern Hemisphere*. CalCOFI Atlas 29. Scripps Institution of Oceanography, University of California, San Diego, 265 pp.

69. Everyone is familiar with oscillations from the evening news. The stock market oscillates between two states called *bull* and *bear*. During the bull market,

most investors buy because everyone else does. During the bear market, there is a tendency to sell, again by entrainment. Reversals occur because enough investors decide that a given run has produced a large number of lemons or bargains, respectively.

70. J. W. Hurrell, *North Atlantic Oscillation (NAO)*.

71. J. Meincke, 2002, *Climate dynamics of the North Atlantic and NW-Europe: An observation-based overview*, in G. Wefer, W. Berger, K.-E. Behre, E. Jansen (eds.), *Climate Development and History of the North Atlantic Realm*. Springer-Verlag, Berlin, pp. 25–40, p. 37.

72. W. S. Broecker, 1997, *Thermohaline circulation, the Achilles heel of our climate system: Will man-made CO_2 upset the current balance?* Science 278, 1582–1588; W. S. Broecker, 1999, *What if the conveyor were to shut down? Reflections on a possible outcome of the great global experiment*. GSA Today 9 (1), 2–7.

73. J. Meincke, *Climate Dynamics*, p. 36, fig. 10.

74. Benjamin Thompson (1753–1814) was at one time a British intelligence officer, and at another the secretary of war for the kingdom of Bavaria.

75. In an outstanding example of applied oceanography, members of the expedition cooled their champagne in the tropics in deep-water samples brought on board for the purpose. C. W. Thomson, 1873, *Challenger* Reports, vol. 11, p. 147.

76. For a formal treatment of this balance, see W. H. Munk, 1966, *Abyssal recipes*. Deep-Sea Research 13, 707–730.

77. This point has been urged in recent years by Scripps oceanographer Lynne Talley, who tracks the origin of thermocline waters to subpolar source regions. For example, L. D. Talley, 1996, *Antarctic Intermediate Water in the South Atlantic*, in G. Wefer, W. H. Berger, G. Siedler, D. J. Webb (eds.), *The South Atlantic: Present and Past Circulation*. Springer-Verlag, Berlin, pp. 125–162. In a conceptualization dominated by horizontal motions (common since the 1970s), the vertical stratification of the ocean more or less reflects the differences in water properties at the surface, from the subtropics to the polar regions.

78. A striking example of midocean sinking was observed directly, in the 1960s, during the MEDOC program. D. Anati, H. Stommel, 1970, *The initial phase of deep water formation in the northwest Mediterranean during MEDOC '69 on the basis of observations made by Atlantis II, January 25, 1969 to February 12, 1969*. Cahiers Océanographiques 22, 343–351; Medoc Group, 1970, *Observation of formation of deep water, 1969*. Nature 227, 1037–1040.

79. Joseph Lee Reid, Scripps's prominent observational physical oceanographer, was a graduate student at Scripps and received his M.S. degree in 1950. A seagoing scientist, he continued at the institution and later joined the faculty. He became professor emeritus in 1992. In 1988, he received the coveted Albatross Award, a stuffed bird for which the honoree becomes host till it is awarded to another deserving oceanographer. The award was given to commend Reid "for insisting that our knowledge of the ocean should bear some resemblance to reality." Much of his career has been dedicated to mapping water-mass properties and producing atlases and profiles that allow the testing of ideas about the origin and destination of water masses.

80. J. L. Reid, 1979, *On the contribution of the Mediterranean Sea outflow to the Norwegian-Greenland Sea*. Deep-Sea Research 26, 1199–1223.

81. The layering of deep waters in the Pacific is rather simple compared with that in the Atlantic. Abyssal waters come in from the Southern Ocean and move north, filling the basin. Intermediate waters are supplied by sources in the Subantarctic and Subarctic regions. The Pacific has no source of deep water in the north. The reasons are not entirely clear. Perhaps the precipitation patterns in the North Pacific prevent a northern deep-water source. There is a net transfer of water vapor from the Caribbean across the Isthmus into the Pacific, by the prevailing winds. In principle, this makes the North Atlantic saltier than the North Pacific. If the Caribbean were colder, less vapor might be transferred and thus increase the chance to make North Pacific deep water. Production of northern deep water during glacial times is a strong possibility.

82. Georg Wüst (1890–1977), who emphasized observation, was joined in this effort by his distinguished Austrian colleague Albert Defant (1884–1974), a theoretician.

83. G. Wüst, 1935, *Schichtung und Zirkulation des Atlantischen Ozeans. Die Stratosphäre*. Wissenschaftliche Ergebnisse der Deutschen Atlantischen Expedition "*Meteor*" 1925–1927, 6, erster Teil, 2, 109–288; G. Wüst, 1936, *Die Vertikalschnitte der Temperatur, des Salzgehaltes und der Dichte, ibid*. Teil A des Atlas zu 6, Beilage II-XLVI; G. Wüst, 1955, *Stromgeschwindigkeiten im Tiefen- und Bodenwasser des Atlantischen Ozeans auf Grund dynamischer Berechnungen der Meteor-Profile der Deutschen Atlantischen Expedition 1925/27*. Papers in Marine Biology and Oceanography. Deep-Sea Research 3 (Supplement), 373–397.

84. A. Defant, 1941, *Die absolute Topographie des physikalischen Meeresniveaus und der Druckflächen, sowie die Wasserbewegungen im Atlantischen Ozean*. Wissenschaftliche Ergebnisse der Deutschen Atlantischen Expedition, "Meteor" 1925–1927, 6, Teil 1, Lfg. 5, 191–260.

85. H. Stommel, 1958, *The Gulf Stream: A Physical and Dynamical Description*. University of California Press, Berkeley, 202 pp.

86. J.C. Swallow, L.V. Worthington, 1961, *An observation of a deep countercurrent in the western North Atlantic.* Deep-Sea Research 8, 1–19.

87. The deployment of deep Swallow floats fit well with the program of "hypothesis testing" urged by Stommel. The motions of the floats suggested that the eddy structure of the sea is pervasive.

88. H. Stommel, A.B. Arons 1960, *On the abyssal circulation of the World Ocean.* Deep-Sea Research 6, 140–154 and 217–233.

89. B.A. Warren, 1981, *Deep circulation of the World Ocean,* in B.A. Warren, C. Wunsch (eds.), *Evolution of Physical Oceanography(Scientific Surveys in Honor of Henry Stommel.* MIT Press, Cambridge, Mass., pp. 6–41.

90. A strong northward current was found in the western South Pacific, in the expected location, that is, the Samoan Passage. J.L. Reid, P.F. Lonsdale, 1974, *On the flow of water through the Samoan Passage.* Journal of Physical Oceanography 4, 58–73.

91. H.U. Sverdrup et al., *The Oceans,* p. 751.

92. J.L. Reid, R.J. Lynn, 1971, *On the influence of the Norwegian-Greenland and Weddell Seas upon the bottom waters of the Indian and Pacific oceans.* Deep-Sea Research 18, 1063–1088; A.W. Mantyla, J.L. Reid, 1983, *Abyssal characteristics of the World Ocean waters.* Deep-Sea Research 30 (8A), 805–833.

93. A.W. Mantyla, J.L. Reid, *Abyssal characteristics,* p. 805.

94. Willard F. Libby (1908–1980), Nobel Prize 1960.

95. Hans Suess (1909–1993). One of the famous papers resulting from Hans Suess's work was the one with Roger Revelle (Revelle and Suess, 1957), which established the rate of exchange of carbon dioxide between ocean and atmosphere. Also, Suess soon realized that the burning of fossil fuels is changing the radiocarbon content of the atmosphere, because the fuel carbon has no radiocarbon. The resulting decrease in the ratio of radiocarbon to normal carbon is now referred to as the Suess effect. In addition, Suess showed that radiocarbon showed fluctuations in the atmosphere, on scales between decades and centuries. These variations are known as "Suess wiggles."

96. The oxygen content is an important qualitative indicator of age for deep water. "Young" deep water, that is, water recently in contact with the atmosphere, has high oxygen content. As the water ages, the oxygen is continuously used up in the oxidation of organic matter, which produces carbon dioxide.

97. The first measurements of radiocarbon from various parts of the ocean became generally available in the last years of the 1950s. A summary of the early phase of this research is in W.S. Broecker, R.D. Gerard, M. Ewing, B.C. Heezen, 1961, *Geochemistry and physics of ocean circulation,* in M. Sears (ed.), *Oceanography,* American Association for the Advancement of Science, Washington, D.C., pp. 301–322. Broecker (the paper, although listing four authors, is written in the first person singular) cites Suess and associates for publishing an apparent age of deep water in the Pacific Ocean (in 1959). Clearly, a race to discover deepwater ages was in progress, and the race was close, as can be read from the statement on page 303: "The application of C^{14} to oceanographic mixing problems was originally proposed by Ewing and Kulp at Lamont." See also W.S. Broecker, R. Gerard, M. Ewing, B.C. Heezen, 1960, *Natural radiocarbon in the Atlantic Ocean.* Journal of Geophysical Research 65 (A), 2903–2931.

98. Scripps geochemist Harmon Craig (1926–2002) employed a somewhat similar model to estimate the exchange rate of carbon dioxide between atmosphere and ocean, in 1957.

99. The overall circulation is strongly reminiscent of Broecker's global conveyor proposed more than 25 years later. See W.S. Broecker, G.H. Denton, 1989, *The role of ocean-atmosphere reorganizations in glacial cycles.* Geochimica et Cosmochimica Acta 53, 2465–2501. See fig. 6.17 (*Upper*).

100. Broecker et al., *Geochemistry and physics of ocean circulation,* p. 317.

101. Sverdrup et al., *The Oceans,* p. 15.

102. One sverdrup is defined as 1 million cubic meters per second. The transport in the Gulf Stream is of the order of 100 sverdrups (abbreviated sv).

103. Deep-sea radiocarbon analyses were performed by Minze Stuiver at the University of Washington and by G. Östlund at the University of Miami.

104. R.M. Key, 2001, *Radiocarbon,* in J.H. Steele, S.A. Thorpe, K.K. Turekian, *Encyclopedia of Ocean Sciences.* Academic Press, San Diego, pp. 2338–2353.

SEVEN

Sardines and the California Current

ON A COLD CURRENT RUNNING SOUTH

The California Current brings cold water south along the sunny parts of the West Coast, and it does so while moving surface waters offshore (fig. 7.1). As a consequence, additional cold water rises from the thermocline and below, producing a narrow cold strip of highly productive waters in sight of the land. The processes involved (referred to as "upwelling in an eastern boundary current") are responsible for the rich sea life supported by the current, including seabirds such as cormorants and pelicans, and mammals such as the California sea lion and the elephant seal.

Productivity varies through the decades. There was a time when the coastal ocean off the shores of California was famous for its rich harvests of sardine (fig. 7.2). John Steinbeck, in his book *Cannery Row* (published in 1945), celebrated the hustle and bustle of the period. "Cannery Row in Monterey in California is a poem, a stink, a grating noise, a quality of light, a tone, a habit, a nostalgia, a dream"—thus the opening line of his book.[1]

At the time Steinbeck wrote the story, which features "Doc," the spirited marine biologist collecting worms, mollusks, and crabs.[2] Cannery Row's 24 canneries were already on the path to extinction. In Monterey, the business of canning had started in 1902, following the example set by San Francisco (1889) and San Pedro (1893). By 1920, the canneries were producing fishmeal and fertilizer at an ever-increasing pace. In the 1936/37 fishing season well over 700,000 tons of sardines were hauled into California harbors, an all-time record (fig. 7.2). This was about one-quarter of the U.S. tonnage of fish. Scientists at the California Department of Fish and Game warned about the dangers of overfishing, but with no appreciable effect on fisheries regulation.[3] The fishery shrank to a shadow of its former glory after 1945 and never recovered.

Scientists at Scripps began systematic studies on the productivity of the California Current in 1938, along offshore transects laid out by the new director, Harald U. Sverdrup, and by Oscar E. Sette of the U.S. Bureau of Fisheries. One important goal was to provide the scientific underpinnings for managing the sardine fishery.[4] Eventually, this effort resulted in great advances in the understanding of coastal upwelling and mixing, processes that govern the productivity of the sea. The California Current became the best-studied high-production region in the world, thanks to the collaborative project arising from these initial efforts, after World War II (the California Cooperative Oceanic Fisheries Investigations [CalCOFI, or CALCOFI] program).

Perhaps the most important insight gained was the realization that this productive system is highly variable, on various scales from a few years to centuries. This variability greatly impacts any attempt to provide for intelligent management of fisheries. In fact, fisheries management anywhere in the world has been greatly frustrated in trying to meet the challenges posed by the combination of ever-increasing fishing pressures and a highly variable physical environment.

FISH IN A COLD OFFSHORE CURRENT

The southward flowing current off California is a poorly organized maze of swirls and eddies with an overall tendency to move southward. Similar so-called eastern boundary currents exist elsewhere: off Peru and Chile, off northwestern Africa, and off southwestern Africa. An important feature of these currents is that cold deep water moves upward to the surface, along the coast. This upwelling of water from below surface layers brings nutrients into the nutrient-starved sunlit zone. Here minute algae use it to grow, in turn feeding all kinds of organisms, including fish.[5] Upwelling off California depends on favorable northerly winds. Whenever winds die down or reverse, production plummets, fish starve, and fisheries suffer. In recent decades, on the whole, upwelling has decreased and the current has become less productive.

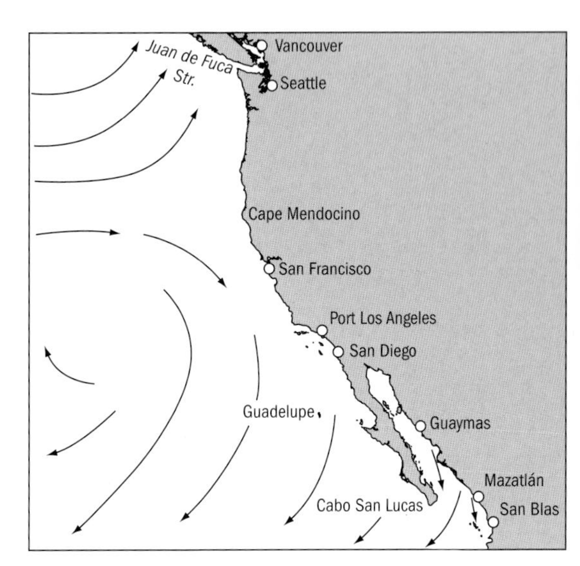

FIGURE 7.1. Currents off California according to H. Thorade, as given in Kruemmel (1907). Note the offshore component of the flow.

As mentioned, the driving force behind the studies initiated during Sverdrup's tenure at Scripps was the sardine fishery, which boomed in the 1930s and went into free fall after 1945. The sardine catch remained quite high until the end of the war, typically yielding between 400,000 and 600,000 tons for each season. But in 1946/47, the number of sardines dropped precipitously to 234,000 tons, that is, to one-third of the landings in the best season 10 years earlier. Fishermen and the entire fishing community were devastated. Canneries were closed—first in Washington and then progressively farther south in Oregon and California—as the sardines disappeared and withdrew southward to their regular breeding areas off the shores of Baja California.

No one knew what had caused the collapse. Overfishing was an obvious candidate for the culprit. Presumably, as fishing pressure mounted, the older and more mature members of the populations were removed, which left these populations without the resilience needed to overcome a period of poor reproductive success.[6] However, as it turned out, this was not the whole story.

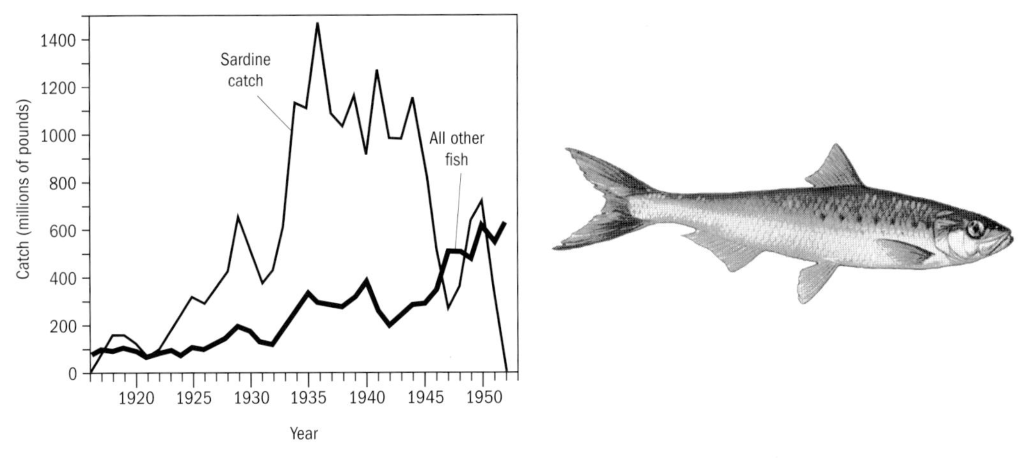

FIGURE 7.2. Rise and fall of the California sardine fishery in the first half of the twentieth century, according to CALCOFI records. Note the more or less steady increase of "all other fish," as a result of increased fishing efforts.

The story of the sardine fishery off California has a number of elements of interest: the cold waters off the West Coast sustain a lot of fish, the fish harvest fluctuates for reasons unknown, and a fishery can collapse from one season to the next, without warning. The first item is a fundamental problem in oceanography: why are fish populations so dense in a few places in the ocean, while the rest of the sea is a desert? The second has to do with climate variability, a factor that ocean scientists started to worry about in the 1970s. The third concerns the results of overexploitations of systems with unknown dynamics of their own. This third item has become a principal concern in environmental studies and management, in the twenty-first century.

The conundrum of sudden collapse reflects a pattern that emerges all over the world in forest ecology, river ecology, estuarine systems, coral reefs, and a large number of major fisheries, such as cod and herring. The complexity of problems greatly increases from one issue to the next: from the simple task of explaining why the fish are abundant in some places, to the hard one of explaining why abundance varies through time, and on to the impossible one of determining the point at which the effects of overfishing trigger collapse.

The central item of interest, as concerns the California Current, is the relationship between cold water and fishery success. Why are there so many fishes in the cold current off California? First of all, it is not the only cold current to sustain large populations of small schooling fishes.[7] Similar conditions prevail off Peru and Chile where cold offshore waters in a subtropical to tropical setting support enormous populations of sardinelike fish, the anchoveta. The Peruvian anchovy fishery greatly expanded after the collapse of the Californian sardine fishery.[8] By 1970, Peru contributed some 20 percent of global landings! The fishery collapsed in 1973. Off Namibia in southwest Africa we also find unusually cold waters near the coast, rich in fish. Great congregations of whales once fed off Namibia and gave Walvis Bay its name (*walvis* stands for "whale-fish").

In each of these instances, as in southern California, arid regions border the coast. The pattern of cold offshore current, high marine productivity, and arid onshore conditions is in fact repeated at all the west coasts of the major continents—North America, South America, North Africa, South Africa, and (to some extent) Australia. Southern California, then, provides a geographic prototype for certain coastal regions thousands of miles away and rarely visited.[9]

What do these regions have in common? They have coastal waters that are colder than is normal for their latitudes. The long-term average sea-surface temperature for latitude 30° is near

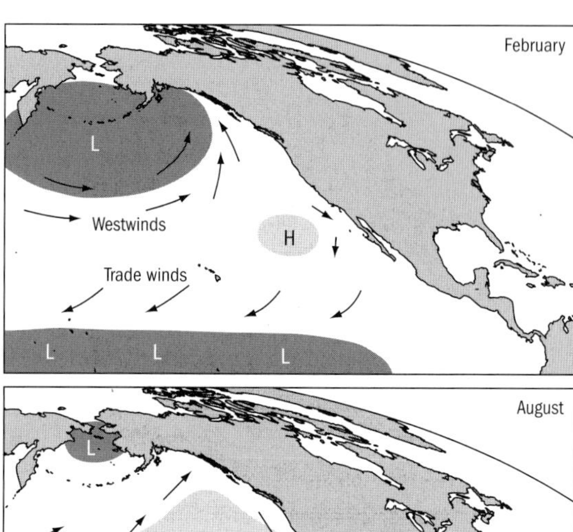

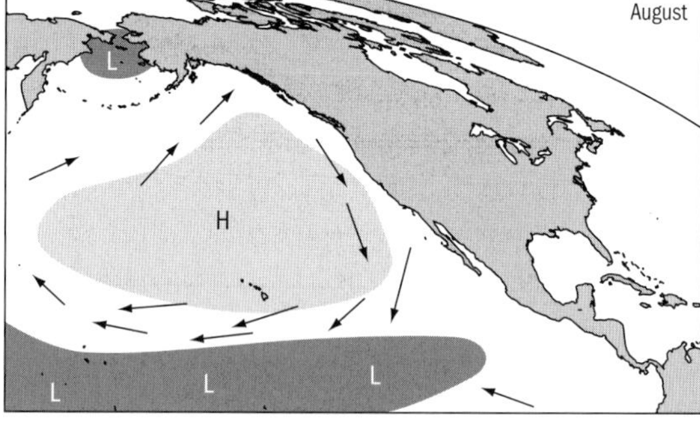

FIGURE 7.3. Seasonal changes in the distribution of high and low atmospheric pressure and the response of the California Current.

20 °C. This corresponds to a reasonable 68 °F, as concerns swimming in the sea. Yet, looking at the temperature record measured at the Scripps Pier since 1916, we find average values closer to 17 °C, that is, 5 to 6 °F less than expected!

Why is the water off San Diego so cold on average? It is cold, normally, for several reasons. First, there is an overall movement of offshore waters in a southeasterly direction, that is, from higher to lower latitudes, in accord with prevailing winds. The southward motion includes surface waters several hundred miles out. This motion is part of the general circulation of the North Pacific, which is driven by westerly winds in a broad zone centered on roughly 45° N, and easterly winds (the trades) centered on about 20° N. Upon finding their path barred by North America, the westerly winds along 45° N adjust their trajectories northward and southward, blowing into the Gulf of Alaska and along the shores of California. In addition, the cold southward-flowing current helps to set up a high-pressure center off California that forms the pivot for westerly winds moving south and later merging into the trades off southern Baja California (fig. 7.3). The offshore high provides for strong winds parallel to the coast (fig. 7.4). Also, it keeps southern California sunny and dry, as it tends to block any storms trying to make landfall. Thus, there is a close relationship between the coldness of the current and the parchedness of the land to the east of it. The rainy season is in the winter months, when the offshore high pressure has shrunk and moved south.

The water is cold because it comes from farther north, but it is also cold for a second reason, as mentioned, especially in the zone close to shore. Cold water is coming up to the surface at many places along the coast of California. The nature of this upward motion—the depth from which the water rises, and how it is made to rise—is quite complicated and is the subject of expert study. What is important about the process in the context of productivity is that the rising cold water brings nutrients. It is because of the fertilization by cold water from below that the coastal ocean off California has an olive tint. The microscopic algae (diatoms, dinoflagellates) at the base of the food chain use the nutrients to grow and thus provide the nourishment

FIGURE 7.4. Surfing in the wind blowing parallel to the coast, north of Point Conception.

to zooplankton that feeds the fishes. Also, sardines and anchovies directly feed on diatoms and dinoflagellates, which greatly increases the amount of food available to them. In addition, since the high-production zone is in view of the coast, there are abundant larvae of nearshore organisms in the water (the so-called meroplankton). All these factors combine to make herringlike fish more abundant in upwelling zones than other types of fish.

The reason cold water rises to the surface has to do with the removal of warm water toward the open ocean, by a general offshore drift of surface waters. The drift is the result of the Coriolis force, mentioned previously. Because the eastward motion of Earth's surface increases toward the equator, a current running south on the Northern Hemisphere (as off California) tends to move waters westward and offshore, thus making room for colder water from below. On the whole, because of the upwelling, the coastal portion of the California Current is the coldest part, and also the one richest in nutrients. That is why the sea is greenest right next to the coast.

REGIME SHIFT

Lately, the waters offshore of San Diego and elsewhere off California have not been as cold, on average, as in the past. This has been good for swimming, but not so good for the growth of kelp beds, or for most types of coastal creatures, including fishes and seabirds. The California Current is warming, roughly since the mid-1970s. Could this be a result of global warming, caused by the addition of greenhouse gases to the atmosphere?

There is general agreement that the world has been warming for more than a hundred years, by somewhat less than 1 °C, that is, between 1 and 2 °F, on average. Just how much was contributed to this warming by human activity cannot be ascertained with any precision, for several reasons. Mainly, the climate a hundred years ago was recovering from a cold spell several centuries long, called the Little Ice Age. The cooling began sometime around AD 1200[10] and accelerated in the fourteenth century, bringing an unusual number of harsh winters and poor summers to northwestern Europe (where records are best),

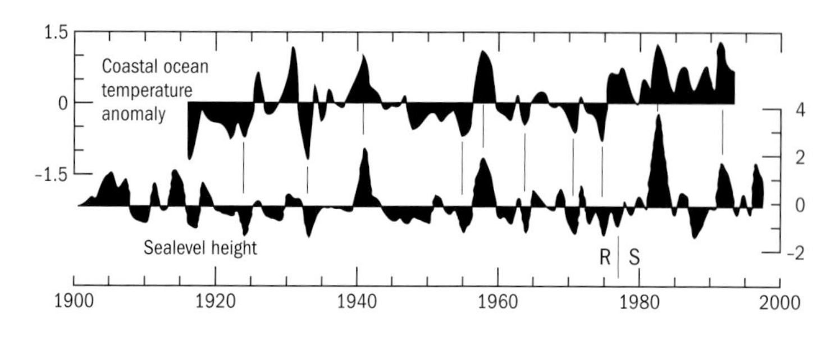

FIGURE 7.5. Comparison of the histories of the coastal temperature anomaly of the California Current with sea level height along the coast. Traditionally, warm anomalies (in degrees Celsius) go together with increased sea level (in centimeters) and vice versa. Since 1974, this correspondence has greatly weakened, for reasons unknown. RS, regime shift.

but also to other areas in the Northern Hemisphere.[11] The observed warming of the past century conceivably could be in part a recovery of the climate system from an anomalously cold condition to a more normal one.[12]

However, the last two decades of the twentieth century have been warmer (as far as this can be ascertained) than any other two decades in the last 1,000 years.[13] The one thing that is clearly different about the 1980s and 1990s is the unusually high content of greenhouse gases, mainly carbon dioxide.[14] Man-made carbon dioxide comes from burning coal, petroleum, and natural gas, and it warms the Earth, as we know from elementary physics. The exact effect cannot be calculated because the climate system responds in ways that are poorly understood. Also, activities related to industry and agriculture increase the air's content of tiny particles. This dust tends to block some of the sunlight from reaching Earth's surface.

Whatever the reasons for the general warming, the California Current is responding to it (fig. 7.5). The water has been getting warmer for the last few decades. Scripps oceanographer Dean Roemmich, who works on the circulation of the ocean, has been studying the matter of ocean warming along California. His database is the unique hydrographic series collected by numerous expeditions since 1948, within the program called the California Cooperative Oceanic Fisheries Investigations (CalCOFI), described below. What he found is that during the past 42 years temperatures have increased by 0.8 °C off the coast of southern California, uniformly in the upper 100 meters, and that significant warming can be detected down to about 300 meters.[15]

A surprising result of Roemmich's study is the finding that the warming is much the same all across the California Current, along a line running offshore between La Jolla and Los Angeles, out to a distance of 500 kilometers. This means that the warming did not simply result from a decrease in the upwelling of cold water near the coast. If that had been the reason, the change would be larger near the coast, tapering out to sea. Instead, the change is pervasive, presumably reflecting general warming through increased overall heating at the top of the sea.

Apparently, the warming had drastic effects on the biology of the sea off California: Scripps biologist John McGowan and colleagues in the Marine Life Research Group report evidence for a dramatic reduction in productivity over the last several decades.[16] They have come to a striking conclusion. It is as though the entire northeastern Pacific ocean-atmosphere system has moved into a new way of operating—a way that leads to much lower levels of biological production. McGowan and associates summarize these findings as follows:

> Beginning sometime in the late 1970s, the mean temperature and salinity shifted over a large area of the California Current. . . . There has been a long-term decline in total commercial, pelagic catch since well before the regime shift, but this decline accelerated after its onset. . . . Kelp forests . . . are now systematically smaller and depauperate, a trend that began in the late 1970s. . . . A large change attributed to climate variation occurred in the intertidal flora and fauna in central California,

with many southern species now dominating the community.[17]

Perhaps the most astonishing result, however, is the large-scale reduction of zooplankton that can be documented in the CalCOFI data, especially the heavily sampled transect running offshore from halfway between San Diego and Los Angeles, called Line 90, which runs some 500 kilometers out to sea, spanning the great Southern California Eddy. Between the 1950s and the years around 1990, there was a reduction by roughly a factor of three, *all along Line 90, regardless of distance to shore*. This is surely surprising.

The "regime shift" of the California Current apparently is embedded in a much larger climate shift involving the entire northern and tropical Pacific. Indeed, it would be surprising were this not so. The California Current is part and parcel of a system comprising the western Pacific warm-water pool from the Solomon Islands to the Philippines and the Indonesian archipelago to the fog-shrouded coastlines of Oregon and Chile, at the eastern margin. It all hangs together, and nothing important happens in the east without the west knowing about it. When correlating sea-surface temperatures off La Jolla with air pressures at sea level around the globe one finds wonderfully strong agreement in monthly fluctuations with such unlikely places as the Seychelles in the Indian Ocean, the sea off Perth in western Australia, the Philippines, the Society Islands in the South Pacific, the sea south of the Aleutian Islands, and the sea off the mouth of the Amazon.[18] All of which goes to prove that the climate system is global.[19]

The first inkling that a regime shift was underway came from a study of chlorophyll distributions in the central North Pacific, by diatom expert and ocean ecologist Elizabeth Venrick and associates.[20] They found a remarkable increase in the amount of chlorophyll at a depth just below the mixed layer, where chlorophyll content reaches a maximum, sometime in the mid-1970s. This increase was tied to a cooling of surface temperatures and a strengthening of winter winds (from a stronger Aleutian Low). Various other regions showed distinct changes in the mid-1970s.[21] Beginning in the mid-1970s, the Oyashio, the cold current moving south along Kamchatka and the Kuril Islands toward Japan, has penetrated farther south, meeting with the northward-flowing Kuroshio at lower latitude. Presumably a strengthened Aleutian Low is responsible. Zooplankton abundance in the Oyashio region dropped sharply at the same time. In the subarctic Pacific, mixed-layer depths became shallower by 20 to 30 percent in the eighties, compared with the previous two decades.[22]

How extensive and how permanent is the regime shift? Does it mainly concern the California Current or the North Pacific or the Pacific—or the entire planet? Is this shift similar to other climatic variations that have occurred previously, or is its nature fundamentally different from "run-of-the-mill" multidecadal variation? Will conditions return to "normal" or will they keep moving off into previously unknown conditions? What is "normal"? When precisely did the shift occur, and why then? Is the cooling of the current during the last few years indicating a shift in reverse?

These are urgent but difficult questions, especially the one about the nature of the shift, which involves the dynamics of climate and air-sea interaction, playing out over many decades. There is one conclusion that we can draw rather readily from these data, though. Without a good understanding of the nature of climate variations and their effects on the California Current we cannot grasp the nature of fluctuations in the sardine populations, or any other fish populations in the productivity centers of the ocean. Furthermore, without the type of historical background delivered by a long-term monitoring program or similar documentation, we have no idea what is normal and what is unusual.

THE CONCEPT OF UPWELLING

The questions listed, regarding the response of coastal productivity to global warming, can be discussed only in light of what we know about coastal upwelling. This phenomenon was discovered sometime within the first two decades of the twentieth century, with attention focused on the California Current.[23] Sverdrup came to struggle with the intricacies of this process in his attempt to understand the nature of the California Current as an environment supporting rich fisheries. He was very much aware of the complexity of the problem, including the role of seasonal variation of winds, gusty winds, and eddies. By the time the text *The Oceans* made its way into print, in 1942, considerable understanding had been gained through fieldwork and theory. The upwelling process was seen as related to Ekman transport of warm waters away from the coast, as a result of winds blowing toward the equator along the eastern shores of the major ocean basins. Also, it had become clear (thanks in part to studies by Sverdrup and associates) that the cold water came from depths no greater than about 300 meters.

The relevant passage in *The Oceans* reads as follows:

> Consider a wind in the Northern Hemisphere which blows parallel to the coast, with the coast on the left hand side. In this case the light surface water will be transported away from the coast and must, owing to the continuity of the system, be replaced near the coast by heavier subsurface water. This process is known as upwelling, and is a conspicuous phenomenon along the coasts of Morocco, Southwest Africa, California, and Peru. The upwelling also leads to changes in the distribution of mass, but now the denser, upwelled water accumulates along the coast and the light surface water is transported away from the coast. This distribution of mass again will give rise to a current that flows in the direction of the wind.[24]

In the last sentence of this paragraph, Sverdrup refers to the notion of geostrophic equilibrium, discussed in connection with Gulf Stream transport. This equilibrium requires that a current should flow roughly parallel to the density gradient set up by the distribution of warm and cold water-masses. Off California, with the cold water being toward the shore and the less-dense warmer water being piled up farther offshore, this flow is southward, parallel to the coast. The warm water well offshore is at a level inches higher than the inshore cold water; thus, gravity encourages downhill flow of the warm water toward the coast. This gravitational force toward the coast is balanced by the opposing Coriolis force, which is generated by the southward flow of the current. Geostrophic equilibrium is a crucial concept when mapping the average flow of surface currents. The concept originated with Scandinavian meteorologists and oceanographers, and Sverdrup was thoroughly familiar with it.

As always in the ocean, simple rules are not enough to describe the complicated motions of its waters. In the upwelling region, the cold water tends to dive below the warmer water offshore, thus forming fronts that are visible as long frothy lines parallel to the shore. In addition, turbulent motions, including eddies, result from interactions of the current with the shoreline, much as may be observed along the banks of a large river. Sverdrup et al. recognized these complications:

> In spite of the added knowledge, it is as yet not possible to discuss quantitatively the process of upwelling or to predict theoretically the velocity and width of the coastal current that develops. It is probable that the tendency of the current is to break up in eddies, and that the forced vertical circulation limits the development of the current. Also, the wind that causes the upwelling does not, as a rule, blow with a steady velocity, and variations of the wind may greatly further the formation of eddies.[25]

From detailed studies of the physical environment supporting the marine life within the California Current, one could hope to make inferences regarding its fish stocks and the harvest that could be sustained. These concepts fueled much of the research at Scripps, beginning in the late 1930s. After the war, these efforts converged in the program labeled the "California Cooperative Oceanic Fisheries Investigation," referred to earlier.[26] Sverdrup's legacy in combining the physics of the ocean with its biology, together with the urgent need of government agencies for a scientific basis supporting fisheries regulation, eventually resulted in the longest-running survey of an eastern boundary current anywhere in the world. Thus, among other things, both the seasonal regularities and the year-to-year variability of coastal upwelling are uniquely documented here. Similar, though less extensive, programs were developed elsewhere. In the 1970s, especially, research on upwelling expanded rapidly, largely as a result of the International Decade of Ocean Exploration.[27]

Upwelling, as a concept, is not difficult to grasp. But real-world upwelling systems, when described in all their complexity by physical oceanographers, are difficult subjects for study, and the implications for marine life are hard to sort out. After several decades of research, physical oceanographer and Scripps alumnus Warren Wooster (Ph.D. 1953) wistfully noted that the upwelling concept had remained static among the marine community at large, despite much progress in the science. He referred to the persistence of "an upwelling mythology," commenting as follows:

> Here, for example, is a statement about upwelling that might have been made 20 years ago without serious challenge and whose equivalent is often repeated today: Upwelling is a coastal process whereby cold, nutrient-rich bottom water is brought to the surface. Vertical speeds are of the order of 10^{-3} cm/s, and the vertical motion is caused by the local winds, as demonstrated by Ekman. Upwelling causes high productivity leading to large quantities of fish. Upwelling research should be supported so that fishermen can catch more fish.[28]

Wooster then proceeds to dismantle the hypothetical statement that upwelling is not restricted to coastal regions but occurs in the open ocean as well, for example, along the equator. The rising water is not necessarily cold and nutrient rich but can be warm and nutrient poor, during El Niño conditions. The rising water does not necessarily come from the bottom (except on the shelf). The vertical transport is not a general "upward oozing" but more likely is concentrated in plumes. The German oceanographer H. Thorade was the first to note that Ekman's equations are useful in explaining coastal upwelling (not Ekman). Finally, tidal action may be very important in promoting upwelling and mixing, in addition to wind forcing. Also, he points out that a direct benefit of upwelling studies to fishing success is highly debatable: while roughly half of the world's fish catch derives from upwelling regions, research there is unlikely to result in more fish caught. While Wooster's assessment seems correct—the collapse of fisheries suggests as much—one might hope that such research could and should help in protecting fisheries in the long run, if the advice of scientists is followed in prudent management.

The central issues are not difficult to identify: what are the ingredients of biologically effective upwelling, what are the controlling factors, and how can emerging insights be applied to management of biological resources? Progress in understanding is hampered by the fact that the various upwelling systems differ greatly from each other and, in addition, show large differences through time, from one decade to another. The complexity of upwelling systems includes subsurface flow of water in the opposite direction of the surface current. Thus, an important element of the eastern-boundary upwelling systems is a poleward flow in subsurface waters. Off California, when this flow comes to the surface, it is known as the Davidson Current. This subsurface flow brings nutrient-rich waters from the eastern tropical regions, where it acquired them on its countergyre journey that started in high latitudes. (We know this because the water is cold, and it must have acquired its low temperature in high latitudes.)

A separation of an inner zone, dominated by a strong influence of coastal upwelling, and an outer one, where filaments of cold nutrient-rich water arrive from the coastal zone, is nicely seen on images taken by satellite (fig. 3.4). It is readily appreciated, when studying such images, that a ship running a series of stations, along a line at right angles to the coast, will encounter a somewhat confusing sequence of patches with high and low nutrient content, and plankton at various stages of response to the nutrient input. It is difficult to extract general rules from such data, unless much replication and long time series are available.

THE LINK TO BIOLOGICAL PRODUCTION

The "mythological" concept so nicely summarized by Wooster may be referred to as a simple two-dimensional model of the upwelling process, whereby we are only allowed to look at a cross section along an offshore transect. The 1 meter per day rise of the water (or 10^{-3} cm/s, which is roughly the same) is readily translated into a corresponding flux of nutrients. Under the assumptions that all nutrients are incorporated into microscopic algae, we obtain an estimate of a typical rate of photosynthesis from this simple scheme.

Ten percent of the photosynthetic rate provides an estimate for the rate of growth of small zooplankton including fish larvae, and 10 percent of this production is the rate of growth of sardinelike fishes. In this fashion, we can make a ballpark estimate of how many fishes the upwelling process can support. For the California Current this comes to several hundred thousand tons of fish. We might call this a bottom-up estimate. A top-down estimate can be obtained from the fishery landings before the fishery collapses. Some fraction of that (say, between one-fourth and one-half) is probably close to the sustainable yield of the stock and describes its growth potential. The details are much more complicated, of course, since each species, and perhaps even each population within a species, has its own life history and responds to variations in upwelling in different ways.

As in theoretical economics, the math can be impressive, while the fit between prediction and actual outcome of the system's behavior can be quite poor.

There are many hurdles preventing us from translating physical conditions of an upwelling system into abundance patterns of phytoplankton, zooplankton and fish. For one, stirring by wind increases vertical mixing, which takes the algal cells out of the sunlight into the dark below, where they cannot photosynthesize. Sverdrup introduced the concept of the "compensation depth" to emphasize this process.[29] It is the depth where photosynthesis just balances respiration; that is, where net algal production goes to zero. Bacteria are active within the entire mixed layer, remobilizing the nutrients fixed within organic matter, releasing ammonia and nitrate, phosphate and silicate.[30] The nutrients released above the compensation depth are available for immediate re-use, while those released below the compensation depth are available for upward mixing to the sunlit zone. Some nutrients are remobilized more readily than others. Nitrate reenters the cycle quite readily, but silicate, which is bound within the glassy shells of diatoms, has a tendency to sink out of the recycling system. As a result, shallow recycling tends to favor dinoflagellates, which do without silica, and puts diatoms at a disadvantage. In contrast, deep-reaching upwelling favors diatom production.

Surprisingly, after almost a century of research, the path from nutrients to phytoplankton growth is still a matter of lively discussion. Only recently, this field of research has been greatly stimulated by the ideas of the marine chemist John Martin, who had his lab at Moss Landing.[31] Martin drew attention to the fact that nutrients in surface waters are not used efficiently everywhere by the growing phytoplankton. He and his collaborators suggested that a lack of trace amounts of iron is what holds back production in large areas rich in nutrients.[32] This idea met with some considerable resistance from biologists who emphasized the effects of grazing by zooplankton. In the grazing scenario, zooplankton reduces phytoplankton sufficiently to prevent it from depleting surface waters of nutrients. In the meantime, a number of open-ocean experiments have shown that addition of trace amounts of iron to nutrient-rich surface waters indeed stimulates the growth of phytoplankton cells.[33]

The fact that plankton researchers were still arguing in the 1990s whether the excess nutrients in surface waters in certain region are due to iron-limitation or to rapid grazing—two fundamentally different mechanisms—says much about where we are in understanding the single most important process in the ocean: that of producing living things from sunlight and carbon.

To be fair, upwelling science has come a long way since back around 1930, when the phenomenon of upwelling was still poorly understood, poorly studied, and poorly appreciated in the oceanographic community. In a 1932 state-of-the-art summary of oceanographic investigations published by the National Research Council of the U.S. National Academy of Sciences, "upwelling" is not even mentioned in the index.[34] The panel included leading figures in North American ocean sciences of the time, such as the marine biologists H. B. Bigelow and A. G. Huntsman (both working on the East Coast), the physical oceanographers C. O'D. Iselin (Woods Hole), and the Scripps scientists G. F. McEwen, Erik Moberg, and T. Wayland Vaughan.[35]

The omission of an index entry for upwelling is surprising, especially since both physical and biological aspects are mentioned in the text. Citing Ekman's (1923) analysis of currents resulting from a wind blowing parallel to the coast in the Northern Hemisphere, McEwen wrote as follows:

> The lower layer of thickness D, called the bottom current, is the bottom part of the gradient current. In order to satisfy the condition of continuity, the flow toward the coast in this layer must equal the flow away from the coast in the upper layer. Thus an upwelling of cold bottom water would result. Such a circulation is found off the Pacific coasts of North and South America and accounts for the abnormally cold water characteristic of these coastal regions.[36]

Thus, the fact that upwelling exists was known and appreciated at least by some oceanographers. Also, even in the early 1930s marine biologists were well aware of the importance of limiting nutrients for the level of production of plankton (and hence of fish), and they were aware that these nutrients—nitrate, phosphate, and some substances so dilute that they were hardly detectable—are rare in the sunlit zone. Huntsman well described the implications of nutrient supply by upwelling:

> On the Pacific coast of North America the phenomenon of upwelling of the deep water is quite pronounced, and furnishes rich surface water for almost uninterrupted growth of the marine plants, and following these there is a wealth of pelagic, as well as littoral crustacea and fishes. The Pacific herring and the Pacific pilchard occur in immense quantities along the coast; and a number of species of salmon of the genus *Oncorhyncus*, feed upon these and upon the pelagic crustacea [copepods and euphausids] and, on return to their rivers, bring within ready reach of man a literally enormous quantity of food from the sea. The rationale of this exceptional production of valuable commercial fishes in the deep, open ocean is as yet very imperfectly understood. A somewhat similar phenomenon is to be found on the Pacific coast of South America.[37]

Huntsman also was aware that strong winds and tidal action can greatly contribute to mixing in shallow water, which can lead to a local decrease in temperature and a stimulation of plankton growth and corresponding benefits to the fisheries. He emphasized, at the same time, the lack of theory and measurements that would allow an understanding of the path from limiting nutrients to phytoplankton to zooplankton and finally to the commercial fishes (which were his main interest). The development of theory would have to await a more intimate marriage of physical and biological concepts, combined with a mathematical approach.

Sverdrup's (1953) work "On conditions for the vernal blooming of phytoplankton"[38] (which built on concepts introduced in the 1930s by his countrymen H. H. Gran and T. Braarud) proved fundamental to this endeavor, but much of the credit for subsequent developments rests with scientists on the East Coast, especially at Woods Hole, notably G. A. Riley and colleagues.[39] One very important development was the quantitative measurement of productivity of the sea, which was greatly advanced through the radiocarbon method introduced by the Danish biological oceanographer Einar Steemann Nielsen, who sailed on the Galathea Expedition (1950–1952). The new method allowed systematic mapping of the patterns of ocean production (see chapter 8).

CALCOFI: LINKING FISHERIES AND PHYSICS

At Scripps, fishes were not a major aspect of marine biology studies in the early years. The one person who concerned himself with fishes was the curator of the aquarium, Percy Spencer Barnhart (1881–1951), who joined the staff in 1914. In 1935 he published a book on the coastal marine fishes of southern California, and he displayed mounted casts of many of them in the institution's museum.

While fishes as objects of marine biology were somewhat neglected, the same was not true for fisheries. As mentioned, shortly after his arrival Sverdrup began to work with the U.S. Bureau of Fisheries. The agency wanted to know where the California sardines spawn. To find out, one must map fish larvae and eggs. The bureau's interest was of a practical sort. The once mighty Chinook salmon fishery had collapsed, with the last salmon cannery on the Sacramento River having closed in 1919. The demise of that fishery had demonstrated that ignorance is not bliss when dealing with the exploitation of natural resources. The people at the bureau wanted to know how the sardine stock maintained itself, with a view to developing ideas about intelligent regulation of the fishery.

Sverdrup saw the opportunity to move the focus of the institution from coastal studies to the systematic investigation of the entire fish-bearing sea off the western shores of North America. As mentioned, he laid out, together with scientists of the bureau, a grid of stations spanning the sea off southern California, from just north of Point Conception to the border with Mexico. The grid consisted of transects running at right angles to the coast straight out to sea, with stations spaced more closely toward the shore, where conditions vary more readily. The idea was to occupy these stations repeatedly and to make measurements of temperature, obtain seawater samples for nutrient and oxygen analysis, and extract filtered samples for the study of the plankton that sustains the fish populations. Results would define the seasonal changes in the activity of the California Current and in upwelling, and the variations in the abundance of plankton that go with these changes. From 1938 to mid-1941 the institution's research vessel *E. W. Scripps* took scientists to the 40-some stations on a tight regular sampling schedule, in coordination with the bureau's surveys.[40] This program represents the intellectual forerunner of a major survey program instituted in the last year of Sverdrup's directorship and subsequently known as CalCOFI. Through this program, running for decades right up to the present, the major distributions of important plankton organisms became well known (fig. 7.6), as well as changes in their abundance patterns through time.

Among the participants in the initial survey cruises was the marine biologist Martin W. Johnson. Among other things, Johnson was looking for the planktonic larvae of a nearshore organism, the common sand crab *(Emerita analoga)*. This small crab is abundant on the beach off Scripps, where it lives in large aggregations buried in the sand for protection, and with antennae protruding to catch food from the wash of the waves. Johnson was intrigued by the life history of this crab, which disperses through its planktonic larvae. He first worked out the complex development of the larval stages of this crustacean. Then he counted the specimens in plankton hauls away from the coast. He found that seaward transport of the larvae was an effective distribution method. While the general movement of the currents was parallel to the coast, eddies moved the plankton inward and outward over a period of weeks, providing an opportunity for dispersal in all directions. His work made clear that the California Current is not some kind of "river in the sea," but a system of complicated swirls, eddies, and countercurrents, whereby coastal waters

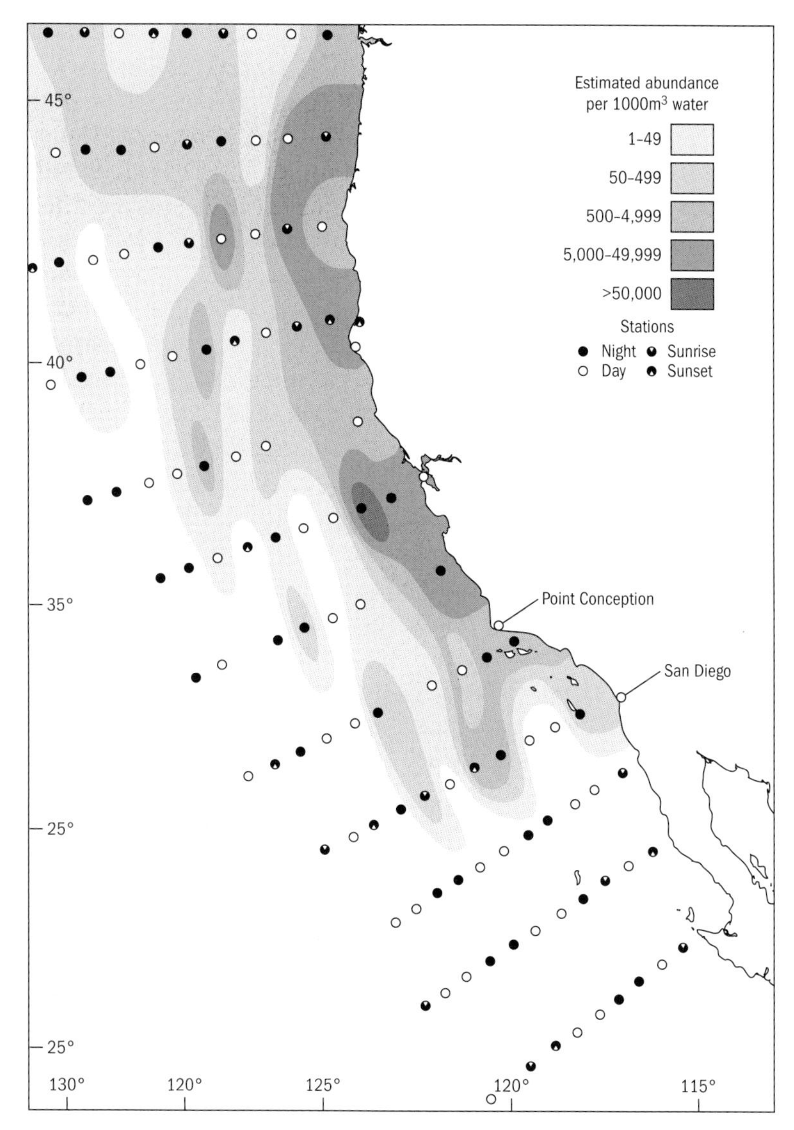

FIGURE 7.6. Example of the abundance distribution of an important planktonic food organism, *Euphansia pacifica*, mapped by the CALCOFI program. Note the break between a northern and a southern region of the California Current, roughly at the latitude of San Diego.

and oceanic waters are mixed in a broad zone along the coast.[41]

In *The Oceans*, Johnson's discoveries are summarized as follows:

> Perhaps the most obvious importance of currents on marine life results from its direct influence on dispersal. It has already been pointed out that nearly all benthic animals in the littoral zone possess free-swimming or floating eggs and larvae. The significance attached to this type of life history in any given species must be that it provides a means of taking advantage of water movements to bring about and maintain as wide distribution as possible, so that all environmental facies favorable to the life requirements of the adult both for spawning and vegetative growth can be inhabited.[42]

The map of currents drawn by Johnson[43] nicely illustrates the fact that the Southern California Bight hosts a large supereddy, with the California Current moving southward outside the bight, but with countercurrents dominating the shoreward side. Johnson and Sverdrup were aware of the implication of these findings for the dispersal of the eggs and larvae of the fishes of the current, as the subsequent discussion makes clear. In essence, these researches showed that the sea off California is richly endowed with seeds (that is, larvae) from both benthic and pelagic organisms. The seeds would settle and grow in favorable places if such are presented in a timely manner.

Fish larvae, the scientists knew from studies elsewhere, have their own strategy of improving their chances to get to favorable places, mainly by changing the depth at which they reside. At different depths the water moves in different directions. Not that the larvae know which direction that is, or why they should prefer one depth over another, but the correct program for their behavior is embedded in their genes, with errors eliminated by failure to grow up and reproduce. Their genes know how the ocean works, at least on average. (This is precisely the reason why there is trouble when the ocean changes its way of operating.)

The natural history of fishes became an important focus of study when Carl Hubbs arrived in La Jolla, in 1944, eager to scour the Pacific for specimens, nearshore, offshore, and even inland.[44] At the time of his arrival, Scripps was without an ocean-going research vessel, as the result of the war.[45] To make matters worse, the rationing of gasoline limited his driving along the coast, impeding his efforts of sampling from the shore. Nevertheless, during his first six months at Scripps, he and his wife Laura, his constant field partner and assistant, managed to collect 107 species of fishes along the local coast, several of them rare and some previously unknown to science.[46] In 1946, Hubbs had occasion to collect fishes from an unusual base: the luxurious yacht of a Hollywood star. The expedition (to Mexico) yielded notable results, including some memorable photo opportunities, showing a dapper Errol Flynn with the eminent ichthyologist.[47]

After the collapse of the sardine fisheries in the late 1940s, fortunes changed. The California legislature voted to provide funds to study the situation (and to institute a tax of 50 cents per ton on sardines). In 1947, the funding to Scripps came to $300,000; it was increased to $400,000 the following year. Other organizations were heavily involved: The U.S. Fish and Wildlife Service concentrated on the spawning, survival and recruitment of the sardines. The California Department of Fish and Game set out to determine the abundance of sardines, and their distribution, migration, and behavior. The California Academy of Sciences began laboratory studies on the behavior and physiology of sardines. Hopkins Marine Laboratory of Stanford University joined in 1951 to study the Monterey Bay area in detail. Established in 1948 under the name "California Cooperative Sardine Study," this project in 1953 became the "California Cooperative Oceanic Fisheries Investigations" (CalCOFI) and from then on took all of the state's fisheries as its charge.[48]

The funds provided to Scripps by the new program, and the availability of surplus military ships after World War II, allowed the institution to obtain three new ships: a harbor minesweeper renamed *Crest*, an ocean-going tug renamed *Horizon*, and a fishing vessel by the name of *Paolina-T*.[49] The U.S. Fish and Wildlife Service also had a research ship, and so did the California Department of Fish and Game. Beginning in 1948 all these vessels participated in the great survey of 670,000 square miles of ocean from the mouth of the Columbia River to halfway down Baja California, Mexico, and extending outward 400 miles. The CalCOFI program has been invaluable in generating a long-term record for judging the response of the sea to human impact.[50]

For the newly born program, the grid of stations that had been established in the earlier Scripps survey (1938–1941) was expanded by placing transects at right angles to an average coastline, at 120-mile intervals, between the end points previously established. Along these lines stations were spotted every 40 miles, and each station was to be occupied once a month by one of the participating ships. At each station a plankton tow, a hydrographic cast, a bathythermograph record,[51] and a phytoplankton cast were taken. Dip-netting for fishes was done at night, notes on marine birds and mammals were logged, and weather observations were sent to the U.S. Weather Bureau four times a day.[52]

For a few years every Scripps graduate student was required to go to sea on at least one CalCOFI cruise, without regard to his or her major subject.[53] For many future oceanographers at the institution, this was their first time at sea—and some left the profession because of the experience. Working the stations during stormy weather on one of the small vessels in the Scripps fleet could be a downright miserable trip. The physical oceanographer Joseph L. Reid, who was on the first CalCOFI cruise (and many cruises after that), later said of the *Paolina-T*[54]: "She was a perfectly seaworthy vessel. She would never sink—though we kept hoping! She just didn't care which end was up." Naturally, severe limits are put on the delicacy of operation of equipment during such times. As Sam Hinton noted, the need is "for a product that can be operated by a seasick technician who must hold on to a stanchion with both feet and one hand while doing his work."[55]

Pills to combat seasickness came into general use about 1949. The British geophysicist Edward C. Bullard, a frequent visitor at Scripps, hailed the pills as "probably the greatest scientific contribution of the twentieth century to oceanography."[56] There is no question that the availability of the pills enabled the great discoveries of the rest of the century to be made in greater comfort.

From the point of view of oceanography, it was fortunate that the motivation for the large postwar program on the dynamics of the California Current was a fish that had a large range and a long spawning season. The result was that scientists could collect an incredible body of information about the physical oceanography of the eastern Pacific Ocean and the life within it, all within the framework of a well-defined goal. The introduction to the first formal report of the program states the aim of the program as follows (p. 7): "To seek out the underlying principles that govern the Pacific sardine's behavior, availability, and total abundance." The report continues (somewhat defensively, perhaps):

> Scientists are investigating the sardine in relation to its physical and chemical environment, its food supply, its predators and its competitors, and attempting to evaluate the findings in terms of the survival of the young and in terms of the distribution and availability of the sardines when they reach commercial size. The pursuit of these studies has occasionally led researchers into some fields that may seem, at first glance, to have little to do with sardines. Investigations have been undertaken that were impracticable a few years ago. But the integrating principle of them all is the patient quest for more knowledge of the Pacific sardine.[57]

Physical oceanographers were able to map the patterns of temperature gradients and mixed-layer depth over a large area and compare them with nutrient and oxygen distributions. This provided insights into the origin of the layer of greatly diminished oxygen content below the mixed layer (the oxygen-minimum layer). Biologists were able to map out the elements of the basic food chain: phytoplankton to zooplankton to small fishes to larger fishes, and they obtained clues to the transfer rates and transfer efficiencies from one step to the next, from variations through time. Thus, the entire California Current became a laboratory where the effects of physics on productivity and fish abundance could be studied in some detail and over sufficiently long periods to capture the dynamics of the system.[58]

As far as the sardine, investigations in the 1950s confirmed and extended earlier suspicions that the home range of the fish is off Baja California, and that it comes north whenever conditions are favorable. The rest of the time its close cousin, the anchovy, dominates the scene. As far as general patterns off California, one striking result was the close correspondence of phosphate concentrations in upper waters with the abundance of zooplankton; that is, the food for fishes.[59] This discovery confirmed a strong link between upwelling, nutrient supply, primary production (that is, growth of microscopic algae), and fisheries success.

THE DISCOVERY OF EL NIÑO NORTE

Routine data gathering is not something people get excited about. The work at each station, in each season, in each year is much the same. Boredom is prevented mainly by mishaps and bad weather, and sometimes by trying out a new instrument, or by unexpected findings such as a tropical fish in cold waters, or the sighting of a rare bird.

The payoff for repeated standardized observations and for doing them right is that patterns emerge that document variations through time, some quite subtle. The first important discovery of such variation was not subtle at all, however. After a decade of gathering routine data, CalCOFI scientists noticed rather unusual oceanic conditions in 1957: the entire California Current had warmed considerably. This was exciting news, and a symposium was promptly called, in June 1958, to discuss the observations.

The meeting was held in Rancho Santa Fe, then a half-hour's drive from La Jolla, in part over dirt roads. Attending were about 30 oceanographers, including fisheries personnel and meteorologists. Among the participants were many prominent scientists, including, from Scripps, Carl Eckart, John Isaacs, Martin Johnson, Walter Munk, June Pattullo, Joseph Reid, Roger Revelle, Milner Schaefer, and Warren Wooster. Fritz Fuglister, Henry Stommel, and William von Arx had come from Woods Hole, and Jules Charney from MIT. John Marr, Garth Murphy, Jerome Namias, and Oscar Sette represented their respective government agencies. From abroad came Nicholas Fofonoff (Canada) and Yositada Takenouti (Japan).[60]

John Isaacs, in the Marine Life Research group at Scripps, gave the introduction.[61] He presented himself as "a naïve, enthusiastic sort of person" and hinted that the range of temperature variation from one year to another, in the California Current, must be an important factor to consider when discussing the dynamics of the current. In particular, he was struck by the fact that the variability was low during the previous decade, which introduced a bias into ideas of how the system works. MIT meteorologist J. G. Charney then introduced the first speaker, the meteorologist Jerome Namias of the U.S. Weather Bureau, with the following words:

> It is appropriate that the first speaker is Mr. Namias, who will discuss the atmospheric events that are presumably responsible for the oceanic changes. I wish to say here that we meteorologists have been particularly at fault in regarding oceans as a kind of passive body which merely responds to the motions of the atmosphere. It is clear, if one considers long-period changes, that this cannot be the case. Indeed, for long-term changes, it is probably more the other way around. Ultimately we shall have to consider the oceans and atmosphere as a coupled dynamical system.

We can dimly discern here, on the horizon, the important consequences that the CalCOFI-driven discoveries would have for Scripps and, indeed, for all of oceanography: a move toward integration of atmospheric sciences and ocean sciences.[62]

The report on the symposium, entitled *The Changing Pacific Ocean in 1957 and 1958*, documents the extent of the strange event in some detail, both the physical changes in the Pacific and the biological response of the California Current. Warren Wooster made the connection to the phenomenon known as "El Niño" off Peru.[63] In that region of strong upwelling and cold offshore waters, the upwelling ceases during certain years, around Christmas (the season of the Christ Child, that is, el Niño). Warm water invades from tropical regions. Fishes disappear, fishermen and their families suffer, and birds die. The recognition that the unusual events in the California Current in 1957/58 were manifestation of large-scale Pacific anomalies, and that they were linked to the Peruvian El Niño, changed the way oceanographers look at the operation of eastern boundary currents, and the way we understand ocean-climate interaction in general.

The changes seen in 1957/58 all through the tropical Pacific and, indeed, in many other regions of the globe had a profound impact on ocean studies. It forced scientists whose subjects are tied to climate to change the framework of their thinking in fundamental ways. Meteorologists, largely concerned with predicting the weather, had to take seriously the proposition that large-scale changes in ocean temperature are possible from one year to another, and that they have ramifications in the distribution of typhoons and rainfall and, undeniably, all kinds of weather patterns. "Air-sea interaction" emerged as the watchword of the new paradigm.

Physical oceanographers now had new reasons to map the changes of ocean properties from one year to the next—things such as surface temperature, depth to the thermocline, density gradients. The variability of these properties, part of the "weather" of the ocean, became the new focus of research, while the geography of averages became background information. Biological oceanographers were faced with drastically new phenomena such as a sudden drop in productivity in an entire eastern boundary system and the invasion of tropical species into regions where this is entirely unexpected. Quite generally, it became clear that measurements and observations on a Thursday afternoon in any part of the ocean did not necessarily yield typical conditions, but that observations have to be systematically repeated to make sure of the range of variations about the average condition and to catch the extremes, if possible.

The sense of surprise, even shock, that initiated the shift in paradigm permeates the Rancho Santa Fe proceedings of June 1958 that summarize the discussions during that first El Niño meeting. The following quotes illustrate the point:

> By the fall of 1957, the coral ring of Canton Island, in the memory of man ever bleak and dry, was lush with the seedlings of countless tropical trees and vines.[64] Two remarkable and unprecedented events gave rise to this transformation, for during 1957 great rafts of sea-borne seeds and heavy rains had visited her barren shores. One is inclined to select the events of this isolated atoll as epitomizing the year, for even here, on the remote edges of the Pacific, vast concerted shifts in the oceans and atmosphere had wrought dramatic change.

And further,

> The meteorology of the North Pacific was most unusual, with intensification of the North Pacific low and slackening of the winds along the California Coast. In regions of the Pacific where intensive oceanographic measurements were being carried out, investigators were sharply aware of changes. Over much of the eastern North Pacific water temperatures were as much as three degrees centigrade higher than normal, and in the California current, more than four times the solar heat actually received, would have been necessary to account for the warming.[65]

The point about the solar heat proved very important in steering the discussion regarding the processes at work. It emphasized the necessity for large-scale transport of warm waters toward the shores of California, either northward along the coast by the countercurrent (as seen in drift bottles) or by spreading out of warm water from the central gyre (as proposed by Walter Munk during the meeting). Presumably both mechanisms are at work during El Niño conditions: tropical marine fauna does move north off California, and there is a rise in sea level with a thickening of the warm-water layer all along the coast, suggesting a slackening in the rotation of the central gyre, which results in spreading.

At the end of the proceedings, the editors of the symposium (Oscar Sette and John Isaacs) summarized the main events in the Pacific as follows:

> The year 1957 terminated a monotonous decade, which involved low temperatures and strong northerly winds along the coast of the eastern North Pacific, a low stand of sea level, warmer conditions in the western Pacific, and possibly cold conditions in the equatorial Pacific and around Hawaii. From the evidence presented in this Symposium, the first indication of the upcoming change occurred in 1956, perhaps as early as July of that year. By

July 1956 the temperatures at Christmas Island, almost on the equator, had started to depart from the previous seasonal record, starting a warming trend that was to be accentuated through 1957/58. In the same month there began rapid and vacillating shifts in the axis of the Kuroshio quite unlike the previous record. In August 1956, the sea temperatures off the Baja California Coast rose sharply without an associated decrease in wind velocity, and by October a conspicuous anomaly existed in the meteorological conditions of the central Pacific, quite unlike previous autumns of the decade. By November 1956, the equatorial crossings showed a strong warming in the north central equatorial Pacific. This trend was well developed by January 1957, and was associated with a drop in the trade winds by this time.

Further, with a reference to weakened trade winds, they note: "It is interesting to conjecture that the sudden decrease in the trades from their high spring level occurred in the summer of 1956, resulting in a sudden decrease in east to west flow, which was associated with the disturbances in the Kuroshio and the California Current, similar to that suggested by events in 1957."[66]

It is evident from this summary that the discussions had brought out the major phenomena that need to be considered when talking about El Niño dynamics: the weakening of the trade winds (which allows warm water to move eastward along the equator and helps slow the gyre rotation), the expansion of the Alaska Low, and the expansion of the warm-water lens that defines the central gyres.

Sette and Isaacs then turned to the biological consequences of the physical changes:

> As early as the spring of 1957, there was evidence of a northward encroachment of warm water species of fish and possibly a retreat of the northern species along the California Coast. This trend intensified over the year and into 1958, with southern fish migrating even into Alaska waters. Among the planktonic forms, warm water phytoplankton made its appearance off southern California early in 1957. The tropical nature of the flora intensified until, in December 1957 and January 1958, the population was essentially tropical, with some tropical species extending to Monterey by March 1958. . . . Although Central Pacific species apparently did not encroach on the California Current from the west, southern species extended at least as far as northern California close to the coast.

Their discussion here highlights the importance of the drifting of plankton organisms in solving the question of whether warm water was brought in mainly from the central Pacific, or from the south along the coast, in a warm countercurrent. While the general rise in sea level suggests that offshore waters moved inward, the presence of tropical plankton shows that warm water was brought in from farther south, close to the coast. Evidence from drift bottles also indicates a strong narrow countercurrent along the California Coast. Sette and Isaacs now add a sentence on sardines—the topic that brought so many organizations together in the first place: "The pattern of spawning of sardines during 1958 was considerably altered, with spawning taking place both farther north and earlier than at any time in the previous decade."

Perhaps the single most important concept in precipitating the revolution in thought was the linking of the Peruvian El Niño to the strange happenings off the shores of California, as suggested by Wooster. In the words of the editors: "One of the most provocative documentations of the Symposium was that of the apparent relationships between oceanic and atmospheric changes between the North and the South Pacific."

The ocean-atmosphere system had provided a fantastic natural experiment, but, as the discussions showed, the theoretical understanding of ocean dynamics was quite inadequate to cope with it or to grasp its implications. The El Niño phenomenon had made its entry. It is still not clear to what degree it is predictable. Its typical period of recurrence is 4 to 6 years, an interval encumbered with great uncertainty.

In the meantime, other similar oscillations in multiyear climate patterns have joined

El Niño in the discussion of trends and fluctuations of prevailing weather patterns, such as the Pacific Decadal Oscillation (ca. 20 years), and the North Atlantic Oscillation (typically 7 to 8 years). These various fluctuations determine, for any series of years, where production is high in the sea and where the rain will fall on land. They are not well understood, as is evident from the fact that the shifts in patterns they produce are not well explained. They may be unpredictable in principle.

EL NIÑO: A NEW PARADIGM FOR INTEGRATED CLIMATE AND OCEAN STUDIES

A quarter century after the meeting in Rancho Santa Fé the superlarge El Niño of 1982/83 was discussed in another such meeting, in a conference called by Warren Wooster (who had left Scripps for the University of Washington) and David Fluharty, in 1984.[67] By this time the phenomenon was well appreciated as a major climatic anomaly affecting the entire tropical Pacific, with implications for the upwelling regions off Peru and California. It is now seen as a manifestation of an enormous oscillation (El Niño–Southern Oscillation, or ENSO).[68]

The oscillation derives from the alternation of the buildup and decay of a continent-size warm-water lens in the western tropical Pacific, the western equatorial Pacific warm-water pool. The warm-water pool is one of the most remarkable and crucially important features in the heat budget of the planet. It is the Earth's climatic "hot spot," a region of intense convection where heat is gathered and from which it is redistributed. Roughly, it is bounded to the west by the most extensive and most diverse reef structures in the ocean, and to the south by Papua New Guinea and Australia. It is open to the north and serves as a breeding ground for typhoons migrating into the Philippines and toward Japan.

The El Niño oscillation can be visualized as a seesaw motion of the warm-water masses between the western and the eastern side of the tropical equatorial Pacific (fig. 7.7). Normally, the Pacific trade winds move warm tropical surface waters to the west, piling them up in the warm-pool region between Indonesia, the Philippines, and New Guinea. The water there has a year-round temperature close to 29 °C (85 °F). In the east, where the westward-moving surface water is replenished by water imported from off South America and from upwelling, the sea surface is relatively cool (less than 25 °C east of 140° W). Warm water throughout the tropical Pacific gives off heat and moisture to the atmosphere, so the western tropical region where moist air and warm water collect becomes the great center for convection and rainfall on the globe. Also, the heat from the warm pool feeds energy to tropical storms there.

For some reason, during certain years, the trade winds weaken. Warm surface water is no longer carried westward. On the contrary, the piled-up warm-pool water starts moving east, taking the convection region with it. Drought strikes Indonesia, forests burn. Unusual rainfall is recorded on the tropical islands between 170° W and 100° W, including Tahiti. Upwelling along the equator is greatly diminished. Corals suffer from the warming and shed their symbionts (which absorb sunlight for photosynthesis); bleaching sets in. The surface waters in the eastern region become warm and deliver warm water southward and northward. Moisture from these warm waters can then power torrential rains and flooding in the coastal regions of the adjacent continents.

In the North Pacific, the low-pressure area associated with the Aleutian Islands (the Aleutian Low) and the high-pressure area off the West Coast (the Northeast Pacific High) characterize normal conditions (fig. 7.3). The California Current runs strong. When the trade winds weaken and the eastern Pacific warms up, the global wind system changes. The Aleutian Low tends to expand southward. The Northeast Pacific High weakens. Winter storms take a more southerly track than normal and impinge on central and southern California. As the California Current slows and warms, fisheries collapse. Kelp forests suffer from the lack of nutrients in

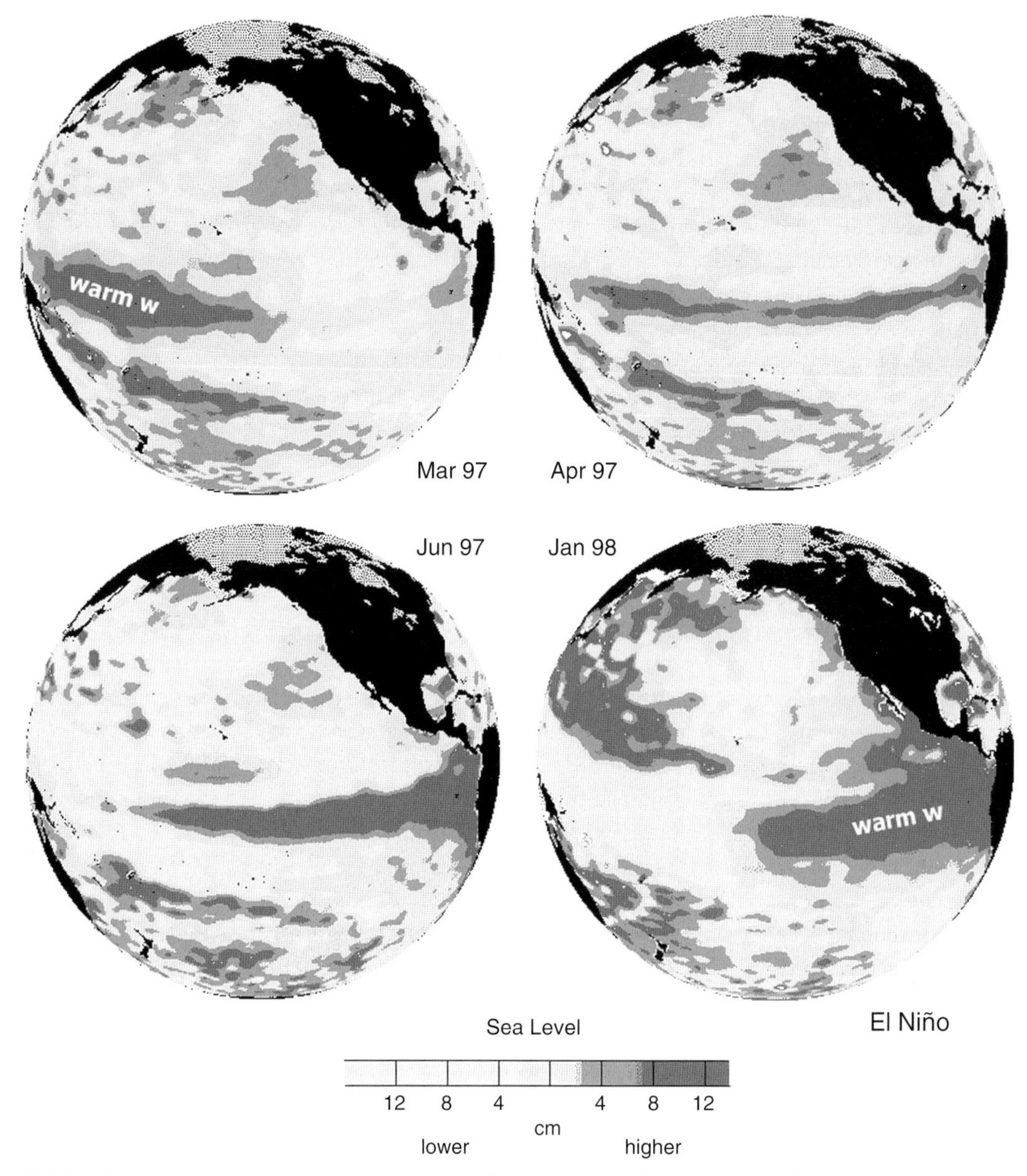

FIGURE 7.7. Sea level changes in response to the change from normal conditions (warm water [warm w] in the west, upwelling in the east) to El Niño conditions, as seen in satellite surveys. Warm water has the higher sea level. Note the west-to-east seesaw motion of the warm-water body in the equatorial Pacific.

the warm surface waters. Baby sea lions starve because the fish are elsewhere. Warm, moist air is widely available, favoring development of storm centers. The jet stream in the upper troposphere and the lower stratosphere, which runs along the boundary of arctic and tropical air, moves farther south. Storms following the jet stream path may pick up tropical moisture-laden air at their southern edge, which makes for especially heavy rains in the Southwest.

In this manner are the unusual events observed in the California Current and in California's weather linked to drought in New Guinea, and even to the monsoon system in India. This tropical control of the strange weather patterns off the West Coast was not appreciated when the phenomenon was first discussed in 1958 at the meeting in Rancho Santa Fe and for some time afterward—although Wooster's talk on the Peruvian El Niño already hinted at the fact that the tropics must be involved. Scripps climatologist Daniel R. Cayan, who studies the links between California's water budget and the changing conditions at the surface of the Pacific Ocean, put it this way: "What Namias missed is that the tropical ocean is the biggest gorilla in the Pacific zoo [of weather-controlling factors]."[69]

A shift to a tropical perspective, which emphasizes the east-to-west air pressure differences in the tropical Pacific, was well established by the early 1980s.[70] By this time, satellite images were available, showing year-to-year basinwide changes in cloud patterns. Also, the systematic comparison of temperature records from the equatorial Pacific with temperature and precipitation records from elsewhere around the globe produced more or less consistent patterns that pointed to underlying tropical control.[71]

In 1972, a strong El Niño episode resulted in the collapse of the largest fishery that ever was, the Peruvian anchoveta fishery. Ten years later, the event of 1982/83 had devastating consequences around the rim of the Pacific and elsewhere in the world and led to extensive coral bleaching in many parts of the ocean. It greatly raised the awareness of the public regarding the importance of the El Niño phenomenon. In California, newspaper headlines proclaimed "El Nino's Weird Effects," "Havoc on Land, Sea—El Nino: The World Turns Topsy-Turvy," "El Niño Decimates State's Kelp Beds," "Brown Pelicans Threatened as El Nino Sends Anchovies Away," "El Nino Blamed for Huge Drop in Commercial Catch," "El Niño: Harbinger of Deadly Weather," "Result of El Nino: Tropical Barracuda Caught inside Moss Landing Harbor," and many other strange and newsworthy items.[72]

The highly unusual El Niño event of 1982/83 did not hold the record for long. The event of 1997/98 was the biggest on record by several measures. Some have labeled it the "El Niño of the century."[73] Damages have been estimated as four times larger—more than $30 billion—than those of the 1982/83 event, with the reinsurance industry absorbing considerable losses.[74]

Only the future will show whether this ominous piling up of large El Niño events toward the end of the twentieth century reflects a new long-term trend, or whether it is a passing wave in the overall climate narrative.

We now know that oscillations of the climate system are common in the ocean, involving both currents and wind fields. The main examples, mentioned earlier, are the (North) Pacific Decadal Oscillation and the North Atlantic Oscillation, on which oceanographers have bestowed much research effort in recent years. (The preeminence of northern oscillations in the collection may be related to the distribution of land and sea, or else to the fact that most oceanographers work north of the equator.) The El Niño phenomenon itself is part of the Southern Oscillation (the *SO* in *ENSO*), which is tropical in nature (rather than truly southern). Typical periods in these oscillations range from 7 years to 50 years or so, but longer cycles also exist, as shown in records of coral growth and in sediments accumulating below upwelling areas.[75] In the context of the attempt to predict how the frequency and amplitude of ENSO events will change in the future, the longer cycles (that is, cycles near 55, and between 90 and 110 years) present great difficulties for computational modeling.

One would hope that eventually it should be possible to predict the next El Niño half a year ahead, with some confidence. But inasmuch as the ENSO events are embedded in larger oscillations, we need to understand these long-term periods of change before we can predict, with some assurance, what the future is likely to bring in terms of ENSO activity.

The ENSO events as such (when not focusing on their extreme manifestations) are not exactly rare events but are part of the regular behavior of the tropical ocean-atmosphere system. In the twentieth century, there were 24 unusual warmings of the eastern tropical Pacific, that is, on average every fourth year had an ENSO event. The prize questions are why El Niño events occur at all, and why they come at intervals of typically four to six years. The answers are not known. Apparently, the buildup of the western warm pool by the trade winds sets up an unstable situation. Once decay of the pool is triggered, the process feeds on itself, since the redistribution of warm water interferes with normal trade wind patterns. It then takes well over a year to get things back to "normal." Strong episodes were recorded for 1891, 1918, 1925, 1931, 1940/41, 1957/58, 1972, 1976, 1982/83, 1986, and 1997; that is, these major events occurred 10 years apart, on average. The reason for such spacing is obscure.

A number of ENSO events resulted in floods in California, but not invariably so. On the whole, the likelihood of flooding is increased during such times. The El Niño winters of 1992/93, 1994/95, and 1997/98 brought heavy storms to southern California, with much flooding and coastal erosion. However, the milder El Niño events during the late 1980s were associated with drought.[76] Non-ENSO years also have produced floods, on occasion. Hydrologists and civil engineers talk about a "50-year flood" and a "100-year flood," referring to exceptionally big events. As a rule, the big events are linked to El Niño, and they do, by far, most of the geologic work.

Floods typically occur during years of high precipitation (greater than 12 inches in a year) during the months of December to March (that is, in the rainy season). Occasionally, an amount equivalent to the average rainfall for one year can come down within a couple of weeks or less, in a localized region. For an extreme example, in 1891 30 inches of rain was measured in two days in an area near today's Wild Animal Park in San Diego County. Much of the rain in San Diego County (and elsewhere in California) falls in the mountains. Thus, floods may occur even if there is little or no rain in the lower reaches of the rivers draining into the sea. In any one region, intervals between floods are highly variable, ranging from a year to decades. After many years of quiet, the ever-present reality of the danger of flooding may be forgotten, an amnesia effect that may help explain the building of major developments in many a flood plain, for example on the San Diego River.

In summary, the dynamics of the California Current is tied to the climate dynamics of the entire Pacific, and in turn helps control the weather and the water budget of the state of California. Extraordinary events of precipitation in our region (which produce high stream flow or flooding) are commonly associated with El Niño conditions, when the eastern Pacific is unusually warm. Winter storms, generated in the North Pacific, move farther south than at "normal" times. Also, they have the opportunity to pick up moist tropical air along the way to California's shores. Without the offshore high (which acts as a road block), the storm systems can invade our region and dump their load.

An interesting corollary of the link between the California Current and the water budget of the state of California is that the water-stressed trees of southern California should contain information about offshore sea-surface temperatures, in their growth record. Indeed, such a record is contained in tree rings. In 1973, a graduate student by the name of Arthur V. Douglas, working with the guidance of the tree-ring expert Harold C. Fritts, presented a master's thesis to his committee at the Faculty of Geosciences at the University of Arizona, which explored this link for the first time.[77] Douglas collected tree-ring cores from the Santa Ana Mountains southeast of Los Angeles, and from these and other core data (San Gabriel Mountains, San Bernadino Mountains, San Jacinto Mountains) obtained a consensus from trees in southern California about the varying climate, especially the supply of precipitation. From Scripps he obtained sea-surface temperature data for Port Hueneme (northwest of Los Angeles), Balboa (southeast of Los Angeles), and La Jolla. He found that he could use the tree record to approximate the variation in sea-surface temperature back to AD 1611.

Thus, A. V. Douglas started a program whereby tree proxy records are used to reconstruct the history of the sea-surface temperature, making it possible, in principle, to detect long-term variations beyond the reach of direct measurement. His results were soon incorporated into ongoing work at Scripps, on finely layered offshore sediments in Santa Barbara basin, enhancing the value of these data for reconstruction of the climate changes in the Santa Barbara region.[78]

Douglas credits Scripps's McEwen with first noting a link between summer sea-surface temperature at La Jolla and the following rainy season at various locations in southern California.[79] The results obtained by Douglas are still relevant today in the context of deciding whether the unusually warm period comprising the final quarter of the last century is extraordinary (and hence likely caused by human impact), or whether periods like this occurred earlier.[80]

A NEW LOOK AT FISHERIES MANAGEMENT

The shift in thinking forced by the Pacific-wide climate anomaly in 1957/58, upon the oceanographers and meteorologists meeting in Rancho Santa Fe, greatly impacted the fishery sciences, as well. Of course, a transfer of physical insights to the problems of biological management of the oceans was precisely what the CalCOFI program was designed for in the first place. But the manner in which this came about was entirely unanticipated, and thoroughly unhelpful to the task of managing fisheries.

At the meeting, two presentations addressed fisheries concerns, one on the redistribution of fishes in the eastern North Pacific in 1957 and 1958,[81] and one on fish spawning during this period.[82] The first of these, by J. Radovich of the California Department of Fish and Game, emphasized the "heavy influx of southern species into the waters of California" during earlier warm years (1926 and 1931) and again during the period in question. The discussion centered on the question of to what degree temperature changes are responsible for the changes in catches of these fishes and others, and to what degree it is the food available to fishes. There is no indication of surprise, but rather a déjà vu response: the event of 1957/58, as far as this fishery biologist was concerned, was much like the ones observed in 1926 and 1931. Isaacs pointed out that the catch record, which goes back some time, might allow reconstruction of the frequency of southern invasion events.[83]

The report on fish spawning during the 1957/58 event focused on the sardine. Striking changes were seen in the spawning behavior of this species. In short, egg production was unusually reduced in 1957 and 1958, in both spawning centers, the Southern California Bight and off central Baja California.[84] These findings reinforced the message that the 1957/58 event had great impact on the biology of the current, including the main focus of the CalCOFI program, the Pacific sardine.

Regarding the abundance of the sardine, large-scale climatic variation now had to be considered in addition to fishing pressure and so-called background fluctuation. Eventually, John Isaacs looked to the history of fish-scale accumulation in the marine basin off Santa Barbara basin for evidence of large variations in sardine abundance through time. He started a coring program in the basin, where sediments preserve a year-by-year record undisturbed by the burrowing of large organisms. (These are absent because of the lack of oxygen at the floor of the basin.) Graduate student Andrew Soutar was put in charge of the program. Soutar identified and counted the numbers of fish scales from the Pacific sardine, the northern anchovy, and the Pacific hake, which together constituted 80 percent of all fish scales in the sediment. The record covered many centuries of change.[85]

The abundance of sardine *(Sardinops sagax)* scales proved to be highly variable, with some periods (e.g., in the sixteenth century) showing no scales at all (fig. 7.8). In contrast, remains of its relative the northern anchovy *(Engraulis mordax)* are much more reliably abundant. Only once previously, about 800 years before the present, did the core record indicate an abundance of

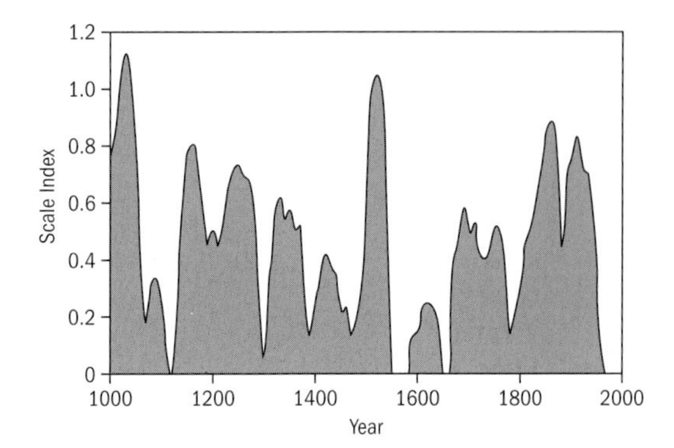

FIGURE 7.8. Changes in abundance of fish scales in the marine sediments off Santa Barbara suggest large natural variations in the abundance of the Pacific sardine.

sardines like that of the 1920s and 1930s when California fishermen exploited the stock with such success.

The conclusion drawn by John Isaacs and Andy Soutar was that natural variation in fish populations off California is very large indeed, in response to natural climatic changes. Sardine populations had come and gone, flourished and collapsed, without any human interference. In accepting the message from the sediment record, and linking it to the evidence from observing the ENSO event of 1957/58, Isaacs went much further than others in pursuing the implications of the El Niño paradigm shift. He realized that the nature of change of fish populations would make it exceedingly difficult to assess effects of fishing, and to measure the success of management.

Another fallout from studying the El Niño phenomenon from the viewpoint of fisheries is this: in terms of value to the fishermen (but not in tonnage), the losses in coastal pelagic fishes and shelf and slope fisheries were almost made up by the gains in the temperate and tropical tuna.[86] It is difficult to escape the impression that a focus on the sardine in this context is beside the point. The scope of the problem being studied had expanded enormously by the early 1980s. The sardine was no longer at the center of attention. It comes and goes and has made a modest comeback in recent years (fig. 7.9). It is unlikely that this welcome comeback is a consequence of good fisheries management.

What then did we learn? We learned that if one would study the dynamics of the weather-making system off California and its effects on biology (and fishes), one needs to engage in long-term monitoring of the environment. It is the analysis of the patterns of change, through the years, that reveals the processes. Without such knowledge, there is no chance whatever to gain a full understanding of the year-to-year variations seen in the distribution and abundance of marine organisms.[87]

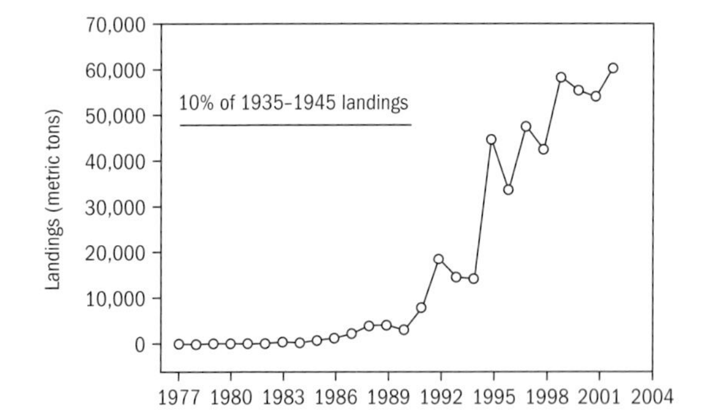

FIGURE 7.9. Comeback of the sardine, since 1990. The level is modest, compared with the landings in the 1930s and 1940s.

NOTES AND REFERENCES

1. J. Steinbeck, 1945, *Cannery Row.* Viking Press, New York. In the preface, Steinbeck paints a picture of frenzied activity revolving around the processing of the sardine harvest from the California Current.

2. The character "Doc" in Steinbeck's novel was inspired by his friend Edward F. Ricketts, marine naturalist and commercial collector who supplied marine specimens to academic institutions. (See chapter 3.)

3. A. F. McEvoy III, 1979, *Economy, Law, and Ecology in the California Fisheries to 1925.* Ph.D. dissertation, University of California, San Diego, 484 pp.

4. From 1938 to mid-1941 scientists on the institution's research vessel, the *E. W. Scripps,* repeatedly occupied a predetermined grid of 40 stations within the California Current system. At each station, plankton samples were taken, temperatures recorded, and seawater samples collected for analysis of phosphate and oxygen.

5. *Fish* and *fishes* both denote a collection of the finned marine vertebrates. *Fish* is the older term and originally inclusive of many types of animals (cf. finfish, cuttlefish, starfish, crayfish, jellyfish). Scientists commonly prefer to use the grammatical plural *fishes,* where appropriate, and reserve the term for vertebrates.

6. G. I. Murphy, 1966, *Population biology of the Pacific Sardine* (Sardinops caerulaea). Proceedings of the California Academy of Sciences, 4th Ser. 34 (1), 1–84.

7. Sardines are in the herring family along with anchovies, anchovetas, and many types of herring proper. There are around 50 genera and some 180 species recognized in this group. Most of them feed on plankton, form schools, and have their habitat near the surface in coastal waters. Among the larger forms is the Pacific sardine, *Sardinops sagax* (or *S. caerulaea*). It is less than 12 inches long and colored a glittering dark green to blue above and iridescent silvery below. Sardines and their kin prefer highly productive waters, as their manner of gathering food, by straining water through gill rakers, requires a high concentration of plankton.

8. Ships and equipment readily travel from a waning fishery to a more promising one.

9. The climatic similarity between southern California and southern Africa is conspicuously illustrated by the fact that South African plants do well in San Diego County. The golden yellow flowers along the highways of San Diego are commonly African daisies, rather than California poppies.

10. A shift in the nature of fluctuations of sardines and anchovy during this time is indicated in the sediments of Santa Barbara basin. W. H. Berger, A. Schimmelmann, C. B. Lange, 2004, *Tidal cycles in the sediments of Santa Barbara basin.* Geology 32 (4), 329–332.

11. Mountain glaciers advanced in the Alps, and extensive permanent snowfields developed in Canada, as though the Earth was getting ready for the next ice age. Harsh winters and bad harvests in northwestern Europe and elsewhere are the hallmarks of the Little Ice Age. See J. M. Grove, 1988, *The Little Ice Age.* Methuen, New York, 498 pp.

12. There is evidence that the Little Ice Age resulted from a dimming of the Sun, and that the Sun has been brighter than average in our time. See J. Lean, 2002, *Solar forcing of climate change in recent millennia,* in G. Wefer, W. H. Berger, K.-E. Behre, E. Jansen (eds.), *Climate Development and History of the North Atlantic Realm.* Springer Verlag, Berlin, pp. 75–88; K. D. Pang, K. K. Yau, 2002, *Ancient observations link changes in Sun's brightness and Earth's climate.* EOS, Transactions of the American Geophysical Union 83 (43), 481–490.

13. Some have proposed, unthinkingly, that the sensitivity of the climate to changes in the brightness of the Sun argues against the importance of human impact on recent warming. To the contrary, since the Sun's variations in brightness are quite small, a climate response documents a high sensitivity of climate to disturbance. For details see chapter 15.

14. The season-to-season and year-to-year change in the atmospheric content of carbon dioxide was first systematically measured by Scripps chemist Charles David Keeling. His series of measurements, beginning in 1956, are the best-known geochemical data series in the world today (referred to as the Keeling Curve). See chapter 15.

15. D. Roemmich, 1992, *Ocean warming and sea level rise along the southwest U.S. coast.* Science 257, 373–375.

16. The question of how the biology of the California Current responds to changes in its physical operation is the focus of the work of a number of biologists at Scripps. Among these are or were Paul Dayton and Mia Tegner (1947–2001), experts on the response of the kelp forests to warming; John McGowan and Tom Hayward, experts on plankton; and Carina Lange, an expert on diatoms. Other scientists working on this problem are physical oceanographer Dean Roemmich, climatologist Daniel Cayan, and geophysicist LeRoy Dorman.

17. J. A. McGowan, D. B. Chelton, A. Conversi, 1996, *Plankton patterns, climate, and change in the California Current.* CalCOFI Reports 37, 45–68; J. A. McGowan, D. R. Cayan, L. M. Dorman, 1998, *Climate-ocean variability and ecosystem response in the northeast Pacific.* Science 281, 210–217. John A. McGowan,

professor emeritus at Scripps, received his Ph.D. at the institution in 1961 and continued on the staff in the Marine Life Research program of CalCOFI. His field of research is the plankton ecology of the North Pacific. He went to sea on many of the early CalCOFI cruises.

18. McGowan et al., *Climate-ocean variability*, p. 210.

19. And that La Jolla is at the center of the action, naturally.

20. E.L. Venrick, J.A. McGowan, D.R. Cayan, T.L. Hayward, 1987, *Climate and chlorophyll a: Long-term trends in the central North Pacific Ocean.* Science 238, 70–72.

21. Thomas Hayward, pers. comm., 2000.

22. The realm of the regime shift is larger than the North Pacific. Off Peru and Chile surface waters warmed after the mid-1970s, and a large decline ensued in the abundance of zooplankton, as well as in the dominant pelagic fish, the anchoveta *(Engraulis ringens)*. At the same time sardines increased there, for reasons poorly understood.

23. Thorade, H., 1909, *Über die Kalifornische Meeresströmung.* Annalen der Hydrographie und Maritimen Meteorologie 37, 17–34, 63–76; G.F. McEwen, 1912, *The distribution of ocean temperatures along the west coast of North America deduced from Ekman's theory of the upwelling of cold water from the adjacent ocean depths.* Internationale Revue der gesamten Hydrobiologie und Hydrographie 5, 243–286.

24. H.U. Sverdrup, M.W. Johnson, R.H. Fleming, 1942, *The Oceans—Their Physics, Chemistry, and General Biology.* Prentice-Hall, Englewood Cliffs, N.J., 1087 pp., p. 501.

25. *Ibid.*, p. 501.

26. The program began in 1947. The collapse of the sardine fishery brought home the urgency of the problem of understanding the fish yield of the California Current. The state government responded by setting up a multiorganizational program including the U.S. Fish and Wildlife Service, the California Department of Fish and Game, the California Academy of Sciences (San Francisco), Scripps Institution of Oceanography, and representatives of the sardine and the tuna industries. The new program brought together agencies whose goals differed markedly: the federal agency had a mission of promoting the fishing industry, while the state agency was concerned with restoring and preserving fish stocks. Such conflict in mission still influences all discussion on fisheries management, at various national and international levels.

27. A summary of this phase of research is in the symposium: F.A. Richards (ed.), 1981, *Coastal Upwelling, Coastal and Estuarine Sciences,* vol. 1. American Geophysical Union, Washington D.C., 529 pp.

28. W.S. Wooster, 1981, *An upwelling mythology,* in F.A. Richards (ed.), 1981, *Coastal Upwelling, Coastal and Estuarine Sciences,* vol. 1. American Geophysical Union, Washington D.C., p. 1.

29. H.U. Sverdrup, 1953, *On conditions for the vernal blooming of phytoplankton.* Journal du Conseil, Conseil Permanent International Pour L'Exploration de la Mer 18, 287–295.

30. The role of bacteria in recycling nutrients has emerged as a high priority in plankton research, in the last 30 years. Scripps biologist Farooq Azam has played an important role in these developments. See, for example, F. Azam, T. Fenchel, J.G. Field, J.S. Gray, L.A. Meyer-Reil, F. Thingstad, 1983, *The ecological role of water column microbes in the sea.* Marine Ecology Progress Series 10, 257–263.

31. John Martin (1935–1993) worked at Moss Landing Marine Laboratory near Monterey, California. His research focused on making very precise measurements of trace elements in seawater, and he related the cycling of such elements to the productivity of the sea.

32. J.H. Martin, S.E. Fitzwater, 1988, *Iron deficiency limits phytoplankton growth in the north-east Pacific subarctic.* Nature 331, 341–343. The iron hypothesis is not entirely novel. The Norwegian oceanographer H.H. Gran expressed the idea that iron is limiting to growth in phytoplankton in the 1920s, tying high production along the coast to availability of iron. Also, the British marine chemist H.W. Harvey had published on iron as a nutrient for diatoms. H.W. Harvey, 1937, *The supply of iron to diatoms.* Journal of the Marine Biology Association of the U.K. 22, 205–219. But the high precision necessary to show the true amount of available iron was not at the command of those earlier workers.

33. Quite a large number of such experiments have been made recently. For example, K.H. Coale and 18 others, 1996, *A massive phytoplankton bloom induced by an ecosystem-scale iron fertilization experiment in the equatorial Pacific Ocean.* Nature 383, 495–501; V. Smetacek, 2001, (Antarctic experiment) U.S. JGOFS News 11 (1), 11; A. Tsuda and 25 others, 2003, *A mesoscale iron enrichment in the western subarctic Pacific induces a large centric diatom bloom.* Science 300, 958–961.

34. N.H. Heck and 17 others, 1932, *Physics of the Earth—V, Oceanography.* Bulletin of the National Research Council 85, 581 pp. The concept of growth-limiting nutrients is mentioned, with reference to the British oceanographer H.W. Harvey and the German biologist K. Brandt.

35. Henry Bryant Bigelow (1879–1967) was the founder-director of Woods Hole Oceanographic Institution. C.O'D Iselin (1904–1971) was its second director. Archibald Gowanlock Huntsman (1883–1972) was the director of the Bedford Institute in

Halifax, Canada. T. Wayland Vaughan was director of Scripps Institution of Oceanography. McEwen and Moberg represented physical and chemical oceanography, respectively. (N. H. Heck was a geophysicist.)

36. G. F. McEwen, 1932, *A summary of basic principles underlying modern methods of dynamical oceanography,* in N. H. Heck and 17 others, 1932, *Physics of the Earth—V, Oceanography.* Bulletin of the National Research Council, 85, p. 340. An earlier exposition by McEwen is in G. F. McEwen, *The distribution of ocean temperatures.*

37. A. G. Huntsman, 1932, *Relation of biology to oceanography,* in N. H. Heck and17 others, 1932, *Physics of the Earth—V, Oceanography.* Bulletin of the National Research Council, 85, p. 525.

38. H. U. Sverdrup, *On conditions for the vernal blooming.*

39. Gordon A. Riley (1911–1985) was the first graduate student of G. Evelyn Hutchinson, the famous limnologist. He joined Yale's Bingham Oceanographic Laboratory in 1938, as a marine biologist, and worked closely with Woods Hole afterward, using the *Atlantis.* G. A. Riley, 1938, *Plankton studies, I.* Journal of Marine Research 1, 335; G. A. Riley, 1939, *Plankton studies, II.* Journal of Marine Research 2, 145–162; G. A. Riley, H. Stommel, D. F. Bumpus, 1949, *Quantitative ecology of the plankton of the western North Atlantic.* Bulletin of the Bingham Oceanographic Collection 12 (3), 1–169. Riley's influence on biological oceanography is emphasized in J. H. Steele, 1974, *The Structure of Marine Ecosystems.* Harvard University Press, Cambridge, Mass., 128 pp., and in E. L. Mills, 1989, *Biological Oceanography—An Early History, 1870–1960.* Cornell University Press, Ithaca, N.Y., 378 pp.

40. The program sponsored by the U.S. Bureau of Fisheries ended with the U.S. entry into World War II. Conservation and fisheries management took a backseat to the needs of the U.S. Navy.

41. Martin W. Johnson (1893–1984); M. W. Johnson, 1939, *The correlation of water movements and dispersal of pelagic larval stages of certain littoral animals, especially the sand crab,* Emerita. Journal of Marine Research 2, 236–245.

42. H. U. Sverdrup et al., *The Oceans,* p. 858.

43. *Ibid.,* p. 860.

44. Carl L. Hubbs (1894–1979) was an expert in ichthyology and many other marine vertebrates besides, with a strong interest in natural history, ecology, and paleoecology.

45. The only research vessel at Scripps had been loaned to the University Division of War Research, and in 1946 it had not yet been returned.

46. Hubbs was fond of taking students along on his trips. But many found it a grueling experience. On trips along the coast the students were expected to spend most of the day collecting fishes, and then sometimes past midnight sorting and preserving them. Hubbs occasionally allowed some specimens to be turned into dinner, with proper attention to possible health hazards. For example, Hubbs knew from earlier experience that eating cabezon *(Scorpaenichthys marmoratus)* was somewhat hazardous. To locate the problem, he obtained a fresh cabezon with ripe roe from a local fisherman (in 1949). Hubbs then enlisted the aid of chemist Arne N. Wick of Scripps Metabolic Clinic, who tested the roe on rats and guinea pigs. The animals became ill and several died. The roe of the cabezon is toxic, but the flesh is not. C. L. Hubbs, A. N. Wick, 1951, *Toxicity of the roe of the cabezon,* Scorpaenichthys marmoratus. California Fish and Game 37, 195–196.

47. Errol Flynn offered the collecting opportunity in a letter to SIO, motivated by the desire of his father, Theodore Thomson Flynn, a biologist in Ireland interested in fisheries, to do studies off Mexico. Hubbs joined the cruise, in hopes of opportunities for collecting around Mexican islands. After stops at several islands off Mexico, Flynn made a sudden decision to go straight to Acapulco. While Flynn partied, Hubbs escaped to collect 160 different species of fishes in the bay and along the beach. Footage from the "expedition" was later incorporated into *The Cruise of the Zaca,* a short movie widely shown around the country. For details, see E. N. Shor, 1981, *Hollywood and the scientist.* Oceans 14 (3), 53–55.

48. E. N. Shor, 1978, *Scripps Institution of Oceanography: Probing the Oceans.* Tofua Press, San Diego, 502 pp. See pp. 43–78. Scripps scientists who attended the 1948 meeting defining the new program were Carl Eckart (director after Sverdrup returned to Norway), Carl L. Hubbs, Martin W. Johnson, Roger R. Revelle, John D. Isaacs, and graduate student J. Laurence McHugh.

49. Having a fleet of ships made Scripps unique among oceanographic institutions. These ships were used for many other expeditions, of course, besides the CalCOFI monthly surveys.

50. At this juncture, early in the twenty-first century, the CalCOFI program has entered a difficult stage because of a shortage of state funding. It is to be hoped that funding will be found to continue this long-running survey, which turned out to be so valuable to understanding the dynamics of the California Current and, indeed, the entire northeastern Pacific.

51. The bathythermograph is an instrument containing a temperature-sensitive and a depth-sensitive part whose deformations are transferred to a stylus that scratches a line into a carbon-covered glass slide, tracing out a temperature-versus-depth line. The apparatus is housed in a solid metal tube, with slits to admit the surrounding water. The tube is lowered down to 100

meters or thereabouts, and the glass slide with the temperature profile is removed and replaced after the tube is on deck again. In the late 1950s, Scripps engineer James M. Snodgrass developed a throw-away unit, the "expendable bathythermograph" (XBT), which records the temperature as a function of depth and sends the information back to the ship via a thin copper wire. It is not retrieved after the observation is made.

52. E. N. Shor, *Scripps Institution of Oceanography*, pp. 49–50.

53. This regulation did not exempt the women students; Joseph L. Reid told Shor that he recalls June Patullo and Margaret Robinson participating in MLR cruises. Some thought it would be difficult having women aboard ship. But it proved not to be.

54. J. L. Reid, 1982, *An oceanographer's perspective.* CalCOFI Reports 23, 39–42.

55. Quoted in E. N. Shor, *Scripps Institution of Oceanography*, p. 58.

56. Quoted in T. F. Gaskell, 1960, *Under the Deep Oceans.* Eyre and Spottiswoode, London, p. 73.

57. Marine Research Committee, State of California, 1950, *California Cooperative Sardine Research Program, Progress Report, 1950.* State Printer, Sacramento, 54 pp.

58. Within the framework of the CalCOFI program, Scripps oceanographers developed a great number of new types of gear, including the Isaacs-Kidd Midwater Trawl, free-fall instruments for deep observations (including a fish trap and a baited camera), and moored traps for catching phytoplankton remains falling toward the seafloor. A training program for marine technicians was another benefit. In large part, these developments were led by John D. Isaacs, director of Marine Life Research unit from 1958. (See below.)

59. J. L. Reid Jr., G. I. Roden, J. G. Wyllie, 1958, *Studies of the California Current system.* CalCOFI Progress Report 1958, 27–56.

60. O. E. Sette, J. D. Isaacs (eds.), 1960, *The Changing Pacific Ocean in 1957 and 1958.* California Cooperative Oceanic Fisheries Investigations (CalCOFI) Symposium, Rancho Santa Fe, California, June 1958. CalCOFI Reports 7, 217 pp.

61. John Dove Isaacs (1913–1980) was a major figure in ocean sciences, an ebullient seaman with unbounded curiosity and little respect for established dogma. An athletic six-foot-plus, and never short of striking ideas, he was a commanding presence in any meeting. He started his sea-going career as a fisherman, during the Depression, when jobs and food were scarce. (His various adventures as a fisherman in Oregon are described in D. Behrman, J. D. Isaacs, 1992, *John Isaacs and His Oceans.* American Geophysical Union, Washington, D.C., 230 pp.) This experience gave Isaacs a very special relationship to the sea, one that included respect for its dangers, and an abiding interest in the living things in it. He came to Scripps in 1948 and took a leading part in the new CalCOFI program, becoming director of the Marine Life Research unit in 1958. Isaacs contributed importantly to ocean instrumentation (his background was in engineering), knowledge of waves and beach dynamics, ENSO dynamics, desalination methods, ocean resources, atmospheric jet streams, the distribution of sardine larvae, vertical migration of plankton and scattering layer, currents near the deep-sea floor, food from the sea, internal tides, meteorology, history of fish populations inferred from fish scales in sediments, distribution of zooplankton in the California Current, the grand patterns of marine ecology, large scavengers on and near the seafloor (perhaps his most important discovery), and the energy contained in salt deposits, among many other items. If it had to do with the ocean it fell into the range of Isaacs's probing questions. (For a list of publications, see D. Behrman, J. D. Isaacs, *John Isaacs and His Oceans*, p. 205 ff.)

62. Jerome Namias (1910–1997), after retiring from the U.S. Weather Bureau, later joined Scripps as the head of a fledgling climate group, which was to become the fastest growing research division at Scripps, and the one dealing with some of the most perplexing and exciting topics in climate change. Namias brought with him a new perspective—that of air-sea interactions and teleconnections, married to the conceptual tools of Scandinavian meteorology, which arose in response to the practical demands of long-term weather prediction. Results of his studies were published in CalCOFI atlases 22, 27, and 29 (1975, 1979, 1981). In his autobiographical notes, he traced the new research paradigm to the meeting in Rancho Santa Fe (as reported in CalCOFI Report 7, p. 23):

> It was not until 1958 that my research took another major turn—this time into the realm of air-sea interaction on time scales of the order of a month to seasons. The stimulation for this was the now-famous Rancho Santa Fe CalCOFI Conference, held by Scripps Institution of Oceanography, at the idyllic Rancho Santa Fe Inn, not far from La Jolla. . . . It struck me that some of the secrets of long-range weather forecasting might lie in the coupled air-sea system. It was especially noteworthy that the mismatch of time scales in the two media, air and sea, could account for the frequently observed long-term memory required for long-range problems. (J. Namias, 1986, *Autobiography*, in J. O. Roads (ed.), *Namias Symposium.* Scripps Institution of Oceanography Ref. Ser. 86-17, 1–59.)

63. Warren S. Wooster, a sea-going physical oceanographer at Scripps studying circulation in the

eastern Pacific and its relationship to fisheries, gave a core presentation at the symposium. Wooster may have realized that studying the anomalies in the North Pacific (the focus of the work of Namias) would not yield the whole story, and that the center of action was farther south, in the tropics. His talk was entitled, simply, "El Niño." He drew attention to the recurrent nature of anomalous warming off Peru and emphasized that, as far as he could see, what happens off Peru and what had just happened off California was pretty much the same thing (*ibid.*, p. 43):

> One of the most celebrated of oceanic disturbances is that known as El Niño, an occurrence of the first half of the year which is reported at irregular intervals from the coast of Northern Peru. Conspicuous outbreaks were reported in 1891 and 1925 and more recently in 1941, 1953, and 1957–58 [refs. here omitted]. Similarities between this phenomenon and conditions observed off the California Coast and in other coastal upwelling zones suggest that the underlying causes of the observed abnormalities are the same in such regions.

We now realize, thanks to an enormous amount of research since the Rancho meeting, that Wooster was correct when linking the Peruvian El Niño to the anomalies observed in the North Pacific and especially off California. However, he understated the importance of the direct link: not only are the causes much the same, *it is the very same cause that underlies both the southern and the northern abnormality*. Wooster subsequently addressed this point, suggesting that the term *El Niño* be reserved for events involving the equatorial Pacific (W. S. Wooster, 1998. Where is El Niño? EOS, 79 [21], 251). In the book titled *El Niño North*, the focus is on El Niño effects in the northeastern Pacific. W. S. Wooster, D. L. Fluharty (eds.), 1985, *El Niño North—Niño Effects in the Eastern Subarctic Pacific Ocean*. University of Washington, Seattle, 312 pp.

64. Canton Island is north of Samoa, just south of the equator.

65. O. Sette, J. D. Isaacs, *The changing Pacific Ocean*, p. 21.

66. *Ibid.*, p. 211; the quote is considerably shortened and paraphrased in places, and references are omitted.

67. W. S. Wooster, D. L. Fluharty (eds.), *El Niño North*.

68. S. G. Philander, 1990, *El Niño, La Niña, and the Southern Oscillation*. Academic Press, San Diego, 293 pp.

69. Talk given on 8 Jan 2003 at the Birch Aquarium at Scripps.

70. The pressure difference between Darwin in Australia and Tahiti in the South Pacific is commonly taken as a measure of the east-to-west difference in air pressure. This difference varies as the air pressure values over Darwin and over Tahiti move in opposite directions, in a seesaw manner. Sir Gilbert Walker referred to the phenomenon producing the negative correlation between the air pressure records in the two regions as "Southern Oscillation," in the 1930s. See S. G. Philander, *El Niño, La Niña*.

71. Studies by the physical oceanographer Klaus Wyrtki (Hawaii; at Scripps from 1961 to 1964), and the meteorologists J. D. Horel, J. M. Wallace, and E. M. Rasmusson were important in these developments. K. Wyrtki, 1975, *El Niño—the dynamic response of the equatorial Pacific Ocean to atmospheric forcing*. Journal of Physical Oceanography 5, 572–584; J. D. Horel, J. M. Wallace, 1981, *Planetary-scale atmospheric phenomena associated with the Southern Oscillation*. Monthly Weather Review 109, 813–829; E. M. Rasmusson, J. M. Wallace, 1983, *Meteorological aspects of the El Niño/Southern Oscillation*. Science 222, 1195–1202.

72. A. Bakun, 1996, *Patterns in the Ocean—Ocean Processes and Marine Population Dynamics*. California Sea Grant College System, NOAA, Washington, D.C., 323 pp. As mentioned, the rendering of *El Niño* as *El Nino* is incorrect. *El Ninyo* is the correct transliteration.

73. K. E. Trenberth, 2001, *El Niño Southern Oscillation (ENSO)*, in J. H. Steele, S. A. Thorpe, K. K. Turekian (eds.), *Encyclopedia of Ocean Sciences*. Academic Press, San Diego, pp. 815–827.

74. Jürgen Krause, manager at Cologne Re, pers. comm., 2001. (The Cologne Re, like other reinsurance companies, insures insurance companies against catastrophic losses.)

75. T. R. Baumgartner, A. Soutar, V. Ferreira-Bartrina, 1992, *Reconstruction of the history of Pacific sardine and northern anchovy populations over the past two millennia from sediments of the Santa Barbara basin, California*. California Cooperative Fisheries Investigation Reports 33, 24–40; F. Biondi, C. B. Lange, M. K. Hughes, W. H. Berger, 1997, *Inter-decadal signals during the last millennium (AD 1117–1992) in the varve record of Santa Barbara basin, California*. Geophysical Research Letters 24 (2), 193–196; W. H. Berger, J. Pätzold, G. Wefer, 2002, *Times of quiet, times of agitation: Sverdrup's conjecture and the Bermuda coral record*, in G. Wefer, W. H. Berger, K.-H. Behre, E. Jansen (eds.), *Climate Development and History of the North Atlantic Realm*. Springer, Berlin, pp. 89–99.

76. K. E. Trenberth, 2001, *El Niño Southern Oscillation*, p. 825.

77. A. V. Douglas, 1973, *Past Air-Sea Interactions off Southern California as Revealed by Coastal Tree-Ring Chronologies*. M. S. thesis, University of Arizona, Tucson, 98 pp.

78. A. Soutar, P. A. Crill, 1977, *Sedimentation and climatic patterns in the Santa Barbara vasin during the 19th and 20th centuries.* Bulletin of the Geological Society of America 88, 1161–1172.

79. Douglas cited G. F. McEwen, 1925, *Ocean temperatures and seasonal rainfall in southern California.* Monthly Weather Review 53 (11), 483–494.

80. Douglas found that his tree-ring records supported earlier conclusions of Hubbs, based on unusual fish occurrences. He reported an anomalously warm period from about 1847 to 1856 (that is, increased rainfall and larger tree rings) and noted that Hubbs postulated unusual warm times in the middle of the nineteenth century, based on the reported abundance of tropical fishes. C. L. Hubbs, 1948, *Changes in the fish fauna of western North America correlated with changes in ocean temperature.* Journal of Marine Research 7 (3), 459–482.

81. J. Radovich, 1960, *Redistribution of fishes in the eastern North Pacific Ocean in 1957 and 1958.* CalCOFI Reports 7, 163–171.

82. E. H. Ahlstrom, 1960, *Fish spawning in 1957 and 1958.* CalCOFI Reports 7, 173–179.

83. Isaacs was keen on finding ways to reconstruct the history of the California Current, realizing that therein was the clue to many intriguing questions.

84. The high concentration of eggs around Punta Eugenia in the 1950s in general (as seen in the maps provided by Ahlstrom (E. H. Ahlstrom, *Fish spawning,* pp. 174 and 175) suggested that upwelling in a subtropical setting is favorable for egg production in the sardine.

85. A. Soutar, J. D. Isaacs, 1969, *History of populations inferred from fish scales in anaerobic sediments off California.* CalCOFI Reports 8, 63–70; A. Soutar, J. D. Isaacs, 1974, *Abundance of pelagic fish during the 19th and 20th centuries as recorded in anaerobic sediment off the Californias.* Fisheries Bulletin 72, 257–273.

86. P. E. Smith, 1985, *A case history of an anti–El Niño to El Niño transition on plankton and nekton distribution and abundance,* in W. S. Wooster, D. L. Fluharty, (eds.), *El Niño North—Niño Effects in the Eastern Subarctic Pacific Ocean.* University of Washington, Seattle, pp. 121–142.

87. The point that the dynamics of the physical environment is paramount in controlling the abundance of fish is no longer in doubt. It was made, most recently, by the biological oceanographer Francisco Chavez of the Monterey Bay Aquarium Research Institute and colleagues. F. P. Chavez, J. Ryan, S. E. Lluch, M. Cota, C. Ñiquen, 2003, *From anchovies to sardines and back: Multidecadal change in the Pacific Ocean.* Science 229, 217–221. They propose a back-and-forth shift of conditions in the entire North Pacific between a cool "anchovy regime" and a warm "sardine regime." The Santa Barbara record does not support this suggestion. For the anchovy scales, the dominant cycle is near 100 years, for the last 800 years or so, but for sardine scales the cycles are distinctly shorter than this value. With such discrepancy in cycling, no seesaw is possible over the long run. In defense of his hypothesis, Chavez wonders whether the basin's record is representative of the ecology of the entire California Current system (pers. comm., August 2004). We do not know. But for the time before instrumental records, for the California Current, it is the information that we actually have in hand.

EIGHT

Meadows and Deserts of the Sea

ON THE ELUSIVE CONCEPT OF OCEAN PRODUCTIVITY

Production varies widely in the sea; it is largely controlled by nutrient supply and the availability of sunlight. The crucially important nutrients are phosphate (fig. 8.1) and nitrate (which has a distribution similar to that of phosphate), as well as silicate (which tends to follow the same distribution patterns but with important exceptions). Silicate is the nutrient that allows diatoms to make shells of glass. Diatoms are an important group of primary producers sometimes referred to as "grass of the sea." Primary production in the sea is carried out by both photosynthetic bacteria and microscopic algae, including diatoms and dinoflagellates. It is measured in terms of grams of carbon fixed per square meter per year (or per day). Typical values are around 100 production units (grams of carbon per year); somewhat lower than the equivalent photosynthetic fixation on land.

The central questions of interest are the controls on production—largely the supply of nutrients wherever sunlight is adequate—and the manner of transfer of the matter generated in primary production into fish and consumers at the end of the food chain (apex consumers), including seabirds, whales and other marine mammals, and, prominently, people. This dual focus informs the discussions about the amount of food available for apex consumers in the context of fishery resources. One major topic concerns the amount of carbon fixed each year per unit area by chlorophyll-bearing organisms. An early global estimate of this production is by R. H. Fleming (the third author of the *Oceans* text).[1] His map nicely shows the major features of the productivity of the sea (fig. 8.2). The other topic centers on the food chain. The "length of the food chain," as emphasized in the 1960s by Woods Hole oceanographer John Ryther, turned out to be the most significant element of this concept.[2]

The ocean's productivity is nothing like the one on land, where trees, brush, and grass

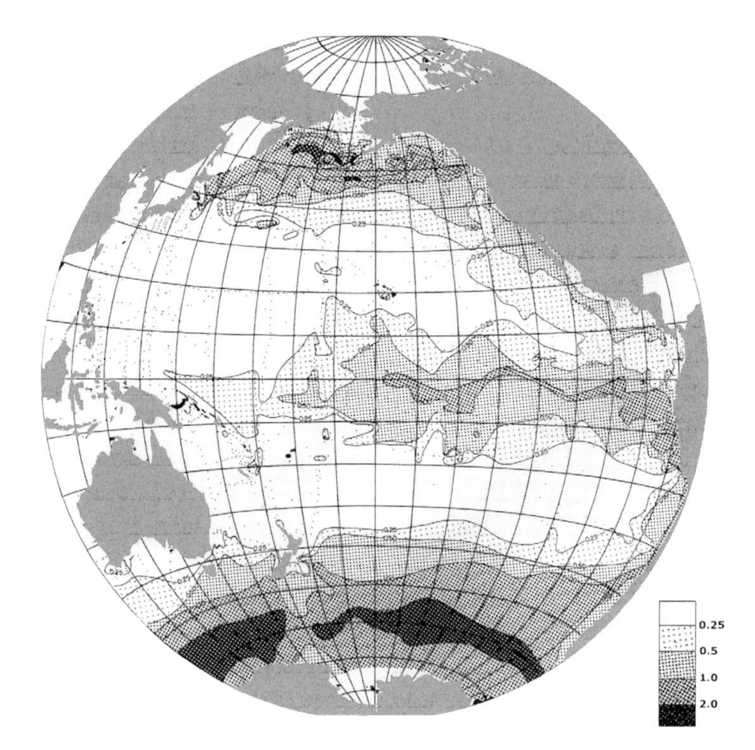

FIGURE 8.1. Phosphate concentrations below surface waters in the Pacific, according to J. L. Reid (1962). Nitrate shows similar patterns. Abundant nutrients make for meadows in the sea, with high abundance of zooplankton.

dominate. The fact is, the green primary producers in the sea, on the whole, are microscopically small and are unavailable for consumption by large animals. Only small creatures can eat the tiny algae. In turn, these small organisms feed small zooplankton and fishes, which serve as food for the larger animals that are of interest to fisheries. In other words, it takes several steps in the "food chain" till we come to the apex consumers. At each step, typically, only 10 to 20 percent of the matter at the lower step is transferred to the next-higher level. After two such transfers, a few percent of the original primary production is left to feed the apex consumer. After four transfers, less than one-thousandth is left.

In the deserts of the sea—the central gyres and other warm-water regions—the chain is long. This is why the desert, in spite of having millions of green microbes in each liter of water, does not support large animals. In contrast, regions of deep mixing and upwelling, where the microbial algae are relatively large and food chains are much shorter therefore, support abundant marine mammals and seabirds, and fisheries. The question that moves to the foreground, then, is what controls the length of food chains. The answer holds the key to understanding the productivity patterns that matter to apex consumers. In the same manner, when asking how global warming will affect the productivity of the sea, we must ask how the warming of the planet affects the length of the food chain.

To anticipate the answer: both the amount of nutrients and the particular mixture of nutrients that is locally available to support primary production help determine the length of the food chain. Specifically, a high supply of silicate shortens the food chain.

THE PROBLEM OF SUSTAINABLE YIELD

Just how large is the potential harvest of the ocean? Perhaps not surprisingly, the answer to this simple question depends on what is meant by "harvest." If whales and large fishes are counted as "harvest," as was true for the nineteenth century and much of the twentieth, there is one answer. If small herringlike fishes are counted, there is another, giving a much greater number. If we count krill (the food of baleen

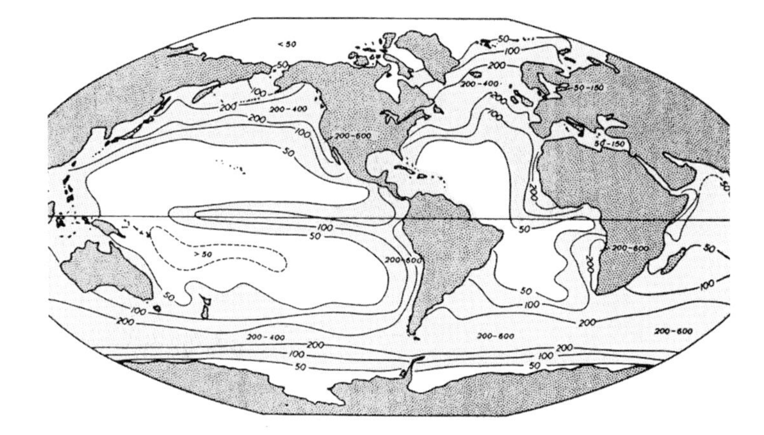

FIGURE 8.2. Primary production of the ocean in grams of carbon per square meter per year, according to R. H. Fleming (1957). The central deserts are shown in white.

whales, penguins, and crabeater seals) and squid (the food of dolphins and other marine mammals), the answer is different again. Fishing at a lower level in the food chain yields the greater harvest. It is a similar story on land: harvesting plants yields more food than harvesting plant-eaters.

Among harvestable fish, herring and their kin occupy a privileged position: they filter the water in high production regions, thus feeding at a relatively low level in the food chain, or better, the food web (fig. 8.3). Much of what was learned about the life history of plankton in the open sea resulted from studies focusing on the food web supporting herring.

The facts regarding global fish catches are reasonably clear (although the reporting system is not fail-safe, of course).[3] The total catch reached 20 million tons in 1950. After that, the rate of increase was about 1.5 million tons each year. The collapse of the Peruvian anchoveta fishery slowed the increase in global tonnage in the 1970s, but it continued nevertheless until 1988, when the total catch stabilized at roughly 90 million tons.

When attempting to interpret this historical trajectory of fishing success in terms of sustainable yield, great difficulties arise. All through the time of increasing yield, there has been a tremendous expansion in the sophistication of fishing equipment, both with regard to detection (echo-sounding, radio communication, helicopter spotting) and to processing and freezing of the catch at sea. Factory ships of advanced northern fishing nations, serving large national or global markets, now greatly out-do the traditional coastal fishing boats that serve local communities and small regional markets. Thus, the rules of the game have changed—and are still changing all the time. Not only is there practically no negative feedback on the fishing activity from the collapse of any one fishery (the effort simply moves elsewhere), but also there is a constant readjustment and evolution of technology for ever-more-efficient forms of exploitation, including a move to lower levels in the food chain.[4]

Some take comfort in the human ingenuity that allows exploitation of dwindling resources, and believe that there is, in principle, a way to manage stocks for maximum yield, like domestic cattle. Thus, in a book edited by the late Julian Simon, the irrepressibly optimistic economist, one can read this: "The bottom line is that tripling or even quadrupling production of human food from the sea remains a reasonable possibility [at 90 million tons]. Achievement of maximum sustained yield will depend on appropriate management of the resources and their fisheries, a goal that so far has proven difficult to achieve."[5] The second part of this statement remains valid, and for good reasons.

Others take a less sanguine view. A publication of the Worldwatch Institute, dedicated to "foster a sustainable society" emphasizes the damage done by wasteful fishing practices:

> Trawling fishers, who drag a large sock-like net through the water, can catch large quantities of non-target species. Shrimp trawlers have the

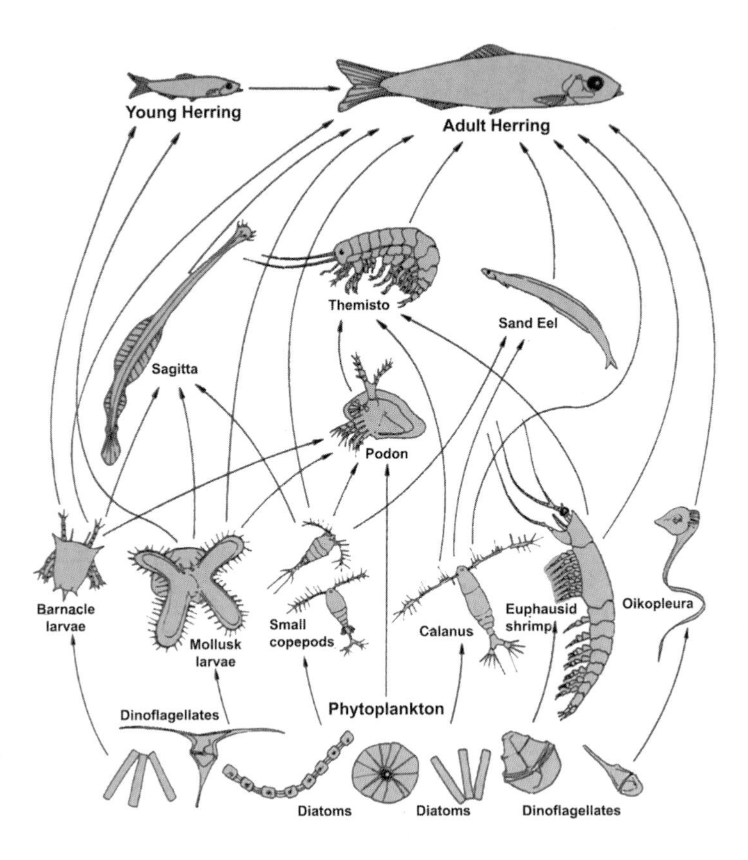

FIGURE 8.3. The food web supporting herring, according to Hardy (1924).

> highest recorded rate of by-catch because their nets have a very small mesh size to capture the tiny shrimp. In tropical waters, they bring in 80 to 90 percent "trash fish" with each haul. "Trash fish," however, is a misnomer. The discards include species that other fishers would want but will not be able to catch. . . . Worldwide, shrimp fishers are estimated to jettison up to 15 million tons of unwanted fish each year, and other fishers are thought to discard at least another 5 million tons.[6]

Comparing this quote with the preceding one, one might rightfully conclude that there is an opportunity to increase the catch by cutting back on waste. However, such an increase would be marginal.[7]

Yet others are entirely unimpressed with the problems facing fisheries in the context of the global economy and global food supply. Thus, B. Lomborg, a Danish political scientist with a background in statistics, noted that saltwater fishes represent less than 1 percent of our total calorie consumption and only 6 percent of the protein intake.[8] His assessment is somewhat of an oversimplification. The problem centers on the meaning of the word *our* as used by Lomborg. Our fish-derived protein intake, for the average human, may be no more than 6 percent, but there are hundreds of millions of people depending on fish for a considerable portion of their sustenance.

Overfishing has long been a topic of intense discussion, first in connection with the collapse of the Californian sardine fishery, then with that of the Peruvian anchoveta fishery, and lately with the collapse of the Atlantic cod and herring fisheries. According to the Food and Agriculture Organization of the United Nations, more than 25 percent of all fish stocks are severely damaged by overfishing. Of the rest, two-thirds are being fished at the biological limit. In the 30 years since 1970, it is estimated, the sum of fish stocks has been halved.[9] An important aspect of the change in fishery success is a shift from highly valued pelagic predatory fish (high on the food chain) to less valuable small plankton-eaters and groundfish (lower on the food chain). This phenomenon is called "fishing down the food web" by marine ecologists.[10] Fishing down the food web increases the available catch. In the extreme, if we eat what the fishes eat, we can greatly increase the take. In the same manner, on land, growing grain yields more food than growing cattle.

Managing the problem of fish mining (that is, removing target stocks and moving on to less-exploited ones) and rogue fishing (that is, destroying resources that approach or exceed the value of the target fish through disposal of by-catch and destruction of bottom ecologies) will require international cooperation and regulation.[11] The problem is a prime example of the "tragedy of the commons," a conflict situation named and first explained in some detail by the economist Garret Hardin.[12] Hardin analyzed the example of a common meadow available to the members of a community. If there are no constraints, every participant in the commons sends another cow to the meadow, until it is ruined. The behavior is entirely rational, because the benefit of having food for the additional cow accrues to the decision maker. The downside, that all other cows now have somewhat less, is a tax levied from all across the board. The gain to the decision maker outweighs the distributed loss, even when that loss is larger than the gain. Therefore, without agreed-on regulation, the problem is unmanageable.

But what principles are going to provide the basis for the type of regulation called for by Hardin? As we have learned from studying herring age distributions off Norway early in the twentieth century, and from the story of the California sardine,[13] large natural fluctuations in marine ecosystems make it impossible to determine that elusive entity, the maximum sustainable yield. It is this yield that the government agencies are supposedly aiming for when regulating exploitation. Unfortunately, the maximum sustainable yield is a moving target, changed both by fishing pressure and by climatic variation. It emerges only upon collapse of a fishery, when everyone agrees that it has been exceeded. It is true that the concept of maximum sustainable yield has given way to optimum sustainable yield, emphasizing a more cautious approach. However, optimum sustainable yield faces the same fundamental problems as its progenitor.

The concept of maximum sustainable yield (which underlies much of fisheries management, no matter what the label) is simple enough in the abstract. We start with the "stock" being exploited. The stock has an age structure, a mixture of younger and older individuals. It has input from new young each year ("recruitment") and from the growth of members already present, which adds weight. It has losses from natural predation and death, and from being exploited by fishing. For a stock that is neither expanding nor shrinking, recruitment and growth are balanced by natural predation and fishing mortality. An increase in fishing mortality has the following effects: 1) stock size decreases, as first the large adults and then the smaller ones are removed; 2) natural mortality decreases, as stock removed is denied to natural predators; 3) growth increases, as the age distribution shifts to the more rapidly growing younger members of the stock; 4) recruitment may stay the same, or it may decrease if too many mature adults are taken, or increase if a smaller stock size favors survival of juveniles.

At which point do we get the best yield from such a system? A mathematical treatment of this problem suggests that the most favorable yield is obtained at the point where the fishing effort is just large enough to reduce the stock to one-half its natural equilibrium size.[14]

On paper, the solution is quite elegant. Unfortunately, in the real world, we do not know the size of the stock, its natural age distribution, or its current age distribution. Since "stock" refers to breeding populations (rather than to the entire species) we would have to first identify such populations—a task that is quite daunting. Furthermore, the size of a stock is difficult to assess, even if there were only one in any one region and even if it were not migrating all the time. Not knowing the equilibrium size, the would-be regulator has problems in estimating half that size as the target to be fished toward. Not knowing the age distribution, inputs regarding growth are mainly guesswork.

Fisheries scientists do have some means of making estimates, of course. They count eggs and larvae in the plankton, for example. These are dispersed by currents and are more evenly distributed than the adults. Also, the scientists may capture, mark, and release members of the adult population. The percentage of fishes caught that are marked then gives an idea about the total number in the population. Finally, there are the fishing operations themselves, which are a kind of sampling program. Presumably, the more fish the fishermen bring in, the more fish there are out there: stock size as a function of fishing luck. Of course, improved technology may constantly shift the baseline for assessing fishing luck—last year's success-to-effort ratio may not be comparable to this year's.

One of the most volatile parameters in assessing maximum sustainable yield or similar hypothetical quantities is recruitment, that is, the entry of young fish into the exploitable stock. Year-to-year changes in food supply for fish about to become recruits, as well as survival of larvae providing recruits, are the factors determining recruitment to the stock. In essence, these factors are tied to climatic variation and therefore unpredictable. The task, thus, is to manage living resources undergoing unpredictable fluctuations. One way to do this, without getting caught in equations with too many unknowns, is to set aside preserves of appropriate extent and close them to fishing. Another is to have fishing simulate natural predation, for example by restricting the use of equipment and the right to switch between different prey species. Without some such regulation, the tragedy of the commons takes its inevitable course.

FOOD FOR FISH: PRIMARY PRODUCTION

Naturally, an estimate of the fish yield of the sea must proceed from the primary production of the ocean. Like all animals, fish ultimately depend on primary producers for sustenance, that is, on photosynthesizing organisms. In the simplest analogy to the terrestrial ecosystem, the problem resembles calculating the number of cattle that can be grown on a given range. After determining how much grass and other edible vegetation the range produces, we can translate the amount growing each year to the corresponding yield in meat, using a conversion factor. In the sea, things are a bit more complicated.

Before we can make a guesstimate about fish production in a manner analogous to that about cattle production, we need to determine the ocean's primary production, that is, how much photosynthesis is going on. We also need to study just how this production is transferred to fish. These two tasks have posed great challenges to marine biologists ever since the marine biologist H. H. Gran took samples off Norway in an attempt to relate plankton to its physical environment.[15] The problem is by no means solved. Recent estimates of the primary production of the sea vary by more than a factor of two. How the carbon is transferred from primary production to the fish of commercial interest is a field of active study and has been for many decades.

The "grass of the sea" consists of minute greenish and yellowish single-celled algae (Fig. 8.4). Some of these, among the most important in the economy of the sea, make a glass house to live in, a shell built on the principle of a pillbox. The shell material is very porous—more like a screen than a wall. These algae are called diatoms, and they are ubiquitous in the ocean, wherever there is light and nutrients for growth, and are dominant in highly productive regions. In the open sea, diatoms float in surface waters as plankton, but near the shore, where light penetrates to the bottom, they also live on the floor. On the beach, they readily cover the wet part of it, nutrient supply permitting. The free-floating forms are the planktonic diatoms, and the ones on the bottom the benthic diatoms. Some benthic species live attached to other organisms.[16]

There are many other types of primary producers besides diatoms, of course, just as there are plants besides grass in a pasture. Another

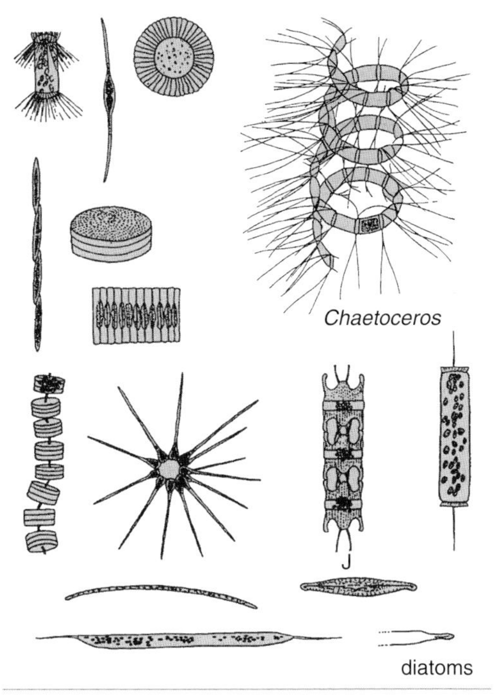

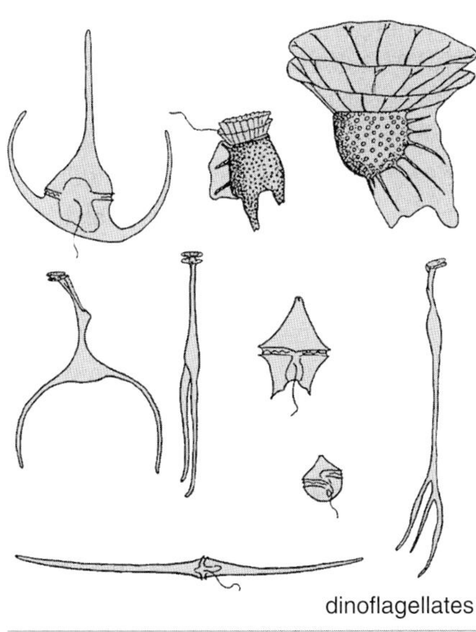

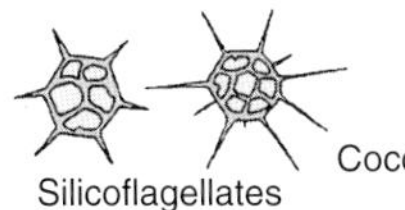
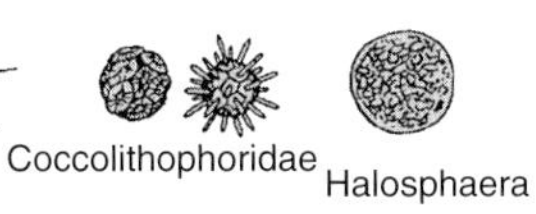

FIGURE 8.4. At the base of the food chain in high-production regions: diatoms *(upper)* and dinoflagellates *(lower)*, as presented in Sverdrup et al. (1942). *Chaetoceros* is a common form making bristly chains in upwelling regions.

important group comprises the dinoflagellates, a group common in the phytoplankton (plant plankton). As their name implies, these organisms have flagellae propelling them through the water.[17] By moving through the water, the cells improve their chances of finding nutrient molecules, since the water surrounding a cell is soon depleted. Some dinoflagellates eat other plankton, and some have lost their chlorophyll, relying entirely on prey for food. Luminescent dinoflagellates are responsible for the phosphorescence frequently observed in tropical waters.

Another group of phytoplankton whose members are extremely abundant are coccolithophorids. Most are quite small compared even to the microscopic diatoms and dinoflagellates; highly specialized microscopes are needed to see them properly. Coccolithophorids cover themselves with calcareous platelets, perhaps as defense against microbial predators. These platelets, as fossils, cover about one-half of the seafloor of the World Ocean, making sediment called nannofossil ooze.[18] In addition to this nanoplankton,[19] there are photosynthesizing bacterioplankton (picoplankton), which are even smaller, and numerically even more abundant.[20]

Fisheries are concentrated in highly productive regions, which implies that diatoms (which are particularly abundant here) are the most important primary producers in a fisheries context. Diatoms, especially the large and robust forms, are conspicuous in the plankton of the more productive regions, while coccolithophorids tend to be prominent in the phytoplankton of the deserts of the sea, the great central gyres. This large-scale production pattern was recognized by the marine biologist Ernst Hentschel when he compiled the plankton survey done during the *Meteor* Expedition (1925–1927).[21]

Diatoms in upwelling regions can produce heavy shells, and some grow quite large by making long chains. In most of the ocean, in contrast, their shells are extremely delicate, presumably reflecting, at least in part, that the raw material for building shells (silicate) is in short supply almost everywhere. There are a great variety of different types of planktonic diatoms; many of them occur in colonies, typically in long chains. An especially abundant form, familiar from upwelling regions, is the genus *Chaetoceros*. Its members bristle with long glass needles sticking out in all directions (fig. 8.4). The pioneers of plankton studies called these bristles "floating spines," identifying a potentially important function for these hairlike appendages. However, we may confidently assume that they are mainly useful in slowing down grazers, by clogging their mouthparts and interfering with swallowing.[22]

Diatom populations can grow extremely rapidly, once nutrients (nitrate, phosphate, silicate) have been injected into the sunlit zone with cold water from below. Diatoms multiply by simple cell division; after 10 such divisions, by plain arithmetic, the population has increased by a factor of 1,000. If this happens quickly, in a population explosion, we have a bloom. A bloom ends when the nutrients are used up, or when grazers reduce the population, or when conditions become otherwise unfavorable. At that point, many diatoms make so-called resting spores, that is, seeds that can start a bloom later, when conditions are favorable again.

So, how fast do the diatoms double? It depends on circumstances. During favorable conditions, a doubling of once a day or so would seem typical, from laboratory experiments and observations on rates of bloom expansion. With this information, if we could count the cells present under each square meter of ocean surface, we could make a rough guess in regard to production. Making such counts is by no means easy. The first plankton scientists towed very fine meshed nets through the water and examined the concentrate at the end of the net. This method is still used, although with the knowledge that most organisms are smaller than the mesh size and will escape. To obtain a more reliable count, one takes a water sample and centrifuges it, or one can add preservative (such as formalin) and wait until the organisms have settled out, for examination of the sediment in a special microscope (looking upward through a transparent container holding the sediment).

We can use such information to make a (very) rough guess as to how the production of grassland compares with the production of highly fertile parts of the ocean. During a bloom, let us say, we find enough diatoms below a square meter of ocean surface to make up 10 grams of organic matter, when summed through a depth of 10 meters (billions of cells, actually). The uppermost part of the water column represents the most productive zone under bloom conditions. A daily doubling would then translate into a carbon production of 4 grams per square meter per day, similar to the production of fertile grasslands. Of course, a bloom lasts only a short time, so in general, the production will be much lower in the sea than on land, where growth, fed by nutrients in the soil, continues as long as water is available.

Measuring is commonly preferred over guessing. One measures primary production by determining the amount of carbon fixed by photosynthesis in a given water sample containing phytoplankton, over a period of several hours. The chemical equation describing photosynthesis says that carbon dioxide and water, in the presence of sunlight, combine to make carbohydrates (and similar organic compounds) and oxygen. The oxygen is released to the environment. That is why, in a home aquarium by the window, we see bubbles emerge from the water plants. In principle, we can measure the amount of oxygen evolved over several hours and readily calculate the corresponding amount of carbon fixed by the plants.

In their text *The Oceans*, Sverdrup, Johnson, and Fleming cite the results from an experiment in the Gulf of Maine, carried out in the 1930s, which nicely illustrates the oxygen method of determining productivity.[23] In the experiment, transparent bottles as well as bottles covered with dark cloth, both kinds full of seawater with diatoms, were suspended along a wire at just below the surface, and in 9-meter intervals below that, down to 45 meters. After 9 hours and 10 minutes of daylight (under a variable sky) the bottles were pulled up and the change in oxygen content determined. All the covered bottles, into which sunlight could not penetrate, had lost oxygen—typically corresponding to about 5 percent of what was present to begin with. The highest losses occurred in the shallower bottles. The bottles whose contents were exposed to sunlight gained oxygen if they hung above 20 meters and lost oxygen if they were put below that depth. Down to 40 meters this loss was less than that of the covered bottles, but below that depth, the losses were the same, whether the bottle admitted light or not.

The loss of oxygen in the dark bottles at all depths is a result of respiration, which is the reverse of photosynthesis. Respiration is necessary to maintain body functions for diatoms as well as any other organisms present in the bottles, such as animal plankton and bacteria. Respiration, in these bottles, decreased with depth because metabolism slows in the deeper water, which is cold. In the deepest bottles, there was no difference between exposed and covered bottles because there was not enough light for photosynthesis. Upward from there, the excess oxygen produced over background respiration (which was equally going on in all bottles, whether covered or not) kept increasing right to the surface. The reason is that more light is available at shallow depths, and light is needed for photosynthesis. At about 20 meters, the amount of oxygen used up in respiration was equal to the amount produced in photosynthesis. The overall net gain of oxygen, summed over the water column, suggests a net fixation rate of 16 grams of carbon per square meter for that day, not counting the loss expected for the night.

Assuming that not much photosynthesis is going on in late fall and winter in this region, and on rainy days, we still get a yearly production of something of the order of 1,500 grams of carbon per square meter—at least 10 times higher than a typical open-ocean value. The Gulf of Maine, then, is not precisely a good analog for the rest of the ocean, mainly because it is very nutrient rich, and these nutrients easily reach the sunlit zone; they cannot escape to great depth because the seafloor is shallow.

An important result of the experiment was the determination of the depth where there is just enough light so that photosynthesis and respiration balance. The light intensity at which this balance is achieved is called the compensation point, and the corresponding depth is the compensation depth.[24] Gordon Riley of Woods Hole early on realized the implications of the compensation principle for the spring bloom in the North Atlantic.[25] He made a concerted attempt at a formal treatment of primary productivity, taking into account all the known factors controlling phytoplankton growth. His 1946 equation for phytoplankton growth contains these factors: incident light and its attenuation with depth, availability of phosphate, the depth of the mixed layer, the surface temperature, and the quantity of zooplankton grazing on the phytoplankton. By choosing appropriate weightings for these different factors, an excellent fit was produced between the calculated seasonal cycle of phytoplankton on Georges Bank (off Maine) and spot checks from plankton tows. His work has been acclaimed as being central to biological oceanography.[26]

In 1953, focusing on the single most important parameter in the mix of factors controlling productivity, Sverdrup subsequently defined the critical depth as the level above which photosynthesis and respiration balance.[27] When the surface mixed layer is greater than this depth, there is no net production, but nutrients are supplied to the sunlit zone through the mixing. When the mixed layer shoals, upon warming in spring in the northern North Atlantic, sunlight and nutrients combine to generate the spring bloom.[28] When the bloom has extracted the nutrients, it declines.

Thus, the optimum time for production is in spring, right after the onset of stable stratification, when nutrients are still abundant from past mixing, but when green cells are no longer mixed down into the dark. Blooms occur in all sorts of phytoplankton: diatoms, dinoflagellates, coccolithophorids, and blue-green bacteria.

Since diatoms support the shorter food chains, their blooms support the largest number of fish, birds, and mammals.

MAPPING THE DISTRIBUTION OF PRIMARY PRODUCTION

By around 1950 it was fairly clear what factors mainly control ocean production: the essential ingredients are light, nutrient supply, and intensity of grazing and of mixing of upper waters and the flip-side of mixing, stratification. The overall dichotomy between the plankton-poor blue ocean and the plankton-rich olive-green coastal ocean, known to every sailor, suggested that the supply of nutrients is crucial to overall patterns of productivity.

Assuming that nutrient concentration in upper waters captures most of the relevant information, Sverdrup was able to give a first crude map of overall patterns of production, from considering the distribution of vertical mixing and upwelling.[29] He identified the storm-stirred regions in high latitudes, the coastal upwelling zones, and the equatorial divergence zones as areas of high production.

What was still missing was a method that would allow rapid determination of the *rate* of primary production, while underway at sea. With the discovery of radiocarbon by Willard Libby in the 1940s,[30] an entirely new way of studying carbon fixation became available: if one provided the phytoplankton with radioactive carbon dioxide, one could measure the rate of carbon uptake by determining the radioactivity of the phytoplankton after the experiment. The Danish chemist and oceanographer Einar Steemann Nielsen revolutionized the field by introducing this new methodology.[31] Steemann Nielsen sailed on the Danish *Galathea* Expedition (1950–1952), perhaps the last great expedition in the tradition of scientific circumnavigation. On this voyage, about 200 radiocarbon uptake measurements were made in various parts of the ocean. Typical uptake values came in between about 30 grams and 150 grams of carbon per square meter per year, indicating a level of productivity comparable to that of deserts and semiarid regions on land.

From Steemann Nielsen's database and the relevant principles tying nutrient supply and sunlight to productivity, R. H. Fleming (as mentioned) constructed the first quantitative global map of phytoplankton production in the sea (fig. 8.2). Fleming's map nicely shows the major patterns, with high production around the continents, and the large deserts associated with the central gyres of the ocean. His map also reflects the high productivity zone in the equatorial Pacific, with the highest values in the east, as is proper. It is a map of fishery success, as well, at least in general outline. Fleming's map is an educated guess, an interpolation and extrapolation of a very slim database, complemented by a general understanding of the processes controlling productivity. Fleming was fully aware of the difficulties introduced by seasonality. Also, he had earlier identified grazing as an important factor in controlling diatom populations.[32]

The Fleming map was much copied and modified and soon ended up as a document with the Food and Agriculture Organization of the United Nations (FAO). It provided a base for estimating ocean productivity until well into the 1970s. The various resulting estimates of the rate of carbon fixation were near 20 billion tons of carbon per year. Assuming that fishes eat the crustaceans that feed on the phytoplankton, and using the conventional factor of 10 reduction between levels up the food chain, we can make the first rough estimate of the global fish production: 20 billion tons of edible phytoplankton carbon make 2 billion tons of zooplankton carbon, make 200 million tons of fish carbon. This is more than 10 times what the fishing industry was removing from the sea at the time. Thus, we can appreciate that there was much room for optimism, based on this sort of simplistic calculation.

Following the *Galathea* work, hundreds of radiocarbon determinations were made for the purpose of estimating productivity in the following decade. Many of these measurements, along with other estimates based on phytoplankton abundance and chlorophyll content of surface waters, were compiled by three Russian marine biologists (O. Koblents-Mishke, V. Volkovinsky, Yu. Kabanova) for a new global map, which they published in 1968.[33] An English translation became available two years later.[34] This map dominated the estimation of primary production for the next two decades, appearing (in simplified form) in the *FAO Atlas of the Living Resources of the Sea,* and subsequently in many textbooks.[35] Twenty years later, a new compilation further refined the patterns mapped by Koblents-Mishke and coworkers.[36] In that newer compilation, the total primary production of the ocean came out just above 30 billion tons of carbon per year.[37] This amount happens to be rather similar to the estimate for primary production on land.[38]

What exactly is the meaning of this number? For an answer we must take a close look at the method used to obtain it.[39] In essence, the radiocarbon method makes a number of assumptions that affect the reliability of the results. The main problem is that grazers are in the sample being tested for carbon fixation. The effect of grazing on the phytoplankton and other sources of die-off during incubation is to lower the estimates for total production. In addition, it is now appreciated that the minute nanno- and picoplankton are responsible for a large portion of total photosynthesis.[40] While such minute organisms would be largely retained on the membrane filters used in radiocarbon studies, the amount caught would be small relative to their productivity, because they quickly fall prey to equally minute microbes within the bottles. As a consequence, radiocarbon is released back into the water, from its fixed state.

In addition to a plethora of problems of measurement, there are questions concerning assumptions made regarding the depth interval where photosynthesis takes place. Considerable carbon fixation may occur at rather low light intensities, where nutrient supply compensates for the lack of light. All these issues are rather complex and have engendered much experimentation and learned discussion. However, as far as the food supply for fish, many of these issues are not crucially important. For example, the minute picoplankton is not available to fishes and must pass through several collection stages up the food chain to become accessible. By that time, through the factor of 10 attenuation at each step, there is little left to feed on. From this viewpoint, a total primary production estimate of 30 billion tons of carbon, counting only the coarser phytoplankton, is just as valid as one of 60 billion, including the picoplankton. Multiplying the earlier estimate by two does not increase the ability of the ocean to support fish.

In essence, an estimate of productivity depends on the purpose for which it is made. Analogous difficulties arise when assessing the productivity on land: In forest production the *wood* is of interest economically, while for atmospheric chemistry the annual growth and decay of *leaves* is equally interesting. To ranchers growing cattle on open range the production of wood and leaves is more or less irrelevant. A number describing total photosynthesis on land, then, has little use unless it is split back into its relevant parts.

Entirely new aspects have been introduced to the mapping of ocean production by the flood of data emerging from remote sensing of color of the ocean's surface, from satellites. Most maps now reproduced in textbooks are from this source. They are attractive because of the detail displayed. Reading the chlorophyll colors (mainly greens but also others) in terms of productivity poses some problems, of course: not all chlorophyll is seen (only that near the surface), and not all chlorophyll is equally productive—much depends on sunlight and the presence of various nutrients. Converting color maps to production is subject to error, therefore. The factor of uncertainty is somewhere near 1.5 and may go up to 2.[41] "Sea-truthing" is necessary to relate the satellite-derived color patterns to what is actually in the water.[42] At Scripps, Richard Eppley and coworkers contributed substantially to converting chlorophyll measurements to estimates of primary production.[43] This type of work is very much going on right now.

An important result of the monitoring of the ocean's productivity from space is the ability to define and map seasonal fluctuations on a global scale.[44] Also, the new information is useful in defining and resolving questions surrounding year-to-year changes in productivity, including long-term trends such as have been observed in the North Pacific.[45] In the Pacific, the largest such fluctuations are those tied to the El Niño–Southern Oscillation phenomenon.

THE WORKINGS OF THE PRODUCTION MACHINE

One reason why coastal waters are productive is because nutrients that are sent toward the bottom in fecal matter and in aggregates of organic debris do not sink far—the seafloor is shallow. Bacterial decay of the organic rain regenerates the nutrients and releases them to the water above the seafloor. From here the nutrients can readily cycle back into the surface layer, through upwelling, eddy mixing, storm action, and tidal turbulence. As a bonus, iron may be added to the mix in cases where iron-rich mud from land is abundant and encounters reducing conditions within the mud, which makes iron mobile. The presence of iron stimulates phytoplankton growth, wherever other nutrients are plentiful.

Besides coastal waters, there are other regions in the ocean that are highly fertile. We can readily identify them by plotting the amount of zooplankton caught in net tows, for example.[46] Zooplankton live on diatoms and other algae, and their abundance reflects the productivity of these algae. From these patterns and from the chlorophyll distribution mapped from space, we see that the basic pattern of fertility can be described as a ring around each ocean basin (the coastal zone), combined with three major belts, one each in midlatitudes, and one on the equator. The latitudinal belts coincide with regions of strong zonal winds and currents.

Let us briefly recapitulate the important factors controlling primary production.

First, let us consider the sunlight. Not surprisingly, the effects of the Sun's mean radiation on primary production vary strongly with latitude. At equinox, for example, when the Sun is overhead at the equator, the amount of sunlight falling on a unit area at 60° N or S latitude is only about one-half that at the equator. The amount of sunlight typically available in tropical areas all year round is delivered only during midsummer at 50° latitude. In midwinter, the available sunlight at 50° is one-tenth of the normal tropical value. Of course, during summer there is plenty of sunlight available even at high latitudes, because nights are short.

Not all of the incoming light enters the water. The proportion of the light reflected at the ocean surface varies with the Sun's angle and with the state of the sea. When the Sun stands high and the sea is calm, only a negligible fraction of the incoming light is reflected—the water looks very dark from an airplane. In contrast, at low angles and with a rough sea, up to one-half and even more of the incoming sunlight can be reflected. Regions in high latitudes are at a disadvantage, therefore, as far as receiving light for growth.

Light that penetrates into the water is quickly attenuated by absorption and scattering, red light the most, blue green light the least. At 100 meters depth, the intensity of the light will be less than 1 percent of the light penetrating the surface, even in the clearest ocean waters. The remaining light is almost all bluish green. In the rather murky waters of the coastal ocean and other fertile regions, the 1 percent level is reached already after about 20 meters depth, or even less. (The layer above the 1 percent level is called the euphotic zone, which means "well-lit.")

Much of the murkiness in coastal waters is due to the green microalgae themselves. Thus, even though the euphotic zone is relatively shallow, production is high in such murky waters. Of course, production in the coastal zone is not high *because* the euphotic zone is shallow. Rather, production is high in the first place, and this makes the water murky. The fact that light is absorbed so quickly in the uppermost few tens of meters is fundamental to understanding the fertility of the ocean. It means that photo-

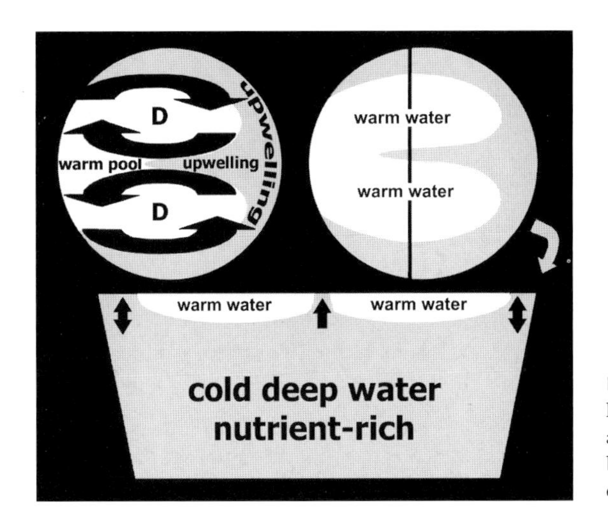

FIGURE 8.5. Schematic representation of high- and low-production regions in the Pacific basin *(upper)* and a vertical section *(lower)*, showing the contrast between nutrient-poor warm-water lenses and nutrient-rich deep waters.

synthesis—the basis of almost all other life in the ocean—is restricted to the surface waters.

On the whole, productivity and abundance of plankton well reflect nutrient distribution. What then is a "nutrient"? Actually, any element or compound that organisms incorporate into their bodies may be considered a nutrient. For plants, carbon dioxide and water are the primary nutrients. In addition, plants use nitrogen and phosphorus to build carbohydrates, lipids, and proteins. Furthermore—with variations from one species to another—organisms need minor amounts of sodium, magnesium, sulfur, chlorine, potassium, calcium, iron, molybdenum, boron, fluorine, silicon, manganese, copper, zinc, and iodine to grow and live. In fact, virtually every element in the ocean serves the needs of at least some organisms. But not all of these nutrients need to be discussed. Only the nutrients that are in short supply in the euphotic zone, where growth occurs, are of interest. These are the "limiting" nutrients.[47]

Limiting nutrients are readily recognized in that they tend to disappear from surface waters entirely, due to extraction by organisms, and they accumulate at depth, where decay and respiration set them free. Thus, the ocean's deserts are identical to the great subtropical gyres: shallow warm-water lenses that have lost their nutrients and have trouble acquiring replacement from the cold waters below (fig. 8.5). The chief nutrients are fixed nitrogen (nitrate and ammonia, mainly), phosphate, dissolved silicate, and certain trace elements, of which iron apparently is the most important. Basically, where these micronutrients are abundant, the ocean is productive, and where they are rare, it is not. When oceanographers talk about nutrients, they mean these biologically active and limiting compounds and elements. The single most important observation about nutrient distribution is that the various nutrients show roughly the same pattern. For example, if we measure the concentration of nitrate and of phosphate or of silicate and phosphate in various regions and at various depths of the ocean, we find that they vary together. This high degree of correlation considerably facilitates the task of studying nutrient distributions: any one of the nutrients tends to show the major patterns for any other.

Dissolved nutrients are constantly taken out of the water and fixed into algal matter within the mixed layer. Some of the organic matter generated within the mixed layer is lost every day to dark waters below the euphotic layer, by sinking of fecal pellets, shells, and discarded skeletal parts such as copepod molts. Below the sunlit zone, respiration and decay dominate: the photosynthetic process now runs in reverse, driven by bacteria. As a result, the nutrients are re-released to the surrounding water. Slowly, these nutrients diffuse back up into the sunlit zone, where they stimulate plant growth. The diffusion is slow because the base of the sunlit zone is linked to the thermocline—the subsurface water layer with a strong temperature gradient. The thermocline produces stable layering (warm water is lighter than cold), and the strong gradient forms a barrier to vertical diffusion. The thermocline barrier is broken wherever there is vigorous mixing and upwelling.

It took a while for oceanographers to fully grasp these basic principles of nutrient cycling—principles that are now readily taught in Oceans 101. The British chemist H. W. Harvey of the Plymouth Laboratory in the United Kingdom was the first to have a clear conception of the ocean's productivity machine, that is, the great loop of production and decay and recycling.[48] The present ocean, in its deviation from equilibrium, is a "Harvey Ocean." The Harvey Ocean differs from an equilibrium ocean in major ways.[49] Mainly, the silicate concentrations are controlled in large part by biological processes rather than by chemical equilibrium, indicating that equilibrium arguments are suspect. Some oceanographers prefer the label "Redfield Ocean" to emphasize the biological control on ocean chemistry, thus honoring another great pioneer of the ocean's nutrient cycles.[50]

The chief corollary of the Harvey nutrient cycle is that, without replenishment of nutrients from below, the process of extraction by phytoplankton growth would soon come to a halt for lack of extractable matter. Replenishment is by upward mixing of nutrient-rich deep waters rather than by river input, which is negligible on a global scale, in this context. (Locally it can be important, however.) Mixing activity is favored by strong winds. Hence, the occurrence of winds and storms is one important factor setting the fertility level of the ocean. This insight has implications for considering the question whether the ongoing global warming will increase or decrease the productivity of the ocean. On the whole, one would expect that a weakening of zonal winds in the Northern Hemisphere (from a weakening of the planetary temperature gradient there) should decrease mixing and vertical circulation, and hence diminish the overall productivity of the sea, at least in northern latitudes. However, the development of strong monsoon winds, from the rise of seasonal temperature differences between land and sea, could compensate for such an effect.

Whatever the future of winds, the supply of silicate to the euphotic zone will be more affected than that of phosphate, because silicate concentrations tend to increase more slowly with depth, compared with phosphate. By the argument that silicate supports a short food chain (through diatom production), a decrease of vertical mixing would lengthen the food chain to fish and, hence, accentuate the loss of production as far as the fisheries harvest, while a strengthening of mixing (for example, around the Antarctic) would have the opposite effect.

Quite generally, the rate at which nutrients are supplied to the sunlit zone from deeper water layers depends not only on the intensity of mixing but also on the concentration of the nutrients in the water below the euphotic zone. For the same amount of mixing, the supply rate of nutrients directly depends on this concentration. What then controls the concentration of the nutrients at depth? As Harvey and his contemporaries explained, respiration and decay produce nutrients. Clearly, the rates at which these reactions proceed are crucial. The faster the decay, the more nutrients will be added per unit time and the more quickly concentrations will rise. Concentrations will keep rising as long as the water is below the sunlit zone and as long as there is oxygen to drive decay and respiration. Below the zone of production, the rate of liberation of nutrients depends largely on the abundance and efficiency of bacteria, which in turn depend on the kind of organic matter delivered from the sunlit zone. Thus, the rate at which decay proceeds—which has to do with the rate of growth of bacterial populations and their equilibrium density and with the rate of diffusion of materials in and out of bacterial cells—is a limiting step in the great production cycle of the ocean. Warming at depth will speed decay, hence increase nutrient concentration in subsurface waters.

Still this is not the whole story: We have to consider both shorter and longer cycles of phytoplankton growth and decay. The shortest cycle occurs within the sunlit zone itself. Most of the nutrients in that zone are locked up within organisms. Within a food web of minute microbes, these nutrients move largely from one organism to another, with only a very short residence time in the water. The cycling within the bacterial plankton and minute microbial grazers is known as the microbial loop, a concept that took center stage only in the last two decades or so. At Scripps, the marine biologist Farooq Azam has made this topic the focus of his research, since the 1980s.[51]

The microbial loop is now recognized as equal in importance to the traditional plankton food web,[52] in the framework of plankton ecology (rather than fisheries, as discussed). The rapid cycling of nutrients and carbon in this loop may have ramifications for various aspects of the traditional food chain from microalgae to zooplankton to fishes, which remain to be worked out. In particular, the (tiny) ciliates feeding on the bacterioplankton may contribute significantly to feeding certain zooplankton. In any case, it is remarkable that the single most abundant (but minute) photosynthesizing organism in the ocean was only recently discovered, within the bacterioplankton.[53] Its name is *Prochlorococcus marinus*.

Also, within the food web of the larger organisms (diatoms, dinoflagellates, copepods, and such) some nutrients recycle directly, without leaving the mixed layer. This cycle is based on grazing and excretion by zooplankton but is also tied to the microbial loop. The production that is due to this kind of internal recycling is referred to as regenerated production and is treated separately from so-called new production, for which nutrients are brought in from the subsurface through mixing and upwelling. The separation of these two types of production processes was an important conceptual advance, which was made in the 1960s.[54]

The concept that production needs to be separated into two types, one involving recycling

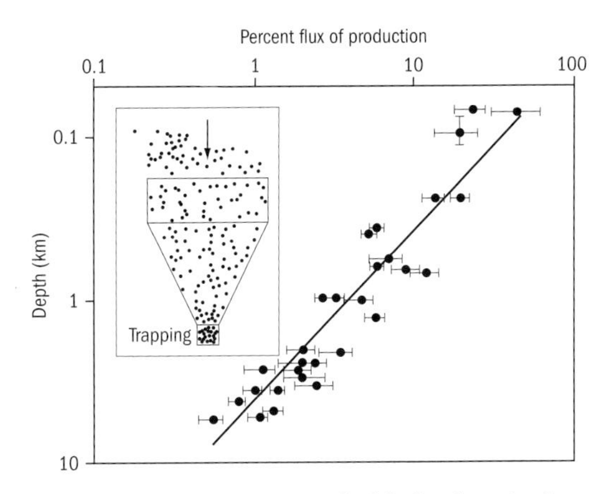

FIGURE 8.6. Export production: food for benthic animals. The supply decreases markedly with depth. Data points represent amount of organic matter found in traps, as percent of overlying production, according to E. Suess, Kiel. *Inset:* Principle of a trap. Notice the logarithmic *x* and *y* scales.

within the productive surface layer and the other including the supply of nutrients from below this zone, rapidly became a central tenet in the study of primary productivity.[55] A crucial aspect of this dichotomy is that new production is closely related to export production (fig. 8.6), since both reflect that part of the nutrient supply that is not recycled. Export production—the organic matter that finds its way to below the productive zone—feeds all living things at depth.

THE GREAT NUTRIENT CYCLES

On longer time scales than those usually considered in marine biology, nutrient control must involve geologic processes. Thus, phosphorus is continually lost from the ocean as it enters sediments, not only within organic matter but also within skeletal parts such as bones of vertebrates and molts of crustaceans, and within calcareous shells. The phosphorus thus locked up sinks to the seafloor (fig. 8.7). To remain available for the ocean's production, it needs to be liberated. Indeed, the skeletal parts largely redissolve, yielding back to the water the phosphorus locked up previously. The rate of dissolution of phosphatic skeletal parts, then, is an important

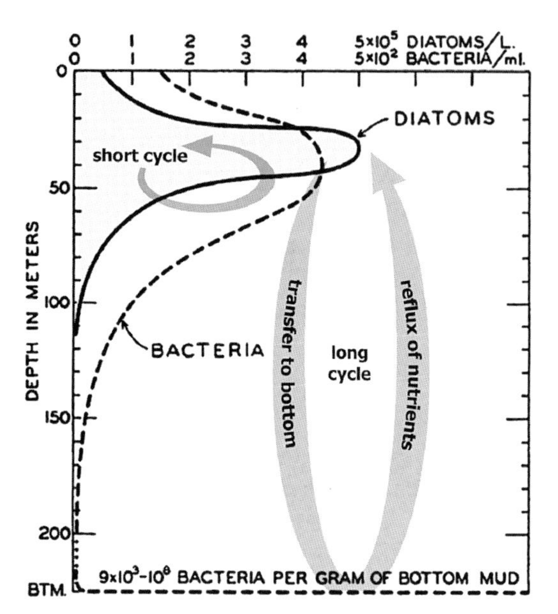

FIGURE 8.7. Nutrient cycles in the sea run between photosynthesis in the sunlit layer to the bacteria in the water and in the bottom mud. The bacteria cause decay, liberating nutrients for renewed production. The illustration is in Sverdrup et al. (1942); cycles are here added.

control on the fertility of the ocean. An even longer cycle involves the erosion on land of marine phosphorites, formed in the sea and incorporated into continents millions of years ago. There is, then, a short nutrient cycle that involves processes in the water, and a number of long nutrient cycles that involve sediments. Of these, the one in the water has been studied quite intensely for the last century. However, as we shall see, fundamental elements of nutrient cycling are still poorly understood, as illustrated by the discussions surrounding the role of iron in phytoplankton production.

So far, we have left open exactly which of the nutrients is the most important. A number of nutrients are taken out of the sunlit zone until they reach values near zero in this uppermost layer. Each one of them, therefore, must be limiting to the growth of at least some algae. The great paradox of nutrient availability is that one of the most abundant nutrient elements—nitrogen—is a limiting factor for growth in the ocean. Most of the air we breathe consists of nitrogen, yet even on land nitrogen is in short supply, as every gardener knows. To add nitrogen to the soil, we grow leguminous plants, or we fertilize the soil using ammonia-rich manure, or nitrate, or urea.[56] Nitrogen in the air is present as a very stable molecule, N_2, with a triple bond. This molecule is loath to react with anything, including organic molecules. Few plants, therefore, can use it to synthesize the amino acids and proteins necessary for growth.

Fortunately for life on Earth, specialized bacteria are able to overcome this problem, applying a number of chemical tricks to break the nitrogen-nitrogen bond. In the sea, blue-green bacteria that produce biologically available nitrogen are ubiquitous. Nitrogen-fixing cyanobacteria have been found in a wide range of environments—from salt marsh, tidal zone, and estuary, to coral reef and central gyre. Nitrogen-fixing bacteria also may be important on the seafloor in shallow water where light reaches the bottom. In addition, atmospheric fixation plays some role: nitrogen oxides are produced within lightning.

With so much fixing of nitrogen going on, why is nitrate is in short supply at all? The answer must be that there is a process that reverses the effects of nitrogen fixation. It exists indeed, and it is called denitrification. It is a biological process involving bacteria that live in anaerobic conditions. Anaerobic conditions occur in areas where all free oxygen has been used up in driving decay reactions. When this happens, certain bacteria strip the nitrate of its oxygen, to use it for oxidation of organic matter. This stripping produces molecular nitrogen, which returns to the air. The air, then, is a product of the denitrifying bacteria, which provide the N_2, and of photosynthetic organisms that make O_2. Our atmosphere, it turns out, is almost entirely a product of life processes. This fact has been well appreciated since the Yale ecologist G. E. Hutchinson elaborated on this theme (and pioneered the field of biogeochemistry in the process).[57] More recently, the British engineer and science writer James Lovelock has incorporated the concept of biological control of atmospheric composition into his Gaia hypothesis, which holds that biological processes keep the Earth habitable. They do, and the pervasive agents are microbes. On land, they make the soil fertile. In the sea, they recycle the nutrients.[58]

From the link between oxygen and the nitrogen nutrient cycle, we might conclude that the present ocean, which is cold and holds plenty of oxygen, restricts the opportunities for denitrification. In turn, this should allow the buildup of nitrate in the system. A well-aerated ocean, it would seem, should be more fertile than a poorly aerated one. On the other hand, if the solubility of phosphate in the ocean is the crucial factor, the ocean's productivity is controlled by the rate at which phosphate is returned to the sea from organic and skeletal matter. A high rate of phosphate recycling will provide for high concentrations, closer to the theoretical saturation value, while a low rate will starve the ocean of phosphate.

In effect, one of these two cycles will be limiting overall, and it will be the one that has the lower yield relative to the needs of the phytoplankton. With nitrate being protected by oxygen in the present well-ventilated ocean, we must assume that phosphorus is limiting. If so, nitrate simply follows the phosphate around within the marine cycle of production and decay, like a dog on a leash that follows his master. The reason is that blue-green bacteria are ready to fix nitrogen up to the level supported by the available phosphate.

Alfred C. Redfield of Woods Hole, who thought deeply about these matters, wrote: "If the argument presented is sound [that phosphate is precipitated from seawater and sequestered in sediments] it may be concluded that the quantity of nitrate in the sea, and the partial pressure of oxygen in the atmosphere are determined through the requirements of the biochemical cycle, by the solubility of phosphate in the ocean."[59] His statement well reflects the cross-links within the biogeochemical cycles.[60]

The ratio between the numbers of phosphate and nitrate molecules dissolved in the ocean is called the Redfield ratio. It is roughly 1 to 16. This is also the ratio in plankton. It was this apparent coincidence that led Redfield to postulate biological control of the nutrient chemistry of the sea, which in turn leads to regulation of the atmospheric composition, because both oxygen and nitrogen are part of the production cycle.

There is yet another nutrient of great importance in biogeochemistry, and that is iron. As red-blooded creatures, we are dependent on its ability to carry oxygen from our lungs into brain and muscles. Being warm-blooded, we need plenty of oxygen to feed the fires within. Iron is equally vital for the processes involving photosynthesis. H. H. Gran, the great pioneer of phytoplankton dynamics in the sea, was puzzled that the spring bloom in the North Atlantic began earlier in coastal waters and only later offshore.[61] He suggested (in the 1920s) that nutrients were enriched at the coast because of runoff from land. When this proved untenable, because the source of nitrate and phosphate was clearly the deep water, he suggested that "iron-containing humus-compounds" from land might provide the missing element. In fact, this proposition was greatly bolstered by experiments showing that phytoplankton responded favorable to the addition of iron compounds and soil extracts, in culture.[62]

As mentioned, Gran's suggestion has experienced a dramatic revival in the last two decades. A puzzling situation of unused nutrients in surface waters is widespread in the northeastern North Pacific, in parts of the equatorial Pacific, and in the Southern Ocean.[63] Could iron be the limiting factor? Measurements seemed to show that there was plenty of iron in the water to satisfy the needs of the phytoplankton. However, when the marine chemist John Martin[64] improved the analytical procedures and made superclean measurements, he found that iron was in fact very rare and did limit phytoplankton growth.[65] Researchers making measurements before that, apparently, had not taken all the precautions necessary when working in an iron-rich environment, such as the vicinity of a research vessel!

Martin's findings, published in the 1980s, had a large impact. If iron is limiting over large parts of the sea, concepts about the plankton dynamics of these waters have to be seriously revised. Naturally, scientists are reluctant to give up familiar explanations for unproven ones, and much discussion of Martin's claims ensued, stalling acceptance of his hypothesis.[66] However, the various implications were readily explored.[67] One corollary of Martin's iron hypothesis is that a relatively modest application of dissolved iron sprayed from aircraft over large areas of the sea will greatly stimulate phytoplankton production in those areas where nitrate and phosphate are abundant in the mixed layer. As a result of fixing carbon into plankton, and removing it from interaction with the atmosphere, the carbon dioxide content of the air could be artificially decreased, in principle.

When Martin proposed that adding iron to the surface waters of the Antarctic could pull down carbon dioxide in the atmosphere, much media attention followed, as well as negative reactions from colleagues in biological oceanography. Martin recounts the beginning of this story as follows:

> I first said this more or less facetiously at a Journal Club lecture at Woods Hole Oceanographic Institution in July 1988. I estimated that, with 300,000 tons of Fe, the Southern Ocean phytoplankton could bloom and remove two billion tons of carbon dioxide. Putting on my best Dr. Strangelove accent, I suggested that with half a shipload of Fe . . . I could give you an ice age. After which we all had beer on the lawn outside Redfield Laboratory.[68]

The idea of applying dissolved iron to the sea surface to stimulate phytoplankton growth was soon tried and was found to produce striking results.[69] Nevertheless, it is questionable whether iron fertilization is going to help with the carbon dioxide problem in a serious manner, for a number of reasons.[70] Also, we must not lose sight of the scale of the problem. The human input of carbon to the atmosphere each year exceeds the entire carbon fallout from the upper waters of the Southern Ocean by a factor of seven or so. So, even doubling the output by Martin's method (if this were possible) would solve but a modest fraction of the problem and only for a limited time.

FOOD CHAINS AND FOOD WEBS

The classical food chain concept, whereby phytoplankton is consumed by herbivorous zooplankters, which are in turn consumed by fishes (that is, the useful part of the ocean's production as far as humans are concerned), was the reigning paradigm well into the 1970s.[71] Even today, it is a good way to think about production dynamics for a first-order description of how fish get their food (fig. 8.3).

At each transfer, most of the carbon is lost and becomes available for recycling, through bacterial decay. Because there are from three to five transfer levels from primary producers to fish, only a very modest portion of the original production makes it to the higher levels of the food chain, such as tunas (which eat other fishes). Thus, the fraction available for human consumption is but a fairly trifling share of the total production of the sea.

Also, in the classical food chain, the primary producers—diatoms, dinoflagellates, and nanoplankton—are at the base, carrying the pyramid, which tapers to the edible fish. A focus on these three types of representative producers—diatoms, dinoflagellates, nannoplankton—immediately captures crucial ingredients of the dynamics: the fact that the dominance of large diatoms signals high nutrient supply, the fact that silicate is limiting to production (not just phosphate and nitrate), and the fact that low productivity is associated with minute producers (nanoplankton) using, largely, nutrients recycled within the mixed layer.

How well does the classical concept match the modern understanding? Apparently not all that well. Richard Barber and Anna Hilting, reviewing the last 50 years of landmark achievements in biological oceanography in the year 2000, state the case as follows:

> Over the last 25 years our vision of the pelagic foodweb structure has changed dramatically. We now view the traditional "diatom-copepod-fish" foodweb as a relatively minor component. The foodweb consistently present in all oceanic habitats is based on pico- and nanoplankton-sized autotrophs and heterotrophs, which are efficiently grazed by flagellates and ciliates. The pelagic foodweb is microbe-centric.[72]

According to Barber and Hilting, this discovery of the centrality of the microbial loop is the foremost achievement in biological oceanography in the last 50 years. The traditional food web, as reflected in the scheme offered by Hardy,[73] is relegated to "special case" status, evidently. Of course, as mentioned, it happens to be the special case that is relevant to the discussion of the ecology of marine mammals and birds, and of fisheries. That is, the special case is the one of general interest.

It is been realized for some time that the so-called food chain is an idealization and that we actually deal with a food web in all marine ecosystems, pelagic and coastal. What is new is that the flow of fixed carbon begins with an assemblage of producers that prominently includes bacterial plankton. Thus, the size-range of producers is expanded downward into the picoplankton, smaller than nanoplankton by a factor of 10 or more. The range in size of the producers in the plankton now is comparable to that between the smallest moss on the forest floor and the giant trees of Sequoia National Park. In principle, if grazing organisms are of equal size or larger than the producer cells, the grazers can eat them, but not if the size difference is too large. Copepods eat diatoms and dinoflagellates but not bacterial plankton. Thus, bacterioplankton fall prey to tiny flagellates and ciliates. Also, many of the small flagellate producers, starved for nutrients, prey on planktonic bacteria. Such minute flagellates and ciliates, in turn, are large enough to be eaten by the regular zooplankton, such as copepods, and this link provides a path from the microbial loop into the traditional food web describing the interaction between the larger organisms (fig. 8.8).

It is clear that adding the microbial loop (or loops) to the web of plankton represents a fundamental revamping of our understanding of how carbon and nutrients move through this system. However, because it is a long way from pico- and nannoplankton to fish, the fundamental arguments concerning the amount of food for fishes remain valid.[74] These arguments have been elegantly laid out by Woods Hole oceanographer John Ryther, in the late 1960s.[75]

Ryther emphasized, first of all, the great difference in primary production from the richest to the most nutrient-poor regions. He put the average near 50 grams of carbon per square meter per year for nine-tenths of the ocean area, based on then-available data. For coastal waters and regions of open ocean upwelling and intense mixing, he proposed 100 grams per square meter per year.[76] He took these productive regions to comprise about one-tenth of the ocean area. In addition, he postulated that 0.1 percent of the ocean surface has superproductive areas comparable to the one off Peru, with values near 300 grams of carbon per square meter per year. In terms of percentage of total production, this comes out to 81.5 percent to 18 percent to 0.5 percent for the respective types of regions. Ryther next pointed out that the bulk of the production in the more highly productive regions is delivered by microplankton, with diameters of 100 microns or more, while the chief producers of the low-production regions are nannoplankton cells between 5 and 25 microns in diameter. He stressed the implications of this marked difference in size: "The larger the plant cells at the beginning of the food chain, the fewer the

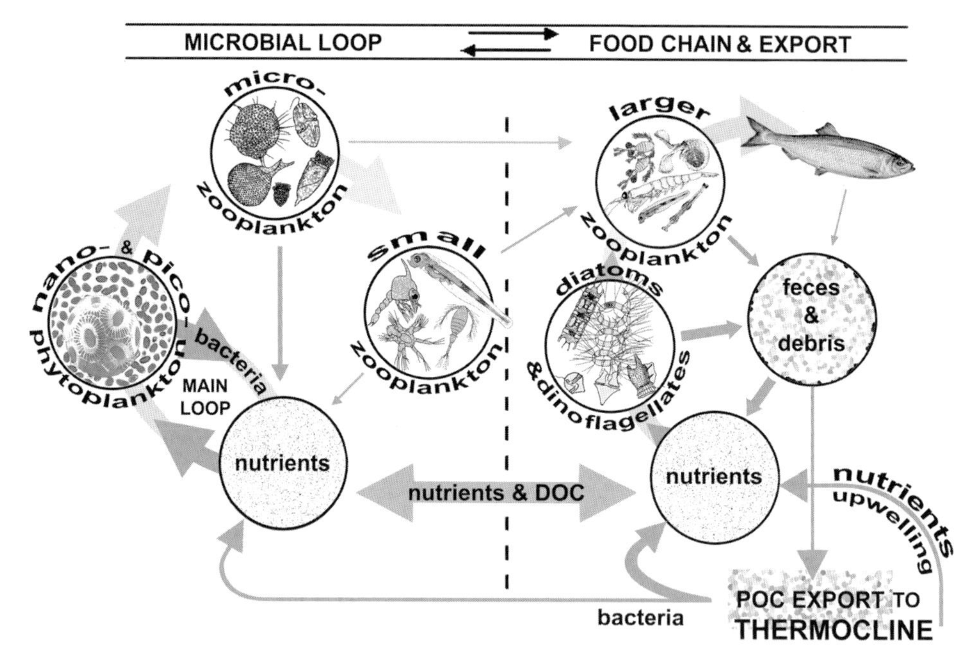

FIGURE 8.8. Relationship between the traditional food chain (diatoms–copepods–fish) *(right)* to the microbial loop that dominates production in the sea over vast regions *(left)*. DOC, dissolved organic carbon; POC, particulate organic carbon.

trophic levels that are required to convert the organic matter to a useful form."[77] The word *useful* is important here. It refers to food for fish. Ryther says, in essence, that a short food chain is the path to fish (fig. 2.18).

Ryther's argument hinges on the number of steps to be climbed, up the food web pyramid, to get from the primary producer to the desired product. He attempted to make sense of the confusing multitude of small organisms that make up the pelagic food chain, in terms of the traditional (terrestrial-based) ecologic series from plant to herbivore to carnivore:

> Intermediate between the nannoplankton and the carnivorous zooplankton are a group of herbivores, the microzooplankton. . . . Representatives of this group include protozoans such as Radiolaria, Foraminifera, and Tintinnidae, and larval nauplii of micro-crustaceans. These organisms, which may occur in concentrations of tens of thousands per cubic meter, are the primary herbivores of the open sea. Feeding upon these tiny animals is a great host of carnivorous zooplankton, many of which have long been thought of as herbivores. Only by a careful study of the mouthparts and feeding habits [is it possible] . . . to show that many common copepods are facultative if not obligate carnivores. Some of these predatory copepods may be no more than a millimeter or two in length.

Ryther speculated that these small copepods are in turn prey for chaetognaths, a group of voracious small zooplankters.[78]

According to Ryther, then, by the time the carbon gets from photosynthetic nannoplankton to animals of sizes approaching an inch or so, it had to move through three to four trophic levels, paying a tax of 90 percent at each transit. The food remaining for fish, after this chain of transfers, is between one-thousandth and one-ten-thousandth of the original production. Considering that at least two more steps are needed to convert this food to top-predator fish such as tuna and mahimahi, the output of the open-ocean realm in terms of desirable dinner fare is exceedingly meager.

For the realm of coastal ocean and intense pelagic mixing, the chain is taken to be only half as long, and the efficiency of transfer from one level to the next is supposed to be somewhat better. An even shorter chain may be postulated

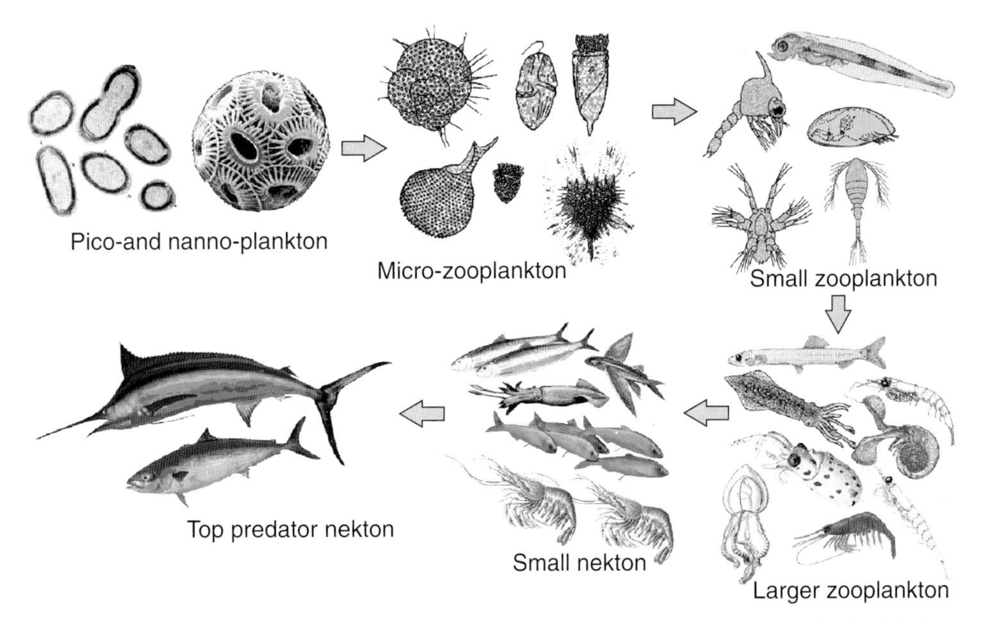

FIGURE 8.9. The long food chain, beginning with pico- and nannoplankton, has many steps and yields little in terms of fish. "Small zooplankton" measured in millimeters, "large zooplankton" in centimeters.

for the regions of intense upwelling. According to Ryther, the reason for this is the large size of diatom colonies:

> Many of the above-mentioned species of phytoplankton [diatoms in upwelling regions] form colonies several millimeters and, in some cases, several centimeters in diameter. Such aggregates of plant material can be readily eaten by large fishes without special feeding adaptation. In addition, however, many of the clupeoid fishes (sardines, anchovies, pilchards, menhaden and so on) that are found most abundantly in upwelling areas and that make up the largest single component of the world's commercial fish landings, do have specially modified gill rakers for removing the larger species of phytoplankton from the water.

From this type of reasoning, Ryther concluded that, in essence, oceanic areas (90 percent of the total) contribute very little to the total potential fish catch, while the high production regions (10 percent of the total) and the intense upwelling regions (one-thousandth of total area) contribute each about one-half of the potential catch. In other words, the long food chain (fig. 8.9), which dominates by area, can be neglected when assessing the harvest of the sea.

Ryther's line of argument makes clear why the 200-mile fishing limit is vital in fisheries management. It also makes clear why whales are abundant around Antarctica. Here an important food item—the relatively large planktonic shrimp, or "krill," *Euphausia superba*—occurs in great concentrations. The Antarctic krill feed directly on the primary producers, largely diatoms. Estimates for total production of krill are in the range of tens to more than a hundred million tons per year.[79]

Ryther's estimate of the total amount of fish that can be taken out of the ocean (without destroying the resource) is of some interest: it is 100 million tons. He predicted, in 1969, that the fishing industry would expand for no more than a decade and then hit the limit. Not a bad guess. Of course, given Ryther's methodology, there is in fact no difference between an estimate of 50 million and 200 million tons, so we must admit he was simply lucky in hitting the target fairly well. In any case, Ryther showed that the food-chain concept is consistent with fishery success. The subsequent ascent of the microbial loop as a parallel production system to the traditional one does not change this outcome. The loop is crucial to maintaining the chemical and biological environment in the pelagic realm, but its contribution to feeding fish is insignificant. In analogy, on land, the concept of grass-to-cow works—even though in using it we neglect the rich microbial activity in the soil.

The success of the concept of the food chain in reconciling the productivity of the ocean with the yield of fish invites the question when it was first discovered. As usual when considering a major ecologic concept, there are a number of converging sources. In the present context, some of these go back to lake research. However, one important figure stands out, the German biologist Victor Hensen.[80] Hensen realized that to understand the yield of fish one had to study the base of the production chain, the plankton. He used finely meshed plankton nets to sample and measure the standing stocks of the different kinds of phyto- and zooplankton, in an effort to make a comprehensive inventory. Thus he hoped to begin to construct the food chain, which he envisioned as a "cone of metabolism" reminiscent of the pyramid of trophic levels invented later.

Hensen failed to link his plankton inventories to nutrient chemistry. It was the nutrient supply that held the answer to the puzzling question of why plankton was sparse in the tropics, but so abundant off the shores of Germany, and in high latitudes in general. The integration of nutrient chemistry with plankton dynamics became the hallmark of the famed Kiel School of plankton ecology, led by the biologist Karl Brandt.[81] The Kiel School flourished in the first two decades of the twentieth century. Among its many accomplishments were a shift of thinking from distributions to bloom dynamics, and the beginnings of a standard model in terms of nutrient-phytoplankton-zooplankton transfer.[82]

There was one severe shortcoming in Hensen's attempt to establish the base of the food chain in the plankton. It was the evidence that the fine-mesh net did not catch most of the phytoplankton. This challenge came from a younger colleague at Kiel, Hans Lohmann.[83] Lohmann had noted that appendicularians[84] filtered the water for extremely small food particles, consisting of minute flagellates, protozoans, and coccospheres. Most of these were too small to be caught by Hensen's net. To show how much was being missed, he filtered seawater from Kiel Bay and from the Mediterranean, using filter paper and centrifugation to concentrate the plankton. He found that all his "nannoplankton" organisms (Lohmann's phrase) were severely underrepresented in Hensen's inventory.

Lohmann (in the years between 1901 and 1905) had taken the first step toward establishing the microbial loop concept, which reemerged 80 years later as a reigning paradigm requiring adjustment of our thinking about plankton dynamics.[85]

ARCHETYPES OF ZOOPLANKTON

Traditionally, the macrozooplankton—the plankton that is readily visible to the unaided eye—has shaped our views of what the world of plankton is like. It has been described in expedition reports for some 150 years. Copepods, shrimps, arrow worms, tunicates, small jellyfish, comb jellies, pelagic snails, segmented worms and the larvae of fishes, squids, and some benthic organisms are in this category. Important aspects of the life-styles of these creatures are the manner of feeding, the rate of survival and the manner and rate of replacement, and the phenomenon of daily vertical migration. Among this collection of organisms, copepods (sometimes called the insects of the sea, although they are not, of course, insects) have been taken as representing typical zooplankton (fig. 8.10).

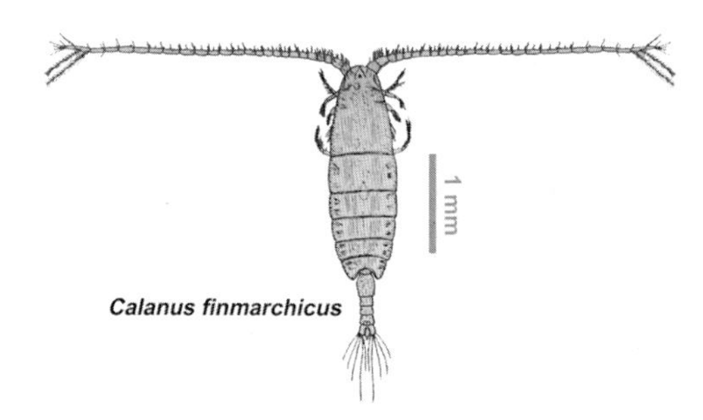

FIGURE 8.10. The copepod Calanus finmarchicus, a classic experimental plankton animal.

In their delightful book *The Fertile Sea*, the marine scientists Andrew P. Orr and Sheina Marshall say this about copepods: "[Copepods] are important because they are so numerous and so widely distributed and act as a link between the phytoplankton and the fish, turning the phytoplankton into a food palatable to these larger carnivorous creatures." and further: "[Copepods] . . . are too small to be conveniently collected and so we leave it to the fish to concentrate them for us."[86]

Copepods are commonly classed as herbivores, although they feed on microzooplankton, as well. They fall prey to carnivores such as arrow worms, jellyfishes, and fishes. Small copepods get eaten by large microplankton, as well. Thus, foraminifers (single-celled shelled amoebae) have been seen to consume small crustaceans. There is in fact no clear path from primary producers (phytoplankton) to herbivorous (plant-eating) zooplankton to carnivores. Instead, there is an expectation that whatever is edible will readily be found by a consumer willing and able to eat it.

Copepods feed at several size levels; their larvae can handle the very fine stuff and the adults the larger prey. Each species has its own special adaptations, which became known through the decades from experiments in the laboratories of the various marine biological institutes all over the world. Pioneering work was performed at the Marine Station in Millport in Scotland, where Sheina Marshall and collaborators worked on the biology of *Calanus finmarchicus* for more than two decades and made this particular copepod the paragon of zooplankton.[87] At Woods Hole, George L. Clarke and collaborators likewise cultured *Calanus* in the 1930s, and R. Conover continued this type of work at the Marine Biological Laboratory. An important finding (among others) was that if given a chance, copepods will feed wastefully, that is, they will excrete a lot of undigested material when phytoplankton is plentiful, reducing the efficiency of transfer of carbon from phytoplankton to zooplankton.[88]

At Scripps, the Food Chain Research Group within the Institute of Marine Resources[89] carried out feeding experiments on copepods in the late 1960s and 1970s, under the leadership of biologist Michael Mullin.[90] Mullin was especially interested in those details of copepod anatomy and life histories that would reveal the organism's strategy in maintaining large populations and competing for food resources. Mullin studied rates of growth and of ingestion of food to determine the gross efficiency of converting food to zooplankton. An important finding was that for the optimum growth of populations, palatable food for the larvae must be provided, food that is not the same as for the adults. The larvae are especially important because they are much more efficient than adults in converting phytoplankton food into copepod mass.[91]

Mullin and his collaborators had the use of a unique facility in experimental marine ecology, the Scripps Deep Tank, a steel cylinder 10 meters deep and 3 meters in diameter, coated internally with an inert plastic and surrounded externally by an insulated layer of polyurethane. The large volume of the tank (70 cubic meters) gave hope that within, ecologic systems would behave more naturally than in the relatively small containers in the lab.[92] When culturing *Calanus* in the Deep Tank, the population bred well, with an egg-to-egg generation of about 20 days.[93] The advantage of the large container was that the animals felt right at home and reproduced readily (which they did not do in the lab). The disadvantage was a certain lack of control: at the end of one experiment (almost 3 months), two other copepod species (both common off La Jolla) were 50 times more abundant than the target species. Perhaps not surprisingly, simulation of natural conditions also implied lack of predictability.

An important aspect of zooplankton research is to establish whether the animals seek out their food purposefully and actively. According to Barber and Hilting (2000), establishing this is in some detail was a landmark achievement of the last 50 years or so.[94] They cite 10 papers between 1977 and 1998 that they consider crucial. Eight of the references are to the feeding behavior of copepods, and more specifically calanoid copepods, that is, *Calanus* or related genera.[95] It turns out that the calanoid copepods can smell their food and adjust their filtering activity accordingly, which can be precisely observed with a high-speed, strobe movie camera focused on tethered animals in the lab. The copepods hunt, they do not just strain water hoping for good luck.

It would be unfair to say that the role of zooplankton in the pelagic food chain is being determined entirely by analogy with marine copepods. Much was learned from culturing the brine shrimp *Artemia* and two small arthropods living in lakes and ponds (and commonly referred to as water fleas), *Daphnia* and *Cyclops*. All of these are easily maintained in aquaria. Lakes have pelagic systems whose size is somewhere between the Deep Tank and the Mediterranean, and they can provide useful models for what happens in the sea.[96] In addition, many other zooplankters have been cultured and checked for their efficiency in converting food into body mass.[97] In low-production regions, the phytoplankton particles, as mentioned, generally are extremely small. There the protozoans (tintinnids, radiolarians, acantharians, foraminifers) play an important role in providing bite-size food for copepods. Such systems are difficult to study in the laboratory.

Size plays the crucial role in determining who eats whom. Being small (such as copepod larvae) has the advantage of finding abundant food in the pico- and nannoplankton, but carries the risk of being eaten by a host of protozoans. Being large (if organisms an inch long or less can be called "large") brings the advantage of being able to search for food by moving over some distance, but carries the risk of being found by sophisticated hunters in the macrozooplankton, or being strained out by the gill rakers of anchovies, sardines, and herring. Given the gauntlet of predation from egg to adult, the best strategy for the larvae and juveniles of macrozooplankton is to grow as fast as possible and to grow protuberances and spines, increasing the effective size at minimum investment of energy. The morphologies of many small plankton organisms nicely reflect this strategy.

In addition, it is advisable to be invisible. Thus, practically all plankton organisms are transparent, unless they are green for catching sunlight, or live at depth in the dark, where it does not matter. Those that live in the dark during much of the day may come up to feed at night. The daily vertical plankton migration, which is pervasive, enormously complicates the original Kiel School version of the food chain that postulates a simple progression from nutrients to phytoplankton to zooplankton.[98]

Despite (or perhaps because of) its striking simplicity, the Kiel model has survived well throughout its 100-year history. It is still alive, transmuted through the equations of quantitative ecologists for the last half-century. Scripps ecologist Peter Franks defends it as follows:

> The nutrient-phytoplankton-zooplankton (NPZ) model is a common tool in oceanographic research. The NPZ model incorporates one of the simplest sets of dynamics that usefully describe oceanic plankton dynamics. . . . Acceptability as a research tool is by no means universal, however. In a review of a recent manuscript in which we used an NPZ model . . . one of our anonymous reviewers commented, "The real world cannot be modeled with a 3-compartment NPZ model that agrees (possibly fortuitously) with satellite images. . . . Even in open ocean systems the scientific community has long abandoned the use of 3-compartment models." This attitude reflects a common bias in modeling—that a more complicated model is necessarily a better model. . . . I would argue that there is no compelling reason to reject the NPZ model until it is clear that it cannot describe the system being studied.[99]

The pelagic biologists Victor Hensen and Karl Brandt, when arguing with their younger colleague Hans Lohmann, must have had thoughts like this. To invoke a terrestrial analogy: If the task is to find out why lynx populations fluctuate, it seems futile to count the ants in the area or the bacteria on the roots of the herbs eaten by the rabbits that the lynx are feeding on. However, if one wants to know just how nitrogen is replenished within the soil, one does need to look at the roots. As Franks puts it, in regard to the sea: "On the other hand, there is a great deal that NPZ models cannot tell us about the workings of the ocean, and it is important to continue to develop and test new models for plankton dynamics."[100] Hans Lohmann would have been pleased with this statement.

NOTES AND REFERENCES

1. R. H. Fleming, 1957, *General features of the ocean*, in J. W. Hedgpeth (ed.), *Treatise on Marine Ecology and Paleoecology*. Geological Society of America Memoir 67, pp. 87–107.

2. J. H. Ryther, 1969, *Photosynthesis and fish production in the sea*. Science 166, 72–76. Ryther was at Woods Hole from 1951 to 1981.

3. It is estimated that Chinese fishermen considerably overreport the catch, for political reasons. R. Watson, D. Pauly, 2001, *Systematic distortions in world fisheries catch trends*. Nature 424, 534–536.

4. D. Pauly, J. MacLean, 2003, *In a Perfect Ocean. The State of Fisheries and Ecosystems in the North Atlantic Ocean*. Island Press, Washington, D.C., 163 pp.

5. J. P. Wise, 1995, Trends in food from the sea, in J. L. Simon (ed.), *The State of Humanity*. Blackwell, Oxford, U.K., pp. 411–415. (Wise is listed as "consultant in marine affairs.")

6. P. Weber, 1994, *Fish, Jobs, and the Marine Environment*. Worldwatch Institute, Washington, D.C., p. 20.

7. Considering the waste described by Weber (which is common knowledge among fisheries biologists), and the fact that close to one-third of the global catch is processed for fertilizer and animal feed, an increase for *human* consumption might be feasible, even with decreased fisheries success. But a tripling or quadrupling (as suggested by the optimistic assessment by Wise, note 5) presumably would call for a serious modification of the ecosystem, and consumption at a significantly lower level of the food chain.

8. B. Lomborg, 2001, *The Skeptical Environmentalist*. Cambridge University Press, Cambridge, UK, 515 pp. (See page 106.) The author is listed as associate professor of statistics in the Department of Political Science at the University of Aarhus, Denmark. His book, an optimistic assessment of the economic future and environmental resources, has generated much discussion. It has been criticized for poor scholarship in the course of an inquiry by the Danish government.

9. Based on U.N. Food and Agriculture Organization and European Union statistics. See M. von Baratta (ed.), 2001, *Der Fischer Weltalmanach*, 2002. Fischer Verlag, Frankfurt am Main. Entry: *Ozeane*.

10. Daniel Pauly, V. Christensen, J. Dakgaard, R. Froese, F. Torres Jr., 1998, *Fishing down marine food webs*. Science 279, 860–863.

11. Carl Safina, marine biologists and conservationist, refers to "scorched-earth fishing" when discussing the effects from bottom trawling. C. Safina, 1998, *Scorched earth fishing: Bottom trawls*. Issues in Science and Technology (Spring), 33–36. See also S. Thrush, P. K. Dayton, 2002, *Disturbance to marine benthic habitats by trawling and dredging: Implications for marine biodiversity*. Annual Review of Ecology and Systematics 33, 449–473. Other examples for ecocide fishing and rogue fishing come to mind in the context of shark finning and long-line fishing, where the scope of the damage to all can readily exceed the gain to the few.

12. G. Hardin, 1968, *The tragedy of the commons*. Science 162 (1), 243–248.

13. The Norwegian fisheries expert Einar Lea showed that a few age classes were dominant in the herring caught off Norway, between 1907 and 1927. Cited by F. S. Russell, C. M. Yonge, 1936, *The Seas, Our Knowledge of Life in the Sea and How It Is Gained*. Frederick Warne, London, 379 pp. (See p. 333.) Great historical fluctuations in sardine populations off California were proposed by A. Soutar, J. D. Isaacs, 1974, *Abundance of pelagic fish during the 19th and 20th centuries as recorded in anaerobic sediment off the Californias*. Fisheries Bulletin 72, 257–273. (See chapter 7.)

14. The fisheries expert Milner Baily Schaefer (1912–1970), director of the Institute of Marine Resources at Scripps (from 1962 to 1970), worked on the theory that underpins such calculations.

15. H. H. Gran, 1902, *Plankton des Norwegischen Nordmeeres von biologischen und hydrografischen Gesichtspunkten behandelt*. Norwegian Fishery and Marine Investigations, Report 2 (5), 1–222. Haakon Hasberg Gran (1870–1955), the leading Norwegian marine biologist of his generation, pioneered the study of plankton dynamics, early in the twentieth century.

16. There has been some discussion in recent years as to whether *planktonic* is a correct derivation of the Greek root word for "wanderer," or whether *planktic* is more appropriate. While learned etymologic and grammatical analysis may suggest otherwise, we use the old-fashioned *planktonic* because it sounds better than *planktic*. Usage counts, in grammar.

17. Singular: *flagellum*; a whiplike appendage used for locomotion. A dinoflagellate cell has two of these, as a rule.

18. Also "coccolith ooze" or simply "calcareous ooze." Where foraminifer shells are abundant in this sediment, the names "foram ooze" or "foraminifer ooze" are commonly used. The older version "*Globigerina* ooze" is obsolete: most foraminifers do not belong to the genus *Globigerina*.

19. From the Greek word for "dwarf." Smaller than "micro" but larger than "pico."

20. Perhaps the most abundant photosynthesizing organism is the bacterium *Prochlorococcus*, discovered in the 1980s by Sallie Chisholm of MIT and collaborators. S. W. Chisholm, R. J. Olson, E. R. Zettler, R. Goericke, J. B. Waterbury, N. A. Welschmeyer, 1988, *A novel free-living prochlorophyte abundant in the oceanic euphotic zone.* Nature 334, 340–343. Its small size makes it unlikely that it contributes to feeding fish. We may safely ignore it in making the fish production estimate, therefore.

21. E. Hentschel, 1926, *Bericht über die biologischen Arbeiten.* Dt. Atlant. Exped. "*Meteor.*" Zeitschrift der Gesellschaft für Erdkunde (Berlin) Jahrg. 26 (1) and Jahrg. 27 (5–6).

22. Diatoms are not the only organisms among the dominant primary producers to use silica for defense. Grasses on land use silica to fortify their blades (this is why the blades can cut fingers). On land, mammalian grazers evolved teeth that are specialized to cope with the glass particles in the grass, by folding hard enamel into the bone structure. In the ocean, crustacean grazers evolved various types of apparatus for crushing glass spines and shells (including siliceous "teeth" in certain copepods). On land, silica is rather readily available, as the elements silicon and oxygen are the main components of the most common rocks and minerals. In the sea, dissolved silica is in low supply in surface waters but is modestly abundant within the colder waters below, just like other nutrients.

23. H. U. Sverdrup, M. W. Johnson, R. H. Fleming, 1942, *The Oceans—Their Physics, Chemistry, and General Biology.* Prentice-Hall, Englewood Cliffs, N.J., 1087 pp., p. 777; the citation is to G. L. Clarke, R. H. Oster, 1934, *The penetration of the blue and red components of daylight into Atlantic coastal waters and its relation to phytoplankton metabolism.* Biological Bulletin 67, 59–75.

24. H. U. Sverdrup et al., *The Oceans,* pp. 778–779. The concepts of compensation point and compensation depth are central to understanding the productivity of the ocean, and hence to all of biological oceanography. They are crucial when explaining the spring bloom in the North Atlantic, one of the most highly visible annual events when studying the productivity of the ocean. It was largely the work of the Norwegian biologist H. H. Gran and his contemporaries, and later of George Leonard Clarke (1905–1987) and colleagues at Woods Hole, that helped these concepts emerge. For example, G. L. Clarke, 1936, *Light penetration in the western North Atlantic and its application to biological problems.* Rapports et Proces-verbaux des Réunions Conseil International pour l'Exploration de la Mer 101 (2), 1–14.

25. Gordon Arthur Riley (1911–1985), a leading figure in plankton dynamics, worked at Woods Hole from 1942 to 1948 and moved to Dalhousie University after that. Important contributions include G. A. Riley, 1942, *The relationship of vertical turbulence and spring diatom flowering.* Journal of Marine Research 5, 67–87; G. A. Riley, 1946, *Factors controlling phytoplankton populations on Georges Bank.* Journal of Marine Research 6, 54–73; G. A. Riley, H. Stommel, D. F. Bumpus, 1949, *Quantitative ecology of the plankton of the western North Atlantic.* Bulletin of the Bingham Oceanographic Collection 12 (3), 1–169.

26. E. Mills, 1989, *Biological Oceanography, an Early History, 1870–1960,* Cornell University Press, Ithaca, N.Y., 378 pp. Eric Mills is at Dalhousie University.

27. H. U. Sverdrup, 1953, *On conditions for the vernal blooming of phytoplankton.* Journal du Conseil International pour l'Exploration de la Mer 18, 287–295.

28. V. Smetacek, U. Passow., 1990, *Spring bloom initiation and Sverdrup's critical depth model.* Limnology and Oceanography 35, 228–233.

29. H. U. Sverdrup, 1955, *The place of physical oceanography in oceanographic research.* Journal of Marine Research 14, 287–294.

30. Willard Libby, 1908–1980, discoverer of radiocarbon (see chapter 6).

31. E. Steemann Nielsen, E. Aabye Jensen, 1957, *Primary oceanic production, the autotrophic production of organic matter in the oceans.* Galathea Report 1, 49–136.

32. R. H. Fleming, 1939, *The control of diatom populations by grazing.* Conseil Permanent International pour l'Exploration de la Mer, Journal du Conseil 14 (2), 210–227.

33. O. I. Koblents-Mishke, V. V. Volkovinskiy, Yu. G. Kabanova, 1968, *Noviie danniie o velichine pervichnoi produktsii mirovogo okeana.* Doklady Akademii Nauk SSSR, 183, 1189–1192.

34. O. I. Koblents-Mishke, V. V. Volkovinsky, J. G. Kabanova, 1970, *Plankton primary production of the World Ocean,* in W. S. Wooster (ed.), *Scientific Exploration of the South Pacific.* National Academy of Sciences, Washington, D.C., pp. 183–193.

35. Department of Fisheries, 1972, *Atlas of the Living Resources of the Seas,* 3rd ed., FAO, United Nations, Rome, vii pp., 62 charts, 19 pp.

36. W. H. Berger, K. Fischer, C. Lai, G. Wu. 1987, *Ocean Productivity and Organic Carbon Flux.* SIO Reference 87-30. University of California, San Diego, 67 pp.

37. The total value is 23 billion according to O. I. Koblentz-Mishke et al., *Plankton primary production,* and 31 billion tons according to T. Platt, D. V. Subba Rao, 1975, *Primary production of marine microphytes,* in J. P. Cooper (ed.), *Photosynthesis and Productivity in Different Environments.* International Biological Programme, vol. 3. Cambridge University Press, Cambridge, pp. 249–279.

38. H. Lieth, R. H. Whittaker, 1975, *Primary Productivity of the Biosphere.* Springer-Verlag, New York, 339 pp.

39. Steemann Nielsen gives a detailed description of the method in A. F. Bruun, S. Greve, H. Mielche, R. Spaerck (eds.), 1956, *The Galathea Deep Sea Expedition 1950–1952.* Macmillan, New York, 296 pp., 53–64.

40. T. Platt, W. K. W. Li (eds.), 1986, *Photosynthetic Picoplankton.* Canadian Bulletin of Fisheries and Aquatic Sciences, vol. 214, 1–583.

41. W. E. Esaias, G. C. Feldman, C. R. McClain, J. A. Elrod, 1986, *Monthly satellite-derived phytoplankton pigment distribution for the North Atlantic Ocean basin.* Eos 67 (44), 835–837.

42. H. R. Gordon, D. K. Clark, J. W. Brown, O. B. Brown, R. H. Evans, W. W. Broenkow, 1983, *Phytoplankton pigment concentrations in the Middle Atlantic Bight: Comparison of ship determinations and CZCS estimates.* Applied Optics 22, 3929–3931.

43. R. W. Eppley, E. Stewart, M. R. Abbott, U. Heyman, 1985, *Estimating ocean primary production from satellite chlorophyll: Introduction to regional differences and statistics for the Southern California Bight.* Journal of Plankton Research 7, 57–70. More recently, Scripps oceanographer Greg Mitchell has focused research on this problem of conversion.

44. J. A. Yoder, C. R. McClain, G. C. Feldman, W. E. Esaias, 1993, *Annual cycles of phytoplankton chlorophyll concentrations in the global ocean: A satellite view.* Global Biogeochemical Cycles 7, 181–193.

45. E. L. Venrick, J. A. McGowan, D. R. Cayan, T. Hayward, 1987, *Climate and chlorophyll a: Long-term trends in the central North Pacific.* Science 238, 70–72. (The trends were established from direct measurements, not from satellite observations.)

46. As did Joseph L. Reid, comparing zooplankton abundance with phosphate concentrations at 100 meters depth. J. L. Reid, 1962, *On the circulation, phosphate-phosphorus content and zooplankton volumes in the upper part of the Pacific Ocean.* Limnology and Oceanography 7, 287–306.

47. The concept of limiting nutrients was introduced by the German chemist Justus von Liebig (1803–1873), pioneer of organic chemistry and the chemistry of plants and animals, in the context of agriculture. The concept was adopted for marine plankton dynamics in the nineteenth century by the marine biologists H. H. Gran and K. Brandt, and their contemporaries.

48. H. W. Harvey (1887–1970) wrote two influential books on the subject of the ocean's productivity: H. W. Harvey, 1928, *Biological Chemistry and Physics of Sea Water.* Cambridge University Press, Cambridge, 194 pp.; H. W. Harvey, 1955, *The Chemistry and Fertility of Sea Waters.* Cambridge University Press, Cambridge, 234 pp.

49. The "equilibrium ocean" is a concept investigated by the Swedish chemist Gunnar Sillén. He calculated the ocean's chemistry from first principles, taking note of the chemistry of the atmosphere and the materials on the seafloor.

50. Alfred C. Redfield (1890–1983), the Woods Hole oceanographer who made fundamental contributions to the nutrient chemistry of the ocean.

51. F. Azam, T. Fenchel, J. G. Field, J. S. Gray, L. A. Meyer-Reil, T. F. Thingstad. 1983, *The ecological role of water-column microbes in the sea.* Marine Ecololgy Progress Series 10, 257–263. Among the early papers on this subject are J. A. Fuhrman, J. W. Ammerman, F. Azam, 1980, *Bacterioplankton in the coastal euphotic zone: Distribution activity and possible relationships with phytoplankton.* Marine Biology 60, 201–207; P. J. LeB. Williams, 1981, *Incorporation of microheterotrophic processes into the classical paradigm of the planktonic food web.* Kieler Meeresforschung 5, 1–28. A review of the concept is in F. Azam, 1998, *Microbial control of oceanic carbon flux: The plot thickens.* Science 280, 694–696.

52. R. T. Barber, A. K. Hilting, 2000, *Achievements in biological oceanography,* in J. Steele and Ocean Studies Board Members (eds.), 2000, *50 Years of Ocean Discovery.* National Academy Press, Washington, D.C., pp. 11–21.

53. By Scripps alumna Sallie Chisholm and collaborators. S. W. Chisholm et al., *A novel free-living prochlorophyte.* Chisholm works at MIT.

54. R. C. Dugdale, J. J. Goering, 1967, *Uptake of new and regenerated forms of nitrogen in primary production.* Limnology and Oceanography 12, 196–206.

55. R. W. Eppley, 1989, *New production: History, methods, problems,* in W. H. Berger, V. S. Smetacek, G. Wefer (eds.), *Productivity of the Ocean: Present and Past.* John Wiley and Sons, Chichester, U.K., pp. 85–97.

56. Appreciable amounts of bioavailable nitrogen—comparable to biological fixation itself—are added to the world's store by industrial fixation, and also by lightning (which produces nitrogen oxides).

57. G. Evelyn Hutchinson (1903–1991)—zoologist, planktologist, nutrient chemist— contributed in fundamental ways to the understanding of the ecology of lakes and of the ocean. His work and his students were very influential in marine ecology of the second half of the twentieth century. Regarding nitrogen: G. E. Hutchinson, 1944, *Nitrogen in the biogeochemistry of the atmosphere.* American Scientist 32, 178–195.

58. Another important pioneer of biogeochemistry was Vladimir Ivanovich Vernadsky (1863–1945), who insisted that Earth's environment was controlled by biological processes. Pioneers in thinking about the marine nitrogen cycle, besides G. Evelyn Hutchinson, were H. W. Harvey and L. H. N. Cooper

of Plymouth, A. C. Redfield at Woods Hole, and R. H. Fleming. At Scripps, chemist Norris W. Rakestraw (1895–1982) devoted much of his career to the marine nitrogen cycle.

59. A. C. Redfield, 1958, *The biological control of chemical factors in the environment.* American Scientist 46 (3), 205–221.

60. A high solubility of phosphate, presumably, would foster burial of organic carbon, which would drive up the oxygen content in the atmosphere.

61. E. Mills, *Biological Oceanography*, p. 165.

62. H. H. Gran, 1931, quoted in E. Mills, *Biological Oceanography*, p. 166.

63. Important papers addressing this problem are R. T. Barber, J. H. Ryther, 1969, *Organic chelators: Factors affecting primary production in the Cromwell Current upwelling.* Journal of Experimental Marine Biology and Ecology 3, 191–199; J. D. H. Strickland, R. W. Eppley, B. Rojas de Mendiola, 1969, *Poblaciones de fitoplancton, nutrientes, y fotosintesis en aguas costeras Peruanas.* Boletín Instituto del Mar del Perú, 2, 4–45; W. H. Thomas, 1979, *Anomalous nutrient-chlorophyll interrelationships in the offshore eastern tropical Pacific Ocean.* Journal of Marine Research 37, 323–335. Richard Barber, a leading figure in productivity research, is at the Marine Laboratory of Duke University, N. C. Strickland, Eppley and Thomas were members of the food chain group at Scripps.

64. John H. Martin (1935–1993), at the Marine Laboratory at Moss Landing, investigated the role in trace metals in ocean productivity. He also was a pioneer in the study of export production, using open-ocean experiments to determine its magnitude.

65. J. H. Martin, S. E. Fitzwater, 1988, *Iron deficiency limits phytoplankton growth in north-east Pacific subarctic.* Nature 331, 341–343; J. H. Martin, R. M. Gordon, S. E. Fitzwater, 1990, *Iron deficiency limits phytoplankton growth in Antarctic waters.* Global Biogeochemistry Cycles 4, 5–12.

66. R. T. Barber, A. K. Hilting, *Achievements in biological oceanography.*

67. For discussions see S. W. Chisholm, F. M. M. Morel (eds.), 1991, *What controls phytoplankton production in nutrient-rich areas of the open sea?* Limnology and Oceanography 36 (8) (Special Volume), 1507–1965.

68. J. M. Martin, 1990, *A new iron age, or a ferric fantasy.* U.S. JGOFS News 1(4), 5, 11.

69. See, for example, K. H. Coale and 18 others, 1996, *A massive phytoplankton bloom induced by an ecosystem-scale iron fertilization experiment in the equatorial eastern Pacific Ocean.* Nature 383, 495–501.

70. One reason is that continuous application of iron would be required to keep the carbon from coming back up from below the mixed layer (where most of the downward moving organic carbon is reoxidized). Another reason, applicable in the Antarctic Ocean, is that where light is limiting, and where the mixed-layer depth is below the critical depth most of the year, adding iron will not have the desired effect except over a very short period in the year. Massive application of iron fertilizer during that time would surely impact the ecology that sustains marine mammals and seabirds in the region.

71. See, for example, J. D. H. Strickland's introduction to the symposium *Marine Food Chains*, edited by J. H. Steele (University of California Press, Berkeley, 1970, p. 3).

72. R. T. Barber, A. K. Hilting, *Achievements in biological oceanography*, p. 17. *Autotroph* is the technical term for "producer." *Heterotroph* means "consumer." The authors define *microbe* as meaning small producers, grazers, and organisms both producing and consuming, and as including both bacteria and eukaryotes.

73. The food web proceeds from primary producers to zooplankton and to herring. Note that young herring are part of the food supply for adult herring (fig. 8.3). Hardy's scheme is reproduced in the text by J. L. Sumich, 1976, *An Introduction to the Biology of Marine Life.* Wm. C. Brown, Dubuque, Iowa, 348 pp.

74. From diatoms to copepods is one step; from bacterioplankton and nannoplankton to copepods is between two and three steps. Thus, equal production in the two systems would imply a contribution of less than 10 percent from the microbial loop to the traditional chain.

75. J. H. Ryther, 1969, *Photosynthesis and fish production in the sea.* Science 166, 72–76. Ryther was at Woods Hole from 1951 to 1981.

76. The contrast in production between meadows and deserts of the sea, as stipulated by Ryther, is in fact conservative: a factor of three would be acceptable. W. H. Berger et al., *Ocean Productivity.*

77. J. H. Ryther, *Photosynthesis and fish production*, p. 73. The *trophic level* is an idealization used by ecologists for projecting the food web into something like a food chain. The number of levels considered typically varies between two and six. The levels are the steps in the food-chain pyramid.

78. Chaetognaths, literally "bristle jaws," by common name "arrow worms," are in a phylum by themselves. They are ubiquitous predators in the plankton, in the millimeter size range.

79. Ryther pointed out (*Photosynthesis and fish production*, p. 74) that there is "considerable interest at present in the possible commercial utilization" of the krill. In the years that followed, a krill fishery indeed developed; it peaked near 400,000 tons in 1990, with Soviet and Japanese factory ships the chief

operators. After 1990, Japan and Poland became the most active krill fishing nations. The take is said to be 100,000 tons or less, with minor participation of Russia and the Ukraine. I. Everson, 2001, *Southern Ocean fisheries,* in J.H. Steele, S.A. Thorpe, K.K. Turekian, *Encyclopedia of Ocean Sciences,* 6 vols. Academic Press, San Diego, pp. 2858–2865, p. 2861. Only a small portion of the catch is for human consumption; most of it goes to aquaculture and pet food.

80. Victor Hensen (1835–1924) was professor of zoology at Kiel University. See E. Mills, *Biological Oceanography* for a thorough discussion of Hensen's many contributions.

81. Karl Brandt (1854–1931) was professor of zoology and director of the Zoological Institute at Kiel from 1888 until 1922.

82. See E. Mills, *Biological Oceanography,* p. 44 ff. and p. 175.

83. Hans Lohmann (1863–1934) went to Kiel in 1886, where he became an expert in plankton ecology and challenged the ideas of his elders, Hensen and Brandt. Eventually he left Kiel to become professor of zoology in Hamburg.

84. Appendicularians are planktonic protochordates related to salps and sea squirts. They are specialists in extracting extremely small particles from the surrounding seawater.

85. J.H. Steele, 1998, *Incorporating the microbial loop in a simple plankton model.* Proceedings of the Royal Society London B 265, 1771–1777.

86. A.P. Orr, S.M. Marshall, 1969, *The Fertile Sea.* Fishing News (Books) Limited, London, p. 68. A.P. Orr was a chemist, Sheina Marshall a biologist; both worked at the Scottish Marine Biological Association, at the Marine Station in Millport, from 1922 and 1923, respectively.

87. S.M. Marshall, A.P. Orr, 1955, *The Biology of a Marine Copepod,* Calanus finmarchicus *(Gunnerus).* Oliver and Boyd, Edinburgh, 188, pp.

88. R.J. Conover, 1966, *Factors affecting the assimilation of organic matter by zooplankton and the question of superfluous feeding.* Limnology and Oceanography 11, 339–345.

89. The Food Chain Research Group was established by marine chemist John D.H. Strickland, in 1963. The Institute of Marine Resources, with branches at several campuses, was headquartered at Scripps. It was created (in 1954) to coordinate research in fisheries and other ocean resources. It is no longer active.

90. Michael M. Mullin (1937–2000) was a recognized leader in the study of copepod ecology.

91. M.M. Mullin, E.R. Brooks, 1970, *Growth and metabolism of two planktonic, marine copepods as influenced by temperature and type of food,* in J.H. Steele (ed.), *Marine Food Chains.* University of California Press, Berkeley, pp. 74–95.

92. J.D.H. Strickland, O. Holm-Hansen, R.W. Eppley, R.J. Linn, 1969, *The use of a deep tank in plankton ecology. I. Studies of the growth and composition of phytoplankton crops at low nutrient levels.* Limnology and Oceanography 14, 23–34.

93. M.M. Mullin, G.-A. Paffenhöfer, 1971, *Mass culture of the copepod,* Calanus helgolandicus. University of California Institute of Marine Resources Report 71-10, 83–91.

94. R.T. Barber, A.K. Hilting, *Achievements in biological oceanography,* p. 16.

95. For example, J.R. Strickler, 1982, *Calanoid copepods, feeding currents, and the role of gravity.* Science 218, 158–160; J.J. Price, G.A. Paffenhöfer, J.R. Strickler, 1983, *Modes of cell capture in calanoid copepods.* Limnology and Oceanography 28, 116–223.

96. The standard work on limnology is the two-volume opus by the ecologist G.E. Hutchinson, pioneer of biogeochemistry. G.E. Hutchinson, 1967, *A Treatise on Limnology,* John Wiley and Sons, New York.

97. For example, see articles in J.H. Steele (ed.), 1970, *Marine Food Chains,* University of California Press, Berkeley, 552 pp.

98. See, for example, A.R. Longhurst, A.W. Bedo, W.G. Harrison, E.J.H. Head, D.D. Sameoto, 1990, *Vertical flux of respiratory carbon by oceanic diel migrant biota.* Deep-Sea Research 37, 685–694.

99. P.J.S. Franks, 2002, *NPZ models of plankton dynamics: Their construction, coupling to physics, and application.* Journal of Oceanography 58, 379.

100. *Ibid.,* p. 386.

NINE

Of Whales and Sharks and Giant Squid

REFLECTIONS ON THE BIG, THE STRANGE, AND THE POWERFUL

W. H. Berger and E. N. Shor

The largest animals on the planet are the great whales. The biggest among these weigh more than 20 big elephants.

Modern whales arose within the last 30 million years or so, as a result of the cooling of the planet, which changed the productivity patterns of the sea in ways favorable for the development of large size. This is true both for whales that hunt (toothed whales) and for whales that filter the water containing small fish and krill and other plankton (baleen whales). Other vertebrate groups also have large animals; pinnipeds have elephant seals and walrus, sharks and their kin include whale sharks, the great white, and manta rays, and bony fishes have swordfish and sturgeon. Various unverifiable observations suggest that there are deep-sea giants yet to be discovered. Among the mollusks, the largest—the giant squid—apparently grow to well over a ton in size. Warm oceans also had large marine animals, especially among reptiles (ichthyosaurs, plesiosaurs, mosasaurs, crocodiles) and mollusks (giant ammonites). It is likely that their food requirements were met in highly productive shelf seas.

THE WORLD OF WHALES

"Thar she blows!" That call coming from the lookout in the masthead high above the deck of a nineteenth-century whaling vessel used to spell doom for the object of attention.[1] Wholesale killing of whales continued well into the second half of the twentieth century, with large and fast metal ships armed with harpoon guns turning whale hunting into a whale removal operation, in a prime example of the "tragedy of the commons."[2] The annual catch peaked near 65,000 large whales in the years around 1960. A moratorium on whale hunting went into effect from 1985, after it was realized that continued whaling would result in the extinction of

FIGURE 9.1. Humpback whale breaching in Hawaiian waters, as part of a social ritual. Males of this species sing. The humpback is a baleen whale.

FIGURE 9.2. Highly intelligent and trainable, the small whales called dolphins entertain thousands of visitors to open-air aquariums. Here, seen on the island of Oahu, Hawaii. Dolphins are toothed whales.

some of the largest animals that ever lived on the planet.

In our time, the call-out for whales serves to alert a throng of tourists to the presence of the animal they had come to see. Vast numbers of tourists swarm to places from which whales can be watched from land or on day trips—on the East Coast, the West Coast, among the islands of Hawaii, and off Baja California. To many people the ocean is first and foremost the habitat of whales.[3]

Cetaceans as a group (whales and dolphins and their kin) have the most intelligent, the most playful, the largest, and the fastest swimming animals in the sea (figs. 9.1 and 9.2). Also, they exhibit complex social behavior, as appropriate for large-brained mammals living in groups. It is now common knowledge that whales and dolphins are mammals rather than fishes, but for centuries it was thought that whales were some kind of fish.[4] Cetaceans have in common a fishlike shape and a tail fluke for propulsion. Their front limbs are steering paddles; they have lost the hind limbs of their four-footed ancestors.[5] The head is large and highly modified compared to other mammals, with elongated jaws and one or two breathing holes near the top of the skull (dolphins have one

FIGURE 9.3. Large whales. Upper: baleen whales (*Balaenoptera*); left: Blue Whale, right: Minke. Lower: toothed whale, Orca.

only). There is no neck (or not much of one), the cervical vertebrae having been much shortened in the course of evolution. Hair has been lost. Insulation from cold water is by a layer of fat, and this blubber also serves to store energy in the absence of food (much like the camel's hump). The fishlike shape is a matter of convergence—the reptilian ichthyosaurs of the Jurassic adopted the same shape 100 million years before the whales did. The shape, the fluke, and a skin that decreases friction when traveling through the water are adaptations forced by the physics of bodies moving through water. These features reduce resistance and save energy.

Both toothed whales and those lacking teeth and having baleen instead have enormously large representatives, but the very largest are baleen whales. The skeletal architecture of all whales is rather similar, except for the skull, which shows large differences between the various types of whales (fig. 9.3).

There is much that cetaceans can tell us about the nature of life in the sea—about the habitat that sustains them and about the forces that drive their evolution. Each species, finely adapted to its surroundings, carries in its evolutionary history a record of its responses to the opportunities offered by the ocean environment. Closely related species tell similar stories. But more distantly related forms have explored different ways of making a living. Together, their life stories reflect the many facets of the ocean as a life habitat.

Some of the elements of this life habitat are unchanging, such as the nature of the resistance

FIGURE 9.4. Baleen-supported feeding, with baleen plates hanging from the roof of the mouth for retaining zooplankton when pressing out large gulps of water. Note the large head for holding seawater. Example shown is a bowhead whale.

of water to a swimming body, or the fact that light does not travel far in water. But some aspects of the life habitat change through geologic time, such as the temperature distribution, and the patterns of productivity. Such changes, ultimately, are responsible for the ascent of the great baleen whales, the largest animals on Earth, now or ever. They are the products of a cooling planet with ice at the poles, as is clear from their manner of feeding. They filter the water by taking big gulps of it and pressing it through the fringed horny plates—called baleen,[6] or whalebone—growing from the roof of their mouths and draping its circumference like a valance curtain (fig. 9.4). For this method to be profitable, there needs to be plenty of food in the water. As we have seen when discussing meadows and deserts in the sea, plankton abundance is controlled by nutrient supply, hence deep mixing and upwelling. Mixing and upwelling, in turn, depend on strong winds. Strong winds characterize a planet with strong temperature differences, that is, a planet that has ice. Baleen whales evolved in adaptation to ice buildup on Earth, starting some 35 million years ago!

Whales and dolphins, on the whole, are prodigious sources of different types of sound, from low-frequency moans to high-frequency whistles and chirps, depending on species and circumstances. The elaborate and hauntingly beautiful songs of humpback whales have been compared to human music.[7] However, much of the sound emitted routinely by cetaceans is of a mechanical nature, consisting of regular clicks or, in the case of sperm whales, of pulsed claps reminiscent of hammering nails into a wall.[8] Research vessels and modern fishing vessels use similar pulses for echolocation, that is, for mapping the seafloor and for finding fish, respectively. Whales and dolphins also use sound for communication, presumably much of it of the kind that says: "I am here. Where are you?" The members of a group, unable to see each other, confirm their presence and relative position by making noise.

The ancestors of modern whales, of course, had ears like all other mammals. But the use of sound for "seeing" in the water arose in the ancestors of modern cetaceans by the late Oligocene, some 30 million years ago. How do we know this? We do not know for sure, of course; sound makes no fossils. But we can infer it from the change in the shape of skulls. The late Oligocene is the time when we first find skulls that suggest adaptations to processing sound. These modifications include a doming and foreshortening of the skull, a pulling out of the snout, and a widening of the skull, presumably making room for a bulbous structure in front of the forehead.[9] Also, increasing the distance between the hearing centers improved the ability to judge where sound comes from.

In modern toothed cetaceans, strange-looking bulbs are carried in front of the forehead (a "melon" or similar protuberance), which enhance the power of echolocation. Outgoing sound, it is now realized, can be focused and beamed at objects to be explored, using a "melon" in front of a reflective forehead. Lens and mirror of a searchlight provide a familiar analogy for this arrangement.[10] The proper analysis of sound, to produce mental images analogous to those made by our brain using input from the eyes, takes large amounts of computing power. Cetaceans have some of the largest brains in the animal kingdom.

The adaptations concerning the use of sound for "seeing" in the ocean are a conse-

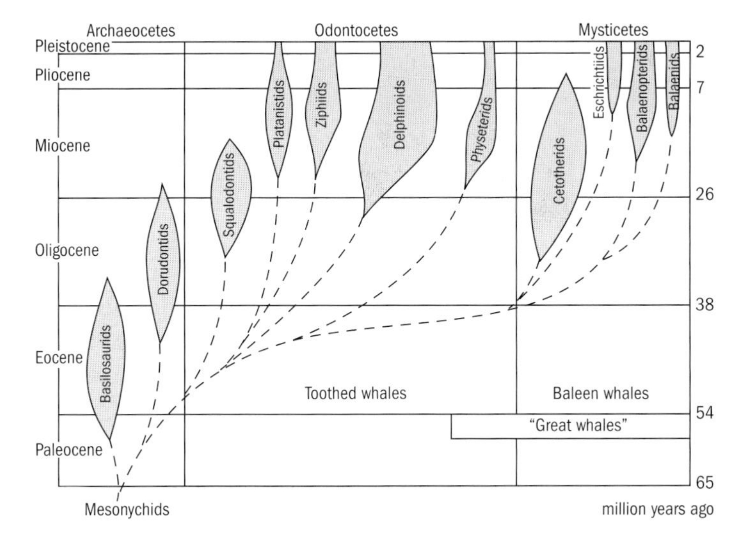

FIGURE 9.5. Origin of the various groups of cetaceans, according to Bonner (1989). Note the development of maximum diversity in the Miocene. Examples for the four groups of modern odontocetes (toothed whales) are Amazon river dolphin (platanistids), beaked whale (ziphiids), common dolphin and orca (delphinoids), and sperm whale (physeterids). Examples for the three groups of modern mysticetes (baleen whales) are the gray whale (eschrichtids), blue whale (balaenopterids) and bowhead whale (balaenids).

quence of the fact that sound travels well in water, in contrast to light, which is quickly scattered and absorbed. Thus, the development of the cetacean sonar capabilities, like their fish-like shape, is driven by the elementary physics of the environment, rather than by a changing environment. However, the effects of a changing climate are not so easily dismissed. Again, we have to focus on changes in productivity. For echo-based hunting to evolve, there must be something to ping on—schools of small fish or squids. Such schools are common in high-production regions. Many of the prey preferred by echo-hunting dolphins engage in vertical migration, hiding at depths below the sunlit zone during the day, and coming up to feed at night. As they ascend, they enter the sonar range of dolphins, which "see" in the dark. Quite likely, echo-hunting evolved in response to vertical migration of prey animals in productive regions. Thus, the same basic driving forces related to the cooling of the planet since the Eocene are at work here as well, just as for the baleen whales. This is fundamentally the reason why the evolution of baleen whales and of echo-hunting whales runs in parallel: the adaptations of both toothed cetaceans (odontocetes) and whalebone whales (mysticetes) are geared toward the increase in productivity in privileged regions of the World Ocean (fig. 9.5).

The split between the two great groupings in the order of cetaceans occurred at a time of major cooling, in the early Oligocene, when some of the toothed whales grew baleen.[11] However, the main opportunities for specialization clearly arose at the end of the Oligocene, with the intensification of upwelling in privileged regions (see fig. 9.5). At that time, echo-hunters (odontocetes) found plenty of prey to ping on, and filtering-grazers (mysticetes) caught plenty of zooplankton in their frayed baleen and could now do without teeth. Their way of life became again more similar to that of their hoofed cousins on land, which fed low on the food chain and depended on a strongly seasonal food supply.

Krill occurs in enormous abundance around Antarctica *(Euphasia superba)* during southern summers. The Southern Ocean, in consequence, provides the traditional feeding grounds for the largest of the mysticetes, and the history of productivity in the sea around Antarctica therefore likely contains important clues to the evolutionary history of the baleen whales. While krill make no fossils, their diatom food does. Diatomaceous sediments greatly expanded in the Miocene (fig. 9.6),[12] and this is when the whales achieved maximum diversity. The seasonal feeding in cold waters favored large size and blubber buildup, and a habit of migration. Large size and blubber are good for retaining

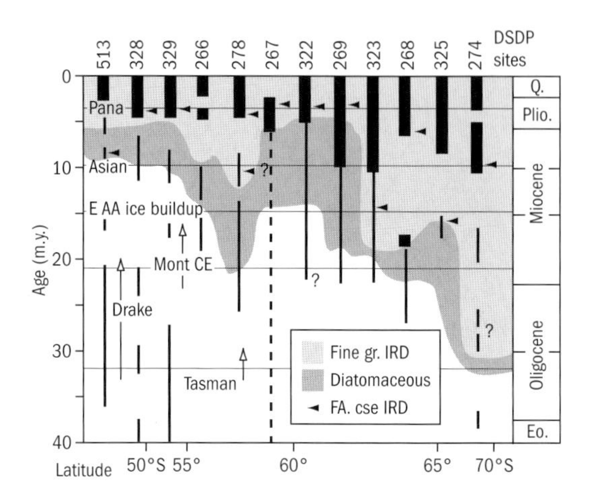

FIGURE 9.6. History of expansion of diatom-rich deposits around the Antarctic in the Miocene (based on drilling results, according to Ciesielski and Weaver [1983]). This provides the chief evidence for the rise of seasonal feeding opportunities for baleen whales. Pana, emergence of Panama Isthmus; Asian, restriction of Indonesian passage from Pacific to Indian Ocean; E AA, East Antarctic; Mont CE, upwelling off California (Monterey Carbon Event); Drake, widening of Drake Passage; Tasman, widening of Tasmanian Passage; fine gr. IRD, fine-grained ice-rafted debris; FA cse IRD, first appearance of coarse IRD; Q., Quaternary; Plio., Pliocene; Eo., Eocene.

heat, and both are especially good for being able to go without much food till the next season. The very largest of the cetaceans, the baleen whales, can go without food for long periods.

In contrast to the baleen whales, the toothed cetaceans, hunters of the sea, have a less predictable food supply. The forces of evolution working on the toothed cetaceans, then, differed greatly, depending on the mix of prey hunted. Thus, there was great opportunity for the development of different hunting skills, and for diversification. The result is that there are many more types of toothed cetaceans than of baleen whales. (The ratio is around four to one.)

A MATTER OF SIZE

The largest of the toothed cetaceans is the sperm whale, *Physeter macrocephalus,* a favorite target of whaling since the late eighteenth century. (Moby Dick was a sperm whale.) *Physeter macrocephalus* (so named by Linné in 1758) has a unique shape, with an enormous bulbous structure in front of the forehead that makes up much of the square head, which in turn represents about a third of the animal. The largest males measured reach a length of almost 20 meters (65 feet) and a weight of 50 tons or more. Females are somewhat smaller. Teeth are mainly restricted to the lower, comparatively narrow jaw. They fit into sockets in the upper jaw. Sperm whales are the diving champions of marine mammals; they can go down more than a mile to hunt, and stay submerged for more than an hour. The primary food items are squid, including giant squid weighing as much as a ton. Males migrate over long distances in search of the best hunting grounds, while females show a lot less wanderlust, being in charge of the young.

The baleen whales as a group, together with the giant sperm whale, are called great whales, because of their impressive bulk. What factors are responsible for their enormous sizes? According to the whale expert Roger Payne, it has to do mainly with the way they gather food. Zooplankton cannot escape a very wide mouth. And for seasonal feasting in far-away places, and fasting for the rest of the year, a large size is of great advantage.[13] If so, seasonal feeding has the chief clue to the enormous size of baleen whales. Thus, the history of diatom deposition around the Antarctic (which records the rise and abundance of seasonal whale food) becomes highly relevant to the story (see fig. 9.6). Diatoms feed krill, and krill feed baleen whales, penguins, and the crabeater seal.

How large are these giants that exceed in bulk all other animals on the planet? Very large indeed.

The maximum size of the largest of the baleen whales, the blue whale *(Balaenoptera musculus),* is not well known because the largest members of the species have long fallen prey to whaling. Scaling the maximum length reported

(33.5 meters, or 110 feet) to present dimensions of typical length and weight (24 meters, or 80 feet, and 100 tons) one readily arrives at a weight of 200 tons—well in excess of the largest dinosaurs.[14] The newborn calf of a blue whale weighs around 7 tons, that is, more than a big elephant. The smallest of the baleen whales, the rarely seen pygmy right whale *(Caperea marginata)* is up to 21 feet long and reaches a weight of 5 tons.

The blue whale and its relatives belong to the group called rorquals.[15] These are whales with enormous feeding pouches anchored at the lower jaw—like a pelican's—and reaching halfway across the body to the navel. When inflated with water (and the plankton within), the pouch expands with its folds (called throat pleats) ballooning outward. This expansion nearly doubles the size of the whale. The water within the pouch is pressed out through the baleen sieve, and the plankton thus trapped goes down a relatively narrow gullet.

The rorquals comprise blue, fin, sei, Bryde's, Minke, and humpback whales. Of these, the humpback *(Megaptera novaeangliae)* is different enough from the rest to be assigned to a separate genus. The others belong to *Balaenoptera*. The smallest of the rorquals is the Minke whale (at about 30 feet and 9 tons). Thenonrorquals —the baleen whales without a pouch with throat pleats—are the right whales, with the right (*sensu stricto,* northern and southern), the bowhead, and the pygmy right whales. (The label "right" is a whaling term: the "right" whales are the good ones to hunt; they move slowly, have lots of fat to make oil, and float after being harpooned.) The gray whale *(Eschrichtius robustus),* the whale best known to San Diegans, is neither a rorqual nor a right whale but is off to the side in its own family. Members of this species reach 50 feet in length and a weight up to 30 tons. They feed and breed in shallow waters and are thought to be closer to the ancestral form of baleen whales (more "primitive") than the others.

If size is largely a matter of feeding, a comparison of feeding habits of the great whales should be instructive. Feeding habits vary substantially between species, even when excluding the sperm whales from the comparison. Blue whales are specialists for the consumption of krill, the small shrimplike zooplankton that occurs in enormous clouds in the Southern Ocean, during the productive season. The whales engulf huge swarms of the crustacean, get rid of the water through the baleen, and swallow the plankton a hundred pounds at a gulp. The blue whale's closest relative, the fin whale, likewise is adept at gulping planktonic crustaceans such as krill and copepods. As well, fin whales go after schools of small fish. A white patch on the right side of the fin whales apparently is useful when herding fishes into a clump, by swimming in circles around the school, clockwise. (Perhaps the fishes react as they would to circling by sharks.) During such herding the fin whales work in pairs.[16] The next smaller of the rorquals, the sei whale, has similar food preferences as its larger relatives but also takes larger fishes such as cod. Thus, in this group of large rorquals, the largest members feed on the smallest prey.

Size, it appears, is an adaptation to feeding close to the base of the food chain, where food is abundant but not particularly concentrated: a ton of plankton-rich water has less meat than a school of fish. A big gaping mouth—like a large plankton net—makes it difficult for the krill to escape before the water is engulfed into the feeding pouch.[17] Interestingly, the largest living fish, the whale shark, also feeds at the base of the food chain. Other factors may drive large whale species to increase their size further: the ability of males to find mates by sounding off, a preference of females to mate with large males, increased immunity from attack by sharks (some of which were much larger several million years ago, in the Pliocene, than now), and by the fearsome killer whales (*Orcinus orca,* dolphins large enough to attack rorquals).

Feeding at the base of the food chain is typical for large mammals on land, as well: elephants, rhinoceros, and hippopotamus are familiar examples. Also, the long-distance travelers among the hoofed animals tend to be rather large.

There are, of course, also disadvantages to being large—the Minke whales do quite as well as the blue whales. At least some of these disadvantages have to do with low reproduction potential, which raises vulnerability to disease and parasites. An individual succumbing to disease presumably is more readily replaced in a rapidly reproducing population with smaller members, than in a herd of leviathans. We cannot, therefore, categorically list the advantages for giant size and assume that they drove evolution. Instead, we must assume that at some point the large size itself drove evolution, by decreasing options and forcing the organism into paths where bigger is better. We also must consider factors that are ecologically irrelevant, such as female choice.

The huge investment of resources in individuals, within the great rorquals, was ultimately the reason that they became the prime target of a new type of predator: whalers and their factory ships. In this new environment, being huge is a ticket to extinction.

FEEDING, BREEDING, AND MIGRATION

The strategies for feeding are crucial in the evolution of cetaceans, as we have seen. This is true not just for the obvious anatomical adaptations that are so conspicuous in the baleen whales—the huge pouches of the rorquals, the enormous heads of the right whales—but also for the development of patterns of behavior, such as echolocation, deep diving, and long-distance migration, with the mix of activities depending on the species.

Of course, the imperatives of reproduction are just as important as the need for food in driving evolution. Adult females, in particular, have to invest in a growing fetus in their wombs, and they have to nurse the newborn till it can fend for itself. During this time of nursing the young are especially vulnerable to predators. Thus, females have an interest in raising their young in places where predators are rare. These are not generally places where food is abundant. Naturally, then, it is the females that seek out sheltered (or warm, or less frequented) breeding areas. The females quit feeding when the time has come to go to the refuge. They start moving. The males follow the females. The result is long-distance migration for the entire herd.

The great whales are long-distance champions, but they are not the only ones making the effort. Migration is a way of life for many of the large swimmers of the open sea. In addition to whales, patterns of migration have been studied in fishes, turtles, penguins, seals, and seabirds. Stay-at-homes among the large swimmers can be found in the kelp forests and in the reefs where fishes (including sharks) stake out territories.

The main reason for long-distance migrations is no big mystery—we know it from watching birds and mammals on land. Animals migrate over large distances because a rich supply of food is seasonally available in otherwise inhospitable regions. Quite generally, the two chief reasons for migration are seasonal change and different requirements for feeding and breeding. Another important aspect is that migrating prey species get some relief from their predators by periodically moving out of their range.[18]

Migration patterns differ greatly between the different types of great whales, and also between populations of the same species. Humpback whales of the Atlantic, for example, spend their feeding time in high latitudes—off Spitsbergen (Svalbard), off Greenland, and off Iceland in the north, and all through the Southern Ocean in the south (fig. 9.7). They move to wintering grounds in temperate and low latitudes in the "off season." Northern populations congregate off New England (hence the name *M. novaeangliae*), off Bermuda, and off the Antilles. Some prefer an eastern route of migration, moving to the British Isles and then south to the upwelling regions off Africa and the Canary Islands. The southern populations of the same species, naturally, feed in the fertile seas around Antarctica during this same time, since northern winter coincides with southern summer. They will come to low latitudes during southern winter. In the Atlantic these are areas off Brazil and off Angola, the latter appropriately named Walvis Bay, using the Dutch word for "whale-fish."[19]

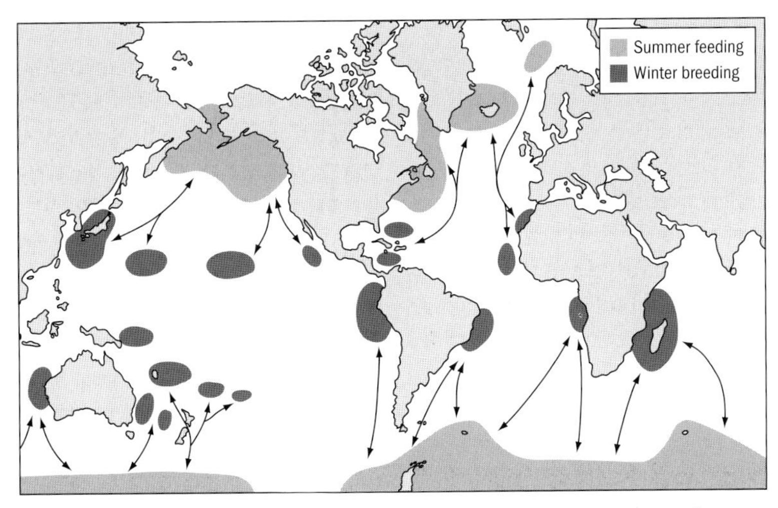

FIGURE 9.7. Migration routes of humpback populations, according to Bonner (1989).

Because of their migration schedules, the southern and northern populations hardly have occasion to meet, and thus there is little gene exchange. They are well on the way to evolve toward different species. Why then are they still so closely related?

Presumably, two factors have played a role in the evolutionary history of the migrating whales and their southern and northern populations. One is that permanent ice in the north is a relatively recent development, geologically speaking. It happened only 3 million years ago. Before that, it would seem the northern high latitudes were much less attractive as winter feeding grounds for baleen whales because winter storms were less fierce and nutrient supply to the sunlit zone was less efficient. In contrast, feeding was good in the south, where productivity has been high for more than 20 million years. We must assume, then, that the baleen whale way of life was a southern affair for a very long time. It is interesting to note, in this context, that the penguin way of life, which depends heavily on krill, always remained southern.[20]

Still, if northern populations started to split off some 3 million years ago we should have distinct northern and southern species by now. This is where the second factor comes in: the enormous expansion of northern ice caps from time to time, which displaced the northern feeding grounds far to the south, to upwelling regions off Portugal and off northwestern Africa. Thus, the need for migration was sporadically greatly reduced for the northern populations. Northern and southern populations could then mingle, and presumably some gene flow was restored, slowing the divergence into different species.

These scenarios are difficult to prove or disprove, of course, but they are not just idle speculation. We have reasonably detailed information on the changes in oceanographic conditions in the Atlantic during the last ice age.[21] For the Pacific, productivity was greatly increased during glacial conditions in the eastern tropical Pacific, off Middle America, and this region would have provided a similarly centrally situated meeting ground suitable for year-round baleen feeding.

The giant rorquals—blue whales and fin whales—apparently have no well-defined breeding regions. When not congregating to feed, according to Roger Payne, they are broadly scattered all over the ocean.[22] Payne suggests they keep in touch with loud regular sound emissions, at the lower end of human hearing. Such signals can carry over thousands of miles in the "sound channel," a layer of minimum sound velocity within the thermocline. Sound made within this layer is trapped and will radiate out horizontally rather than being dispersed in three dimensions.[23] For the fin whales, the sounds made are exceedingly stereotyped and of a pure tone—exactly what a scientist would send if he wanted another one to know, a thousand miles away, that the sound is a signal and not just some random rumble from an undersea volcano or from storm waves.

The discovery of the potential for long-distance communication in the great whales was a breathtaking step forward in the integration of physical and biological oceanography. It is vital for the understanding of the life history of the great rorquals. Revolutionary ideas, by definition, interfere with conventional wisdom. Thus, the idea of long-range communication in whales was not taken seriously when first proposed, in the 1970s. Evidence that such communication could lead to whales keeping track of each other's whereabouts became available when it was shown that a blue whale could be tracked for more than a month, from a distance of a thousand miles. The feat was accomplished by the marine biologist Chris Clark.[24]

Why would the whales wish to be dispersed, and why would they be calling to each other? Payne thinks they disperse to improve the chances of finding food, and they emit sound to advertise success in finding it. Presumably, if there is a congregation someplace, a single animal would realize that "something is going on" in that area and join the group to feed. Payne suggests that the patchiness of krill clouds (which may follow storm patterns and eddies) makes a scouting strategy pay off for the population, provided the find is shared.[25] This cannot be the whole story, of course. Something must be in it for the caller who advertises any new food sources: evolution works by rewarding useful behavior. The currency of the reward is offspring.

Perhaps the best known of the long-distance marine migrations, certainly to Californians, is the annual migration of the gray whale, a baleen whale of intermediate size. The gray whale's feeding grounds are in the subarctic North Pacific, where productivity is high during the long days of the northern summer. The lagoons in Baja California, which the whales seek in winter, are sheltered places that are readily defended against predators. Mothers have brought provisions for their newborn from the feeding grounds, stored as blubber. They now turn the blubber into milk. In the sheltered lagoons, the young are kept from the fierce and cunning *Orca* for the most vulnerable period of their life.

But *Orca* is no longer the biggest threat to baleen whales. Once whalers discovered the lagoons as points of assembly, in the middle of the nineteenth century, the gray whale populations were rapidly reduced. For some years, starting in the 1890s, no one reported any whales, and it was feared that they had gone extinct. Fortunately, this was not the case. In 1945, Hubbs[26] learned that gray whale spouts had been seen off Point Loma, the promontory protecting the harbor of San Diego. He promptly began a whale watch in the winter months, when the whales pass southward to their calving lagoons.[27] The habits and the breeding grounds of the gray whales were poorly known then, in the 1940s, and Hubbs did much to shed light on the issue, visiting the regions by airplane, to make counts, for the years from 1952 to 1964.[28] The tally of gray whales rose every year, from the initial count of 4,000, to more than 26,000 in the 1990s for an overall growth rate of 4 percent per year. This is slightly higher than human population growth at its fastest (about 3.5 percent per year).

Scammon's Lagoon (Ojo de Liebre) in Baja California, where enormous numbers of gray whales congregate to breed, is now a favorite tourist attraction where people observe female whales and their calves at close range. Most of the visitors who reach out to pat the gentle creatures that approach the inflatable boats do not know that whalers once called them Devil Fish. When whalers killed a calf in order to keep the mother within the range of the harpoon, they exposed themselves to ferocious attack by furious mother animals.[29]

The gray whale *(Eschrichtius robustus)* has a black skin but appears gray because it is usually encrusted with white barnacles and with small crustaceans called whale lice. The females are slightly larger than the males, reaching about

45 feet (14 meters) in length and about 35 tons in weight. Their migration follows a highly predictable course and time and is one of the longest such treks recorded for any mammal, equaling or exceeding those of humpback populations. Apparently, the gray whales do not eat much during their journey of 6,000 miles south in December and January, nor do they feed on the return voyage in March to May. The whales fatten up in summer, in the Arctic seas (Bering, Chukchi, and Beaufort seas). They feed on bottom-dwelling amphipods and other benthic organisms, which they gather by scraping their open jaws along the bottom, stirring up the mud and pouring it out the sides of the mouth as they filter out the edible animals, through the baleen.

Did baleen whales have a bottom-feeding ancestor? Or did the bottom-feeding arise as a supplement to plankton-feeding? We do not know. Presumably, bottom-feeding requires less skill in finding food than plankton-feeding, so the filtering of mud may have marked the beginning of a baleen-based feeding strategy.

The toothed cetaceans do not have the travel urges of the baleen whales, except perhaps for the sperm whale.[30] In this species females congregate in groups with their calves and juveniles, while mature males are off wandering (as in African elephants). The mixed matrilineal herds stay in tropical and subtropical waters, covering territories up to 1,000 kilometers in length.[31] The males explore the highly productive subpolar waters until they are ready to mate.[32] The fact that juveniles stay in the tropics supports the hypothesis that the fertile high latitudes are dangerous hunting grounds, presumably largely because of the abundance of killer whales. The various smaller toothed cetaceans—the dolphins and porpoises and their relatives—are certainly capable of ranging widely (as shown by tagging and tracking), but many of them seem to have more or less well defined home ranges. Among the killer whales (the largest members of the dolphin family), biologists have distinguished "residents" feeding on fishes, and "transients" hunting mammals.[33] Tagged individuals have been observed to range widely.

As mentioned, migrating oceanic mammals, at some point in the past, made a choice of whether to visit northern or southern feeding grounds. Thus, when the northern high-productivity regions emerged, as a result of northern cooling several million years ago, sibling species evolved. Such a north-south split is also observed for the elephant seal (fig. 9.8), which has closely related southern and northern species. Recent tagging of elephant seals hauled out on Año Nuevo off California revealed migrations over more than a thousand miles into the Gulf of Alaska, to the shores of the Aleutian Islands, and (for some proportion of the females) into the open ocean within the Westwind Drift of the North Pacific.[34]

DIVING FOR FOOD

> For though other species of whales find their food above water and may be seen by man in the act of feeding, the spermaceti whale obtains his whole food in unknown zones below the surface; and only by inference is it that any one can tell of what, precisely, that food consists. At times, when closely pursued, he will disgorge what are supposed to be the detached arms of the Squid; some of them thus exhibited exceeding 20 and 30 feet in length.[35]

The above lines, written by Herman Melville, refer to the sperm whale, largest of the toothed cetaceans and of all deep-diving mammals. Many questions come to mind when reading these lines. How deep do sperm whales go? How do they find their prey down there? How did they get there in the first place, in the course of evolution? What modifications to the physiology were necessary to make dives of greater than 2,000 meters depth possible?[36] Do sperm whales restrict the distribution of giant squid? Mostly, we do not know the answer to any of these questions, but there are glimpses of clues for some of them.

There are a number of major obstacles in the way of deep-diving hunters.[37] First of all, oxygen is available only at the surface for an air-breathing animal, so it has to be stored somehow for diving. Second, hydrostatic pressure increases rapidly and compresses whatever air is in the lungs: the lungs collapse. Third, it is hard to

FIGURE 9.8. Elephant seals hauled out at Piedras Blancas, California; mainly females and juveniles, molting. The two immature bulls are sparring in mock fight. The animal in the foreground has completed the molt.

see very far, making a visual hunt ineffective. Fourth, it is very cold at depth, which means loss of energy, for a warm-blooded animal.

The reason why deep-diving mammals found ways to overcome these obstacles is simple: there is food available at depth. By increasing the supply of oxygen taken in by the lungs and stored in blood and muscles, by developing ways to favor the brain over other organs in receiving it, and by developing mechanisms for building up oxygen-debt in the less-sensitive organs (liver, kidneys, gastrointestinal tract), mammalian deep-sea hunters could hold their breath longer and longer—up to over one hour. A key element is a high content of the oxygen-binding protein myoglobin within the muscle tissue of diving mammals. In addition, seals and sea lions have a greatly enlarged spleen for storing red blood cells. Also, diving mammals markedly slow down their heart rate as the dive begins. Instead of swimming toward the target depth, they glide downward, using the fact that below a certain depth, with lungs compressed, the bone-carrying body is heavier than the surrounding water. Under these circumstances, only the brain still needs substantial oxygen—all other organs are on minimum supply.[38]

The enormous compression of air within the lungs of diving animals poses very special physiological problems in the history of deep-diving mammals. In human divers, the nitrogen gas (which makes up 80 percent of the air) enters into the bloodstream, and this causes dangerous degassing upon decompression. At Scripps, the physiologist Per Scholander was greatly interested in the question of how diving animals cope with this conundrum. He persuaded the National Science Foundation to provide him with a special laboratory, the Physiological Research Laboratory, or PRL, with a large ring-shaped pool suited for continuous swimming of marine mammals.[39]

Among those invited to work with Scholander at the new laboratory was Gerald Kooyman, who designed special instruments to record the diving behavior of marine mammals and birds.[40] Kooyman began by using recorders attached to pinnipeds in the 1960s. He found that the sperm whale dives deeper than any other mammal, more than 2,000 meters, with the northern elephant seal the runner-up at around 1,500 meters. Diving periods vary between half an hour and more than one hour for extra deep and long dives.[41]

Special adaptations within the lungs of diving mammals allow compressed air to collect in inert pockets, where there is no exchange with the blood. Thus, the nitrogen is not forced into the blood, as it would be in deep-diving humans. This avoids the bubbling out of the gas during rapid ascent, an effect that is known to produce the dreaded "bends" that human divers suffer. Also, on surfacing, the diving mammals are able to exhale and inhale unusually large volumes of air, by the standards of terrestrial mammals.

In the dim light at depth, visual hunting demands unusually large eyes and an unusually high density of light-sensitive rods in the retina. Like cats, dogs, and deer, the seals (and many other vertebrates) have a reflective layer behind the retina, doubling the chance of capturing a photon entering the eye. But by developing sophisticated sonar systems (similar to those in bats) the "vision" of the deep-diving sperm whales and their smaller teeth-bearing relatives became based on sound, rather than light. Armed with such a detection system, hunting at night and at depth becomes not only feasible but brings great advantage, as the predators can "see" while their prey cannot. Prey species had to adjust their anatomy and their behavior to the echo-hunting predators. Their swim bladders have become shining white spots in the acoustic flashlights. One defense is to swim in schools, to confuse the probing enemy with the acoustic glitter of concerted but unpredictable motion. Another is to do without gas in the swim bladder and substitute fatty substances. As so often in evolution, a race was on—evasive behavior and concealment against sophisticated detection.

For millions of years, the pelagic ecosystem has been characterized by daily vertical migration of much of the zooplankton, involving the upper several 100 meters of the water column. Like deer at the edge of the forest, the migrating zooplankton hide in the dark during the day, and come out at night to feed in the meadows of the surface layer. We may assume that the darkness at depth protects the migrating organisms from visually hunting fishes. In addition, many fishes may not like to enter the cold water below the warm surface layer, an aversion that would provide additional protection for their intended prey.

On a global scale, a strong contrast between warm water above and cold water below, at a depth of as little as 100 meters, developed only around 25 to 20 million years ago. Thus, the present pelagic ecosystem with its peculiar vertical migration patterns evolved since then. It is geologically young.[42]

Naturally, predators follow their prey. Many of the deep-living fishes and squids that feed on the zooplankton mimic the vertical zooplankton migration, some by simply just staying with their prey, others going somewhat deeper than their prey. Emerging at twilight, the night-hunters come up to feed in the teeming surface waters till the stars fade. What these dark-hunters lack in speed, they make up through stealth, superior vision, and superior sound detection. It stands to reason that acoustic-hunters, such as dolphins and porpoises, know about this vertical migration and time their hunting of lantern fishes and squids accordingly. One big advantage to feeding in the evening or at night is that the small predators that are the prey of the aural-hunters congregate in greater density as they come up against the surface boundary. During the day, they would have an open-ended range of depths for dispersal. It would be interesting to learn how the prey animals react to being hit by the sound probe sent out from a hunting dolphin. Presumably they dart to the side.[43]

Interestingly, it is after a shallow thermocline developed 25 million years ago that dolphins diversified and flourished. Presumably, the behavior of their prey, and the dolphin's hunting habits, evolved along with a strong thermocline.

One of the remarkable diving champions among mammals is the ungainly looking elephant seal (*Mirounga angustirostris* in the north, *M. leonina* in the south). These true seals are familiar from nature films showing the massive bulls stabbing at each other with their long tusks while battling for status as harem master. The northern elephant seal can be viewed at several beaches of California (notably in a reserve near Santa Cruz), and also on offshore islands. At a length of 14 to 16 feet (largest 22 feet), the males are much larger and heavier than the females, which rarely reach 10 feet. A reigning bull dominates a harem of as many as

10 to 40 cows, fighting constantly with other males during the breeding season.

The northern elephant seals were once greatly reduced by whalers. Carl Hubbs first saw these behemoth seals on Guadalupe Island off Mexico in 1946 while traveling on Errol Flynn's yacht. He repeatedly counted the animals in the following years, seeing the small colony grow to thousands of individuals within the following two decades.[44] Another marine mammal brought to the brink of extinction by sealers was the Guadalupe fur seal *(Arctocephalus townsendi)*, a 6-foot graceful animal with a beautiful coat of soft thick hair. Hubbs found a group of them in 1954, in a remote cave. His tallies document the recovery of this species, as well. The government of Mexico, in 1922, had declared the island a wildlife sanctuary, making these recoveries possible.

In the United States, marine mammals have been protected by the Marine Mammal Protection Act since 1972 (modified 1995). Protection has had the desired effect: the populations of various types of seals have recovered, even to the point where there is grumbling among some fishermen, who feel that seals compete for fish.[45]

Another diving mammal that has benefited from protection is the playful sea otter *(Enhydra lutris)*.[46] The smallest of the marine mammals, it is larger than its terrestrial relatives, up to 6 feet in length. Sea otters dive for food on the bottom not far offshore, for clams, abalone, sea urchins, crabs, and octopus. Their diving range is modest compared with that of other marine mammals—typically several tens of meters, maximally around 100 meters. Sea otters use rocks for anvil and hammer to open shells, while lying on their back using the stomach as butcher table. Their role in maintaining the health of kelp forests has been studied intensely over the last several decades. By keeping down the populations of sea urchins, they prevent overgrazing by these kelp-eating echinoderms, thus protecting the kelp. If the predators of sea urchins are removed, through hunting and overfishing, "sea urchin barrens" are the likely result: the kelp system collapses.[47]

Mammals, it should be mentioned, are not the only deep-diving air breathers. Large penguins have been tracked to depths around 500 meters, and a diving depth of more than 750 meters has been reported for the leatherback turtle, hunter of jellyfish.[48]

INTERLUDE: STRANGE TALES ABOUT STRANGE CREATURES

In the time of sailing vessels, all sailors were convinced that there were monsters lurking in the sea, and some swore to have seen some. Cartographers added their likenesses to the annoyingly large empty spaces on their maps. For thousands of years, setting sail for the open sea was indeed one of the more dangerous things one could do, and the idea of sea monsters made this fact palpable.

It is practically impossible to prove the absence of anything. This is the secret behind the theory that visitors from space make strange markings in British crop fields. So some of the markings are proven frauds. So anyone with some skill and string and boards on his feet can make them. But how does that disprove the little-green-men theory? It does not, of course, although it does raise doubts.

In the same manner, we can explain some of the sightings of sea monsters as optical illusions, or as confusing a line of jumping dolphins with an undulating snake, or as mistaking a half-rotten whale carcass for a strange fish. But this does nothing to dispel the notion that sea monsters do exist. What is more, they do exist, of course. It depends on one's definition of "sea monster"!

One famous legendary creature is the Loch Ness monster, affectionately known as "Nessie."[49] Sightings are exceedingly rare but reported with some enthusiasm. Nessie's immediate ancestors must have come from the deep sea: the loch is a fjord (now cut off from the sea) and was filled with ice not so long ago. This would suggest that Nessie's cousins are still alive in the deep ocean. Or perhaps it is the dragon in the Beowulf Saga, now become marine: "The

FIGURE 9.9. The oarfish, up to 20 feet long, might be confused with a large sea serpent.

burning one who hunts out barrows, the slick-skinned dragon, threatening the night sky with streamers of fire."[50] In any case, Nessie is an asset to the tourist industry, which presumably favors her survival in the media.

Many or most of the early monster sightings presumably were about various kinds of whales. "The monsters that plough its waters have been hunted by man till the race is well nigh exhausted; for the leviathan which frightened the ancients is the whale which has illuminated the moderns."[51] There is no question, in addition, that some of the sightings reported are authentic and concern strange creatures, or perhaps known creatures behaving strangely. In truth, the ocean is dark and mysterious, with vast depths that have room for many things unknown, perhaps even very large organisms. Even in modern times striking discoveries have been made concerning strange creatures, such as the "megamouth," a huge shark, and *Latimeria,* a fairly big fish once thought to have suffered extinction more than 60 million years ago.

An obvious representative of the "sea serpent" category is the oarfish *(Regalecus glesne).* It is a most unusual fish, oddly attractive, only rarely seen. The body is silver colored, flattened vertically, with a crested red dorsal fin along the top, which extends onto the head as a plume that can be raised (fig. 9.9). It has rodlike fins that end in a blade shape (hence the name). The largest specimen measured was 21 feet, but there are reports to 35 feet and 500 pounds.[52] Most oarfishes have been found washed ashore or dying at the sea surface. They have been observed throughout the world's oceans.[53] The oarfish comes to mind when reading the account by a sea captain that starts as follows: "Sir, in the latter end of August, in the year 1746, as I was on a voyage, on my return from Trondhjem, . . . I heard a kind [of] murmuring voice from amongst the men at the oars, who were eight in number, . . . and was informed that there was a sea-serpent ahead of us. . . . The head of this sea-serpent, which it held more than two feet above the surface of the water, resembled that of a horse." The captain goes on, listing a number of features including grayish color, a large black mouth, large black eyes, and a long white mane hanging down to the surface of the water. He suggests that the length of the undulating body, following head and neck, was more than 8 fathoms, that is, some 50 feet.[54]

The length of this creature, then, was more than 50 feet. A monster roughly fitting the description is pictured in the town square of Bergen in Norway (fig. 9.10). According to the inscription, it was seen in the eighteenth century, and it exceeded the length of a 50-foot fishing vessel.

If the monster was an oarfish, it weighed a ton or more, by simple scaling of observed specimens. An even larger snakelike animal was seen off Gloucester, Massachusetts, in this case at least 50 feet long, perhaps even 100 feet, bearing a head as large as a horse, and at times swimming at a speed of more than 15 knots. The sightings were by experienced fishermen, both from boats and from land, on 10, 12, 13, 14, 15, 17, 18, and 28 August 1817.[55] After 1817, there were additional sea serpent sightings every summer through 1827 in the area of Massachusetts Bay, and these were reported by hundreds of people, on land and on boats at sea. Such reports continued at irregular intervals for more than a century, at least as verbal descriptions. The

FIGURE 9.10. Reconstruction of a sea monster seen in Bergen harbor, as shown in Bergen's town square. The inscription says "eighteenth century."

observations, from Norse to modern, have almost always been in fine calm weather, the best of times for clear viewing.[56]

What the drawings did for ancient sea maps, that is, make sea monsters real, can be done nowadays using modern scientific nomenclature. In 1995, E. L. Bousfield (of the Royal Ontario Museum) and P. H. LeBlond (of the University of British Columbia) attached the name *Cadborosaurus willsi* to an uncaught large creature, said to have been sighted at numerous occasions off western Canada and from southeastern Alaska to Monterey Bay, California. They consider it reptilian. For "material examined" they offered three photographs of a "juvenile" specimen removed from the stomach of a sperm whale in 1937. They also comment on a number of reported sightings of adults. The creature was said to have a head shaped like a horse or camel, anterior flippers without claws, a large posterior flipper webbed to the tail, and spinelike processes on the side and trailing edges of the tail. From testimony of observers, the authors estimated that an adult would be 15 to 20 meters long, with large luminous eyes, a long neck, a mane of hairlike filaments, and two to five or more vertical humps on the body.[57]

So: are there sea serpents out there? Might they have been sighted less frequently since ships with noisy engines replaced sailing ships? Heuvelmans analyzed 587 reports (from 1639 to 1966) of unknown sea animals that he considered significant, grouped them into nine categories, and concluded that there are large creatures in the ocean not yet identified.[58] From the reported behavior of sea serpents it appears that, in any case, they have been misnamed. Identified members of the snake class do not undulate in vertical motion and neither do other kinds of reptiles. Fishes undulate sideways, excepting rays and flatfish and a few others swimming on their side. This leaves mammals as candidates for many sightings, or perhaps the arms and tail of the giant squid *Architeuthis*.[59]

The giant squid is a cephalopod, that is, a relative of the common octopus (and of the extinct ammonites, which produced giant forms back in the Cretaceous). The enormous kraken, a legendary Scandinavian creature that goes back to prehistory, apparently is a cephalopod, as well. According to available drawings from the Middle Ages, it is big enough to threaten a ship (anyway, a smallish Viking ship), and it is armed with many tentacles, as befits an octopus or squid. There is no reason, of course, why giant squids should not have washed up on Norwegian shores for centuries before records were kept, just as they did (and still do, occasionally) elsewhere. An early description of a dead giant squid (referred to as a sea monster but readily recognizable as a cephalopod) is in fact from a stranding in Iceland, in 1639.[60]

Giant squids are impressive creatures, with lengths of 25 to 30 feet (7.5 to 9 meters) being quite common. The maximum size is unknown, of course, since the animals live at great depth, hidden from view. One specimen that washed ashore in New Zealand in 1887 was 55 feet (16.5 meters) long, overall. Another was reported as 65 feet, also from New Zealand, washed up in 1880.[61] This makes the giant squid the largest invertebrate ever known.

Greater sizes than those reported have been calculated by scaling of squid beaks and sucker marks on the skin of sperm whales. The reliability of such scaling exercises is unknown. Weights of beached specimens have been given as around a ton or so. Unconfirmed estimates of maximum weights vary from 3 to well over 10 tons. We suggest 3 tons as a reasonable guess for such a maximum, based on a statistical evaluation of stranded animals.[62]

The remains of giant squids have been described for some time, at least as far back as the seventeenth century.[63] Through the 1900s about 200 of these creatures were recorded in the literature as stranded on beaches or tangled in deep fishing nets. Whalers often found chunks of squid tentacles vomited by sperm whales that they had harpooned. In 1895, near the Azores, a badly wounded sperm whale that had been injured by a nearby whaling crew attacked the research vessel of Prince Albert of Monaco. As it died, the whale regurgitated its last meal of "fragments of gigantic cuttle-fishes hitherto unknown to science."[64] The Prince published the find. Sperm whales carry scars of squid suckers on their skin as large as 18 inches in diameter.[65] The largest such suckers so far measured on squid specimens in hand are only about 2 inches. Pieces of squid arms found in the stomachs of whales suggest the existence of creatures as long as 80 or 90 feet, or perhaps even more than 100 feet, by extrapolation.

Giant squids may be the stuff of monster stories but they are objects of serious research as well. Clyde F. E. Roper of the National Museum of Natural History at Smithsonian Institution has spent much of his career studying these intriguing animals; he would very much like to observe them where they live instead of as occasional decaying masses on the beach.[66] These creatures seem to be present in almost all parts of the ocean and may migrate and congregate at times, perhaps for spawning, as is common with smaller squids. Presumably the sperm whales would know where they are, as they prey on them. Thus, tracking the whales (which have to come up for air) might be a way to locate them.

Clearly, the giant squids are deep-water predators—perhaps they live at or near the bottom, in cold waters at depths of about a thousand meters or more, among steep slopes. Looking upward with their giant eyes (the largest on the planet, as much as a foot in diameter) they would see the silhouettes of prey against a sky a million times dimmer than ours. Midwater fishes and smaller squids could be among their prey. (Stomach contents of beached specimens have not been very informative.) Alternatively, or in addition, they might scare up prey from nooks and crannies along a steep-walled seafloor.

Judging from the hunting success of sperm whales, giant squids must be reasonably abundant. The number of different species is not known; the number of names given various specimens or parts of specimens is about 20. There may only be a few, in fact, or just one.[67]

Are giant squids dangerous to humans? Not usually (since we do not much meet them), but yes, they might take us as prey when it is convenient, after a shipwreck for example. Heuvelmans describes an attack on survivors of a shipwreck.[68] The attack occurred at night, the time when squids are active in surface waters. The report, if true, would suggest that at least some giant squids hunt in shallow waters at night.

SHARK!

Sharks are among the most ancient of marine vertebrates. Thus, their adaptations reflect the ocean environment and the development of marine ecosystems over the last several hundred million years, unlike the adaptations of the marine mammals, which entered the sea but 50 million years ago. The range of adaptations of sharks is truly remarkable. As a group, they

FIGURE 9.11. Mako shark, among the largest of predatory fishes.

have mastered all major habitats, from the sunlit open sea to the seafloor in shallow waters and at great depth. Their niches (or life-styles) range from fearsome predator swallowing large fishes, squids, and even seals, to plankton-straining grazer, and to bottom-living detector of worms and mollusks in the mud.

As in cetaceans, the very largest among the selachians are the species that feed on plankton: the huge whale shark *(Rhincodon typus)* and the somewhat smaller basking shark *(Cetorhinus maximus)*.[69] These behemoths represent the largest of all fishes, with lengths of 14 meters and more, and weights of 15 tons and greater.[70] The largest known predator shark, the great white shark *(Carcharodon carcharias)* is distinctly smaller but still impressive, with a length exceeding 6 meters (20 feet) and a weight greater than 2 tons.[71] The largest among the plankton-straining sharks compare in size with the smaller of the baleen whales, while the great white shark overlaps in size with some of the larger toothed cetaceans, excepting the sperm whales.

Sharks belong to a group of fishes whose skeletons are made of cartilage rather than of bone, and they have no swim bladder. Fertilization of the female's eggs is internal. Most shark species bear live young; some produce large eggs, which they deposit in egg cases that protect the embryo. The young look like small adults; there is no larval stage. Sharks have a keen sense of smell, and they detect vibrations and sound. They use vision for close-up orientation.

Most of the sharks are predators, with an exquisite sensory apparatus to find prey. In the hammerhead, the need to move paired detectors (smell, electric field) far apart from each other has resulted in extreme modification of the skull. Some sharks are quite large and have large teeth with serrated edges for slashing and cutting (fig. 9.11). To become the target of attention of one of these predators is clearly very dangerous. However, most sharks are entirely harmless to swimmers, and even the big ones (which can and do cause serious harm) tend to be shy, unless provoked. Diving pioneers who attempted to take pictures of big and toothy sharks quickly learned that even sharks much larger than the divers would flee when approached. However, one way to attract them is to have blood in the water.[72]

As pointed out by the pioneering diving ecologist Irenäus Eibl-Eibesfeld, every predator

FIGURE 9.12. Manta ray, a plankton-eater common in coastal warm waters.

has stored in its memory a template for prey. If the object fits the template, an attack is appropriate. If not, there is no trigger for attack. For large predatory sharks, blood in the water lowers the threshold for the trigger, obviously.[73]

One group of sharks, early during their long evolutionary history, became specialized in preying on small organisms on the bottom. Their bodies became flat, and propulsion moved from the tail to broadened pectoral fins that merged with the flat bodies. These sharks became the ancestors of the rays. Some of the rays discovered plankton as a food source, in preference to benthic organisms. They became the ancestors of the giant manta rays (*Mobula* spp.), peaceful plankton-eaters that have left the seafloor and now wing their way through sunlit waters unlike any other fish (fig. 9.12).[74] Manta rays attain a width of close to 7 meters, with a corresponding weight of around 1.3 tons.

For both rays and sharks, then, the plankton-eating species are among the giants of the cartilaginous fishes—indeed of all the fishes. Surprisingly, one of these giants had been undiscovered by science until quite recently. In November 1976, northeast of Kahuku Point, Oahu, a large plankton-feeding shark was entangled in a parachute being used by a research vessel of the Naval Undersea Center in Kaneohe, Hawaii. The ichthyologist Leighton Taylor named it *Megachasma pelagios* and erected a new family for this species.[75] The animal, now known as megamouth shark, was an adult male 4.5 meters long (14.6 feet) and weighed 750 kilograms (1,650 pounds).

The contrast between the peaceful grazers among the large sharks and the fearsome large predators is striking. The whale shark moves slowly and purposefully through plankton-rich waters, basically unperturbed by other activities in the water (boats, rafts, or divers). The great white shark is "curious"—checking out what moves in the water. It attacks and swallows other large fishes, as well as smallish seals and even entire sea lions.[76] In addition, remains of sea turtles have been found in its stomach. The great white is considered a warm-water denizen, but it has been found as far north as Washington off the West Coast, and Cape Cod off the East Coast.

The raylike fishes also evolved a giant predator: the sawfish *(Pristis pectinatus),* said to reach up to 10 meters in length. The bladelike protuberance from its head with the symmetrically arranged rows of skin teeth makes up about one-fourth of that length, so this corresponds roughly to the size of the great white. The weight of one female (more than 6 meters long) was given as 2.4 tons! The "saw" reportedly is used for digging in the seafloor and for slashing motions when swimming through a school of fishes, which leaves hurt individuals that can be picked up at leisure. The teeth on the two edges of the saw are formed from denticles, structures originally developed in ancient fishes as armor plates on their skin, but now largely confined to sharks. In sharks and rays the origin of teeth from skin armor plates is still very obvious. Our own teeth (as in other vertebrates) have this same origin, but it is less evident after many modifications along the path from fish to mammal.[77]

Besides the great white, the mako shark *(Isurus oxyrhinchus)* and the hammerhead *(Sphyrna zygaena)* are considered dangerous, along with the tiger shark *(Galeocerdo cuvieri)*, the blue shark *(Prionace glauca)*, the nurse shark *(Carcharias arenarius* and *C. tricuspidatus)*, the bull shark *(Hexanchus griseus)*, and the thresher shark (*Alopias* spp.). All of these are very large and very mobile hunting sharks with occasional giant members, with maximum lengths between 5 and 6 meters, typically. The weight of the largest members of each species or genus approaches 1 ton.

Do sharks eat people? In the 64 years from 1926 to 1990 a total of 114 attacks worldwide by great whites *(Carcharodon)* were documented, roughly two each year on average. Twenty-five of these attacks were fatal.[78] Almost all of the victims died from massive loss of blood, not from being devoured. The great white is found throughout the world in temperate waters, usually where seals and sea lions are common. Presumably, to the great white, a human swimming at the surface appears as just another type of seal. However, unlike seals, the victim is usually released after a vicious bite. The areas where most attacks have occurred are off southeastern Australia and northern California.[79] Great whites have been known to seize small boats and humans on surfboards, as well as swimmers and scuba divers. However, the great white shark may not be the most dangerous of the lot. It is likely that more attacks on humans have been from a closely related shark *(Carcharinus leucas)*, which is more common than the great white, is found in warm-temperate and tropical shallow waters (including some freshwater lakes), and has an indiscriminate appetite. It reaches only 10 (maybe 11) feet and 400 pounds but has massive jaws and large teeth.[80]

Parenthetically, the danger to humans from sharks is commonly greatly exaggerated. To put the danger posed by sharks (several deaths per year) into perspective, more than 10 times the number die from lightning strikes each year, and the same is true for insect stings (ants, wasps, bees).[81]

The question of why shark attacks involve a pattern of bite and release has intrigued two marine biologists at the University of California, Peter Klimley at Davis and Burney Le Boeuf at Santa Cruz. Their experiments indicate that the great white seizes seals or sea lions viciously, but it grabs an unfamiliar object in the water tentatively and often spits it out. They think that the shark may be testing for the fat of blubber. Humans do not pass the test. Great whites apparently are somewhat territorial but also migrate between familiar haunts over considerable distances. Le Boeuf tagged white sharks to follow their movements, around Año Nuevo Island off northern California. Of six adult sharks he found that they stayed for some weeks in the area, feeding intermittently, and in winter moved well offshore eastward. One male reached Hawaii, returned to California, and repeated the jaunt the following year.[82]

What should be our attitude toward "man-eating" sharks? A response to the 1970's movie *Jaws* was to reel in trophy-sized whites, and some even suggested that these sharks should be destroyed wherever they threatened swimmers. By 1994 the tide had turned. Apparently, the numbers of white sharks were declining. They reproduce but slowly, as they are viviparous like most other sharks. So, California banned the taking of white sharks, as South Africa had done earlier. Presumably, white sharks, like other large predators, have an important role in structuring the food chain below their level. One might surmise that its presence keeps the seal and sea lion populations from expanding in ways that has unfavorable impacts on their health from crowding and lack of food.

While shark attacks on humans are rare, humans have become a significant source of mortality for sharks. Shark meat has become part of the standard menu. The demand for shark-fin soup in Asian countries represents a major problem. Fins are much more valuable than shark meat. The problem is that a single fishing operator can do enormous damage in a short time, because it takes a large number of sharks to fill a boat with fins. The fins are slashed off, and the rest of the big fish is discarded (commonly alive).[83]

What we know about sharks is mainly from observing these fishes in shallow waters. That large sharks are present at great depth was dramatically illustrated by moving pictures taken by a deep-sea camera built by John Isaacs and his team of engineers. Isaacs was eager to learn what creatures would show up if he put dead meat on the deep-sea floor. The results were most gratifying. The pictures showed that

> a deep-sea population of large, active animals thrives in what was generally assumed to be a province inhabited mainly by small, feeble creatures such as worms, snails and sponges. . . . Much of the deep-sea floor teems with numerous species of scavengers: vigorous invertebrates and fishes, including some gigantic sharks, that are supported by a marine food web whose extent and complexity is only beginning to be perceived.[84]

Perhaps the most surprising results came from deploying Isaacs's "monster camera" below the central gyre of the North Pacific, one of the great deserts of the ocean. One would not expect many large creatures near the bottom there, at depths below 5,000 meters, where practically nothing edible arrives in the organic rain from above. However, surprisingly, "more than 40 large fish and shrimps were attracted by the bait within a few hours."[85] In contrast, below highly productive regions, no such activity was evident, and in the most productive region, off Peru, the scientists found "furious masses of amphipods that stripped the bait in a few hours." We must assume that large fishes, including big sharks, patrol the low-production regions like vultures circling over the desert. Presumably, the high-production areas sustain a high density of local scavengers; there would be nothing left by the time the vultures arrive. They do not even try to compete with the flocks of crows, as it were.

Large sharks, then, search the deep seafloor for fallen chunks of meat, be it fish or mammal. Interestingly, if hagfishes are present they keep other fishes from feeding on the remains by extruding a slimy substance that clogs their gills. Apparently, sharks are not impressed by this threat—the hagfishes depart when sharks arrive. Perhaps the sharks have ways to protect their gills from the noxious stuff, and perhaps they do not mind eating hagfish. Sharks and hagfishes have evolved together for several hundred million years, and they know each other's tricks. The memory of innumerable encounters is in their genes.

MORE ABOUT BIG ANIMALS IN THE SEA

While baleen whales and whale sharks hold the world records in terms of size of living animals, there are some fairly large bony fishes in the sea, as well.[86] Some of these are among the fastest swimmers in the sea (fig. 9.13). Just as a sampling, we have marlin species (*Makaira* spp.), which can attain a length of 15 feet (4.5 meters) and a weight of 2,000 pounds (900 kilograms). Marlins are high-speed specialists; velocities in excess of 40 miles per hour have been estimated. Other high-speed species are members of the same clan, in the mackerel family. The bluefin tuna *(Thunnus thynnus)* has lengths of more than 10 feet (3.5 meters) and weighs up to 1,500 pounds (700 kilograms). Atlantic bluefin migrate over long distances; they go north in summer, even to Norwegian waters, to feed on smaller fishes. They manage to keep their body temperature above that of the ambient water in cool regions, based on a circulation system that conserves heat. Tunas spawn in warmer waters in the subtropics, producing enormous numbers of small eggs that are released into the plankton—a strategy quite different from that of large pelagic sharks.[87]

Another remarkably large fish among the fast predatory ray-finned fishes is the swordfish *(Xiphias gladius)*, said to attain almost 6 meters in length (4 meters, or 13 feet, without the

FIGURE 9.13. Profile of speed: blue marlin. Note the large, thin, forked tail.

sword) and a weight of around 1,000 pounds (450 kilograms). It is known to be aggressive on occasion; its sword has penetrated the hull of wooden boats.[88] The disk-shaped sunfish *(Mola mola)* is neither particularly long (up to 10 feet, 3 meters) nor aggressive, but it is quite heavy (weight estimated at 2,000 pounds, or 900 kilograms) and very visible. The sunfish can be seen close to the sea surface, lying on its side and lazily feeding on small animals near the water-air interface, including jellyfish.

The large marlin, tuna, and swordfish are modern ray-finned hunters chasing other fish; they are large and fast, and their way of life is comparable to that of the larger dolphins. In contrast, their plankton-eating relatives are comparatively small; they follow a strategy quite different from that of baleen whales and whale sharks. The typical representatives of plankton-eating ray-finned fishes belong to the clan of clupeids; that is, herringlike fishes (herring, *Clupea harengus*; sardine, *Sardina* spp., *Sardinopsis* spp.; anchovy, *Engraulis* spp.) They form large schools, in a sense simulating a large organism. The wide-open mouths are close together while feeding by straining the water, forming an enormous filtering device that is slowly pushed through the water. Such schools, of course, provide tempting targets for echo-sounding-hunters. Interestingly, all the clupeids have special hearing aids: a connection between swim bladder (a sound-intercepting organ) and the ears.[89] Of course, the increased sensitivity to sound may well serve the needs for communication in a school of fishes, but it is likely useful, as well, in detecting the search-beam of a dolphin.

The British ichthyologist N. B. Marshall thought that modern ray-finned fishes among the bony fishes (teleosts, with more than 30,000 species) are much smaller than sharks, on average, because sharks are more ancient and have been able to fill available niches for large animals. In contrast, the more recently evolved teleosts had to make due with nooks and crannies, literally hiding out between the rocks, and more lately (geologically speaking) working their way up into the world of large fishes.[90] If this is so, perhaps the most ancient types among the living bony fishes should also have the largest members.

The largest of the bony fishes in fact belong to a very ancient tribe. They are bottom-feeding animals, feeling around in the mud for crustaceans, worms, and mollusks. Sturgeons (genera *Acipenser, Huso, Polypterus*) represent an archaic clan of bonefishes that have retained many primitive characteristics of the ancestral forms. Sturgeons have rows of bony armor plates on their sides, reminiscent of Devonian ancestors. Some have retained the ability of using their swim bladder to breathe air when necessary (*P. birchir*, which lives in the Nile). Like sharks, sturgeons have a fishtail whose upper lobe is supported by the tail end of the spine. Sturgeons are found both in freshwater and in saltwater. Marine forms enter rivers for spawning—reminding us that the ancestors were freshwater fishes. The beluga *(Huso huso)* is the largest species; it is found in the Caspian, Adriatic, and Black seas. According to the ichthyologist Marshall, this species reaches a length of 28 feet (7.5 meters) and a weight of 2,860 pounds (1.3 tons).[91]

However, sturgeons have become rare because of fishing pressure and changes in their river habitats, so such prime specimens are unlikely to be seen today. The type sturgeon *A. sturio* is the main source of caviar (sturgeon roe), which brings a high price in the market. Market demands generate unsustainable fishing: as the fish becomes rare, the price of the product increases. Poaching undercuts protection.

Perhaps the longest of the ray-finned fishes is the oarfish *(Regalecus glesne)*, mentioned in connection with sea serpents. It is reported at up to 33 feet long (10 meters) with a weight of 600 pounds (275 kilograms). (Or 35 feet and 500 pounds, depending on reports.) The confirmed length is 20 feet (6 meters). It belongs to a relatively small group of teleosts characterized by an unusual configuration of the jaws. Its close relatives are the ribbonfishes (with three genera and seven species). In some ways similar in form, but belonging to a different order, are the eels (Apodes), the largest members of which are among the moray eels (*Thyrsoidea macrura*, more than 3 meters in length) and the conger eels (*Conger* spp.), said to attain a length between 2.5 and 3 meters (perhaps 6 meters, depending on source). For the conger eel, a maximum weight of 71.6 kilograms has been reported.[92] These eels are predators and scavengers. Moray eels have poison glands at the base of their teeth.

A number of eel species (among the hundreds that are strictly marine) leave the sea to spend much of their lives in rivers. The European river eel *(Anguilla anguilla)* arrives at the estuary as a juvenile (a 3-inch "glass eel" a few millimeters thick) and leaves the river as an adult several years old, to return to the Sargasso Sea to spawn. The larvae, which look more like a miniature ribbonfish than an eel, take 3 years for the trip back to Europe.[93] The trip of the larvae of the North American eel *(A. rostrata, A. bostoniensis)* is much shorter, about 1 year. In essence, these river eels on both sides of the North Atlantic probably have their origin in deep-sea fishes, initially adapted to live and feed in the oxygen-poor subsurface waters of the early Cretaceous sea, and hence able to "hold their breath" for a long time. These versatile hunters discovered food-rich brackish and freshwater environments along the coast, closed their gills to avoid osmotic stress while feeding, and eventually made the physiological adjustments to invade the waterways of the continents. As the Atlantic widened, different genetic instructions had to evolve for the migrations of European and American eels.

We earlier mentioned the leatherback turtle, an expert diver in a reptilian group long adapted to the sea. At one time, reptiles were the largest predators both on land and in the sea; some of the sea lizards had cetacean proportions. Even today, well after the age of reptiles, some of them are quite large. Sea turtles and sea snakes are (or *were*, in the case of turtles, which are being fished out) abundant in the warm waters of today's ocean. The various sea snakes, commonly between 1 and 3 meters long as adults, are closely related (family Hydrophidae). Their cousins on land are cobras and mambas. The tails of sea snakes are commonly flattened at the end, for improved eel-like propulsion. Most species bear their young in the water, where they spend all their lives; some lay eggs on land. The yellow-bellied sea snake *(Pelamis platurus)*, colorful in yellow and black, is widespread in the warmer waters of the Pacific and Indian oceans, sometimes forming milewide rafts of moving snakes, perhaps mating.[94] Fortunately for humans, this creature has to chew its highly toxic venom into its victim, so a glancing bite is not usually fatal.[95] Interestingly, the Atlantic has no sea snakes; the reason is not known.

Unlike the sea snakes (which have plenty of closely related species) sea turtles are quite diverse on a fundamental level, in keeping with their ancient pedigree. They have remembered their terrestrial ancestry for more than a hundred million years, coming on shore to lay their eggs in the sand. The hatchlings emerge together and make straight for the sea and so avoid the many predators ready to feed on turtle babies. The distance between brood beaches and feeding areas at sea can be a thousand miles or more, engendering extensive migrations across the open ocean. Perhaps the most impressive migrations are those of the loggerhead turtle in the Pacific. This carnivorous species *(Caretta caretta)* is at home throughout the warm regions of the ocean, in deep and shallow waters. Loggerhead turtles nesting on islands off Japan actually spent their youth in Mexican waters, 10,000 kilometers away! Studies showed that these turtles can sense the magnetic field, and it is thought that this ability, together with a memory of the smell (or taste) of the natal beach, makes it possible to find the way "home."[96]

The afore-mentioned leatherback (*Dermochelys coracea*) is the largest of the marine turtles. It grows to a length of over 8 feet (2.5 meters), has a "wing-span" of 10 feet (3 meters), and attains a weight of up to a ton (2,000 pounds). It feeds on jellyfish and is able to dive to great depths (3,000 feet, or 1 kilometer) to find food.[97] The largest turtles are long gone—life has become short for these ancient denizens of the sea. Turtles are being rapidly reduced in abundance, by fishing, through drowning in drift nets, from egg hunting, and from other adverse human impact.

While the locals along any beach are usually a great source of information about dangerous marine animals in the area, it is wise to use discretion. In Darwin, Australia, the local jokers tell visitors not to worry about the sharks when swimming—pause—because the crocodiles eat the sharks. A saltwater crocodile 15 feet long may well hold its own against any shark, but this ability is not exactly reassuring to a swimmer, of course. Crocodiles are by nature amphibious, that is, they spend much of their lives in water but come on land to lay eggs. Females commonly guard their eggs. Propulsion is by undulation of a flattened tail, and speeds are modest. Attack of prey is done from ambush. The saltwater crocodile in southeastern Asia and northern Australia *(Crocodilus porosus)* is quite large—it grows to a length of at least 23 feet (7 meters).[98]

The one species among today's lizards that is mainly aquatic is the iguana of the Galapagos Islands *(Amblyrhynchus cristatus)*. It feeds on seaweed and hauls out to warm up on the basaltic rocks along the shore. Like the rocks it rests on, it is black. It uses the tail for propulsion and routinely dives to a depth of more than 10 meters (30 feet), holding its breath for up to an hour. The largest ones attain about 5 feet in length. Eggs are laid on land, within beach sand. To cope with lack of food during El Niño years, marine iguanas use up body reserves, actually shrinking in the process.[99]

Some very large organisms in the sea belong neither to the vertebrates nor to the mollusks: they are members of the ancient marine phylum of cnidarians, animals with stinging cells. Certain cold-water jellyfishes (*Aurelia* sp.) have an enormous umbrella up to 7 feet (2 meters) in diameter and have trailing tentacles 100 feet long. Likewise, the lion's mane jellyfish *(Cyanea capillata)* has a belled sac up to 8 feet across, and tentacles around 100 feet long. The Portuguese man-of-war *(Physalia physalis)*, a siphonophore (a colonial cnidarian), trails tentacles up to 50 feet long. Clearly, such large predators, as well as their numerous smaller cousins, are an important source of mortality for fish larvae and other small animals. Cnidarians subdue their prey using poison, and contact with their tentacles can be extremely painful to humans, even fatal in certain cases.[100]

Poison, for subduing prey and for defense, is used across the major taxonomic categories in marine animals. Thus, there are poisonous snails *(Conus geographus)*, octopus *(Octopus maculosus)*, sea urchins *(Toxopneustes pileolus)*, seastars *(Acanthaster planci)*, certain annelids, and various fishes. In addition, there are toxic protists (dinoflagellates, diatoms), some of which kill fishes (or can make sea otters and people sick when they eat bivalves that fed on toxic dinoflagellates).[101] Much recent research in marine resources is focused on unraveling the chemistry and physiological effects of such poisons, for medical purposes.[102]

Interestingly, the most poisonous animals tend to be in warm, tropical waters. Here there has been much time, more than a hundred million years, to perfect poison chemistry in biological interactions. Cold-water environments, in contrast, are geologically young, and the newly adapted forms had to evolve mainly to meet the physical challenges that arose when invading a comparatively inhospitable habitat. Evolving poisons takes time.

NEKTON OF ANCIENT SEAS

Many of the more charismatic large animals in the sea, especially among the mammals and birds, evolved in response to a cooling planet. Whales and dolphins, seals and sea otters, penguins and puffins evolved in taking advantage of a new habitat characterized by seasonal upwelling and deep mixing on a planet with strong winds. What did large or speedy or smart nekton look like on a warm planet? Did such categories apply at all? Quite generally, what were the factors driving changes in the nekton? Does such information have lessons for the future?

As with so many things in the history of life, reconstruction is difficult and a source of contention between experts. Of course, it is impossible to extract general rules from reconstructions not agreed on. What we know for sure is that predators and prey invariably evolve together, in a kind of arms race. If such a race involves speed and sophisticated behavior, it is bound to culminate in highly agile, intelligent forms, regardless of whether the ocean is warm or cold.[103] However, in a cold ocean we would expect warm-blooded mammals and birds to have the edge on fishes and reptiles, which set their temperature to that of the environment.

The story begins with fossils—marine fossils—more than 600 million years old. By the nature of preservation, practically all such fossils from the ancient part of fossil-bearing rocks, the Paleozoic, are from shallow water. We have lots of deep-water fossils only for the last 100 million years. But these deep-water fossils are shells of single-celled organisms: microplankton, not nekton. The microplankton can tell us much about the course of environmental change in the sea (as discussed in chapters 13 and 14) but not about the response of the large, the fast, and the smart (if any). Thus, our knowledge of the history of nekton will always be highly fragmentary.

What we know is that already in the Paleozoic, all three of the dominant animal phyla—mollusks, arthropods, and chordates—had impressively mobile representatives. The most advanced mollusks were the orthoceratids, cephalopods related to *Nautilus*, a "living fossil" of today's ocean. However, for a time, in the early Paleozoic, the largest and strongest animals were arthropods. They were the sea scorpions or eurypterids, which were up to 6 feet in length. Some of these apparently were agile swimmers, judging from the shape of their hind legs.[104] Eurypterids were armed with powerful claws bearing spikes for impaling prey. Large compound eyes scanned the scene for food. The ancestral fishes at the time were freshwater forms, and they stayed in contact with the bottom and carried heavy armor. They did not rule the seas, as far as we can tell. Only in the Devonian do we see a flowering of vertebrates, that is, a great diversification in fishes. The first sharks appeared, and many of them became fast-swimming predators, such as *Cladoselache*, a late Devonian fish with a streamlined shape and a modern-looking symmetrical tail for efficient propulsion.

None of the Paleozoic forms developed a large brain. Neither arthropods nor mollusks ever went far in this direction (although octopuses can be trained to recognize differently shaped objects). Intelligence evolved in the vertebrates, especially in the land-dwelling descendants of the ancestor of the small-brained crossopterygian fishes. Of this group, the coelacanth has survived (fig. 9.14), possibly giving us a glimpse of the life-style of members in this ancient clan.

The discovery of a living coelacanth, in the 1930s, was a major event in marine biology and paleontology, and for natural history in general. The find was announced in the scientific journal *Nature* in March 1939, by the ichthyologist J. Smith.[105] The 5-foot, 127-pound fish had heavy scales, a stubby tail, and fins that look like clubs. When still alive it was deep blue in color. By coincidence and good luck, the curator of the East London Museum in South Africa, M. Courtenay-

FIGURE 9.14. The modern lobe-finned fish *Latimeria* closely resembles crossopterygian fishes that lived in the Devonian, some 400 million years ago.

Latimer, had been able to salvage it from the catch of a trawler working off eastern South Africa. She wrote for advice to Smith, a professor of chemistry at Rhodes University in Grahamstown. Her sketch greatly puzzled Smith. It looked like the drawings of a kind of fish known from fossils but presumed extinct since the end of the Cretaceous. It looked as though Courtenay-Latimer had chanced upon a "living fossil" of major importance: a fish that was of special interest because it resembled the ancestor of terrestrial vertebrates, back in the Devonian.

Since then at least 200 more coelacanths have been retrieved for science, mostly around two of the Comoros Islands and a few from South Africa, Mozambique, and Madagascar. None has stayed alive after capture for as much as a day. Examinations of preserved specimens have shown that the coelacanth preys on smaller fishes and on squids, and that the females bear their young alive.[106] The fish is not restricted to the western Indian Ocean: one specimen showed up in the outdoor market of Manado on the Indonesian island of Sulawesi.[107]

The discovery of the coelacanth was a sensation because it was large, yet unknown to science, and it did not change much through geologic time, as far as this can be determined. It represents a group that flourished in the distant past, and its taxonomic position is close to the freshwater fishes that gave rise to amphibians and also to the modern lungfish. The lung of the lobe-finned ancestral fishes became a swim bladder. (In *Latimeria*, for unknown reasons, it is filled with a fatty substance.) At some point in the Paleozoic the freshwater ancestors of this fish invaded the sea. Perhaps, at that time, having a lung was of advantage since there was little oxygen in the water, as reflected in widespread black shale deposits rich in sulfide minerals.

During the Mesozoic[108] various kinds of terrestrial reptiles returned to the water from where their lobe-finned ancestors had invaded the land. The most remarkable forms to evolve in this process were the ichthyosaurs, sleek efficient swimmers with a relatively large brain (fig. 9.15). Ichthyosaurs (the name means "fish-lizard") were fishlike in their appearance. To raise air-breathing young in the water from birth may have required special parental behavior. Perhaps the ichthyosaurs protected their young against sharks and other predators as the dolphins do today. The best-preserved specimens of ichthyosaurs come from the lower Jurassic of southern Germany.

Other large marine reptiles, familiar from exhibits and reconstructions in most large natural history museums, are the plesiosaurs. The original ichthyosaur and plesiosaur finds, from the lower Jurassic in England, can be seen in the Natural History Museum of the British Museum in London.[109] Plesiosaurs apparently were very efficient swimmers, judging from their attributes.[110] However, speed may not have been the main issue in catching prey. Instead, the long neck may have been important in striking, snakelike, at fishes and cephalopods.

FIGURE 9.15. Reconstruction of an ichthyosaur, perhaps the fastest and smartest of the Mesozoic swimmers.

In North America, the chalk beds in western Kansas—hundreds of square miles in extent—are a rich source of fossils of oceanic animals of the Cretaceous period, when shallow seas covered almost half of the continent. Here, too, we find ichthyosaurs and plesiosaurs, as little modified descendants of Jurassic ancestors. In addition, we find fossils of mosasaurs, enormous lizards with paddles for feet and with mouths studded with large teeth. These predators were as long as 40 feet (12 meters) and must have weighed over 6 tons. Among the reptiles, they were the most fearsome predators at sea, ever.[111] They had jointed jaws that allowed them to swallow large prey. Crushing and swallowing a 3-foot ammonite may have been routine.[112]

The impressive maximum length of plesiosaurs, some 50 feet (15 meters), owes much to their elongated necks (to 20 feet), which had as many as 76 vertebrae.[113] We do not know what the long neck was useful for. Other species of plesiosaurs had short necks, with as few as 13 vertebrae. How fast the plesiosaurs swam and how they hunted is a matter of speculation, of course.[114]

Marine crocodiles, turtles and sea snakes also belong in this group of ancient marine reptiles—and they are the ones that survived the end of the Cretaceous. At that time, about 65 million years ago, most large animals species then in existence became extinct, as a result of bombardment from space. Some crocodilians in Jurassic time were 20 feet long (7 meters). In the Cretaceous, the largest sea turtle (probably) was the 12-foot (3.6 meters) *Archelon,* which had a parrotlike beak and a leathery shell like that of the slightly smaller leatherback turtles of the present day.

Sharks, mammals, and birds evolved to fill many of the niches of the extinct reptilian predators, after the catastrophe that ended the Mesozoic. In the following geologic era, the Tertiary period produced its own giants, matching and exceeding those of the Cretaceous.

A true giant even among big sharks was the megatooth shark *(Carcharodon megalodon),* which may have grown to more than 80 feet long, according to some estimates, implying a weight of well over 50 tons. Sharks replace their teeth frequently, so thousands of teeth of this shark, including impressive 6-inch ones, have been found in marine deposits on land, well fossilized. In contrast, the cartilage that forms the skeleton does not fossilize readily, and so nothing much else is commonly found. Just how big these sharks really were is not known. Perhaps they were half the size commonly estimated but had especially large teeth. Still, a shark the size of a killer whale would have been a troublesome presence offshore of seal breeding colonies, with consequences for the evolution of marine mammals.

With the great marine reptiles extinct, the cetaceans—the brainiest creatures on Earth—evolved very rapidly in early Tertiary time, from coastal mammals looking superficially like some of the mosasaurs long gone. By the Eocene, when we find their fossils *(Eocetus, Basilosaurus),* they were already very large and had lost their hind limbs. Their conical teeth betray the fish hunter. As we have seen, modern offspring of the ancient cetaceans include both the toothed cetaceans (dolphins, porpoises, sperm whales) and the baleen whales.

Mammals now dominate the sea, as did the reptiles back in the Cretaceous, before the catastrophe that ended the Mesozoic. On the whole, the few aquatic modern reptiles do well in warm waters, as did their ancestors. The marine mammals, in contrast, are especially adapted to a cold ocean. It was not just the disaster at the end of the Cretaceous that changed life in the sea. The disaster gave mammals a chance to dominate, but the grand cooling in the last 40 million years ensured that they reached the top and stayed there.

NOTES AND REFERENCES

1. On occasion, it might also spell doom for at least some of the crew. Whale hunting was dangerous business, as summarized in the adage "A dead whale or a stove boat." The fluke of a sperm whale could readily damage a boat and fling the crew into cold waters. The deliberate sinking of a whale ship (the Essex from Nantucket, in 1820 in the eastern equatorial Pacific) by an enraged sperm whale underlies the novel *Moby Dick* (published 1851) by Herman Melville (1819–1891), seafarer and writer. The same event (the sinking of the Essex by an 85-foot male sperm whale with a weight estimated at 80 tons) is the subject in a detailed account by Nantucket historian Nathaniel Philbrick (N. Philbrick, 2000, *In the Heart of the Sea*. Penguin Books, New York, 302 pp.

2. As outlined by Garrett Hardin in 1968 (see chapter 8). Eventually, the discovery of petroleum ("rock oil") eliminated the need for whale oil in industry, making whale mining economically less attractive. Even so, more than 2 million large whales were killed in the twentieth century. (According to an exhibit on whale hunting, Stralsund Schiffahrt Museum, 2005.)

3. What is a "whale"? In everyday English a distinction is commonly made between whales on one hand and dolphins and porpoises on the other, with the first term referring to large cetaceans and the other two to relatively small toothed cetaceans typically weighing several hundred pounds. The big ones, then, are "whales," with the largest denoted as "great whales." Dolphins (Delphinidae) comprise the dominant family of cetaceans as well as the one containing small freshwater forms, while the porpoises (Phocoenidae) denote a separate family of dolphinlike animals with special anatomical features. When talking about small, toothed cetaceans of the open sea we deal almost exclusively with true dolphins. Their family also includes some animals not usually referred to as "dolphins": the melon-headed whale, the pilot whale, and the killer whale.

4. *The Oxford English Dictionary* (Compact Edition, 1971, Oxford University Press, Glasgow, p. 3746) quotes an article in the *Philosophical Transactions* of 1725 (vol. 33, p. 256): "The Right or Whalebone Whale is a large Fish, measuring sixty or seventy Feet in Length."

5. Hind legs have been reported on rare occasions—imperfect rudimentary freak appendages that remind us of the four-legged ancestry of the cetaceans. (See V. B. Scheffer, 1969, *The Year of the Whale*. Charles Scribner's Sons, New York, 244 pp., p. 62). Rudimentary bones *inside* the animal, at the place were the legs once connected to the pelvis, are commonplace.

6. The word *baleen* is derived from the Latin word for whale, *balæna*. The syllable *bal* occurs in both Latin and Sanskrit denoting *big and strong*. It is related to *val* as in valor. The Norse word for whale is *hval*, and it is clearly the immediate ancestor of *whale*. The connection of *hval* to *big, strong, terrible* is made in the name of the ice giant *Hvalur*. The word *whale* itself has come to denote "very large" in many contexts, thus returning to its origins. A relationship of *whale* to the word *wheel*, as suggested by one distinguished whale expert, with reference to circling while feeding, is not evident.

7. By the marine ecologist Roger Payne. He pioneered the recording of vocalizations of large whales. See R. Payne, 1995, *Among Whales*, Dell Publishing, New York, 431 pp. Payne suggests that the Greek myth of the Siren's song may derive from hearing humpback whales sing below small boats. *Ibid.*, p. 160.

8. A herd of sperm whales, according to Minasian et al. sounds like a team of carpenters nailing up a wall. S. M. Minasian, K. C. Balcomb, L. Foster, 1984, *The World's Whales*. Smithsonian Books, Washington, D.C., p. 82. Philbrick says whalemen referred to the sperm whale as "carpenter fish." He suggests that it was the hammering associated with repair of a whaleboat that attracted the sperm whale that sank the *Essex*. N. Philbrick, *In the Heart of the Sea*, p. 87.

9. Skull shapes are shown in A. S. Romer, 1966, *Vertebrate Paleontology*, 3rd ed. University of Chicago Press, Chicago, 468 pp., p. 298.

10. The power to focus sound is elaborated in K. D. Norris, B. Möhl, 1983, *Can odontocetes debilitate prey with sound?* American Naturalist 122, 85–103.

11. N. Bonner, 1989, *Whales of the World*. Facts on File, New York, 191 pp.

12. Ciesielski, P. F., F. M. Weaver, 1983, *Neogene and Quaternary paleoenvironmental history of Deep Sea Drilling Project Leg 71 sediments, Southwest Atlantic Ocean*. Initial Reports DSDP 71 (1), 461–477.

13. R. Payne, *Among Whales*. Payne emphasizes the ability of the great whales to go without food for a long time, as "masters of fasting." *Ibid.*, p. 35.

14. The data are in Minasian et al., *The World's Whales*, p. 40. *The Oxford English Dictionary* (p. 3746) cites a report from 1769 where it says that "whales are still seen one hundred and sixty feet long." If these are modern feet (12 inches, 0.3 meters), we get 48

meters, twice the length reported for today's average blue whale, which seems unlikely. If the length of the whale in question was exaggerated by 25 percent relative to present standards, we still get 120 feet, or 36 meters, which translates to a weight of well over 250 tons. Colbert (E. H. Colbert, 1955, *Evolution of the Vertebrates.* John Wiley and Sons, New York, p. 308) states that blue whales may reach weights of 150 tons.

15. Apparently from the Norwegian *roeyr-kval, roeyr* being a type of fish. Presumably, the Norwegian word goes back to Old Norse *reythar-hval,* which might refer to driving and pushing ("push whale"). The latter would be descriptive of the feeding habits of the rorquals.

16. *Fide* Minasian et al., *The World's Whales,* p. 38 and p. 46.

17. R. Payne (*Among Whales,* p. 33 ff.) emphasizes the width of the gape in reducing the success of escape responses of krill.

18. Ectoparasites—animals clinging to whales and digging into the skin and hitching a free ride—are another source of stress. Changing the environment during migrations may help to get rid of them.

19. Humpback migration routes in the Atlantic according to U. Sommer, 1998, *Biologische Meereskunde.* Springer-Verlag, Berlin, p. 218.

20. Fossil penguin remains are known only from the Southern Hemisphere: New Zealand, Australia, South America, South Africa, and islands off the Antarctic Peninsula. L. S. Davis, 2001, *Sphenisciformes,* in J. H. Steele, S. A. Thorpe, K. K. Turekian, *Encyclopedia of Ocean Sciences.* Academic Press, San Diego, pp. 2872–2880, p. 2872. In the north, the Great Auk, a flightless diving bird, had a life-style similar to that of penguins. The Great Auk was hunted to extinction in the nineteenth century. The last breeding pair was taken in June 1844, on the island of Eidey near Island, by a man named Jon Brandsson and his companions. (*Fide* Natural History Museum, Braunschweig.)

21. Glacial-ocean environmental patterns were first explored in the 1960s. In the 1970s, comprehensive studies for the global ocean were initiated and undertaken by the CLIMAP group, led by John Imbrie, James Hays, and T. C. Moore. See articles and references in W. H. Berger, V. Smetacek, G. Wefer (eds.), 1989, *Productivity of the Ocean: Present and Past.* Dahlem Konferenzen. John Wiley, Chichester, 471 pp.

22. R. Payne, *Among Whales,* p. 184.

23. The sound channel is discussed in chapter 11.

24. *Ibid.,* p. 180.

25. *Ibid.,* p. 189.

26. Carl Leavitt Hubbs (1894–1979) became professor of biology at Scripps Institution in 1944. (See chapter 7.)

27. "Willing associates and drafted graduate students" joined him in taking turns on a rooftop platform on two-story Ritter Hall on the Scripps campus, for 15-minute watches through the daylight hours, gazing through a pair of high-powered binoculars to count the whales. The gray whales turned out to be alive and well, though in the 1940s the population was estimated at fewer than 4,000 individuals. See E. N. Shor, R. H. Rosenblatt, J. D. Isaacs, 1987, *Carl Leavitt Hubbs, 1894–1979.* Biographical Memoirs of the National Academy of Sciences, 56, 215–249.

28. The airplane was piloted by Scripps physical oceanographer Gifford C. Ewing, "a superb pilot" who "knew Baja California as few others ever have," according to Hubbs. Hubbs was concerned with protecting the gray whale, and he conferred with officials in Mexico about hazards to the whales in the lagoons during their calving. Chiefly because of his entreaties, in which Ewing participated, Mexico declared Laguna Ojo de Liebre (Scammon's Lagoon) a sanctuary in 1972.

29. It was the whaling captain Charles Scammon who discovered the breeding grounds of the gray whales. He initiated the slaughter of whales within the lagoon and wrote about the aggressive reaction of the victims. C. M. Scammon, 1874, *The Marine Mammals of the Northwestern Coast of North America.* G. P. Putnam's Sons, New York, pp. 28 and 29.

30. P. J. Corkeron, S. M. Van Parijs, 2001, *Marine mammal migrations and movement patterns,* in J. H. Steele, S. A. Thorpe, K. K. Turekian, *Encyclopedia of Ocean Sciences.* Academic Press, San Diego, pp. 1603–1611, p. 1603.

31. Members may switch groups. S. K. Hooker, 2001, *Sperm whales and beaked whales,* in J. H. Steele, S. A. Thorpe, K. K. Turekian, *Encyclopedia of Ocean Sciences.* Academic Press, San Diego, pp. 2865–2872, p. 2869.

32. P. J. Corkeron, S. M. Van Parijs, *Marine mammal migrations,* p. 1605.

33. *Ibid.,* p. 1606.

34. B. J. Le Boeuf, D. E. Crocker, D. P. Costa, S. B. Blackwell, P. M. Webb, D. S. Houser, 2000, *Foraging ecology of northern elephant seals.* Ecological Monographs 70 (3), 353–382.

35. H. Melville, 1851, *Moby-Dick.* Harper and Brothers, New York, 538 pp.

36. Typical diving depths for sperm whales are between 500 and 1,000 meters. W. A. Watkins, M. A. Daher, K. M. Fristrup, T. J. Howald, 1993, Sperm whales tagged with transponders and tracked by underwater sonar. Marine Mammal Science 9, 55–67.

37. G. L. Kooyman, 2001, *Marine mammal diving physiology,* in J. H. Steele, S. A. Thorpe, K. K. Turekian, *Encyclopedia of Ocean Sciences.* Academic Press, San Diego, pp. 1589–1595.

38. *Ibid.,* p. 1592.

39. Per F. Scholander (1905–1980) joined the Scripps faculty in 1958. In addition to a laboratory he also obtained a ship from the National Science Foundation, the *Alpha Helix*, to use as a floating base for his physiological research.

40. Gerald L. Kooyman joined Scripps in 1967. Most recently he has worked on migration in the life history of penguins. His preferred study area has been McMurdo Sound in the Antarctic. Over the years his subjects have included sperm whales, elephant seals, bottlenose dolphins, Weddell seals, leatherback sea turtles, emperor penguins, and sea otters.

41. G. L. Kooyman, P. J. Ponganis, 1997, *The challenges of diving to depth.* American Scientist 85, 530–539.

42. We do not mean to imply that there was no vertical plankton migration before that. However, the nature of the migration must have changed with the shallowing of the thermocline, a barrier that became much more distinct as the planet cooled.

43. Moths avoid sound-emitting bats in this fashion.

44. Historically, according to Scammon (*The Marine Mammals*, p. 115), the elephant seal was found over a wide range of latitudes, from Cabo de San Lazaro in Baja California (near 25° N) to Point Reyes in Alta California (near 38° N). Wherever the numbers of prime target whales had been greatly reduced, whalers turned to other marine mammals to capture and process. By the beginning of the nineteenth century they began taking elephant seals, especially the southern variety around Kerguelen Island and Heard Island in the southern Indian Ocean. They soon found the elephant seals on the California coast and islands, which they attacked on land with rifles and clubs during mating and breeding seasons. The herds were quickly decimated. In 1892 Charles H. Townsend of the U.S. Fish Commission learned about eight animals on Guadalupe Island. He collected seven of them as museum specimens.

45. Commercial sealing, much like whaling, has declined since the 1960s in any case, although there are some remarkable exceptions. Harp seals in eastern Canada and in western Greenland remain a prime target of hunting. An estimated 350,000 were taken in 1998, the great majority by shooting, but with thousands of molting pups being clubbed to death (to prevent damage to the pelts). R. R. Reeves, 2001, *Marine mammals, history of exploitation*, in J. H. Steele, S. A. Thorpe, K. K. Turekian, *Encyclopedia of Ocean Sciences.* Academic Press, San Diego, pp. 1633–1641, p. 1637. Besides Canadian commercial seal hunting, Norwegian and Russian sealing continues in the Greenland Sea and the Barents Sea, with a take of tens of thousands each year. A new factor in the seal products market is a demand in Asia for the reproductive parts of seals. The high prices paid for such items has provided an incentive for poaching in certain marine reserves.

46. After extensive hunting for its fur, beginning in the early eighteenth century, the sea otter was thought extinct by 1925. A 1999 census along the central California coast by the U.S. Geological Survey totaled 1,970 individuals. Today there are more than 100,000 sea otters spread out between Baja California and the Kuril Islands in Japan. Not everyone responds with enthusiasm. To those who fish for the increasingly rare abalones off the central California coast, the sea otter is seen as a menace to their livelihood. (For present-day distribution, see J. L. Bodkin, 2001, *Sea otters*, in J. H. Steele, S. A. Thorpe, K. K. Turekian, *Encyclopedia of Ocean Sciences.* Academic Press, San Diego, pp. 2614–2641, p. 2614.)

47. J. A. Estes, D. O. Duggins, 1995, *Sea otters and kelp forests in Alaska: Generality and variation in a community ecology paradigm.* Ecological Monographs 65, 75–100. Animals other than sea otters are important also as predators on sea urchins (e.g., lobsters and certain fishes). Thus, removal of sea otters by itself may not automatically produce sea urchin barrens.

48. F. V. Paladino, S. J. Morreale, 2001, *Sea turtles*, in J. H. Steele, S. A. Thorpe, K. K. Turekian, *Encyclopedia of Ocean Sciences.* Academic Press, San Diego, pp. 2622–2629, p. 2624.

49. For a summary of Loch Ness lore, as well as other monster stories, see R. Ellis, 1994, *Monsters of the Sea.* Robert Hale, London, 429 pp.

50. S. Heaney, 2000, *Beowulf, a New Verse Translation.* W. W. Norton and Co., New York, p. 155. Alternatively, Nessie is an offspring of the slimy companions of the awful swamp creature Grendel, also described in the Beowulf Saga, and which later morphed into something like a plesiosaur, perhaps as a result of the spectacular discoveries of the fossil collector Mary Anning, in the first half of the nineteenth century. (The Anning story is recounted in C. McGowan, 2001, *The Dragon Seekers.* Perseus Publishing, Cambridge, Mass., p. 70.)

51. F. B. Goodrich, 1873, *The History of the Sea.* Donohue, Chicago, p. 25. The reference to illumination concerns the use of blubber in oil lamps, not any intellectual preoccupations.

52. A length of 35 feet, or 10 meters, would normally suggest a weight much greater than 500 pounds, except that the fish is much like a ribbon.

53. The flesh is soft and rather tasteless, as one of us, ENS, confirmed when a fresh specimen was given to the Scripps Institution fish collection, decades ago. (The fish was too large to store whole.)

54. Quoted in B. Heuvelmans, 1968, *In the Wake of the Sea-Serpents.* Hill and Wang, New York, 645 pp., p. 103.

55. Upon request, the observers' verbal accounts were delivered to a committee of the Linnaean Society of New England. In its published report the society included a sea captain's description of a similar creature in June 1815, off Plymouth, Massachusetts, and observations from Norway by Erik Ludvigsen Pontoppidan, Bishop of Bergen, of a very large snake-like animal that was said to come often to the surface in July and August for spawning. B. Heuvelmans, *In the Wake*, p. 162. (Hubbs's library had a copy of the following report: Anonymous, 1817, *Report of a Committee of the Linnaean Society of New England, Relative to a Large Marine Animal, Supposed to Be a Serpent, Seen near Cape Ann, Massachusetts, in August 1817*. Cummings and Hilliard, Boston, 52 pp.)

56. Heuvelmans *(In the Wake)* reported a sighting by two Icelandic fishermen, in February 1963, of a two-humped creature that rose 15 feet above the sea surface and was unlike anything they had previously seen. The crew of a Nantucket fishing vessel, in May 1964, saw a 50-foot black creature that had several humps and a "blowhole in the top of the alligator-like head" (B. Heuvelmans, *In the Wake*, p. 527). The blowhole would indicate a reptile or a mammal rather than a fish, of course, unless the observers saw the funnel of a giant squid.

57. The name given by these authors was in recognition of Archie H. B. Wills, editor of the British Columbia newspaper *Victoria Daily Times* in the 1930s, who suggested the name "Cadborosaurus" and the nickname "Caddy" for such creatures in his region, from nearby Cadboro Bay. While the presentation of the paper by Bousfield and LeBlond is in standard taxonomic format, its scientific validity is in doubt. Without a type specimen, Cadborosaurus has no anchor in reality. The technical expression for such a name without a recognized basis is *nomen nudum*. (The report was published in April 1995. E. L. Bousfield, P. H. LeBlond, 1995, *An account of* Cadborosaurus willsi, *new genus, new species, a large aquatic reptile from the Pacific coast of North America*. Amphipacifica 1, suppl. 1, 3–25.)

58. B. Heuvelmans, *In the Wake*, pp. 537–588.

59. The giant-squid connection is the one favored by Richard Ellis (*Monsters of the Sea*, p. 365 ff.). In agreement with Henry Lee (cited as "1884, *Sea Monsters Unmasked*, London"), he thinks that in many sea serpent sightings what people saw is one of the long arms of the giant *Architeuthis*, along with a glimpse of the tail (at the other end), with the main body of the animal submerged, and moving by jet propulsion, rather than being propelled by undulating motion. He criticizes Heuvelmans for missing the possible connection of giant squids to sea serpents.

60. R. Ellis, *Monsters of the Sea*, p. 125 ff. (The label kolkrabbe means squid or cuttlefish.)

61. *Ibid.*, p. 128. and C. F. E. Roper, 1998, *Tracking the giant squid: Mythology and science meet beneath the sea*. Wings (Xerces Society) 21 (1), 12–17.

62. Weights of around 1 ton are well confirmed. R. Ellis, 1998, *The Search for the Giant Squid*. Penguin Putnam, New York, 322 pp., p. 106. Ellis quotes Heuvelmans's weight estimates of between 16 and 216 tons and deems them utterly ridiculous. While this conclusion is defensible, we do not follow the logic of his argument, which invokes compositional detail. Instead, we use statistics. If we take as "typical" a giant squid of 30-foot length and a weight of 450 pounds (200 kilograms) (*Ibid.*, p. 99), we get eight times that weight for a 60-foot squid, provided the proportions stay the same. Thus, the weight would be around 1,600 kilograms, or 1.6 tons. By the same token, an extra-large giant squid with a length of 120 feet (as postulated by some) would weigh around 13 tons. A squid 80 feet long, in this scaling scheme, would weigh just a little less than 4 tons. If we use the data on a 34-foot squid weighing 150 kilograms (*Ibid.*, p. 155), we get a weight of 2 tons for an animal 80 feet long. An estimate of a maximum weight of 3 tons for the giant squid would seem reasonable, given the information readily available, since the lengths of beached squid reported by Ellis extrapolate statistically to a maximum size near 80 feet.

63. The Danish zoologist Johan Japetus Steenstrup (1813–1897) described a giant squid in 1857, giving it the name *Architeuthis monachus*, from remains found on a beach in Jutland in 1853. (The description apparently was not properly published, however.) During the 1870s a number of similar creatures were washed up on beaches of Labrador and Newfoundland and died there. Scientists took note and confirmed lengths of 30 to 60 feet for *Architeuthis*, including the extended tentacles (which can be stretched). Professor Addison E. Verrill (1839–1926) of Yale noted in 1879 that there were apparently two different species of large squids, one with long arms and another with short arms. (B. Heuvelmans, *In the Wake*, p. 52.) Many more have been found since. (A list of strandings is in R. Ellis, *Monsters of the Sea*.)

64. W. A. Herdman, 1923, *Founders of Oceanography and Their Work: An Introduction to the Science of the Sea*. Edward Arnold, London, 340 pp., p. 121.

65. I. T. Sanderson, 1956, *Follow the Whale*. Bramhall House, New York, p. 208. Roper (C. F. E. Roper, *Tracking the giant squid*.) believes that the scars on sperm whales are not trustworthy as a measure of the size of the squids that left the mark. Extraordinary squid sizes have been reported as seen from afar only. One whaler (perhaps in 1875) described a moonlight battle between a sperm whale and an enormous squid. Through night glasses he saw the very large

whale, which was encased in the writhing arms of an almost equally large squid; the whale seemed to be gnawing through the cephalopod's "tail," while sharks hung about the battle scene. (B. Heuvelmans, *In the Wake*, p. 68.)

66. C. F. E. Roper, K. J. Boss, 1982, *The Giant Squid.* Scientific American 246 (4), 96–105; C. F. E. Roper, *Tracking the giant squid.*

67. For discussion of expert opinions on this point, see R. Ellis, *Monsters of the Sea*, p. 72.

68. According to Heuvelmans (*In the Wake*, p. 78), 12 men were left adrift on a small raft, after the *Britannia* was sunk on 25 March 1941, in the middle of the tropical Atlantic. They took turns at hanging on to the raft in the water, because it was so small. Then: "One night a huge squid hauled one of the men off the raft and he was not seen again. Soon afterwards a tentacle attacked Lieutenant R. E. G. Cox, fortunately letting go again, but removing discs of skin and flesh the size of a penny (1-1/4 inches diameter). This implies that the squid was 23 feet long."

69. *Selachii* is the scientific name for sharks (selachians). Together with the rays (Batoidea or Rajiformes) they make up the elasmobranchs (Elasmobranchii). The elasmobranchs are the dominant group within the fishes with a skeleton made of cartilage.

70. Sizes and weights of sharks commonly get much attention. As a rule, 5-meter-long sharks of the typical fish shape weigh around 800 to 900 kilograms, while lengths of 10 meters and 15 meters correspond to weights of roughly 6 tons and 18 tons, respectively, with error bars of about 30 percent. Discrepancies arise because many weight numbers reported in the literature are but rough guesses. A scale for fishes weighing more than a ton is rarely available!

71. An oft-quoted 36.5 feet (11 meters) was thought by shark expert Perry Gilbert to be a printer's error for 16.5 feet; the jaws of the specimen in question are in the British Museum. Other discrepancies in maximum lengths given in the literature may result from including or omitting the tail length; ichthyologists measure to the base of the tail. A length of 20 feet (6 meters) is reliably recorded; the maximum would be greater than this. Occasional reports of much larger animals exist, based on unconfirmed sightings.

72. Jacques-Yves Cousteau, who did more than anyone to bring the living sea into the homes of people, described his first encounter with a large *Carcharodon carcharias*, a fearsome brute 25 feet in length, in his book *The Silent World*. When it saw Cousteau and his diving companion, it turned and fled. As well it should—an earlier encounter of Cousteau's with a somewhat smaller shark had left the shark with a harpoon in its head. The same flight reaction was observed in many other types of large sharks. Cousteau's associate subsequently killed a small whale, and after the blood was in the water they obtained good shark pictures. One large shark came close enough to be bumped on the nose with the camera. Others were circling. J. Y. Cousteau, F. Dumas, 1953, *The Silent World.* Harper, New York, p. 181 ff.)

73. For experiences with sharks by pioneer divers and expert observers, we consulted H. Hass, I. Eibl-Eibesfeldt, 1977, *Der Hai, Legende eines Mörders.* C. Bertelsmann, München, 258 pp.

74. For some reason, perhaps related to their fearsome size and strange appearance, they are also known as "giant devilfish" or "giant devil ray."

75. L. R. Taylor, L. J. V. Compagno, P. J. Struhsaker, 1983, *Megamouth—A new species, genus, and family of lamnoid shark (*Megachasma pelagios, *family Megachasmidae) from the Hawaiian Islands.* Proceedings of the California Academy of Sciences 43 (8), 87–110. Leighton Taylor, then director of Waikiki Aquarium in Honolulu, received his Ph.D. from SIO in 1972. Discovering a member of a previously unknown vertebrate family was a major event in taxonomy; discovering one that weighs close to a ton was amazing news.

76. Reported to have been found in the stomach of *Carcharias.* A. Remane, V. Storch, U. Welsch, 1976, *Systematische Zoologie.* Gustav Fischer Verlag, Stuttgart, 678 pp.

77. We do not mean to imply that sharks were in the line of human ancestry, of course. Rather, the sharks and some rays retain this relationship between armor and teeth from an ancient common ancestor.

78. R. Ellis, J. E. McCosker, 1991, *Great White Shark.* Harper Collins, N.Y., 270 pp., p. 140–142.

79. The 120-mile coastline between Monterey Bay and Tomales Point has been called the White Shark Attack Capital of the World, according to Ellis and McCosker (*Ibid.*, p. 156). This stretch of coast has seen more than half of all such attacks.

80. R. Ellis, J. E. McCosker, *Great White Shark*, pp. 146–147.

81. According to the National Geographic News web site (lightning) and an estimate by a Dr. Scott Camazone of Cornell University, quoted on the APIS site of the University of Florida (insect stings).

82. B. J. LeBoeuf, 2004, *Hunting and migratory movements of white sharks in the eastern North Pacific.* Memoirs of the National Institute for Polar Research, Special Issue 58, 91–102.

83. It is unlikely that the ecocidal finning industry will last long. Scripps ichthyologist Jeff Graham estimates that the death rate of sharks is in the tens of millions. Sharks have a slow rate of reproduction; the populations will not cope with this kind of fishing pressure. In recent years taking sharks solely for the

fins has been forbidden in waters of the United States. Shark fin soup, however, can still be ordered in certain types of restaurants, right here in San Diego.

84. J. D. Isaacs, R. A. Schwartzlose, 1975, *Active animals of the deep-sea floor.* Scientific American 233 (4), 84–91.

85. *Ibid.,* p. 86.

86. The bony fishes (Osteichthyes) are on the level of a class in the vertebrates. The bulk of modern bony fishes are ray-finned and belong to the teleosts (Teleostei). The flesh-finned fishes include the lobe-finned fishes (crossopterygians) and the related lungfishes (dipnoi). The crossopterygians are considered the ancestors of the first amphibians, from which all land vertebrates take their origin. A. S. Romer, 1959, *The Vertebrate Story,* 4th ed. University of Chicago Press, Chicago, 437 pp.

87. The strategy of producing thousands of small eggs, common in pelagic fishes, is radically different from that of the sharks, which early during evolution took the option to invest in a few large eggs at a time. In pelagic shark species, the eggs are not released; rather, they are hatched within the mother, or else the embryos are nourished inside the mother, in a fashion reminiscent of mammalian reproduction. The two strategies (called "r" strategy and "K" strategy, respectively) both have advantages and disadvantages. A highly unpredictable environment would seem to favor the many-eggs strategy.

88. Swordfish have long been reported to attack boats on occasion, with their viciously sharp upper jaw that can be 5 feet long on a fish of 15 feet overall length. One thousand pounds of weight can put some considerable emphasis on the pointed business end. Presumably, the sword is useful in slashing through schools of fish, after which the attacker swims back to pick up the injured. An 8-foot swordfish jammed its sword into Woods Hole's research submarine *Alvin,* in 1967, at a depth of 1985 feet (600 meters). The weapon entered the gap between the vessel and its fiberglass skin. The fish thrashed about as the vehicle rose to the surface and there broke from its firmly stuck sword. The escort swimmers grabbed it near the surface as it died. All were relieved that it had not struck the porthole and that it had missed—just barely—the electrical cables. The attacker was served for dinner. V. A. Kaharl, 1990, *Water Baby: The Story of Alvin.* Oxford University Press, New York, 356 pp.

89. N. B. Marshall, 1966, *The Life of Fishes.* Universe Books, New York, 402 pp.

90. *Ibid,.* p. 363.

91. *Ibid.,* p. 360. Not to be confused with the white toothed whale in Arctic waters, also known as "beluga."

92. J. R. Norman, F.C. Fraser, 1963. *Riesenfische, Wale, und Delphine.* Verlag Paul Parey, Hamburg, Germany, 341 pp. [Translated and edited by G. Krefft and Kurt Schubert, from *Giant Fishes, Whales, and Dolphins.* Putnam, London] p. 106. There may be larger eels still to be discovered. B. Heuvelmans (*In the Wake,* p. 484) reports that in 1930 the Danish biologist Anton Bruun (1901–1961) collected an eel larva 6 feet long, which would imply an adult having a length of more than 100 feet (30 meters), assuming the usual scaling.

93. The amazing story of finding the spawning area of the European river eel was worked out by the Danish ichthyologist Johannes Schmidt (1877–1933). J. Schmidt, 1906, *Contributions to the life-history of the eel* (Anguilla vulgaris *Flem.*), Rapports et Proces-verboux des Réunions Conseil International pour l'Exploration de la Mer 5, 137–264; J. Schmidt, 1922, *The breeding places of the eel.* Philosophical Transactions of the Royal Society B 211, 179–208.

94. The observation commonly cited is that of Willoughby Lowe, who saw a huge aggregation of snakes in the Malacca Straits, in 1932, "many intertwined with each other." See R. Ellis, 2001, *Aquagenesis.* Viking Penguin, New York, p. 158.

95. On one Scripps expedition the chief scientist collected some of these sea snakes on the Pacific side of Panama, for the San Diego Zoo. They were easy to catch by dip net, but not easy to confine to a container full of seawater. The ship's doctor, on seeing them slithering around the deck, shook his head sadly and commented: "There is no antidote. And no time. Death occurs within ten minutes." Fortunately, no one was bitten.

96. F.V. Paladino, S. J. Morreale, *Sea turtles,* p. 2628.

97. Sources differ: A. Remane et al. (*Systematische Zoologie,* p. 551) give 500 kilograms; R. Ellis (*Aquagenesis,* p. 145) gives "over a ton."

98. R. Ellis, *Aquagenesis,* p. 164; A. Remane et al. (*Systematische Zoologie,* p. 553) give 8.5 meters (28 feet) for the largest crocodiles.

99. M. Wikelski, C. Thom, 2000, *Marine iguanas shrink to survive El Niño.* Nature 403, 37.

100. For example, contact with the sea wasp *(Chironex fleckeri),* a small jellyfish occasionally found in great numbers on some Australian coasts, can be deadly. The venom in its trailing tentacles is extremely toxic and acts directly upon the heart muscle. B. W. Halstead, D. D. Danielson, 1970, *Death from the depths.* Oceans 8 (6), 14–25.

101. A popular summary of dangerous marine animals is given in T. A. Dozier, 1977, *Dangerous Sea Creatures,* Time-Life Films, New York, 128 pp.

102. For example, the deadly poisons of cone snails. L. Nelson, 2004, *One slip, and you're dead . . .* Nature 429, 798–799 [news feature].

103. For an exposition of the race between armor-making and armor-breaking organisms see G. J. Vermeij, 1987, *Evolution and Escalation, an Ecological History of Life.* Princeton University Press, Princeton, N.J., 527 pp.

104. A. Remane et al., *Systematische Zoologie,* p. 229. The closest living relatives are the bottom-living slow-moving horseshoe crabs (East Coast: *Limulus polyphemus*).

105. James Leonard Brierly Smith (1897–1968) named the fish *Latimera* after the discoverer, Marjorie Courtenay-Latimer (1907–2004). J. L. B. Smith, 1939, *A living fish of Mesozoic type.* Nature 143 (3620), 455–456.

106. Scripps eventually obtained a coelacanth from the Comoros for its collections, as a gift from John E. McCosker, in 1975. An earlier attempt by John D. Isaacs and Scripps ichthyologist Richard H. Rosenblatt to procure a specimen of this marvelous fish, in 1971, was thwarted when French authorities refused access to the Comoros.

107. The Indonesian specimen of coelacanth was discovered by Mark and Arnaz Mehta Erdmann, in 1997. They learned that local fishermen caught two or three of these fish every year in deep-water shark nets. *Washington Post,* 19 November 1998. Subsequently, in October 2000, Pieter Venter, a recreational diver who descended to the remarkable depth (for a free-swimming diver) of 320 feet, saw three coelacanths off the coast of South Africa. He took film footage showing that the fishes were sometimes standing on their heads. *San Diego Union-Tribune,* from Reuters News, 2 December 2000. Presumably they do so to minimize their silhouette for upward-looking (and upward-swimming) potential prey. Earlier, the German zoologist Hans Fricke had made observations on coelacanths off the Comoros. H. Fricke, O. Reinicke, H. Hofer, W. Nachtigall, 1987, *Locomotion of the coelacanth* Latimeria chalumnae *in its natural environment.* Nature 329, 331–333; H. Fricke, K. Hissmann, 1991, *Coelacanths—fate of a famous fish.* Oceanus 34, 44–45.

108. The Mesozoic era lasted from 225 million years ago to 65 million years ago and includes the Triassic, the Jurassic, and the Cretaceous periods. It starts and ends with great catastrophes producing widespread extinction. (See appendix.)

109. Some of these reptilian fossils were reconstructed with great skill by Thomas Hawkins (1810–1889) before being sold to the museum as perfect specimens. See C. McGowan, *The Dragon Seekers,* p. 9.

110. J. Braun, W. E. Reif, 1985, *A survey of aquatic locomotion in fishes and tetrapods.* Neues Jahrbuch für Geologie und Paläontologie, Abhandlungen 169, 307–332.

111. For comparison, today's largest lizard, the predatory Komodo Dragon, attains about 10 feet in length (3 meters).

112. The presence of shell-crushing ammonite eaters is evident from the increased abundance of shells with strong ribs and knobs, throughout the first two-thirds of the Cretaceous. At some point, the crushers had become so efficient that the strategy of stronger shells lost appeal. Presumably, speed and better ink bags, or better sensory devices and changes in behavior were then more effective.

113. That oh-so-long neck got paleontologist Edward Drinker Cope, 1840–1897, into trouble in his early collecting days of 1869. He assumed that the vertebrate pieces were the tail of the plesiosaur *Elasmosaurus* and promptly arranged the specimen with the head at the wrong end. O. C. Marsh (1831–1899), who soon became Cope's most energetic rival, was delighted to point out the error. E. N. Shor, 1974, *The Fossil Feud.* Exposition Press, New York, 340 pp.

114. Samuel Wendell Williston (1851–1918), who spent considerable time in those chalk beds in the mid-1870s, concluded that these reptiles were not fast swimmers and that only the forward part of the neck was flexible, but not capable of snakelike twisting. E. N. Shor, 1971, *Fossils and Flies.* University of Oklahoma Press, Norman, Okla., 285 pp.

TEN

The Deep, the Cold, the Dark

LIFE AT THE END OF THE LINE

Living in the dark is the normal state of affairs on our planet: the dark and cold waters below the sunlit surface layer of the sea constitute the largest life habitat on Earth. This habitat is largely a desert, except at the margins, including the outer shelf and upper continental slope. The animals seen in the dark and cold habitat are on the whole quite closely related to those near the surface, but many of them have special adaptations for seeing at low light, for making light, and for stalking and attracting prey in the dark.

Different depth levels bear somewhat different kinds of animals, both in the water and on the seafloor. In the first several hundred meters below the sunlit zone, daylight still is present, albeit extremely dim. Large eyes are useful here, as well as biologically produced light for hunting, hiding, and signaling. The detection of sound over a large frequency range is useful for hunting and for avoidance of bigger hunters. The dark-adapted grazers and hunters come up into the surface layer at night to feed in shallow water, where both phytoplankton and zooplankton are highly concentrated compared with subsurface depths. This daily vertical depth migration is the largest coherent motion of animals on the planet and governs all marine ecosystems away from the shore to a degree commonly not fully appreciated in textbooks.

Below the uppermost subsurface layer—the midwater zone—depth zonation continues, largely because food supply keeps on decreasing with depth. Animals living in the dark and cold waters, in general, tend to be relatively small or, if large, commonly have a disproportionate amount of watery gelatin in their bodies. They move slowly to conserve energy, and many have large mouths to enable them to swallow prey their own size, to compensate for the rarity of prey. The exploration of the dark and cold realm of the sea has benefited greatly from developments in deep-sea photography and deep diving since the middle of the twentieth century. It is truly the "inner space" of life habitats on Earth, at the edge of the unknown.

EXPLORING THE LARGEST HABITAT ON EARTH

When dealing with the largest life habitat on Earth, the bathypelagic realm, we are in unfamiliar territory.[1] Our experience as creatures living on a two-dimensional interface makes it difficult for us to comprehend a way of life in three dimensions. Also, deep-sea creatures live in the dark, and we do not. Our language reflects our preferences for a dry and well-lit habitat. We call our water-dominated planet "the Earth" and say "I see" when we believe we understand. The contrast in feelings we have for a sunlit versus a dark world may be gauged by comparing the words *light, bright, brilliant, luminous, radiant,* and *clear* with the words *dark, dim, shadowy, murky, gloomy,* and *opaque.*

Taking the living space on land as 20 meters thick, to include treetops and insect clouds, the sunlit part of the sea (the uppermost 100 meters or so) may be estimated as having 10 times the terrestrial living space. The same rough comparison makes the deep, dark part of the ocean more than 300 times larger than the terrestrial life habitat! Our ecologic intuition then, fundamentally built on living on a solid surface lit from above, and informed by a childhood of watching ants, catching tadpoles, gathering mushrooms, and climbing trees,[2] is likely to be of limited use in relating to 99 percent of the Earth's life zones. Here light is scarce or absent, and there is no surface—only a black expanse in all directions.

When invading that realm, a wonderful surprise awaits the explorer, one that is comparable perhaps with seeing, for the first time, a star cluster in the sky through a powerful telescope. We are talking about the abundance of living lights at depth, an experience often recounted by divers in moving words.

One well-known ambassador of the deep, who went there to report not only on what he saw but also how he felt about it, was William Beebe.[3] In his widely read book *Half Mile Down,* he recalled the feeling of estrangement and wonder when underwater in a diving helmet, saying that "adventuring under the sea is an unearthly experience."[4] In 1930 Beebe joined forces with engineer Otis Barton. With advice from Captain John J. Butler, they built a steel sphere that could be lowered into the sea from a large barge. Beebe named it the bathysphere.[5] The bathysphere introduced deep diving to the study of the ocean. The diving feats were well publicized in *National Geographic* and other scientific and general publications, as well as in Beebe's book. A new world never seen before started to open to public awareness.

From his tiny probe penetrating a vast and unfamiliar space, Beebe made what observations he could, through the small quartz portholes. He noted how the color of the water faded from a brilliant blue into dark gray and black (at 1,700 feet, or about 500 meters), reflecting both the fact that blue light penetrates the water farthest, and that our eyes see no color in dim light. He recorded that animals that he knew from net hauls to be red in daylight appear black at depth; there is no red light to reflect and make the bodies visible. And he was greatly impressed by the numerous sparks and flashes of bioluminescent animals in total darkness—10 to 50 at a time, "ghostly things in every direction."[6] He caught glimpses of unnamed living things at the edge of his range of vision. He observed creatures that had not been caught in the many deep net hauls in that area and understood why the nets came up empty when he saw the animals dart and twist and turn away from the vehicle's light. Some of these he recognized, but others he thought were unknown species. He described them in some detail and even assigned scientific names to some fishes.[7]

Others soon followed Beebe's example; ocean engineers and physicists devised equipment to invade and explore the deep sea, directly or indirectly.[8] The age of scientific manned submersibles had started, including tethered capsules and freely moving research submarines.

In this line of development of deep submersibles also belong the exploits of the Swiss adventurer-explorer Auguste Piccard.[9] Piccard was bent on setting vertical records. After going higher than 15 kilometers in a balloon (in the Alps, in 1932) he created the bathyscaph *Trieste*, a free-diving metal sphere with a gasoline-filled float. This underwater balloon took him and his son Jacques down to a depth of 10,168 feet off the Cape Verde Islands, a record at the time.[10] In 1960, the *Trieste* set a depth record in the Mariana's Trench on a U.S. Navy dive in the Pacific, at 35,800 feet (near 10.9 kilometers).[11]

Woods Hole's *Alvin* is the most famous and most productive in a long line of manned diving vehicles that followed the exploits of the pioneers. It brought us, among other things, a view of a life habitat never before dreamed of, that is, the environment of the strange community growing around volcanic hot springs on the seafloor, in the eternal dark.

While direct observation has much to recommend it besides the sense of adventure that stirs public interest, it is not invariably the best way to explore the abyss. Many measurements can be made profitably from unmanned vehicles, in rapid survey fashion and on a large scale, or with attention to local detail, and yet avoiding the large costs associated with making operations fail-safe. Marine engineers have designed remotely operated vehicles of various sizes for the purpose. More recently, autonomous underwater vehicles are being used to survey the deep ocean, that is, unmanned small submarines with a resident computer programmed to gather and store information, and to communicate via satellite as necessary.

The various vehicles, manned and unmanned, free or tethered, are used everywhere in the water column down to the seafloor, with a preference for a zone just above the seafloor. Right above the seafloor there are things to discover, while information density is low in the water column below the faintly lit zone. The reason for the low abundance of animals here is lack of food. As far as settling organic matter, less than 20 percent of the original amount that entered the midwater, or *mesopelagic*, zone[12] remains for downward transfer to the realm below, and about half of that will make it to the seafloor.[13] Thus, the rain of edible particles at depth is very thin, except perhaps right at the equator, below the region of high production.

We must assume that much of the food at great depth is coming in by active downward transfer, that is, by predation of deeper living forms on those above, including predation on the daily migrants when resting at their deepest hideouts, by stealth-hunters looking upward. Also, we must consider horizontal transfer, that is, predation by bathypelagic organisms at the edge of the broad ribbon of increased animal density on the upper continental slope and within the curtain below the equator (fig. 10.1). Upward transfer occurs at the seafloor, through predation by wholly pelagic forms on animals that live closer to the seafloor, at the continental slope, and which make a living cropping benthic animals. The logic of this process is a kind of diffusion of biocarbon through chains of predation, from regions of high abundance into the thinly stocked vast spaces of the interior.

The seafloor, where it intersects the boundary between the mesopelagic and the bathypelagic depth zone (900 meters or so) still supports a relatively rich fauna. One suggestion is that this is the zone where the deep scattering layer has its maximum depth; thus every day there is a congregation of migrants impeded by the seafloor and then available as prey for the locals.[14] Also, at around 1,000 meters on the continental slope, much organic matter is still delivered from the coastal zone of high productivity, both by a pulsed rain of organic matter and by transfer from the shelf, along the bottom. Again, it seems we must envision a diffusion of food supply into the central bathypelagic zone from this richer boundary—a relay of predation starting with benthopelagic forms living at the continental margin (or, away from continents, on the slopes of islands).

From the simple principle that food must come from regions where it is abundant and move to the otherwise barren interior by a relay of

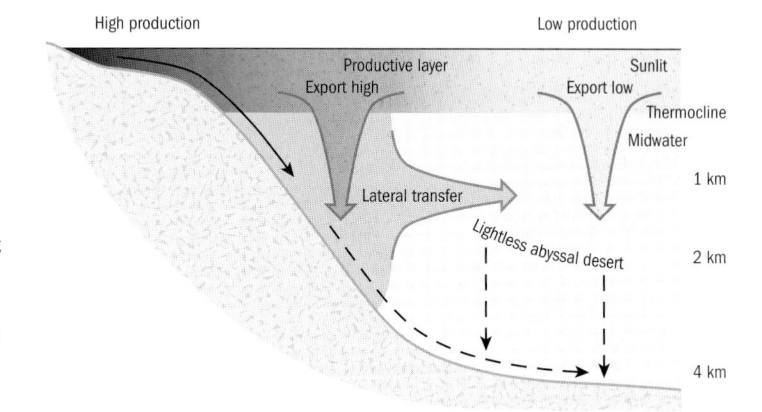

FIGURE 10.1. Food for those living in the dark: the main pathways of transfer of organic matter into the deep interior of the ocean *(arrows)*. Level of export depends on level of overlying production. The abyss itself is a desert.

predation, we can appreciate the ecologic dynamics that sustain the bathypelagic realm: predation on migratory animals resting at their maximum depth level, predation on the fauna living below privileged regions such as equatorial upwelling, and predation on benthopelagic forms living on elevated seafloor including the continental slope and the top of submerged seamounts. Also, the deep bottom of the ocean acts as the ultimate trap for settling organics and thus can produce prey in some modest abundance.

Consider the opportunity to prey on fish and squid engaged in daily migration, resting at their lowermost depth range. What kind of sensory apparatus would it take to detect them? How might the intended prey defend itself?

Between 500 and 1,000 meters depth, in the extremely clear waters of the open ocean, the light is now attenuated so much that only highly sensitive eyes—whether of fishes, squids, or shrimps—can detect anything at all using background illumination. Sensitivity is generally enhanced in the blue green part of the spectrum (since both residual sunlight and bioluminescence are located there). In fishes, rods proliferate on the retina, in preference to cones. Some have two or three layers of rods. Also, there are special adaptations that allow the concentration of light onto sensory cells. An extreme case of a specialized system for vision is that of the fish *Dolichopteryx*, which has night glasses mounted on top of its head, looking upward.[15] The fish is less than a foot long. We may infer that it looks for silhouettes above its cruising range. Other hunters apparently use flashlights to detect prey: the organs sit close to the eye and they have a reflective back shaped to beam light forward. This method should work even if the prey is dark rather than silver sided, because of eye shine, familiar from animals on the road, in the high beam of a car. Some fishes produce "invisible" light, in the red and infrared—the prey may be unaware that it is being illuminated and cannot immediately react to the predator's approach, therefore.[16]

The main danger, presumably, is to present a silhouette against the sky to predators cruising below one's depth level. Defense against being detected can be achieved by matching the naturally downwelling light with carefully adjusted light output from light organs (photophores) on the underside of the body (fig. 10.2). Defense against searchlight detection is dark coloration, and escape response upon being flashed at. In

FIGURE 10.2. Photophores on the underside provide for countershading, preventing the fish from being seen against the sky, from below.

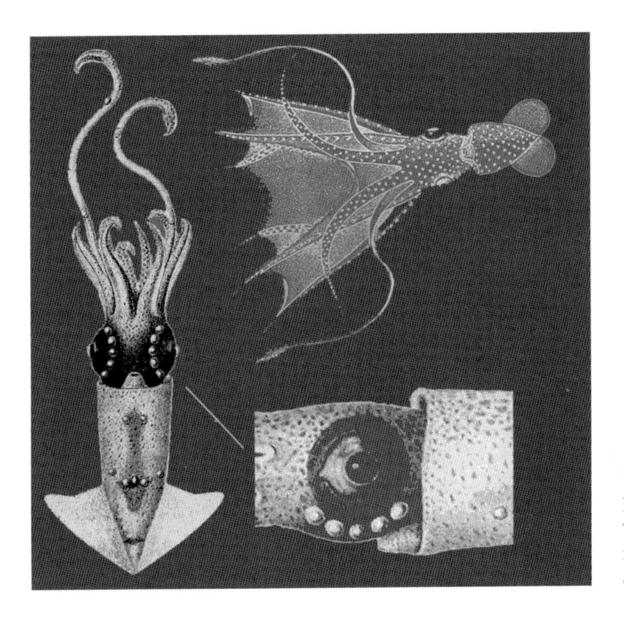

FIGURE 10.3. Left-right asymmetry in eyes and photophores in squid *(upper right)* suggests sideways swimming. Up-down asymmetry *(bottom)* is normal (that is, photophores pointing downward on both eyes).

addition, flashing back vigorously will momentarily disturb the dark-adapted seeing of the predator, which then has to rely on other senses (hearing, touch) to pursue the game.[17] Silver-mirror sides (in hatchet fishes) might serve a similar purpose: the attacker gets the flash right back into its eyes, while the target darts away.

All hunters also are potential prey. The small squid *Calliteuthis* nicely exemplifies the combination of upward search and downward defense typical for the mesopelagic and upper bathypelagic environment. One eye is small and has a ring of photophores, while the enlarged upward looking eye is free of photophores (fig. 10.3, upper right). We can imagine the animal swimming on its side looking up for prey against a dim sky, with its large eye, while using its smaller eye for camouflage and to flash confusion into the face of a predator attacking from below. Of course, we can only speculate about the habits of this little deep-water squid: it would be difficult to follow it around to observe its natural behavior.

With bioluminescence so central to the ecology of the deep, a few words are necessary here about this highly technical topic.[18] Basically, the light is produced by a chemical reaction. It is the oxidation of the main fuel (called luciferin) by means of a catalyst (called luciferase). Luciferin comes in many types of compounds, some specific to certain groups of organisms, some occurring in several phyla. Mostly, organisms generate their own luciferin, but some (among fishes, squids, and tunicates) host bacteria to produce the reaction—and the light—for them. The method of making light other than blue green involves the funneling of the energy from the luciferin reaction into a fluorescent compound. It is important to realize that bioluminescence is not just an adaptation to deep living: it is important in the ecology of plankton throughout the water column. At depth, though, it plays the more conspicuous role.

In the Stygian interior,[19] many animals have lost their ability to "see"—that is, form images using eyes—and make do with simplified light organs allowing them to sense the presence of bioluminescent light. Any light shone on them would be from a flashlight by a predator, most likely. Their rudimentary eyes allow them to kick in with an escape response, or a return flash, or both.

For the interior, two aspects of ecology are of paramount importance: the rarity of encounters, and the virtual absence of stimuli useful for sensory input. Obviously, there is here an enormous premium on being quiet, both when stalking and when escaping. Slow, wakeless and vibrationless motion is called for. Detection of sound and low-frequency pressure variations, in fishes, is by hearing and by highly developed lateral systems along the entire body, including the head. In deep-living forms, these senses are especially well developed. Gentle, undulating movements provide for propulsion. Some of the deep-living squids have very broad fins, presumably to glide through the water with a minimum of noise. Others have one pair of exceedingly long arms, analogous to the thin long feelers of deep-sea prawns and the extended appendages of various deep-sea fishes.

If there is a "silent world," as once announced by Jacques Cousteau when first diving just below the surface of the ocean,[20] this is it, deep down in the interior. We know very little about this type of world. Can a fish use its lateral line to "visualize" prey that is trying to escape? Or use its tail to set up vibrations that reflect from potential prey? Can some fishes shed a part of their very long body, to provide a wiggling detraction for a predator, as lizards do? Can a squid wiggle the end of a tentacle to lure the curious by vibration? All we can say is, with all these opportunities and adaptations concerning sound and vibration, there must be interesting interactions in this dark world.

Perhaps the best-appreciated adaptation for the vast silent world of the Stygian interior is the fact that many deep-sea predators have enormous jaws (compared to their small bodies) and extensible stomachs that allow them to swallow prey bigger than themselves (fig. 10.4). Who eats and who gets eaten in one of the rare encounters may depend on who has the bigger mouth. The winner may then work on digestion for weeks or even months, analogous to the feeding in snakes. Deep-sea squids with webs between their tentacles are hard to beat for size of gape: the webbing produces, functionally, a mouth several times larger than the body.

Finding a mate can be a problem in the thinly populated bathyal environment. Hermaphrodism, the ability to produce both eggs and sperm, is widespread. Every encounter between members of the same species is then potentially a male-female double encounter. In some fishes, with separate sexes, the male makes sure he stays with his female, once found. Juvenile males of bathypelagic anglerfish swim about till they find a female, then lock on to her with their jaws. The successful males become parasites on their female, with which they fuse permanently. Being very small, the parasitic male can be cared for with little discomfort to his mate. In turn, she has no problem getting her eggs fertilized, especially if she hosts several of the little freeloaders.

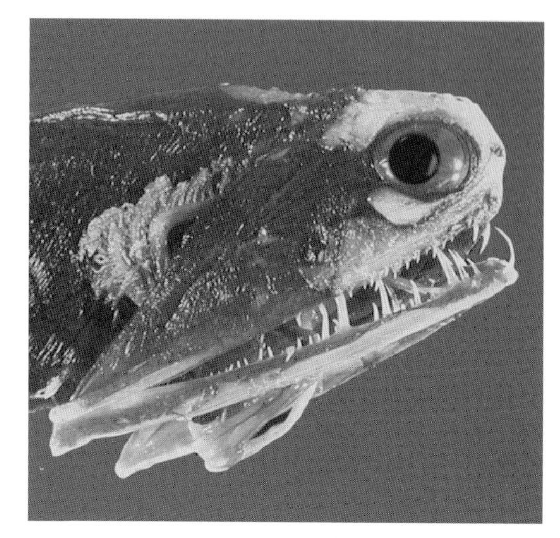

FIGURE 10.4. Extensible jaws and an expandable stomach help take advantage of rare encounters with larger prey.

How a male finds his target female in the first place is a mystery. Perhaps he recognizes the flashes on her lure. The large olfactory organs in some males suggest that their females release chemicals—sex pheromones—to attract them, much as female butterflies do on land. In any case, if the male makes a mistake in picking his partner, he probably will not get another chance. In fact, a female anglerfish may well lure males to flesh out the menu. They are living traps, with a huge mouth and with lures dangling from the jaws. Once the need for one or two males is filled, it would seem reasonable to eat any latecomers.[21]

LIFE IN THE STYGIAN DESERT

Most of the action in the ocean, as far as living things, is in the sunlit zone of the sea and on the seafloor in shallow waters. The reason is obvious: it is here that green organisms produce the food on which all others depend. As we go to greater depths, the abundance of organisms diminishes rapidly. Very soon, below a depth of half a kilometer or so, animals are sufficiently rare so that some nineteenth-century biologists, such as Edward Forbes, had reason to believe there were none at all.[22]

Forbes reasoned that algae need sunlight and disappear at some depth because of the attenuation of light, and he supposed that animals must do the same because they depend, ultimately, on the food from algae. While this describes the situation correctly to a first approximation, he neglected to consider that edible leftovers would find their way into the deep sea, where they would support at least some animal life. Thus, even on the abyssal seafloor several miles down, there is everywhere evidence for animal life, as was established beyond doubt by the *Challenger* Expedition in the 1870s.[23] Both the pelagic fauna at depth and the benthic fauna on the deep-sea floor are shaped by the lack of food. But the seafloor, as the place that collects whatever crumbs remain from the rich table of the sunlit zone, after transit through the water column, is privileged in comparison with the dark cold abyss above it. Interestingly, the scarcity of food does not impede evolution. On the contrary, the deep-sea environment has an extremely diverse fauna, at least on the continental slope, as was discovered in the 1960s.[24]

That food supply is the crucial factor at depths below the sunlit zone is well reflected in the fact that benthic animal abundance decreases away from continents, in marked fashion. Production is high in coastal waters, and the seafloor of the upper continental slope receives sufficient organic matter, therefore, to support a rich benthic fauna. Also, the rain of organic matter has but a short distance to the seafloor, on the upper continental margin. In contrast, at great depth, the overlying waters commonly have low production, and the path to the seafloor is long, that is, most of the settling organic matter is used in the water column. The result of the contrast between regions of high production with a short path to the seafloor and regions of low production with a long path is well appreciated by fishermen, who trawl the seafloor for fish around continents but not in the deep sea.

The pervasive lack of food that is the hallmark of the abyss favors adaptations that deal with saving energy when searching for sustenance. One answer to this need, through evolution, is the rise of gelatinous organisms. Representatives of this type of adaptation occur in many different phyla, from protists to chordates.[25] Gelatinous organisms are most abundant in the sunlit zone (like other types of plankton) but are also uniquely suited to exploit environments with low food supply. Their main asset is the fact that much of the body consists of water-rich jellylike material, which costs little energy to make or to maintain. Making jelly puts the organism in a larger size category in the cheapest way possible, and its larger size increases the potential for capturing prey and decreases the number of predators able to attack.

The mesopelagic and bathypelagic depth zones of the ocean—that is, the twilight zone and the completely dark zones—host especially large and delicate representatives of gelatinous organisms.[26] This fact emerged only in recent decades, with increased investigation using cameras and submersibles. The traditional methods of sampling animals living at depth—towing nets—results in the recovery of bits and pieces of slimy substance of uncertain origin. In fact, gelatinous zooplankton turns out to be quite important in the ecology of the pelagic realm.

The best-known jelly-makers are, of course, the jellyfish. These are medusae—relatives of the corals, the anthozoans—characterized by the possession of stinging cells, which they use to stun prey. The typical medusa has a flexible umbrella, which serves for propulsion as well as a scaffold to bear the organs of digestion and reproduction, and the tentacles. In the sunlit zone, medusae are transparent, to stay inconspicuous to both prey and predators. But in the dark, at depth, this is not necessary. As a consequence, deep-living medusae are commonly quite colorful. Also belonging to the stinging kind are the siphonophores, animals that might be considered as medusa colonies, but where each individual has a specialized task: propulsion, digestion, or reproduction. Their size range is astonishing: from the length of an ant to the height of a tall cedar. The most familiar of the siphonophores is the Portuguese man-of-war, which has a float at the surface from which long tentacles with vicious stinging cells hang deep into the water, trapping fishes and squids. Some siphonophores, denizens of the mesopelagic, undertake large vertical migration every day, coming up at night and disappearing into the twilight zone during the day. Functionally somewhat similar to medusae, but not closely related, are the comb jellies (ctenophores), predators that use paired tentacles with sticky cells, rather than stinging cells. Some specialize in eating medusae and also other comb jellies.

Mollusks have many representatives in the gelatinous plankton, including both gastropods and cephalopods. One might not be surprised to have slow-moving representatives of gastropods. But it is surprising, perhaps, that there are gelatinous squids—one would usually associate a highly active life-style with these dynamic predators. Deep-living holothurians presumably have extremely low energy requirements to begin with, gelatinous or not. But in this group also there are some gelatinous forms, among swimming or drifting forms. Commonly such species stay quite close to the bottom, where the particle content of the water is high enough to warrant filtering it for food. Tunicates include the sea squirts, abundant on the seafloor in shallow waters. Pelagic forms, all gelatinous, display great diversity and range from tiny barrel-shaped doliolids to 20-meter-long colonies in one species of pyrosome. In salps, also, chains can be several meters long.[27]

The twilight zone (from below the sunlit zone to around 900 meters depth) hosts many organisms that take part in a daily vertical migration, as mentioned. That is, the migrating animals come up to feed and go down to hide. The fact that the hiding place is only dimly lit (discouraging visual predators) and also very cold is of some advantage. Fishes that hunt in warm waters are reluctant to dive below the thermocline (the boundary zone between warm and cold water).[28] Fishes of the twilight zone (that is, midwater, or mesopelagic, fishes) are prominent participants in the daily migration between the upper food-rich zone (epipelagic) and the thinly populated mesopelagic resting place. The mesopelagic fishes belong to some 30 families with more than 100 genera.[29] Most familiar are the lanternfishes (family Myctophidae) and the hatchet fishes (family Sternoptychidae). They have in common rather small size, large eyes, dark backs and silvery sides, and light-emitting organs on various parts of the body, especially the belly. These organs, as mentioned, are called photophores. Also, many other mesopelagic fishes have these adaptations (fig. 10.5). The photophores on the underside are designed to emit light just strong enough to match the background light in the eye of an upward-looking predator. This works, during the day, from about 400 meters on down, where sunlight is sufficiently attenuated that it can be matched by bioluminescence. A disproportionately large mouth with small teeth, to grasp large prey, is another adaptation common to many mesopelagic fishes.

The diversity of mesopelagic fishes is remarkable, as mentioned, especially in the myctophids (lanternfishes). For a number of years Robert L. Wisner, a fish specialist at Scripps Institution working with Carl Hubbs, studied the taxonomy of this group. It is highly diverse, with at least 32 genera and more than 200 species. Once, in mock despair, he was heard to remark, "I think this group is evolving faster than I can get them identified." We suspect that the proliferation of species may have something to do with the signaling between the sexes, using easily modified light patterns. Except for those points of light on the body of the fishes, they look much alike (fig. 10.6). Why should fishes within a given population have an interest in changing the light patterns? Because it is a password code, and confusion must be avoided.

FIGURE 10.5. Various small midwater fishes, including one hatchet fish *(top, center)*. Some are clearly adapted to twilight conditions (large eyes, efficient propulsion, countershading). Others are eel-like stealth-hunters with extensible jaws, suggesting rarity of prey.

The species living at the shallower depths are the better swimmers; they need to be agile because they face visual predation. Their larvae grow up in the sunlit zone; they are transparent. The deeper-living forms tend to have less gas in their swim bladders, to the point where, at great depth, swim bladders may be lost entirely. Gas, presumably, is useful in vertical migration, provided the swim-bladder content can be regulated for upward and downward swimming. It may also be important in trapping sound, thus improving auditory sensing. However, gas-filled bladders also make fish more "visible" to echo-hunting dolphins.

Among the mollusks, squid are prominent members of the mesopelagic and bathypelagic fauna (fig. 10.3). In fact, they are the mesopelagic predators par excellence: more squids are below the thermocline than above it. We have mentioned jelly squid, deep-living, slow-moving forms. But the twilight zone is rich in fast-moving species, which also take part in vertical migration. They are agile predators, moving forward and backward with equal ease. At night on a research vessel, when on station, squid can be seen hunting for small fish in the light cone of a shipboard lantern. Their tentacles form a forward net, extended at the end of a sudden forward

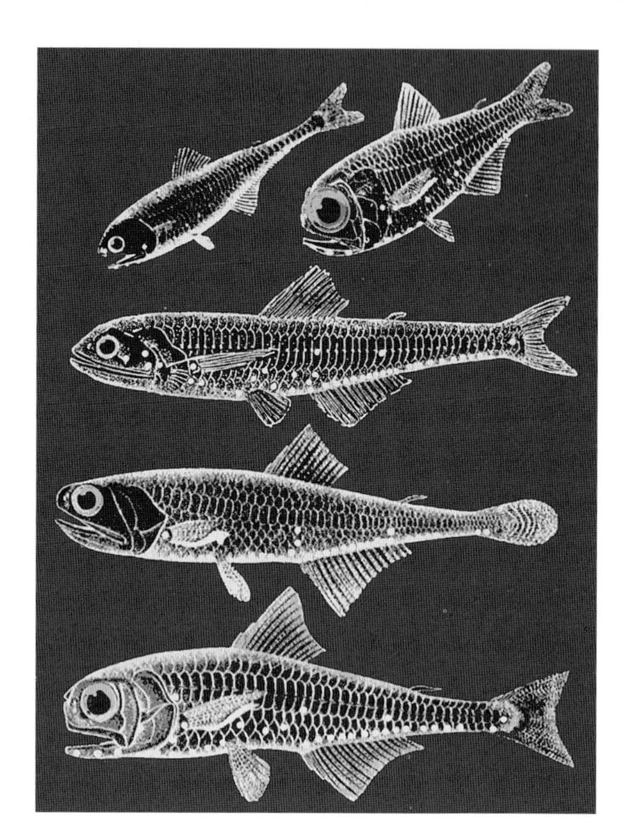

FIGURE 10.6. Lanternfishes are look-alikes except for the photophore patterns. Presumably the patterns are important in signaling between members of the same population.

dash. Attack from larger predators triggers a powerful escape response—a jet blast from the mantle tube moving the squid backward at high speed. In addition, a squid can eject ink, to confuse the predator as to its whereabouts. In deep-living forms, the ink is luminous. Also, many mesopelagic species have light-emitting organs, just like the fishes with which they share the deep and dark environment.

The evolution of light-emitting organs, undoubtedly, proceeded independently in animals with an entirely different pedigree, such as fishes, squids, and shrimps. Thus, tremendous advantage must have accrued to organisms that developed bioluminescence. Clearly, the adaptation must be useful in food procurement, escape from predators, and reproduction. For obvious reasons, details of the role of bioluminescence in controlling the behavior of animals in the twilight zone have remained largely obscure.

Little is known about the life history of deep ocean squid—most of what we know is from observations in the coastal ocean. We do know that sperm whale stomachs can contain thousands of squid beaks, and that squids are a favorite prey of smaller toothed whales, as well. Consumption by whales alone is estimated as well in excess of 100 million tons,[30] comparable to the entire fish landings by human predation. The human take of squid is reported as around 3 million tons per year.

A host of other animals participate in the complex ecosystem that is the mesopelagic realm, and that merges into the enormous bathypelagic zone. All the major marine phyla are represented. As in the gelatinous zooplankton, strange forms characterized by unusual size (large worms) or by striking pigmentation (red prawns) or by interesting applications of light-emitting organs (as in the lures of angler fish) invite speculation about the special adaptations that evolved for life in the deep and dark ocean. The deep ocean, the most common life environment on the planet, remains the least known.

Much has been learned from deep-sea photography starting in the 1950s and from deep-diving vehicles such as Woods Hole's *Alvin*, during the last several decades. The seafloor turned out to be a highly diverse environment, with many opportunities for a highly diverse fauna. The biggest surprise, which emerged in the late 1970s, was the existence of a unique fauna flourishing around hot springs, which exist wherever there is active volcanism on the seafloor.[31]

The fact that attention is drawn so irresistibly to the drama of the hot vents and their fauna reflects the circumstance that elsewhere the action is much less spectacular. There simply is not much going on that can be photographed (unless one provides bait, as John Isaacs discovered). The reason is the general scarcity of food in the deep ocean. At the vents, specialized bacteria make a living from oxidizing sulfur in the hot water emanating from active volcanic springs. They provide food for animals, partly as symbionts in worms and mollusks, partly as bacterial mats than can be grazed. If we were to look on land for an analog of deep-sea hot vents, we might consider localized artesian springs in a desert: a rare but interesting ecosystem.

FIGURE 10.7. Benthic foraminifers: pinhead-size shelled microbes on the seafloor, indicators of depth zones.

THE RULES OF DEPTH ZONATION

As one moves upward on a high mountain in low latitudes, from sea level, one crosses many zones of entirely different plants and animals, until one finally reaches the snow line. If we could dive down along the seafloor in analogous fashion, we would see something similar. In Hawaii or Florida, we would start with coral reefs thriving in sunlit warm waters and end up in the lightless, icy cold abyss.

Just as on land when crossing elevations, temperature changes with depth in the sea. The warmest zone is at sea level. At the same time, with increasing depth of water and increasing distance from shore, abundances of large organisms on the seafloor decrease. By and by, we see fewer fishes, fewer crabs and lobsters, fewer clams and snails, fewer sea urchins.[32] On land, the zonation by elevation is clearly a function of temperature (and also of seasonal temperature extremes) and of precipitation. Plants need a growing season and they need water. In the ocean, depth zonation is a function of the penetration of light suitable for photosynthesis, the depth of the thermocline, and the decreasing food supply with depth and with distance from the coast.[33]

Even before these factors were clearly recognized, the reality of depth zonation was appreciated.[34] In the 1930s and 1940s, geologists discovered the depth zonation of benthic foraminifers (shell-bearing protists that are ubiquitous on the seafloor; fig. 10.7), showing that different species have different depth ranges.[35] The geologists were especially interested in the matter because foraminifers are practically the only way to determine the depths of deposition of older marine sediments, a knowledge that is relevant in oil exploration. It turned out that the zoning is quite distinct in shallow water but becomes very fuzzy with depth, below the shelf break. Given that all sorts of environmental depth gradients are strongest in shallow water, this is hardly surprising.

In the 1950s, a nomenclature was proposed that became the basis for subsequent depth zonation schemes.[36] More or less, these schemes are arbitrary, since there are no sharp boundaries. However, it is useful, in going from

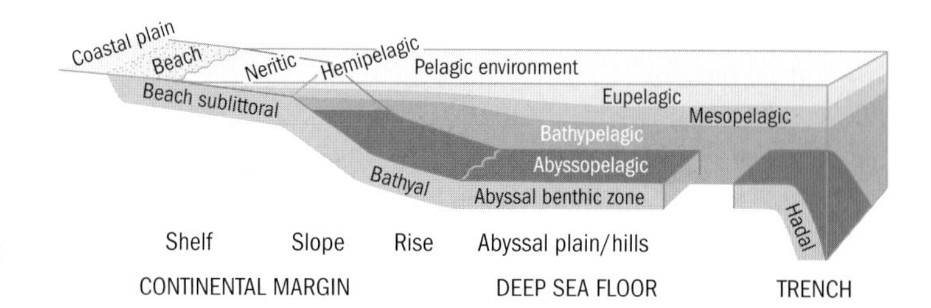

FIGURE 10.8. Common names of oceanic environments, with respect to depth and distance from land.

the littoral to the abyss, to recognize the oceanographic categories *mixed layer, thermocline, intermediate water,* and *deep water;* and the geologic categories *inner shelf, outer shelf, upper slope, lower slope,* and *abyssal.* Ecologists use *euphotic*" (meaning "well-lit") for the productive zone and the terms *epipelagic* (upper layer), *mesopelagic* (within the main thermocline to about 1,000 meters depth), and *bathypelagic* (everything deeper than 1,000 meters).[37] Other common terms are *bathyal* (deep bottom), *abyssal* (very deep bottom), and *hadal* (trench depths).[38] *Abyssopelagic* and *hadopelagic* are sometimes used when talking about the very deep environments[39] (fig. 10.8).

Depth zones based on light, on temperature, and on food are not congruent, of course. But their interactions are of considerable interest. Thus, a comparison between mixed-layer thickness and depth of the sunlit zone is at the heart of the compensation depth concept of Sverdrup.[40] Likewise, a comparison of the depth of the thermocline and the depth of the shelf break determines how easily nutrient-rich waters move across the shelf toward the coast, where upwelling is concentrated. Clearly, when the thermocline is distinctly deeper than the shelf break (when the warm water layer is thick), upwelling will not pull in nutrient-rich waters. In such situations upwelling becomes ineffective as a source of nutrients, and coastal productivity suffers.[41]

We can readily understand depth zonation, when taking food supply, light, and temperature as the controlling factors. However, one more factor needs to be considered, that is, turbulence. The reason that turbulence is important is that it keeps organic-rich debris in suspension. Suspended material is easily transported away by currents. Thus, turbulence tends to help remove organic matter from the inner shelf toward the outer shelf and upper slope. As a consequence, there is less benthic production on the shelf than could be, and more on the upper slope. In places where strong currents move along the upper slope, the organic matter keeps moving even farther down the slope, feeding feather stars and sea pens on the lower slope at the expense of the animals on the upper slope.

Just how much food is coming into deep waters and onto the seafloor? This question has been the topic of intensive studies for more than three decades. We shall return to this subject and highlight the important discoveries. First, let us now turn to the next important factor, light. What do we mean by sunlit zone, or euphotic zone? How deep does sunlight penetrate into the sea? What are the consequences of the progressive dimming of the light with depth?

A convenient way to think of the sunlit or euphotic zone is as the layer where there is a net contribution to primary production throughout. It does not mean that there is no light below this zone; there clearly is. However, the intensity is insufficient to balance the surrounding respiration, and there is net uptake rather than production of oxygen. The thickness of the euphotic zone very much depends on the clarity of the water. Sunlight penetrating the water column downward is both absorbed and scattered, and these two processes together determine how fast the light dims downward. Particles in the water greatly contribute to the absorption and to the scatter (particles are dark but can be seen to glitter from the side). Thus, the more particles there are in the water, the thinner is the euphotic zone. Commonly, the particles are photosynthesizing organisms; the more abundant the algae, the less light remains for those deeper down.

Among the first systematic measurements of the transmission of sunlight into deeper waters were those of N. G. Jerlov, a member of the Swedish Deep-Sea Expedition (1947–1949). His report on the expedition results concerning light transmission in the sea, and later his book on the subject, entitled *Marine Optics*, became the standard reference in the field.[42] Jerlov, among others, showed that water affects light in various ways, depending on the wavelength of the light. Long wavelengths (red, yellow) and short ones (violet) are much more readily attenuated during transit in seawater than blue light. Thus, as any diver knows, blue green light is what remains even just beyond a few tens of meters down. If one would capture colorful scenes at depth, a flash must be used for photography; otherwise everything looks bluish gray.

Within clear water, the intensity of the light dims by one-half about every 25 meters. After 250 meters, then, the light is down to one-thousandth of its original intensity, and after another 250 meters, to one-millionth. At that point, the remaining sunlight is roughly equivalent to that of a moonlit night in shallow water.[43] Because of the rapid attenuation of sunlight with depth, midwater organisms "know" where they are by the strength of the faint glow of the "sky," looking up. They know when it is time to come up to feed, or go down to hide and rest, simply by keeping the skyward light intensity at the same level, by vertical migration. Adjusting their level to surrounding light intensity is a good way to make sure they are at the "right" depth. Indeed, the sensitivity of vertically migrating organisms to changing intensity of skylight can be observed during echo profiling at high sound frequencies in productive waters, as a change of the depth in the midwater scattering layer in response to clouds passing over the location.[44]

There are good reasons to stay at the "right" background level of illumination, and thus use light intensity as a guide for vertical migration. These reasons have to do with the relative intensity of background lighting and bioluminescence. As bioluminescence expert P. J. Herring of the Southampton Oceanography Center points out, the camouflage requirements against flashlight-hunters are fundamentally different from those appropriate to a dimly lit scene where predators look for silhouettes.[45] Presumably, once a fish or squid species has made a commitment to solving the problem of hiding where there is no hiding place, the particular solution adopted works best at a certain level of light, under given circumstances of predation pressure. This light level becomes the optimum sought out, with the additional proviso that migrating takes energy and should be minimized. If, like in the hatchet fishes, the solution for hiding is to make an ultrathin body, the payoff is in staying in dimly lit waters, where the silhouette disappears for the view from below. In addition, since the shape allows for straight sides, it becomes advantageous to make those sides mirrorlike, to reflect any flashlight attack into the eyes of the predator. This works in dark waters. The combined payoff from the two adaptations sets a "most-safe" light level at the boundary between dimly lit and dark.

Much of the deep plankton is typically red, bright orange, or black, colorations that are absent from the sunlit zone. It makes sense to be black in an environment of near-total dark, if one wants to remain unseen. Red and orange have the same effect; these colors are neither in the background light nor in the bioluminescent light, thus, they appear black. The copepods, ostracods, and shrimplike creatures of the deep more often have red pigment rather than black, perhaps because of their body chemistry. Fishes prefer to make black pigments. The result is the same.

We may ask why not more creatures are transparent. This should work just as well in dim light and against detection from bioluminous flashes as it does with the plain sunlight in the epipelagic zone. Actually, some transparency does occur, especially among the gelatinous plankton in the upper reaches of the midwater zone. Clearly, there must be some cost in relying on being transparent; presumably a loss of agility is one of these, with no iron in the blood to feed muscles. Also, it is actually quite difficult to make a body transparent: the index of refraction of light has to be exactly the same in the body and in the water. This equality is never quite achieved. Thus, a transparent organism that is backlit from above can be seen in outline, a fact much used in plankton photography. In addition, whenever there is food in the stomach, an organism loses any transparency it had.

Naturally, the typical fishes of the midwater zone, the lanternfishes and the hatchet fishes, have specially adapted eyes that can see at very low light intensities. One such adaptation is a reflector behind the eye's retina, much like those in cat, dog, and deer eyes. A great number of crustaceans also have this reflector device, which must have developed entirely independently. These highly specialized eyes, one assumes, are not only used to process the exceedingly attenuated sunlight that arrives at a depth of several hundred meters. As we have seen, the world these creatures live in is one of moving points of light and random flashes with potent meaning in terms of encounters with friend and foe, and food. Bioluminescence is a vital part of their visual environment.

We must assume that in most deep-living animals other senses besides eyes are highly developed, for detecting prey, or predator, or potential mate. Sound and vibrations, chiefly, carry messages, and we have evidence (e.g., the lateral lines on fishes) that many deep-living creatures are highly sensitive to their acoustic environment. Other senses (touch, electric field perception) are employed as well. Each organism, in essence, lives in a different world, depending on which senses it depends on for adjusting its behavior to the environment. At great depth, where even bioluminescence wanes, the sensing of sound and related pressure waves must become increasingly dominant over the sensing of light.

We have briefly touched on the two major controls of depth zonation, food supply and light. But what about temperature? Clearly, it is important also, especially in providing a major barrier for those organisms that prefer to stay warm or cold, rather than coping with two entirely different temperature regimes. The main temperature gradient happens to coincide, over large areas, with the depth where sunlight becomes insufficient to support net production. This is only in part a coincidence: while one can readily envisage a situation where the thermocline is weaker and deeper, it is difficult to move the thermocline closer to the surface. The reason is that the existence of a strong thermocline implies a strong planetary temperature gradient, which in turn calls for winds strong enough to keep the thermocline from rising further, by vigorous mixing of surface waters. In the present ocean, the thermocline is about as shallow as it can get. In the distant past some 30 million years ago, when it was weaker and deeper, the mixed layer did not coincide with the sunlit zone, and the open ocean was less productive and supported much less life at depth.

THE UPS AND DOWNS OF DEEP-SEA LIFE

We have seen that vertical migration is an integral part of life in the mesopelagic zone. Many, perhaps the overwhelming majority of the animals in the midwater zone, take part in a daily routine of going up when the lights get dimmer, going down when it gets brighter.

Also, within the life history of deep-living animals (including those living on or near the bottom at intermediate depth) there is a good chance that their larvae and young live much higher up than the adults; larvae are small and depend on high prey density, they cannot wait long for food or go search for it. The juveniles of many bottom-feeding deep-sea fishes on the continental slope live at shallower depths than the adults.[46]

Let us briefly look at some of the more important aspects of daily vertical migration and of life-cycle vertical migration, which in effect merge the ecosystems of the sunlit zone with those in the twilight of the mesopelagic realm. Looking at this system from afar, we can appreciate that the sunlit zone is the place of maximum activity, and that the mesopelagic realm has grazers and predators that take a toll at night. While the toll may not be large compared with the internal cycling of the productive zone, there are bound to be ramifications for the carbon cycle, for life at greater depths, and for the evolution of organisms that never leave the sunlit zone.

We already appreciate that the diurnal vertical migration is related to feeding strategies in several ways. First, to be seen is to be eaten, for small creatures. Thus, it is a viable strategy to stay out of sight during the day and to come up at night only, for feeding. It is true that most of the zooplankton is transparent and is hard to see even in plain daylight. However, like H. G. Well's Invisible Man upon eating cheese, the zooplankton becomes visible when feeding on phytoplankton. There is some advantage in digesting the food in the dark, below the sunlit layer.[47] Second, during upward migration, there is a chance to hit upon a water layer enriched in phytoplankton, perhaps a diatom meadow in the very top of the thermocline in the zone of the chlorophyll maximum. Here, suspended matter (including phytoplankton) tends to congregate. An upward migrating copepod could stop at such a meadow and start filtering the water for food.[48]

Another reason for vertical migration has to do with getting around in the vast ocean, as a small organism. By migrating to various depth levels, an animal has a chance to travel to unexplored regions, or roughly back to where it came from, because deep currents and shallow currents move in different directions, which may be opposite in fact. Thus, by going down and coming up later, a migrating copepod has changed its scenery. The best places to feed are those with lots of phytoplankton in the sunlit zone. For a copepod drifting along with the subsurface current, a good place to prospect for food would look dark from below. Noting a dark sky, the copepod would automatically start to migrate up. Thus its chance of finding a "good" place would be improved. We know that sudden darkness does indeed trigger upward migration, regardless of the time of day.[49] This has been demonstrated nicely by showing the response of migrating zooplankton to an eclipse.[50]

Grazers hiding at depth, then, have a strategy analogous to that of deer who leave the edge of the forest to graze at dusk, when they are difficult to see. In the sea, the migrating grazers feed on phytoplankton while avoiding predators. Predators have a similar strategy, but in addition they use the night for stalking the unwary. They do not need much light to find their prey—a bit of moonlight will do fine. Predators of the mesopelagic zone are adapted to life in the dark. Like owls at the edge of the forest, the night-hunters come out after the Sun sets to feed on smaller grazers and predators in the moonlit waters till early in the morning. Coming out of the dark, these hunters are experts in stealthy ambush, guided by superior vision and detection of motion.

The daily migration of mesopelagic predators was first discovered as a result of high-frequency echo sounding developed during World War II for detecting submarines. The deep scattering layer is seen on the echo recorder during the day, between 250 and 800 meters. It tends to be diffuse, indicating that the sound scatterers are spaced through some considerable depth interval. Quite commonly there are several distinct layers. At sunset, the deep layers begin to rise and merge into a thick and strongly reflecting unit that comes right up into the epipelagic zone. At daybreak, the sound scatterers descend again to their original depths.

It has proved difficult to catalog the organisms responsible for the sound scattering. Prime candidates are those that have gas in their bodies, notably fish with swim bladders (mainly among the lanternfishes). Sound travels much more slowly in gas than in water. Thus, a layer of water where swim bladders are abundant is "seen" acoustically as a large flat object with a lowered inherent velocity, as the sound energy is reflected and scattered from such an object. Siphonophores, with their gas floats, scatter sound very efficiently. Their importance in this context was realized only after directly observing the deep scattering layer from diving vehicles, because, as mentioned, net tows damage these delicate gelatinous forms. Thus, the sampling of the scattering layer using the "midwater trawl" developed at Scripps by John D. Isaacs and Lewis Kidd, in 1950, while resulting in the discovery of many of the scattering organisms, had this drawback.[51]

Research since the discovery of the deep scattering layers (discussed in the next chapter) revealed that many different types of organisms scatter sound in many different ways.[52] In addition to gas-filled swim bladders of bony fishes, sound is scattered by bladders filled with organic substances. Many fishes have no swim bladder at all. Sharks have large livers, which differ from water in their acoustic properties, and therefore also scatter sound. Among the zooplankton, there are those that bear gas in their bodies, such as siphonophores, and others with hard shells, such as pteropods. Yet others, for example the euphausiid shrimps, have neither gas nor hard shells but contain fluids whose properties differ from water sufficiently to present an obstacle to sound transmission.

From an ecologic and evolutionary standpoint, the question is how the fact that so many creatures undergo vertical migration affects the general ecology of the ocean, and how the creatures cope with the fact that they are "visible" in a noisy sea, especially when being pinged at by echo-hunters.

Of special interest in this context are the deep-sea lanternfishes, the myctophids, which are among the most abundant fishes on the planet. Generally about 3 to 4 inches long, they have bioluminescent photophores on the underside, in patterns distinct for each species, as mentioned. At night lanternfishes come to the surface to feed and, in turn, become targets for fast-moving squid.

Observations from submersibles off San Diego, in the 1950s, had suggested that lanternfishes are important participants in the migrating deep scattering layers. Some 10 years later the role of this kind of fish emerged during *Alvin* dives, with Woods Hole biologists Richard Backus and Jim Craddock using *Alvin's* sonar to investigate the layer. As ocean explorer Robert Ballard tells it:

> With the lights turned off they homed in on a good-sized blob on the scope. When they turned on the lights, they were surrounded by thousands and thousands of lanternfish. The fish were pointed and moving in all directions—"a fantastic aggregation not a school." The biologists had thought that photophores, which were usually on the underside of the fish, were meant to shine down and blind predators. But the photophores of these lanternfish were aimed in every conceivable direction.[53]

We can guess at aspects of the life history of some of these fishes from simple principles. By hiding at depth during the day, they not only avoid visual predation, but also they become less "visible" in an acoustic search beam, because their swim bladder shrinks as pressure increases.[54] At some point, even a 100-kilohertz sound beam (at the upper end of a dolphin's hearing) has waves too long (one-half inch) to produce much of an echo from the shrinking bladder of a fish. Thus, from the point of view of a dolphin, the best time to hunt such a fish is at dusk or dawn, when it is out of the "safe" depth and becomes acoustically visible, but not yet immersed in all the other scatterers in surface waters, where it is difficult to separate one from the other. From the point of view of the fish, with its well-adapted eyes, the best time to hunt is when it is not too dark to find prey, but yet too dark to be seen well by the visually hunting squid. There are many more squid than dolphins, so that avoiding squid presumably has priority in influencing the evolution of behavior.

Myctophids (that is, lanternfishes) are just one of the families making up a large community of midwater fishes, which have their daytime depths somewhere between 200 and 900 meters and live in great abundance below the more productive regions of the sea. The migrating species among these mesopelagic fishes are the more agile swimmers; on the whole they tend to live at the shallower depths in the overall range.[55] A number of these fishes exhibit microscopic daily increments in the growth of their

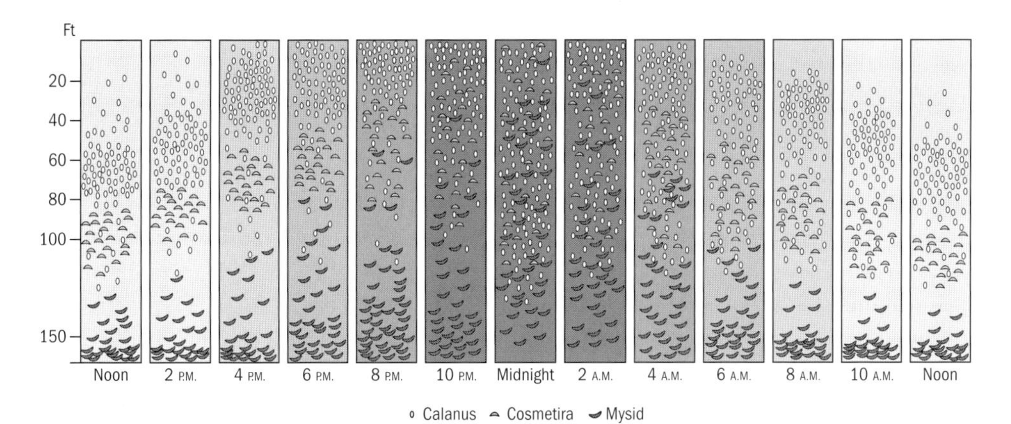

FIGURE 10.9. Ups and downs of different types of zooplankton somewhere off the shores of Britain, as reported in Russell and Yonge (1936). Depth is in feet: this is the uppermost portion of daily vertical migration.

otoliths, suggesting a daily ration of food, such as would result from migration. The lanternfishes and their midwater colleagues are exceedingly difficult to keep alive in aquaria and so do not make good experimental subjects.[56] In the sea, members of the same species may be vertically segregated (the young ones tend to live at shallower depths), and the same is true for the sexes. In this fashion, a species can hedge the various risks arising.[57] Presumably, the bigger animals find it less difficult to migrate through a large depth range than the smaller ones.

What is ultimately the effect of these daily migrations on the pelagic environment? Vertical migration helps to transfer food from the sunlit layer of algal production downward to greater depths. Thus, the pelagic living space is greatly increased, and a large web of predator-prey interactions—a vertically tiered food web—is established. Different species can employ different migration strategies: full or partial migration, early or tardy ascent or descent, or even countermigration (up during the day). Such variations in migration patterns would translate into variations in feeding and escape strategies. By doing things a little differently, competing species can coevolve into different niches. In this manner, vertical migration contributes both to living space and to diversity. Migration is one way the seemingly uniform environment of the open ocean is made highly complex (fig. 10.9).

Of course, the general effect of vertical migration—increase of living space and diversity—must not be seen as a direct cause for migration. Individual species have no grand strategy of increasing diversity: they simply adapt to survive and to thrive. By exerting control over their environment, through migration, they automatically increase opportunities for different life-styles. One type of explanation for vertical migration that has been put forward can be readily dismissed: the observation that animals follow certain light intensities. Animals do not migrate *because* they follow light intensity, or pressure levels, or density levels, or whatever. Light, pressure, and water density are the cues by which animals find their way. They are programmed to follow such cues because it means more food, fewer predators, finding a mate, or a good place to put eggs.

The distinction between cause and cue may seem rather subtle. It spells the difference, however, between real understanding based on evolution and mere description of animal behavior.

EXPORT PRODUCTION: FOOD FOR THE ABYSS

We have seen that vertical migration results in downward transfer of organic matter. The copepods that graze in the sunlit zone and void below the thermocline, or become food while in their resting place, have effectively transferred edible carbon to the next deeper layer. The night-hunters that take their gut contents back to mid-water depths likewise have transferred carbon downward. Bathypelagic predators feeding on myctophids and hatchet fishes continue the transfer to depth.[58]

Just how important this transfer may be is difficult to estimate. The biomass of squid below the sunlit zone has been estimated at 300 million tons.[59] Doubling this value for fishes and doubling twice again for other types of organisms, we arrive at a value of 2.4 billion tons of deep-sea predators. We assume that less than one-third of this mass is carbon, which gives us an upper limit for the amount of solid living carbon: a billion tons. Assuming that this mass has to be replaced once a year, we get an upper limit for the transfer of biocarbon necessary to support the deep-living predators—a billion tons per year. This number, of course, is embarrassingly uncertain, but useful as a ballpark estimate. The point is that the transfer of carbon by the settling of biodetritus is almost 10 times greater, as far as this can be determined.[60] Thus, the "active" transfer by predation apparently is quite modest compared to the "passive" transfer by general settling.

Much of the transfer of biocarbon from the sunlit zone to depths takes place in the productive regions of the sea, whether of the active or passive kind. Thus, most of the transfer feeds life in the dark below the coastal ocean, especially also the benthic ecosystems there. Benthic organisms, therefore, are abundant around the continents, where the supply of organic carbon from overlying waters is high. In addition, there are large differences in the structure of benthic communities, depending on the overall supply and the seasonal pulsing of the supply of food to the bottom.

Where the supply is high enough to support large mobile animals on and in the mud, the bottom sediments are stirred and mixed through a depth of several inches, readily re-releasing carbon and nutrients to the waters in contact with the seafloor, for recycling within the ocean's production machine. Intense mixing, of course, is unfavorable for sessile organisms that prefer a stable floor. Where supply of food is greatly reduced (below the deserts of the sea) small organisms predominate, and mixing is slow and inefficient, involving but the uppermost inch of sediment. Foraminifers, shelled protozoans, are an important component of this type of benthic community in the deep deserts of the sea.

The export flux reaching intermediate waters is much greater than that reaching the deep waters of the ocean, as has been ascertained by trapping experiments.[61] We can readily make a rough estimate of the organic fallout into the upper half of the water column, that is, the upper 2 kilometers, where this fallout is almost all oxidized by bacteria. From radiocarbon measurements we know that the typical residence time of this layer is around 500 years.[62] From this renewal time and the content of phosphate, we get a rate of upwelling of phosphate, and hence a fixation rate for the carbon, assuming that one phosphate atom is needed to fix a hundred carbon atoms, the so-called Redfield ratio.[63] We get an export of 3 billion tons of carbon by this route, about 10 percent of the standard radiocarbon productivity of the sunlit layer.[64] This is the amount of carbon flux that is needed to keep the upper half of the water column at steady state, under the assumptions stated. The actual export (say, at 100 meters) has to be considerably greater than this value (at least twice as much), because so much of the downward flux is intercepted at very shallow depths and is not available in deeper waters.[65]

The type of back-of-the-envelope calculation that allows us to make a rough guess of fluxes of carbon to the midwater environment is widely employed in standard geochemical box models, with additional complexity. One of the early estimates of export production was by Hank Postma, a Dutch oceanographer and chemist who pioneered the method. His multibox model gave a value for the total export from the sunlit zone of approximately 6 billion tons of carbon per year.[66] The estimate is not much different from subsequent ones, achieved by similar box models constrained by better data.

However, it is now considered to be on the low side of the probable range.

In principle, what is exported from the surface layer can be caught in traps moored at depth. In this manner, the export production from the sunlit zone has been studied intensely since the early 1970s. Among the pioneers in this field are W. G. Deuser and Susumu Honjo at Woods Hole, Bernd Zeitzschel in Kiel, the late John Martin and his colleagues at Moss Landing on the Monterey Bay, and the Bremen trapping group led by Gerold Wefer.[67] These experiments established that output from the productive zone varies greatly, in terms of types of particles, of chemistry, and of amounts falling downward, depending on region, season, and year of observation. For biogeochemistry, the proportions of organic matter to opal and to carbonate are of central interest. Because of the considerable variation in these ratios, typical values are difficult to extract from the data.

Export production may be seen as a tax on the output of the sunlit zone. The tax is progressive: the more productive areas deliver proportionally more to deeper waters.[68] This is another crucial factor (besides high production itself) behind the great contrast in the output from the coastal ocean versus that from the open sea, and the reason why there is so much more life associated with the seafloor on continental margins than in the deep sea. The origin of the oxygen minimum as a coastal-ocean phenomenon is closely tied to this pattern, as well. The rapid decline of the amount of falling organic matter with depth, owing to the activity of bacteria, nicely accounts for the presence of the oxygen minimum in the thermocline.[69]

In a workshop on ocean productivity in Berlin, in 1988, marine biologists, geochemists, and geologists gathered "to assess the state of knowledge and ignorance concerning the processes which lead to the export of organic matter from the photic zone, its transit to the seafloor, and its burial within the sedimentary record."[70] The conveners started with the expectation that the problems of export production would somehow show analogies to economic activity.[71] Instead, they ended up being surprised by the importance of "unscheduled" events, much as the pioneers of biological oceanography were puzzled by the spring bloom:

> One outstanding problem which arose during discussions was this: can the record [on the seafloor] be regarded as the remainder of export from the euphotic factory, diminished by road tolls on the way down and by import duties at the seafloor? Or should we adopt an entirely different analogy: that of trash left over from (seasonal) Oktoberfest events and from (sporadic) rock festivals? If the latter analogy is better, what chance is there to reconstruct the typical or average productivity (as seen on global maps) from the sediment which largely contains the memories of happenings?

And: "The factory analogy guided our planning while the event analogy emerged as a unifying theme of the conference."[72]

A brief survey of the field at the time suggested a strong seasonality in the export, for all regions studied.[73] This came as no surprise—plankton growth is seasonal, and the plankton supplies the material for export, as well as the means of transport, in the shape of aggregates and fecal pellets. In the longest such data set, from the Sargasso Sea (a result of the efforts of Werner Deuser and colleagues at Woods Hole) the variations of the amounts of carbonate, opal, and organic matter arriving in the traps is closely linked to the annual progression of productivity, dominated by the spring bloom.[74] Elsewhere also, blooms of diatoms, dinoflagellates, and coccolithophorids are crucially important in producing materials for export. Depending on circumstances, admixture of detritus from rivers and shelf sediments can be a significant component of the rain to the seafloor.[75]

When setting traps—large conical containers with the opening looking upward—one attaches them at various depths along a mooring (a line with an anchor at the bottom and a float at the top). Such vertical series typically show the greatest amount trapped in the upper traps and much less material in the lower ones. In the uppermost kilometers (that is, within the mesopelagic realm), the organic flux roughly decreases as the inverse of depth (that is, a doubling of the depth yields half as much flux), as a result of bacterial decay.[76] At greater depths, the more resistant organic carbon remains, and the rate of loss slows. Somewhere near 0.3 grams of carbon per year are delivered to the seafloor in the bathyal realm, from each square meter of a typical open ocean production zone.

Traps set out in the Southern Ocean have demonstrated that all around the Antarctic there is an enormous flux of diatom debris downward.[77] In fact, it seems likely that so much silicate is being precipitated in the Antarctic that it affects diatom production all over the world, starving the diatoms of their most vital nutrient and thus favoring the production of dinoflagellates and other nonsiliceous phytoplankton in the global ocean. If so, the Antarctic production system is a crucial element in controlling all of the ocean's fertility. As discussed in chapter 8, diatom production is the basis for a short food chain, which in turn supports marine mammals and birds. Increasing silica piracy by the Antarctic diatom production system, in the last 25 million years, must have severely affected the geographic production patterns of the sea. While the Antarctic became reliably productive, all other regions suffered shortages, at least occasionally.

The hallmark of Antarctic production is the short growing season, which results in a sharp spike of export production. More than 90 percent of what is trapped comes down in two or three summer months, as was shown by experiments launched from the research ice breaker *Polarstern*, of the Alfred-Wegener Institute in Bremerhaven.[78] Of course, when pulses of production are so short and strong, much more of the production can go into the export than when production is spread out throughout the year: pulsing overloads the recycling system.

Before there were highly pulsed upwelling systems (which profited greatly from the general cooling in the last 10 million years) there was much less food for the deep-living organisms on the continental slopes. Thus, we should expect that these environments were invaded, within the last several million years, from shallower depths, where food supply is greater and organisms are already adapted to pulsed high food supply.[79]

THE HABITAT OF THE DEEP-SEA FLOOR

The size of the marine benthic habitat, the seafloor, is twice that of the corresponding area on land. Only a small portion is bathed by sunlight; the rest is dark and also very cold. The shallow portion of the seafloor is influenced by seasons (especially outside of the tropics), by run-off from land, by storm waves, and by strong currents. In general, the deep portion of the seafloor, beyond the shelf break, feels much less of such disturbances. The deeper one descends, the more uniform the environment becomes.[80]

Far off the coast and below a depth of 1,000 meters or so, the main sources of ecologic variation are the fluctuating supply of food and the nature of the substrate. Since food supply diminishes with distance from the coast, the abundance of benthic organisms—the benthic biomass—likewise diminishes as we leave the coast and enter ever-greater depths. It is useful, as indicated above, to distinguish the shelf environment (neritic zone) from upper continental slope and bathyal environment (and then down to abyssal and hadal). Neritic and bathyal seafloor environments are comparatively rich in large mobile organisms; the abyssal and hadal environments are thinly populated with sessile and slow-moving creatures, on the whole.

As for the ways of life of the various benthic creatures in the deep sea far below the range of daylight and scuba, we know very little about their habits and life histories. We have no reason to believe that they differ fundamentally from those of their relatives living in and on soft bottom in shallow waters. By location, we can distinguish animals living above, on and within the sediment, some sessile and others mobile. By feeding habit, we have deposit feeders, suspension feeders, scavengers, and hunting predators. All this is known from shallow water.

When historians review the discovery of the deep-sea fauna, they commonly focus on the

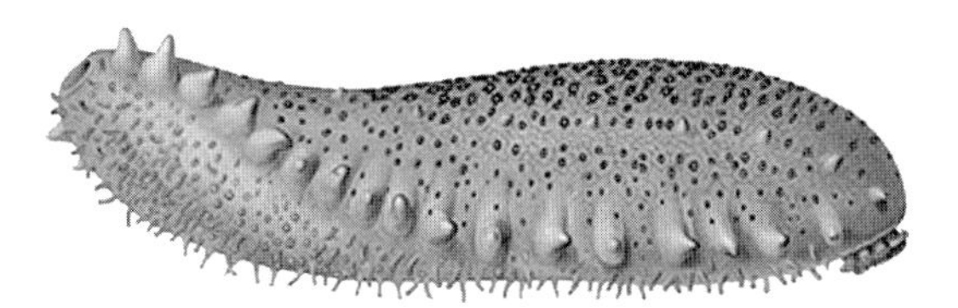

FIGURE 10.10. A sea cucumber, representative of the most abundant type of large animal on much of the deep-sea floor. It eats surface mud, and processes the organic matter within it. The cleaned mud emerges in long fecal strings at the opposite end.

collections made during the nineteenth century.[81] But what such collections brought up are more or less damaged specimens, usually quite dead. It was not until the second half of the twentieth century that deep-sea cameras brought back a view of the bathyal seafloor environment, with a promise to let us see living animals in their environment, even if briefly.

Underwater photography goes back to the nineteenth century.[82] In the 1930s, the geophysicist Maurice Ewing (founding director of Lamont Geological Observatory) became interested in looking at processes on the seafloor using photography. He developed an underwater camera (around 1938), which could be dropped free fall, take pictures, release its ballast, and return to the surface. He also used cameras lowered by wire. Development of cameras by Ewing and associates continued through the 1940s and 1950s, at Woods Hole, mainly for the study of geologic processes.[83] British and Russian scientists soon followed their example. In the United States, Harold E. Edgerton at MIT designed a camera that set new standards for depth and performance.[84]

Intriguing results of looking at the deep-sea floor, a few square meters at a time, are superbly illustrated in the volume entitled *The Face of the Deep*, by the geologists Bruce C. Heezen and Charles D. Hollister, based on work in the 1960s.[85] What is striking, from the point of view of deep-sea biology, is that a selection of the most interesting photos shows so few animals: the occasional filter-feeding sea pen or sponge is typical, beautiful but lonely. In a large proportion of the photos, tracks, trails, and burrow mounds are what one sees, along with the rare sea cucumber feeding on the mud.[86] As far as large mobile animals, sea cucumbers are in fact among the more important representatives in the abyssal fauna (fig. 10.10): the mud they eat is plentiful, and their energy requirements are exceedingly modest. They feed on the vacuum-cleaner principle, preferring the very surface of the mud, which has some freshly fallen material.

The first really dramatic photographs taken at the deep-sea floor were from the baited cameras set by John Isaacs and his group.[87] But the most spectacular photos ever are those taken from inside Woods Hole's *Alvin*, when exploring the hot vents. The discovery of the hot vent community was a major event in biological oceanography, an event that has many contributors deserving credit. The very first indication of the existence of a warm-water community in the midst of the vast icy spaces of the deep sea was in photographs taken by the Deep-Tow instrument package of the Marine Physical Laboratory at Scripps.[88] But the significance of the aggregation of bivalve shells seen in the photos emerged only after *Alvin* had taken observers to the hot springs on the seafloor. We shall return to the exciting story of the hot vents when discussing the volcanism associated with seafloor spreading, in chapter 11.

Back to the 99.99 percent of the deep seafloor that is icy cold. As mentioned, most deep-sea photographs that show any animal tend to emphasize organisms attached to the seafloor, or moving very slowly over it. Mainly, they show sponges, stalked coelenterates, and the occasional sea cucumber.[89] These organisms, while large, have very low energy requirements (and little to offer to potential predators, in the way of digestible material). Heezen and Hollister, after studying hundreds of photographs, came to the following conclusion:

> "The deep sea is ruled by the spiny echinoderms, which are by far the most important animals decorating and marking the abyssal

FIGURE 10.11. Feathery crinoid, related to sea stars, with a long geologic pedigree. Abyssal representatives enrich the deep-sea benthic fauna.

> landscape. The feathery crinoids, although the least abundant, are perhaps the most photogenic. Starfish, sea urchins, and sea cucumbers, which constitute the vast majority of the living animals seen in deep-sea photographs, are found in all seas, at all latitudes and to the greatest depths. As deep-sea animals go, all are relatively large.[90]

Interestingly, many of the sponges and sea pens and crinoids (fig. 10.11) separate themselves from the seafloor by means of a stalk. Even some fishes do this, using pectorals and tail as a high tripod. Presumably, the more nutritious particles and small plankton are higher up; in the layer directly adjacent to the floor they are diluted with inedible mud.

Many of the photographs show tracks and trails and fecal strings—deformations of the mud that mark the erstwhile presence of sizable animals, such as sea cucumbers, sea stars, sea urchins, crabs, and various shelled mollusks and worms. The tracks and trails seen in the photos do not necessarily document much activity: in analogy with the desert on land, we must presume that tracks and trails stay around for some considerable time. Tracks and trails and burrows and the fecal strings of worms and sea cucumbers provide many special environments, offering a diversity of habitats to support a diversity of small benthic species not seen in photographs. In addition, the fluctuations in food supply (which reflect the changes in export production through the seasons and also between years) introduce opportunities for different life-styles—opportunities for those that do well when food is plentiful, and those that survive the extended periods of dearth and hunger.

Occasionally, large clumps of organic matter fall to the seafloor, such as a large dead fish or marine mammal. Like carrion attracting vultures and other scavengers from afar, the rotting meat attracts a host of large organisms, from hand-sized isopods (related to the familiar pill bugs on land) to rattail fishes and sharks. As mentioned in the previous chapter, the discovery of this community of scavengers came with baiting a film camera. Isaacs called his contraption the "monster camera" because it gave evidence of large sharks cruising the depths for dead meat.

The organic matter that reaches the seafloor fuels the activities of benthic organisms, presumably largely through a pathway involving bacteria and associated microbes, which dominate the chemical processes in the sediment.[91] Bacteria grow on the organic flocs and are in turn consumed by protists and nematodes. As practically everywhere in the sea, small copepods are an important component of the fauna in the food web. That the abundance of the mass of benthic organisms reflects the supply of organic matter is evident from biomass distributions. Generally, such supply is high below the coastal ocean regions, as mentioned, and comparatively low elsewhere.[92] Already the *Challenger* scientists realized (by 1874) that dredges were full of benthic animals at stations close to continents and had little to offer at stations in the open ocean far from land.[93] In addition to overall abundance, food supply also controls the general makeup of benthic communities. Both species composition and size distribution are affected. Recognition of these patterns emerged in the 1960s and 1970s, in some detail.[94]

As microbes attack the organic matter arriving on the seafloor, they consume oxygen. This loss represents a substantial portion of the overall oxygen consumption below a depth of 1,000 meters, that is, roughly one-third of the total. Along the continental margins the proportion is larger and greatly contributes to the strong oxygen minimum in many coastal regions. The patterns of oxygen content in the ocean are part of the overall circulation and the marine carbon cycle, topics that are still very much under discussion. In any case, measurements of respiration of the seafloor in the last several decades demonstrate that slope and abyssal environments differ greatly, as expected. The blue-ocean-versus-green-ocean contrast that permeates the entire production system of the sea naturally also emerges in the patterns of seafloor respiration, and in benthic abundances.[95]

With a high supply of food, below the coastal ocean (the green ocean), it is here that relatively large burrowing animals can thrive. Their activities mix the uppermost sediment, down to several inches. Thus, the zone for very active bacterial growth is thickened. As the seafloor builds upward, because of sediment accumulation, bacterial activity persists to several meters depth below the floor. After the free oxygen is used up within the sediment, the bacteria strip oxygen first from nitrate and then from the sulfate ions in the interstitial water. Different types of bacteria are involved in these chemical processes.[96] The conversion of sulfate to hydrogen sulfide produces the rotten egg smell of stinking mud. It also provides for the precipitation of iron sulfide—the familiar fool's gold (pyrite) seen in many marine sedimentary rocks on land.

As mentioned, the resuspension of detritus arriving at the seafloor is important in delivering food for benthic organisms to the lower reaches of the continental slope, and perhaps to the continental rise and adjacent abyssal plains. Direct observations, made in the 1970s and 1980s, suggest that the material arriving as a consequence of the spring bloom forms a detrital carpet, which moves over the sediment surface due to bottom currents.[97] Slowly settling matter is then available for downslope redistribution within the zone of turbulence called the benthic boundary layer. This layer is several meters thick and owes its existence to slow currents—in part tidal—running across a bumpy seafloor. The layer is rich in materials recycled from the seafloor and provides a habitat for benthopelagic organisms interacting with the bottom and with the suspended particles above it.

Practically all these various insights about the environment controlling the patterns of benthic life were obtained through a systematic probing of the deep sea, in the second half of the twentieth century. By the 1980s, the old concept that benthic patterns are somehow tied to depth zones was interpreted as reflecting their sensitivity to food supply, which is a function of depth, as elaborated above.

For a long time it was believed that the seemingly inhospitable environment of the deep sea would not harbor a very diverse fauna. Compared with the variety of species known from the shelves, the number of species recovered from the abyss was small indeed, despite the extensive dredgings of HMS *Challenger* and many other expeditions that followed. This general perception changed substantially when attention shifted to the smaller members of the deep-sea fauna. In the 1960s, using a new type of sampling device, a sled gliding over the seafloor and cutting off a slice of mud, marine biologists discovered that a rich variety of organisms exists even at abyssal depths, although the organisms are quite small.[98]

Interestingly, the earlier belief, that there are few species in the abyssal setting, had readily been "explained" by the uniformity of the deep-sea environment. Presumably, a uniform environment offers few opportunities for diversification. The new results were embarrassing: this seemingly reasonable explanation was now without a phenomenon to explain. Also, the new fact of high diversity proved much more difficult to explain. Perhaps the idea that a uniform environment necessarily means few species is incorrect. Or perhaps there is something wrong with the concept of deep-sea uniformity. No one has tried to measure this particular factor, so it is hard to apply to a scientific hypothesis.

How would we define the degree of uniformity of an environment, independent from the diversity of the fauna? How could we be sure that the organisms, through evolution, react to the particular conditions we refer to in our definition? The truth is, we do not know exactly what kind of measurements would best resolve the question. We know so little about the life history of any of the deep sea creatures—what they eat, who eats them, how they reproduce, how they defend themselves, and what kind of interspecific relationships there are—that all we can do is speculate in rather abstract terms.

How far down must we go to the depths of the sea till the environment becomes unfavorable for living things? One task of the Danish *Galathea* Expedition (1950–1952) was to find out by sampling the very greatest depths in the ocean.[99] Even the deepest of trenches were found to host sizable animals (that is, recognizable as animals by the unaided eye). Likewise, bacteria are ubiquitous on the seafloor, as the microbiologists Claude ZoBell and Richard Morita discovered when they joined the expedition near the Philippines. They brought living bacteria to the surface from 10,060 meters in the Philippine Trench, kept them alive aboard ship, and then for many years longer in their laboratory at Scripps.[100]

ON THE ANTIQUITY OF DEEP-SEA LIFE

A century ago and earlier, marine naturalists thought that the deep sea was definitely a bad place to live: icy cold waters, eternal night, and enormous pressures. Also, there is not much food: it has to come a long ways from the surface. (The feeling persisted, in a general way, into the 1960s.)

Because of this negative feeling about the place, Forbes's hypothesis of an azoic zone (a zone without animals below a depth of 800 meters or so) gained some acceptance early in the nineteenth century.[101] As mentioned, this azoic zone idea was proven wrong by a number of findings in the middle of the nineteenth century, but especially by the *Challenger* Expedition, which dredged a variety of hitherto unknown creatures even from the greatest depths.

Subsequent expeditions brought up additional organisms, all of which turned out to be much the same from ocean to ocean. The types (although not necessarily the species) of creatures found were more or less familiar, and any hope that the deep-sea floor provided a haven for ancient life forms—so-called living fossils—was not fulfilled. In fact, later research suggested that the closest relatives of many of the deep-sea benthic animals live at shallower depths in high latitudes. These organisms are primarily cold-adapted forms rather than depth-adapted forms. Icy poles and a cold abyss have appeared quite recently on Earth (30 million years ago rather than 100 million), and the deep cold abyss, therefore, is as an environment as modern as the rest of the world. The strong aspect of cold adaptation of the abyssal fauna explains why there is very little overlap, on the species level, between deep sea and shallow waters.[102]

The early ideas of an azoic zone and of living fossils have been defunct for more than a hundred years. At issue is the rate of evolution and the ease with which organisms invade adjacent habitats. Rates of evolution are best studied in organisms that make fossils. In benthic foraminifers, for example, the rate is such that after 15 million years one-half of the species have been replaced. Rates of invasion express themselves in the similarity of faunas in adjacent depth zones. Since there is a considerable degree of endemism within the major depth zones, it appears that the rate of evolution within each province is fast, compared with the rate of invasion.

The first encyclopedic work on foraminifers—still in use by the specialists—is the report on the *Challenger* dredge samples by H. B. Brady (1884).[103] He illustrated more than 900 species,

280 of them from depths greater than 1,000 meters. This work by itself put to rest the concept of a barren abyss and foreshadowed the discovery, a century later, of high biodiversity in the deep sea.

Through deep-sea drilling, the main evolutionary history of the deep-sea benthic foraminifers became accessible. Whether this history is typical for deep-sea life in general we do not know, but it is the best information available. Robert Douglas and Fay Woodruff of the University of Southern California summarized the state of affairs after 20 years of drilling and study.[104] They identified three major events: one in the mid-Cretaceous, one in the Maastrichtian (latest Cretaceous), and one in the Eocene. Of these three, the first and the last had the most profound effect on changes in the deep-sea fauna, so that a threefold division emerges: older than mid-Cretaceous, from mid-Cretaceous to Eocene, and post-Eocene (that is, modern). In millions of years, these are older than about 90 million years ago, about 90 to 37 million years ago, and younger than 37 million years ago, respectively.

So, what happened in the Eocene? Douglas and Woodruff find "that the transition from the 'warm' ocean conditions of the Cretaceous and early Tertiary to those of the 'cold' ocean of the later Tertiary was a time of major reorganization in deep-sea benthic Foraminifera. At the species level, similar evolutionary changes occurred in middle Miocene time, about 14 million years ago, when modern oceanic conditions began."[105]

We might look for clues especially from the genera that diversified after the cooling. Many of them belong to the group specializing in taking advantage of pulsed high production. With the overall cooling, opportunities related to pulsed food supply opened up with the intensification of winds and upwelling. As the environment changed, the fauna changed. According to Douglas and Woodruff, most of the living species of benthic foraminifers evolved in the Miocene, with but a small percentage of the total fauna having an origin in the Oligocene or earlier.[106]

Regarding the deep-sea macrofauna (that is, all the animals brought up by dredging or seen in photographs) an assessment of "geologic age" is not nearly as straightforward as for the foraminifers. The deep dredge hauls of the *Challenger* proved that life is diversified and abundant everywhere in the deep ocean, even down to 5,000 meters depth. Moreover, the cold temperatures (a few degrees above freezing), and the enormous pressures (1 atmosphere for every 10 meters depth) proved no obstacle to successful adaptation to the environment. We know that the pressure is unaltered through time, but the temperature changed after the Eocene. Thus, the cold-adapted forms have to be younger than Eocene. We can also appreciate that the dark realm is being invaded from zones where eyes are useful: at the greatest depth, creatures have no eyes, even creatures whose relatives at intermediate depths have large eyes.[107]

The concepts of *diversity* and of *antiquity* are closely linked—both relate intimately to the rate of evolution. *Diversity* may be roughly defined as the number of different types of animals, where "types" is usually taken to mean species, but quite commonly also genera or families, depending on the depth of knowledge available or the sense of the question. There is a surprisingly large number of species of benthic animals, many more than in the plankton. The foraminifers provide a typical example: benthic species outnumber pelagic ones about 50 to 1 (2,000 species versus 40). Among the thousands of cnidarian (or coelenterate) species, the benthic forms (stinging corals, sea anemones, true corals, and gorgonians) vastly outnumber the pelagic ones (siphonophores and jellyfishes). The same is true for the flatworms, ribbon worms, and roundworms. The several thousand species of colonial moss animals, the bryozoans, are entirely benthic, as are their distant relatives, the brachiopods. The several thousand polychaete species (cousins of the familiar earthworm) are almost all benthic.[108]

Among the crustaceans, the ostracods (more than 2,000 species), the cirripeds (about 500), the copepods (about 6,000), the amphipods (about 3,000), the isopods (more than 3,000) and the decapods (crabs, lobsters, shrimps: more than 8,000 species) all have many more benthic forms than pelagic ones. Yes, even the copepods, which play such a dominant role in the zooplankton! It is true there are some major groups that have no benthic representatives: radiolarians (1,500 species) and arrow worms and comb jellies (each less than 100 species). When considering the 40,000 species of gastropods (snails, slugs, etc.) and some 10,000 species of bivalves, however, this imbalance in favor of the plankton fades to insignificance. The sea cucumbers, sea stars, brittle stars, sea urchins, and sea lilies (all echinoderms) also add their 6,000 species to the benthos. The 2,000 species of tunicates are mostly benthic sea squirts; only a few are pelagic.

The marine vertebrates, especially the more than 10,000 species of marine fishes, may seem to redress the imbalance by adding their lot to the nekton. However, thousands of the species of fish live in the coral gardens of the tropics. Many others also live in close contact with the seafloor, although not necessarily as intimately as the 500-plus species of flatfishes. Among the 600 species making up the sharks, rays, and chimaeras, again benthic forms dominate over pelagic ones.

Why this extraordinary diversity of the benthic life realm, compared with the pelagic one? Why this particular ratio of 50 to 1? The second question is easily answered: we have no idea. We simply do not know why there are just this many animal species around, no more and no less. In fact, we are quite ignorant about why the pelagic realm has several thousand species of animals rather than several hundreds or several tens of thousands. Likewise, there is at present no accepted theory from which we could derive that there should be on the order of 200,000 benthic animal species rather than 10,000 or 1 million. In fact, we do not know whether perhaps our count of species is sufficiently off, so that there *are* a million.

Evidently, with a ratio of 50 to 1, the difference between benthic and pelagic species diversity is not at all subtle, and we should be able to make some educated guesses as to why it exists, despite our ignorance about the absolute levels of species diversity. We might reasonably argue that a physical environment that creates more opportunities for making a living in different ways will have more species than another, more uniform environment. This is an old idea, linking species diversity to habitat diversity.

Benthic habitats—and especially shallow habitats—are indeed very diverse: exposure to light, waves, currents, and supply of sediment and food can vary substantially over short distances. Also, there are a great number of different ways an animal can make itself at home on the seafloor. It can live on top of the bottom as part of the epifauna, or within it as infauna. The ratio of epifaunal species to infaunal ones is about 4 to 1, on average. In polar areas it is nearer 1 to 1; in the shallow tropical areas, especially the coral reefs, the ratio is shifted greatly in favor of the epifauna, but in the deep sea, to the infauna.

When compiling the types of animals that live together on sandy or muddy seafloor in various parts of the world's shelves, we notice something very interesting: infaunal soft-bottom communities from various shelves in mid- and high latitudes show a certain similarity. This similarity has given rise to the concept of parallel communities.[109] As the Danish ecologist G. Thorson has urged, species are different from one faunal region to the other, but the genera tend to be the same. We have here one clue to the proliferation of benthic species: provincialism in their distribution. Instead of having one community settling all appropriate environments, we have a number of similar ones developing independently. The reason is a certain degree of isolation between the different shelf regions, which allows separate evolution of populations of a species.

In the pelagic realm, isolation is much more difficult to provide than in the benthic realm. Thus, there is much less opportunity for separate development of closely related species. This chain of reasoning gives rise to the provincialism hypothesis of species diversity.[110] It is, clearly, difficult to apply to the deep sea, where barriers are much less in evidence than on the shelves of the world. However, for species restricted to depths above 3 kilometers or so, islands and undersea mountains would presumably constitute something akin to provinces, which could foster isolation and speciation.

Besides the habitat and the provincialism hypotheses, there are what may be called "biological interaction" and "disturbance" hypotheses. One calls on evolutionary pressures for diversification from interaction between organisms in the same community—opportunities for symbiosis and the "arms race" effect advocated by the evolutionary biologist G. J. Vermeij come to mind.[111] The other, which has been explored in the connection with fire ecology, calls on episodic disturbance to keep dominant species from taking over, and to provide opportunities for a great variety of forms and for new adaptations.[112]

The various hypotheses attempting to explain diversity patterns are quite generally poorly tested. The result (diversity) is known (sort of), but the forces driving evolution are of a historical nature and are basically impossible to measure. We do not live long enough for that.[113]

Diversity reflects the balance between the rise of new species and the demise of existing ones, integrated over millions of years. It is a result of history. Present patterns and the associated physical and ecologic conditions contain insufficient information to explain its distribution. This aspect, that the history of the ocean is crucial in understanding biodiversity of the abyss, or in reefs, or anywhere else, tends to get insufficient attention in many discussions.

NOTES AND REFERENCES

1. The prefix *bathy-* means "very deep." Its root is of Greek origin, as is that of *abyss*, used for much the same purpose, and emphasizing dark and cold.

2. Or else, more recently, watching TV shows on handling snakes and crocodiles. In the same vein, conventional theoretical ecology owes much to observations of birds and insects, and nothing to midwater fishes.

3. William Beebe (1877–1962) was a biologist with the New York Zoological Society. After some zoological adventures in the tropical jungle, he turned to exploring tropical marine life using a diving helmet. During more than a decade he entered the ocean "hundreds of times," usually to not much more than 40 feet. Among the corals of the West Indies Beebe became entirely absorbed and entranced by observing underwater life, especially in night dives when the waters surrounding a moving diver light up with phosphorescence.

4. W. Beebe, 1934, *Half Mile Down.* Harcourt Brace, New York, p. 7.

5. The bathysphere was 4 feet 9 inches in diameter, with steel walls more than 1 inch thick; it had three windows of fused quartz, each 3 inches thick and 8 inches in diameter, and it weighed 5,000 pounds. Oxygen tanks supplied air for breathing, and a soda lime scrubber prevented buildup of carbon dioxide. A heavy electric cable provided for lighting, telephone, and a spotlight. In the summer of 1930 Beebe and Barton, crammed together in their pressure capsule and established a deep diving record by being lowered to 1,426 feet off Bermuda; in 1934 they went to 3,028 feet. Maximum depths reached at that time by other means were 306 feet by U.S. Navy divers in full suits, and 383 feet by a submarine; see W. Beebe, *Half Mile Down*, p. 100.

6. W. Beebe, *Half Mile Down*, p. 285.

7. The ichthyologist Carl L. Hubbs, then at the University of Michigan, reviewed Beebe's book and scolded him for naming creatures from a vision "faintly seen through the bathysphere windows." C. L. Hubbs, 1935, *Review of* Half Mile Down. Copeia 2, 105. Hubbs, an indefatigable observer himself, undoubtedly envied Beebe's opportunity for making observations at depth, but he was right in insisting on proper procedure when naming organisms.

8. E. Shenton, 1970, *Where have all the submersibles gone?* Oceans 3 (6), 39–56.

9. Auguste Piccard (1884–1962) was a physicist by background.

10. The precision of the maximum depth reported is illusionary; it reflects instrument readings, not necessarily reality.

11. The *Trieste* was purchased by the U.S. Navy in 1958 and located at the Naval Electronics Laboratory in San Diego, where it was modified and used for experimental dives. That modified vessel, carrying Jacques Piccard and Lieutenant Don Walsh of the U.S. Navy, sank into and returned from the deepest spot in the ocean, the Mariana's Trench in the Pacific Ocean. On 23 January 1960 it went straight down, with no maneuverability, and, to everyone's relief, came straight back up after reaching its target. The exercises involving the *Trieste* established that humans could reach the deepest part of the sea. However, the circumstances of the experiment did not

suggest the procedure as a routine method of exploration. For details see J. Piccard, R.S. Dietz, 1961, *Seven Miles Down*. G.P. Putnam, New York, 294 pp.

12. Numerous definitions are available for the depth zones in the sea, commonly tied to the decimal system (for example, taking 1,000 meters as a boundary). Here we recognize these zones: the sunlit zone of the mixed layer, the zone of rapidly changing temperature below the mixed layer (thermocline), the zone of dim light (midwater, mesopelagial), and the abyssal zone of eternal dark. Very roughly, this corresponds to depths of 0 to 100 (or 50) meters, then down to 300 meters, then to 900 meters, and everything below. In addition, the shelf edge is important (near 100 meters) and so are the various depth limits of vertical migration in zooplankton and nekton within the mesopelagic zone.

13. See articles in W.H. Berger, V.S. Smetacek, G. Wefer (eds.), 1989, *Productivity of the Ocean: Present and Past*. Wiley-Interscience, Chichester, 471 pp.

14. J.D.M. Gordon, 2001, *Deep-sea fishes*, in J.H. Steele, S.A. Thorpe, K.K. Turekian, *Encyclopedia of Ocean Sciences*, 6 vols. Academic Press, San Diego, pp. 687–693, p. 690.

15. P.J. Herring, 1996, *Light, colour, and vision in the ocean*, in C.P. Summerhayes, S.A. Thorpe, *Oceanography, an Illustrated Guide*, Manson, London, pp. 212–227, fig. 14.17, p. 221.

16. R.H. Douglas, 2001, *Fish vision*, in J.H. Steele, S.A. Thorpe, K.K. Turekian, *Encyclopedia of Ocean Sciences*, 6 vols. Academic Press, San Diego, pp. 987–1000, p. 996.

17. Flashing back by bioluminescent animals, upon being flashed at, is commonly observed in aquaria keeping the right kind of fish.

18. See P.J. Herring, *Light, colour, and vision*, p. 222; P.J. Herring, E.A. Widder, 2001, *Bioluminescence*, in J.H. Steele, S.A. Thorpe, K.K. Turekian, *Encyclopedia of Ocean Sciences*, 6 vols. Academic Press, San Diego, pp. 308–317, p. 308–317.

19. Referring to the river Styx, which runs in the dark underworld of Greek mythology.

20. J.Y. Cousteau, F. Dumas, 1953, *The Silent World*. Harper, New York, 225 pp.

21. Such a pattern is familiar from certain spiders. Also, certain firefly females lure males of related species to eat them.

22. In 1844, the British marine biologist Edward Forbes (1815–1854) proposed an "azoic zone" below 600 meters or so, reflecting his experience that nothing of note was caught at depth in his sampling gear. See M. Deacon, 1971, *Scientists and the Sea, 1650–1900, a Study of Marine Science*, Academic Press, London, p. 281.

23. It has been claimed that the "azoic zone" was finally laid to rest when living bacteria were discovered in the Philippine Trench during the *Galathea* Expedition, in 1951. D.J. McGraw, 2002, *Claude Zobell, hadal bacteria, and the "azoic zone,"* in K.R. Benson, P.F. Rehbock, *Oceanographic History, The Pacific and Beyond*. University of Washington Press, Seattle, pp. 259–270. This assertion is questionable. Naturally, finding bacteria at depths greater than 10,000 meters was a major discovery, but it had little or nothing to do with Forbes. He was talking about animals. The word is *azoic*, not *abiotic*.

24. H.L. Sanders, R.R. Hessler, 1969, *Ecology of the deep-sea benthos*. Science 163, 1419–1424. Robert R. Hessler joined SIO in 1969 and continued here his studies on various aspects of the life histories and life-styles of deep-sea animals, including hot-vent organisms.

25. A.L. Alldredge, 1989, *The quantitative significance of gelatinous zooplankton as pelagic consumers*, in M.J. Fasham (ed.), *Flows of Energy and Materials in Marine Ecosystems*. Plenum Press, New York, pp. 407–433; L.P. Madin, G.R. Harbison, 2001, *Gelatinous zooplankton*, in J.H. Steele, S.A. Thorpe, K.K. Turekian, *Encyclopedia of Ocean Sciences*, 6 vols. Academic Press, San Diego, pp. 1120–1130.

26. L.P. Madin, G.R. Harbison, *Gelatinous zooplankton*, p. 1129.

27. *Ibid.*, p. 1129.

28. This reluctance of escaping downward has been found useful when setting purse-seine nets on pelagic fishes.

29. A.G.V. Salvanes, J.B. Kristofferson, 2001, *Mesopelagic fishes*, in J.H. Steele, S.A. Thorpe, K.K. Turekian, *Encyclopedia of Ocean Sciences*, 6 vols. Academic Press, San Diego, pp. 1711–1716.

30. P. Boyle, 2001, *Cephalopods*, in J.H. Steele, S.A. Thorpe, K.K. Turekian, *Encyclopedia of Ocean Sciences*, 6 vols. Academic Press, San Diego, pp. 436–442.

31. It was a momentous discovery. The impression is sometimes created, when looking at the marvelous *Alvin* pictures from the deep sea, that there is not much else. We may safely assume, however, that the hot-spring environment comprises less than 0.01 percent of the deep ocean floor.

32. As mentioned, however, the microfauna remains very diverse. H.L. Sanders, R.R. Hessler, *Ecology of the deep-sea benthos*.

33. Pressure is another factor, formerly much stressed but apparently of subordinate importance.

34. The concept of depth zonation was pioneered by Edward Forbes, as mentioned. C.W. Thompson, who did much to dispel the notion of azoic depths, gave Forbes abundant credit for this important work. C.W. Thompson, 1873, *The Depths of the Sea*. Macmillan and Co., London, 527 pp.

35. M.L. Natland, 1933, *Temperature and depth ranges of some Recent and fossil Foraminifera in the southern California region*. Scripps Institution of Oceanography Bulletin, Tech. Ser., 3 (10); F.L. Parker, 1948, *Foraminifera of the continental shelf from the Gulf of Maine to Maryland*. Bulletin of the Museum of Comparative Zoology, Harvard 100 (2), 213–241. Parker joined Scripps in 1950.

36. See J.W. Hedgpeth (ed.), 1957. *Treatise on Marine Ecology and Paleoecology*. Geological Society of America Memoir 67, vol. 1, 1–1296.

37. Textbooks tend to give rather precise depth ranges, where there are none, actually. It is better to follow the use of practicing ecologists. The British ichthyologist N.B. Marshall categorized the more than 2,000 species of deep-sea fishes living at low light levels and in the eternal dark by distinguishing a *mesopelagic* fauna, with centers of abundance between levels of 200 and 1,000 meters, and a *bathypelagic* one, in the eternal night. We use the English equivalents *midwater* and *deep water*. In addition, Marshall recognized a *benthopelagic* and a *benthic* fauna, with fishes that habitually swim near the bottom and others that habitually rest on the bottom, respectively. We use *near-bottom* and *bottom* fauna as equivalent to these technical terms. See N.B. Marshall, 1966, *The Life of Fishes*. Universe Books, New York, 402 pp.; and N.B. Marshall, 1954, *Aspects of Deep Sea Biology*. Hutchinson, London, 380 pp.

38. *Hadal* derives from the Greek *Hades*, realm of shadows.

39. As mentioned, none of the deep boundaries are sharp. The sharp boundaries are the sea level, the seafloor, and (in many places) the bottom of the mixed layer, where a very strong temperature gradient can develop. The depth of this boundary (the bottom of the mixed layer or the top of the thermocline) varies greatly. In areas of active upwelling it may be only a few meters down, but in the subtropical central gyres, it can be as deep as 150 meters. For a thorough survey of depth zonations up to 1970, see R.J. Menzies, R.Y. George, G.T. Rowe, 1973, *Abyssal Environment and Ecology of the World Oceans*. John Wiley and Sons, New York, 488 pp.

40. That is, the relationship between the two zones determines the likelihood that a photosynthesizing organism in the mixed layer has sufficient sunlight for net production.

41. The recent decrease in the productivity of the California Current may be a result of such dynamics, in part.

42. N.G. Jerlov, 1976, *Marine Optics*. Elsevier, Amsterdam, 231 pp.

43. Since particles are more abundant in the sunlit zone than in deeper waters, light is less readily absorbed at depth than in surface waters. In clear water, early experiments demonstrated that some light does penetrate even down to 1,000 meters, where it can still blacken photographic plates, given an hour-long exposure, that is (experiments by Helland-Hansen published in 1912, cited in H.U. Sverdrup, M.W. Johnson, R.H. Fleming, 1942, *The Oceans—Their Physics, Chemistry, and General Biology*. Prentice-Hall, Englewood Cliffs, N.J., 1087 pp., p. 82). Taking the highest transmission possible (half the light energy at 50 meters, for pure water and greenish blue color), the light intensity at 1,000 meters must be much less than one-half multiplied by itself some 20 times, that is, well below a millionth of what it was near the surface. Below that depth, then, there simply are not enough photons left to "see" by—at that depth, all seeing is related to bioluminescence.

44. Or so watch-standers have claimed, when watching the echo-sounder.

45. P.J. Herring, *Light, colour and vision*.

46. J.D.M. Gordon, *Deep-sea fishes*, p. 693. The information is based on sampling and fishing experience. A downward migration in the grenadier fish, off California, from juvenile to adult could be demonstrated by analyzing the isotope chemistry of ear stones, which record the temperature of growth. S.A. Mulcahy, J.S. Killingley, C.F. Phleger, W.H. Berger, 1979, *Isotopic composition of otoliths from a benthopelagic fish,* Coryphaenoides acrolepis, *Macrouridae: Gadiformes*. Oceanologica Acta 2, 423–427.

47. Feeding on bioluminescent dinoflagellates has another hazard: the guts might light up! Some fishes have black guts so the bioluminescence of their food does not show through, in the dark. Elizabeth Venrick, SIO, pers. comm.

48. That copepods react to an increase in particle density by filtering more rapidly was demonstrated in the laboratory of M.M. Mullin at SIO, in the 1970s.

49. Scripps scientists Brian Boden and his wife Elizabeth Kampa collected evidence on this point. B.P. Boden, E.M. Kampa, 1967, *The influence of natural light on the vertical migrations of an animal community in the sea*. Symposium of the Zoological Society of London 19, 15–26.

50. E.M. Kampa, 1975, *Observations of a scattering layer during the total solar eclipse, 30 June 1973*. Deep-Sea Research 22, 417–423.

51. The Isaacs-Kidd trawl, in essence, is a large-frame coarse-mesh net with a wing-shaped depressor that, when towed, takes the net downward, acting like a kite but moving downward rather than up.

52. K.G. Foote, 2001, *Acoustic scattering by marine organisms* in J.H. Steele, S.A. Thorpe, K.K. Turekian, *Encyclopedia of Ocean Sciences*, 6 vols. Academic Press, San Diego, pp. 44–53, p. 44 ff.

53. R. D. Ballard, 2000. *The history of Woods Hole's deep submergence program,* in J. Steele and Ocean Studies Board Members (eds.), *50 Years of Ocean Discovery, National Science Foundation 1950–2000.* National Academy Press, Washington D.C., pp. 67–84. The aggregation of the lanternfishes is an important discovery. The comment about the photophores is interesting. As shown by experiment, ventral photophores are designed to simulate downwelling light to make a fish invisible from below. Thus the normal orientation of lanternfishes in the water is indeed the conventional one: dorsal side up. The curious chaotic orientation of the fish observed may have been influenced by the approaching *Alvin*: perhaps the myctophids were getting ready to flee in all directions.

54. According to Boyle's law, doubling the pressure halves the volume of a given volume of gas. A fish with a gas-filled bladder at 10 meters depth, at night, would end up with a bladder 30 times smaller by the time it reached a depth of 300 meters in the morning.

55. A. G. V. Salvanes, J. B. Kristoffersen, 2001, *Mesopelagic fishes,* in J. H. Steele, S. A. Thorpe, K. K. Turekian, *Encyclopedia of Ocean Sciences,* 6 vols. Academic Press, San Diego, pp. a1711–1717, p. 1711 ff.

56. *Ibid.*, p. 1713.

57. The strategy of changing life-styles and habitat as a function of maturity and sex is a common one—we might compare it with the strategy of an investor with a diversified portfolio.

58. Some of this loss is reversed by dolphins feeding on myctophids and squid: they bring the organic matter back up to the photic zone, where they spend most of their time and presumably also void their gut contents.

59. P. Boyle, 2001, *Cephalopods* in J. H. Steele, S. A. Thorpe, K. K. Turekian, *Encyclopedia of Ocean Sciences,* 6 vols. Academic Press, San Diego, pp. 436–442, p. 442.

60. See flux estimates in W. H. Berger et al., *Productivity of the Ocean.*

61. For review and references see G. Wefer, *Particle flux in the ocean: Effect of episodic production,* in W. H. Berger, V. S. Smetacek, G. Wefer (eds.), *Productivity of the Ocean: Present and Past.* Wiley-Interscience, Chichester, pp. 139–154.

62. W. S. Broecker, T.-H. Peng, G. Ostlund, M. Stuiver, 1985, *The distribution of bomb radiocarbon in the ocean.* Journal of Geophysical Research 90, 6953–6970.

63. The ratios of phosphate and nitrogen to carbon were found to be more or less the same for much of the ocean, according to Alfred C. Redfield (1890–1983), of Woods Hole.

64. The volume of the layer in question is about 50 percent of that of the total ocean, so the flux is 650 million cubic kilometers in 500 years, that is, 1.3 million cubic kilometers per year. The phosphate content is typically near 2 microgram-atom per liter. H. U. Sverdrup et al., *The Oceans,* p. 239.

65. Estimates of export at 100 meters vary around a value of 10 billion tons of carbon, with a factor of 1.5.

66. H. Postma, 1971, *Distribution of nutrients in the sea and the oceanic nutrient cycle,* in J. D. Costlow (ed.), *Fertility of the Sea.* Gordon and Breach, New York, pp. 337–349.

67. At Scripps, the geologist Andrew Soutar, working with John Isaacs, began trapping experiments in the 1960s. His method helped determine the rate of reproduction of planktonic foraminifers. W. H. Berger, A. Soutar, 1967, *Planktonic foraminifera: Field experiment on production rate.* Science 156, 1495–1497.

68. The "progressive tax" relationship is implied in the "new production" scheme of R. W. Eppley, B. J. Peterson, 1979, *Particulate organic matter flux and planktonic new production in the deep ocean.* Nature 282, 677–680.

69. E. Suess, 1980, *Particulate organic carbon flux in the oceans: Surface productivity and oxygen utilization.* Nature 288, 260–263.

70. W. H. Berger et al., *Productivity of the Ocean,* from the preface.

71. In fact, such analogy is the origin of the term *export production.*

72. W. H. Berger et al., *Productivity of the Ocean,* p. 4.

73. G. Wefer, 1989, *Particle flux in the ocean: Effects of episodic production,* in W. H. Berger, V. S. Smetacek, G. Wefer (eds.), *Productivity of the Ocean: Present and Past.* Wiley-Interscience, Chichester, pp. 139–153.

74. See, for example, W. G. Deuser, E. H. Ross, 1980, *Seasonal changes in the flux of organic carbon to the deep Sargasso Sea.* Nature 283, 364–365.

75. Where such detritus is important in surface waters, it can get incorporated into the fecal matter of filter feeders and accelerate the sinking of the fecal particles to the seafloor. R. B. Dunbar, W. H. Berger, 1981, *Fecal pellet flux to modern bottom sediment of Santa Barbara basin (California) based on sediment trapping.* Bulletin of the Geological Society of America 92, 212–218.

76. E. Suess, *Particulate organic carbon flux.*

77. G. Wefer, *Particle flux in the ocean.*

78. G. Wefer, E. Suess, W. Balzer, G. Liebezeit, P. J. Müller, C. A. Ungerer, W. Zenk, 1982, *Fluxes of biogenic components from sediment trap deployment in circumpolar waters of the Drake Passage.* Nature 299, 145–147; G. Fischer, D. Fütterer, R. Gersonde, S. Honjo, D. Ostermann, G. Wefer, 1988, *Seasonal variability of particle flux in the Weddell Sea and its relation to ice cover.* Nature 335, 426–428.

79. The pulsed supply of food is a major area of study in marine biology and biogeochemistry.

R. S. Lampitt, 1996, *Snow falls in the open ocean*, in C. P. Summerhayes, S. A. Thorpe, *Oceanography, an Illustrated Guide*, Manson, London, pp. 96–112.

80. This is not to say that the environment is entirely predictable. Locally, the arrival of carcasses of fish or whales provides for rare and sporadic disturbance. Over large regions, the occasional arrival of muddy water (turbidity currents), or the turbulence from tsunamis, can bring abrupt changes.

81. See, for example, L. Saldanha, 2002, *The discovery of the deep-sea Atlantic fauna*, in K. R. Benson, P. F. Rehbock, *Oceanographic History, The Pacific and Beyond*. University of Washington Press, Seattle, pp. 235–247.

82. L. Boutan, 1893, *Mémoire sur la photographie sous-marine*. Archives de Zoologie Expérimentale et Générale 3 (1), 281–324.

83. M. Ewing, A. C. Vine, J. L. Worzel, 1946, *Photography of the ocean bottom*. Journal of the Optical Society of America 36, 307–321.

84. For a history of deep-sea photography, see J. B. Hersey (ed.), 1967, *Deep sea photography*. The Johns Hopkins Oceanographic Studies 3, 1–310. For a general account in the context of instrumentation, see A. McConnell, 1982, *No Sea Too Deep: The History of Oceanographic Instruments*. Adam Hilger, Bristol, 162 pp.

85. B. C. Heezen, C. D. Hollister, 1971, *The Face of the Deep*. Oxford University Press, London, 659 pp.

86. In San Diego, Carl Shipek of the Navy Electronics Laboratory used a wire-lowered camera on the NEL-Scripps MidPac Expedition in 1950. These early efforts showed mostly interesting but unoccupied seafloor rocks and then-unexpected ripple marks on the deep-sea floor.

87. J. D. Isaacs, R. A. Schwartzlose, 1975, *Active animals of the deep-sea floor*. Scientific American 233 (4), 84–91.

88. The deep-tow package was developed under the direction of Fred N. Spiess, then director of the Marine Physical Laboratory at Scripps. A cluster of bivalve shells was seen in a photo taken in 1976, during the initial deep-tow search for hot vents along the Galapagos Rift by the Scripps research vessel *Melville*. P. Lonsdale, 1977, *Clustering of suspension-feeding macrobenthos near abyssal hydrothermal vents at oceanic spreading centers*. Deep-Sea Research 24, 857–863.

89. See, for example, the photos in P. A. Tyler, A.L. Rice, C. M. Young, and A. Gebruk, 1996. *A walk on the deep side: Animals in the deep sea*, in C. P. Summerhayes, S. A. Thorpe, *Oceanography, an Illustrated Guide*, Manson, London, 352 pp.

90. B. C. Heezen, C. D. Hollister, *The Face of the Deep*, p. 53.

91. B. B. Joergensen, 2000, *Bacteria and marine biogeochemistry*, in H. D. Schulz, M. Zabel (eds.), *Marine Geochemistry*. Springer-Verlag, Berlin, pp. 173–207.

92. G. T. Rowe, 1983, *Biomass and production of the deep-sea macrobenthos*, in G. T. Rowe (ed.), *Deep-Sea Biology, The Sea*, vol. 8. Wiley Interscience, New York, pp. 97–121.

93. J. Murray, J. Hjort, 1912, *The Depths of the Ocean*. Macmillan, London, 821 pp.

94. G. T. Rowe, 1972, *Benthic biomass and surface productivity*, in J. Costlow (ed.), *Fertility of the Sea*, vol. 2. Gordon and Breach, New York, pp. 441–454; H. Thiel, 1983, *Meiobenthos and nanobenthos of the deep sea*, in G. T. Rowe (ed.), *Deep-Sea Biology, The Sea*, vol. 8. Wiley Interscience, New York, pp. 167–230.

95. The respiration patterns emerged in the 1970s. Kenneth Smith, who came from Woods Hole to Scripps, was one of the most active pioneers in this field. K. L. Smith, K. R. Hinga, 1983, *Sediment community respiration in the deep sea* in G. T. Rowe (ed.), *Deep-Sea Biology, The Sea*, vol. 8. Wiley Interscience, New York, pp. 331–371.

96. M. L. Bender, D. T. Heggie, 1984, *Fate of organic carbon reaching the deep-sea floor: A status report*. Geochimica et Cosmochimica Acta 48, 977–986; C. Hensen, M. Zabel, 2000, *Early diagenesis at the benthic boundary layer: Oxygen and nitrate in marine sediments*, in H. D. Schulz, M. Zabel (eds.), *Marine Geochemistry*. Springer-Verlag, Berlin, pp. 209–231; S. Kasten, B. B. Joergensen, 2000, *Sulfate reduction in marine sediments*, in H. D. Schulz, M. Zabel (eds.), *Marine Geochemistry*. Springer-Verlag, Berlin, pp. 263–281.

97. R. S. Lampitt, 1985, *Evidence for the seasonal deposition of detritus to the deep seafloor and its subsequent resuspension*. Deep-Sea Research 32, 885–897; H. Thiel, O. Pfannkuche, G. Schriever, K. Lochte, A. J. Gooday, Ch. Hemleben, R. F. G. Mantoura, C. M. Turley, J. W. Patching, F. Rieman, 1989, *Phytodetritus on the deep seafloor in a central oceanic region of the northeast Atlantic*. Biological Oceanography 6, 203–239.

98. Major contributors to this revolution were Robert Hessler and Howard Sanders, at Woods Hole. R. R. Hessler, H. L. Sanders, 1967, *Faunal diversity in the deep-sea*. Deep-Sea Research 14, 65–78. At Scripps, Robert Hessler worked with Paul Dayton, continuing the study of mechanisms maintaining diversity in the deep sea. P. K. Dayton, R. R. Hessler, 1972, *Role of biological disturbance in maintaining diversity in the deep sea*. Deep-Sea Research 19, 199–208. He later worked with others, as well. M. A. Rex, C. T. Stuart, R. R. Hessler, J. A. Allen, H. L. Sanders, G.D.F. Wilson, 1993, *Global scale latitudinal patterns of species diversity in the deep-sea benthos*. Nature 365, 636–639.

99. A. F. Bruun, S. Greve, H. Mielche, R. Spärck, 1956, *The* Galathea *Deep Sea Expedition 1950–1952*,

Described by the Members of the Expedition. Macmillan, New York, 296 pp. (first published in Danish, in 1953).

100. Claude E. ZoBell (1904–1989), pioneer in marine microbiology, joined Scripps in 1932. The recovery of bacteria from a depth exceeding 10 kilometers and their culturing on board ship is described in C. E. ZoBell, R. Y. Morita, 1956, *Bacteria in the deep sea*, in A. F. Bruun, S. Greve, H. Mielche, R. Spärck, *The* Galathea *Deep Sea Expedition 1950–1952, Described by the Members of the Expedition*. Macmillan, New York, pp. 202–210. The article contains the remarkable sentence, "In spite of their small size, bacteria are so numerous that they may constitute an appreciable part of the volume or total weight of living organisms in the sea." It is now thought that the biomass within the sediments of the sea is roughly equivalent to that of all other organisms on the planet, mainly because bacterial activity persists deep below the seafloor.

101. M. Deacon, 1996, *How the science of oceanography developed*, in C. P. Summerhayes, S. A. Thorpe, *Oceanography, an Illustrated Guide*, Manson, London, pp. 9–26.

102. Those species that span a wide depth range, incidentally, are called eurybathyal, while the ones with a narrow range are labeled stenobathyal.

103. H. R. Brady, 1884, *Report on the Foraminifera dredged by H.M.S.* Challenger *during the years 1873–1876*. Challenger Reports, Zoology vol. 9, 814 pp.

104. R. Douglas, F. Woodruff, 1981, *Deep sea benthic Foraminifera*, in C. Emiliani (ed.), *The Oceanic Lithosphere*, vol. 7 of *The Sea*, John Wiley, New York, pp. 1233–1327.

105. *Ibid.*, p. 1309.

106. *Ibid.*, p. 1311.

107. The direction of invasion, evidently, is marked by loss of eyes, rather than by gaining eyes. Presumably, when spreading into the starved and barren environments at great depth, the invaders find that bioluminescence is no longer a viable strategy for making a living.

108. M. V. Angel, 1996, *Ocean diversity*, in C. P. Summerhayes, S. A. Thorpe, *Oceanography, an Illustrated Guide*, Manson, London, pp. 228–258.

109. G. Thorson, 1957, *Bottom communities (sublittoral or shallow shelf)*. Geological Society of America Memoir 67 (1), 461–534.

110. A detailed exposition of various theories of originating new species through isolation is in Ernst Mayr's book *What Evolution Is* (Basic Books, New York, 2001, 318 pp.).

111. G. J. Vermeij, 1987, *Evolution and Escalation, an Ecological History of Life*. Princeton University Press, Princeton, N.J., 527 pp.

112. P. K. Dayton, R. R. Hessler, *Role of biological disturbance*.

113. Geologists studying both fossils and the history of environmental change would seem to have an advantage in guessing the right answers. However, the level of trust in their methods may be limited. As one geochemist put it when defining what geology is about: "I give geologists a rock and they tell me a story about it" (Ed Goldberg, 1970s, pers. comm.).

ELEVEN

Seeing in the Dark

A SOUND APPROACH TO EXPLORATION

W. H. Berger and E. N. Shor

Sensing sound in the sea is a strategy for survival that is geologically ancient, going back to the time when fish evolved a lateral-line system for detecting pressure waves, several hundred million years ago. The most ancient of marine mammals, the toothed whales, have lived by their skills as echo-hunters for millions of years. Within the last 25 million years, they have perfected their echo system to an astonishing degree, including the growth of ever-larger brains, to process acoustic information.

The science of sound in the sea started with echo sounding for icebergs early in the twentieth century. By the middle of the century, marine geologists were collecting information about the topography of the seafloor at a rapid rate, resulting in a new appreciation for the diversity of major subsea landforms, including enormous mountain chains, vast abyssal plains, and continent-size rolling-hill provinces. In the process, they stumbled on the "deep scattering layer," acoustic evidence of daily migrations in a diverse fauna living below the surface waters, a phenomenon of fundamental interest in the biology of the ocean. After the 1950s, interest in biological sounds and their meaning greatly increased, in response both to the need of the U.S. Navy to become familiar with the acoustic environment in the sea and to the desire of marine biologists to understand the role of sound transmission in the life histories of a host of organisms, mainly mammals and fishes.

Also in the 1950s, geophysicists were using powerful sound sources including dynamite explosions to gather echoes from sediments and rocks deep below the ocean bottom, to ascertain the nature of the oceanic crust that forms the ocean basins. As the exploration for new oil fields went farther offshore, into regions on the shelf and continental slope likely to hold hydrocarbons, this type of surveying became routine, using sudden release of compressed air for a

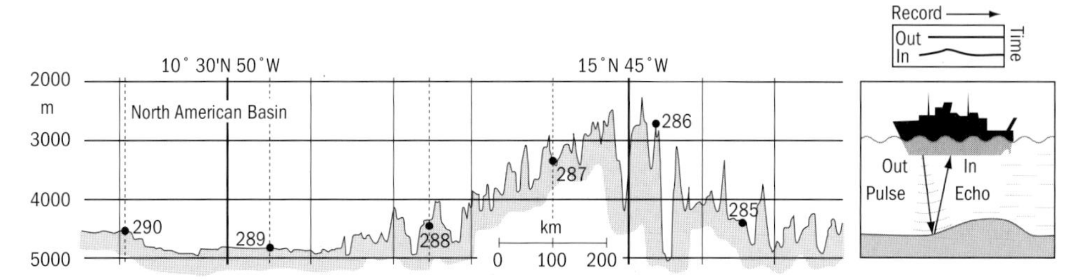

FIGURE 11.1. Discovery of the ruggedness of the Mid-Atlantic Ridge, by the *Meteor* Expedition, 1925–1927. *Left:* Topographic profile and stations reported. *Right:* Principle of echo sounding.

sound source and listening for the echoes from the bottom and below with a towed array of microphones. Since the 1970s, concern has been expressed that this type of noise, or any other type of loud noise, may injure or otherwise harm marine deep-diving mammals. The U.S. Marine Mammal Protection Act (1972) makes it mandatory to consider such concerns. Thus, much research has been done since on the actual and possible impacts of human-generated sound in the sea on marine mammals. Such research has further expanded our understanding of the sound-filled ocean, in the last 30 years or so.[1]

TO SEE WHERE LIGHT IS DIM

Light does not penetrate far in the ocean—even in clear tropical waters a scuba diver will see her buddy disappear within a few tens of meters. But sound carries over miles in any water, clear or murky. Using sound, dolphins have managed to "see" where light is dim for millions of years. Early in the twentieth century humans caught on and developed instruments that would allow the detection of objects from listening to the sound they reflect, and the determination of the depth of the sea at the location of a ship, by counting the seconds between an outgoing sound pulse and the arrival of its echo.

The initial trigger for developing sound-detection systems was a fatal encounter. On 15 April 1912, the passenger steamship *Titanic* struck an iceberg, and the ship sank, with a loss of 1,500 lives. Engineers in several countries began experimenting with underwater sound as a means to detect obstacles.[2] The beginning of World War I greatly stimulated additional efforts. The new technology was soon used in oceanographic research. The geologist and oceanographer Edward Smith, an officer with the ice patrol of the U.S. Coast Guard,[3] used echo sounding while searching for icebergs. "Iceberg Smith," as he was known among colleagues, was especially interested in the origin of the icebergs in Greenland and keen to follow their meandering path from their place of birth into the traffic lanes of the North Atlantic.

"Sonar," that is, echo sounding, is now used routinely on all ships to track the depth of the bottom or to find fish.[4] Navies of all countries use it to detect enemy submarines. Scientifically, the most familiar result of early echo sounding probably is the discovery of the ruggedness of the Mid-Atlantic Ridge, from the echolocation profiles made across the central and southern Atlantic by the *Meteor* Expedition (1925–1927) (fig. 11.1). The discovery put to rest the notion of a continuous barrierlike swell in the middle of the Atlantic basin, a picture that had emerged from interpolating the scattered soundings of the *Challenger* Expedition, half a century earlier.

Echo sounding is used to measure the depth of the ocean by keeping track of the time it takes sound to travel from the research vessel down to the seafloor and back (fig. 11.2).[5] From routine application of the method, maps emerged that revealed a wondrously dynamic landscape hidden underneath the sea. The first such maps that entered public awareness are the "physiographic diagrams" drawn by the Lamont geologists Marie Tharp and Bruce Heezen. These maps contained much guesswork but provided

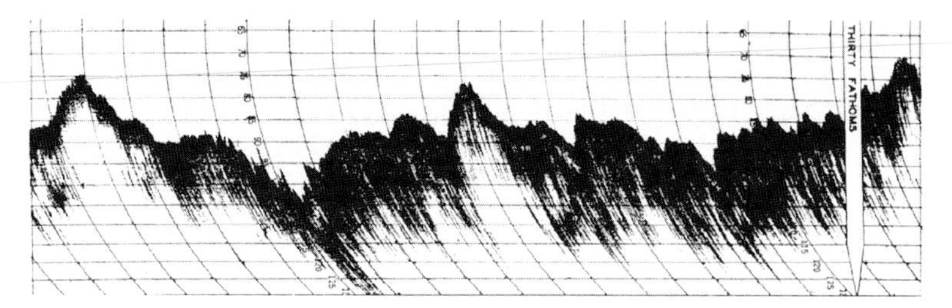

FIGURE 11.2. An echo signal from the seafloor, obtained in the 1930s, by a U.S. vessel using an early type of echo recorder.

an excellent general appreciation of the subsea landscape.[6] "Seeing" the ocean floor in the light of sound, along with other new information, eventually changed our views of the home planet in fundamental ways.

During World War II, a great need arose to detect hostile submarines—German U-boats were taking a painful toll on Allied shipping. Roger Revelle later recounted the early history of this effort. He said that a group of physicists arrived in San Diego early in 1941 to test the idea that light was the best way to detect submarines. They built a powerful underwater searchlight, created a black canvas cylinder to represent a submarine, and then turned on the light. They were able to recognize the object when it was about 100 feet away, but no farther. A new set of physicists, familiar with acoustics, then arrived and turned to sound for probing the environment.[7]

As with light, sound becomes weaker away from the source, because the energy is distributed over a larger region as the distance to the source increases. Yet, just as light, sound can be focused, and thus there is no problem in sending out a sound pulse to a sufficiently large target hundreds of meters away and receiving enough of an echo to estimate the distance from the time difference between sending and receiving. For a difference in time of 1 second, the object is 750 meters away: in water, sound waves travel close to 1,500 meters per second.[8]

LIFE IN A NOISY SEA

Jacques Yves Cousteau spoke of a "silent world" when talking about his adventures.[9] If this world is so silent, why do fish have highly developed ears and lateral lines? The ocean is, in fact, a noisy world—but the noises may be hard to pick up.

Of course, the fishes and other vertebrates in the ocean have long lived with the fact that in the sea, sound carries as much information as light, or more. Many of them have developed a keen sense of hearing. Or perhaps we should say a sense for detecting sound and pressure waves in the water (which is really the same thing). Fishes have the required senses in their ears and in the lateral-line system, which picks up low-frequency sound. Whales and dolphins also *make* sound and listen to its returns (fig. 11.3). While our brain is exquisitely programmed to analyze optical images, the brains of odontocetes (toothed whales) do an equally amazing job analyzing acoustic signals. Also, they use sound to drive fish. For example, when making bubble curtains by circling and exhaling air below a fish school to trap it, they apparently trick their quarry into thinking there is no escape, because the fish sense the vibrations

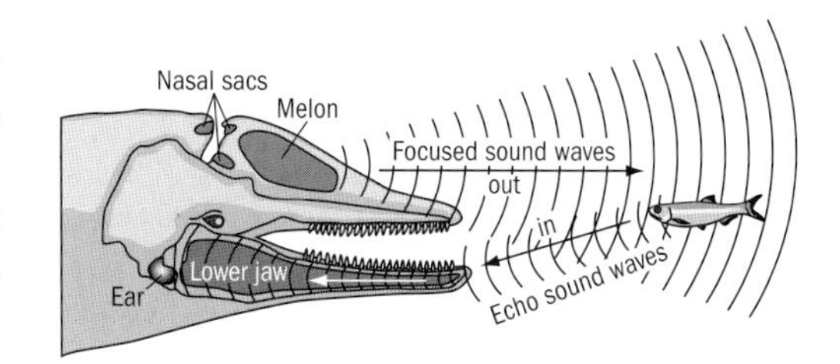

FIGURE 11.3. Echo-hunting in dolphins. The sound is made in the nose (nasal sacs) and is emitted through the melon, which acts as an acoustic searchlight. The echo returning from the target is guided to the ear through the lower jaw. Swim-bladders make a strong echo; thus a dolphin can readily distinguish schools of fish from aggregations of squid.

from the moving bubbles, which then "look" like a wall to them.

Not all the noise in the sea, perhaps not even most of it, happens to be in our range of hearing. Some of it is, for sure. Fishermen have long been familiar with the fact that some ocean creatures are noisy, and they gave such expressive names to fishes as "croaker," "drumfish," and "grunt." We speak of "snapping shrimp," because their crackling noises are audible even through the water surface. A time-honored way to find noisy fish was to put one's ear to an oar submerged in the sea. N. B. Marshall, in his classic book on the *Life of Fishes*, reports that fishermen in Ghana catching herringlike fish "use three-pronged paddles about 5-feet long to listen for schools of fishes. The paddle is lowered into the water over the stern of the canoe and rotated slowly, the fisherman holding his ear to the end. Those skilled in this art are not only able to judge the direction of the school, but also to get an impression of how far away it may be."[10]

Many teleost fishes (the bulk of modern bony fishes dominating the seas since Cretaceous times) make sounds deliberately. The use of the swim bladder for this purpose results in a range of noises described by Marie Poland Fish as "thumps, grunts, groans, growls, knocks, thuds, clucks, boops and barks. Typically they have a hollow quality, like the sound of a distant tomtom or of hammering on a wooden wall or like the sound produced when a wet finger is rubbed along the surface of an inflated balloon."[11] According to Marshall, such sounds have frequencies between 50 and 1,500 cycles per second (middle C on a piano being 256 cycles per second, or hertz) and most of the energy is concentrated below 300 hertz.

Noise equals food, not just for fishermen. Making noise attracts predators. An approaching predator makes pressure waves that can be sensed by the target of the attack. Thus, whether as prey or predator, it makes sense to be able to detect noise over a wide range of wavelengths. The inner ear of fishes resembles that of other vertebrates. Differences in vibrations of a calcareous ear-stone (called otolith) and its support structure, in response to disturbance by sound, are recorded by sensory hair cells that transmit the information to the brain through the appropriate nerves. As in other vertebrates, the ear's duties include keeping the balance and giving feedback on accelerations, besides the detection of sound and a preliminary analysis of it.

A large number of teleost fishes use their gas-filled swim bladder as a hydrophone, sensing its vibrations through a coupling with the inner ear. The membrane of the bladder, at the interface between gas and water, is an excellent collector of acoustic energy within the familiar range of hearing and beyond to higher frequencies. (Of course, sound waves larger than the bladder cannot be intercepted.) Such a fish knows very well, therefore, when a mammalian hunter is nearby acoustically "looking" for food! And it can analyze its approach and use its powerful escape mechanism—a sudden flip of the fin with several g of acceleration—to take action at the right moment.[12]

Just how well do fish hear? By offering food to fishes for responding to sound (a method well known to kids with an aquarium), one can determine the range of hearing and the sensitivity at different frequencies. Generally speaking, many of the fishes tested hear more or less as humans would, although with less ability in the lowest and highest portion of the range.[13] Members of the genus *Alosa* (alewives, shads) hear surprisingly well at very high frequencies. Presumably, they have evolved this ability to detect the echolocation sounds of their mammalian predators.

The purposes of the lateral-line system of fishes, for a long time, were something of a mystery. However, it became clear some 50 years ago that this sensing device is capable of detecting and locating disturbances in the water that result in pressure waves or pulses.[14] The mechanical aspects of sensing are similar to that of the ear: in both cases sensory hair cells are the key receptor of a signal. Sensors are distributed both on the head and along the body. The "forward-looking" sensors, presumably, are useful in detecting obstacles, or rapidly approaching predators, or evasive prey. The long dimension of the lateral-line system suggests that it is capable of sensing long sound waves and pressure waves, such as made by the swimming motion of other fishes. The sensory apparatus that is to detect such waves must be larger than the wavelength of the signal. According to A. N. Popper and D. M. Higgs, the lateral-line system is important in schooling behavior: "The lateral line tells the fish where the other fish are in the school, and helps the fish maintain a constant distance from its nearest neighbor. In experiments where the lateral line is temporarily disabled, the ability of fish to school is disrupted and fish tend to swim more closely together."[15]

An important consideration is that to locate prey or predator, the fish needs to know the direction from which the sound comes. We terrestrial vertebrates use two ears for this purpose; the brain analyzes the differences in arrival time of sound. In the water this is more difficult. Sound travels five times faster in water than in air, so the ears would have to be five times farther apart for the same effect. By having widely spaced sensors along the body, the direction of a sound source can be more readily verified, especially sound at the higher end of the frequency range.

Marine bioacoustics—the study of organisms making and detecting sound in the sea—is a relatively young field of investigation. In the 1960s many of the researchers in this field were brought together, with the encouragement of the Office of Naval Research, in two symposia (one at the Lerner Marine Laboratory in Bimini, in 1963, the other at the American Museum of Natural History in New York, in 1966). The various contributions to these meetings provide an excellent overview of the early efforts in the field.[16]

Of central interest is the fact that fishes produce sound in the first place. The most efficient way for them to make noise deliberately is to somehow rub the swim bladder to make it vibrate. Making noise attracts predators, so it is probably a good idea to stay quiet within sunlit waters, where escape from fast visual predators may be difficult. The bottom fishes, on the whole, do not have a swim bladder and cannot make noise with it, therefore. Many or most of the midwater fishes do have one, but they keep it quiet, judging from the lack of vibrating muscles. The deepwater fishes, as a rule, do not have gas in their swim bladder, and thus they too are quiet. Drummers are most common among fishes that live near the bottom but do not rest on it habitually. In sum, it seems that fish that make noise are most common in shallow waters and live close to the bottom.

Why would many fishes living near the bottom make noise, since it presumably can attract predators? Perhaps for the same reason that birds sing—to mark their territory or to advertise their presence to potential mates. In a slope-dwelling group of rattails, the male fish has large drumming muscles on the bladder but not usually the female,[17] which supports the analogy with male birds claiming territory and attracting a female partner. The noise-making species, it turns out, have the larger otoliths among the rattail fishes, suggesting better hearing. This supports the idea that the noise is for communication between members of the same species.

Apparently, in shallow-water fishes the ones living close to the bottom or whose home is in rocks, corals, and weeds are more likely to make noise than others. Marshall (very tentatively) suggested that claiming territory could be an important aspect of this pattern.[18] In shallow-water fishes, also, there is a tendency for cycles of sound activity during a day.[19] The greatest noise-making usually occurs at dusk, at night, and at dawn, and less so during daylight hours—at depths where daylight penetrates. Singing at dusk and dawn is a pattern familiar to bird watchers. Recently, it was found that noise waves can move for hundreds of miles along the continental margin, as the noise made by one fish stimulates a neighbor to sound off.[20] Martin Johnson, in 1948, already reported a "regular and pronounced seasonal and diurnal habit of certain *Sciaenida* to form localized choruses."[21]

We know that male humpback whales sing complicated songs that have become legendary for their strange beauty. Presumably the intended listeners are nearby, females to be

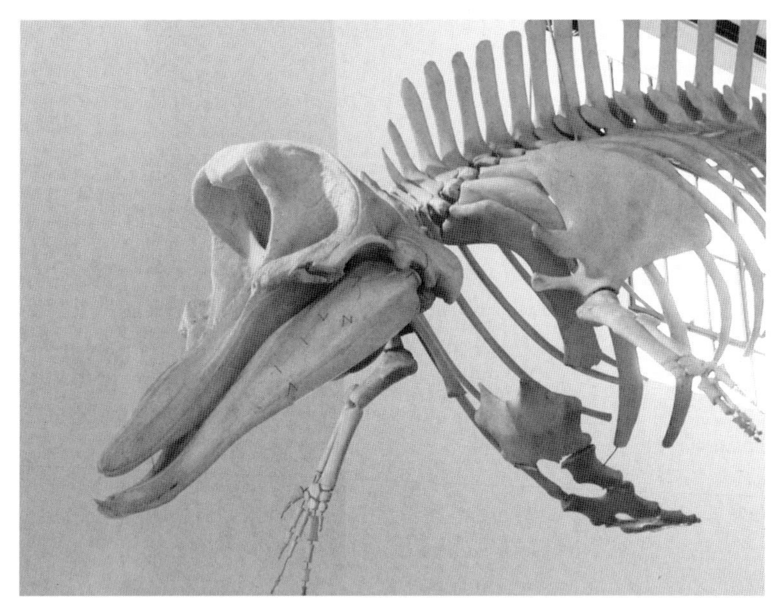

FIGURE 11.4. The strongly modified skull of *Ziphius*, a beaked whale, accommodates a large melon, for the focusing of sound in echo-hunting.

impressed, rival males to be warned off. Do some large whales also communicate with each other over long distances? When they jump out of the water, 30 tons momentarily airborne, and then fall back on the surface (an activity called breaching), are they impressing their group members by the large noise they make? Humpbacks, right whales, gray whales, and sperm whales are famous for breaching—they do so during the times of congregation for mating and breeding. The marine ecologist Roger Payne suggests that breaching is stimulated when a whale hears another one breach, some miles away.[22] It is a pattern reminiscent of the me-too chorus of dogs barking in the neighborhood in response to one individual sounding off, for whatever reason.

To interpret the various roles of sound in communication, obviously, we would have to know far more about the patterns of behavior of whales than we do. The most easily understood function of marine mammalian noise is the method of echolocation; it can be treated as an engineering problem. Already around 1960, through the work of Kenneth Norris and associates, it became clear that dolphins and porpoises direct their acoustic search beam forward, concentrating the energy in the direction they are heading.[23] The shape of the skulls of toothed whales, and the "melon" on the upper jaw, are thought responsible for directing the sound forward (fig. 11.4). (This is analogous to having a lens in front of a flashlight, to make a directed beam.) Norris and Evans found that one species (the rough-tooth porpoise, *Steno bredanensis*) produced sounds of extremely high frequency.[24] Presumably, this would enable it to pick out even small fishes quite distinctly. Experiments on other species, as well, show that the range of hearing of dolphins and porpoises reaches much higher than ours. It may be too high for many fishes, making the sonar beam inaudible to prey and thus giving the mammals a great advantage.

Since the mid-1970s, the principles of echo-hunting worked out by dolphins and their kin over millions of years are being studied and applied in fisheries management. Fisheries scientists now find themselves in the position of an echo-hunter trying to estimate the abundance of fish. The new echo-sounding technology opened up opportunities for doing this, making it possible to supplement the information received directly from the fishing industry. The fisheries biologist D. H. Cushing pioneered the adaptation of the new acoustic methods for this purpose.[25] Before long, of course, the fishermen themselves realized that echo surveying would improve success of fishing, and it became a routine tool for hunting fish. The fact that 100-kilohertz sound can discriminate individual herring[26] shows that the dolphins know what they are doing when using high-frequency sound. Routine surveys of fish stocks (using somewhat lower frequencies) have been made since 1975.[27]

In summary, we see that vertebrates, including fish, mammals, and people, make and detect noise, to good purpose. This also includes seals, "true" and "eared," whose emissions are recorded as grunts, snorts, buzzes, clicks, yelps, roars, groans, creaks, knocks, trills, barks, whinnies, and bleats, in various proportions for various species, as well as other sounds with more technical descriptors.[28] But what about invertebrates? Do mollusks and arthropods react to sound? They do. This may be one reason why large plankton avoid deep-towed nets—they hear them coming. Specialized organs in crustaceans and in cephalopods, useful in keeping balance and in sensing currents, suggest the ability to pick up sound waves. We know that some insects on land can hear quite well: crickets chirp to attract a mate; some moths take evasive action when bats approach. It would be surprising if deep-sea shrimp and deep-sea squid were deaf to the sound of approaching predators.

That crustaceans make sounds became widely appreciated when the crackling sounds of snapping shrimp interfered with submarine war research. During World War II, U.S. Navy scientists worked with Scripps physicists in developing sensitive listening equipment. When submerging the instruments in the sea, an intermittent crackling sound was heard at times. Was there something wrong with the design of the instruments? The biologist Martin Johnson came to the rescue.[29] He spent many nights on the end of the Scripps pier, with equipment for recording animal sounds, and then worked for days relating them to specific animals. The source of the crackling sound proved to be snapping shrimp (genus *Alpheus*, formerly *Crangon*). It was strictly confined to certain depths and rocky areas, where this shrimp is abundant. Clusters of these crustaceans, each snapping its large claw, make quite an underwater clatter.[30] In the nearshore waters off California, shrimp noise can exceed that from ships, wind, and even waves.[31]

Animals that deliberately make noise usually do so to communicate. This works only if they also *hear* the noise they make. The fact that this shrimp (and other crustaceans) makes noise indicates that crustaceans can hear.

DISCOVERY OF THE SCATTERING LAYER

The standard reaction to puzzling and entirely unexpected findings from a new technology is the thought that there must be a problem in the equipment. This was a common situation with the many new instruments that were being created for ocean exploration, during and right after World War II.

Thus when the physicist Russell Raitt and colleagues[32] used their newly developed echo-sounding equipment, they were most gratified to get a good echo of the seafloor for much of the time, but they were greatly puzzled by mysterious indications of shallow seafloor, where the water was known to be deep. For example, on one occasion the echo-sounder indicated a depth of 1,500 feet where the bottom was known from other traverses to be 12,000 feet down. They satisfied themselves that the instruments worked fine. Something—some layer in the sea 1,500 feet down off San Diego—was reflecting sound. In ignorance of its nature, the physicists named it the "deep scattering layer," which correctly describes what they observed.

The acoustic obstacle appeared as several fuzzy horizontal layers on the echo-sounder, layers that seemed to move upward to the surface at the end of the day and become indistinct, and reform on moving down again at dawn (fig. 11.5).[33]

When called in for consultation, the biologist Martin Johnson decided that the pattern

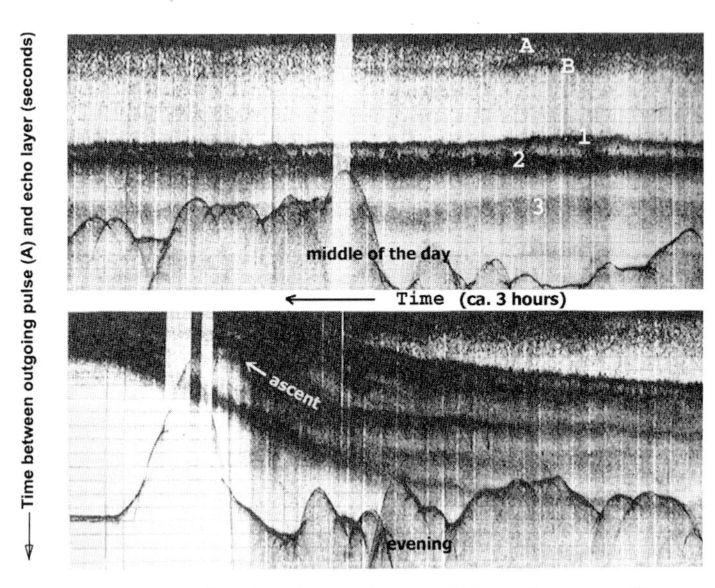

FIGURE 11.5. Output from the echo recorder on an SIO vessel, as reported by Robert Dietz, in 1962. The fuzzy heavy traces are echoes from the deep scattering layer. The sharp record at the bottom is the echo trace from the seafloor. This deeper record is on a different listening cycle, for which the instrument waits several seconds before recording. A, outgoing pulse; B, scattering organisms near the bottom of the warm-water layer; 1–3, different groups of deep-scattering organisms. The depth is roughly 1,000 feet for echo 2.

looked much like animal behavior, so he spent 24 hours in June 1945 watching the echo-sounder on the *E. W. Scripps,* with the physicists at his side. He concluded that the layers must be large clusters of small animals engaged in vertical migration. It was a major discovery, whose impact on pelagic ecosystem research is still reverberating.[34]

Johnson undoubtedly knew about sound echoes from fish schools, a phenomenon discovered in the 1930s. Also, he knew about vertical migration of plankton from earlier observations at Scripps and elsewhere. Under the directorship of Ritter, repeated collections had been made at specific offshore stations, supported by measurements of water temperature and chemistry. One abundant form in these collections is a small arrow worm, *Sagitta bipunctata,* a fierce predator in the world of microplankton, with bristles for teeth and a finned pencil-shaped body for sudden movement.[35] From 5 years of records, biologist Ellis Michael recognized that this arrow worm was most abundant at the surface in the time between an hour after sunset and an hour before sunrise, and that it spent the day 15 to 20 fathoms down (ca. 100 feet). He realized that the animal performed a daily vertical migration, following a preferred level of light intensity. The same is true for most planktonic copepods.[36] In their popular book *The Seas, Our Knowledge of Life in the Sea and How It Is Gained,* which went through two editions and many printings, the prominent British marine biologists F. S. Russell and C. M. Yonge (1936) described vertical migration and noted that herring follow their migrating prey: "We all know that the herring fishermen only shoot their drift nets at night. This is because at night, the herring, like the plankton animals, also come to the surface."[37] Their observation does not involve the deep scattering layer, but it does draw attention to the importance of vertical migration in the uppermost waters.

The nature of the deep scattering layer became a question of central interest to the Navy, which aimed at understanding the acoustic environment of the sea. In the 1950s, a concerted effort was made, funded by the Navy, to better understand the biology behind the readily apparent physical phenomenon. It was soon realized that deep scattering layers occur throughout the ocean, and that they are better developed in the more productive regions. Usually several layers are involved, with somewhat different migration ranges. As Johnson had hinted in his 1948 paper, the details proved difficult to work out because the larger animals (which do much or perhaps most of the scattering) are the ones that more easily avoid capture. Several important aspects emerged after more than a decade of exploration: the density of scattering organisms did not have to be high for the scattering layer to be noticeable, in many cases fishes are involved (even lanternfishes, with a fat-filled swim bladder), and apparently also comparatively large shrimps. The range of migration of the main layer was found as typically between a thousand feet down and the base of the mixed layer.

The strongest effects, not surprisingly, are from schools of fishes with gas-filled swim bladders.[38] In the decades following, the demands of the fishing industry and fishery science drove improvements in instrumentation for detection of schools and even individual fishes of certain sizes (and species). From the 1970s on, not just fishes but all kinds of zooplankton were found to be acoustically recognizable, including euphausids, pteropods, siphonophores, jellyfishes, and copepods. Even animals with bodies having acoustic properties close to that of the surrounding water, such as squids, scatter sound, and some do so surprisingly strongly.[39]

The nature of the reflected sound waves shows that the scattering is caused by a change in density in the path of the sound, such that the density suddenly decreases downward. This can be achieved, statistically, by an increase in the number of swim bladders and oil droplets in the water—devices used by fish and plankton for buoyancy. Some of the swim bladders of deep-living fishes are not filled by gas but by organic substances (e.g., wax esters). In 1953 Scripps physicist Victor C. Anderson showed, in his Ph.D. thesis, that the scattering organisms caused a complete sound inversion (a negative reflection) of the signal, which established the nature of the density change at the level of the scattering layer. Direct observations by Eric Barham at the Naval Undersea Center in the 1950s, from submersibles, suggested that clusters of fishes and invertebrates are involved in creating the scattering layers. Anderson's results indicated that these organisms carried buoyant matter, such as low-density fluid, fat, and gas.

The search for the sound-scattering organisms stimulated much research concerning the layered distribution of plankton and of small swimmers below the sunlit zone (mesopelagic fauna), and their daily migrations. At Scripps, John D. Isaacs and Lewis W. Kidd developed a "midwater trawl" to explore this relatively unknown region with a coarse-mesh net, as mentioned in the previous chapter. The net, 12 feet across and 150 feet long, engulfed specimens that had rarely or never been seen before.[40]

Early models of the Isaacs-Kidd Midwater Trawl had no closing control, but later ones did, so that the depth of capture could be set. The use of opening-closing nets for plankton (with a finer mesh than used by the trawl) became routine in the 1960s. In the 1960s, Scripps biologist John A. McGowan created a double net called Bongo Net, because of its characteristic shape. It was used to sample zooplankton down to several hundred meters depth. The double catch provided interesting data for sampling statistics. Surveys using the bongo net greatly contributed to sorting out the vertical zonations and migrations of plankton within the upper kilometer of the water column.

Vertical migration turned out to be a general phenomenon involving both zooplankton and the swimmers feeding on the plankton animals, much as hinted at by Russell and Yonge when explaining why fishermen set on herring at night. The overall effect of the overlapping migrations of plankton, shallow-living nekton, and deep-living lanternfishes and scarlet prawns, as reflected in the migrating scattering layers, is to introduce enormous complexity into the food web of the upper ocean.

THE MYSTERY OF THE "AFTERNOON EFFECT"

Early in World War II, German submarine commanders discovered how to use local oceanographic conditions to try to escape from the acoustic searchlight of echo-hunting destroyers of the Allies. The commanders made use of a principle of sound transmission that had been recognized by the U.S. Navy, in the context of a puzzling "afternoon effect." The observation was that, during testing, acoustic targets were present in the morning and virtually disappeared in the afternoon.

To understand how this works, we need to review a bit of physical oceanography and some elementary principles of sound transmission in the sea. Over most of the ocean, except at high latitudes, there is a comparatively thin layer of warm water on top of the water column, and temperature decreases with depth. Sound velocity is relatively high in warm water and decreases within the thermocline below the warm mixed layer (that is, in the transition layer between warm and cold water).

When sound is sent sideways from a surface ship, in search of a submarine, sound waves will be bent, or "refracted," within the thermocline, into the medium that has the lower sound velocity, that is, downward. Fundamentally, in the propagation of a sound wave, the wave front within the slower medium lags behind the wave front in the faster medium, and a rotation ensues, as in a marching band on a turn, where one side makes shorter steps.

In effect then, a beam of sound aimed at a submarine situated below the warm layer in the upper thermocline is bent downward before it can reach its target. Some of the sound aimed higher will reach the target, and some will return, but it will be difficult to tell just where the target is located. It is a bit like locating a fish while looking sideways at the water surface. A spear aimed directly at the image of the submerged fish will not reach it.

Obviously, the nature of the boundary of the warm surface waters and the cool waters below—the thermocline—had to be of great interest to naval forces in all major seagoing nations. The existence of the thermocline was not the question; the knowledge of its presence goes back to oceanographic antiquity. What was less well known, right into the 1930s, was how variable the thermocline could be from season to season, and how it could change depending on local circumstances. That question required the development of a new instrument: the bathythermograph, soon called simply the BT. First developed by Athelstan Spilhaus in 1936 at the request of Carl-Gustaf Rossby at the Massachusetts Institute of Technology, this device recorded temperature versus depth in surface layers.[41]

The Spilhaus BT was small and rugged, with a metal bellows, a straight bimetal strip that curved with the temperature, and a stylus that marked a smoked microscope slide.[42] For a few years skunk oil was used for a coating before smoking the slides.[43] Subsequently this coating was replaced by a single molecule layer coating of gold. Several oceanographers at Woods Hole made various improvements, especially Allyn Vine.[44] The BT recorder was housed in a torpedo-shaped cylinder, for protection. The device, light enough to be handled by one person, was fastened to a wire at the top and dropped from a ship under way to a depth below the mixed layer, typically to between 100 and 200 meters down.

Spilhaus presented a paper on his instrument in 1939, and it immediately attracted keen interest by the Navy (and everything about it went secret). It helped the Navy to explore the thermocline, the layer of rapid change in temperature at the heart of the "afternoon effect," that is, the fact that test runs produced no echoes from a target in the afternoon, except at very close range. Were sailors getting sleepy in the afternoon? Not so—they were as diligent in the afternoon as in the morning. Much of the testing was carried out in the Caribbean near Cuba. Iselin suggested that the cause might be in the water itself, and, with Maurice Ewing, he used the *Atlantis* to explore the problem. Sure enough, measurements showed that the uppermost layer of the water, a layer several meters thick, warmed under the tropical sun by as much as 2 °C. Thus, by the afternoon the sound waves were bent downward, hiding the target from the search.[45]

During the 1950s, oceanographic research ships usually took BT records every hour, whatever the weather and condition of the sea. Recording the ocean temperature with a bathythermograph while underway required constant attention as it was dropped, braked, and recovered. Once the operator had started the record, he had to stay at the winch to completion, in spite of any waves breaking over the vessel (and over the person). Recovery could be dangerous, as the instrument came in quite swiftly and often swung insolently in circles while waiting to be caught, threatening to hit whatever was nearby. In the 1960s the procedure was abandoned in favor of "expendable" bathythermographs—small probes that were left in the sea after deployment.[46]

The "afternoon effect," then, was not caused by drowsy sailors, but by changes in temperature affecting the velocity structure of the water column. As mentioned, the sound velocity in seawater is close to 1,500 meters per second, almost five times faster than in the air.[47] The velocity varies according to temperature (several meters per second per degree), with salinity (relatively little), and with depth (1 to 2 meters per second for each 100 meters).[48] Generalizing the afternoon effect to the summer season, we can predict that sound transmission on the shelf should be poor, owing to a surface layer of warm water with a high sound velocity. This is indeed the case. A beam of sound aimed horizontally will bend down toward the bottom, be reflected from it and go upward, bend to the horizontal and go downward again, and so on, traveling a path like a bounding deer, but with rapidly decreasing energy because of absorption at the seafloor.[49]

Unlike on the shelf, in deep water sound is able to travel unobstructed over hundreds, even thousands of miles. In a typical sound profile for the Pacific, a velocity minimum appears between 600 and 1,000 meters depth, with speeds near 1,480 meters per second. In the overlying warmer waters, the velocity increases to a maximum between 1,530 and 1,540 meters per second in surface waters. In the underlying part of the water column, the velocity steadily increases with the increasing compression of the water to near 1,540 meters per second at a depth of 5 kilometers.[50] The fact that there is a layer, in midwater depths, which has velocities lower than the water above or below means that a beam of sound moving at a slight angle to the horizontal will be bent back into the layer when attempting to leave it, whether it be through the roof or the floor of the layer. This effect produces a "sound channel," where sound travels long distances with but modest attenuation.[51]

Maurice Ewing, working at Woods Hole Oceanographic Institution during World War II, explored the ocean's sound channel and proposed that it might be used for long-range transmission of sound. In early tests in the Atlantic Ocean in 1944, he set off 4-pound explosive charges at depths to 4,000 feet, which were recorded on the seafloor as far as 900 miles away. One 6-pound charge was recorded 3,100 miles away.[52] A system for detecting the origin of sound traveling within the channel was developed, using triangulation from several listening stations. It was named SOFAR (for *sound fixing and ranging*). In 1950, on the MidPac Expedition, Scripps scientists set off SOFAR bombs for a Navy project related to air-sea rescue. One of the 4-pound bombs was recorded 3,500 miles away and proclaimed "the shot heard round the Pacific." The sound channel had trapped the energy in a thin layer and decreased the attenuation accordingly.[53]

Whether the sound channel has implications for the biology of the ocean is not known. Sperm whales can readily dive to the depth of the sound channel, as can elephant seals. Do they use it for long-distance phone calls? No one really knows. However, we do know that the channel does not work where cold water is right at the surface, that is, in polar regions. If mammals use the sound channel for calling, they go easily out of range when in their feeding grounds.[54]

One use of the sound channel is for monitoring changes in ocean properties, especially temperature. The project taking advantage of this opportunity is a multiinstitutional ongoing experiment named Acoustic Thermometry of Ocean Climate, or ATOC for short.[55] The goal is to measure global ocean temperature with sound waves transmitted at very low frequencies from deep-water stations (to 900 meters) and recorded at various long distances. After traveling for thousands of miles, the sound waves integrate subtle changes in the ocean that would go unnoticed if only point measurements were made. Early tests in the Pacific Ocean demonstrated that this technique provides temperature data that compare favorably with information accumulated over many years from measurements by ships and submarines.[56] An important object is to provide a baseline and continuing measurements for following the effects from the rise in global surface temperatures that accompanies the burning of fossil fuels. How fast do such changes produce noticeable effects in the deeper parts of the ocean?

The ATOC program (among others using sound sources for ocean exploration) has raised concerns about the possible effects on marine mammals, regarding the schedule of sound emissions, the locations of the emissions, the frequencies used, the duration of individual sound trains, and the level of the energy of sound emission. For ATOC, we deal with low-frequency sounds emitted at a noise level lower than that of many ship screws. However, sounds of such frequencies are indeed used by baleen whales, presumably for communication. The need is for observations on how marine mammals react to different types of noise in their vicinity.[57]

The potential impact of using human-generated sound in the ocean is not known. Questions arising must be seen in the context of the general acoustic background and the associated behavior of marine mammals.[58] Some regions of the sea are naturally quite noisy, especially in high latitudes, where icebergs and pack ice grind into each other with constant crackling and occasionally loud explosive sounds, long described by many explorers. The whales and seals feeding there seem to manage somehow with this aspect of their environment.

EYES FOR THE NAVY AND THE PAYOFF TO SCIENCE

On 2 September 1914, a German U-boat torpedoed and sank the cruiser HMS *Pathfinder* off the Firth of Forth in Scotland.[59] From then on during World War I, considerable Allied effort was devoted to find methods to hunt the elusive killer boats.[60] The development of seeing by sound—sound navigation ranging, or sonar—was to provide, eventually, powerful eyes for the U.S. Navy for its operations, and it was to have enormous consequences also for economic activities and for science. Thus, as is common in the history of science, military needs resulted in technology that proved immensely useful in industry and scientific exploration.

As noted, ocean scientists and engineers in San Diego became engaged in matters relating to submarine warfare when the Navy's research agency initiated the University of California Division of War Research (UCDWR) on Point Loma. The focus of research was on the acoustic environment of the sea.[61] The tasks of UCDWR included compiling oceanographic information relating to underwater sound, devising offensive and defensive equipment for submarine warfare, and training Navy personnel in using the equipment. With a large contingent of scientists.[62] and support personnel, the staff built up to 600 in all, impressive for a research institute by any standards. Scripps's only research ship, the *E. W. Scripps*, was taken over to work for the laboratory during the time of war. The director of the facility, Vern O. Knudsen, recruited the physicist and mathematician Carl Eckart at the University of Chicago and appointed him director of research.[63]

The work of the scientists and engineers at UCDWR was eminently successful in contributing to the war effort—and pushing the frontiers of science in the process. Along with other laboratories in the United States and Great Britain,

San Diego in fact contributed to ending the menace of U-boats. In May 1943, the German submarine commander Karl Doenitz (a U-boat veteran of World War I) warned his captains that the Allies were depriving the U-boats of their invisibility.[64] After the war, U.S. Navy Admiral Jonas H. Ingram stated: "I'd say that what we won on was the ability of the American boys to learn faster than the Germans how to become expert in using the stuff scientists put out."[65] In this, he confirmed the wisdom of having scientists work with navy personnel when developing technology to be used at sea.

The war effort had catapulted ocean sciences into new territories, but much of what was gained was inaccessible, for obvious reasons. The transition from war research to "regular" research came in 1946, when the University of California Division of War Research ended and the Marine Physical Laboratory (MPL) was established by the University of California. This was chiefly at the urging of Roger Revelle, who persuaded the Navy to establish long-term research funding. The scientists and engineers at MPL have since created a host of novel and highly successful equipment for the acoustic exploration of the sea.

The research programs at the MPL at Scripps and Point Loma have dealt with ambient noise (the ever-present background noise in the water, from creatures and from ships), self-noise (generated by the equipment itself and the ship carrying it), and the scattering of the sound waves (from solid objects such as whales and schools of fishes and from peaks and hills on the seafloor). Other topics of investigation include the sediments and rocks beneath the seafloor, the magnetic field over the subsea landscape, and the amount of heat emanating from the seafloor. The central figure in guiding the research in these various fields of marine physics was the late Fred Spiess, physicist, consummate oceanographer, and inventive engineer. He joined MPL in 1952 and was its director from 1958 to 1980. His leadership has resulted in a host of important discoveries and in a large number of well-trained young scientists who became prominent in their own right.[66] Among seagoing scientists, Spiess's record is unique: after the age of 80 he was still leading expeditions exploring the ocean.

Among the outstanding achievements of MPL are those resulting from two unique platforms of observation: the FLIP (floating instrument platform), a tube-shaped vessel that rotates into a vertical position by flooding the bottom end, and the Deep-Tow, a tethered instrument package lowered to near the seafloor and carrying side-looking sonar, cameras, and other instruments powered from the ship. The package is taken close to targets near the seafloor, while underway. It is in constant communication with the ship, allowing the operator to maneuver the vehicle, that is, to "fly" it through the undersea landscape.

FLIP is a stationary platform for precise measurements of underwater sound (fig. 11.6). It is also useful in measuring waves, and especially internal waves, because it hardly partakes in vertical motions of the sea surface. Its design grew from an idea that a submarine positioned vertically, and with a platform on top, should make a fine stable observatory in the sea. Since its construction in 1962, FLIP has contributed to studies of long-range, low-frequency sound measurements, short-range, high-frequency sound, and the scattering of sound waves down from the surface. It has also been used for seismic studies, the motions of internal waves in the ocean, storm waves, upper ocean currents, air-sea interactions, and more.

The long tube-vessel is towed to its location, where its ballast tanks are filled, making it rotate into a vertical position (in about 20 minutes). The crew ends up standing on the former forward-facing walls, now horizontal. All internal equipment items are mounted on gimbals, so that they swivel to a usable position (even the stove). Upright, FLIP exposes a tower 55 feet high and extends 300 feet below the surface. Its ocean motion is about 5 percent of the height of passing waves.[67]

In one early project, FLIP was being used as a receiving station off Hawaii, while sound sources were activated at depth off the shores of the

FIGURE 11.6. The towed instrument platform FLIP in upright position, and ready to work.

Aleutian Islands. Spiess was content to take care of the receiving end. As he recalls it: "I can remember sitting out on FLIP a hundred miles or so north of Hawaii while other people ran sound sources up and down, and I sat there in lovely tropical weather, looking out at the surroundings and you knew that the people running the sound source were up there toward the Aleutians and the weather was not so nice, and you could sort of feel that you owned the ocean."[68]

The development and deployment of the Deep-Tow instrument package, conceived by Spiess, is one of the great success stories in deep-sea exploration. The purpose was to have "eyes and ears" at depth, sending down all sorts of instruments capable of recording important properties of the environment, and an arm for grabbing things. Spiess's group already employed a host of geophysical remote-sensing techniques, and Spiess now wished to adapt this technology for having a closer look, by bringing it next to the seafloor.[69] In particular, sonar equipment had progressed to a stage where it could light up the seafloor, acoustically, with high-frequency sound. Such sound does not carry far, but its echo delivers incomparable detail. Spiess knew that the costs of operating an unmanned deep-towed vehicle would be very modest when compared with the expense of sending a manned submersible to the seafloor. As an instrument platform, it would provide great flexibility in accepting different types of devices, and it could be readily moved and used on various research ships.

For those to whom "seeing is believing" it could produce close-up photographs, albeit without the sense of adventure linked to personal observation. In Spiess's words:

> A real strength of the vehicle is that you make multiple kinds of measurements all at the same time, and so they're all there in the same navigated reference frame. So you map the fine scale magnetics, the fine scale topography, the seafloor images, acoustic images using side-looking sonars, and all of this is there at once. And if you really need to document something in a little more detail, you can fly down a little closer to the bottom and use film cameras or video cameras to determine even more precisely what's down there.[70]

The depth accessible to Deep-Tow was set for at least 6 kilometers, using as much as 9 kilometers of wire. The vehicle position was to be determined within 1 to 2 meters, by means of triangulation among acoustic transponders.[71] The idea to use transponders was new and promising; it subsequently became standard in deep-sea exploration, for precise positioning of submersibles and drilling platforms.

Deep-Tow began with a towed echo-sounding device in 1961. An early addition was a proton precession magnetometer, to record the seafloor magnetic anomalies as closely as possible. Then came side-looking sonar, which sends out sound waves at an angle of 45 degrees and is useful in searching for specific objects on the seafloor. Other equipment has included a stereo photographic system to take photos of the seafloor geology and its life. A strobe light can be used for high resolution. Water samplers can be opened and closed at specified depths. Sensors for water temperature, clarity, and conductivity are part of the package. For safety, forward-looking sonar warns of obstacles in the way of the vehicle.

The Deep-Tow system (or "fish" as it is called at sea, like other towed devices) can be lowered, operated, and retrieved from any of the larger oceanographic ships, even in heavy seas. It weighs about a ton in air. While being towed, it can be operated by remote control of the main winch by the scientific party, from the ship's laboratory. As a general rule the unit is operated about 50 meters above the seafloor, to avoid accidents. (However, Spiess, when on board, did reserve the right to go a little closer if he chose.) On one early Deep-Tow operation, the towing wire broke and the equipment was left on the seafloor. But, from transponder records they knew exactly where it was, so Spiess arranged to have the instrument recovered on a scheduled trip 6 months later. The fish was retrieved.

The steady stream of results from Deep-Tow observations contributed greatly to the growing knowledge of the seafloor since the 1960s, on scales that matter in pushing the frontiers of exploration in that remote environment. Few areas of study in marine geology and geophysics—seafloor spreading, volcanism, fracture zone dynamics—have remained untouched, and the scope of results extends to abyssal circulation and abyssal biology. But perhaps the crowning achievement was the role Deep-Tow played in the discovery of the hot vents in the Galapagos Rift and elsewhere.[72]

LANDSCAPES UNDER THE SEA

The single most important discovery in marine geology concerns the nature of the Mid-Ocean Ridge. It is the mountain range that surrounds the entire planet, the place where the Earth cracks open and mantle basalt comes to the surface. The forces shaping the face of our planet were understood only when it was realized that this range marks the spreading center where new seafloor is being formed. Once the nature of the Mid-Atlantic Ridge was established, beginning with the survey of the *Meteor* Expedition (1925–1957) and culminating in the visually striking diagrams of Marie Tharp and Bruce Heezen, in the 1950s and 1960s, the door was open to speculation of what might be the cause of this ocean-dominating feature. But the significance of this jagged mountain range in the Atlantic remained a mystery for more than 30 years after it was discovered. The reason was not necessarily a dearth of ideas about its origin.[73] The reason was that the crucial observations regarding the magnetism of the deep seafloor were not made till the 1960s.

Surveying deep ocean landscapes using large ships for a long time is expensive and possible only with substantial government support. Seagoing nations everywhere are especially interested in acquiring detailed information about the nature of the seafloor in coastal waters, the site of much of the economic and military activity of a country. Thus, there was plenty of opportunity for surveying in coastal waters, beginning in the 1920s.

Off California, surveying by echo sounding started in the early 1920s, with equipment developed by the engineer Harvey Hayes.[74] The U.S. Navy ships *Hull* and *Corry* gathered information all along the Pacific coast, for the Carnegie Institution of Washington, where scientists worked on the causes of earthquakes. The resulting chart from San Diego to Santa Rosa Island, published in 1923, was the first bathymetric chart compiled from sonic soundings, with about 5,000 control points. It illustrated that the floor of the ocean from the coast out to 2,000 fathoms (3,700 meters) had plenty of topographic variation. Near the coast there were submarine canyons, and farther offshore one found banks, basins, and shoals, before the seafloor dropped off into deep water. Undersea landscape mapping had started.

By 1939 the U.S. Coast and Geodetic Survey had compiled a map of the intricate topography of the continental slopes of the United States. By the same year, the International Hydrographic Bureau had accumulated more than 300,000 soundings from oceanographic ships, cable ships, and commercial steamships, for their *General Bathymetric Chart of the Oceans*. The face of the deep was beginning to emerge on a large scale.

At Scripps, in the 1930s, seafloor research was being pursued on a shoestring, on the continental margin, just offshore. The results turned out to be of great interest to the geologic community in general.

One prominent question at the time was the origin of "submarine canyons."[75] With the systematic mapping of offshore regions, it was realized that wherever one looked, shelves and continental slopes bear deeply incised valleys. Their origin was the subject of intense discussion and emphatic disagreements (see chapter 5).

Off Scripps, we have the La Jolla Canyon, which ends in the deep San Diego Trough, off Point Loma. This canyon has two smaller tributary canyons, the Sumner branch and the La Jolla branch. Conveniently, one starts at the beach less than a mile north of the Scripps pier, the other about a mile south of it. Not surprisingly, the La Jolla Canyon and its tributaries are perhaps the most-studied and best-mapped canyon on the planet, with geologist Francis Shepard starting the project.[76] By default, these features have become paradigmatic in the explanation of ancient sediments, with the "typical" canyon system serving as a conduit for funneling mud and sand from the nearshore environment to the trough offshore, as discussed in chapter 5.

Initial studies were by rowboat and sounding lead, but in 1936, the *E. W. Scripps* had obtained an intriguing new piece of equipment—the fathometer.[77] The labors of Shepard's graduate students K. O. Emery and Robert Dietz were greatly simplified and enhanced by the new instrument. With it, they could record the depth every 2 minutes.[78] They now gathered more data in a day than they could earlier hope for in a week of good weather! The result was a great increase in data density.[79]

While the fathometer was nice, it was but the beginning. There followed continuous echo recording and automatic charting.[80] Continuous two-dimensional sounding combined with real-time graphic mapping arrived in the 1980s. In July 1992, Chris de Moustier, Peter Lonsdale, and Jacqueline Mammerickx watched the floor of the ocean appear as a map on the computer screen while the Scripps ship *Melville* cruised the waters off Acapulco, Mexico, at about 10 knots. The test of the new equipment, called Seabeam™, was going well, and a swath of ocean floor 120 degrees wide showed as humps and hollows, as detailed as a topographic map of San Diego County. The three scientists were elated. Mapping the seafloor had reached a new level of sophistication.[81]

The detailed mapping of the deep-sea floor is a vast undertaking and is by no means completed. In recent times, satellite gravity maps have given us a coherent view of seafloor topography, albeit not a detailed one.[82] Back in the 1930s, when echo sounding was in its infancy, the enormous expanses of the Pacific Ocean still had only scattered soundings. During World War II U.S. Navy ships were outfitted with echo-sounders. These were primarily intended to help avoid underwater hazards in shallow water, and the Navy advised against using them on the open ocean, because their pinging could reveal the ship's location. When the United States entered the war, Harry Hess,[83] an officer in the Naval Reserve, was called to active duty; he became captain of the assault transport *Cape Johnson*. Against the recommendations, he used the echo-sounder during regular transits in the Pacific. The reward for his initiative was the discovery of 20 flat-topped seamounts between Hawaii and Guam. He named them guyots. He explained them as ancient sunken islands—islands that had their top planed off by wave action at one time. The discovery and Hess's proposition became the focus of much discussion regarding the origin and nature of the Pacific basin (see chapter 12).

To some geologists, Hess's findings and his provocative interpretation were the single most intriguing problem in deep-sea geology after the war in the 1940s. Roger Revelle, Robert Dietz, K. O. Emery, Bill Menard, Russ Raitt, and a graduate student by the name of Ed Hamilton, set out to find answers in the central Pacific west of Hawaii.[84] The scientists of the MidPac Expedition, SIO's first major postwar venture, left La Jolla in July of 1950. Their goal was to explore the nature of the Pacific seafloor—its deep structure, its heat flow—and the nature of the guyots. Keeping the echo-sounder running, they found additional guyots, confirming the abundance of these features in the region west of Hawaii. More significantly, they discovered a major undersea mountain range, 2,000 miles long and 60 miles wide, entirely new to science. They called the region the Mid-Pacific Mountains.

The most striking discovery of the expedition, in regard to the guyot problem, was the finding that at least some of the flat tops of the deep seamounts are covered by shallow-water reef material of mid-Cretaceous age.[85] The proof was in the material obtained by dredging on top of the guyots. Obtaining and describing such stuff was to form the basis for the doctoral thesis for Ed Hamilton, a point established with Robert Dietz, before the expedition.[86] Hamilton's write-up, "Sunken Islands of the Mid-Pacific Mountains," was published (in 1956) by the Geological Society of America and gave him instant prominence. Hess's proposition that the guyots were sunken islands had proved correct. However, the concept of great age had to be abandoned—these seamounts had been at the surface only a hundred million years ago.

What made them sink? Had sea level risen since? Most geologists familiar with Cretaceous marine sediments on land would have sworn that sea level had fallen since the Cretaceous, not risen. The geologists on board were creative thinkers, and there was no shortage of hypotheses. But the time had not yet come for synthesis—that was still some 15 years into the future.[87]

On the way home from the MidPac Expedition, the three remaining geologists in the scientific crew[88] took the *Horizon* on a northerly path to try to intersect the Mendocino Escarpment, a long linear feature on the seafloor running almost straight west from off Cape Mendocino in California, and which could be tracked on echo-sounding records of the Coast and Geodetic Survey some 300 miles out from the coast. The idea was to find how far west this scarp could be detected.[89]

What emerged from this and a follow-up expedition with additional crossings (Northern Holiday, in 1951) was the existence of an incredibly long east-west-running linear fault scarp, representing a persistent jump in seafloor elevation and spanning some 40 degrees of longitude.[90] Nothing like this had ever been seen previously by any geologist, anywhere on the planet. And this scarp is not the only one—similar features run parallel to it farther to the south. In subsequent expeditions, again using simple echo-sounding equipment, Menard mapped out these great fracture zones of the eastern North Pacific: Mendocino, Murray, Molokai, Clarion, and Clipperton (fig. 11.7). Their nature was a complete mystery and remained so till Jason Morgan explained them in a talk given at the American Geophysical Union, in April of 1967. He proposed that they describe the tracks of large rotating pieces of real estate on the surface of the planet. His talk signaled a new way of looking at the planet: plate tectonics started its ascent to the dominant concept in the Earth sciences.[91]

Thus, two major discoveries relating to the nature and history of the Pacific basin—the presence of flat-topped seamounts in the Mid-Pacific mountains, and the enormous linear features called fracture zones—resulted from echo sounding. And they both are linked to that first trans-Pacific expedition on the *Horizon*, exploring the unknown floor of the basin that covers almost half of the planet. Menard called the *Horizon* a "fat, slow, dirty, uncomfortable, but happy little ship." It displaced about 900 tons and sailed more than 600,000 miles in its travels between 1948 and 1969.[92]

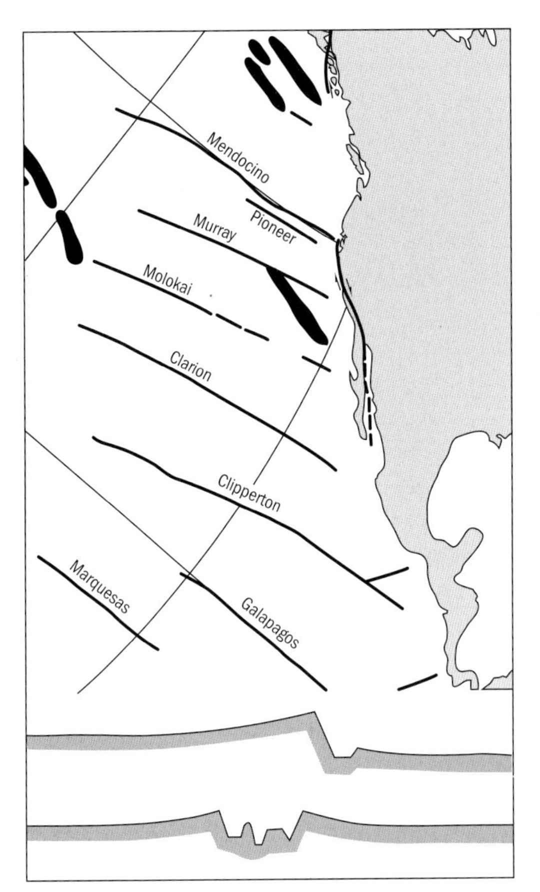

FIGURE 11.7. Menard's great fracture zones of the northeastern Pacific, as published in 1964. The features were discovered by echo sounding. Their explanation (in 1967) marks the beginning of "plate tectonics."

The years that followed brought a flood of soundings and feverish mapping activity, along with geophysical exploration. In his book *The Ocean of Truth* Menard refers to a flood of new soundings from numerous expeditions that threatened to drown the investigators in data. Once, he notes, it took him and Bob Fisher three months to plot and reduce the soundings from a single expedition. When either he or Bob Fisher were part of an expedition, they saw to it that the navigation was adjusted and the soundings plotted before they stepped ashore, but with six Scripps ships it was impossible to be on board all of them. Menard recalls: "I kept hiring more technicians to plot data and sending students out to do research on marine topography. Still the problem grew. In the early 1960s Bruce Heezen and I attended several meetings at the International Hydrographic Bureau at Monaco. . . . By the end of the first meeting Bruce and I realized that Lamont and Scripps were collecting more and better soundings than any navy in the world."[93]

Within two decades after the war, the maps and profiles resulting from echo sounding and seismic exploration would reveal how the planet really works, and a few years later all earlier geology textbooks would become obsolete.

LOOKING BELOW THE SEAFLOOR

In echo sounding, sound penetrates the water column and returns from the seafloor. But if the sound is powerful enough, and the seafloor has mud on it rather than very firm sediment or hard rock, much of the sound will penetrate into the seafloor, and some of it then will be reflected from layers below. Suddenly, we are looking *inside* the seafloor. We are dealing with a form of "seismic" exploration. Such exploration, when using high frequencies, can yield detailed information about layering of sediments, down to several hundred feet.

The point is, the step from seeing the seafloor to seeing below the seafloor, using sound, is not very great in principle. What is needed is for the sound to penetrate into the seafloor, and for some of it to come back with sufficient power to be recorded. In the example shown in fig. 11.8, we see the layering in sediments off southwestern Africa that accompanies the great climate cycles of the last million years.

In making the step from echo sounding to deep seismic profiling one turns to the use of longer wavelengths, that is, to sound of lower frequency. A "pinging" on the seafloor, at a frequency of several kilohertz, will show an echo from the bottom, and below, depending on conditions. But "booming" on the seafloor in a frequency close to 100 hertz, will produce echoes from well below the bottom.[94] The lower the frequency, the better the penetration deep into the seafloor. But there is a trade-off in this quest for penetration. Resolution suffers when one switches from high to low frequencies: at a

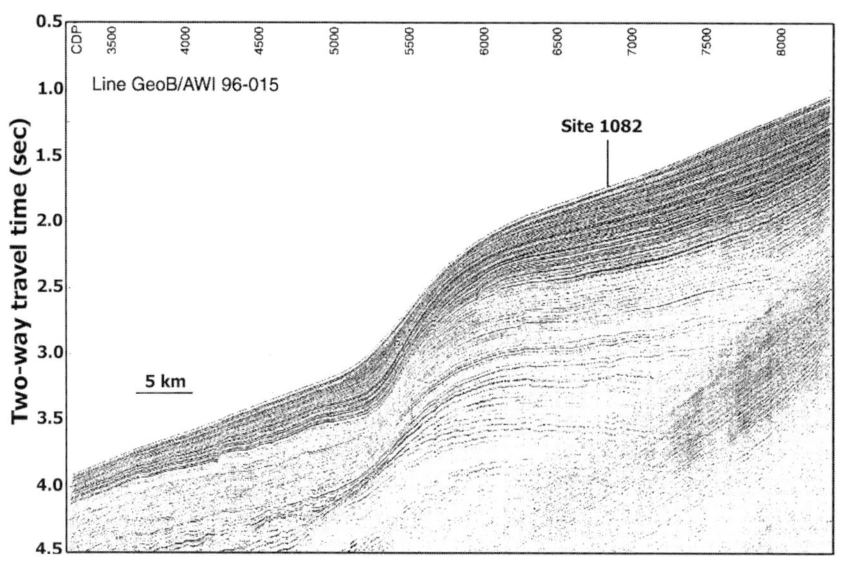

FIGURE 11.8. Sediment layers on the continental slope off southwestern Africa, obtained in preparation for drilling site 1082 of the Ocean Drilling Program, using a sound source of intermediate strength and frequency. GeoB, Geo Bremen; AWI, Alfred-Wegener-Institut.

velocity of 2,000 meters per second (normal for upper sediments), a 2-kilohertz sound has 1-meter resolution, but a 100-hertz sound has a wavelength of 20 meters, thus bypassing structural details that are smaller than that. For many purposes, such as mapping sediment layers on a continental slope, a sound source intermediate between a ping and a boom yields good results (fig. 11.8).

The use of sound to look inside the Earth has a long tradition in the investigation of the transmission of earthquake energy. One type of earthquake wave (there are several) is the one described as compressional, which is the same as sound waves. From the way these waves travel through the Earth, and from the way shear waves (transverse waves) are transmitted, the internal structure of the Earth can be determined.[95] Of central importance to the understanding of the origin of the seafloor is the nature of a jump in sound velocity some 10 kilometers below the seafloor, a transition known as the Mohorovičić discontinuity, M discontinuity, or (in a mixture of jargon and vernacular) the Moho.[96]

The layer between the surface of the solid Earth and the M discontinuity is known as the crust. Thus, the crust is foremost a sound velocity unit, and its composition is a question that remains open till materials are sampled and their nature determined. (This is not always clearly stated in textbooks.) The thickness of the oceanic crust was established in the 1950s by seismic exploration. It turned out to be several times thinner than the crust on the continents.

The fact that the ocean's crust is much thinner than the continental crust was a major discovery in the Earth sciences. It established that the processes shaping the ocean bottom must be fundamentally different from those shaping the continents.

The foremost application of the technology involving seismic exploration (that is, listening to returns from self-generated loud noise with a strong low-frequency component) has been the search for petroleum and gas reservoirs. The development of the appropriate sound-making and sound-recording instruments took off after World War I, as demand for petroleum increased, and again after World War II, with the expansion of the automobile market thirsting for oil, and the advent of computing devices that facilitated the analysis of seismic data. The classic task in early oil exploration was to find an upward bulge of layers with some sort of "closure," within which hydrocarbons could be trapped.[97] Oil and gas are trapped below a bulging layer, provided it is impermeable, because hydrocarbons rise, being lighter than the water filling the spaces between solids in the sediment.[98] The new methods for finding oil were readily adapted to studying the structure of the oceanic crust, at sea.

Studies using seismic refraction are about following the path of sound when source and recorder are separated over a considerable horizontal distance. Since sound travels faster at greater depth below the seafloor, the various arrival times from different layers in the sediment will vary in characteristic fashion as the horizontal distance is varied. For a larger distance, the deeper-traveling sound waves gain an advantage over the shallow-traveling ones in the race to the recorder. The result of this race can be analyzed and the nature of the layering (or better, the speed within the race tracks below the seafloor, which are tied to the layers) can be inferred. The output is information about the thickness of the layers and the various velocities of sound therein. In contrast, seismic reflection is about counting seconds from sending out a pulse downward, to hearing its echo. In the reflection mode of exploration, there is no way of extracting the speed of the sound in its path from sound source and back. Nevertheless, the various layers are nicely "reflected" in the record (fig. 11.8). We just do not know how thick they are, unless there is independent information on sound velocity as a function of depth below the seafloor.

The refraction method was the first one to achieve success, both in finding oil and in measuring seafloor structure, but the reflection method is the one now more commonly used, because it is more amenable to routine application.

The first to use field seismic techniques to look below the seafloor was the physicist and geologist Maurice Ewing, when he was an instructor at Lehigh University in 1935.[99] While a student at Rice University, he had worked during the summers with an oil-exploration crew in Louisiana, which used gravity and seismic-refraction methods. He subsequently applied these methods to exploring the nature of the continental margin off the East Coast.[100] From the research vessel *Atlantis* of Woods Hole Oceanographic Institution, he lowered a seismograph to the seafloor to record the arrivals from explosive charges going off at various distances from the ship. He put the charges on the seafloor, so the energy could immediately penetrate into the sediment below.[101] To the surprise of the few who were interested in these experiments, he recorded sediments some 3,800 meters thick (more than 2 miles) off the coast of Massachusetts on the continental shelf.

Ewing concluded that the continental slope was not a fault separating a rising continent from a sinking seafloor. For the shelf to collect so much sediment, it had to sink. The continent-ocean boundary was hidden below a thick stack of sediments! In the next few years following, he extended this program to deeper water beyond the shelf whenever he could find an available ship, until World War II made such research impossible.[102]

Ewing's studies opened new avenues to study fundamental questions about the nature of the seafloor: How thick are the sediments in the margins? How thick in the deep sea? How much of Earth's history do they contain? What is the nature of the rock below the sediments? What is the depth to the M discontinuity that marks the bottom of the ocean's crust and the top of the mantle?

Generally, for those thinking about these matters, the assumption was that the deposition of sediments had been going on, in the deep sea, for all of Earth's history, and that an undisturbed record going back into the Precambrian would be found. This expectation included the presence of a very thick layer of deep-sea sediment. Reality proved to be quite different, eventually. The average age of the seafloor turned out to be near 60 million years—only 10 percent of the time since we have rocks on land bearing visible fossils. But that discovery was still several decades away when Ewing started his seismic experiments.

After World War II Ewing resumed his quest. He joined the faculty of Columbia University and a few years later (in 1948) obtained support to found a new institute within the university: Lamont Geological Observatory.[103] In 1949, Ewing's associate J. Lamar Worzel and Woods Hole physicist J. Brackett Hersey, analyzing a long seismic-refraction line they had run, determined that the M discontinuity was only 5 kilometers beneath the ocean floor, much shallower than on land.[104] In 1953 Lamont acquired the schooner *Vema,* and with growing funding from the Navy, Ewing was able to expand the exploration of the deep ocean rapidly, while developing and adapting new technology to do so.[105]

Lamont, from there on, became a major force in advancing knowledge about the structure and history of the continental margins and the deep-sea floor, especially in the Atlantic Ocean. The geophysical data collected at Lamont in the following years eventually became a crucial ingredient in the plate-tectonics revolution. The sediment cores collected by the *Vema* from the seafloor helped revolutionize our understanding of the history and origin of the ice ages.

SEISMIC EXPLORATION OF THE PACIFIC

In the Pacific, the deep ocean floor likewise became the subject of intense scrutiny, after World War II. Comprising nearly half of the planet, it was still entirely terra incognita. It was scientists from Scripps and their colleagues at the nearby Navy research facility (NEL) on Point Loma whose efforts began to change this situation after the war.[106]

The leader of the effort was the physicist Russ Raitt[107] (fig. 11.9). Raitt was one of the first scientists to be appointed to the newly founded Marine Physical Laboratory. He immediately set out to devise new equipment for seismic exploration. This included developing the technique of floating the hydrophone array at nearly neutral buoyancy at depths of 100 to 200 feet. He tested new ideas on many trips using Scripps ships just offshore, in the Southern California Borderland, a region made up of islands, banks,

FIGURE 11.9. Geophysicist Russ Raitt at sea, collecting data on the nature of the ocean bottom in the central Pacific. Capricorn Expedition, 1952–1953.

and basins. Knowing exactly where one was located when doing experiments during these short expeditions "left a great deal to be desired," he said in an interview, commenting on the means of navigation and the quality of maps available at the time.[108]

Seismic refraction at sea requires setting off a series of explosive charges along a straight line, while increasing the size of the shots as the distance to the recording station increases (to overcome the loss of energy with increased time of travel). The recording station can be a moored listening device or a second ship. The moving "shooting" ship sets off the explosives and the stationary "receiving" ship, knowing the shooting time by radio, records the time of the returns. Initially, Raitt tried the two-ship method with one person in a whaleboat to set off the explosives from a single point while the recording ship sailed off to stop at predetermined points along a line to record each explosion. His assistant then was Art Raff.[109] Raff, sitting alone in a small boat full of explosives, recalls an "eerie, lonely feeling" when observing the mother ship disappearing over the horizon. He did have radio contact, of course, which came in handy when he had to ask to be picked up one time when a sudden wind made things highly uncomfortable.[110] This particular variety of the two-ship method was soon given up.

From his experience with previous experiments, Raitt was ready for the MidPac Expedition in 1950. Besides the *Horizon*, the research ship of the Navy Electronics Laboratory *EPCER-857* participated, which gave Raitt a second ship for refraction work. The size of explosive shots was 5 pounds close in and as much as 480 pounds at the far end of a line, which could be as distant as 100 miles. The big shots and the long distances allowed the sound to dive deeply below the seafloor to probe the M discontinuity, identified by an increase of velocity from 7 kilometers per second or slightly less to about 8 kilometers per second or slightly more.[111]

The work by Raitt and his associated in the deep Pacific gave indications that the ocean sediment layers are only a few hundred meters thick, rather than thousands of meters, and that they are even thinner near the crests of ridges. So much for the expectation of a thick layer, accumulating since the beginnings of Earth history. Where was all the sediment that supposedly collected on the bottom of the ocean since the Earth was young? Were the sediments there but had become so solid that they looked like hard rock when measuring sound velocities? Or were they covered up with thick layers of lava? No one had the faintest notion.

In addition, the depth to the M discontinuity, that is, the thickness of the crust, turned out to be only 10 kilometers or thereabouts. This was a great surprise, since it is typically about 30 kilometers beneath land (and more beneath mountain ranges). Raitt also discovered on MidPac that the crust was layered in a rather simple fashion. Menard recalled:

> Moreover, the crust consisted of only three layers that Russ with typical clarity called the First, Second, and Third layers. The first was clearly sedimentary, because we cored the top, and the seismic-wave velocity was typical for sediment. The second, at station M1, was 0.9 km thick and had a velocity of 5.9 ± 0.2 km/sec. After Midpac, Russ gave a talk about his results at a geological meeting in Los Angeles. In the question period, an impatient geologist asked what kind of rocks he was talking about, basalts or whatever. Russ is sometimes painfully deliberate but he answered directly, "These are rocks that transmit seismic waves at about 5.9 and 7.0 km/sec." He brought down the house.[112]

We know more about the nature of these layers today, largely thanks to drilling. The first layer is sediment, as Menard stated. The second layer consists of basaltic dikes and pillow basalt, and the third layer consists of gabbro, that is, coarsely crystalline basalt. Thus, layers two and three are both volcanic. They originate at the spreading center, the upper one from intrusions and extrusions of basalt rising from the central magma chamber that is located underneath the spreading center, and the lower one from the walls of the magma chamber. It took another 30 years or so for Raitt's discoveries to become the stuff of exams in Geology 101. By now, of course, they are entirely integrated into the new worldview based on the theories comprising continental drift, seafloor spreading, and plate tectonics. The nature of the oceanic crust is fundamental knowledge, and the need to explain Raitt's results was a crucial ingredient for inventing seafloor spreading.

Raitt continued to go to sea on Scripps expeditions throughout the 1950s, often with his colleagues, geophysicist George Shor[113] and geologist Robert Fisher,[114] and by the end of that decade it was determined that the sediments are quite thin in practically all parts of the Pacific, that the velocity in "layer 2" is quite variable, and that the primary crustal layer at the bottom ("layer 3") has a very uniform sound velocity of 6.8 kilometers per second. Below the crest of the East Pacific Rise the sound velocity in the mantle is comparatively low, the M discontinuity is deeper beneath the axis of trenches, and it is found at "continental" depths under the continental shelves.[115]

Raitt was known to change from an amiable colleague, somewhat hesitant in conversation as he pondered a point, to a drill sergeant at sea, when he was directing his seismic runs. When handling seismic charges, mistakes are forbidden.

The runs could last for 12 hours. In total concentration, Raitt firmly snapped out brief specific commands. Over the radio to those on the firing ship, he would say: "I want another 200-pound shot in 9 minutes. Tell me as soon as it's ready," and to those around him in the lab: "Get me a position from the bridge; turn the volume up on that speaker." Everyone accepted his leadership and jumped to do his bidding. Raitt enjoyed going to sea: both the intensive work aboard ship and the exotic ports, where he took jaunts into the hinterland and relished the local cuisine.

Doing it, and doing it professionally, was everything to Raitt. Publishing and showing off his results were much less important. When he knew the results of a survey, he was eager to go on and find out more. It took 6 years to get the MidPac results published. In the meantime, Ewing and his associates at Lamont started to explore the crustal profiles in the northwestern deep Atlantic, with comparable results regarding the nature of the oceanic crust.[116] The fact that Raitt had data for a large area of the Pacific makes it fair to say that he and his team established the typical thickness of the oceanic crust and what it looked like, and that it looked the same over a wide region.[117]

By the 1970s, the seismic-refraction method lost appeal, partly because of the cost of running two ships simultaneously. Methods were now needed that would map the subsurface in finer detail, and that would work well on inclined and thickly covered continental slopes. Geophysicists working on land looking for oil had developed seismic-reflection techniques to a high art, using single explosive shots and multiple receivers. Seismic reflection, which is essentially echo sounding using powerful low-frequency waves, gives much more detail than refraction does about the shape of the interfaces between layers, but requires many more explosive shots than refraction work and usually gives less depth penetration.

Ways to make loud noise for use in seismic exploration had proliferated quickly once this type of survey became important for finding oil. Explosives were replaced by propane-oxygen mixtures, electric sparks, steam bubbles, compressed air, hydraulically actuated plungers, mechanical vibrators, and other ingenious schemes. One of the more spectacular systems was the Rayflex Arcer, used for some time at Scripps. It ran on electric energy stored at 20,000 volts in a tremendous capacitor bank and discharged as a short circuit into the water. It lit up the sea with a blue green flash. On one occasion at night in the Indian Ocean, the Arcer got a beautiful response from bioluminescent organisms on each flash.[118] During surveys off the Pacific entrance to the Panama Canal, passing ships diverted to watch the light show; their curiosity was a hazard for the long hydrophone streamer towed behind the ship.

Generating sound by electric sparking had its drawbacks—not everyone was born to handle high-voltage electricity with the ease of the unusually dry-skinned technician Sammuli.[119] A much safer and more easily handled instrument for making noise proved to be the air gun, a steel cylinder that releases high-pressure air into the sea through a fast-opening valve. Air guns come in various sizes, and the larger ones make the bigger noise, with the lower frequency, for deep penetration. The smaller ones are good for high resolution, but their sound does not penetrate as deeply into the sediment. The air gun is towed not far behind the ship, and the receiving hydrophones are towed farther back, away from the ship's noise.

Air guns have been used routinely, on Scripps ships and those of other institutions, to map the thickness of sediments in the deep ocean and on the margins. Such surveys are important in preparing for drilling, because they reveal where the sediment stack is likely to be complete and where it is disturbed (fig. 11.8).

When compiling the data from the central and eastern equatorial Pacific, an important discovery emerged: the sediments just north of the equator were found to be unusually thick.[120] Drilling revealed that the sediments accumulated under the high-productivity zone of the equator. The sediment stack kept migrating northward, across the equator. It took the insights from seafloor spreading and plate tectonics to explain the pattern.[121]

For mapping the structure under the edge of the continental shelf, the traditional methods of seismic refraction and seismic profiling fell short, because of the many side reflections from the rugged slopes. A more elaborate system, which had been developed for the oil industry, came into use: multichannel seismic reflection, with air guns as the sound source. In these systems, the streamer containing the hydrophones became ever longer, with arrays 7 kilometers long not unusual and holding a thousand or more hydrophones. The signals from the many hydrophones are digitized, combined, and computer processed in many ways to obtain velocity data and a cross section of the structure below the seafloor. These very expensive systems are widely used for commercial mapping, and only occasionally for oceanographic research.[122]

A new development in marine seismic exploration is the use of ocean-bottom seismometers, which passively record vibrations from earthquakes. Traditionally, seismometers have been used on land, on continents and islands. This situation has changed rapidly in recent years, as researchers in Japan, Germany, and the United States have emplaced considerable numbers of seismometers on the seafloor.[123] Results bear on the workings of the volcanically and tectonically active machinery at the Mid-Ocean Ridge. In addition, since 1993, the U.S. Navy's declassified underwater sound and surveillance system has been employed to monitor earthquakes, and also for other purposes such as monitoring changes in current structure and temperature (through acoustic tomography), and biological sound production.[124]

Sound, in the deep sea, is the magic key to opening the treasure chest of Earth knowledge—many fundamental discoveries are yet ahead.

NOTES AND REFERENCES

1. A recent survey of findings of acoustic research relevant to the Marine Mammal Protection Act is in W.J. Richardson, C.R. Greene, C.I. Malme, D.H. Thomson, 1995, *Marine mammals and noise.* Academic Press, San Diego, 576 pp. The U.S. Office of Naval Research provided funds toward completion of this review.

2. Reginald Aubrey Fessenden (1866–1932), a former assistant of Thomas Edison, invented a powerful sound generator that proved successful: in a field experiment, steaming away from a visible iceberg, at a distance of 2 miles the ship was still recording echoes from an oscillator suspended in the water. Also, a clear echo came back from the seafloor, 1 mile deep. S. Schlee, 1973, *The Edge of an Unfamiliar World.* E.P. Durron, New York, 398 pp. Fessenden had found a way to detect objects in the water, and to measure distance to the seafloor.

3. Edward Hanson Smith (1889–1961) worked with the International Ice Patrol, which was established in the wake of the *Titanic* disaster. The U.S. Coast Guard's ice patrol operated within that framework. Smith became director of Woods Hole Oceanographic Institution in 1950.

4. The name is an acronym, from *sound navigation ranging.*

5. The words *sound* (noise) and *sounding* (finding depth) are not related. The first has a Latin origin (*sonus;* French *son*). The second is related to depth and is of Anglo-Saxon origin. To fishermen around the North Sea, a "sound" (Dutch *sont,* German *Sund*) is a safe path for ships through shallows. Thus, "sounding" has to do with finding the right depth for proceeding in treacherous waters. For centuries, the technique for measuring beyond the depth of a hand-held pole from a boat was with a rope marked at intervals (usually the spread of one's outstretched arms, a "fathom" in English) with knots or tags that could be counted, and with a weighted sinker called a sounding lead. It is amusing that sounding for depth is now determined by sound waves, giving "echo sounding" a double meaning.

6. Lamont (today's Lamont-Doherty Earth Observatory) is part of Columbia University. B.C. Heezen (1924–1977) obtained his Ph.D. from Columbia University; he began his work in collaboration with Maurice Ewing, Lamont's founder director. The physiographic diagrams, based largely on Lamont data but also others, were published in the 1960s, the earliest such effort being B.C. Heezen, M. Tharp, and M. Ewing, 1959, *The Floors of the Ocean I.* The North Atlantic Geological Society of America Special Paper 65, 83–104. A photo of Marie Tharp next to such a map (on p. 616), and several examples of the maps are in B.C. Heezen, C.D. Hollister, 1971, *The Face of the Deep.* Oxford University Press, New York, 659 pp.

7. The distance of sight discovered by the first set of experimenters was not news to the many oceanographers who had used a Secchi disk onboard a research vessel. The Secchi disk is a solid white wheel

of 30 centimeters diameter. By noting the depth to which it can be seen when lowered overboard on a weighted line, a rough estimate of absorption of light can be made. The method was used routinely in the early part of the twentieth century to estimate the clarity of the water. Hence, the distance to which objects can be seen—in plain sunlight—was known, in principle.

8. The precise value depends on temperature, pressure, and salinity.

9. J. Y. Cousteau, F. Dumas, 1953, *The Silent World.* Harper, New York 225 pp.

10. N. B. Marshall, 1966, *Life of Fishes.* Universe Books, New York, 402 pp, citing J. M. Mouton at Woods Hole. Marshall was the senior principal scientific officer in the Department of Zoology at the British Museum at the time of publication of the book *Life of Fishes.* He began his career in 1937, with research on plankton.

11. M. P. Fish, quoted in Marshall, *Life of Fishes,* p. 166.

12. The term g stands for Earth's gravity, that is, for an acceleration near 10 meters per second per second. At 3g, for example, given one-third of a second to evade an attack, a fish will travel at 20 miles per hour.

13. A. N. Popper, D. M. Higgs, 2001, *Fish hearing, lateral lines,* in J. H. Steele, S. A. Thorpe, K. K. Turekian, 2001, *Encyclopedia of Ocean Sciences,* 6 vols. Academic Press, San Diego, pp. 922–928, p. 923.

14. N. B. Marshall, *Life of Fishes,* p. 145.

15. A. N. Popper, D. M. Higgs, *Fish hearing, lateral lines,* p. 926.

16. W. N. Tavolga (ed.), 1967, *Marine Bio-Acoustics,* vol. 2. Pergamon Press, Oxford, 353 pp.

17. N. B. Marshall, 1967. *Sound-producing mechanisms and the biology of deep-sea fishes,* in W. N. Travolga (ed.), *Marine Bio-Acoustics,* vol. 2. Pergamon Press, Oxford, pp. 123–133 p. 126.

18. *Ibid.*, p. 131. A link to territory would explain why rattail populations below 1,000 meters depth apparently are not sound-producers. Food supply decreases rapidly with depth, away from the fertile coastal zone. The territories necessary to support a pair of fish may be too large to make noisy claims worthwhile.

19. H. Schneider, 1967. *Morphology and physiology of sound-producing mechanisms in teleost fishes,* in W. N. Tavolga (ed.), 1967, *Marine Bio-Acoustics,* vol. 2. Pergamon Press, Oxford, pp. 135–158, p. 149.

20. Jules Jaffe, pers. comm., 2005. Jaffe is at Scripps; he studies marine bioacoustics.

21. M. Johnson, 1948, *Sound as a tool in marine ecology, from data on biological noises and the deep scattering layer.* Journal of Marine Research 7, 443–458. Johnson was talking about croakers.

22. Roger Payne, pers. comm., 2003. Payne wrote the book *Among Whales* (1995, Bantam Doubleday, New York, 431 pp.). He is with the World Wildlife Fund.

23. Kenneth Stafford Norris (1924–1998) received his Ph.D. at Scripps in 1959. He did research on mammals in several places, last at the Santa Cruz campus of the University of California, with an emphasis on conservation.

24. K S. Norris, W. E. Evans, *Directionality of echolocation clicks in the rough-tooth porpoise,* Steno bredanensis (Lesson), in W. N. Tavolga (ed.), *Marine Bio-Acoustics,* vol. 2. Pergamon Press, Oxford, pp. 305–316, p. 305.

25. D. H. Cushing, 1951, *Echosurveys of fish.* Journal du Conseil International pour l'Exploration de la Mer 18, 45–60. Cushing worked at the Fisheries Laboratory, Lowestoft, Suffolk, England.

26. D. H. Cushing, 1967. *The acoustic estimation of fish abundance,* in W. N. Tavolga (ed.), *Marine Bio-Acoustics,* vol. 2. Pergamon Press, Oxford, pp. 75–91, fig. 5, p. 83.

27. Standard-target calibration has made the surveys quantitative, and the ability to use sound beams like giant flashlights and to detect changes in the pitch of returning sound (Doppler shift) makes it possible to follow the motions of individual schools or even individual fish.

28. W. J. Richardson et al., *Marine mammals and noise,* pp. 190–191.

29. Scripps geologist Francis P. Shepard and his physicist colleague Russell Raitt had for some while noted crackling sounds at times in La Jolla cove, their favorite spot for a daily swim. One of them commented on it to Martin W. Johnson, who took up the challenge to figure out the source of the sounds.

30. F. A. Everest, R. W. Young, M. W. Johnson, 1948, *Acoustical characteristics of noise produced by snapping shrimp.* Journal of the Acoustical Society of America 20 (2), 137–142.

31. W. J. Richardson et al., *Marine mammals and noise,* p. 347.

32. Russell W. Raitt (1907–1995), with colleagues Ralph J. Christensen and Carl F. Eyring, worked on the development of high-resolution downward-directed echo-sounders at a lab of the University of California on Point Loma, San Diego. Raitt later joined the Scripps faculty, in 1946.

33. On occasion such shallow echo readings had been entered on charts as shoals, providing useless warnings to cautious captains, and misleading clues for geologists.

34. While we like to think of the discovery as one made by Scripps scientists (since both Russ Raitt and Martin Johnson were on the faculty at Scripps for many decades), reality is a bit more complicated.

Early in 1941, under the U.S. Navy's Office of Scientific Research and Development, a laboratory was established in San Diego called the University of California Division of War Research. It found a home at the navy's Radio and Sound Laboratory on Point Loma, the peninsula protecting the bay and port of San Diego. Headed by physicist Vern O. Knudsen from the University of California Los Angeles, it was set up to develop ways to locate enemy submarines, which turned out to be a problem in ocean acoustics. It was this setting, at the end of the war, that produced the discovery of the deep scattering layer. Early results were published in a special issue of the *Journal of Marine Research* dedicated to H. U. Sverdrup, on the occasion of his sixtieth birthday, in 1948. R. W. Raitt, 1948, *Sound scatterers in the sea.* Journal of Marine Research 7, 393–409; M. W. Johnson, *Sound as a tool;* R. S. Dietz, 1948, *Deep scattering layer in the Pacific and Antarctic Oceans.* Journal of Marine Research 7, 430–442. Dietz was at Point Loma, as well, and worked closely with colleagues at Scripps.

35. *Sagitta* belongs to the phylum Chaetognatha, bristle jaws, a rather isolated and smallish group with the general name arrow worms.

36. Michael's study is identified by Sir Alister Hardy as the first of its kind. A. Hardy, 1956, *The Open Sea: Its Natural History. Part I, The World of Plankton.* Houghton Mifflin, Boston, 335 pp., pp. 202–203. Also, Martin knew about the work on the vertical migration of copepods by F. S. Russell, in Britain. F. S. Russell, 1927, *The vertical distribution of plankton in the sea.* Biological Reviews 2, 213–262. F. S. Russell, 1928, *The vertical distribution of macroplankton. VII. Observations on the behavior of* Calanus finmarchicus. Journal of the Marine Biological Assoc. of the U.K., N.S. 15, 429–454. (This is the paper cited by M. Johnson, *Sound as a tool.*) Hardy also refers to important work on vertical migration by A. G. Nicholls, in the Clyde Sea, and by George Clarke, off the U.S. East Coast, both in the 1930s. Thus, the phenomenon of the scattering layer fit well with known or suspected plankton behavior at the time Martin Johnson considered the matter.

37. F. S. Russell, C. M. Yonge, 1936, *The Seas, Our Knowledge of Life in the Sea and How It Is Gained,* 2nd ed. Frederick Warne, London, 379 pp., p. 128.

38. J. B. Hersey, R. H. Backus, 1962, *Sound scattering by marine organisms,* in M. N. Hill (ed.), *The Sea, Ideas and Observations on Progress in the Study of the Seas.* Vol. 1, *Physical Oceanography.* John Wiley and Sons, New York, pp. 498–539.

39. K. G. Foote, 2001, *Acoustic scattering by marine organisms,* in J. H. Steele, S. A. Thorpe, K. K. Turekian, 2001, *Encyclopedia of Ocean Sciences,* 6 vols. Academic Press, San Diego, pp. 44–53.

40. D. Behrman, 1992, *John Isaacs and His Oceans.* American Geophysical Union, Washington, D.C., 230 pp., p. 154.

41. Athelstan Frederick Spilhaus (1911–1998), meteorologist and oceanographer, came to the Unites States from South Africa in 1931. He was associated with Woods Hole Oceanographic Institution from 1936, and dean at the University of Minnesota from 1949. Carl-Gustaf Arvid Rossby (1889–1957) came to the United States from Sweden. He was professor of meteorology at MIT from 1931 to 1939 and then had appointments at the U.S. Weather Bureau and the University of Chicago.

42. A. Spilhaus, 1987, *On reaching 50: An early history of the bathythermograph.* Sea Technology 28, 19–28.

43. The slides were greased before smoking, at first simply by rubbing a forefinger along the side of one's nose and then onto the slide. When greater numbers were needed, skunk oil was recommended by C. O'D. Iselin of Woods Hole Oceanographic Institution, and that advice was followed for several years. The reasons that drew Iselin to skunk oil had to do with the oil's resistance to being washed off. However, the Iselin method did involve some olfactory discomfort.

44. Woods Hole's Allyn Collins Vine (1914–1994), engineer and oceanographer, is best known for designing the research vessel *Alvin,* which carries his name, albeit in somewhat cryptic form.

45. J. B. Hersey, 1977, *A chronicle of man's use of ocean acoustics.* Oceanus 20 (1), 8–21.

46. Scripps engineer James M. Snodgrass devised the instrument. It is a free-falling probe tethered to very fine copper wire and sinking at 20 feet per second while sending information electrically to a chart recorder. A single launching takes less than a minute on deck and about five minutes to record in the ship's laboratory. The ship never slows down.

47. In 1826, the Swiss physicist Daniel Colladon and the French mathematician Charles Stunn measured the velocity of sound in water by ringing a bell in Lake Geneva and listening some distance away, underwater. They announced the speed as 1,435 meters per second at 8 °C. Not a bad guess for a first attempt.

48. F. B. Jensen, 2001, *Acoustics, shallow water,* in J. H. Steele, S. A. Thorpe, K. K. Turekian, 2001, *Encyclopedia of Ocean Sciences,* 6 vols. Academic Press, San Diego, pp. 89–96, p. 89 ff.

49. At high frequencies, in addition, there is much scattering. The physics of sound propagation in shallow waters has been intensely studied since the 1950s, and it is quite well understood. However, the environment of the sea above shelves is highly variable, and thus prediction of what sound will do under given circumstances is impossible without detailed regional and seasonal knowledge.

50. W. A. Kuperman, 2001, *Acoustics, deep ocean,* in J. H. Steele, S. A. Thorpe, K. K. Turekian, 2001, *Encyclopedia of Ocean Sciences,* 6 vols. Academic Press, San Diego, pp. 61–72, p. 61 ff.

51. Within this channel, the size of the front of a wave generated by a point source increases much less than would be expected in a three-dimensional setting, where the energy is distributed over a spherical (or elliptical) wave front, expanding with the square of the distance.

52. M. Ewing, J. L. Worzel, 1948, *Long-range sound transmission.* Geological Society of America Memoir 27, 1–35.

53. To the U.S. Navy the channel was of interest for detecting submarines, so a network of hydrophones was placed at many locations in the ocean, where the low-velocity layer intersects the continental slope. The array was called Sound Surveillance System, or SOSUS. In 1999 it became available to scientists for undersea studies.

54. The sound channel is not the only way whale-to-whale communication could be achieved acoustically. For arguments advocating such communication and the necessary background in acoustics, see R. Payne, 1995, *Among Whales,* Bantam Doubleday, New York, 431 pp.

55. The project was initiated by Walter H. Munk of the Institute of Geophysics and Planetary Physics at Scripps, and Carl Wunsch of the Massachusetts Institute of Technology, in the early 1990s.

56. W. Munk, P. Worcester, C. Wunsch, 1995, *Ocean Acoustic Tomography.* Cambridge University Press, Cambridge, 433 pp.

57. What can be studied is whether animals react negatively to the type of sound source employed by the ATOC project. Biologists have long known that some marine animals are attracted to unusual underwater sounds and even vocalize nearby, while others flee. One way to warn off animals that show a negative response is to provide them with advance notice by gradually increasing the intensity of the sound, over a time span of several minutes. Mammals then have time to move away to where the sound is weak. Whether such remedies (and others contemplated) will be judged satisfactory is an open question.

58. For details on natural background noise, see W. J. Richardson et al., *Marine mammals and noise,* p. 87 ff.

59. The development of submarines started well before 1900. The first successful wartime destruction by a submarine was in Charleston harbor on 17 February 1864, during the Civil War. The human-powered vessel, *H. L. Hunley,* 40 feet long, sank the Union ship *Housatonic* with a torpedo. But the *Hunley* and its crew of nine men did not survive the action. (The *Hunley* was lifted from the seafloor in August 2000, for restoration.) The most significant advance in submarine design was the introduction of the diesel engine, combined with battery power. Another important advance was the invention of the periscope, by the British Captain R. H. S. Bacon in 1904.

60. The methods of detection went by the names of Asdic (after the Anti-Submarine Development Investigation Committee) and Sonar (sound navigation ranging) by the British and the Americans, respectively.

61. The U.S. Navy's research agency was the Office of Scientific Research and Development. It became the U.S. Office of Naval Research, or ONR, after the war ended. ONR supported the rapid and sustained postwar expansion of ocean sciences at Scripps and Woods Hole, and subsequently of many other seagoing research institutions, including those at Lamont, Miami, Texas, Oregon, Washington, Hawaii, and Rhode Island.

62. Attached to the laboratory from Scripps were Harald U. Sverdrup, Martin W. Johnson, Richard H. Fleming, Eugene C. LaFond, Walter H. Munk, Francis P. Shepard, and Roger R. Revelle (who soon went to Washington, D.C., on Navy duty). New people who came to it included Carl H. Eckart, Jeffery D. Frautschy, and Russell W. Raitt.

63. Carl H. Eckart (1902–1973) was the first director of the Marine Physical Laboratory, which followed on UCDWR, and he also served as interim director of Scripps, from 1948 to 1950. Eckart's mastery of physics and math were invaluable, and he eventually edited and wrote much of the UCDWR report *Principles and Applications of Underwater Sound* (1946). This document was a landmark resource for those who had access to it. C. H. Eckart, 1946, *Principles and Applications of Underwater Sound.* Summary Technical Report of Division 6, vol. 7. Natl. Defense Research Council, Washington, D.C. For details, see W. H. Munk, R. W. Preisendorfer, 1976, *Carl Henry Eckart.* Biographical Memoirs of the National Academy of Sciences 48, 195–219.

64. R. Garrett, 1977, *Submarines.* Little, Brown and Co., Boston, 143 pp., p. 95.

65. *San Diego Union,* 30 September 1946.

66. Fred Noel Spiess (1919–2006) became a reserve officer in the navy after graduation (Berkeley, 1941). He went to submarine school, influenced in part by his father having served on a submarine chaser in World War I. He was on submarine duty at Manila when Pearl Harbor was attacked and subsequently served on a dozen submarine patrols in the Pacific during the war. After the war he returned to Berkeley for his Ph.D. in physics. Spiess was acting director for Scripps from 1961 to 1963 and director from 1964 to 1965.

67. Physicist Philip Rudnick (1904–1982), at MPL, worked out a design to minimize the craft's motion, and Frederick H. Fisher, at MPL, became the project officer while naval architect Lawrence M. Glosten completed the final design. Spiess worked closely with all of them.

68. F. N. Spiess, 2000, SIO Reference Series, no. 004, 81 pp., p. 49 [oral history].

69. F. N. Spiess, 1980, *Some origins and perspectives in deep-ocean instrumentation development,* in M. Sears, D. Merriman, *Oceanography: The Past.* Springer-Verlag, Berlin, pp. 226–239.

70. F. N. Spiess, 2000, SIO Reference Series, no. 004, 81 pp., p. 58 [oral history].

71. A transponder is a noise-making device sitting on the seafloor that transmits an outgoing acoustic signal as soon as it receives an incoming one from the ship that seeks to determine its position.

72. As described in chapter 12.

73. The British geologist Arthur Holmes (1890–1965) had the right idea in postulating upward-moving mantle material under the ridge—but he failed to realize the nature of the ridge itself. A. Holmes, 1929, *Radioactivity and Earth movements.* Transactions of the Geological Society of Glasgow 18, 559–606.

74. Harvey Cornelius Hayes (1877–1968) was then with the Naval Experimental Station in New London, Connecticut, and later had a distinguished career in underwater sound as superintendent of the Sound Division of the Naval Research Laboratory in Washington, D.C.

75. *Submarine canyon* was the term preferred by Scripps geologist Francis P. Shepard (1897–1985), who spent much of his career on research regarding the origin of these incisions in the continental margins. There is no connection to submarines.

76. Shepard started the mapping of the La Jolla canyon system during his visits at Scripps from the University of Illinois in the 1930s. In 1936, he brought along his two graduate students, Robert S. Dietz (1914–1995) and Kenneth O. Emery (1914–1998). Both eventually gained great prominence.

77. The name *fathometer* is trade-marked by Submarine Signal Company.

78. R. S. Dietz, K. O. Emery, 1976, *Early days of marine geology.* Oceanus 19 (4), 19–22.

79. Also, some of the time thus gained could be used for collecting abalone for dinner, for example.

80. With baby-sitting the echo sounder and noting depths in a logbook every 5 minutes, in the 1960s and 1970s.

81. The equipment was created by General Instruments, in Massachusetts. Jacqueline Mammerickx, who mapped large areas of the Pacific in some detail, received her geology training in Belgium. She was a strong advocate for using acoustic swath mapping to explore the seafloor.

82. D. T. Sandwell, W. H. F. Smith, 1996, *Global Seafloor Gravity.* University of California, Berkeley. Sandwell is at Scripps, Smith at the National Oceanic and Atmospheric Administration.

83. Harry Hammond Hess (1906–1969) was professor of geology at Princeton University and also in the Naval Reserve. He had earlier worked with Navy submarines to measure Earth's gravity anomalies.

84. Roger R. Revelle, the director of SIO, led the Joint University of California–U.S. Navy Electronics Laboratory Mid-Pacific Expedition. H. William Menard, Robert S. Dietz, and Edwin Lee Hamilton (1914–1998) were at the Navy Electronics Lab; Hamilton also was working on his degree at Stanford University.

85. As noted in chapter 4, when discussing coral reefs.

86. This was a wise move, as the discovery turned out to be eye-popping, with a concomitant temptation for senior scientists to get in on it.

87. In fact, in 1961 Robert Dietz gave the answer in a famous paper published in the science magazine *Nature*: seafloor spreading. In this scheme, new seafloor is created at the center of a spreading ridge, moves away from its origin, and sinks as it cools. In the process, it carries the flattened seamounts to greater depths. But Dietz's priority in identifying the mechanism was challenged by Hess, the original discoverer of the guyots, on the basis that he (Hess) had circulated a typed manuscript in 1960, proposing the production of new seafloor at the Mid-Ocean Ridge. Hess's paper was published in 1962. Neither paper had much of an impact on the marine geology community at the time. See chapter 12.

88. H. W. Menard, Jeff Frautschy, Lou Garrison; the others had left the ship and flown back.

89. H. W. Menard, 1986, *The Ocean of Truth.* Princeton University Press, Princeton, N.J., 353 pp., p. 58.

90. H. W. Menard, R. S. Dietz, 1952, *Mendocino submarine escarpment.* Journal of Geology 60, 266–278.

91. W. J. Morgan, 1968, *Rises, trenches, great faults, and crustal blocks.* Journal of Geophysical Research 73, 1959–1982. Morgan was beaten into print by Dan McKenzie and Robert Parker, who published in *Nature* later in the year of Morgan's AGU presentation. D. P. McKenzie, R. L. Parker, 1967, *The North Pacific; An example of tectonics on a sphere.* Nature 216, 1276–1280.

92. H. W. Menard, *The Ocean of Truth,* p. 40. As W.H.B. recalls from student days, the *Horizon* was pervaded by a sickening smell of a mixture of diesel fumes and coffee, which became hardly bearable when one's sleep was severely curtailed by the pounding of breaking bow waves against the hull next to the

bunk in bad weather. Happiness, such as it was, resulted from the proper working of the echo sounder and other instruments and a feeling of mission achieved in spite of adversity.

93. H.W. Menard, *The Ocean of Truth,* p. 40. The reference to "Bob Fisher" is to Scripps geologist Robert L. Fisher.

94. *Hertz,* abbreviated *Hz,* is read as "cycles per second."

95. In 1896, the German physicist Emil Wiechert (1861–1928) concluded, based on seismic data, that a rocky mantle 1,200 kilometers thick surrounds an iron core. While the basic idea was correct (Earth consists mainly of mantle and core), a thickness of 2,900 kilometers for the mantle would have agreed with modern values. By 1913 the geophysicist Beno Gutenberg (1889–1960) (then at Strasbourg, later at the California Institute of Technology) obtained results for the depth of mantle-core boundary that are close to those now accepted.

96. In 1900, the Croatian physicist and meteorologist Andrija Mohorovičić (1857–1936) established a Center for Meteorology and Geodynamics in Zagreb, Croatia. In 1909, Mohorovičić analyzed the recordings from an earthquake near Zagreb and determined that there was a discontinuity in seismic velocity within the Earth at a depth of about 60 kilometers. Below this depth, he thought, sound moved a lot faster. The jump in velocity he inferred became known as the "Mohorovičić discontinuity." The correct pronunciation is Mo-ho-ro-vichik.

97. Such a bulge looks much like a stack of wet napkins draped over an egg, and much enlarged. Gas and oil collect around the top of the egg.

98. Many fields were found in this rather elementary fashion. Subsequently it was discovered that a host of other types of traps are worth looking for, and the science of finding petroleum now fills libraries.

99. (William) Maurice Ewing (1906–1974), founding director of Lamont Geological Observatory of Columbia University, and a central figure in the geophysical exploration of the deep ocean floor and deep-sea geology in general.

100. In doing so, Ewing was responding to suggestions from senior colleagues. He was approached in 1934 by Princeton geologist Richard M. Field and by William Bowie of the Division of Geodesy of the Coast and Geodetic Survey, who urged him (and offered funding) to use seismic refraction to determine the nature of the continental shelf off the East Coast. The purpose was to determine the true edge of the continent. Ewing readily agreed.

101. It is said that the way Ewing handled explosives on deck at the time added much to a sense of adventure in geophysical exploration at sea.

102. M. Ewing, A.P. Crary, H.M. Rutherford, 1937, *Geophysical investigations in the emerged and submerged Atlantic coastal plain.* Bulletin of the Geological Society of America 48, 753–802.

103. Lamont (now Lamont-Doherty Earth Observatory) had no ship, but Ewing was able to use Woods Hole's *Atlantis* on occasion.

104. M. Ewing, J. L. Worzel, J. B. Hersey, F. Press, G. R. Hamilton, 1950, *Seismic refraction measurements in the Atlantic ocean basin, Part I.* Bulletin of the Seismological Society of America 40, 233–242.

105. E.C. Bullard, 1980, *William Maurice Ewing.* Biographical Memoirs of the National Academy of Sciences 51, 119–193.

106. Both the Navy Electronics Laboratory on Point Loma and the Marine Physical Laboratory at Scripps were offspring of the erstwhile University of California Division of War Research, which was discontinued with the end of the war.

107. Russell Watson Raitt (1907–1995), a founding member of the Marine Physical Laboratory of the University of California, had a Ph.D. in physics from the California Institute of Technology (1935). For several years, with two colleagues, he searched for oil in the Los Angeles basin and the San Joaquin Valley, using seismic-reflection techniques. In 1941 he was invited to join the University of California Division of War Research. When the wartime laboratory closed in 1946, Raitt accepted the invitation to stay and continue his researches at MPL.

108. Bathymetric charts created by Francis P. Shepard and Kenneth O. Emery were available but as yet based on spotty information.

109. Arthur D. Raff (1917–1999) later had a role in the discovery of magnetic stripes on the seafloor.

110. A.D. Raff, 1997, *Early days of seismic and magnetic programs at MPL,* in *Seeking Signals in the Sea.* Marine Physical Laboratory, SIO Reference Series 97-5, 10–22.

111. A. Harding, 2001, *Seismic structure,* in J.H. Steele, S.A. Thorpe, K.K. Turekian, 2001, *Encyclopedia of Ocean Sciences,* 6 vols. Academic Press, San Diego, pp. 2731–2737, p. 2732.

112. H.W. Menard, *The Ocean of Truth,* p. 50.

113. George Shor, geophysicist, joined Scripps in 1953, after receiving his Ph.D. from CalTech, where Charles Richter was his advisor.

114. Robert L. Fisher obtained his Ph.D. from Scripps in 1957 and stayed on in the Geological Research Division. He mapped large areas of the deep-sea floor, in the Pacific and especially in the Indian Ocean, and contributed greatly to the knowledge of deep-sea trenches.

115. Raitt's shipboard techniques for seismic work at sea became a standard in this sort of work. G. Shor

developed a safety protocol for handling the explosives; there were no accidents.

116. The stations that reached the mantle were few in number, and they comprised a relatively small area, according to H. W. Menard, *The Ocean of Truth*, p. 51. In any case, the fact that results in the Atlantic agreed with those in the Pacific, regarding the thickness of the crust, was a most exciting result of these pioneering experiments.

117. The realization that the M discontinuity is not so far below the seafloor in some regions encouraged some geophysicists to think about drilling into it. A project, dubbed "Mohole" was invented and funded and ultimately failed to achieve its goals: the idea to drill through the entire crust before having explored the crust proved unrealistic. (See chapter 12.)

118. As reported by participants on board.

119. The marine technician Harold Sammuli, originally from Finland, thoroughly enjoyed operating the Arcer. Watching him work on a wet deck among the high-voltage equipment, George Shor concluded that Sammuli was among the few people who have unusually high resistance to electricity, which was confirmed with an ohmmeter.

120. J. Ewing, M. Ewing, T. Aitken, W. J. Ludwig, 1968, *North Pacific sediment layers measured by seismic profiling*. Geophysical Monographs 12, 147–173.

121. W. H. Berger, E. L. Winterer, 1974, *Plate stratigraphy and the fluctuating carbonate line*, in K. J. Hsü, H. Jenkyns (eds.), *Pelagic Sediments on Land and under the Sea*. Special Publications of the International Association for Sedimentology 1, 11–48; Tj. H. van Andel, G. R. Heath, T. C. Moore, 1975, *Cenozoic history and paleoceanography of the central equatorial Pacific Ocean*. Geological Society of America Memoir 143, 1–134.

122. Questions have been raised in recent years about the use of sound-producing equipment to determine geologic structure beneath the seafloor. Is this type of sound harmful to marine creatures—specifically marine mammals? This is being investigated by teams of physicists and biologists, as discussed in W. J. Richardson et al., *Marine mammals and noise*.

123. At Scripps, Leroy Dorman and John Orcutt have been leaders in the deployment of seismometers at great depths in the sea. L. M. Dorman, 2001, *Seismology sensors*, in J. H. Steele, S. A. Thorpe, K. K. Turekian, 2001, *Encyclopedia of Ocean Sciences*, 6 vols. Academic Press, San Diego, pp. 2737–2744.

124. J. Orcutt, C. deGroot-Hedlin, W. Hodgkiss, W. Kuperman, W. Munk, F. Vernon, R. Worcester, E. Bernard, R. Dziak, C. Fox, C. Chiu, C. Collins, J. Mercer, R. Odom, M. Park, D. Soukup, R. Spindel, 2000, *Long-term observations in acoustics: The ocean acoustic observatory federation*. Oceanography 13 (2), 57–63.

TWELVE

Mountains, Trenches, Sunken Islands

THE GREAT REVOLUTION IN EARTH SCIENCE

What do the mountains of the Sierra Nevada have in common with the San Andreas Fault System of California and with the island chain of Hawaii (fig. 12.1)? They are among the large linear features that characterize the surface of our planet, and that result from the motions of large pieces of real estate called plates. There are about a dozen plates. The largest comprises much of the Pacific basin, the next largest includes much of the Indian Ocean and both India and Australia, the next the Americas and the western half of the Atlantic basin, and so on down to the smallish Cocos and Caribbean plates flanking the Panama Isthmus (fig. 12.2). Their existence and their motions are a result of convection within the mantle. The basaltic mantle rock, when under pressure and hot, can flow slowly and does so at a rate of roughly an inch per year. Volcanism, earthquakes, tsunamis—all these phenomena on the surface of Earth are linked to the convection-driven motions of the plates in recognizable patterns.

The discoveries that led to a new all-encompassing theory of the Earth were made primarily in the 1950s and early 1960s; the theory itself has roots older than that but emerged in modern fashion in the 1970s. The core statement of the theory is that seafloor is created all the time, at spreading centers making up the Mid-Ocean Ridge, and is destroyed in the subduction zones, whose position is marked by trenches and strong deep earthquakes. Such earthquakes, in some cases, are the source of devastating great waves spreading across the sea, waves known as tsunamis. The subsea volcanism at the spreading centers produces hot springs, and the chemical reactions of hot seawater with hot basalt stimulate bacterial activity, which in turn supports a hot vent fauna with unusual animals.

FIGURE 12.1. Basalt forms the cliffs of the Hawaiian Islands, and provides the main building material for all other islands in the deep sea.

SO YOUNG A SCIENCE

Continental drift, seafloor spreading, and plate tectonics comprise the concepts that allow us to explain the major features on the surface of the home planet. These concepts entered the geology textbooks only in the 1970s. All older textbooks suddenly became obsolete. Like writings from a distant century, they give a glimpse of past thinking, offering a host of observations but lacking what matters most: an appreciation of the role of mantle convection in shaping the planet.[1]

The revolution started with the concept of continental drift, the brainchild of the German meteorologist Alfred Wegener. He introduced the concept as a way to explain the origin of the Atlantic Ocean as a great rift opening within the last 200 million years.[2] The idea of seafloor spreading, which lagged the idea of drift by more than three decades, had a number of sources. In 1960, the marine geologist Bruce Heezen (Lamont) narrowed Wegener's broad rift between the Americas and Europe-plus-Africa and put it into the middle of the Atlantic, making it coincident with the Mid-Atlantic Ridge. Shortly after, Harry Hess, of Princeton, made the rift a narrow zone of upwelling basalt at the center of the ridge, basalt that builds new seafloor while interacting with seawater. In 1961, Robert Dietz published a paper exploring

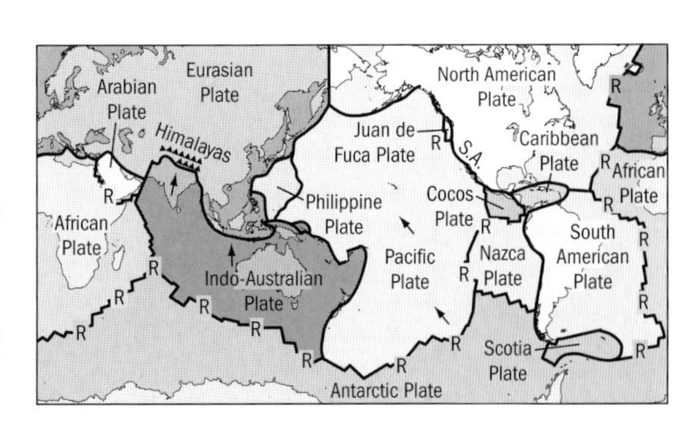

FIGURE 12.2. The great tectonic provinces on the surface of Earth, called plates. Plates (roughly 100 km thick) move as rigid bodies. High mountains such as the Himalayas (Himal.) mark continent-continent collision. Earthquakes and volcanoes mark all boundaries (e.g., Indonesia, Indon.; San Andreas, S. A.). R, crest of the Mid-Ocean Ridge, site of seafloor spreading.

the various corollaries associated with generating new seafloor in the middle of the ocean, while emphasizing the youth of the ocean bottom. He was the one who invented the term *seafloor spreading*.[3] The new concept elegantly explained a number of recently discovered features on the seafloor, including high heat flow and earthquakes associated with a continuous Mid-Ocean Ridge, and the fact that the oceanic crust is thin and carries but little sediment. Plate tectonics, finally, was a multiparented attempt to explain the complicated geologic history of the Pacific Ocean, which includes a rich assortment of volcanic activity, deep earthquakes, and vigorous mountain building along its shores, in addition to seafloor spreading on the East Pacific Rise.[4] Plate tectonics made its debut in 1967, in a public talk by the geophysicist Jason Morgan, at a meeting of the American Geophysical Union.

With plate tectonics, it became feasible to address the origin of the most striking features of the surface of Earth, such as the following:

1. Land areas and ocean regions are large and continuous. We do not have a water-covered planet dotted with small islands adding up to 30 percent land area.
2. The major mountain ranges tend to make chains. Both volcanic and nonvolcanic mountains do this. We do not have a disorganized patchwork of highlands with a more or less random distribution.
3. Some of the highest mountains are not far from the rim of the Pacific Ocean. Also, the mountains are not far from narrow elongated deeps showing the greatest depths in the sea, at the very edge of the basin. Great heights and great depths go together.
4. Large regions of the Pacific are dotted with sunken islands, and with sinking atolls that stay at sea level only because of coral growth. How can seafloor sink everywhere?

It came as a great surprise when all of these puzzling features turned out to be intimately connected. They are the results of the same

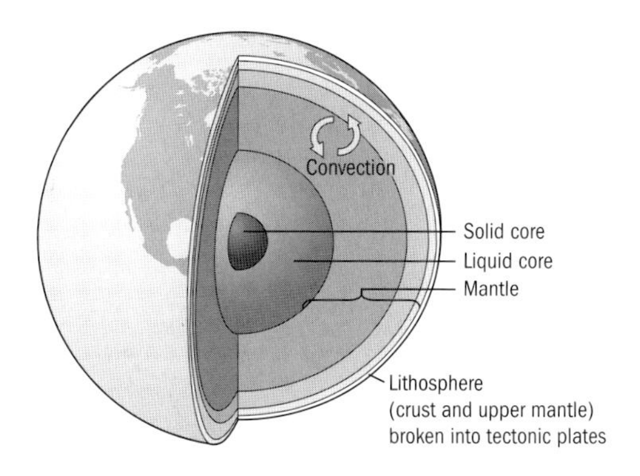

FIGURE 12.3. Layered structure of the Earth. Convection of the mantle holds the clue to the origin of mountains chains, lines of volcanoes, trenches, and earthquakes.

large-scale motions produced by convection deep within the Earth (fig. 12.3). These motions are now seen as the interplay of the creation, displacement, and destruction of relatively few large pieces of the rigid outer shell of Earth, the "plates," responding to mantle convection.

The principles of plate tectonics were established in the late 1960s and early 1970s of the twentieth century. Before that, none of the four fundamental observations and questions listed above could be answered.

In this sense, geology is a very young science. Ironically, geologists had neglected the third dimension of their subject: the processes deep within the Earth. Many of the older geologists alive today took freshman courses when their teachers were clueless about the processes shaping the major features of Earth—the mountains, the trenches, the volcanoes. The reason was simple: little was known in academia about the nature of the seafloor even into the late 1950s. As soon as the observations became available—within less than a decade—a global pattern emerged, and the geophysics of the Earth's surface came to be understood. After the 1960s, the home planet was much less mysterious than before. Continental drift and seafloor spreading became commonplace concepts, even household words.

To be sure, the idea of continental drift had been around for some time, as mentioned. Alfred Wegener proposed it in 1912; his final word on the subject was published in 1929.[5] He envisaged the opening of the Atlantic as a consequence of the Americas drifting away from Eurasia and Africa, since Permian time, some 200 million years ago. He marshaled two major but seemingly unrelated facts: an excellent fit between the continental margins on both sides of the Atlantic, and the similarity of plant and animal fossils on the continents in the Paleozoic. Opponents of the theory readily discounted the fit of coastlines, as coincidence. The similarity-of-fossils problem was taken care of by postulating ancient land bridges that once connected the continents and had since sunken below the sea, in an Atlantis-type disappearance act. A scenario rich in drama, albeit short of evidence.

One of the foremost proponents of the land-bridge scenario was Yale's geologist and paleogeographer Charles Schuchert. Around 1930, he was asked by the National Research Council to prepare a brief outline of the principles of paleogeography, with emphasis on the former distribution of land and sea, and of the explanations proposed for such changes. The views of this distinguished geologist provide a window on where things stood at the time: the permanency of continents and oceans was generally accepted, and large vertical motions, but not horizontal ones, governed the mobility of the crust.[6]

Twenty years after Schuchert wrote his summary article on transgressions and mountain building, the Dutch geologist Philip H. Kuenen, then the world's leading scholar in marine geology, wrote: "The formation of the ocean basins is a subject on which practically no direct evidence is obtainable and speculation finds little sure ground to go on."[7] He emphasized the difference in the nature of volcanism in the interior of the Pacific and at the rim of the basin and credited Hess with drawing an "andesite line" separating the two provinces.[8] The andesite line, we now know, runs along plate boundaries. It separates the ring of volcanism around the Pacific—the Ring of Fire—from the oceanic plates. Concerning continental drift, Kuenen provided some discussion and then commented: "For the time being most geologists appear to have lost faith in continental drift as a sound working hypothesis."[9]

The amazing postwar increase in information regarding the nature of the deep seafloor (which holds the clues to the origin of mountains, trenches, and sunken islands) readily emerges from comparing the chapter on "deep ocean floor topography" in the second edition of Shepard's book *Submarine Geology*, published in 1963, with the earlier textbook of Kuenen's, published in 1950.[10] Besides great advances in the mapping of topographic features (including the world-girdling Mid-Ocean Ridge, the fracture zones of the North Pacific, and a great number of sunken islands), we note a quantum step increase in geophysical information. The latter is especially well reflected in H. W. Menard's book entitled *Marine Geology of the Pacific* (published in 1964), a major summary of geologic research at Scripps in the 1950s. At the time Menard prepared the manuscript for this book (early 1960s), the flood of new information was still entirely undigested. He wrote:

> The origin of the [Pacific] basin . . . is open to unconfined speculations, and it is quite difficult to eliminate—or support—hypotheses at present. It may be anticipated that very deep drilling of the type visualized in the Mohole project during the next ten years will tell us a great deal about the early history and even the origin of ocean basins. Meanwhile it seems profitless to summarize the suggestions, largely unsupported by evidence, that have been presented in the literature.[11]

Menard's prediction that deep drilling would solve the problems concerning the major features of the Earth's surface proved incorrect. The problem of the origin of ocean basins, in essence, was solved before drilling began in earnest (in 1968). It was solved through the interpretation of the peculiar patterns of seafloor magnetism, first mapped by Scripps scientists in the northeastern Pacific.[12]

The magnetic patterns, when colored black and white according to above- and below-average intensity of the magnetic field on the seafloor, become reminiscent of the stripes on a

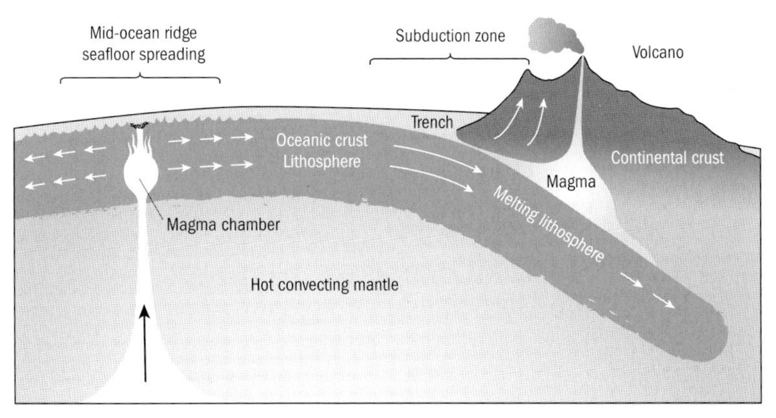

FIGURE 12.4. The chief elements of plate tectonics: seafloor spreading and subduction.

zebra. When first discovered, the origin of the zebra patterns was a complete mystery. The discovery of the answer came with the recognition of symmetry in the patterns. After 1966, the evidence presented left little doubt that the patterns are indeed symmetric about privileged lines centered on the Mid-Ocean Ridge.[13] These are the spreading centers where new seafloor is created from upwelled mantle material, precisely as suggested by H. Hess and R. Dietz. The cooling basalt takes on the ambient Earth magnetism, which changes with time. The result is bands of differing magnetic properties in the seafloor, running parallel to the spreading center.

According to Menard, seafloor spreading was generally accepted by the working experts by 1966.[14] The results of deep drilling simply confirmed what the geophysicists knew already. But it did help to convince the geologists that a new era of Earth science had begun.

Continental drift, which had fallen into disrepute when a British physicist showed that continents cannot possibly move,[15] now leaned on seafloor spreading for support. The new hypothesis of Dietz and Hess, put forward in 1961 and 1962, postulated that new seafloor is continuously being formed at the Mid-Ocean Ridge, and moves out from there, carrying the continents along, as on a conveyor belt. This concept is now fully integrated into plate tectonic theory. A plate is defined as a large piece of rigid real estate on the face of the planet. It may contain both ocean floor and continental areas. Its thickness is roughly 100 kilometers (comprising the "lithosphere"). Earthquakes mark the boundaries of the moving plates.

Suddenly, in the last few years of the 1960s, a host of long-standing questions found answers. The lining up of earthquakes, and the existence of all the linear features on the planet that had been so puzzling—the mountain chains, the trenches, the Mid-Atlantic Ridge, the enormous fracture zones in the northeastern Pacific, even the San Andreas fault—all could now be seen as manifestations of the same fundamental mechanisms, related to mantle convection[16] (fig. 12.4).

Plate motions are the clues to all these major features. The plates (lithosphere and attached crust) move away from each other (diverge) by seafloor spreading, creating the Mid-Ocean Ridge. Here the plates grow at their edge. The plates move toward each other (converge), where joined by subduction zones, usually featuring deep trenches. One of the plate edges (the one at the ocean side) bends down to descend into the mantle. As it heats up on first descending, it injects material into the opposing lithosphere that now overlies it, generating magma for volcanoes and roots for rising mountains above. Plates move past each other where separated by active horizontal faults (transform faults). In some cases transforms are "leaky," that is, the plates move apart slightly as they glide past each other, and new mantle material wells up between them. The Azores mark the

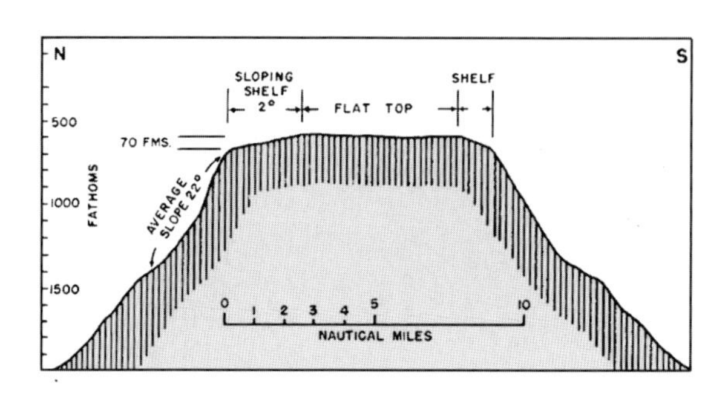

FIGURE 12.5. A typical guyot, as redrawn by Hamilton (in 1956) from Hess (1946).

site of such a leaky transform fault. Each plate may carry both oceanic and continental crust in various proportions, and there is no necessity for plate boundaries to coincide with the border between continents and oceans. Continents do not descend into the mantle, upon collision of plates. If two continents collide, things pile up, and we get very high mountains, as in the Himalayas. Interestingly, plate boundaries are not fixed forever but can change their relative positions through time, perhaps even abruptly (that is, in terms of geologic time scales).

SO YOUNG A SEAFLOOR—AND SINKING EVERYWHERE

The revolution of thought that is plate tectonics rests on three pillars—magnetic properties of the seafloor, as mapped in the magnetic anomaly bands (or "stripes"), the distribution patterns of earthquake centers, and the topography of the seafloor.

It is common, when reviewing the history of the geologic revolution, to concentrate on the first two, which are indeed crucial. But it was the nature and distribution of geographic features that started the process, and that called out most conspicuously for explanation. In the Atlantic, the dominant feature is the Mid-Atlantic Ridge, the paradigmatic expression of seafloor spreading. In the Pacific, there is the ring of trenches associated with active volcanism landward, and there are the fracture zones in the east and the sunken islands over vast regions in the west.

Of the various geographic features in the Pacific—all discovered by simple echo sounding—the flat-topped seamounts picked up by Hess's supply ship during World War II were perhaps the most influential in shaping the activities and the thinking of marine geologists working at Scripps[17] (fig. 12.5; see also chapter 11). Of course, the concept that the seafloor is sinking over very large areas goes back to Darwin's theory of atolls. Hess's proposition was that his flat-topped seamounts, which he named guyots,[18] were once islands at the surface of the sea and had sunk by a mile or more. This greatly expanded the scale of the process proposed by Darwin, and introduced a major conceptual difficulty: how could the seafloor be sinking everywhere?

Clearly, this was a problem worth exploring, and clearly the single most important issue was to find out how much time had elapsed since the islands had sunk—*if* they had sunk from the surface. A sinking island in the tropics, according to Darwin, goes through an atoll stage. Thus, the clue to the time since leaving the surface had to be the age of reef fossils on top of the seamounts. As recounted in the previous chapter, the question was clinched during the Mid-Pac Expedition, in 1950. Spectacular results emerged from using the most primitive means available to marine geologists, that is, dredging. To this expedition, and to the Stanford geology student Ed Hamilton, who published the results, goes the credit for proving Hess both right (the guyots are indeed sunken islands) and wrong (the guyots are 10 times younger than he thought).[19]

Why was this discovery so crucial? The fact that the guyots are sunken islands showed that enormous changes in depth had taken place over vast regions. The fact that their tops are no older than 100 million years showed that the Pacific seafloor—a large portion of the surface of the planet—is vigorously active now, geologically speaking. It became apparent that large-scale processes are right now shaping and changing the planet in major ways. For the next 15 years, after many more dredgings, nothing older than the reef rocks recovered during the MidPac Expedition was brought up. The youthfulness of the seafloor was truly astonishing.

To appreciate the enigma of seafloor youth, Hamilton's discovery has to be set against the prevailing wisdom of the time, the reigning orthodoxy, which emphasized stability. In fact, to understand the revolution at all, we must not just consider the struggle between "fixists" and "mobilists," but rather the lack of dynamics. Geology was largely being taught as facts, and as a history of facts—the sea comes in, the sea goes out from the continents, and mountains rise and are worn down. Where explanations were offered as to the processes ultimately responsible for having the sea go in and out, or for having mountains in the first place, they seemed ad hoc. So, serious geologists ignored them, took up their hammer (or equivalent instrument), and made some more observations.

The founding feat of geology, the publication of the *Theory of the Earth* by the Scotsman James Hutton some 200 years ago, actually did have dynamics in it. In fact, the point of the book was process.[20] Hutton introduced what is now known as the rock cycle, whereby mountains rise and are eroded, and the resulting sediment accumulates in low places. Eventually the sediment load builds up enough so the lowermost layers, under tremendous pressure and with heat from below, are cooked into solid rock that can rise to make mountains. Hutton's concept—a kind of steady-state geochemical cycle expressed on Earth's surface as mountains and basins—was a fundamental alternative to the one-way development envisaged by the German mining expert Abraham Werner. Werner created a mythology of Earth history that moved from primordial conditions (as seen in metamorphic and igneous rocks) to conditions closer to present ones (as seen in sedimentary rocks), in well-defined steps. Werner saw Earth as an evolving system, with early stages being totally different from later ones, and with each stage set off from its successor by catastrophe.

The main problem with Hutton's theory was its vagueness on the hot end of the rock cycle. The upper part of Hutton's cycle is no great mystery. It yields to boots, hammer, and hand lens and emerges from making geologic maps. However, the portion of the rock cycle that concerns making mountains (and especially mountain roots) remained obscure for another century and a half. Hutton knew that there are mighty forces at work below the great mountains of the world, generating earthquakes and volcanism, but he did not—could not, in his days—have the faintest notion what was actually happening deep within the Earth. To solve the riddle of the missing half of Hutton's rock cycle, one had to find out where the sediments, the products of erosion, ended up and how they were stuffed back into the Earth's interior. For this process no theory existed until after the 1950s.

Plate tectonics, then, provides a way to close the Huttonian rock cycle, by subduction. Thus, the theory shows its roots as a child of the steady-state thinking introduced by Hutton.[21]

How fast does Hutton's cycle actually run? Clearly, the answer is closely linked to the youth of the seafloor. The very oldest pieces, we now know, are no older than Jurassic in age (150 million years or so). On land, in contrast, rocks more than 1 billion years old are quite common, especially where the crust has been deeply eroded in the central parts of continents. In fact, the record extends back to more than 3 billion years, in a few places. Thus, the record on land spans a time some 20 times greater than the record of the deep sea. It comes as no surprise, therefore, that the ocean record is much like a snapshot of the workings of the planet, supporting a steady-state view of the dynamics of Earth. The dearth of information about long-term changes in the conditions on the planet gives the geophysical ocean record a strong Huttonian flavor. The new theory is about dynamics, not about history.

With this background, it should be clear that the most important discoveries leading into the new theory were those that forced geologists to accept a highly dynamic Earth. Thus, the 1950 MidPac Expedition's confirmation of Hess's sunken island hypothesis and the discovery of their youth were the findings that were crucial in this respect. Another important discovery signaling a dynamic geologic environment was the evidence for unexpectedly high heat flow from the seafloor. Basalt (the rock underlying the seafloor) is comparatively poor in radioactive elements, which are, however, concentrated in the granitic rocks of the continental crust. Thus, other things being equal, the continental crust should have high heat flow and the oceanic crust low heat flow. This turned out to be not so. Edward C. Bullard, the famous British geophysicist, was very much interested in doing a proper comparison of heat-flow values.[22] A frequent guest at Scripps, he worked with graduate student Art Maxwell to design and build equipment that would be useful in measuring temperature gradients within the uppermost sediments of the seafloor.[23] The first successful measurements were made on MidPac, with an instrument modified from that earlier design. Results were surprising. The heat flow from the seafloor was similar to that on continents![24] This suggested that there must be additional sources of heat, besides radioactivity, in the deep ocean.[25] Could heat be transported upward from the Earth's interior by convection currents?

By the same criterion of a focus on Earth dynamics, the meaning of the fracture zones started to emerge only with the evidence for large-scale horizontal displacement. Before that, they were impressive oddities. Likewise, the chasms of the trenches, and the gravity deficit associated with them, gain meaning only when it is realized that these features have to be maintained by active downward motion. The distribution of earthquakes becomes important when tied to the magnitude and direction of motion. And the magnetic properties of the seafloor achieve their potential when the patterns are linked to a time scale, and therefore to motion and velocity.

Depending on background and inclination, experts will differ on choosing the single most important clue that led to a new theory of the Earth. But by the criteria that result from the dynamics-versus-history paradigm, we would have to insist that two discoveries were fundamental: the ocean floor moves, and the ocean floor is young. More specifically, it is not just the vertical motion of the ocean floor that is at issue here—it is the large-scale horizontal motion. Initial attempts to solve the enigma of the sunken islands did not—could not—call on horizontal motion. It was not in the repertoire of conceptual tools.

The leader of the expedition that made the momentous discovery concerning the nature of the guyots, Roger Revelle, looked to an increase in the volume of water in the ocean for a possible answer.[26] Perhaps the delivery of water from the mantle to the ocean had increased in the last 100 million years, compared with earlier rates. Menard turned to large-scale vertical motion of the seafloor. He suggested that there had been an enormous bump in the seafloor 100 million years ago, studded with islands. For some reason this submarine highland region (which he called Darwin Rise) collapsed and took the islands with it.[27] The sinking of the Darwin Rise, incidentally, could explain the overall retreat of the sea from the continents since the middle of the Cretaceous period, when the continents were widely flooded.

What was still missing was an appreciation for large-scale horizontal motions. Seafloor spreading would eventually posit such motions, in the 1960s. But without some kind of proof it would be just another idea, one of many such ideas floating about in search of evidence.

As mentioned, the evidence came from unexpected quarters, and from an entirely new kind of information, based on the most sophisticated instruments then in the service of marine geology: magnetometers. Before we turn to the topic of paleomagnetism, let us briefly review the relevant evidence that needed explaining by the end of the 1950s.

DISCOVERIES THAT NEEDED EXPLAINING

The general sinking of large parts of the Pacific seafloor (as indicated by atolls and flat-topped seamounts), and the very existence and perfectly central position of the Mid-Atlantic Ridge with respect to the continental margins rimming the Atlantic—those were certainly unexplained features of the planet that loudly advertised the general lack of knowledge about the workings of Earth. However, there were other enigmas, in the 1950s, some of long standing, and some just emerging.

In fact, practically every important geologic discovery presented a new mystery, conundrum, or paradox. In a sense, in this phase of exploration, new results increased the awareness of lack of understanding rather than just adding to knowledge. There were those, such as Raitt and Hamilton, who navigated the sea of ignorance with great confidence, resolutely collecting the information that would eventually provide the breakthrough. And there were others, such as Dietz, whose minds were constantly racing to guess at solutions even when only a few of the necessary puzzle pieces were on the table. Guessing right has its rewards. Guessing wrong can be embarrassing. The majority of scientists are not content to let the facts speak for themselves. They rarely do!

At the end of the reconnaissance period, in the late 1950s, enough data had been amassed by an impressive array of instruments to make guessing worthwhile. The search was on for some sort of synthesis that would explain several major seafloor patterns at once, rather than providing yet another ad hoc explanation for any one of them. The remarkable thing is that such a synthesis did emerge within one decade (the 1960s). Here are some of the major features that needed to be considered in building it.[28]

1. *Rock types.* Extensive dredging had produced basaltic rocks, but not granitic ones; that is, there are no continent-type igneous rocks on the seafloor (excepting some ice-rafted debris in high latitudes). Thus, Wegener had been right in surmising, from the distribution of elevations in the sea and on land, that the ocean floor was essentially basaltic.

2. *Mid-Ocean Ridge system.* Routine use of precision echo sounding on oceanographic expeditions and by navy ships produced greatly improved topographic maps. The well-known Mid-Atlantic Ridge was recognized as part of a globe-encircling midocean rise system. A rift in the very center of the Mid-Atlantic Ridge strongly suggested expansion of the seafloor and upwelling of fresh basalt along the ridge. Characteristically, shallow earthquakes were found to be associated with the ridge, further strengthening this idea.

3. *Trenches and the Ring of Fire.* By the 1940s, a system of deep trenches—with depths exceeding 10 kilometers in places—was emerging, whose pattern was most intriguing.[29] An important contribution was a map produced by Hess,[30] which interpreted the large-scale topographic features between Japan and New Guinea in terms of tectonic structure. He boldly connected the trenches and assigned to them the term *tectogene*, signifying "related to mountain building." Without exception, the deep trenches are associated with volcanic activity and deep earthquakes on their landward side. This zone of volcanic activity and deep earthquakes spans the entire rim of the Pacific, from the southern tip of South America to Alaska and the Aleutian Islands and on to Japan, the Marianas, and the Tonga Islands. Together, the volcanoes and the associated mountains and trenches form one large linear dynamic feature on the Earth, its length exceeding one-half of the circumference of the planet. Only the world-girdling rift system of the Mid-Ocean Ridge is longer.

4. *Fracture zones.* In the Pacific, four great fracture zones were discovered, stretching from off North America almost due west for several thousand miles into the center of the ocean.[31] These zones looked suspiciously like superlarge wrench faults, somewhat like the San Andreas Fault, which runs through San Francisco before turning seaward. Thus, the seafloor appeared to be highly mobile horizontally, and to undergo deformation as rigid blocks, rather than as a plastic mush. A

mantle "current" rising below oceanic rises and sinking near the continental margin was invoked by H. W. Menard to produce a stress pattern resulting in the fracture zones.[32] Though incorrect in detail, this explanation contained important elements of the coming revolution.

5. *Flat-topped seamounts.* As mentioned, routine echo sounding by Harry Hess during World War II led to the discovery of flat-topped mountains in the Pacific, at great depth.[33] Hess thought they were Precambrian features. He invoked sinking of islands, once eroded at sea level. When some of the guyots were dredged, during the MidPac Expedition in 1950, they were found to be drowned islands that had been at sea level only 100 million years ago.

6. *Fossils.* Not only were there no pre-Cretaceous fossils on the seamounts, but also fossils of pre-Cretaceous age were conspicuously absent anywhere in samples from the seafloor.

7. *Layering of the oceanic crust.* In contrast to the thick, deformed, and generally messy continental crust, the deep ocean floor was found to consist of three relatively thin layers covering the mantle proper. Counted from the top, they are layer 1, only a few hundred meters thick and consisting of sediment; layer 2, somewhat less than 2 kilometers thick and with a sound velocity appropriate either for volcanic material or consolidated sediment; and layer 3, about 5 kilometers thick and consisting of basalt.[34] The lower boundary of the crust, the Mohorovičić (sound velocity) discontinuity, which occurs at some 30 to 50 kilometers under the continents, turned out to be at less than 10 kilometers below the deep ocean floor. Also, if layer 2 is assumed to be volcanic rock (as we now know is the case, from drilling), sediment thickness is then inadequate to account for the debris expected from erosion of continents through hundreds of millions of years.

8. *Heat flow through the ocean floor.* Earth radiates its own heat into space, besides that received from the Sun. Granite produces some 10 times more heat per unit mass and unit time than does basalt (from decay of radioactive elements), and continental crust, therefore, is expected to produce much more heat than the basaltic crust below the seafloor. Surprisingly, however, it was found that heat was flowing through the seafloor at about the same rate as through the surface of continents.[35] Perhaps the higher temperatures below the oceanic crust were due to rising mantle currents below the oceans?[36] Significantly, the highest heat-flow values on the ocean floor were found to be associated with the midocean ridges.[37] Could it be that mantle material was rising here at the ridges?

9. *Polar wandering curves.* Parallel to the exciting new discoveries by ocean-going geologists and geophysicists, land-based studies on magnetic properties of rocks of all ages contributed new evidence for the mobility of continents. It had become clear that the magnetic field of Earth changes substantially through geologic time but remains linked to the rotational axis. The prevalent field is frozen into cooling lava from volcanoes and also is locked into newly formed sediments, by alignment of magnetic particles. Thus, the history of the field can be reconstructed. When the ancient pole positions are plotted for rocks from any one continent, they show displacement along a path, along which the pole wandered, relative to the continent. If continents are fixed, they all should show the same polar wander path. If the continents moved relative to each other, the paths should differ. It was found that the paths differ for North America and for Europe—and in such a fashion that only a drifting apart of the continents could explain the difference.[38]

At the First International Oceanographic Congress held in New York, in September 1959, these various observations were presented and discussed. It was to host the last international gathering of marine geologists and geophysicists in midcentury in which the concepts of seafloor spreading and continental drift did *not* play a central role but were mentioned in passing as possible hypotheses, among others. Some of the participants went home in a very thoughtful mood indeed.

One of them, Harry Hess, soon put it all together, explaining every one of these puzzling

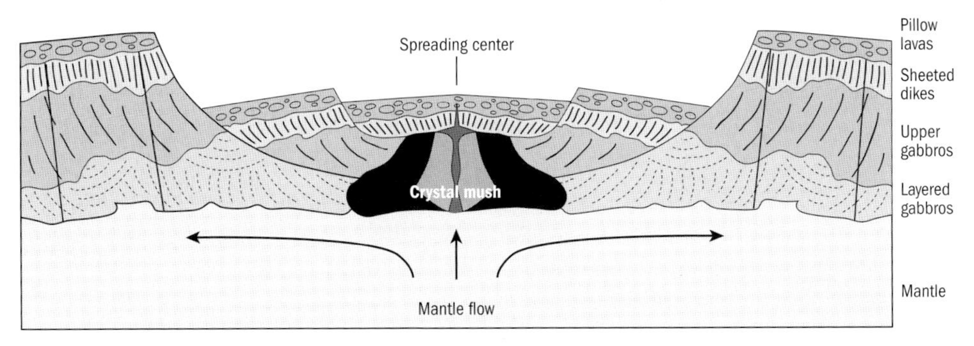

FIGURE 12.6. Seafloor spreading as a result of mantle flow, in a modern version of the concept. Image depicts central zone of Mid-Ocean Ridge. Note the layered structure of the basaltic crust. Sediments not shown.

facts as resulting from mantle convection.[39] His hypothesis turned out to be correct: in essence, the entire deep ocean floor, to a thickness of about 100 kilometers, consists of mantle material moving from the regions of upwelling (the Mid-Ocean Ridge) to the regions of downwelling (the subduction zones). In the Pacific, the voyage takes around 100 million years, hence the youth of the seafloor. Where the seafloor disappears back into the mantle, the sinking slab carrying the ocean floor gives up those portions in its body that are more easily melted, in the presence of heated and pressurized water. This material rises and makes volcanoes and mountains.[40] And central to the concept was the generation of new seafloor at a narrow zone along the Mid-Ocean Ridge (fig. 12.6).

The proof of the new view—if there is such a thing as proof in a historical science—was soon to come, in the shape of the famous "zebra stripes."

MAGNETIC STRIPES HOLD THE ANSWER

The missing link in the dynamics that would explain both the youth of the seafloor and its sinking everywhere was the power of large-scale horizontal motions. *Seafloor spreading*, the term proposed by Robert Dietz in 1961, became the key word for such motions.[41] The concept of seafloor spreading is based on mantle convection, a mechanism proposed earlier by Arthur Holmes to explain continental drift.[42]

Dietz laid out the new concept and named it. His paper on the subject was published in 1961, in the British science magazine *Nature*. Harry Hess had thought along similar lines. His article was published in 1962, in a symposium volume issued by the Geological Society of America. The article had been widely circulated in 1960, in manuscript form, a circumstance that has supported Hess's claim to priority in the creation of the concept. Hess's proposal included localized upwelling of mantle material to make the Mid-Ocean Ridge, and the outward migration and sinking of the seafloor carried on top of convection currents in the mantle. In his model, the seafloor is young, the crust is basalt (altered by reaction with seawater), and the sediment is thin, in excellent agreement with observations.

Dietz summarized the new concept of convection-dominated seafloor morphology with great clarity and economy[43]: "The gross structures of the seafloor are direct expressions of this convection. The median rises mark the upwelling sites or divergences; the trenches are associated with the convergence of downwelling sites; and the fracture zones mark shears between regions of slow and fast creep."

The concept of mantle convection was not exactly new; it has a considerable pedigree in the geologic literature. In a paper published by the Geological Society of Glasgow, in 1929, Holmes explained the breakup of Wegener's Pangaea by upwelling and lateral motion of mantle material, carrying the continental blocks away from each other. In addition, in his scheme, trenches and mountains form above the down-going limbs of the convection.[44] However, without observations from the seafloor supporting such seemingly outlandish concepts, Holmes's ideas withered into oblivion.

For a very few years, a similar fate seemed inevitable for the ideas of Hess and Dietz: no one paid much attention. Their hypotheses simply were contributions to an ongoing discussion. In his book on the geology of the Pacific, published in 1964, H. W. Menard makes no use of the concept of seafloor spreading.[45] He (very) tentatively accepts convection as the process creating oceanic rises, trenches, and fracture zones, but uses its forces to *deform* a rigid surface layer (including the crust), rather than to carry it along over long horizontal distances. Clearly, at the time the manuscript for his book was readied for submission (1962/63), Menard was not convinced that the case for seafloor spreading was strong enough to warrant employing it in explaining the morphology of the seafloor.

Yet, in the same book, Menard provides a succinct summary of the type of clues that were to be instrumental in solving the puzzle. The magnetic properties of the seafloor in the northeastern Pacific, mapped in the mid-1950s to early 1960s, show large-scale regular patterns.[46] Although the origin of these patterns was entirely obscure, it was obvious that they suggested large-scale lateral motion of the seafloor, from the enormous offsets of similar patterns observed across major faults.[47]

Recalling the circumstances that led to the discovery of magnetic lineations on the seafloor, in the 1950s, Menard commented in his book *The Ocean of Truth* that federal funding agencies had shown no interest in conducting a magnetic survey during the Pioneer work, so what he considered "one of the most significant geophysical surveys ever made" was paid for by the small discretionary funds of the director of the Scripps Institution.[48] The director was Roger Revelle. The results of the survey proved a major milestone in the Earth science revolution.

What the *Pioneer* survey showed, and the related work of Scripps geophysicist Victor Vacquier and associates, is that the seafloor carries unique magnetic markers of unknown origin, which are nevertheless useful in providing clues about large-scale horizontal displacement. Unfortunately, the northeastern Pacific is among the most complicated of tectonic provinces of the deep ocean. The solution to the puzzle of the magnetic patterns, and how they record the motion of the seafloor, was to be found elsewhere, where the patterns were simple.

The turning point came when the hypothesis of seafloor spreading was applied to the magnetic patterns on the seafloor, together with new knowledge about the history of the Earth's magnetic field through geologic time. The match between magnetic patterns and seafloor spreading was made by Fred J. Vine and Drummond H. Matthews, in 1963, and was elaborated by Vine and J. Tuzo Wilson, in 1965.[49] It was this match that forced the acceptance of the Hess-Dietz hypothesis as a general theory of the motions of the Earth's crust. From a handful of believers, the new faith suddenly spread, and by 1966, it included essentially all practicing marine geologists.

What was the nature of the remarkable discovery that turned "seafloor spreading" from another clever hypothesis into a basic framework for doing research? Frederick John Vine, then a graduate student at Cambridge, and his advisor, Drummond Hoyle Matthews, combined the idea of seafloor spreading with new evidence that the Earth's magnetic field reverses every once in a while, on a time scale of 100,000 years. They theorized that new seafloor being generated at the ridge crest is magnetized in the prevailing direction of the Earth's field, as the basalt cools below the Curie point (about 500 °C). The seafloor takes this magnetization with it as it moves down the flanks. Upon reversal of the field, new seafloor is now magnetized in the opposite direction. Thus, strips of crust with alternately normal and reverse magnetization drift away from the center of the ridge and produce linear magnetic anomalies parallel to the ridge crest: the zebra stripes were a direct result of seafloor spreading.[50]

For another year or two after being first announced, the new hypothesis did not attract much favorable attention. There was at least one exception. Several months after the Vine-Matthews hypothesis appeared in print, Scripps geophysicist George Backus published a note in *Nature,* outlining how the idea could be tested: one should find anomalies symmetrical about the center of the Mid-Atlantic Ridge, he suggested.[51] Subsequently, Vine and Wilson were able to demonstrate that the sequence of anomalies in various parts of the ocean is indeed symmetric about the crest of the Mid-Ocean Ridge, and that the pattern is the same for each location along the crest, allowing for different spreading rates.[52] The sequence of anomalies exactly agrees with the sequence of magnetic reversals worked out on land by Allan Cox and collaborators. At Lamont, geophysicists James Heirtzler and Walter Pitman started mapping the age of the seafloor, counting off the reversals as seen in the anomalies[53] (fig. 12.7).

A proposal by Backus to the National Science Foundation, to test the Vine-Matthews hypothesis by looking for symmetries around presumed spreading centers, was rejected as "too speculative," according to Menard.[54] A year and a half later, the point was moot. The evidence was overwhelming, even without new tests.

In 1966, coming back from a conference at Columbia University, where Fred Vine and the Lamont scientists had presented the evidence for symmetry of magnetism about spreading centers, Menard wrote a letter to the director of Scripps, William A. Nierenberg, advising him of the new developments. Menard urged the director to pay close attention to the ongoing geological revolution. In his letter, he noted that seafloor spreading is a demonstrated fact (although the mechanism was not entirely clear). He warned that Lamont was moving fast in applying the new paradigm, which "will clearly revolutionize all thinking on the history of the earth." And he insisted that the mapping of magnetic anomalies needed to be taken up at Scripps, in a routine manner, just like echo sounding became standard procedure for doing marine geology.[55]

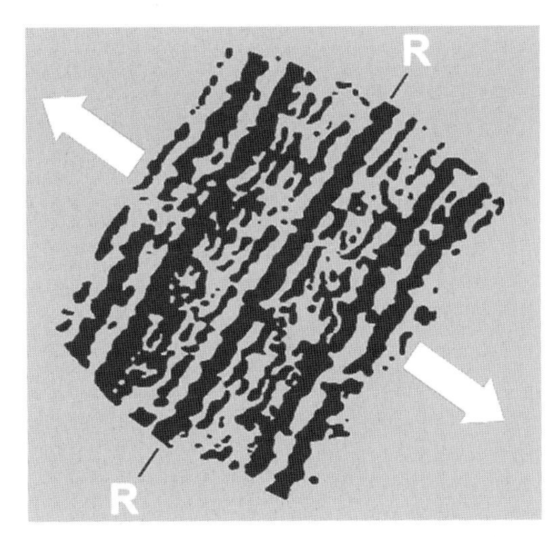

FIGURE 12.7. The symmetry of magnetic lineations about the ridge axis (R) was the crucial evidence for seafloor spreading. Lineations from a magnetic survey of Rejkjanes Ridge (J. R. Heirtzler et al. 1966). Arrows here added. R, ridge crest.

The transition had been made. The mapping of the seafloor now had an entirely new dimension and a new significance. The new program was the detection and reconstruction of large-scale lateral motion. To some of the graduate students at Scripps, having earlier listened to inspired presentations by Robert Dietz, the new program came as no surprise.

COROLLARIES OF THE NEW THEORY

The rate of acceptance of the new theory differed greatly between the working marine scientists and the Earth science professionals on the outside, whose contact with the new ideas was mainly through the literature. Inside, the corollaries to the revolutionary theory came like an avalanche. Outside, skepticism reigned well into the 1970s.

Skepticism came in many guises. Some argued that the new theory could not explain their particular problem (say, the geology of certain Caribbean islands); others had problems with the mechanism for large-scale horizontal motion. The latter attitude was well reflected in the cautious words of the *Encyclopedia Britannica* of 1974, "Until the question of mechanism is

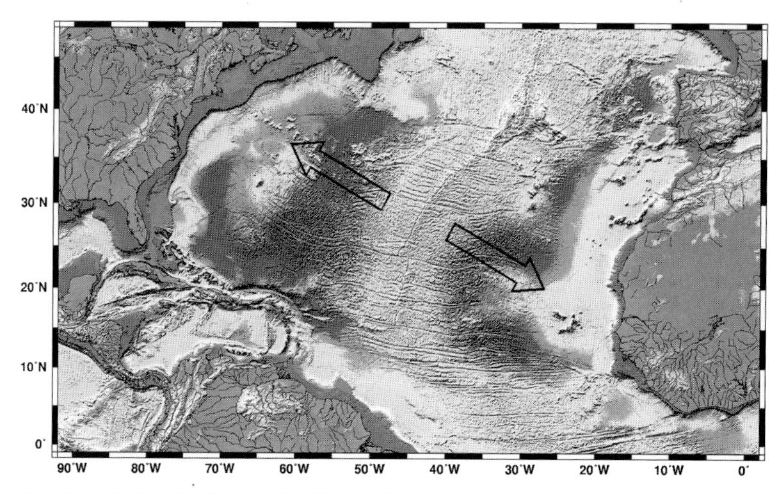

FIGURE 12.8. The fit of North Atlantic coastlines as a result of seafloor spreading, as seen in a modern gravity-based topographic reconstruction of the seafloor. Arrows added.

solved to general satisfaction, it appears premature, in spite of current enthusiasm, to regard continental drift as established."[56] This kind of statement, of course, looks suspiciously like the old Kelvin fallacy. The famous physicist averred that there is no mechanism for the Sun to burn so long, to accommodate geologic time. To this the proper answer is: it did, never mind the mechanism. The distinguished British physicist Harold Jeffreys fell into the Kelvin trap when he pronounced that continental drift is impossible. Obviously, if something happened, it is possible. And the outline of the African bulge does fit quite nicely into the great bay that is the East Coast (fig. 12.8).

The working scientists could not afford skepticism—it was now or never. The race was on to get a piece of the action and to get into print. In the 3 years following the paper by Vine and Wilson (1965) well over a hundred contributions appeared in the English-speaking scientific literature, using the new vantage point in interpreting Earth processes. In 1965, the Canadian geophysicist J. Tuzo Wilson suggested that mobile zones—ridge crests, trenches, and fracture zones—divide the surface of the Earth into several large rigid plates. In the same year, Edward Bullard and associates showed how to best describe the movement of crustal pieces on a spherical Earth, as rotations around a pole.[57] Three years later *plate tectonics* and *the new global tectonics* were on their way to becoming household words among geologists, thanks to several global syntheses of plate motions, which appeared in 1968.[58]

Subsequent and ongoing research has been concerned with refinements of the plate tectonic concepts, and their application to other fields in the Earth sciences. Also, marine geologists and seismologists began an intensive search for processes in the upper mantle, where the driving forces of plate tectonics are to be found.

We have already seen how the moving seafloor acts like a huge tape recorder, memorizing the ambient magnetic field as the newly upwelled magma cools, at the ridge crest. Since the patterns of the magnetic reversals are dated (on land, in volcanic layers), the age of the spreading seafloor can be estimated from the distribution of magnetic anomalies. When we now map the age distributions inferred from the magnetics, the results strikingly illustrate the growth and motion of the plates: there is a belt of young, shallow seafloor associated with the Mid-Ocean Ridge (fig. 12.9). It is relatively narrow in the Atlantic, where spreading is slow, and it is wide in the East Pacific, where spreading is fast. Ages are offset at the fracture zones. Remarkably, the position of the trenches appears to be rather independent of the age

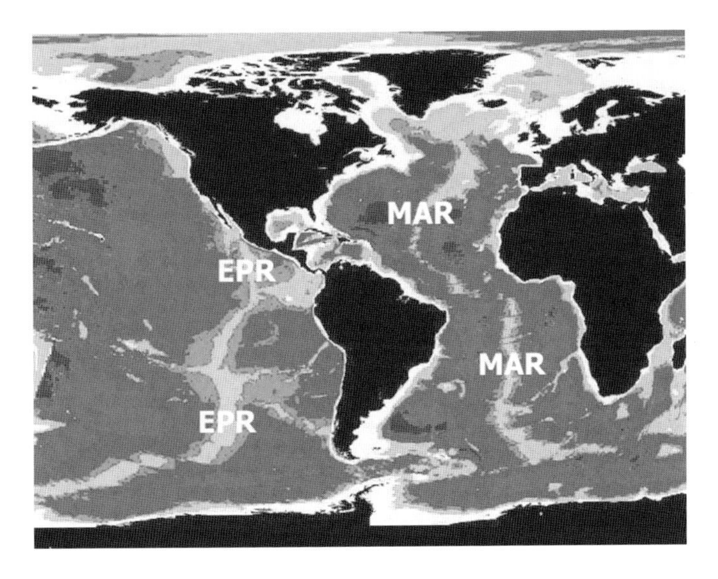

FIGURE 12.9. Youthful seafloor (light gray) marks the shallow Mid-Ocean Ridge in all ocean basins. The East Pacific Rise (EPR) is broad, because of fast spreading; while the Mid-Atlantic Ridge (MAR) tends to be narrow and rugged.

distribution of the seafloor: both young and old seafloors are being destroyed quite impartially.

As an extra bonus, plate tectonic theory delivers an elegant explanation for the existence of the Hawaiian island chain (fig. 12.10). The rigid plates that carry the seafloor are roughly 100 kilometers thick. They move on top of a plastic layer of the upper mantle, called the asthenosphere, which does not share the plate motions. J. T. Wilson, who did so much to further the causes of plate tectonics (after his conversion from a fixist position), proposed that many island chains originate on a plate moving over so-called hot spots.[59] These are large sources of rising magma within the mantle, from which basaltic material makes its way to the seafloor, to produce volcanoes. The origin of magma plumes is uncertain. They may originate at the core-mantle boundary of the Earth.

Finally, then, within the framework of plate tectonics, the answers emerged to the long-standing puzzles in geology—how to make mountain chains, trenches, and sunken islands, why earthquakes and trenches go together, and why trenches are flanked by volcanoes on land, not far from the coast.

Subduction of the seafloor thickens the crust below the mountains—and a thickened crust rises. Intrusions of newly melted rock into the crust, from the downgoing slab, provide for formation of igneous rocks and for volcanism. An enormous mixing machine is at work, with erosion products being delivered to the inner-slope trench, by gravity-driven flow and by slumping. This material is entrained, and some of it joins again the mountain roots at depth, only to be uplifted again as the entire complex keeps rising. The mountains so made are the Andes and

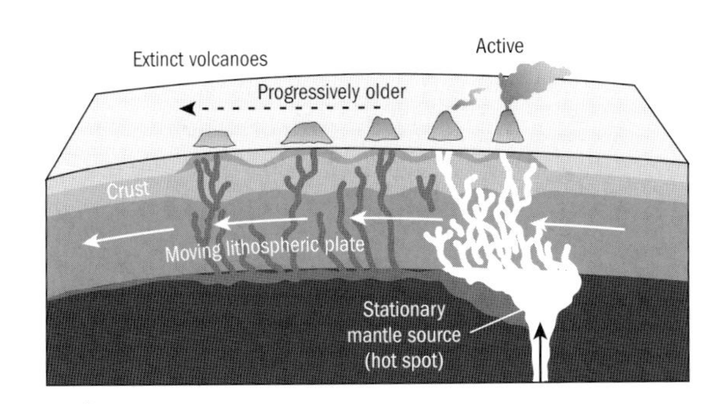

FIGURE 12.10. Origin of the Hawaiian island chain, based on the hypothesis of J. Tuzo Wilson, 1963.

their siblings elsewhere around the Pacific, including the Sierra Nevada. The composition is "andesitic," that is, intermediate between granite and basalt.

Thus, the mountains next to trenches are produced by collision of one plate with another. If one of the plates carries a continent, we get the Andes. If neither carries a continent, we get the Japanese island chain or the Aleutians. If both carry a continent, we get very high mountains: the Himalayas. The history of the collision of the Indian subcontinent with Asia—the uplift of Tibet—is recorded in the history of sedimentation in the Bay of Bengal.[60] Much or most of the uplift occurred in the last 10 million years or so: the Himalayas are geologically young, and rising rapidly.

When understanding came, it created a new kind of geology, destroying and redefining the most familiar and obstinate problems, which had stimulated learned discussion for many decades. Now, suddenly, it became obvious where and how mountains are made and, therefore, how to make continental crust. The continents are made of ancient mountain roots. Now we know, finally, the origin of the Vishnu Schist at the bottom of the Grand Canyon, and similar rocks in the continental crust everywhere. The schist was made within a subduction zone. More likely than not, it had a trench at its seaward edge. Now we know where to find the modern habitat of granitic plutons that dominate the landscape in the Sierras and in the mountains of San Diego and Baja California. They rise as enormous magma balloons within Andean-type mountain-building complexes, eating their way through the rubble of a metamorphosed mixture of sedimentary and igneous rocks, tumbled by a great mixing machine.

Invariably, regions of active mountain building are home to earthquakes. We now understand why deep and strong earthquakes ring the Pacific: they are part of subduction. Earthquakes occur both on the upper side of the downgoing lithosphere slab and on its underside. Earthquakes cease below a depth of about 700 kilometers. Presumably the temperature becomes high enough at that level to accommodate plastic flow, so that the strain (whose release makes quakes) can no longer build up.

The most extensive mountain chain on Earth, the Mid-Ocean Ridge, is seen as a result of processes fundamentally different from those responsible for the Andes and the Japanese island arc. The ridge is the site of upwelling of magma derived from rising mantle materials. The crust is thin and volcanic in origin and is, in essence, a window to the basaltic mantle, just as envisioned by Wegener. The ridge is more than a mile high above the deep seafloor because the new lithosphere forming there is hot, and its density is slightly reduced, therefore. As the new lithosphere cools, it sinks. It does so as it moves away from the spreading center, thus producing the flanks of the ridge.[61]

The crest (and in the Atlantic the central rift) has shallow earthquakes, with centers less than about 60 kilometers deep. The elevation of the crest, along much of the Mid-Ocean Ridge, is roughly the same (–2,500 meters to –3,000 meters), except in the North Atlantic. This suggests that the temperature of the upwelling material is roughly the same all around the globe. The fact that the basalt is hot when emplaced has a most interesting corollary: the heat drives a circulation within the fractured crest, whereby cold water enters into cracks in the upper flank, gets heated within the crest, and exits as hot springs and warm seeps. The hydrothermal vents set up chemical gradients, which are used by bacteria to make a living. In turn, the bacteria support a rich vent fauna, including worms, hydrozoans, mollusks, arthropods, and fishes, as described below.

The hydrothermal circulation leads to chemical alteration of the basalt, thus affecting the salinity of seawater. It also accelerates cooling of the new crust in the initial stages of spreading. The cooling of the plate leads to sinking. It takes about 10 million years for the new seafloor to sink by 1,000 meters. The next 1,000 meters of sinking takes about 26 million years. After 60 million years, the seafloor sinks to about 5,000 meters, which is the average depth of the deep seafloor, away from the Mid-Ocean Ridge. Sixty million years is close to the average age of the ocean floor. It is less than one-tenth the average age of exposed rocks on the continents. Not only is the ocean floor geologically young, it is extremely young.

Ever since the HMS *Challenger* discovered the Mariana Trench, this type of elongate, narrow, and deep depression posed an insoluble enigma, until the late 1960s. The trenches in the western Pacific are associated with zones of large negative gravity anomalies, a fact first discovered by the distinguished Dutch geophysicist Vening-Meinesz, in the 1920s.[62] This means that the crust is in the grip of powerful forces within the mantle, pulling it down, as was realized by Vening-Meinesz and others who followed.[63] On the continental side of the trenches, there are island arcs. Invariably, these are convex toward the oceanic side, that is, toward the trench. The arcs are almost entirely built of volcanic material (except for a certain amount of reef rock in places), including volcanic rock recycled within sediments. Earthquakes are common. These trench-arc systems can be understood only within the new global tectonics. In fact, even with these conceptual tools at hand, the western Pacific island arc systems remain difficult to explain. Behind the arc, on the landward side, there is typically a marginal basin with actively spreading seafloor.[64]

The realization that trenches are but one element in a complex tectonic system opened the way toward reinterpretation of the geology of the West Coast, from Washington to southern California. Of special interest is the nature of the "accreting wedge," which has a mixture of volcanic products from mountains at the landward side, marine sediments delivered in the normal fashion, and scraped-off seafloor materials from below, including altered basalt. Such mixtures (French: *mélange*) are open to inspection in the coastal ranges of California, from Morro Bay northward. These rock masses are tumbled chaotic masses on the seaward edge of a growing continent.[65]

HOT SPOTS AND THE TRACKING OF PLATE MOTION

Darwin's theory of atoll formation and Hess's theory of the origin of flat-topped seamounts in the central Pacific provided the portal to the enigma of the origin of the Pacific basin. Within the new framework of seafloor spreading, the widespread sinking of the seafloor indicated by islands and guyots became a simple corollary of volcanoes riding down the flanks of a spreading ridge. If the reef organisms kept up, the result is an atoll, a structure made of reef rocks on top of a volcano. If the corals and algae failed to keep up, the result is a flat-topped seamount.

In reality, things are rarely quite that simple, however.[66] In the case of the Hawaiian volcanoes, for example, the volcanoes are not produced anywhere near a ridge crest. Instead, a deep magma source keeps loading volcanoes onto the passing plate overhead, to make a chain. Away from the source, the volcanic edifices are reduced by vigorous erosion, they sink, and they acquire reef platforms. Thus we go from snow-covered Mauna Loa on the island of Hawaii (the biggest of the chain), westward to low-lying Nihau, and finally to the Emperor Seamounts, along a north-trending chain. As mentioned, this scenario was first laid out by J. Tuzo Wilson, in the earliest stage of the revolution (in 1963). Jason Morgan set it into the framework of plate tectonics, in 1971 and 1972.[67] The concept has since been widely used to link island chains with the motion of the Pacific plate through time.[68]

The Hawaiian Islands are nicely arranged from high to low, and from active to inactive, along a chain some 1,600 miles long (2,500 kilometers). At the active eastern end, the largest volcano on Earth—Mauna Loa—intermittently produces copious amounts of lava, creating new real estate by invading the sea (fig. 12.11). At the quiet western end, drowned craters tell a story of former volcanic glory long since subdued, its products assimilated into the realm of coral reefs. The Polynesian seafarers who discovered the islands around AD 400, coming from the

FIGURE 12.11. Scenes of Hawaii (Volcano Park). *Left:* Lava making its way to the sea. *Right:* Waves attacking the basaltic cliffs.

South Pacific, understood this sequence. They believed that Pelé, their mythological goddess of volcanism, made her initial home in Nihau, to the west, and then fled from her jealous sister in the sea, from one island to the next, until she reached her present home on the Big Island of Hawaii where (they say) she resides today.

Geologists attempting to explain the origin of the Hawaiian Islands invented a different story: they postulated a deep crack disrupting the surface of Earth, a crack that kept opening toward the southeast, allowing hot magma to rise from below, along its tip.

Both stories correctly reflect the sequence of events. Both stories reflect perplexity about the origin of the Hawaiian island chain.

According to the mechanism proposed by Wilson, the chain precisely records the direction of motion of the seafloor. According to Morgan, the source for the magma below the volcano at the tip of the chain is very deep in the mantle. Its position is fixed (more or less). It has been active for a very long time. For 43 million years, the plate above it has moved in the same direction. However, before that time the chain was lining up more nearly north to south, with the active tip at the southern end. The seafloor apparently moved northward. We see this in the Emperor seamount chain, now submerged below sea level.

What happened? At that time, at 43 million years ago, the entire seafloor carried on the Pacific plate must have changed its direction of motion. This was not a minor event in the history of the planet. Apparently, mountain building intensified, and the sea retreated from the shallow regions of the continent.[69] The widespread shallow-water carbonate deposits on the exposed shelves began to be eroded, and the materials were reprecipitated by calcareous plankton in the open ocean and ended up as new carbonate deposits on the deep seafloor. The climate changed. The polar areas cooled and winds became stronger. Winds generate upwelling. The zones of high productivity of the ocean—the regions of vigorous diatom growth—became concentrated in narrow belts along the equator and around the continents, in the coastal ocean. Food became abundant enough in these zones to stimulate the evolution of high-energy consumers: whales and seabirds. By 40 million years ago, the world had changed profoundly and was well on the way toward its present state, as a planet with snow and ice at its poles.[70]

Thus the corollaries to the bend in the Hawaiian-Emperor chain. Nothing on this planet happens in isolation. Even if we cannot explain the nature of the change, we do know that the bend was part of a large-scale reorganization of tectonics on the planet. It is not clear why the seafloor's motion would change direction. Perhaps slamming the Indian subcontinent into Asia had something to do with it. A major zone of subduction became clogged at that time, conceivably forcing a rearrangement of global convection patterns.

The Hawaiian chain is not the only hot-spot trace; there are others in the South Pacific, as is evident on the maps produced by radar altimetry from satellite surveying.[71] Hot-spot chains, of course, must be parallel (within their framework of plate motion), as was recognized by Wilson (1963) when he proposed the mechanism for creating chains. A prime example of a South Pacific hot-spot track is the Louisville seamount chain, which runs from its origin near the Pacific-Antarctic Ridge (at 53° S) in a northwesterly direction into the Kermadec-Tonga Trench. The chain has more than 60 seamounts and is 4,300 kilometers long. It represents 66 million years of hot-spot activity (comparable to the Hawaiian-Emperor age of 70 million years).[72]

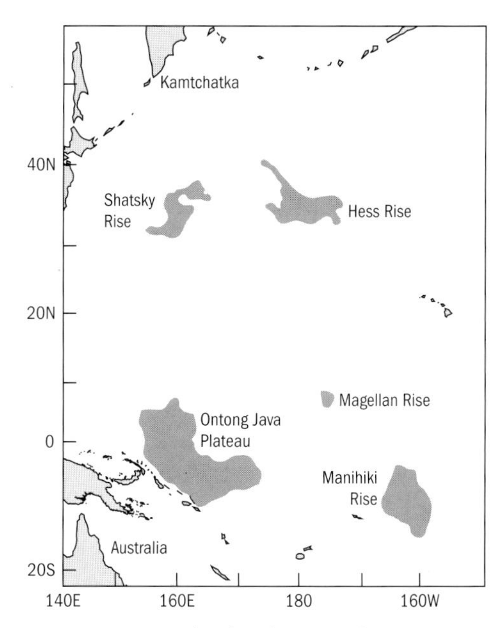

FIGURE 12.12. Large basaltic plateaus in the western Pacific.

How are we to visualize the source for the Hawaiian chain and its look-alikes? A common model is a vertical pipe deep down into the mantle, left over from a blob of hot melt rising from the mantle-core boundary. The hot stuff rises as in a chimney, pooling below the hot spot and feeding the magma chambers there. But how does the pipe get established and maintained? Speculations abound. Presumably, from time to time, superheated material piles up at the core-mantle boundary, becomes unstable, and melts its way up through the mantle as a plume head. On exiting at the surface it produces a "flood basalt." It leaves a pipe as its legacy. The pipe is maintained by continuing ascent of hot material.

The largest flood basalt edifice on the ocean floor is the Texas-size Ontong Java Plateau, in the western equatorial Pacific (fig. 12.12). It is more than 30 kilometers thick—a nice big sample of mantle material. Incidentally, it also serves as a large platform for the deposition of pelagic carbonate, with an outstanding record of ocean history (climate and productivity) for the last 100 million years. On ODP Leg 130, the *Joides Resolution* drilled through the entire sequence, and well into the basalt (see chapter 14). Hawaii's petrologist John Mahoney has discussed the possible sources for this plateau, the choice being between deep mantle plumes and ridge-crest basalts. He favors an origin as a plume in the vicinity of the Mid-Ocean Ridge, a situation analogous to today's Iceland Plateau.[73] Some of Iceland's basaltic edifices, of course, are exposed and can be directly inspected. In North America we have a pocket version of a major flood basalt in Washington: the Columbia Plateau, which takes up a substantial portion of the state. Its basaltic outpourings lasted about 10 million years, beginning in the lower Middle Miocene, 17 million years ago. Today's manifestation of this hot spot, it is thought, is the volcanic activity in Yellowstone Park, including the numerous hot springs there.

The Ontong Java Plateau is not the only large basaltic plateau in the western Pacific. Others are Shatsky Rise, Hess Rise, and Manihiki Plateau (fig. 12.12). As far as is known (all these structures are poorly sampled), Shatsky Rise is the oldest, and Hess Rise the youngest of the plateaus. However, all except perhaps Shatsky were active in the middle Cretaceous, which makes this period a time of incredibly intense basaltic outpourings from the mantle. It is generally agreed that such large-scale volcanism must have had considerable impact on the ocean environment, as well as on the atmosphere.

Long after the basaltic flooding has ceased, it is thought, the tail that trailed the plume now provides the conduit for the hot-spot material. The region around the hot spot is elevated, presumably because of the heating of the lithosphere from below, which makes it expand and float higher on the mantle that carries it. In the case of Hawaii, the swell associated with the hot spot has a dimension of roughly 1,000 kilometers across the trend of the islands, near Oahu. The height above the surrounding seafloor is 1.4 kilometers.[74] Of course, as the islands move off the swell, the lithosphere they ride on will sink back into normal depths, and in addition, the volcanoes will sink a little deeper, because of their own weight.

The fact that a hot spot produces a swell is noteworthy and helps explain the formation of seamounts along the Hawaiian-Emperor chain. But what about the much larger regions in the South Pacific that are anomalously shallow?

The question about the anomalously shallow deep seafloor in the South Pacific gets us into the discussions about *superswells*, a term coined by the marine geophysicist Marcia McNutt.[75] The superswell that carries the islands in the central Pacific south of the equator is reminiscent of (and might be the modern counterpart to) Menard's reconstructed Darwin Rise, a vast midplate swell, the site of active volcanism in the early to middle Cretaceous. Menard, in his book *Marine Geology of the Pacific* (1964), had reconstructed a virtual rise by connecting the flat tops of the sunken islands of the Mid-Pacific Mountains, assuming the resulting surface once was the sea level.[76] Thus, in his view, the entire seafloor over a vast region in the western Pacific was shallow at one time, not just individual islands.

The Darwin Rise, as envisaged by Menard, occupies the center of the Pacific basin, within a triangle formed by Hawaii, New Guinea, and Tahiti—a region the size of North America. If real, the Darwin Rise is a discovery of major proportions, a crowning achievement of the sunken island saga, from Darwin to Hess to Hamilton. If wrong, it ranks with the "Terra Australis" that used to occupy the Southern Ocean, before Captain Cook reduced it to a modest-size continent now known as Australia.

The crucial point, of course, is whether the tops of the various seamounts in the formerly elevated Realm of Islands were in fact at sea level at the same time. If they were not, a vast Darwin Rise is not necessary to explain the pattern. We could then envisage local swells or temporary swells to do the job of lifting the seafloor to the required elevation, one restricted area at a time. Information on this point comes from the intensive dredging and drilling that has taken place since Menard proposed the concept. But the available data apparently are not sufficient to decide the issue. One reading is that the ages of seamounts in the region show a geographic trend, from young in the southeast Pacific (including active volcanoes) to progressively older toward the northwest Pacific. If true, Menard's assumption that the tops of the seamounts are roughly of equal age cannot hold. This is the view taken by the geologists E. L. Winterer and W. W. Sager.[77] Looking at the same data, Marcia McNutt comes to a seemingly opposite conclusion: "Dredging, dating, and drilling of the guyots have now established conclusively that the guyots are closely grouped between 80 and 120 Ma in age."[78] In this view, where plus-or-minus 20 million years counts as "same," Menard's Darwin Rise remains a viable proposition, in the revised version put forward in 1984.[79] The hypothetical Darwin Rise may have a modern analog: the large region of elevated seafloor in French Polynesia, centered on Tahiti, which is rich in young volcanoes.[80] Discussions continue.

HOT VENTS AND THE STABILITY OF SEAWATER

If new seafloor forms from hot basalt at spreading centers—as is not in doubt—the hot rock

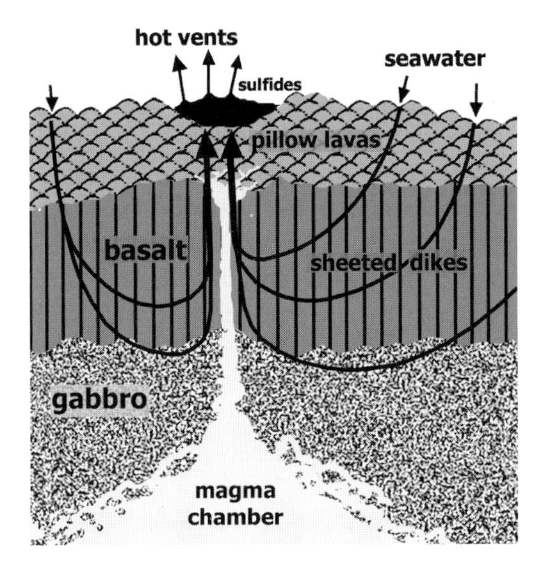

FIGURE 12.13. Modern concept of the circulation above a magma chamber on the seafloor.

cannot avoid reacting with the surrounding seawater, near the crest of the Mid-Ocean Ridge (fig. 12.13). Indeed, among the most exciting discoveries of the last quarter of the twentieth century are the various observations associated with hot vents. For geologists, the most important implications of these findings are those that relate to the chemistry of seawater in general, and to the formation of metal ores in the vicinity of hot vents. For biologists, the discovery of the strange ecosystem surrounding the vents was a momentous event demanding a recalibration of our thinking about the conditions that can support life.

That volcanic activity on the seafloor is associated with hot vents should perhaps not have come as such a surprise. In fact, more than a hundred years ago the Bavarian geologist Carl Wilhelm von Gümbel proposed that since the seafloor is largely of volcanic origin, hot springs should occur practically everywhere.[81] In the meantime, we know that hot springs are indeed found wherever there is hot basalt coming to the seafloor.[82] Most of the volcanically active areas of the seafloor are found at the Mid-Ocean Ridge, at the spreading center, and thus this region has been explored most actively and has yielded most of the vents. Hydrothermal deposits tend to congregate where faults and cracks provide an exit for hot water, laden with chemicals from the interior.[83]

Melvin N. A. Peterson, geologist at Scripps from 1960 till the 1990s, was keen on exploring the implications of the new seafloor-spreading paradigm for the alteration of basalts at the spreading center and any associated formation of minerals.[84] He believed that the enrichment of sediments with iron and manganese, on the flanks of the Pacific section of the Mid-Ocean Ridge, must be due to the release of these metals from hot basalt into seawater.[85] He advised graduate student Jack Corliss, who did his Ph.D. thesis on basalt-seawater interaction, at Scripps. Further pursuing this topic after moving to Oregon State University, Corliss discovered the Galapagos hot vents when diving in Woods Hole's submersible *Alvin*, in 1977, on a site of anomalously warm water, identified during previous exploration by the Scripps Deep-Tow group.[86]

The successful search for hot springs on the seafloor in the Galapagos Rift had been preceded by exploration of young basaltic seafloor in the Atlantic. In 1974, an initial series of detailed surveys of a section of the center of the Mid-Atlantic Ridge in the North Atlantic, using three different submersibles, yielded much interesting information about the nature of the spreading center, but no hot springs. In fact, a search for hydrothermal activity was not prominently on the list of sought-after phenomena. The French-American project—confidently named FAMOUS[87]—provided a detailed look at the rifting and faulting at the center, at the steep scarps that mark the rift, with talus fans at their base, and at the lava that is extruded at the center and looks like submarine pahoehoe. However, the water in contact with the youngest lava was just as cold as the rest of the deep sea.

The subsequent discovery of the warm springs and hot vents in the Galapagos Rift, and the sampling of the water that had reacted with hot basalt, revolutionized our understanding of the origin and history of seawater. The first analyses of seawater issuing from the Galapagos hot vents showed great enrichment with silica and almost total removal of magnesium and sulfate in the reaction.[88] Soon many more vents were found on the East Pacific Rise and elsewhere, and a coherent picture started to emerge, on how seawater chemistry is controlled by reactions with hot basalt. On the whole, these reactions have a stabilizing effect, since the chemistry of basalt, in essence, is constant. However, the results between sites vary, because the reactions depend on temperature.[89]

One aspect of the reactions that attracted great interest from the beginning is the enrichment of the seawater with metals, which results in the production of ores upon precipitation when the warm water enters the cold bottom water. The precipitation of minerals from hot water forms the chimneys seen so prominently on *Alvin* photographs, and also produces clouds, white or black, making "smokers"[90] (see fig. 2.3 in chapter 2).

Even before the discovery of the vents, experiments had shown that seawater should become enriched in iron and manganese, and other trace metals, when percolating through hot basalt. Analysis of sediments deposited near spreading centers had demonstrated that they are, indeed, rich in iron and other metals. Iron-rich sediments also were found in the cores raised by the Deep-Sea Drilling Project, before hot vents were known. These peculiar sediments occur exactly at the boundary between basalt and overlying sediment column, showing that they were formed when the basalt was still young. Thus, the principle of metal buildup by hydrothermal circulation had been outlined before the hot-vent deposits were discovered.

By 1990, at least 10 hydrothermally active sites had been visited and their fluids sampled by deep diving.[91] The great majority of these sites are in the eastern Pacific—off Ecuador, off Middle America, in the Gulf of California, and off Oregon and Washington. A wide range of environments is involved, and a large range of temperatures. Water depths cluster around 2,500 meters, the typical depth of the crest of the Mid-Ocean Ridge.

The chemistry of hydrothermal fluids differs markedly between sites, depending on temperature history and local circumstances. Perhaps the most striking feature is the almost complete removal of magnesium from the heated seawater, upon reaction with the basalt. It is a result of the creation of magnesium-rich silicate minerals within the reaction zone. The process is called serpentinization. Harry Hess had emphasized its importance when proposing seafloor spreading in 1962. Serpentinite, the greenish alteration product of basalt, is common in the coastal ranges of northern California. It is altered basaltic seafloor scraped off into the growing continent. The shiny smooth surface of the serpentinite rock is the reason for the name, invoking the appearance of snakeskin. Serpentinites are part of "ophiolite" sequences—both words translate to "snake rock."[92]

The removal of magnesium by alteration of basalt has major implications for the chemistry of seawater. The process reduces the alkalinity of the ocean and keeps magnesium from building up to excess. Before the discovery of the hydrothermal magnesium sink, the geochemical cycle of magnesium had been a major conundrum. At the same time as magnesium is removed, silica is added to the circulating water. Silicate is a diatom nutrient; its addition to seawater at the hot vents stimulates diatom production in the ocean. Neither the chemistry of seawater nor the long-term controls on the productivity of the ocean could be understood without the knowledge about the reactions of seawater with basalt, at the Mid-Ocean Ridge and elsewhere in active volcanic regions.

The new insights regarding basalt-seawater reactions also impact applied geology: such reactions have profound implications for the origin of many metal deposits now exploited on land. The fluids issuing from hot vents commonly carry high contents of iron and manganese, as well as of zinc, copper, and lead.

Why should this be so? The fundamental reason is that hot basalt removes oxygen from the seawater. When there is no oxygen left in the water, many metals that normally exist as oxides become soluble and can now be moved out from the basaltic crust into the ocean. Hot basalt, in its hunger for oxygen, also rips it out from the sulfate in the seawater. This reaction produces hydrogen sulfide, the poisonous substance that smells like rotten eggs. The rotten-egg smell is easily detected by human noses (at extremely low concentrations of hydrogen sulfide). It was reported from the very first samples taken from hydrothermal vents.

The combination of dissolved metal and dissolved hydrogen sulfide (much of the sulfur is from the basalt itself) results in the precipitation of metal sulfides at the exit of hot vents. The metal sulfides build up walls, for example, within the calcium-sulfate chimneys forming on top of vents. Iron sulfides, zinc sulfides, and copper sulfides are the most common. Lead sulfides have been found as well. As every prospector knows, where there is copper, there is invariably some silver and gold as well, because of somewhat similar chemical behavior of these three elements.

THE DISCOVERY OF A NEW ECOSYSTEM

While geologists led the way to the hot springs, in the search for the physical and chemical processes that must govern the interaction between hot basalt and seawater, it was the biologists who profited most thoroughly from the discovery, when it came. For sheer excitement, the sighting of the strange and wonderful creatures aggregating around the hot vents has had no equal in decades. And what a sight it was—cloudy springs of water boiling out of the rocks in the seafloor, and supporting a rich fauna never before seen by humans, from enormous tubeworms to huge clams, with an abundance of agile crabs, shrimp, fish, and squid in the vicinity, all new to science.

The life-supporting hot springs were first seen in the Pacific, near the Galapagos Islands in the eastern equatorial Pacific (in 1977), near 21° N and 12° N on the East Pacific Rise, and somewhat later in the Gulf of California. Most conspicuous among the vent fauna (and very photogenic) are the large (up to 6 feet long) white and red tubeworms forming colonies, and foot-long clams whose shells litter extinct vent areas. The tubeworms have no close relatives among the known phyla—they were eventually placed into their own phylum, the Vestimentifera.[93]

The discovery is rightly celebrated as a major achievement. The marine biologists R. T. Barber and A. K. Hilting, in a review of the most important events in the last five decades, comment as follows: "When geologists discovered the hydrothermal vents, biological oceanography received a much-appreciated jolt of intellectual stimulation (Corliss et al. 1979). The existence of a new kind of ecosystem with dramatic new biochemical adaptation fueled the imagination of everyone. . . . This work also provides a rational organizing paradigm for the search for life on other celestial bodies."[94]

Why are these creatures gathering around the vents? What is the source of food supply? What kinds of organisms are there, and what are their life histories? How old (geologically) are these communities? These are questions that have aroused intense interest in marine biologists. They have been a hot topic of discussion in deep-sea biology for the years since the discovery.[95]

Of all the wonderful exploits of Woods Hole's famed submersible *Alvin*, exploring and sampling the hot vents surely have been the most thrilling—and certainly the most prominent in terms of public attention. In some ways, *Alvin* and hot vents have become inseparable in perception.[96] The striking photographs are familiar by now—black smokelike plumes belching from light-colored and worm-encrusted chimneys several feet high, white clouds of bacterial debris, white crabs and fish, and always forests of tube worms, several feet tall, with the reddish valves at the top.

It is well to remember that before something can be discovered by a submersible with a very limited range of action, one has to know where to dive. The discussion started in earnest with the concerns of geophysicists, who noted, when measuring heat flow through the seafloor, that not enough heat was coming out of the

Mid-Ocean Ridge for all the hot basalt that was being emplaced there. They postulated that circulating seawater was responsible for cooling the freshly intruded basalt. This implied that hot water issued in springs on the seafloor, in areas of active volcanism.[97]

There was abundant evidence of volcanic activity near the Galapagos Islands in the eastern equatorial Pacific, from heat-flow measurements.[98] Thus, this region became a target of interest in the search for hot springs. The 1976 Deep-Tow work from *Melville* at the Galapagos Rift confirmed the presence of hydrothermal fluxes, from elevated temperatures and from chemical signals. The information could be precisely located because of transponder-assisted navigation of the deep-towed package. A number of scientists participated in these discoveries; Ray Weiss and Peter Lonsdale led the collaborative effort.[99]

One instrument on the deep-towed package was a camera armed with black-and-white film. On developing the film, Lonsdale was surprised to note clusters of empty clamshells in certain areas, not far from where the hot vents were suspected. It was the first tangible indication of the presence of a strange aggregation of large animals in this marine volcanic setting.[100] Early in the following year, in February 1977, *Alvin* arrived at the scene. Before its first dive, hundreds of photographs were taken by the deep-towed ANGUS instrument package (Acoustically Navigated Geological Undersea Surveyor) of Woods Hole. These again showed fields of large clams, presaging the odd encounters that were to come. Jack Corliss and Tjeerd van Andel (both from Oregon State University) made the first dive, piloted by Woods Hole's Jack Donelly. They promptly burst upon the scene that has so captivated marine scientists since.[101]

The geologists were totally surprised. Most of the creatures seen through the window were new to science. On the third dive, water collected from the vent area stunk of hydrogen sulfide, and on the fifth dive clams taken from the bottom also reeked of it. Later it was learned that the warm water and the aggregation of organisms are part of an ecosystem that derives its sustenance from bacteria that "feed" on hydrogen sulfide. But for now, for instant communication, some names were needed for what was being observed, and talked about. They named one area the "Garden of Eden" for its rich fauna, which included crabs, "dandelions," tube worms, "spaghetti," and pink fish. The "dandelions" proved to be a kind of siphonophore, which collapsed into mush on deck. The "spaghetti" was a mass of acorn worms, remotely related to primitive vertebrates.

The hot vent fauna had been discovered, and marine biology had a new paradigm, as did marine geochemistry. The geophysicists calling for hydrothermal activity to explain their heat flow data had been correct.

Additional dives and discoveries followed, both in the Galapagos Rift,[102] and subsequently on the East Pacific Rise. Many exciting discoveries were made elsewhere in many different parts of the ocean—not necessarily near spreading centers.[103] The combination of scouting for hot vents with a deep-towed package and then diving on the vents to study their geology, chemistry, and biology proved its worth. Diving on hot vents has its dangers. In some cases the water is hot enough to damage the submersible's skin, requiring great caution by the pilot in charge.

Bacteria, or more generally, microbes, are at the base of the food chain in the hot-vent environment. The key is the high content of hydrogen sulfide in the water issuing from the basalt. In the reaction with hot basalt, sulfate ions in the seawater were robbed of their oxygen, to oxidize iron-rich minerals. The resulting hydrogen sulfide can then be used as fuel, by appropriately endowed bacteria and archea. Using the oxygen in the surrounding seawater, they oxidize the hydrogen sulfide to obtain the energy for building and maintaining tissue. Just as corals have green symbionts, to allow them to directly benefit from the sunlight, so the vestimentiferan worms have symbiotic bacteria within special organs, to allow them to benefit from the energy source that is hydrogen sulfide in an oxygen-rich environment. The worms and their symbionts make a complete production system: the worms have no mouth, no digestive system, and no anus. Special adaptations allow the transport of normally poisonous hydrogen sulfide through the blood system of the worms to the site where bacteria do the processing.[104]

Besides the tubeworms, clams and mussels also have symbionts for tapping into the sulfidic energy source. Bivalves are commonly settled in the cracks surrounding vents. The water they are in is quite cold but presumably has some admixture from below, enriched with hydrogen sulfide.[105] Microbes also flourish outside of specialized hosts, of course, and their contribution to the food chain is made in the usual manner, through ingestion by grazers (limpets, crabs) and filter feeders (serpulid polychaetes), which then are prey to fishes and other predators.[106]

The origin of the vent fauna is complex.[107] Some of the species, such as the vestimentiferan worms, appear to be ancient. Others apparently switched from cold vents on the continental slopes (called seeps) to a warmer environment. Chemosynthesis also occurs during the decay of large bodies on the seafloor, such as whale carcasses. Possibly, species specialized to take advantage of such an environment found their way into the vent environment. Vents, of course, are geologically older than whales, but dead sharks and other large carrion could have served as well, over long geologic time. The point is that the vent fauna is not quite as isolated from other environments as it might seem, at first glance.

How do new vent fields develop? How do they get colonized? How long, on average, do they exist? What is the biogeographic extent of vent species? Discovery continues.

NOTES AND REFERENCES

1. Textbooks published up into the late 1960s fit that category.

2. Alfred Lothar Wegener (1880–1930) published his concept of an opening Atlantic Ocean from the drifting apart of continents in a provocative book called *The Origin of Continents and Oceans*, in 1915. His proposition became known as the hypothesis of continental drift. He postulated a supercontinent for the time before the drifting apart, which he called Pangaea ("all-Earth").

3. See the previous chapter for discoveries by Heezen, Hess, and Dietz. Bruce Heezen (1924–1977) led the effort to map the deep ocean floor in the Atlantic, with emphasis on the nature of the Mid-Atlantic Ridge; Harry Hess (1906–1969), during World War II, discovered numerous flat-topped seamounts between Hawaii and Guam, which he thought were sunken islands, indicating widespread subsidence of the seafloor. Robert Dietz (1914–1995) was a leading figure in the exploration of marine geology during the Earth science revolution of the 1950s and 1960s. A graduate student (brought from the University of Illinois) with Francis Shepard at Scripps, he maintained a close affiliation with Scripps geologists during his professional career at the nearby Navy Electronics Laboratory in San Diego. He coined the term *seafloor spreading*, in 1961.

4. For a brief review, see E. Seibold, W. H. Berger, 1993, *The Sea Floor, An Introduction to Marine Geology*, 2nd, revised, and updated ed. Springer Verlag, Heidelberg, 356 pp. For individual recollections of many of the major players, see N. Oreskes (ed.), 2001, *Plate Tectonics. An Insider's History of the Modern Theory of the Earth*. Westview, Boulder, Colo.

5. A. Wegener, 1929, *Die Entstehung der Kontinente und Ozeane*. Vieweg, Braunschweig, 231 pp.

6. Charles Schuchert (1858–1942) was one of the leading geologists of his time, a professor at Yale and a prominent author of geology textbooks widely used in America. In his report he opined as follows: "The school holding to the permanency of the continents and oceans is now almost universally in ascendancy"; and "the earth's surface in all of its parts moves either up or down," with the continents, on the whole, moving up, and the ocean bottom moving down. Nothing is said about horizontal motions. He cites the "study of coral islands" as having "demonstrated that the oceanic bottoms have moved at least locally up or down and as much as 5,000 feet." C. Schuchert, 1932, *The periodicity of oceanic spreading, mountain-making, and paleography*. Bulletin of the National Research Council 85, 537–557, p. 538.

7. Ph.H. Kuenen, 1950, *Marine Geology*. John Wiley and Sons, New York, 551 pp., p. 115. Philip Henry Kuenen (1902–1976) was professor of geology at the University of Gröningen, Netherlands, and a member of the Dutch *Snellius* Expedition to the Moluccas, 1929–1930.

8. H. H. Hess, 1948, *Major structural features of the western North Pacific*. Bulletin of the Geololgical Society of America 59, 417–446. The name *andesite* is applied to volcanic rocks similar to those in the Andes Mountains of South America.

9. Ph.H. Kuenen, *Marine Geology*, p. 129.

10. F.P. Shepard, 1963, *Submarine Geology*, 2nd ed. Harper and Row, New York, 557 pp.

11. H.W. Menard, 1964, *Marine Geology of the Pacific*. McGraw-Hill, New York, 271 pp., p. 233.

12. The magnetic anomaly patterns of the northeastern Pacific, mapped in the late 1950s, were known to Menard by the time he wrote the book. However, the patterns were unexplained.

13. J.R. Heirtzler, X. LePichon, J.G. Baron, 1966, *Magnetic anomalies over the Reykjanes Ridge*. Deep-Sea Research 13, 427–443; F.J. Vine, 1966, *Spreading of the ocean floor, new evidence*. Science 154, 1405–1415.

14. H.W. Menard, 1986, *The Ocean of Truth*. Princeton University Press, Princeton, N.J., 353 pp.

15. It is good to remember that the mechanism of continental drift was in fact not the issue. The issue raised by Wegener was whether the continents move or not. Pulling the discussion away from observation into the realm of mechanism put Wegener's hypothesis on hold till convection was discovered and accepted. The British scientist was Harold Jeffreys, one of the leading geophysicists of his time.

16. The complicated geology of California also became accessible—Tanya Atwater (Scripps Ph.D. 1972) was influential in applying plate tectonics to its elucidation.

17. H.W. Menard, *The Ocean of Truth*, p. 73.

18. The name is that of the Swiss geologist of the nineteenth century (Arnold Henry Guyot, 1807–1884), who founded the geology department at Princeton. In consequence, the discoverer of "guyots" had his office in "Guyot Hall" in Princeton.

19. Edwin Hamilton (see previous chapter) came to San Diego in 1949 to see about a thesis project, with advice from Robert Dietz at the Navy Electronics Laboratory. Hamilton later joined the NEL and worked closely with Scripps colleagues in the decades that followed. The results of the dredgings are in E.L. Hamilton, 1956, *Sunken islands of the Mid-Pacific Mountains*. Geological Society of America Memoirs 64, 1–97.

20. James Hutton (1726–1797), author of *Theory of the Earth* (Edinburgh, 1795), is commonly recognized as the founding father of classical geology. His rock cycle concept was a grand vision of mythological appeal and power—cycles upon cycles through infinite time, "no vestige of a beginning, no prospect of an end." Hutton's theory was opposed by those who believed the Saxonian mining engineer Abraham Gottlob Werner (1750–1817). Werner saw the world evolving from a hot ancestral ocean precipitating "primary" crystalline rocks (exposed in mountains, in places) to the present cool state of affairs, dominated by the water cycle and its "tertiary" and "quaternary" sediments. Hutton introduced dynamic steady state to geologic thought, a concept fundamental to modern geochemistry. Werner stood for the idea that Earth has a beginning and an end, and a history punctuated by catastrophe in between. Today's geologic worldview allows room for both views.

21. On a scale of a hundred million years, the steady-state concept yields good results. On longer time scales, things get complicated because of sporadic eruption of large flood basalts, from plumes that rise through the mantle from the core-mantle boundary. The plate tectonic machinery is more or less predictable, a fact used by the French geophysicist Xavier LePichon to calculate the first global map of motions, in 1968. X. LePichon, 1968, *Sea floor spreading and continental drift*. Journal of Geophysical Research 73, 3661–3697. But the spacing of plume events is a different matter altogether.

22. Edward Crisp Bullard (1907–1980) was a frequently seen visitor at Scripps. He had considerable impact on the development of plate tectonics. Bullard was department head at Cambridge University and later director of the National Physical Laboratory in England. He had measured the flow of heat on land in England and in South Africa in the 1930s and was eager to learn about heat flow from the seafloor. See E.N. Shor, 1984, *Edward C. Bullard 1907–1980*. Eos, American Geophysical Union 65, 74–75.

23. Arthur E. Maxwell obtained his Ph.D. degree with a thesis on heat flow studies, in 1959. He went on to a distinguished career in geophysics at the University of Texas. He was cochief on Leg 3 of the Deep Sea Drilling Project, the leg that is commonly credited with proving seafloor spreading, by showing that the age distributions of sediments overlying the basaltic "basement" of the seafloor agreed with expectations.

24. In building the instrument for the MidPac cruise, and in making the measurements, Maxwell worked with Scripps engineer James M. Snodgrass. For a detailed account of developments leading up to the heat-flow measurements, see E.N. Shor, 1984, *E.C. Bullard's first heat-probe*. Eos, American Geophysical Union 65 (9), 73–74.

25. R.R. Revelle, 1955, *On the history of the oceans*. Journal of Marine Research 14, 446–461.

26. Ibid.

27. H.W. Menard, *Marine Geology of the Pacific*, p. 138 ff.

28. Similar lists are readily found in textbooks, for example, see E. Seibold, W.H. Berger, 1996, *The Sea Floor*, 3rd ed. Springer-Verlag, Berlin, 356 pp.

29. The deepest place in the World Ocean is the *Challenger* Deep in the Mariana Trench near the island of Guam. It is close to 11 kilometers deep. On 23 Jan

1960, the U.S. Navy bathyscaphe *Trieste* reached the bottom, given as 35,813 feet deep (10,916 meters, or 5,969 fathoms). Other extreme depths occur in these trenches: the Tonga Trench (10,800 meters [35,433 feet]), the Kermadec Trench (10,050 meters [32,972 feet]), the Philippine Trench (10,055 meters [32,988 feet]), and the Japan Trench (9,700 meters [31,824 feet]). The deepest trenches have in common that they are starved of sediment. Many other trenches, similar in their narrow width and great length, are partially (or even wholly) filled with sediment, which greatly decreases their depth.

30. H.H. Hess, *Major structural features.*

31. H.W. Menard, 1955, *Deformation of the northeastern Pacific basin and the west coast of North America.* Bulletin of the Geological Society of America 66, 1149–1198.

32. H.W. Menard, *Marine Geology of the Pacific,* p. 149 ff.

33. H.H. Hess, 1946, *Drowned ancient islands of the Pacific basin.* American Journal of Science 244, 772–791.

34. R.W. Raitt, 1956, *Seismic-refraction studies of the Pacific Ocean basin.* Bulletin of the Geological Society of America 67, 1623–1640.

35. R.R. Revelle, A.E. Maxwell, 1952, *Heat flow through the floor of the eastern North Pacific Ocean.* Nature 170, 199–202; A.E. Maxwell, R.R. Revelle, 1954, *Heat flow through the Pacific Ocean floor.* Publications Bureau Central de Séismologie International Série A 19, 395–405; E.C. Bullard, A.E. Maxwell, R. Revelle, 1956, *Heat flow through the deep sea floor.* Advances in Geophysics 3, 153–181.

36. A number of geologists, such as A. Holmes, F.A. Vening-Meinesz, and D. Griggs, had earlier suggested that mantle currents could explain continental drift, and also the low gravity values and deep earthquakes associated with the trenches rimming the Pacific.

37. This was shown by measurements made by Scripps graduate student Richard von Herzen, during the Downwind Expedition to the East Pacific Rise, in 1958. He obtained his Ph.D. in 1960 and continued his career at Woods Hole. R.P. von Herzen, 1959, *Heat-flow values from the southeastern Pacific.* Nature 183, 882–883.

38. S.K. Runcorn, 1956, *Paleomagnetism, polar wandering and continental drift.* Geologie en Mijnbouw 18, 253–256.

39. Hess's paper was published in 1962. H.H. Hess, 1962, *History of ocean basins,* in A.E.J. Engel, H.L. James, B.F. Leonard (eds.), *Petrologic Studies: A Volume in Honor of A. F. Buddington.* Geological Society of America, Boulder, Colo., pp. 599–620. He circulated a preprint of the article in 1960. Hess referred to his hypothesis as "geopoetry," but it turned out to be more than that.

40. The chemistry of the volcanic rock produced in the Andes, overall, is intermediate between basalt and granite—hence, this type of rock is called andesite. Hess had earlier discovered a strict separation between "andesitic" and "basaltic" volcanoes, and proposed an "andesite line" tied to trenches. H.H. Hess, *Major structural features.*

41. In 1961, R.S. Dietz published his paper on seafloor spreading, which addressed the same problems as did Hess's manuscript, and solved them with similar dynamics. R.S. Dietz, 1961, *Continent and ocean basin evolution by spreading of the sea floor.* Nature 190, 854–857.

42. A. Holmes, 1944, *Principles of Physical Geology.* The Ronald Press, New York 532 pp. The fact that Holmes proposed ideas that parallel those of Hess and Dietz, although without the benefit of supporting data, makes the question as to whose work deserves priority somewhat less urgent from a historian's point of view. Also, Bruce Heezen, in 1960, published his idea of seafloor generation at the Mid-Atlantic Ridge, from upwelling magma, a concept he proposed at the International Oceanographic Congress in 1959. Furthermore, Menard published a proposition, in 1960, linking mantle upwelling to the elevation of the East Pacific Rise. B.C. Heezen, 1960, *The rift in the ocean floor.* Scientific American 203, 98–110; H.W. Menard, 1960, *The East Pacific Rise.* Science 132, 1737–1746. For Menard's view on the priority question, see H.W. Menard, 1986, *The Ocean of Truth,* Princeton University Press, Princeton, N.J., 353 pp., p. 155 ff.

43. R. S. Dietz, 1961, *Continent and ocean basin evolution,* p. 854.

44. A. Holmes, 1929, *Radioactivity and Earth movements.* Transactions of the Geological Society of Glasgow 18, 559–606.

45. H.W. Menard, *Marine Geology of the Pacific.*

46. The work was carried out by Ronald G. Mason, with the assistance of Arthur D. Raff. R.G. Mason, 1958, *A magnetic survey off the west coast of the United States.* Geophysical Journal 1, 320–329; R.G. Mason, A.D. Raff, 1961, *Magnetic survey off the west coast of North America, 32° N. latitude to 42° N. latitude.* Bulletin of the Geological Society of America 72, 1259–1266; A.D. Raff, R.G. Mason, 1961, *Magnetic survey off the west coast of North America, 40° N. latitude to 52° N. latitude.* Bulletin of the Geological Society of America 72, 1267–1270. Mason, who was at Imperial College in London, came to Scripps to take part in the Capricorn Expedition, in 1952, to make magnetic measurements along the ship's track. Several years later an opportunity arose to work with the U.S. Coast and Geodetic Survey, which was planning a multiyear

gridded hydrographic survey off the west coast, in 1955. At the suggestion of Menard, Roger Revelle arranged for Mason to participate for the purpose of a magnetic survey. Raff built the necessary equipment and helped Mason with the fieldwork. The papers resulting from this joint effort were the first to demonstrate the presence of regular magnetic anomaly patterns (the "stripes") on the seafloor.

47. Victor Vacquier joined Scripps in 1957, to continue magnetic studies of the seafloor. His results indicated very large offsets along fracture zones. V. Vacquier, 1959, *Measurements of horizontal displacement along faults in the ocean floor.* Nature 183, 452–453; V. Vacquier, A. D. Raff, R. E. Warren, 1961, *Horizontal displacements in the floor of the northeastern Pacific Ocean.* Bulletin of the Geological Society of America 72, 1251–1258.

48. H. W. Menard, *The Ocean of Truth,* p. 73.

49. F. J. Vine, D. H. Matthews, 1963, *Magnetic anomalies over ocean ridges.* Nature 199, 947–949; F. J. Vine, J. T. Wilson, 1965, *Magnetic anomalies over a young ocean ridge off Vancouver Island.* Science 150, 485–489.

50. The time scale of magnetic reversals was being established by geologists at the U.S. Geological Survey. A. Cox, R. R. Doell, G. B. Dalrymple, 1963, *Geomagnetic polarity epochs and Pleistocene geochronometry.* Nature 198, 1049–1051; A. Cox, R. R. Doell, G. B. Dalrymple, 1964, *Reversals of the Earth's magnetic field.* Science 144, 1537–1543.

51. G. E. Backus, 1964, *Magnetic anomalies over oceanic ridges.* Nature 201, 591–592.

52. F. J. Vine, J. T. Wilson, *Magnetic anomalies*; F. J. Vine, 1966, *Spreading of the ocean floor: New evidence.* Science 154, 1405–1415.

53. Presenting the famous *Eltanin*-19 record, with its nearly perfect symmetry: W. C. Pitman, J. R. Heirtzler, 1966, *Magnetic anomalies over the Pacific-Antarctic Ridge.* Science 154, 1164–1171; and the spectacular survey in the North Atlantic: J. R. Heirtzler, X. Le Pichon, J. G. Baron, 1966, *Magnetic anomalies over the Reykjanes Ridge.* Deep-Sea Research 13, 427–443.

54. H. W. Menard, *The Ocean of Truth,* p. 222.

55. H. W. Menard, *The Ocean of Truth,* p. 277.

56. *Encyclopedia Britannica,* 1974, vol. 6, p. 41.

57. J. T. Wilson, 1965, *A new class of faults and their bearing on continental drift.* Nature 207, 343–347; J. T. Wilson, *Transform faults, oceanic ridges and magnetic anomalies southwest of Vancouver Island.* Science 150, 482–485; E. C. Bullard, J. E. Everett, A. G. Smith, 1965, *Fit of continents around Atlantic,* in P. M. S. Blackett, E. C. Bullard, S. K. Runcorn (eds.), *A Symposium on Continental Drift.* Royal Society of London Philosophical Transactions, Series A 258, 41–75.

58. Cambridge's Dan McKenzie and Scripps's Robert Parker worked out the mathematics of motions of rigid plates on a sphere and showed that the ocean floor of the North Pacific moved as one. Their paper was published in 1967, in the last issue of *Nature* of that year. Other fundamental papers of the new global tectonics were by Jason Morgan at Princeton, and the Lamont scientists B. Isacks, Jack Oliver, L. R. Sykes, and Xavier Le Pichon (visiting from France), as well as others. D. P. McKenzie, R. L. Parker, 1967, *The North Pacific, an example of tectonics on a sphere.* Nature 216, 1276–1280; W. J. Morgan, 1968, *Rises, trenches, great faults, and crustal blocks.* Journal of Geophysical Research 73, 1959–1982; B. Isacks, J. Oliver, L. R. Sykes, 1968, *Seismology and the new global tectonics.* Journal of Geophysical Research 73, 5855–5899; X. LePichon, 1968, *Sea floor spreading and continental drift.* Journal of Geophysical Research 73, 3661–3697; J. R. Heirtzler, G. O. Dickson, E. M. Herron, W. C. Pitman, X. LePichon, 1968, *Marine magnetic anomalies, geomagnetic field reversals, and motions of the ocean floor and continents.* Journal of Geophysical Research 73, 2119–2136.

59. J. T. Wilson, 1963, *Evidence from islands on the spreading of the ocean floor.* Nature 197, 536–538; J. T. Wilson, 1963, *A possible origin of the Hawaiian Islands.* Canadian Journal of Physics 41, 863–870.

60. Scripps geologist Joseph R. Curray spent many years studying and mapping the enormous Bengal Fan deposits, witness to the rapid erosion in the Himalayas.

61. The relationship between depth and age of seafloor and the associated heat flow has been the subject of studies by the geophysicist John Sclater. See, for example, J. G. Sclater, L. Wixon, 1986, *The relationship between depth and age and heat flow and age in the Western North Atlantic Region,* in *The Geology of North America,* vol. M. Geological Society of America, Boulder, Colo. pp. 257–270.

62. Felix Andries Vening-Meinesz (1887–1966), pioneer of gravity measurements at sea. F. A. Vening-Meinesz, 1930, *Maritime gravity surveys in the Netherlands East Indies.* Proceedings of the Koninklijke Akademie van Wetenschappen 33, 566–577; F. A. Vening-Meinesz, J. H. F. Umbgrove, Ph. H. Kuenen, 1934, *Gravity Expeditions at Sea, 1923–1932,* vol. 2. Waltman, Delft, 208 pp.

63. Including Harry Hess, who at times accompanied Vening-Meinesz on submarines carrying out gravimetric measurements.

64. Daniel E. Karig, a Scripps graduate student in the late 1960s (Ph.D. in 1970), went there on the *Melville* and started to work out the dynamics of back-arc spreading, a hitherto unknown process. D. E. Karig, 1971, *Origin and development of marginal basins in the western Pacific.* Journal of Geophysical Research 76, 2542–2560; D. E. Karig, 1974, *Evolution of arc systems*

in the western Pacific. Annual Review of Earth and Planetary Sciences 2, 51–75.

65. K. J. Hsü, 1971, *Franciscan mélange as a model for eugeosynclinal sedimentation and underthrusting tectonics*. Journal of Geophysical Research 76, 1162–1170.

66. The history of western Pacific guyots is complicated by large-scale vertical motions that are independent of seafloor spreading. See E. L. Winterer, J. H. Natland, R. J. van Waasbergen, R. A. Duncan, M. K. McNutt, C. J. Wolfe, I. Premoli Silva, W. S. Sager, W. V. Sliter, 1993, *Cretaceous guyots in the northwest Pacific: An overview of their geology and geophysics*, in M. S. Pringle, W. W. Sager, M. V. Sliter, S. Stein (eds.), *The Mesozoic Pacific: Geology, Tectonics and Volcanism*. American Geophysical Union Geophysical Monograph 77, pp. 307–334; R. J. van Waasbergen, E. L. Winterer, 1993, *Summit geomorphology of western Pacific guyots*, in M. S. Pringle, W. W. Sager, M. V. Sliter, S. Stein (eds.), *The Mesozoic Pacific: Geology, Tectonics and Volcanism*. American Geophysical Union Geophysical Monograph 77, pp. 335–366.

67. John Tuzo Wilson (1908–1993) invented the hot-spot theory of the origin of the Hawaiian Islands. J. T. Wilson, 1963, *A possible origin of the Hawaiian Islands*. Canadian Journal of Physics 41, 863–870. Morgan elaborated on the nature of the hot spots and their tracks on the surface of the planet. W. J. Morgan, 1971, *Convection plumes in the lower mantle*. Nature 230, 42–43; W. J. Morgan, 1972, *Plate motions and deep mantle convection*. Geological Society of America Memoirs 132, 7–22.

68. R. A. Duncan, D. A. Clague, 1985, *Pacific plate motions recorded by linear volcanic chains*, in A. E. M. Nairn, F. G. Stehli, S. Uyeda (eds.), *Ocean Basins and Margins*. Plenum, New York, pp. 89–121.

69. A shift in global geochemistry can be seen in a distinct change in the ratio of strontium isotopes incorporated into shallow-water carbonates. At the time, in the late Eocene, the type of strontium that comes from mountains on land started to increase in relative abundance, and the ratio started to move toward modern values.

70. A fundamental shift in the ecosystem toward modern times around 40 million years ago is reflected in deep-sea sediment patterns. W. H. Berger, G. Wefer, 1996, *Expeditions into the past: Paleoceanographic studies in the South Atlantic*, in G. Wefer, W. H. Berger, G. Siedler, D. J. Webb (eds.), *The South Atlantic: Present and Past Circulation*. Springer-Verlag, Berlin, pp. 363–410.

71. Scripps geophysicist David Sandwell has recently compiled the results from such surveying, producing (inferred) topographic information where few ship tracks exist.

72. According to Scripps geologist James Hawkins, the composition of the source material remained the same, suggesting a steady supply from a large source.

73. J. J. Mahoney, 1987, *An isotopic survey of Pacific oceanic plateaus: Implications for their nature and origin*, in B. H. Keating, P. Fryer, R. Batiza, G. W. Boehlert (eds.), *Seamounts, Islands, and Atolls*. American Geophysical Union Geophysical Monograph 43, pp. 207–220. Substantial off-axis emplacement is suggested in C. R. Neal, J. J. Mahoney, L. W. Kroenke, R. A. Duncan, M. G. Petterson, 1997, *The Ontong Java Plateau*, in J. J. Mahoney, M. F. Coffin (eds.), *Large Igneous Provinces—Continental, Oceanic, and Planetary Flood Volcanism*. American Geophysical Union Geophysical Monograph 100, pp. 183–216. A double-pulse origin is suggested (125 million years and 90 million years) in J. J. Mahoney, M. Storey, R. A. Duncan, K. J. Spencer, M. Pringle, 1993, *Geochemistry and age of the Ontong Java Plateau*, in M. S. Pringle, W. W. Sager, M. V. Sliter, S. Stein (eds.), *The Mesozoic Pacific: Geology, Tectonics and Volcanism*. American Geophysical Union Geophysical Monograph 77, pp. 233–261.

74. N. H. Sleep, 1992, *Hotspot volcanism and mantle plumes*. Annual Review of Earth and Planetary Sciences 20, 19–43.

75. M. McNutt, 1998, *Superswells*. Reviews of Geophysics 36, 211–244. McNutt was once Menard's student and collaborator and is now director of the Monterey Bay Research Institute.

76. H. W. Menard, *Marine Geology of the Pacific*, p. 140.

77. E. L. Winterer, W. W. Sager, 1995, *Synthesis of drilling results from the Mid-Pacific Mountains: Regional context and implications*. Proceedings of the Ocean Drilling Program, Scientific Results 143, 497–535, p. 498.

78. M. McNutt, *Superswells*, p. 212.

79. H. W. Menard, 1984, *Darwin reprise*. Journal of Geophysical Research 89, 9960–9968.

80. In essence, the Tahitian "superswell" is a factory for volcanic islands and seamounts. These are commonly aligned along chains as a result of the lithosphere motion across the region of mantle updraft and magma injection. The story is greatly complicated by the fact that individual seamounts have histories that include rising well above sea level, after a period of subsidence. As drilling shows, soils have formed on reef rocks, before the final sinking into the abyss.

81. Von Gümbel (1823–1898) knew about hot springs because that is where he went on doctor's advice. He thought manganese nodules were a product of hot springs on the seafloor. W. H. Berger, 1986, *Gümbel's hypothesis regarding the origin of manganese*

nodules. Eos, American Geophysical Union 67, 169, 175, 179.

82. R.A. Lutz, 2001, *Hydrothermal vent biota,* in Steele, J.H., S.A. Thorpe K.K. Turekian, *Encyclopedia of Ocean Sciences,* 6 vols. Academic Press, San Diego, pp. 1217–1227.

83. R.M. Haymon, 2001, *Hydrothermal vent deposits,* in Steele, J.H., S.A. Thorpe K.K. Turekian, 2001, *Encyclopedia of Ocean Sciences,* 6 vols. Academic Press, San Diego, pp. 1228–1234.

84. Melvin N.A. Peterson (1929–1995) joined Scripps in 1960. He was in charge of the scientific management of the Deep-Sea Drilling Project, for its duration.

85. K. Bostrom, M.N.A. Peterson, 1966, *Precipitates from hydrothermal exhalations on the East Pacific Rise.* Economic Geology 61, 1258–1265.

86. The Deep-Tow group is part of the Marine Physical Laboratory at Scripps, then led by the physicist and engineer Fred N. Spiess. (See chapter 11.) The hydrothermal area was located by measurements of temperature and chemical properties of the water, and by photography. See W.S. Fyfe, P. Lonsdale, 1981, *Ocean floor hydrothermal activity,* in C. Emiliani (ed.), *The Oceanic Lithosphere, The Sea,* vol. 7. Wiley-Interscience, New York, pp. 589–638. For the discovery of the hot vents by *Alvin* see J.B. Corliss, J. Dymond, L.I. Gordon, J.M. Edmond, R.P. von Herzen, R.D. Ballard, K. Green, D. Williams, A. Bainbridge, K. Crane, Tj. H. van Andel, 1979, *Submarine thermal springs on the Galápagos Rift.* Science 203, 1073–1083. A preliminary report is in *National Geographic* magazine, vol. 152 (4), 441–453, published in 1977 and authored by J.B. Corliss and R.D. Ballard.

87. Project FAMOUS, a collaborative effort between French and American scientists led by James Heirtzler and Xavier LePichon. For a summary of goals and results see R.D. Ballard in J.H. Steele and Ocean Studies Board Members (eds.), 2000, *50 Years of Ocean Discovery.* National Research Council, National Academy Press, Washington D.C., 269 pp., p. 72 ff.

88. Pioneering work in chemistry was done by John M. Edmond, MIT, who had completed his Ph.D. in geochemistry at Scripps, in 1970. J.M. Edmond, C. Measures, R. McDuff, L.H. Chan, R. Collier, B. Grant, L.I. Gordon, J.B. Corliss, 1979, *Ridge-crest hydrothermal activity and the balances of the major and minor elements in the ocean: The Galápagos Data.* Earth and Planetary Science Letters 46, 1–18.

89. For details on the effects of hot vents on seawater chemistry see H. Elderfield, A. Schultz, 1996, *Mid-ocean ridge hydrothermal fluxes and the chemical composition of the ocean.* Annual Review of Earth and Planetary Science 24, 191–224. The geologist William Normark (Scripps Ph.D., 1970), with the U.S. Geological Survey, in 1978 found a temperature of 32 °C (91 °F) on the East Pacific Rise, where chimneylike structures on the seafloor poured out black "smoke." Samples showed that the smoke consisted of particles of iron sulfide and zinc sulfide. White smokers had clear or milky particulates of silica and barium sulfate. Some were skeptical of Normark's report of high temperatures, but the following dives found the near-unbelievable temperature of 350 °C (662 °F).

90. For mineral precipitation, see, for example, K.L. von Damm, 1990, *Seafloor hydrothermal activity: Black smoker chemistry and chimneys.* Annual Review of Earth and Planetary Science 18, 173–204. Initially, it seemed that the ores in ridge crest areas might be too scattered to be of economic interest. However, oceanographers from the National Oceanic and Atmospheric Administration (NOAA) have reported deposits from off Ecuador and from off the West Coast that may be commercially important. A host of different minerals are involved. At Scripps, geochemist Miriam Kastner has studied such hot-vent mineral associations for many years.

91. K.L. von Damm, *Seafloor hydrothermal activity.*

92. *Serpentinite* derives from Latin, *ophiolite* from the Greek word for "snake."

93. R.A. Lutz, *Hydrothermal vent biota,* p. 1217. The giant tube worms have no digestive tract but live on organic matter synthesized by symbiotic microbes. C.M. Cavanaugh, S.L. Gardiner, M.L. Jones, H.W. Jannasch, J.B. Waterbury, 1981, *Prokaryotic cells in the hydrothermal vent tube worm Riftia pachyptila Jones: Possible chemoautotrophic symbionts.* Science 213, 340–342.

94. R.T. Barber, A.K. Hilting, 2000, *Achievements in biological oceanography,* in J. Steele and Ocean Studies Board Members (eds.), *50 Years of Ocean Discovery,* National Academy Press, Washington, D.C., pp. 11–21, p. 12. The reference is to J.B. Corliss et al., *Submarine thermal springs.* The reference to life on other planets has an assumption that the bacterial flora associated with the vents may throw light on the origin of life.

95. Although less stirring than questions about the origin of life, the questions about the origin of the vent organisms have the advantage of being answerable, in principle.

96. A recent encyclopedia article entitled "Science of Deep Submergence" by Woods Hole scientist D.J. Fornari (in Steele et al., *Encyclopedia of Ocean Sciences,* p. 643–658) has 11 photographs taken by *Alvin.* All are from hot vents. No photographs are shown of any other deep-sea subject (other than machinery). The other major link of *Alvin* into public consciousness is the exploration of the wreck of the passenger ship

Titanic. For an account of that venture, see R. D. Ballard, M. McConnell, 1995, *Explorations.* Hyperion, New York, 407 pp.

97. A prominent voice among those postulating the presence of hot springs was C. R. B. Lister, at the University of Washington. C. R. B. Lister, 1972, *On the thermal balance of a mid-ocean ridge.* Geophysical Journal International 26, 515–535; C. R. B. Lister, 1974, *On the penetration of water into hot rock.* Geophysical Journal International 39, 465–509.

98. J. G. Sclater, K. D. Klitgord, 1973, *A detailed heat flow, topographic and magnetic survey across the Galápagos spreading center at 86° W.* Journal of Geophysical Research 78, 6951–6975; D. L. Williams, R. P. Von Herzen, J. G. Sclater, R. N. Anderson, 1974, *The Galápagos spreading centre: Lithospheric cooling and hydrothermal circulation.* Geophysical Journal International 38, 587–608.

99. P. Lonsdale, 1977, *Deep tow observations at the mounds abyssal hydrothermal field, Galápagos Rift.* Geology 5, 147–152; R. F. Weiss, P. F. Lonsdale, J. E. Lupton, A. E. Bainbridge, H. Craig, 1977, *Hydrothermal plumes in the Galapagos Rift.* Nature 267, 600–603. The Deep-Tow system of the Marine Physical Laboratory at Scripps, developed by Fred N. Spiess and colleagues, proved crucial for the task of locating the hot-water emissions with sufficient precision for later diving.

100. P. Lonsdale, 1977, *Clustering of suspension-feeding macrobenthos near abyssal hydrothermal vents at oceanic spreading centers.* Deep-Sea Research 24, 852–863. Lonsdale speculated that these organisms made their living assisted by the hydrothermal activity, perhaps through currents bringing food, or perhaps by filtering bacteria from the water. A good guess indeed.

101. Rarely was a seemingly esoteric interest (basalt-seawater interaction) rewarded more richly (discovery of a major new habitat of exotic marine organisms). For a lively description of the scene, see Robert Ballard's personal account in J. H. Steele et al., *50 Years of Ocean Discovery*, pp. 67–84.

102. Scripps biologist Robert Hessler went to 8,159 feet in *Alvin* in the same area in 1979, not his first dive. He had been down to 5,850 feet on one of the earliest dives in *Alvin* in 1966 off Bermuda, when he was on the staff at Woods Hole.

103. R. A. Lutz, 2001, *Hydrothermal vent biota,* fig. 2.

104. C. L. van Dover, 2001, *Hydrothermal vent ecology,* in Steele, J. H., S. A. Thorpe K. K. Turekian, *Encyclopedia of Ocean Sciences,* 6 vols. Academic Press, San Diego, pp. 1234–1241; A. J. Arp, 2001, *Physiology of hydrothermal vent fauna,* in Steele, J. H., S. A. Thorpe K. K. Turekian, *Encyclopedia of Ocean Sciences,* 6 vols. Academic Press, San Diego, pp. 1242–1246; C. L. van Dover, 2000, *The Ecology of Deep-Sea Hydrothermal Vents.* Princeton University Press, Princeton, N.J., 424 pp.

105. That the large clam *Calyptogena magnifica* grows at temperatures only slightly above ambient was verified by oxygen isotope analysis. J. S. Killingley, W. H. Berger, K. C. Macdonald, W. A. Newman, 1980, *$^{18}O/^{16}O$ variations in deep-sea carbonate shells from the RISE hydrothermal field.* Nature 287, 218–221.

106. Many authors on the subject make the point that chemosynthesis is an additional way of life, independent from photosynthesis. On some level this is true, but in the case under discussion the free oxygen must be accounted for, which allows the oxidation of the hydrogen sulfide. Free oxygen is a result, presumably, of photosynthesis and photodissociation of water; both depend on sunlight.

107. They are not globally the same. C. L. van Dover, *Hydrothermal vent ecology,* p. 1239. The oxygen content of deep waters was extremely low during parts of the middle Cretaceous (around 100 million years ago), so the bacterial oxidation of hydrogen sulfide at the exit of vents in the deep sea would have been impeded at that time. In this sense, vents at the Mid-Ocean Ridge are not an ancient environment.

THIRTEEN

The Ocean's Memory of the Ice Ages

THE ENDLESS CYCLES OF CLIMATE CHANGE

We live in an ice age, geologically speaking. Some considerable portion of the water on this planet is locked up in ice, at high latitudes (fig. 13.1).[1] More precisely, we live in a warm period within a long series of ice age fluctuations.

Over the last million years, the sea level was about 200 feet (60 meters) lower than today, on average. It was higher than today only for a few percent of this time span. The reason the sea level was normally much lower was the presence of large ice sheets that covered much of Canada and Scandinavia, and of thickened ice on Greenland, as well as increases in ice volume elsewhere. During times of maximum ice volume on land (glacial maxima) ice fields covered the region of the Great Lakes in North America, and the Baltic Sea in northern Europe. Besides, many elevated regions in temperate to high latitudes had substantial mountain glaciers: the Rocky Mountains, the High Sierras, the Alps, and the Himalayas all had enormous ice fields. In the Northern Hemisphere, much of the glacial ice disappeared between 16,000 and 8,000 years ago; but not from the Antarctic continent, which has been in a glacial condition for more than 15 million years (fig. 13.2). Ice age climate fluctuations are prominently a matter of the Northern Hemisphere, the Southern Hemisphere's climate being stabilized by the permanent ice cap on Antarctica.

Modern landscapes—just about all of them, including coastal landscapes—carry a legacy from the last glacial period and cannot be understood when only present-day processes are considered. By the same token, neither the present distribution of plants on the surface of Earth, nor that of coral reefs and atolls can be understood without reference to ice age history. Furthermore, the same is true for the zoogeography of the sea, inasmuch as the evolution of marine birds and mammals is involved. The evolution of present-day warm-blooded vertebrates,

FIGURE 13.1. A glacier calving into a fjord in Svalbard, Norway, near 78 °N. We live in the warm phase of an ice age that began 3 million years ago, when Greenland acquired an ice cap (1).

FIGURE 13.2. Icebergs in the waters off the Antarctic Peninsula, Bransfield Strait, near 64 °S. The Antarctic has been in an ice age for some 15 million years. The sea surface is ice free in summer, during the present warm period.

obviously, was closely linked to the history of ice ages and the associated changes in the ocean environment for the last 3 million years.

The climate future of the planet hinges largely on the response of ice on Greenland and in Antarctica to ongoing warming. As the Egyptian priest told the Greek traveler Solon—if you do not think on a time scale of many thousands of years, your knowledge and understanding remains that of a child.[2] The priest had a point, but his time scale (the last 10,000 years) fell short. It is the last million years that we need to consider. And perhaps more.

OF FJORDS AND REEFS AND PIONEERS

The fjords of western Norway are among the most stunningly beautiful coastal landscapes on

FIGURE 13.3. A fjord in western Norway, filling an ancient glacial valley. The seawater penetrates deeply into the land, bringing heat (and tourist ships).

our planet (fig. 13.3), while the coral gardens in the Great Barrier Reef are among the most alluring scenes imaginable, below the sea surface. Surprisingly, both environments are shaped by the action of ice. The fjords are glacial valleys, carved by the powerful grinding action of enormous ice tongues forcing a path to the sea from the mountains to the east. The reefs are heaps of rubble made by rapidly growing coral forests, forests that responded to the large fluctuations of sea level that accompanied the buildup and decay of northern ice sheets. An ocean with a steady-state sea level would produce carbonate platforms, not the kinds of coral reefs we observe (see chapter 4).

The link between the fjords of Norway and the Great Barrier Reef of Australia is quite tight, geologically. Both are a legacy of the ice ages of the last million years or so. Beginning a million years ago, small ice age fluctuations gave way to big ones, with amplitudes sufficient to lower sea level by more than 100 meters.[3] Since then, the large and stately climatic cycles of the ice ages have come and gone in a rhythm of about 100,000 years, roughly corresponding to the regular changes in eccentricity of Earth's orbit around the Sun. These grand astronomically driven waves of change have ruled life on the planet, from the fish in the fjords to the corals in the South Seas, for the last million years, and they also helped shape the history of our own species. It bears repeating in view of the new climate awareness now arising in the public: our landscapes from the mountains to the shores, the life around us, and the human species itself—all have been sculptured by the coming and going of the vast changes in climate associated with the ice age oscillations.

Whenever large continental ice masses build up in Scandinavia, great streams of ice move from the inland glaciers toward the ocean (fig. 13.4). Then, in the mountainous regions of western Norway, thick glacier tongues fill the valleys and start deepening and widening them right out to the shelf, at the same time carving out the inner passage parallel to the coast. The ice tongues work as so many bulldozers, clearing the valleys of debris, and as giant planes shaving material off their floors and walls. When the ice melts, eventually, the ocean invades the coastal valleys deep into the continent. Thus, fjords exist at one time as steep-walled seaways and at

FIGURE 13.4. Polished rocks in the fjords of Norway bear testimony to the former presence of enormous glaciers.

another as giant ice-river beds, depending on the climatic state of the planet.

The presence of the fjords changes the regional climate, because fjords allow seawater and heat from the ocean to penetrate deeply into the continent. Within the grand cycles, fjords are deepest right after the ice melts, when the continent has not yet risen in response to removal of the vast ice load that pushed the crust down into the mantle like a raft weighted down by cargo. Slowly, the fjords become shallower as the land rises. But the crucial effect of this rise is at the entrance of the fjord. Depending on circumstances, connection of a fjord to the ocean is likely to become much more tenuous as the submerged moraine near the entrance comes closer to the surface. The supply of seawater and of heat to the fjord from the ocean then decreases. Eventually, the fjord can freeze over during strong winters, especially in its inner parts. At the same time, elevations are now higher everywhere on the land surface. Conditions then are favorable for ice buildup again, provided the "stars are right," that is, the factors related to orbital forcing of climate are favorable as well. The resulting cycles are well beyond human experience and can be reconstructed only from the Earth's memory.

In the tropical regions, during the warm periods similar to our own time, corals and coralline algae grow vigorously in the sunlit zone, building a wedge-shaped flat-topped structure next to land in tropical seas, with the blunt edge of the wedge seaward. The wedge is largely made of rubble, with living reef on top. The rubble produced by growing patches of reef extends the margin of the land seaward, and new growth takes place on the surface of this pile of debris, broadening the platform. The fall of sea level that accompanies the buildup of ice in North America and Scandinavia obviously must lead to a widespread dying off of reefs in the South Pacific. Whenever the ocean feeds the growth of ice sheets, sending its water into the frozen wastelands in polar regions, the tropical zones suffer.

The reef platforms close to the sea surface, where so many organisms thrive in the sunlit zone, fall dry as the ice builds. Erosion takes over and carves out valleys and passages. After sea level rises again, these valleys persist because their depth (and consequently a lack of sunlight) prevents rapid growth within them, and because tidal currents tend to keep them clear of the fine debris that is constantly delivered from the reef flats. The passages now form conduits for the sea, which can deeply penetrate into the reef, bringing nutrients and taking out debris. Thus, the intermittent erosion ultimately stimulates reef growth and increases the overall rate of making rubble, which expands the reef structures outward.

Captain James Cook, in the most memorable passage in all of Pacific exploration, was wending his way through narrow seaways that owe their existence to the repeated rise and fall of sea level caused by the same multiple cycles of glacial buildup and decay that created the fjords of Norway, Scotland, Canada, Chile, and New Zealand.[4]

The nature of the multiple glaciations and their origin as a consequence of variations in Earth's orbit and rotation are topics that were intensely discussed during the past century. Modern "astronomic" or "orbital" ice age theory is the brainchild of the Serbian engineer and astronomer Milutin Milankovitch, who worked on it in the 1920s and 1930s.[5] The theory was the subject of intense controversy for many decades and is being debated still.

Milankovitch theory claims that the ice ages are strictly cyclic, driven by orbital factors that can be calculated in detail. The question of whether ice age fluctuations are indeed cyclic was decided well after Milankovitch introduced his proposition, based on evidence from long deep-sea cores. Such cores were retrieved by a coring machine invented by the engineer and oceanographer Börje Kullenberg for use on the Swedish Deep-Sea Expedition (1947–1948).[6]

The Kullenberg piston corer is a clever modification of the plain steel-tube method (called a "gravity corer"). It has a piston that sits at the entrance to the tube on the way down, preventing water from entering the tube at the business end, but admitting sediment when the device hits the bottom and the tube slides past the piston (which is fastened to the wire). During this expedition, the coring machine allowed the taking of cores that were typically 7 meters long.[7] Using this method, the *Albatross* retrieved a multitude of sedimentary sequences from all ocean basins, with memories reaching back 500,000 years and more.

Many cores collected on the *Albatross* revealed cyclic deposition both in sediments and in different types of fossils. Of special significance were the carbonate cycles in the eastern equatorial Pacific, which show a striking alternation of white and dark layers. Other cores from the equator had more subtle cycles—alternations of different types of diatoms, indicating periodic fluctuations of productivity. The member of the expedition who was to give an initial description of these discoveries was a young Swedish chemist by name of Gustaf Arrhenius, then working on his doctorate.[8] The discoveries were important for at least three reasons: first of all, they indicated that the ice age fluctuations are recorded as cyclic sedimentation in the deep sea; second, they suggested that the response of the equatorial system to the ice age cycles could be read from the record; and third, it looked as though productivity was increased during glacial periods, indicating stronger equatorial upwelling and hence stronger trade winds at that time.[9]

Important support for the new concepts regarding the ice age memories of the ocean came from the analysis of oxygen isotopes in the calcareous shells of foraminiferan fossils within the sediments recovered by the *Albatross*. These analyses, made in the 1950s, showed that Arrhenius had correctly assigned the carbonate-rich sediments to the glacial periods, and the carbonate-poor ones to the interglacial ones.[10]

Samples from the Swedish Deep-Sea Expedition were soon analyzed for oxygen isotopes in the shells of foraminifers. This new method (described below) allowed assignment of shells to glacial or interglacial conditions. The first such measurements were made by the Italian-American paleontologist and chemist Cesare Emiliani, working at the University of Chicago in the laboratory of Harold Urey. He obtained the samples from Hans Pettersson, leader of the expedition. On the basis of this work, and also based on the analysis of cores taken on Lamont's ship *Vema*, Emiliani strongly argued for ice age cycles in support of Milankovitch's theory.[11]

But the argument had just begun: Emiliani's time scale was soon questioned, and 20 years later, in the 1970s, it was shown to be incorrect. Once a good time scale was in hand (from the mid-1980s), Milankovitch theory indeed became the basis for understanding the grand climate fluctuations that govern our time in geologic history—the last several million years.

Other marine geologists working on the Swedish cores were the American paleontologists Fred B. Phleger (briefly on board the *Albatross* during the early portion of the expedition), Frances L. Parker, and the Swedish geochemist Eric Olausson, in Gothenburg.[12] Their studies in the 1950s, along with those mentioned, established the new field of ocean history, based on reading the record in deep-sea sediments, a field now commonly called paleoceanography.

Many fundamental concepts were introduced by these pioneers, chiefly with regard to glacial-interglacial fluctuations in oceanic and climatic conditions. Arrhenius produced evidence for great changes in productivity in the eastern tropical Pacific; Emiliani established the correlation of sediment cycles with ice age cycles; Olausson proposed that carbonate preservation patterns contained clues to changes in deep-sea circulation; Phleger and Parker produced evidence for large-scale shifts in climatic zones in the central Atlantic for the last several hundred thousand years.

After the *Albatross* expedition, the fundamental questions regarding the ice age memories of the deep sea were on the table. The decades that followed were dedicated to searching for answers, right to the present time. Much effort was devoted to the reconstruction of sea-surface temperatures during glacial time, the effects of fossil preservation on such reconstruction, the meaning of changes in isotopic composition of fossils (both oxygen and carbon isotopes), the nature of abrupt changes in climate at the times of the melting of ice, and especially the nature of the driving forces generating the grand climate cycles.

It is worth remembering that the discovery of the ice age fluctuations established the field of climate research. Climate change first became an issue when it emerged, sometime in the second half of the nineteenth century, that ice had covered much of the real estate now bearing meadows and forests. What made the ice expand? And what made it leave? A host of tentative answers soon appeared, along with the notion of positive feedback. Snow and ice are white and reflect sunlight, once ice expands it tends to stabilize cold conditions. Snow-free ground and ice-free ocean are dark. Once snow and ice are gone, the Sun can warm the ground and the sea. Modern climate models running on large computers have a long pedigree!

DISCOVERY OF THE GREAT ICE AGE

The discovery of the Great Ice Age was driven, in part, by the large boulders dotting the landscape in several countries in northern Europe, from the British Isles to the lowlands south of the Baltic Sea. Such boulders, weighing many tons, were used by Celtic people in building grave mounds and other megalithic structures thousands of years ago (fig. 13.5). They are called erratics, and they are too large to have been brought by anything but ice from their regions of origin in Scandinavia.

While the erratics posed the conundrum of their origin, back in the nineteenth century, the discovery of the mammoth as an extinct elephant suggested an answer. Even today, the mammoth is the paradigmatic animal of the ice ages, the iconic giant everyone is familiar with. It is a type of large elephant, with a remarkable coat of hair. Various kinds of elephants once populated the steppes of both North America and Eurasia, including several species of mammoth and mastodon. These became extinct, along with many other large mammals, at the end of the last glacial period, for reasons yet unknown.[13]

Not very far from the Scripps campus is abundant evidence of the Pleistocene mammalian fauna—at the La Brea tar pits in Los Angeles.[14] The tar pits are among the richest fossil sites for ice age animals: a mass grave with the skulls and jumbled bones of animals

FIGURE 13.5. Erratics in a forest south of Bremen, remnants of an ancient grave mound.

that roamed the coastal plains and valleys of southern California not so long ago, during the last ice age. As Paul Dayton, Scripps's peripatetic ecologist who is equally at home in the kelp offshore, the saguaro forests of Arizona, and the sponge reefs of Antarctica, points out to his students: "What we see today in the state parks and wildlife reserves of Southern California is an impoverished fauna and a flora that evolved in circumstances quite different from today's, in response to browsing by elephants, giant sloths, horses and camels."[15] In fact, with but slight changes in emphasis, the same is true for all of North America, and for much of Europe and Asia as well.

Our world, then, is strangely deficient, without the elephant herds that roamed the northern continents for a million years and more, and without the camels and the wild horses and their powerful predators. Why did they perish? Was it a profound change in climate that made conditions unfavorable? Was it the arrival of proficient human hunters gradually depleting the herds until the few survivors succumbed to regional calamities? Was it some unknown virus that swept the countryside, jumping from one herbivore species to another and leaving the predators with an empty landscape? Or was it some celestial event that removed the ozone shield, resulting in the blinding of all those animals too large to hide in shadows? The answer is not known. Presumably, as information on climate history becomes more detailed, and as we reconstruct in some detail the activities of ancient hunters, we shall get closer to an answer.

Before the great extinction, we must envisage faunas in North America and in much of Eurasia that resembled the mammal-dominated ecosystems of the grasslands of Africa. After the great extinction, there were still bison, musk oxen, caribou, and horses in Eurasia, but most of the large mammals were gone. Prairie dogs and coyotes, rabbits and foxes, were now prominent. An obvious reaction to the discovery of this striking change was to assume that climate change was responsible. Perhaps the large animals (including elephants and rhinoceroses) were tropical in nature, and perhaps they succumbed to a global cooling. It was the nineteenth century Swiss naturalist Louis Agassiz who promulgated this scenario, after his erstwhile mentor, the French naturalist George Cuvier, had established the mass extinction of a host of large mammals that lived not so long ago in Europe and Asia.

Cuvier had set the stage by calling on severe and sudden cooling as a cause of the demise of the mammoth and other large mammals, based on the reports of frozen carcasses found in Siberia.[16] Soon this idea found enormous amplification in the theory of the Great Ice Age, conceived by Louis Agassiz. To Agassiz, the extinct megafauna was clearly tropical in nature. The giants of the past had been wiped out by the sudden arrival of a worldwide Siberian winter, which stayed on to reign as the Ice Age.[17]

Louis Agassiz confidently proclaimed the new hypothesis, writing forcefully and memorably[18]:

> The gigantic quadrupeds, the Mastodons, Elephants, Tigers, Lions, Hyenas, Bears, whose remains are found in Europe from its southern promontories to the northernmost limits of Siberia and Scandinavia . . . may indeed be said to have possessed the earth in those days. But their reign was over. A sudden intense winter, that was also to last for ages, fell upon our globe; it spread over the very countries where these tropical animals had their homes, and so suddenly did it come upon them that they were embalmed beneath masses of snow and ice, without time even for the decay which follows death.

Thus, Agassiz built his Great Ice Age on the same frozen carcasses in Siberia that had so intrigued Cuvier. In Switzerland (where he held a professorship in Neuchâtel) Agassiz could readily ascertain the evidence for more extended Alpine glaciers in the recent past, as did others.[19] On a much larger scale, northern Europe showed the traces of former glaciation over vast regions, including, as mentioned, the enigmatic erratic blocks strewn about in the flatlands bordering the North Sea and the Baltic, many of which looked like rocks from Scandinavia, that is, entirely out of place. Which they are, actually; that is why they are called erratics.

In 1848, Louis Agassiz moved to Harvard. North America yielded even more evidence of former glaciation, and with an even larger extent of ice. Agassiz's concept of a Great Ice Age became firmly established despite early opposition. Its invention was a major step forward in understanding the history of the planet. It brought great fame to its creator.

The Great Ice Age was a big improvement over the then-current ideas on landscape modification, which included the Great Flood of the Bible and Lyell's "glacial drift," a process whereby erratic blocks from Scandinavia were delivered by icebergs to a submerged northern Europe. In a sense, Agassiz turned out to be right, while Lyell's proposition proved wrong. Much of North America and all of Scandinavia including the Baltic Sea and the North Sea were indeed deeply buried under a huge ice cover that attained a thickness of several kilometers in places. Much of the northern North Atlantic was covered with ice, either with sea ice or with icebergs calved from the surrounding glaciers. Frozen ground was extensive in middle Europe and deep into the Russian plains. The Alps were covered with ice, sending glacier tongues well into Bavaria, France, and Italy.

Agassiz's concept of the Great Ice Age survives in modern popular understanding as the Last Great Ice Age. Scientifically, it has morphed into the Last Glacial Maximum, the period of maximum extent of the last glaciation.

One can certainly appreciate Agassiz's nightmarish vision of a long-lasting Siberian winter and his doubts that a rich megafauna of elephants, rhinos, and bison could be supported under these circumstances. Nevertheless, strictly speaking, the Great Ice Age as envisaged by Agassiz is a figment of the imagination. It never happened the way he thought it did, as a sudden change from a tropical climate to a frozen world, with a more recent return to warmer climes. Also, the onset of this Ice Age had nothing whatever to do with the extinction of the woolly mammoth and the woolly rhinoceros. On the contrary: these giant mammals were creatures of the ice ages, not its victims. They died out at the *end* of the last ice age.

The one great idea that stood the test of time was Agassiz's insistence that there had been a lot of ice around, not too long ago. With this, he set the stage for climate reconstruction in Earth history.

MULTIPLE ICE AGES AND ASTRONOMY

The concept of the Great Ice Age of Agassiz soon gave way to the discovery that there had been several ice ages, separated by warm periods. From this finding arose the theory of astronomic control of ice age cycles.

Once it became clear, from moraines and other ice age deposits, that the Great Ice Age was in fact a long period of repeated glaciations separated by warm intervals, the task was to identify these different glaciations, establish whether they were cyclic phenomena, and attempt to correlate them between the various regions on the same continent and even between continents. As the famous Alpine geologists Albrecht Penck and Eduard Brückner noted, in 1909, the collection of such data must precede the search for causes.[20]

Penck and Brückner uttered their truism in response to the ideas put forward by James Croll, pioneer of ice age studies. Croll, in turn shopkeeper and philosopher, and caretaker and handyman at the Andersonian Museum in Glasgow, was a member of the Geological Survey of Scotland.[21] Croll had proposed that the changing eccentricity of the Earth's orbit would lead to changes in seasonal contrast. Times when the orbit deviated most from a circle, he thought, would be most favorable for ice buildup, through amplification of astronomic effects by changes in circulation, specifically the delivery of heat across the equator and the transport of heat to high latitudes by the Gulf Stream.

When Croll's fellow Scotsman, the distinguished geologist James Geikie proposed (in 1874) multiple glaciations, he was intimately familiar with Croll's work and supported it.[22] Croll was painfully aware that many geologists were not ready to consider the physics behind the glaciations. "No amount of description, arrangement, and classification, however perfect or accurate, of the facts which come under the eye of the geologist can ever constitute a science of geology any more than a description and classification of the effects of heat could constitute a science of heat," he wrote.[23]

The response of geologists to this irksome statement was that his theory was useless in explaining facts. This very effective retort, in 1916, was from the pen of the master-describer and master-classifier Gustaf Steinmann, professor of geology in Bonn, known for being somewhat acerbic, and respected as a leading alpine geologist of his time. Steinmann wrote:

> When we can show, as is in fact the case, that the events were contemporaneous not only for the various regions of the northern hemisphere but also for the southern one, it is then clear that the climatic problem of the ice-age period is a general one [comprising the Earth as a whole] and we can eliminate thereby all those theories that explain glaciations from an alternating unfavorable effect on the two hemispheres by astronomical processes such as the changing eccentricity of the Earth's orbit, etc.[24]

After it became clear, through fieldwork, that the last ice age was too recent to fit Croll's scheme, geologists followed the advice of Penck and Brückner, to look first and theorize later. We may safely assume, therefore, that the brilliant speculations of Croll had no further impact on the working geologists in the field. Croll's approach was too far ahead of his time, and too far ahead of the evidence. In fact, Croll gave an explanation for 100,000-year climate cycles before such cycles were known or suspected!

In hindsight, we can see that Croll argued well, but he was misled in at least one fundamental way—he took the present to be typical. Thus, since at present the Northern Hemisphere is almost ice free, he defined the problem in terms of making ice on the Northern Hemisphere. However, the problem is *not* how to make ice in a warm world, the problem is how to *get rid of the ice in an unusually cold world.* The Earth is already programmed, as it were, to have its high northern latitudes covered by ice—snow falls every winter—the real question is how to get rid of the snow and ice to make and keep an interglacial on a cold planet. The effect of this shift in emphasis about what it is that needs explaining is profound. Instead of focusing on the amount of sunshine in winter, which Croll thought important, we now concentrate attention on whether northern summers are warm enough to melt the snow and ice that the winters readily provide. Concentrating the argument on summer insolation (that is, melting power) turns out to give the correct answer.[25]

In the Darwinian world of scientific ideas, where survival matters, being right is better than arguing brilliantly. By adjusting a few seemingly minor details in Croll's scheme—focusing solely on ice growth and decay in the Northern Hemisphere and identifying summer insolation in high latitudes as the crucial guide to climatic change—Milankovitch succeeded where Croll had failed. Also, where Croll had attempted to master oceanography and climatology on his own, Milankovitch took the reasonable course to ask the experts. He asked the best.

In 1924, the German climatologist Wladimir Köppen and his son-in-law Alfred Wegener, a meteorologist who had made a stir by proposing that continents move, published a book entitled *The Climates of the Geologic Past.*[26] In this volume they included a graph showing the summer radiation for 65° N for the last 600,000 years. Milankovitch had sent the graph to Köppen, for comment. Milankovitch was convinced that the ice ages come in cycles and that long-term changes in the seasonal distribution of sunlight arriving at high northern latitudes are at the heart of the ice age story. The incoming sunlight changes because of subtle cyclical shifts in the tilt of the axis of rotation of the Earth and in the configuration of Earth's orbit about the Sun.

In the graph, Milankovitch had marked the periods of especially cool summers as the ones containing the maximum glaciations. Köppen liked his choice of glaciation events—as far as he knew, the ages agreed with the crude guesses of the contemporary glacial geologists. Milankovitch was thrilled to have the famous man support his theory and agreed to have the graph published, with proper attribution. Publishing Milankovitch's hypothesis was strong support by well-known scientists for an approach that had been sharply criticized by prominent geologists such as A. Penck and G. Steinmann. (But then, Wegener, at least, was used to plenty of criticism and quite willing to discount the opinion of experts.)

Milankovitch's diagram shows two time series of northern summer insolation for the last 600,000 years, calculated from astronomical principles, along with a guess that ice ages were linked to cool summers, that is, minimum seasonal contrast in northern latitudes.[27] The assumed link between ice buildup and cool summers made the ice ages cyclic, naturally, since the astronomical forcing of seasonal contrast is strictly cyclic. It was a bold move. No one in fact knew whether the ice ages are cyclic or not, or whether they were spaced more or less according to the intervals prescribed by Milankovitch. In the absence of methods for dating the glacial deposits, all age assignments were rather crude guesswork.

Using the Milankovitch scale for assigning dates to glacial deposits led into a circular argument. Thus, without independent dates, the hypothesis stood on weak feet indeed. The main strength of Milankovitch's hypothesis derived from the prestige of Wladimir Köppen, who said it was a good idea. Köppen turned out to be correct.

At the heart of Milankovitch theory is the amount of sunlight received during summer, in high northern latitudes. The principles governing the variations in this parameter are quite easily understood, and may be summarized as follows.

We assume that the output of light from the Sun is constant over the time scale considered. Then, at any one place on Earth, the amount of light that is available to warm the surface depends on how high the Sun is above the horizon, how large the Sun appears to the observer, and how much of the sunlight is reflected back into space.

FIGURE 13.6. Ice and snow reflect sunlight, while open water and snow-free land absorb it, being dark. Old ice appears gray. Reflectivity (albedo) determines uptake of sunlight and conversion to heat. Clouds play a more complex role: they reflect sunlight, but they also radiate heat downward, acting as a blanket. (Hinlopen Strait, Svalbard, July 2007.)

The reason Milankovitch chose high northern latitudes (65° N) for his calculations is contained in the third item, the one about reflectivity or "albedo" (fig. 13.6). At this high latitude (central Alaska, northern Canada, central Scandinavia, central Siberia) the sensitivity to modest change is very large: A little cooling will help retain snow and ice, and greatly increase the albedo. A high albedo denies much of the arriving summer sunlight a chance to warm the surface.[28] Conversely, a little warming will tend to remove snow, darken the surface, and increase warming by the sunlight. The process is called positive albedo feedback and is extremely important in increasing the amplitudes of climate change, given relatively weak causes.[29]

The maximum height of the Sun above the horizon, in high latitudes, depends on the tilt (obliquity) of the Earth's axis relative to the plane of its orbit around the Sun. As was shown by the French astronomer Pierre Simon Laplace,[30] the obliquity varies very slowly and cyclically over a range of less than 3° to about a mean of somewhat less than 23.5°. Whenever the obliquity is at maximum—this happens every 41,000 years—the summer Sun is higher above the horizon in high latitudes, at noon, and more sunlight comes in, therefore. Whenever the obliquity is at minimum, the high latitudes get less sunlight and the low latitudes get correspondingly more. At present, the obliquity is 23.45°; it was slightly greater 9,500 years ago, the time of the last maximum, and slightly less 29,000 years ago, the time of the last minimum. Thus, the high latitudes received maximum sunlight 9,500 years ago in regard to the position of the Sun above the horizon. It was like having a few extra summer days of sunlight each year.

The apparent size of the Sun as seen from Earth depends on how close we are to it. As first shown by Johannes Kepler, the orbit is elliptical, and the Sun sits in one of the foci of the ellipse.[31] At present, Earth is closest to the Sun on 3 January and farthest away on 4 July; thus, the Sun appears largest in the midst of winter and smallest in summer, in northern latitudes. The difference in distance is almost 3 percent.

The result is a decrease of seasonal contrast: northern summers are relatively cool, winters mild. According to Milankovitch (and his climate advisor Köppen) this is a good situation, in principle, to make glaciers. If this is correct, why are we not building ice sheets right now, when the Sun appears smallest on 4 July? The reason is that the Sun-distance effect is just one of the factors that are important. The long-term state of the climate system plays a role as well. Importantly, the land surfaces that hold the winter snow have not sufficiently rebounded from the depressed position they acquired when covered by ice. They need to rise to be able to keep that snow through the following summer. Canada and Scandinavia are still rising. When elevations are high enough, snow can stay and the positive albedo feedback can then begin to work its magic.[32]

The Earth is not always farthest from the Sun on 4 July. Each year, the seasons come about 25 minutes earlier than the previous year, relative to the points farthest or nearest to the Sun. (These points are labeled *aphelion* and *perihelion* by the astronomers.) So, some 11,000 years from now, July will have moved all the way into the perihelion position, which will result in extra warm summers in the northern latitudes. The migration of the seasons along the Earth's orbit is called precession and is a result of the Earth's axis wobbling, in space.[33]

By adding the effect of the changing maximum height of the Sun above the horizon to that of the changing apparent size of the Sun, in northern summer, Milankovitch produced his graph of summer insolation as a function of time. When the Sun was low and small in summer (low obliquity, Sun close in winter), the radiation received at 65° N was less than that received today at 70° N. Conversely, the radiation received when the Sun stood high and was large in summer (high obliquity, Sun close in summer) was as high as that for today's 60° N; that is, Milankovitch discovered a difference in summer insolation corresponding to a latitudinal shift of at least 5°, about the latitude of 65° N.[34]

If Milankovitch was correct, ice age deposits should show a regular predictable sequence, and ages could then be assigned to all events with high precision. He was, indeed, basically correct. But it would take another half century to prove it, and to obtain the correct time scales for both the orbital elements and the sequence of climate states of the past.

THE ENCHANTING WORLD OF MICROFOSSILS

Strong support for Milankovitch's theory came from an unexpected source: sequences of microscopic fossils on the deep-sea floor. Two types of information proved to be important: clues as to the changes in the conditions of surface waters in response to climate change (temperature and productivity, mainly), and clues to the changes in the chemistry of seawater, with respect to isotopic composition.

The most intensely studied types of microfossils belong to the foraminifers, amoeboid singe-celled organisms with a calcareous shell, commonly coiled and chambered.[35] The other groups that proved extremely useful in reconstructing ocean history well beyond the ice ages are coccolithophorids or "nannofossils," diatoms, and radiolarians. Much of the seafloor is covered with the remains of these organisms, most commonly nannofossils and foraminifers (about one-half of the seafloor) (fig. 13.7). Below highly productive regions, we find abundant diatoms and radiolarians. All of these fossils are from single-celled organisms that are eukaryotes, that is, cells with a well-defined nucleus carrying the genetic code. Coccolithophores and diatoms are primary producers bearing chlorophyll. Many of the planktonic foraminifers and surface-water radiolarians live in symbiosis with photosynthesizing organisms and thus take part in primary production, as well.

First glimpses of the fact that small fossils (microfossils) are ubiquitous emerged from working up the sediment samples from the *Challenger* Expedition (1872–1876). Both John Murray and Ernst Haeckel (leading naturalists of the nineteenth century) were well aware of the oceanwide abundance of organisms making microfossils. Since then, steel tubes have been used to penetrate the uppermost sediments deep

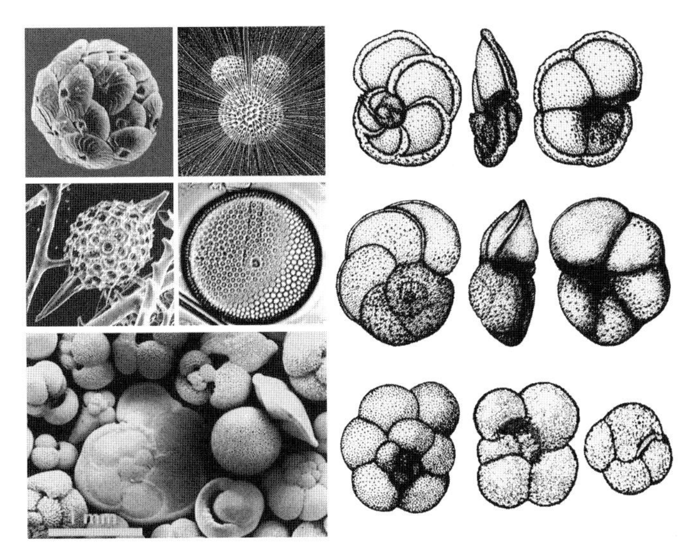

FIGURE 13.7. Microfossils. *Upper left:* coccoliths, spined foraminifer, radiolarian, diatom. *Lower left:* planktonic foraminifers, scanning electron microscope image of washed sediment. *Right:* drawings of planktonic foraminifers.

into the Quaternary (the last 2 million years), beginning with the Swedish *Albatross* Expedition (1947–1948). One by one, almost each core that was raised brought a long and thin cylinder of sediment rich in microfossils to the surface for study.

After recovery, such cores are sampled from the top down, commonly in 10-centimeter intervals. The smallish plugs of sediment are then "washed" to extract the coarser shells (that is the foraminifers), or treated with acid (to retrieve the radiolarians), or examined directly under a powerful microscope as a smear sample on a glass slide (to detect the species that are the smallest: the diatoms and nannofossils). For some purposes (but not for routine shipboard examination) electron microscopes are now used to obtain precise high-resolution images of the fossils, but such equipment did not become available until the 1970s.

Following the Swedish Deep-Sea Expedition, and with Lamont and Scripps recovering deep-sea cores at a rapid clip,[36] vast opportunities opened for mapping fossil distributions on the seafloor, for examining the fossil record for climate change, and for establishing the taxonomy and the evolution of previously unknown or poorly known groups. At Scripps, such studies began in the 1950s, when a number of outstanding experts joined the faculty: Frances Parker and Fred Phleger from Woods Hole and Amherst, for foraminifers; Milton Bramlette from the U.S. Geological Survey, for nannofossils; and William Riedel from Australia, to study radiolarians.[37]

Frances Parker, alumna of MIT, worked on the then best known of the microfossils, the foraminifers, which are the size of sand grains. Parker started her career assisting the paleontologist Joseph Cushman (1881–1949), who established these fossils as a prime tool in oil exploration. By the time she came to Scripps, in 1950, she had already identified, counted, and registered a vast number of fossils. Her favorite place of work was at the microscope; she was highly skilled as an observer and artist. (She drew the figures on the right in fig. 13.7.) It is probable that she saw and identified more fossils than any other paleontologist ever, in a career that spanned more than half a century. Her record-setting performance of inspecting well over a million fossils began even before her studies on shells from the deep-sea floor, on which she started working when the cores from the Swedish Deep-Sea Expedition became available, analyzing samples from both the central North Atlantic and the Mediterranean.[38]

Both benthic and planktonic foraminifers are quite abundant in calcareous deep-sea sediments, although usually planktonic forms are roughly 100 times more common than benthic ones. There is an enormous difference in diversity between the two groups. Where five or six species comprise the bulk of the plankton assemblage, at least five times more species need to be identified to characterize the benthic fauna. Full counts require familiarity with some 40 species of planktonic habitat, and some 200 taxa living on the seafloor, and even then some will end up in the bin for the "unidentified." The presence and absence, and the abundance and rarity, of the various species offer rich sources of environmental information for the changing conditions in the sea. On the time scales of tens of thousands of years, such changes are closely linked to ice age fluctuations in climate. Parker pioneered the division of planktonic foraminifers into warm-loving and cold-loving species. She plotted their down-core abundances for samples taken by the *Albatross*, demonstrating that ice age temperature fluctuations characterized the environment of the eastern Mediterranean in recent geologic history.[39]

Nannofossils are even more abundant than the foraminifers: ours is the nannofossil planet, as far as the most abundant cover on the solid surface of Earth. The nannofossils are so small that they are exceedingly difficult to study in all but the most high-powered and specialized light microscopes. The role of these fossils in unlocking the mysteries of the ice age fluctuations was limited, but they became extremely important in the exploration of deep-sea sediments by drilling (as discussed in the next chapter).

Another important group of microfossils, albeit less widely distributed than the foraminifers and nannofossils, is the radiolarians. They have skeletons made of silica.[40] The radiolarians are typically somewhat smaller than the sand-sized foraminifers. They look like delicate filigree ornaments suitable as jewelry or for Christmas-tree decoration. Their diversity is stunning. Radiolarians first became well known among geologists through the beautifully illustrated work of Ernst Haeckel, based on the study of samples from the *Challenger* Expedition.[41] William Riedel mapped their abundance patterns, which turned out to trace out the high-production regions in the sea.[42] Thus, changing abundances of radiolarians in a core suggests changes in the productivity of the ocean for the time period sampled.

The various types of fossils, through their different responses to a changing environment, have a wealth of information regarding temperature and productivity changes in surface waters, food supply to the seafloor, and even silicate content of subsurface waters. They are the reporters we need to debrief when reconstructing the ocean's past.[43] Fortunately, in more than half of the deep-sea sediments, there usually are plenty of microfossils available for the purpose. However, almost one-half of the seafloor has sediments that are essentially barren of fossils: the vast regions of Red Clay that predominate in the deepest regions far from continents. Both calcareous and siliceous fossils have been dissolved from these brown red deposits comprised largely of extremely fine-grained windblown particles. What information they carry is largely in physical properties, such as the content of ferromagnetic minerals. Because Red Clay is dominant over vast areas of the North Pacific and the central South Pacific, while the Atlantic seafloor has plenty of carbonate, the ice age history of the Pacific is on the whole less well known than that of the Atlantic.

DISCOVERY OF THE ICE AGE CYCLES

While the interpretation of ice age deposits on land was stalled for lack of dating, two new developments brought great progress in the 1950s: the careful study of the deep-sea record, and the radiometric dating of raised coral deposits, which gave a precise date for the last interglacial.

An early attempt to decode fossil assemblages from deep-sea sediments in terms of ice age fluctuations in the sea, by Wolfgang Schott of the *Meteor* Expedition (1925–1927), yielded some interesting results but was greatly limited by the shortness of the cores taken.[44] Nevertheless, the *Meteor* cores did penetrate into the sediments of the last glacial period, and the foraminifer distributions for that time do have information about changes in conditions in the central Atlantic, including increased upwelling of cold water off North Africa. But, as mentioned, the breakthrough regarding the study of ice age cycles came with the Swedish Deep-Sea Expedition (1947–1948) and the new kind of coring device invented by the oceanographer Börje Kullenberg. To the great delight of the geologists aboard the *Albatross,* Kullenberg's contraption worked, and the routine taking of long cores could begin. By the end of the expedition, the *Albatross* had collected close to 300 cores with a combined length of one entire mile.

Nothing on an expedition is ever entirely routine, especially not with a new gadget. Many times during the voyage Kullenberg and his colleagues had to cope with the depressing sight of bent or broken coring equipment returning on deck from its excursion to the deep seafloor. Unfailingly they found spare parts, cut, mended, and welded, to provide a replacement. Any time the corer hit hard bottom, or landed on a steep slope, chances for success were greatly diminished. Vigilance and skill in using the echo-sounder in positioning the vessel, dexterity and deftness in controlling the enormous winch, and correct timing of the release of the corer bouncing up and down while hanging on the wire—these precautions, and good luck, were the ingredients of success.

As mentioned, perhaps the single most important discovery of this expedition, in hindsight, was the observation that the patterns of color, chemistry, and fossil content of sedimentary sequences in the *Albatross* cores show a cyclic nature. Gustaf Arrhenius, youngest of the scientific members of the expedition, proposed an increase in the strength of the trade winds during glacial periods, which led to an increase in productivity. This hypothesis has since received abundant support.[45] In fact, the evidence suggests that productivity and upwelling increased everywhere in the tropics in productive regions during the last glacial period, as far as this can be ascertained.[46]

The first report on the subject of fluctuating ice age conditions that emerged from the expedition was a note by Fred Phleger, who joined the expedition at Port-de-France, in Martinique, and soon returned to land after crossing the Caribbean Sea.[47] A few years later, he and Frances Parker and Phleger's assistant Jean Peirson produced one of the classic reports of the expedition, on the changes in climate in the central Atlantic.[48] The report illustrates the foraminifers of the deep Atlantic in six plates and provides numerous tables describing their stratigraphic abundances. Their method of counting and tabulating the fossils became the standard in the field.[49] Paleontology by numbers was a relatively new idea; it was the founding feat of quantitative paleoceanography. The method lends itself to comparing warm-water and cold-water planktonic species and to using their abundances in the fossil record on the seafloor to reconstruct a generalized temperature curve for the overlying sunlit waters.

Not everyone found cycles when studying deep-sea cores. David Ericson worked at Lamont in the 1950s on the many hundreds of cores filling the rapidly growing Lamont repository. Ericson analyzed the cores for foraminifer content. Based on Schott's discovery that a certain tropical species was absent during the last glacial, but reappeared at the end of the glacial, Ericson used this species *(Globorotalia menardii)* to define glacial and interglacial periods going back in time. The result was a stratigraphy not unlike the one in the geologic textbooks of the time, with four large glacial periods separated by warm intervals.[50] This was generally interpreted, naturally, as support for the traditional view of ice age history—no evidence for cycles, and no points for Milankovitch.

The work that was to revolutionize the field of ice age studies, and indeed our understanding of the forces of climate change on time scales of thousands to hundreds of thousands of years, was linked to isotope chemistry. Cesare Emiliani, while at the University of Chicago, initiated this line of research, at the suggestion of Harold Clayton Urey, his postdoctoral mentor. Urey, with his collaborators Samuel Epstein and Heinz A. Lowenstam, had already established a way to use isotope chemistry in reconstructing past temperatures in the sea. The team had shown that the ratio between two different isotopes of oxygen atoms (oxygen-16 and oxygen-18) changes with the temperature of growth of the shell-forming organism.[51] Emiliani now set out to apply the new method to deep-sea sediments.

In the early 1950s, Emiliani started to analyze the oxygen-isotope composition of planktonic foraminifers from samples of the Swedish Deep-Sea Expedition. Urey was delighted to see that Emiliani's measurements established regular oxygen-isotope cycles in the deep-sea sediments, and especially that these cycles could be correlated over large distances from one region to another. He and his colleagues knew, however, that Emiliani's data could not be simply interpreted in terms of temperature alone. Measurements on snow and rain had already shown that the lighter isotope, oxygen-16, was more abundant, relative to seawater, within the precipitation at high latitudes. This meant that when ice sheets grew, during the glacial periods, oxygen-16 would be preferentially extracted with the water leaving the ocean and destined for the ice. Correspondingly, oxygen-18 would be enriched in glacial-time seawater. How was one to separate the effect from ice growth and decay on the isotope ratios from that of the temperature changes? It turned out to be a difficult problem.[52]

Emiliani reasoned that the great North American ice sheet would have had roughly the same composition, in terms of the oxygen-16 to oxygen-18 ratio, as the snow falling in Chicago during wintertime. From this convenient backyard guess he calculated the change of the isotopic ratio in the ocean, in going from conditions with glacial-period ice sheets to conditions without such sheets. As it turned out, his calculation was off quite a bit—he had seriously underestimated the difference between the composition of ice sheets and that of the glacial ocean. The result was that he attributed most of the change he saw in the foraminifers to the effects from temperature, which meant that his estimates of the range of temperature changes were too large. Subsequently, as better estimates for the ice effect emerged—mainly from drilling into Greenland ice, by Danish scientists[53]—the interpretation of the changing oxygen isotope values in planktonic foraminifers greatly improved.

When first plotting the emerging data from the deep-sea cores, Emiliani was struck by the well-developed cyclicity of the oxygen-isotope record (as was almost everyone else). He immediately and forcefully argued that these cycles provided strong support for Milankovitch's "astronomical theory of the ice ages." In fact, he used a graph much like that of Milankovitch (summer insolation at high latitudes) to assign precise dates to the oxygen-isotope cycles he had discovered. Thus, he invented the dating of deep-sea sediments by referring the oxygen-isotope record to orbital forcing, a method now justly celebrated as the foundation of high-resolution deep-sea stratigraphy.[54]

Unfortunately, two major problems stood in the way of enthusiastic general acceptance of Emiliani's amazing support for Milankovitch theory, offered with Galilean self-assurance. First, it was not clear to what degree the isotopic cycles reflected ice variation. This was after all the problem that Milankovitch had proposed to solve, not merely one of temperature variation in the sea. Second, the estimates of sedimentation rates (which provided the time scale) were of insufficient quality to verify that the cycles matched those of Milankovitch. It took another 20 years to solve these two problems, with the evidence going against Emiliani's preferred (and freely urged) answers.

Eventually, Emiliani had to relinquish the claim of having proved Milankovitch right. He did so reluctantly, knowing that his claim really had been correct, even if it was not justified by his treatment of the data. He did indeed demonstrate the cyclic nature of the ice ages, which is the crucial link between Earth's climate and the external forcing by the thoroughly cyclic celestial mechanics. And he introduced the single most important measure of global change in ice mass through time: the oxygen-isotope composition of the ocean, as recorded in deep-sea sediments. The path was now clear toward obtaining a reliable history of the ice-mass history of Earth, against which the performance of Milankovitch theory (or any other theory) could be tested.

A TIME SCALE AND ORBITAL PACING

The discovery of cyclic sedimentation on the deep seafloor, emerging in the 1950s, gave a strong boost to Milankovitch's theory, which says that the ice ages are a result of periodic changes in the orbit of Earth and its axis of rotation. But acceptance of the link between these observations and the theory was yet slow to materialize. The chief reason for the prevailing skepticism was the lack of a precise time scale, which would make it possible to test the match between the cycles seen in the sediments and the calculated cycles of sunshine intensity in summer in the far north.

From ongoing discussions between the experts one might have gained the impression that Milankovitch's theory was well supported in the 1960s. The two chief protagonists in the game of testing the theory, Cesare Emiliani (who had moved to the University of Miami) and Wallace Broecker at Lamont,[55] agreed that the theory is correct. Unfortunately, they used different time scales to make that argument. This perplexing circumstance suggested to the uninitiated that the acceptance of Milankovitch's theory was a matter of belief rather than one of evidence. Logically, the contestants could not both be right—but they could both be wrong.[56]

Emiliani's original time scale (which he later defended using additional evidence) was based on radiocarbon dating of samples in a deep-sea core. At the time, dating by measuring the beta-radiation from decaying carbon-14 ("radiocarbon") had just been introduced to the scientific community at large—it was *the* new tool for achieving age control in carbon-bearing samples younger than 30,000 years.[57] The radiocarbon dates (which only reached down to the last glacial) allowed calculating a sedimentation rate for the uppermost portion of a core. Taking this rate as typical, one can assign ages to all samples taken farther down-core. Assuming further that Milankovitch cycles govern the oxygen-isotope record, one can refine this age scale by correlating warm peaks within the core sequence to increased summer insolation in high northern latitudes.

In this fashion Emiliani derived his time scale.[58] The method resulted in assigning an age of less than 100,000 years to the last period as warm as today (the "Eemian" in northwestern Europe, later labeled "Stage 5e" in deep-sea sediments). Within a decade or so, in the 1970s, most scientists working in the field believed that the ages Emiliani had assigned to the glacial and interglacial periods were much too young.

In the late 1950s and early 1960s, new dating methods emerged, based on the radioactive decay of thorium and uranium. These could be applied to sediments much older than those datable by radiocarbon. The new methods proved useful in assigning ages to uplifted corals in islands in the Caribbean and the Pacific. On tropical islands that experience uplift, as do Barbados and New Guinea, for example, older coral reefs are exposed on the rising slopes. Most of the reef masses thus exposed grew during high stands of sea level, when the ice-sheet volumes were at a minimum. The corresponding reef terraces can be identified on the seaward slopes of islands and can be sampled and dated. This yields a series of ages corresponding to times of high sea level. Such work was undertaken at Lamont (by Wallace Broecker and his student David Thurber) and at Scripps (by Hans Herbert Veeh, a student of the geochemist Ed Goldberg).[59] Results showed that the last maximally high sea level stand was somewhere between 120,000 and 124,000 years ago—about one-fourth as much again as the age claimed by Emiliani.

Broecker judged the uranium-based results as much more reliable than the extrapolation of radiocarbon dating in deep-sea cores. He proposed a new time scale (in 1966),[60] based on the extrapolation of a 120,000-year date for the 5e warm peak in the oxygen-isotope record, and proclaimed a close match to Milankovitch summer insolation, provided that the effects from the tilt of the Earth received less weight than the ones from precession (and eccentricity). He also (in the 1966 article) introduced the notion of a "mode switch" of ocean circulation, from a warm to a cold state. This intriguing idea was to spawn many corollaries.

A few years later, elaborating on the mode switch theme, Broecker and his student Jan van Donk (in 1970) postulated rapid transitions from periods of maximum glaciation to the following warm periods.[61] They identified six such events for the last 440,000 years, which they called terminations. With this work, Broecker introduced the notion that glaciations grew gradually and ended abruptly, describing "sawtooth" cycles. This new concept was to prove extremely fruitful.[62]

Eventually, Broecker's time scale turned out to be more nearly correct than that of Emiliani. Strong support for a 125,000-year age of the last major interglacial period came from additional dating of exposed reef material on Barbados, in the late 1960s. In addition, independent evidence for a "long" time scale came from dating a deep-sea core from the western tropical Pacific, by magnetic stratigraphy. Magnetic stratigraphy provided a new tool for dating deep-sea sediments, based on the fact that the sediments record the direction of the Earth's magnetic field at the time they were deposited. The field reversed direction some time ago, an event that is now known to have happened around 780,000 years ago.[63] The fact that sediments contain this information had been established in 1964, by the geophysicist Christopher Harrison and the paleontologist Brian Funnell, working on Scripps cores.[64]

Neil Opdyke, Lamont's pioneer paleomagnetician, put this discovery to good use. He dated a large suite of cores from the Lamont core library. Soon after, his colleague James Hays, paleontologist and ocean historian, had the splendid idea to tie oxygen-isotope fluctuations to magnetic-reversal stratigraphy.[65] He sampled the now-famous core V28-238, taken on the twenty-eighth cruise of the research vessel *Vema*, in the western tropical Pacific. Hays sent the samples off to England, to the physicist Nicholas Shackleton, at Cambridge, for isotopic analysis.[66]

The result could not have been more significant. The first distinct magnetic reversal going down in the core (the Brunhes-Matuyama reversal) was found to occur just before isotope stage 19, that is, the nineteenth isotope peak downward, counting the present peak as number 1, and assigning odd numbers to interglacials, even numbers to glacial periods.[67] The reversal had been dated (on land) as 700,000 years before present (or thereabout). It was now possible to assign ages to each isotope stage, by interpolation. The last major interglacial came out near 123,500 years. Bingo! The long time scale was upheld.[68]

With the uranium-based time scale supported by the magnetic reversal dating, an acceptable time scale was now available for dating the various signals emerging from the deep-sea cores. The dated signals now became "time series," analogous to the tidal records of the physical oceanographers. Such time series are routinely subjected to spectral analysis by engineers and physicists. Standard spectral analysis or Fourier analysis[69] represents any time series as a sum of sinusoidal fluctuations. This method allows identification of those cycles within a series that are especially strongly represented.

When Fourier's methodology was applied to deep-sea records (by James Hays, John Imbrie, and Nicholas Shackleton, and independently by Nicholas Pisias, then a graduate student at Oregon State), it emerged that the oxygen-isotope series contained strong cycles with periods near 100,000 years, 41,000 years, and 23,000 years.[70] These are precisely the periods expected if Earth's orbital elements (eccentricity, obliquity, and precession) govern ice age climates, as proposed by Milankovitch theory. Thus, the 1976 article by Hays and colleagues ("Variations in the Earth's Orbit: Pacemaker of the Ice Ages") marks an important milestone in the search for a mechanism responsible for the ice ages. The main cycles discovered in the sediment are identical to the cycles characterizing the orbit and rotation of the Earth. There could be no more doubt that orbital elements had to be considered as important elements of climate change on long time scales.[71]

The following years saw a profusion of information and discussion on the subject. In rapid succession, there were reports by astronomers refining the calculation of orbital elements and the resulting insolation history for various latitudes and months during the year,[72] articles by marine geologists on cyclic sedimentation in both ice age records and in earlier sequences, essays by geologists on cycles in ancient rocks, and equation-rich papers by geophysicists and applied mathematicians on the results of modeling the response of Earth's climate system to cyclic forcing. A revolution in thought was occurring in those years, the 1970s, and our understanding of climate history in particular, and Earth history in general, would never be the same.

The orbit-climate connection puts Earth's climate squarely into the context of the fluctuating gravitational field of the solar system. The planets wandering across the heavens do not control our individual lives, as once postulated during the childhood of astronomy in ancient Babylon,[73] but they do control Earth's orbital elements, which in turn influence the course of climate fluctuations that form the backdrop for the evolution of the human species.[74]

In 1982 almost everybody who was somebody in the climate cycle business met at Lamont, to take stock. Progress had been overwhelmingly fast. Could Milankovitch theory now serve as the ruling paradigm? Why is the 100,000-year cycle dominant in the ice age cycles, when in fact forcing at this frequency is negligible? What about the big change from 41,000-year cycles to 100,000-year cycles a million years ago? How far back could Milankovitch theory be used for precise dating of sediments?

The proceedings from this conference (Milankovitch and Climate) consist of two information-packed volumes with almost 900 pages of results and discussions. These pages represent one great celebration of the role of Milankovitch theory in solving the century-old question about the nature of the ice age cycles.[75] At the same time, these discussions opened a door to the study of cyclicity of the geologic record in general, and greatly enhanced the prospects for a pervasive role for Milankovitch-type mechanisms of climatic change, throughout Earth history.

The revolution was complete. Although many questions remained unanswered, Milankovitch theory entered the textbook stage: the basic premise would no longer be questioned. Or, expressed more precisely, such questioning would largely be ignored from now on. Also, in the proceedings, a standard sequence of oxygen isotopes was offered comprising the last 800,000 years. It turned out to be correct for the last 650,000 years. A modern updated version for the entire period of Pleistocene ice age cycles is given here in figure 13.8. The series shown is an extract from a compilation by paleceanographer James Zachos and colleagues.[76] The data are based on the isotopic analysis of benthic foraminifers from deep-sea cores. To the extent that the deepest waters do not change their temperature very much through the cycles (staying at temperatures just above freezing), the fluctuations in the isotopic index can be read as sea level variations, with the range from the last glacial maximum to the present being close to 120 meters (400 feet).

ON THE ORIGIN OF THE 100,000-YEAR CYCLE

The fact that the deep-sea record contains the same cycles as the Earth's orbit is exciting and illuminating. It does not, however, prove anything about Milankovitch theory. The presence of cycles alone does not address the specific tenet of the Milankovitch mechanism: variations in summer insolation at high-northern

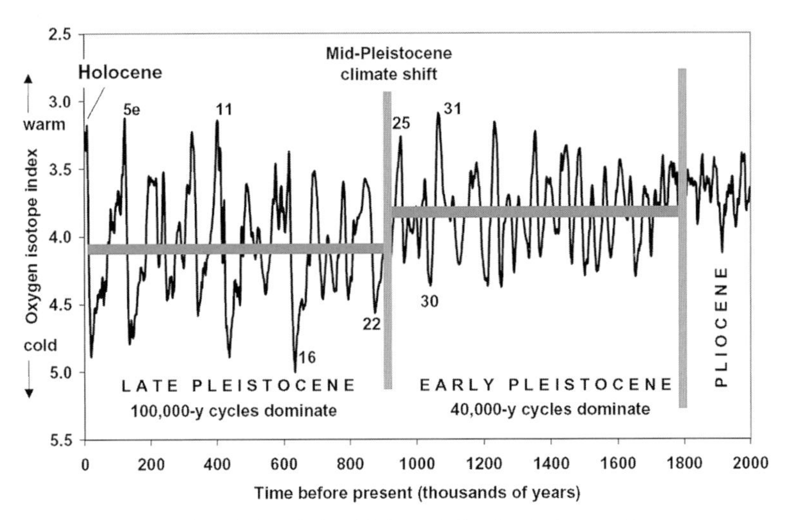

FIGURE 13.8. Isotopic variations in benthic foraminifers for the last 2 million years, comprising the Pleistocene (1.8 million years ago to 0.01 million years ago) and the Holocene, (the last 11,000 years). Note the shift in mean position of sea level near 0.9 million years ago, and the predominance of large cycles after this Mid-Pleistocene climate event.

latitudes are ultimately responsible for the ice-mass fluctuations. How could one show that Milankovitch theory is an adequate explanation for the ice ages?

An early answer to this question came from an entirely unexpected source. In 1974, the British science writer Nigel Calder was writing a book on weather and climate, for which he reviewed the development of the ice age time scale. Calder visited the ice age expert Nicholas Shackleton in Cambridge, where he learned about the project on the time scale of the ice ages that Shackleton and Neil Opdyke of Lamont had been working on. Calder realized that the new ice-mass history provided by core V28-238, which had just been published (in 1973), provided an opportunity for testing Milankovitch theory. Could one invent a set of rules that would transform Milankovitch insolation history into ice-mass history?

Calder thought the attempt was worth a try. He proceeded to write such a set of rules. To his surprise, the rules resulted in a quite remarkable reproduction of the ice-mass series based on core V28-238.[77] Here are Calder's rules:

1. For input we take the variations in summer sunshine at latitude 50° N.
2. Whenever the summer sunshine is (at least) 2 percent stronger than at present, ice melts at a proportional rate, up to a certain limit.
3. Whenever the ice does not melt, it builds up at a rate proportional to the shortfall (below the threshold sunshine), at 22 percent of the melting rate.

Calder's set of rules postulate that melting is four to five times faster than ice accumulation. This stipulation represents an important feature of the ice age record, that is, the asymmetry of growth and decay of ice. The asymmetry is central to finding the rules that will convert Milankovitch *input* into a wiggly line—an *output* series—resembling the oxygen isotope series from the deep-sea record. Calder's output line, derived from orbital input, gave a fit with the isotope data that was decidedly unimpressive. However, it did give positions for the glacial maxima that lined up very nicely with those of the target, that is, the isotopic record of the deep sea. In fact, it turns out that Calder's time scale for glacial periods is more accurate than that published for core V28-238 itself, according to our present understanding.

Calder's playful experiment showed the way for a rigorous test of Milankovitch theory: find a set of physically plausible rules that convert Milankovitch input (summer sunshine in high northern latitudes) to ice-mass history, and check if the resulting output from the rule machine matches the "target," that is, the oxygen-isotope record of the deep sea. Calder's rules were too simple to make a good template, and his target was distorted and dated incorrectly in the older portion. Nevertheless, the results were highly encouraging and definitely supportive of Milankovitch theory, as early as 1974.

A few years later, in 1980, the paleontologist John Imbrie and his mathematician son J. Zeller Imbrie made a more serious effort at simulating the ice-mass history of the late Pleistocene by applying a set of rules to summer insolation in high latitudes.[78] They used a computer rather than a calculator, and a more formal set of equations specifying the response of ice growth and of ice decay to the changing intensity of summer sunshine. Their template is much better than that produced by Calder. However, their set of rules (which also incorporated the asymmetry for growth and decay rates) was unable to simulate the occurrence of a large interglacial near 400,000 years ago. At that period in the Pleistocene, the eccentricity of Earth's orbit was small, so the apparent size of the Sun was about the same for summer and winter. It is not clear, therefore, how summers could have been warm enough to melt the ice to make the prominent interglacial period labeled "stage 11" (fig. 13.8).

The difficulty in simulating the 400,000-year-old warm period became known as the "stage 11 problem." It represents a major stumbling block to unquestioned acceptance of "straight" Milankovitch theory. The stage-11 problem demonstrates that Earth's climate does not react in a simple fashion to the changes in northern latitude sunshine. The reason for the complicated response is ice dynamics combined with the response of the crust to loading and unloading, involving processes dubbed "slow physics." That these responses must be considered was already well recognized by a number of geophysicists and glaciologists.[79] In fact, the necessity to invoke slow physics was early on realized by Emiliani, when discussing Milankovitch theory.[80]

Two elements of the physics behind the ice age cycles are paramount: the stabilization effect of the fact that snow is white and an ice cap makes a high mountain, and the destabilizing effect associated with the buildup of large ice masses that flow under their own weight. When ice builds up, it whitens the ground and increases the reflection of sunlight from the planet. The reflected light goes back into space without warming the ground, or anything else. This albedo response, in essence, is instantaneous. Whitening cools a region, stabilizing snow and ice fields. Thus, ice fields have an inbuilt tendency to keep growing. As the elevation increases, the icy ground experiences lower temperatures, further stabilizing the snow on it. However, there is an opposing effect, from slow physics. As the ice thickens, it presses down on the ground, which starts sinking under the weight. Eventually, the ground may sink below sea level. This makes the ice sheet vulnerable to seawater incursion at its bottom, seawater that can bring heat and take away large chunks of ice as icebergs. Once this happens, the movement of glaciers may accelerate, forced by gravity, with the motion producing more heat inside than is readily dissipated, which can lead to a runaway situation. Finally, if the ice sheet is removed in some way, the ground stays at low elevation where snow is less likely to survive the summer. This effect introduces a long delay in renewed glacial buildup. In summary, the expectation is slow buildup of instability, collapse, and a wait for renewed buildup.

The basic dynamics of large ice sheets just outlined can set up an oscillation. When the planet first cools sufficiently for the temperature to fall below freezing over substantial regions, the positive feedback kicks in, through the whitening effect. Ice keeps growing. At the same time, the ice builds up instability at its base (high pressure, low elevation) and it builds up gravitational energy within its own body. When the time is ripe—when enough potential energy is available and instability reaches a

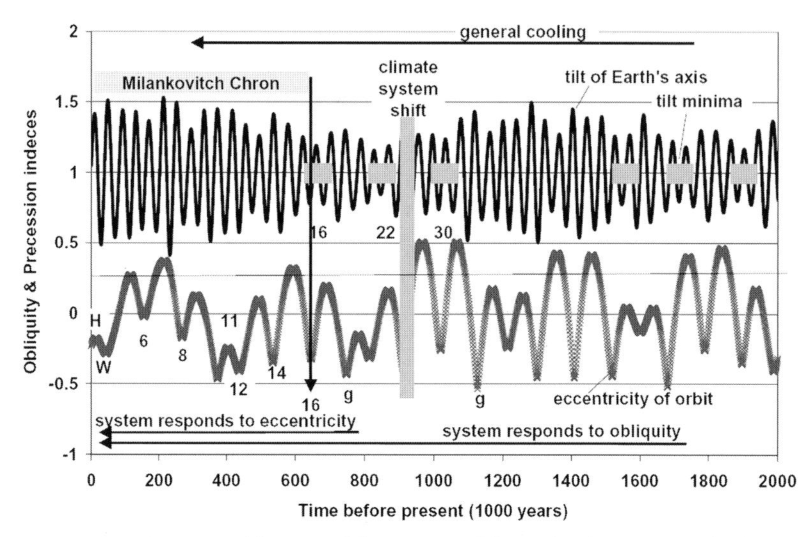

FIGURE 13.9. Astronomical forcing and the response of the Earth's climate system for the last two million years. Shown are the variations in tilt (or obliquity) of Earth's axis and the variations of eccentricity (deviation of orbit from a circle) through time. Even numbers denote major ice age stages. They are common only after the climate shift and tend to occur during periods of reduced eccentricity. However, stage 11, a major warm period, occurs during a time of low eccentricity, as does the Holocene (H), contrary to expectations from Milankovitch theory. This suggests that the external forcing does not determine climate directly but modifies climate oscillations inherent to the system. The system listens to those elements of the forcing that agree with its own preferences. Milankovitch Chron, the period studied by Milankovitch; g, positon of glaciations; H, Holocene (last 11,000 years); W, last glaciation ("Wuerm" in Europe, "Wisconsin" in North America).

threshold—any warming can trigger a collapse of the ice sheet. We are dealing with a process called delayed negative feedback, tied to the release of pent-up energy. A familiar example of oscillations produced by delayed negative feedback is the rattling cover on a pot of boiling water. When the cover is on tight, pressure builds up inside. When the cover rises above the rim, pressure is suddenly released, allowing the cover to fall back and close the pot. Thus the rattling: a cyclic response to steady heating. Old Faithful, the hot geyser in Yellowstone Park, is a well-known natural example for oscillations resulting from steady heat input into a system with a threshold for release of heat buildup.

The warming that triggers the collapse of the "ripe" ice sheets is provided by unusually warm summers in the Northern Hemisphere, in precise agreement with Milankovitch prediction. Summers, for maximum warmth, depend on high eccentricity (changing apparent size of the Sun) and on high obliquity (Sun high over the horizon at noon). One the whole, ice will build up at times when there are no warm summers (fig. 13.9). However, in the case of stage 11, the summers are only slightly warmer than in the previous thousands of years. The effect is more like a tap on the system than a push. But the preceding strong buildup of ice, during stage 12, has readied the system for collapse, and the tap is sufficient to initiate major melting.[81] The fact that stage 11 exists proves that ice buildup alone is sufficient to generate a large climate cycle, by internal dynamics. The shape of stage 11—rapid decay of ice, and a brief period of maximum warmth followed by slow buildup of ice—shows us what the oscillation looks like if it is largely undisturbed by orbital interference. The role of Milankovitch's summer insolation, it is now clear, is to set the pace for rapid melting within those periods for which the ice is ready to go. That is the lesson from stage 11.

Why did the climate system change from a state with dominant 40,000-year cycles to one with 100,000-year cycles? And why was the change so sudden? The answer to the first question is not so difficult. For the late Quaternary, it so happens that the internal dynamics favor an oscillation near 100,000 years (as seen for stage 11), and this works well in concert with the eccentricity changes in Earth's orbit, which are close to this period. Recall that large eccentricity provides the opportunity for having an unusually large Sun in the sky. If that happens in summer (as it must on the precessional cycle), and if ice is ready to collapse, an interglacial period is initiated. The crucial element is the length of time of the buildup of instability, so that the ice will respond with collapse when the summer Sun appears large. The relevant time span is between 50,000 and 60,000 years.[82] After the climate has suffered an ice age for this length of time, it is ready to respond to a thousand years of warm summers, by major melting. Conversely, when warm summers are rare—during times of low eccentricity—there is unobstructed opportunity for a full buildup of ice to the point of instability. This was the case for stage 16, when not only was eccentricity low, but also there was not much interference in ice buildup from variations in tilt (fig. 13.9). (Recall that tilt of the axis, or obliquity, controls how high the noon Sun will appear in the sky, in high latitudes.) Lacking the opportunity for having unusually warm summers, stage 16 saw the greatest northern ice buildup ever. It was centered near 650,000 years ago. It ended abruptly, demonstrating the buildup of instability.

After stage 16, the ice age cycles locked firmly into eccentricity, demonstrating their preference for 100,000-year oscillations over, say, 80,000-year or 40,000-year oscillations. These earlier oscillations did occur further back in time, but their glacial conditions were less severe, and the ingredient of major instability was missing.

The change in the nature of cycles was not really all that sudden; what was rather sudden was the expansion of ice after 900,000 years ago. It suggests that ice-bearing areas, such as shelves, expanded to a larger size by a critical drop in sea level. This expansion, naturally, created more vulnerable ice (that is, ice readily removed when sea level rises). In the early Quaternary, the internal ice age cycles preferred a shorter period, there being less ice, on average. Thus, the oscillations ended up responding to obliquity variation, which is the next shorter period in the options offered by the Earth's motions in orbit. With the increase in ice mass at the Mid-Pleistocene shift, the periods of rapid decay (the terminations) become part of the record, which shows that ice-mass instability became a major player within the climate system at that time.

What clues might we extract for future climate developments from ice age studies? The fact that ice ages are cyclic makes future developments somewhat predictable, as long as the same rules apply as in the past. The problem is, these rules no longer apply: we now have unprecedented amounts of greenhouse gases in the atmosphere. There is no experience within the last million years that we could fall back on for analogy. Ignoring the increased greenhouse effect, we can identify the situation that is closest to ours in the past: it is stage 11, some 410,000 years ago.[83] According to this analogy, we have just passed the warm peak and are on a slow downward slide toward a cooler climate. This trend would normally result, in about 40,000 to 50,000 years, in a new full glaciation that would last at least 20,000 years. What is of some interest in terms of future climate is that the high mark of the sea level during stage 11 was apparently somewhat higher than is that of our time in the Holocene (according to the oxygen isotope data).[84] If so, it suggests that the present sea level is not at a natural maximum but can respond quite readily to warming by moving upward by several meters. A rate of change of 2 or 3 feet per century, for the next several hundred years, seems within the range of possibilities, as suggested by rate of change of isotopes toward the end of the rise in stage 11.

How trustworthy is such reasoning by analogy? How good is the analogy between the Holocene and stage 11 anyway? Nothing stays quite the same in geologic history, and there are no exact analogs. What one needs to consider is that the very landscape has changed through time as a result of loading and unloading with enormous ice masses through many cycles. As the fjords became deeper, the ice collapsed more readily and was removed more thoroughly. This change and many others related to erosion in the heartland of the ice and deposition of material around the periphery is constantly altering the framework within which the climate system operates. Such long-term trends limit the generality of the rules that we assume govern the ice age cycles. Overall, these changes have worked toward greater buildup of ice, along with greater buildup of instability, through time.

THE LAST ICE AGE AND HOW IT ENDED

The last ice age, which culminated about 20,000 years ago, has served—naturally—as the prime example of a glacial period, just as the present has served as an example for a typical warm interval. To what degree this is justified is an open question, but it cannot be entirely off the mark: the sea level changes ranged between more or less fixed limits for the last 600,000 years or so, suggesting that, indeed, there is a pattern defining and limiting the extremes of warm and cold on this time scale.

There is one thing about the recent past that is surely *not* typical, however. It is the *transition* from the last glacial to postglacial conditions. While it was fast, just like the previous ones, it proceeded in two major steps with a distinct pause in between, a feature much less well expressed in older transitions, if at all.

First of all, of course, we would like to know how the ice age ocean was different from the present ocean, beyond the simple statement that it was colder. From the increased upwelling activity in tropical regions (based on a higher supply of organic matter to the seafloor), it seems that winds and therefore surface currents must have been stronger.[85] This is entirely as expected—when the polar front in the Northern Hemisphere presses southward, temperature gradients increase and zonal winds benefit.[86] The increased upwelling, overall, must have extracted nutrients more efficiently from the sea as a whole and thereby depleted, to some extent, the upper water layers of the phosphate, silicate, and nitrate dissolved therein, in the glacial ocean. Thus, with the increase of the winds, at the onset of glacial conditions, there was then a privileged period with both strong upwelling and high nutrient content, resulting in a pulse, a few thousand years long, of very high productivity. This is readily seen in deep-sea sediments in many parts of the ocean. The increased transfer of organic carbon into deep waters also caused a carbonate dissolution pulse over much of the Pacific during the onset of glacial periods.

Of course, the sea was cooler on the whole during glacial times, but the effects are strongest in high and midlatitudes, where the migration of the polar front toward the equator and the associated seasonal expansion of sea ice bring great change. In the tropics, and especially in the western equatorial Pacific, there is but little evidence for cooling in excess of 1 °C or so (2 °F), although the equatorial Atlantic seems to have suffered a somewhat greater drop in temperature. Also, in the central gyres, judging from the lack of much change of the plankton fauna, cooling must have been very modest. The overall message from these findings (made in the 1960s and 1970s) is that the tropical and subtropical regions are quite resistant to change, as far as surface temperatures. The action is in the areas of migrating boundaries of climatic zones in temperate and high latitudes.

From first principles, in addition, we should expect the intertropical convergence zone to move closer to the equator, from its present position north of the equator. A somewhat cooler ocean (by even just a few degrees) would have transferred less water vapor to the atmosphere, and a somewhat cooler atmosphere, especially beyond the horse latitudes, would have carried less water vapor than today. Thus, during glacial periods the heat transfer to high latitudes became less efficient, further stabilizing the ice fields, beyond the effects of the high ice albedo mentioned earlier. A greatly expanded sea-ice cover, in the northern regions, also would have worked in that direction.[87]

A systematic effort toward mapping the glacial ocean was made in the late 1960s and early 1970s, by a large group of paleoceanographers led by Andrew McIntyre, James Hays, and Neil Opdyke of Lamont, Theodore C. Moore[88] of Oregon State University, and John Imbrie of Brown University. Dubbed the CLIMAP project (an acronym memorable for containing allusions to climate and mapping), the well-coordinated interinstitutional collaboration produced a semi-quantitative map for surface-water temperatures of the ice age ocean.[89] To reconstruct the surface temperatures, the fossil assemblages on top of the seafloor were first calibrated against present temperature distributions; that is, in essence, the assemblages found on the present bottom were tagged with the present conditions. In a second step, the assemblages that accumulated in the past—some 20,000 years ago—were then assigned temperatures that seemed appropriate, given the match between distributions of fossils on the top of the seafloor and present sea-surface conditions.[90]

The map that the CLIMAP group produced shows the various features just summarized in striking detail: northern polar fronts much farther south, all high-latitude systems telescoped toward the equator, general cooling of surface waters between 2 and 3 °C on average, increased upwelling along the equator, increased coastal upwelling and strengthened eastern boundary currents, and only very modest change in the central gyres. The map subsequently served as a basis for modeling the climate of the last glacial maximum, using integrated general circulation models, in an effort to simulate patterns of temperature, winds, and precipitation during ice age time.[91]

The end of the ice age came as expected from Milankovitch theory (within 2,000 years or so) but in essence without particular warning. The ice started melting in earnest around 16,000 years ago, and sea level rose rapidly, within 2,000 to 3,000 years, by about 60 meters.[92] It then stood still for a dozen centuries, while a severe cold spell held the North Atlantic realm in its grip. Then, again without warning, extremely rapid warming set in, and the remainder of the vulnerable ice melted within another 2,000 to 3,000 years, to raise the sea level another 60 meters. The reasons for the steepness of the two steps, the reasons why there are steps at all, or why there are two steps rather than three or four, and the reasons for the duration of the cold spell are entirely obscure.

From this circumstance, one might conclude that we do not understand the climate machine very well, especially where there are large ice masses involved.[93] While understanding is limited, there are some clues as to what happened, as follows.

The first of the two great meltwater pulses can be ascribed to Milankovitch forcing, since it is centered near 14,000 years ago, when July insolation in high northern latitudes was at its maximum. The implication is that ice sheets (or rather *some* ice sheets) were ready to go, and they did so when the summer sunlight was of the right intensity. The ice to go first, presumably, was that which was farthest from the poles and most vulnerable to being invaded and lifted off by the sea, that is, the ice that was grounded below sea level.[94] When this rather vulnerable ice was used up, the process stopped. Recall that the eccentricity of Earth's orbit is not particularly pronounced in the last 20,000 years. Thus, the "large-solar-disk" effect in July 14,000 years ago was weak. It needed some time to complete its job of warming up the ice by sending summer melt down the cracks for refreezing or for lubricating the base, or both.[95] During this pause, while the remaining glacial ice warmed in its interior, the climate went back to the state it had left during the first melting step.[96] After all, large ice sheets still dominated the scene with their whiteness and high elevation, so this is perhaps not surprising.

Thus, the central question is what made the ice collapse in such a short time, making sea level rise as fast as it did, at a rate greater than 2 meters per century for a thousand years or more. A likely answer is that ice that is

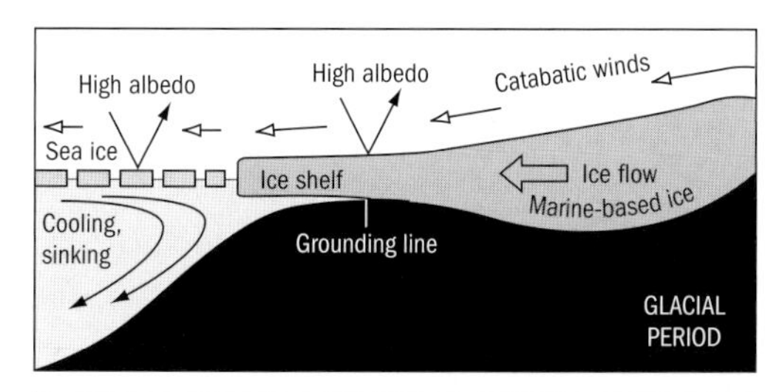

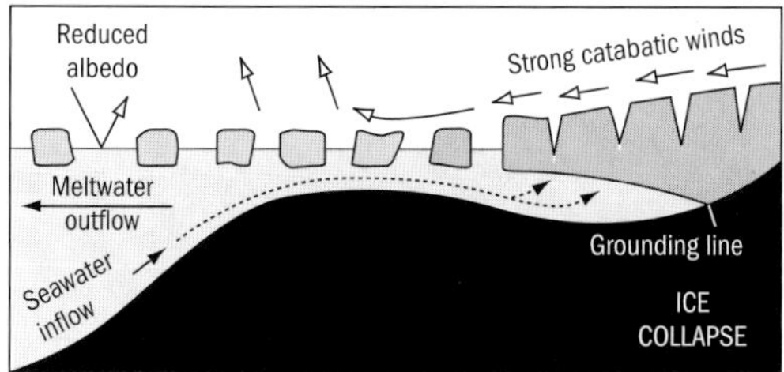

FIGURE 13.10. Hypothesis of collapse of marine-based ice sheets, triggered by a rise in sea level.

grounded below sea level is especially vulnerable to collapse, as pointed out by the geologist T. J. Hughes, in the 1970s (fig. 13.10).

The question of why the system, after the first melting step, went back into a state reminiscent of the glacial period has dominated discussions about the end of the last glacial for some time. What exactly is the origin of the "big chill" that separates the two steps of ice collapse? The origin of the big chill, the Younger Dryas, is no secret, actually. It is the presence of large ice sheets, which set up a wind system that transports Arctic air to temperate latitudes. The result is permafrost south of the ice sheets, for which there is abundant evidence. Thus, there is no mystery here. The really important question is how one could start and sustain the warming of the first melting step, in the presence of enormous ice sheets, during the warm period called the Alleroed that preceded the Younger Dryas and lasting more than 2,000 years.[97] The warmth of the Alleroed, in addition from the slightly warmer summer Sun, must have somehow derived from the melting of the ice itself, since there was nothing else going on that was different in the climate system a thousand years before the Alleroed.

Presumably, the circulation set up by the enormous meltwater input into the northern North Atlantic—a circulation similar to that in a fjord during spring melt—brought heat into high latitudes. In addition, the presence of a thin meltwater layer on the surface of the sea, in high latitudes, trapped summer heat in this layer, making the sea uncommonly warm at the surface. Conversely, in winter this layer would rapidly cool and grow sea ice, greatly slowing heat loss from the water below.[98]

In summary, we have to look to the ice for answers regarding the abrupt changes seen in the transition from the last glacial period to the Holocene. It seems reasonable to assume, from the reasoning given for the entire course of ice age cycles, that instability and ice collapse are at the heart of the matter of stepwise transition, and that ocean and atmosphere follow that lead.[99] Ocean and atmosphere have the role of synchronizing the action in north and south, and indeed globally, by dint of changes in sea level and in greenhouse gas. Furthermore, there is no question that the massive influx of meltwater during the transition from glacial to postglacial time fundamentally affected the deep circulation of the sea, with implications for heat transport.[100]

Detailed studies of Greenland ice cores show that climate change was extremely rapid in cases, with profound and sustained warming and cooling occurring through a time span of a decade or less.[101] Did abrupt climatic change within the transition kill the mammoth and other large mammals, about 12,000 years ago? Everyone agrees that abrupt climatic change is bad for the survival of large mammals. Sudden warming can wreak havoc on rates of reproduction of large mammals (as is well known by cattle farmers). Both sudden warming and sudden cooling can affect the food supply or the habitability of a region. Starving animals do not reproduce well and readily fall prey to sabertooth cats, hyenas, and wolves, and to disease. But for extinction, every single population of a species has to be eradicated. Some large species survived (for example, musk oxen, reindeer, and polar bears). Many ice age experts are convinced that human hunters are responsible for the extinction of the others.[102] Paul Martin of the Desert Institute in Arizona insists that in North America the mammoth and all the rest had no chance; after a millennium or two of being pursued by skilled hunters arriving over the Bering land bridge, they were done for. The big-game predators, the sabertooths, the short-faced bears, the great lions, the hyenas—all then starved. Martin calls it the overkill hypothesis and refers to the process as "blitzkrieg."

On the scale of the lifetime of a species—a million years or so—the annihilation of the Pleistocene megafauna was lightning fast. On the scale of the lifetime of a typical hunter, it would have been unnoticeable.

The question whether natural catastrophe or human overkill was responsible for the demise of the megafauna is unresolved. It may never be resolved, because both processes were involved, climate and people, working together. The fact that the large marine mammals of the Arctic survived the abrupt climate changes associated with the massive melting steps is significant. It suggests that on land, human hunting was indeed a factor, or else that long-distance migration in the sea posed fewer obstacles than on land.[103]

NOTES AND REFERENCES

1. The ice on Antarctica is equivalent to a change in sea level of roughly 70 meters (230 feet), and that on Greenland of roughly 7 meters. The fact that we have open waters in summer around Svalbard (near 80° N, fig. 13.1) confirms that we are in a warm phase of the present ice age.

2. According to Plato, the priest told Solon that the Greeks do not value knowledge that is ancient, but are caught up in current affairs. Thus their knowledge is very limited. Platon (427–327 BC) Timaios (Steph. 22c).

3. The Mid-Pleistocene climate shift was rather sudden. It is dated at 0.92 million years before present. W. H. Berger, G. Wefer, 1992, *Klimageschichte aus Tiefseesedimenten—Neues vom Ontong-Java-Plateau (Westpazifik).* Naturwissenschaften 79, 541–550; W. H. Berger, E. Jansen, 1994, *Mid-Pleistocene climate shift: The Nansen connection,* in O. M. Johanessen, R. D. Muench, J. E. Overland (eds.), *The Polar Oceans and Their Role in Shaping the Global Environment: The Nansen Centennial Volume.* Geophysical Monograph 84. American Geophysical Union, Washington, D.C., pp. 295–311.

4. Another famous navigator has important links to the ice age: Magellan's strait was shaped by glacial ice.

5. Milutin Milankovitch (1879–1958), Serbian engineer and astronomer, is the most prominent among ice age theorists. He clarified the issues associated with the "astronomical theory of climatic changes" (with some help from the climatologist Wladimir Köppen). In particular, he introduced the concept that the receipt of solar radiation in high northern latitudes, in summer, is the single most important factor in driving the ice age fluctuations. This tenet is the heart of Milankovitch theory. Milankovitch taught engineering, physics, and celestial mechanics at the University of Belgrade, from 1909 to 1955. His major work was *Canon of Insolation of the Earth and Its Application to the Problem of Ice Ages,* published by the Serbian Academy of Sciences in 1941, in German. V. Milankovitch, 1984, *The memory of my father,* in A. L. Berger, J. Imbrie, J. Hays, G. Kukla, B. Saltzman (eds.), *Milankovitch and Climate, Part I.* D. Reidel, Dordrecht, pp. xxiii–xxxiv.

6. The *Albatross* Expedition (1947–1948) is to paleoceanography what the *Challenger* Expedition was to oceanography. A total of 299 cores, with a combined length of 1.6 kilometers, were collected by the *Albatross,* using the new piston corer invented by Börje Kullenberg (1906–1991). The expedition was led by the geochemist and oceanographer Hans Pettersson (1888–1966). H. Pettersson, 1953, *Westward Ho with the Albatross.* E. P. Dutton, New York, 218 pp. The long cores obtained during the Swedish Deep-Sea Expedition contained, for the first time, information spanning hundreds of thousands of years.

7. Later, the same principle would be used by scientists at Lamont Geological Observatory and at Scripps.

8. Gustaf O. S. Arrhenius, after writing up his report (published in 1952), went on to a distinguished career as an oceanographer and planetary scientist at Scripps. He joined Scripps in 1953 and has served on its faculty for more than half a century.

9. G. O. S. Arrhenius, 1952, *Sediment cores from the east Pacific.* Reports of the Swedish Deep-Sea Expedition, 1947–1948 5, 1–288.

10. The evidence consisted in a difference in oxygen isotope chemistry of the shells, such that those of glacial age had relatively more of the heavier type (oxygen-18) than the others, because ice caps preferentially incorporate water with the lighter type (oxygen-16).

11. Cesare Emiliani (1922–1995) pioneered isotope stratigraphy, which became the master signal for deciphering the deep-sea record. C. Emiliani, C., 1955, *Pleistocene temperatures.* Journal of Geology 63, 538–578. For a summary of Emiliani's many

contributions, see W. H. Berger, 2002, *Cesare Emiliani (1922–1995), pioneer of ice age studies and oxygen isotope stratigraphy.* Comptes Rendus Palevol 1, 479–487.

12. Fred B. Phleger (1909–1993) and Frances L. Parker (1906–2002) joined Scripps in 1950 and established the Marine Foraminifera Laboratory. Eric Olausson, at the University of Gothenburg, led a long-term systematic study of the hundreds of cores raised by the *Albatross.*

13. Except for a population of dwarfed Eurasian mammoth on Wrangell Island in the Arctic Ocean, which died out several thousand years later. U. Jogel, C. Kamcke, 2005, *Mammut.* Staatliches Naturhistorisches Museum, Braunschweig, Cargo Verlag.

14. To his everlasting credit, the oil tycoon G. Allan Hancock gave the 23-acre site to the county of Los Angeles in 1916, after excavations had proved it to be one of the most remarkable repositories of fossils found anywhere on the planet.

15. Paul Dayton, pers. comm. 2003.

16. George Cuvier (1769–1832), the leading vertebrate paleontologist and zoologist of his time, believed that the "frozen carcasses of the great quadrupeds" discovered in Siberia showed that the cold had come suddenly and catastrophically, killing the animals and preserving them ever since.

17. Jean Louis Rodolphe Agassiz (1807–1873), distinguished Swiss naturalist, was a professor of zoology at Harvard from 1848, and the founding director of Harvard's Museum of Comparative Zoology.

18. L. Agassiz, 1866, *Geological Sketches.* Ticknor and Fields, Boston, 311 pp., p. 208.

19. In fact, Hutton's friend John Playfair had already speculated on the subject, as had Jean-Pierre Perraudin, a Swiss mountaineer, Ignace Venetz, a highway engineer from Wallis, and Jean de Charpentier, director of the salt mines at Bex and a mentor of Agassiz. See J. Imbrie, K. P. Imbrie, 1979, *Ice Ages, Solving the Mystery.* Enslow, Hillside, N.J., 224 pp. A review of the early history of ice age research from a contemporary perspective is in Penck's introduction to the Penck-Brückner treatise on alpine glaciations. (See next note.)

20. Albrecht Penck (1858–1945) and Eduard Brückner (1862–1927), Berlin and Vienna, respectively, are best known for their three-volume treatise entitled *The Alps in the Ice Age,* published in 1909. For decades, it was the standard reference for researchers working on the ice ages. A. Penck, E. Brückner, 1909, *Die Alpen im Eiszeitalter,* 3 vols. C. H. Tauchnitz, Leipzig, 1199 pp.

21. James Croll (1821–1890), a geologist with the Geological Survey of Scotland, was the first to combine knowledge from oceanography, climatology, and glaciology into a coherent and intelligent framework for the explanation of multiple glaciations. His book contains striking physical insights, as when he argues for heat transfer from the Southern to the Northern Hemisphere by ocean currents. His theory of the ice ages turned out to be wrong, however. J. Croll, 1875, *Climate and Time in their Geological Relations. A Theory of Secular Changes of the Earth's Climate.* Daldy, Isbister, and Co., London, 577 pp.

22. James Geikie (1839–1915) was one of the leading geologists of his time. Penck credits him with the concept of multiple glaciations.

23. J. Croll, *Climate and Time,* p. 5.

24. Gustav Steinmann (1856–1929) is best known for having argued that deep-sea sediments do occur in the Alps and in the Appenines, contrary to opinions current at the time. His reply to Croll (and supporters) is in G. Steinmann, 1916, *Die Eiszeit und der vorgeschichtliche Mensch,* 2nd ed. B. G. Teubner, Leipzig, 105 pp., p. 78. The statement was correct as far as Croll's scheme of alternating ice ages on the two hemispheres. However, strictly speaking, Steinmann's criticism turned out to be irrelevant to the argument. As we now know, the glaciation cycles on the Northern Hemisphere entrained changes in the southern one, making them more or less synchronous.

25. In fairness to Croll, even today there is no clear consensus about what is more in need of explanation when tackling the ice age cycles, building northern ice caps or demolishing them. The preference I express here follows Milankovitch.

26. Wladimir Köppen (1846–1940) is best known for his scheme of defining climatic zones on the planet based on vegetation cover, temperature, and precipitation. He was a founder and pioneer of modern climatology. Alfred Wegener (1880–1930) is best known for the theory of continental drift. In their book, the authors explore the shifting of climate zones in the past. W. Köppen, A. Wegener, 1924, *Die Klimate der geologischen Vorzeit.* Bornträger, Berlin, 256 pp.

27. M. Milankovitch (1930) *Mathematische Klimalehre und astronomische Theorie der Klimaschwankungen.* Handbuch der Klimatologie, Bd. 1, Teil A. Bornträger, Berlin, 176 pp.

28. Alfred Russell Wallace, the cocreator with Darwin of the theory of natural selection, argued the importance of positive albedo feedback in high latitudes, when considering ice age cycles, in 1895. He was interested in the effects of past climate change on the present distribution of plants and animals. A. R. Wallace, 1895, *Island Life or the Phenomena and Causes of Insular Faunas and Floras, Including a Revision and Attempted Solution of the Problem of Geological Climates,* 2nd ed. Macmillan, London, 563 pp.

29. Positive feedback is the reason why the present global warming is most strongly expressed in high northern latitudes.

30. Pierre Simon Laplace (1749–1827), one of the great mathematicians of all time. He produced a five-volume treatise on celestial mechanics.

31. Johannes Kepler (1571–1630) discovered the eccentricity of Earth's orbit, as well as that of other planets, along with other rules governing their orbits. Kepler's justly celebrated discoveries were a by-product of applied science: he provided horoscopes to his employer.

32. However, for the next several thousand years, in all probability, man-made global warming will seriously interfere with the regular climate response to orbital forcing.

33. The Earth's axis is not fixed with respect to the stars, but wobbles like a children's top spinning down. This wobble produces precession of the equinox, which means that each vernal equinox (the beginning of the astronomical year) precedes the one of the previous year. Looked at from the position of the vernal equinox, then, the Sun appears shifted toward an earlier sign in the zodiac, each year. It takes about 2,000 years for the Sun to migrate from one sign to the next. Greek astronomers knew this, and it is likely that their knowledge built on that of Babylonian observers.

34. Read from the Milankovitch diagram shown in W. Köppen, 1931, *Klimate der Erde,* 2nd. ed. Walter de Gruyter, Berlin, 388 pp., p. 42.

35. This morphology led the pioneer of foraminifer studies, the French naturalist Alcide d'Orbigny (1802–1857), to class these shells with cephalopods, initially.

36. At Lamont, cores were taken on a fixed schedule, in response to the wishes of the director, Maurice Ewing. At Scripps, it was left to individual chief scientists to decide whether to take cores and where. Lamont obtained the bigger collection.

37. Frances Lawrence Parker joined Scripps at the invitation of Fred B. Phleger, who accepted a faculty position in 1950. Parker made important contributions to the knowledge of benthic and planktonic foraminifers in regard to taxonomy, biogeography, and stratigraphy. She and Phleger helped lay the foundations of modern paleoceanography. Milton Nunn Bramlette (1896–1977) pioneered the use of nannofossils in stratigraphy. William R. Riedel became the world's leading expert in Cenozoic radiolarians, and a major contributor to the planning of the Deep-Sea Drilling Project, one of the most important international ventures in the Earth sciences, ever.

38. F. B Phleger, F. L. Parker, J. F. Peirson, 1953, *North Atlantic Foraminifera,* Reports of the Swedish Deep-Sea Expedition, 1947–1948 7, 1–122, 20 tables, 6 plates; F. L. Parker, 1958, *Eastern Mediterranean Foraminifera.* Reports of the Swedish Deep-Sea Expedition, 1947–1948 8 (4), 217–283.

39. F. L. Parker, *Eastern Mediterranean Foraminifera.*

40. Silica is a substance related to flint and quartz (silicon oxide) but containing substantial amounts of water within a disordered lattice. Many organisms extract silicate from seawater to make skeletons of silica glass: diatoms, silicoflagellates, radiolarians, siliceous sponges. In surface waters, silicate is in short supply and thus limits the productivity of diatoms.

41. Ernst Heinrich Haeckel (1834–1919) was one of the leading zoologists of his time, and the world's expert on radiolarians, with considerable expertise in other zooplankton besides. He is best known for his vigorous defense of Darwin's ideas, to which he added some of his own, regarding the recapitulation of evolution (phylogeny) within individual development (ontogeny). He wrote and lectured forcefully and with great erudition and gladly demolished opposing views in public.

42. W. R. Riedel, 1959, *Siliceous organic remains in pelagic sediments.* Society of Economic Paleontologists and Mineralogists, Special Publication 7, 80–91.

43. The debriefing is not always straightforward. The preserved record is biased throughout against the more delicate forms, for all microfossils. W. H. Berger, 1976, *Biogenous deep-sea sediments: production, preservation and interpretation,* in J. P. Riley, R. Chester (eds.), *Treatise on Chemical Oceanography,* vol. 5. Academic Press, London, pp. 265–388.

44. W. Schott, 1935, *Die Foraminiferen in dem äquatorialen Teil des Atlantischen Ozeans.* Wissenschaftliche Ergebnisse der Deutschen Atlantischen Expedition. *Meteor,* 1925–1927 3, 43–134.

45. G. O. S. Arrhenius, *Sediment cores.* The support comes from the analysis of deep-sea cores for changes in organic matter content, and for barium sulfate content. Both these measures show increased equatorial productivity in the Pacific during glacial periods. A. Paytan, M. Kastner, 1996, *Benthic Ba fluxes in the central equatorial Pacific: Implications for the oceanic Ba cycle.* Earth and Planetary Science Letters 142, 439–450; H. M. Perks, R. F Keeling, 1998, *A 400 kyr record of combustion oxygen demand in the western equatorial Pacific: Evidence for a precessionally forced climate response.* Paleoceanography 13, 63–69.

46. W. H. Berger, J. C. Herguera, 1992, *Reading the sedimentary record of the ocean's productivity,* in P. G. Falkowski, A. D. Woodhead (eds.), *Primary Productivity and Biogeochemical Cycles in the Sea.* Plenum Press, New York, pp. 455–486; G. Wefer, W. H. Berger, 2001, *Reconstruction of productivity from the sedimentary record,* in J. H. Steele, S. A. Thorpe, K. K. Turekian (eds.), *Encyclopedia of Ocean Sciences,* 6 vols. Academic Press, San Diego, pp. 2713–2724.

47. F. B. Phleger, 1948, *Foraminifera of a submarine core from the Caribbean Sea.* Meddelande frin

Oceanografiska Institut Göteborg 16, 1–9. Hans Pettersson, leader of the expedition, published these results in his popular book *Westward Ho with the Albatross,* with addition of the contemporary ice age stratigraphy by Cameron Ovey at the British Museum.

48. F. B. Phleger et al., *North Atlantic Foraminifera.*

49. The method had been introduced by the geologist Wolfgang Schott of the *Meteor* Expedition, in the 1930s, but no one had paid much attention.

50. D. B. Ericson, G. Wollin, 1956, *Micropaleontological and isotopic determinations of Pleistocene climates.* Micropaleontology 2, 257–270; D. B. Ericson, G. Wollin, 1956, *Correlation of six cores from the Equatorial Atlantic and the Caribbean.* Deep-Sea Research 3, 104–125.

51. The two types of atoms both have the chemical properties of oxygen but slightly different atomic weight: 16 and 18. The slower-moving oxygen-18 is more readily precipitated when making carbonate shells, but this advantage decreases with increasing temperature. Urey's insight produced a major tool for reconstructing temperature in the geologic past. H. C. Urey, H. A. Lowenstam, S. Epstein, C. R. McKinney, 1951, *Measurements of paleotemperatures and temperatures of the Upper Cretaceous of England, Denmark and the Southeastern United States.* Bulletin of the Geological Society of America 62, 399–416; S. Epstein, H. A. Lowenstam, 1953, *Temperature-shell-growth relations of recent and interglacial Pleistocene shoal-water biota from Bermuda.* Journal of Geology 61, 424–438; S. Epstein, R. Buchsbaum, H. A. Lowenstam, H. C. Urey, 1953, *Revised carbonate-water isotopic temperature scale.* Bulletin of the Geological Society of America 64, 1315–1325.

52. Other problems arise as well, when attempting to tie changes in isotopic ratios to changes in temperature. Some of these were recognized early on by the Chicago group, some emerged later. See G. Wefer, W. H. Berger, 1991, *Isotope paleontology: Growth and composition of extant calcareous species.* Marine Geology 100, 207–248.

53. W. Dansgaard, H. Tauber, 1969, *Glacier oxygen-18 content and Pleistocene ocean temperatures.* Science 166, 499–502.

54. C. Emiliani, *Pleistocene temperatures.*

55. Lamont oceanographer Wallace S. Broecker, geochemist, geologist, and ice age historian, helped establish a chronology for the late Pleistocene, by the dating of fossil coral reefs. He is best known for the idea that the ocean's "conveyor belt" might shut down with global warming, arguing from analogy with a shutdown of North Atlantic deepwater production during the transition from glacial to postglacial time.

56. In fact, both the time scales of Emiliani and of Broecker were incorrect, as far as the timing of cycles of the last 400,000 years. However, Broecker's date for the last interglacial was on the mark.

57. The dating relies on the (reasonable) assumption that living organisms incorporate carbon-14 and carbon-12 in the ratio given by abundances in their environment, and that carbon-14 decays in the dead shell or skeleton at a constant exponential rate, such that one-half of the carbon-14 present is lost in 6,000 years. Thus, a piece of wood or shell 30,000 years old (five times the half-life) would retain one-half multiplied by itself five times, or one-thirty-second of the original carbon-14, that is, about 3 percent. At this point, a few percent of contamination from the surroundings could greatly falsify results. The method was introduced by Willard Frank Libby (1908–1980) of the University of Chicago (1945–1954), and later at University of California, Los Angeles. His work was recognized with a Nobel Prize in chemistry (1960); it revolutionized archeology and found ample application in oceanography and ice age studies.

58. The method was emulated decades later and hailed as "tuning" to "Milankovitch forcing." It is now in general use.

59. H. H. Veeh received his Ph.D. in 1965. He determined that the last interglacial peaked near 124,000 years ago, in agreement with the Lamont dates obtained at the same time.

60. W. S. Broecker, 1966, *Absolute dating and the astronomical theory of glaciation.* Science 151, 299–304; see also W. S. Broecker, D. L. Thurber, J. Goddard, T. L. Ku, R. K. Matthews, K. J. Mesolella, 1968, *Milankovitch hypothesis supported by precise dating of coral reefs and deep-sea sediment.* Science 159, 297–300.

61. W. S. Broecker, J. Van Donk, 1970, *Insolation changes, ice volumes, and the O-18 record in deep-sea cores.* Review of Geophysics and Space Physics 8, 169–197.

62. The fact that there are terminations, that is, sudden collapse of ice masses and rapid rise of sea level, points to the importance of internal geophysical mechanisms for buildup of instability in the ice age climate system.

63. The commonly assumed age for the event is 780,000 years. In deep-sea sediments, the reversal event is recorded below the sediment-mixing zone, some 10 centimeters down. Thus, in sediments, the boundary must appear several thousand years older than in basalt flows on land.

64. C. G. A. Harrison, B. M. Funnell, 1964, *Relationship of paleomagnetic reversals and micropaleontology in two late Cenozoic cores from the Pacific Ocean.* Nature 204, 566.

65. *Fide* J. Imbrie, K. P. Imbrie, *Ice Ages,* p. 163 ff.

66. Nicholas J. Shackleton (1937–2006), British physicist and paleoceanographer, thus established the first nearly correct time scale for the time span studied by Milankovitch, together with Neil Opdyke of Lamont. N. J. Shackleton, N. D. Opdyke, 1973, *Oxygen*

isotope and paleomagnetic stratigraphy of equatorial Pacific Core V28–238: Oxygen isotope temperatures and ice volume on a 10^5 year and 10^6 year scale. Quaternary Research 3, 39–55. Shackleton had developed an instrument (mass spectrometer) that was modified from the original Chicago design to make high-precision isotopic measurements on small samples. He contributed importantly to the reconstruction of the glacial ocean (the so-called 18-k map), to the confirmation of Milankovitch theory, and to the interpretation of carbon-isotope records. He led the effort to extend orbital dating far back into the Tertiary.

67. The numbering scheme was introduced by Emiliani, and the marine isotope stages are also known as Emiliani stages, therefore. Stage 1 is the Holocene, stage 2 the last glacial period.

68. Later it emerged that the age of the magnetic reversal had been underestimated by 13 percent. By great luck, the core studied was distorted in such a fashion as to absorb the error in the assigned reversal date in the oldest part of the section and thus to yield a nearly correct result for stage 5e. Never has serendipity been more helpful in advancing ice age science.

69. Spectral analysis was invented by the eminent French mathematician Jean Baptiste Joseph Fourier (1768–1830), who worked for Napoleon as an administrator in his spare time.

70. J.D. Hays, J. Imbrie, N.J. Shackleton, 1976, *Variations in the Earth's orbit: Pacemaker of the ice ages.* Science 194, 1121–1132. Pisias's results are mentioned in J. Imbrie, K.P. Imbrie, *Ice Ages*, p. 170.

71. On occasion it is implied that the paper by Hays et al. (1976) clinched the veracity of Milankovitch theory (e.g., J. Imbrie, K.P. Imbrie, *Ice Ages*, p. 172). Such a conclusion would be incorrect. What the study showed, very convincingly, is that the spectral elements expected for Milankovitch forcing are present. But this presence does not specify the manner of converting orbital forcing into climate cycles, that is, the importance of summer insolation in high northern latitudes. The presence of the 100,000-year cycle in particular is not readily ascribed to Milankovitch theory.

72. The Belgian astronomer André Berger is a leading figure in the study of Milankovitch-driven climate changes. He revised the astronomically determined forcing functions and contributed to the dynamics of the climate response. A. Berger, 1978, *Long-term variations of caloric insolation resulting from the Earth's orbital elements.* Quaternary Research 9, 139–167; A. Berger, M.F. Loutre, 1991, *Insolation values for the climate of the last 10 million years.* Quaternary Science Review 10, 297–317.

73. And up into the sixteenth century. Johannes Kepler still delivered horoscopes, around AD 1600, on demand from his employer. However, he knew what he was doing and did it to make a living. Scientists no longer believe that astrology makes useful predictions.

74. There is no question that ice age cycles helped shape human evolution by posing various challenges for survival in a changing environment. However, the central question of human evolution is the growth of the brain. Presumably, once we were walking, evolution was mainly about talking. An influence of climate cycles on this history is not obvious. Cats and dogs and horses in Africa went through the same climate fluctuations and did not emerge with an unusually large brain.

75. A. Berger, J. Imbrie, J. Hays, G. Kukla, B. Saltzman (eds.), 1984, *Milankovitch and Climate*, 2 vols. Reidel, Dordrecht, 895 pp.

76. J. Zachos, M. Pagani, L. Sloan, E. Thomas, K. Billups, 2001, *Trends, rhythms, and aberrations in global climate 65 Ma to present.* Science 292, 686–693.

77. N. Calder, 1974, *Arithmetic of ice ages.* Nature 252, 216–218.

78. J. Imbrie, J.Z. Imbrie, 1980, *Modeling the climatic response to orbital variations.* Science 207, 943–953.

79. J. Weertman, 1976, *Milankovitch solar radiation variation and ice age sheet sizes.* Nature 261, 17–20; G.E. Birchfield, 1977, *A study of the stability of a model continental ice sheet subject to periodic variations in heat input.* Journal of Geophysical Research 82, 4909–4913; J. Oerlemans, 1980, *Model experiments on the 100,000 year glacial cycle.* Nature 287, 430–432.

80. C. Emiliani, J. Geiss, 1959, *On glaciations and their causes.* Geologische Rundschau 46 (2), 576–601.

81. W.H. Berger, G. Wefer, 2003, *On the dynamics of the ice ages: Stage-11 paradox, Mid-Brunhes climate shift, and 100-ky cycle.* American Geophysical Union, Geophysical Monograph 137, 41–59.

82. W.H. Berger, 1997, *Experimenting with ice-age cycles in a spreadsheet.* Journal of Geoscience Education 45, 428–439; A. Paul, W.H. Berger, 1997, *Modellierung der Eiszeiten: Klimazyklen und Klimasprünge.* Geowissenschaften 15 (1), 20–27.

83. The time span of 410,000 years is rather nicely divisible through the long-term eccentricity period of 413,000 years, through the 100,000-year eccentricity period, through the 41,000-year obliquity cycle, and through the 23,000-year precession cycle, at multiples of 1, 4, 10, and approximately 18. Thus, 410,000 years ago, the various cycles were roughly in the same phase as now.

84. The available data show a difference in oxygen-isotope index of around 0.1 per thousand, which allows for a difference in sea level of 5 meters or more, depending on what is assumed for the temperature effect contained in the isotopic difference.

85. As initially proposed by G.O.S. Arrhenius, *Sediment cores.*

86. H. Flohn, 1985, *Das Problem der Klimaänderungen in Vergangenheit und Zukunft*. Wissenschaftliche Buchgesellschaft, Darmstadt, 228 pp.

87. Such sea-ice limits can now be recognized in sediments, because it has been discovered, by scientists at the Alfred Wegener Institute in Bremerhaven, that certain diatom species are very typical for this environment.

88. T. C. Moore Jr. obtained his Ph.D. at Scripps, in 1968.

89. CLIMAP Project Members, 1976, *The surface of the ice-age Earth*. Science 191, 1131–1144.

90. All this was done with a mathematical "transfer" technique introduced by John Imbrie, and involving statistical techniques borrowed from sociology. J. Imbrie, N. G. Kipp, 1971, *A new micropaleontological method for quantitative paleoclimatology: Application to a late Pleistocene Caribbean core*, in K. K. Turekian (ed.), *The Late Cenozoic Glacial Ages*. Yale University Press, New Haven Conn., pp. 71–179.

91. The reliability of the map varies for different parts of the temperature range of surface waters. At the high end, the warm-water species do not care much about small differences in temperature, and it is difficult therefore to extract the precise meaning of changes in abundance. At the low end, the few species present at temperatures approaching freezing provide insufficient information for discriminating small temperature differences. The best resolution for temperature reconstruction is at intermediate temperatures.

92. R. G. Fairbanks, 1989, *A 17,000-year glacio-eustatic sea level record: Influence of glacial melting rates on the Younger Dryas event and deep-ocean circulation*. Nature 342, 637–642.

93. Ignorance is not bliss in this context: the problem is serious. In essence it means that we cannot predict how continental ice will respond to global warming.

94. The pause between melting periods may simply reflect the fact that marine-grounded ice is highly vulnerable to a small rise in sea level, and when it is gone there needs to be an appropriate waiting time till the less vulnerable land-grounded ice has warmed up. W. H. Berger, E. Jansen, 1995, *Younger Dryas episode: Ice collapse and super-fjord heat pump*, in S. R. Troelstra, J. E. van Hinte, G. M. Ganssen (eds.), *The Younger Dryas*. North-Holland, Amsterdam, pp. 61–105.

95. The flux of meltwater down into the glacier, and refreezing there, transfers heat into the interior of land-based ice. Any water that makes it to the bottom will diminish friction there or ice pressure, or both. (The role of water at the base of sliding solids is described in the geologic literature on mass wasting and landslides, beginning in the 1950s.)

96. The cold spell is called the Younger Dryas by geologists. The Younger Dryas period is named for the remains of the arctic flower *Dryas octopetala* in the periglacial deposits of Denmark. Obviously, there is an older period also, with remains of that same flower: the Older Dryas.

97. The Alleroed period is named after a place in Denmark where the time interval is represented by a layer of ancient soil containing remains from temperate plants.

98. An alternative mechanism to produce the observed two-step deglaciation was put forward by Wallace Broecker and George Denton, in 1989. W. S. Broecker, G. H. Denton, 1989, *The role of ocean-atmosphere reorganizations in glacial cycles*. Geochimica Cosmochimica Acta 53, 2465–2501. In their scenario the North Atlantic deepwater production is turned off by meltwater input, and the lack of deepwater production in northern latitudes then fails to attract warm surface waters from the south and results in a cold spell—the Younger Dryas. The hypothesis has a serious flaw, as pointed out by Berger and Jansen. W. H. Berger, E. Jansen *Younger Dryas episode*. The conveyor mechanism constitutes instantaneous negative feedback on melting. If active, the mechanism would stop the melting rather than produce a cold spell hundreds of years after the initial warming.

99. Ice collapse as an important feature of deglaciation was urged by T. J. Hughes in 1977, and many others since. T. J. Hughes, 1977, *West Antarctic ice streams*. Review of Geophysics and Space Physics 15, 1–46.

100. E. Olausson, 1965, *Evidence of climatic changes in North Atlantic deep-sea cores, with remarks on isotopic paleotemperature analysis*. Progress in Oceanography 3, 221–252; also, see articles in W. H. Berger, L. Labeyrie (eds.), *Abrupt Climatic Change: Evidence and Implications*. D. Reidel, Dordrecht, 425 pp. and especially W. S. Broecker, G. H. Denton, *The role of ocean-atmosphere reorganizations*.

101. Jeff Severinghaus, SIO, pers. comm. Severinghaus studies the evidence for abrupt climate change in ice cores.

102. P. S. Martin, R. G. Klein (eds.), 1984, *Quaternary Extinctions: A Prehistoric Revolution*. The University of Arizona Press, Tucson, 892 pp.

103. It stands to reason that the ancient extinction play from 12,000 years ago may be having a re-run, but now involving widespread habitat destruction of estuaries, as well as overfishing and overuse of reefs. The human-induced global climate change in progress will continue to add stress to habitat damage and prevent easy recovery when conservation measures are implemented.

FOURTEEN

Abyssal Memories

A THOUSAND HOLES IN THE BOTTOM OF THE SEA

The deep ocean has memories going back 100 million years, and more. Ramming a steel tube into the seafloor is fine for getting samples for the last 1 percent of that. But to get the whole story one needs to use a floating drilling platform, a ship with a huge derrick over a hole in the center of the vessel, to handle a string of pipes armed with a drill bit that eats its way deep into the bottom (fig. 14.1). Sending a steel tube through the central hole of the bit, we can then sample the sediment well below the seafloor. The memories of the ocean are stored within fossil-bearing deposits dominated by calcareous forms: nannofossils and foraminifers. In many places, one also finds siliceous fossils: radiolarians and diatoms. Both planktonic and benthic fossils are represented, and they tell the stories of climate-driven changes in surface waters and in the waters in contact with the bottom.

There are four great themes that emerge from the study of deposits on land and on the seafloor, as follows. Furthest back there is the ancient warm ocean, whose oxygen content was low and whose climatic fluctuations were subdued. Shallow seas were widespread on the continents. In North America, for example, an enormous seaway reached from the Gulf of Mexico into the Arctic. Great marine reptiles dominated the scene in these seas—ichthyosaurs, plesiosaurs, mosasaurs, turtles, and crocodiles. Mollusks were represented by the usual assortment of clams and snails, as well as a rich variety of ammonites and belemnites, with some bivalves building substantial reefs. Then suddenly, about 65 million years ago, came death from the heavens, in the shape of an asteroid, an enormous chunk of rock on an erratic path in the solar system. Instead of falling into the Sun (the most common fate of such rocks) or into Jupiter's atmosphere (the second most common), it hit Earth, setting off calamitous consequences. Devastation was abrupt, global, and thorough. After this horrifying scene came a

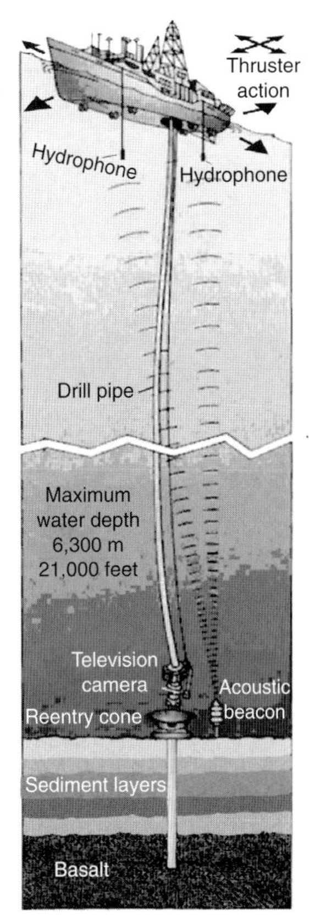

FIGURE 14.1. The recovery of the memories in the deep seafloor requires modern drilling technology.

period of recovery, lasting many millions of years. Again the ocean was warm, even unusually warm in one brief period. Finally there arrived the time of the great cooling, starting 40 million years ago. The shelf seas retreated. Ice started building up at the South Pole. Deep waters became cold. The Drake Passage opened, and strong winds circling the Antarctic drove a great ring of cold water around it as in a merry-go-round. Productivity of the sea responded, and life flourished in the southern ring current. Continued cooling generated strong trade winds and coastal upwelling where these winds blew along the shores. Marine mammals and birds responded with bursts of diversification. Finally, the cooling brought ice to the northern landmasses that surround the Arctic sea, beginning 7 million years ago. Being so far from the pole, however, this ice buildup was highly unstable, and its coming and going helped generate enormous climate fluctuations. The great cooling from 40 million years ago to the ice ages with their waxing and waning northern ice shields: that is the central theme of the Cenozoic (fig. 14.2). It is the period referred to as the Age of Mammals, and it produced the largest organisms on the planet, the great whales.

A NEW WAY OF DOING GEOLOGY

In the deep sea the sediments consist largely of very small fossils, fossils that carry the vital information that allows for the reconstruction of the ocean's history. In contrast, on land, mineral matter dominates the geologic record—we have to "hunt" for fossils. Paleontologists studying deep-sea sediments *process* fossils; they do not hunt them. This fact greatly influences the way they work with the fossil record. Thus, the paleontology of deep-sea sediments tends to have a technical flavor, an aspect that might make it look a bit pedestrian compared with some of the more inspired efforts of traditional paleontology. Large databases are not readily the stuff of brilliant generalizations—there are too many things to be considered.

Among the major feats of discovery resulting from probing the memory of the deep sea, none is more important than tracing the intimate connection between climate history and the development of the biosphere. The historical narrative of relationships between climate change and biological response can now be reconstructed on a scale never before envisioned and with unprecedented resolution in time, for the last 100 million years. We can now find out what a warm ocean really looks like, from studying the minute remains of organisms that once lived in the plankton and on the seafloor. And we can use this information to interpret the fossil record of shallow seas, which has been collected for two centuries and more. The apocalypse that ended life-as-usual some 65 million years ago, as a result of Earth being hit with a 10-kilometer-diameter rock hurtling through space,[1] can be reconstructed in some detail. The great experiment of the Cenozoic

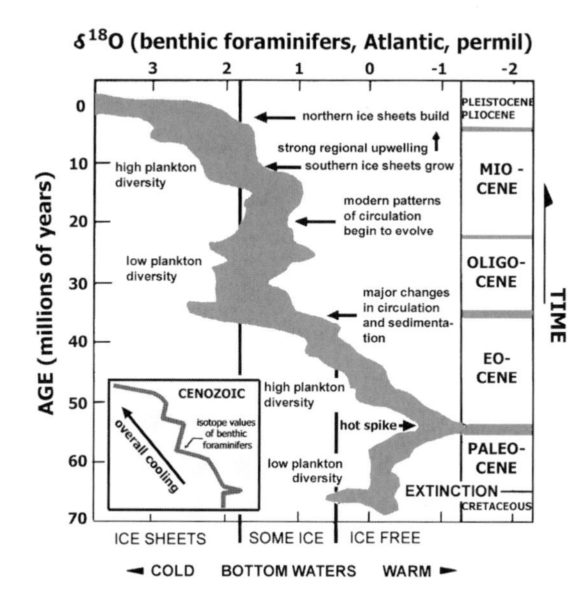

FIGURE 14.2. The great cooling of the Cenozoic as seen in the trend and the steps of the oxygen isotope history of benthic foraminifers on the deep-sea floor (as compiled by Ken Miller in the mid-1980s), based on results of deep-sea drilling.

era—the buildup of ice at the poles—is laid out before us in some detail, as well. Much of the new knowledge gained regarding this experiment has not yet been integrated with the marine record from shelf seas and from land.

The discoveries stemming from drilling go well beyond the realm of fossils, of course. They include a *redefinition of the concept of biosphere.* This redefinition concerns microbes, long appreciated as ubiquitous symbionts within larger organisms and as geochemical agents everywhere in the soil and in seawater, playing a major role in practically all low-temperature reactions, such as making rust, or gas from sewage. We now know that the world of microbes extends many hundreds of meters below the seafloor, in regions with thick sediments, and especially where sediments are rich in organic matter.

Concerning the old question of whether the present is the key to the past or vice versa, we have learned that "our" time, that is the time of human evolution, is entirely unique. The last million years were nothing like any period in the previous 100 million years, ever. The enormous environmental fluctuations associated with the latter part of the Quaternary, ice ages following each other like vast waves on the sea of climate history, driven by the winds of Earth's orbit—they simply have no equal in earlier times. Not in the deep-sea record, that is. Thus, much that we see happening around us is unusual, and not particularly well suited for drawing analogies to explain the more distant geologic past.[2]

What are we learning that is relevant to predicting the future? It depends what one means by "prediction." Since the time of James Watt, inventor of the steam engine, we have entered a situation that is unprecedented in regard to the rate of greenhouse gas addition to the atmosphere and other changes concerning the environment. Thus, we cannot go back and find analogous situations in the geologic record. What we can do is to study abrupt climate change and its effects on the biosphere, and we can reconstruct living conditions on an ice-free planet. Possible effects from abrupt climate change and conditions on a warm planet are issues that are very much under discussion regarding future developments. In this sense, yes, there are lessons to be learned from the ocean's past. But applying them in discussions involving the future is fraught with troubling uncertainty.

As noted when discussing ice age history in the preceding chapter, abrupt change within the more recent geologic past was unfailingly associated with rapid melting or sudden cooling, and involved ice and snow. Collapse of large ice masses, or rapid growth of sea ice and snow cover, are the most likely players in this type of drama. Some abrupt change is "predictable" on the time scale of thousands of years, just like earthquakes are in principle predictable, from their overall patterns, if one does not insist on a precision a hundred times greater than that offered by geologists. Also, we have seen in the last chapter that ice is inherently unpredictable, because its presence creates conditions favorable for buildup and persistence on short time scales, but unfavorable conditions on long time scales. Hence there are opposing trends of buildup and collapse, and we do not know, for any one large ice mass, where the ice mass is positioned in respect to these trends. As the climate warms and the sea level rises, answers will emerge. Many of these answers are likely to be unwelcome.

Abrupt change, while of great interest, is an elusive and difficult subject. We are on firmer grounds when asking how a warm planet differs from a cold one. Mainly, the record shows that the machinery responsible for the present productivity patterns of the ocean is unlikely to persist in the face of general warming. The outlook is for decreased coastal upwelling in the Northern Hemisphere. Furthermore, we know for certain that the amount of oxygen within a warm ocean is much less than in a cold one. The immediate conclusion that can be drawn, regarding deep ecosystems, is that animals that can do with less oxygen will be favored over those that have high energy use: scavengers over hunters, drifting predators over fast-swimming fishes.

The promise of exploring ocean history, then, is to gain insights into the chief elements of different *states* of the ocean environment, as well as the major mechanisms of rapid *change*. It is well to remember, when contemplating such insights, that the time scale studied by geologists is typically measured in thousands of years, rather than in decades.

The great enterprise—the systematic exploration of that last huge white area on the planet in terms of geologic mapping—started in the 1960s. It was formally initiated as the Deep Sea Drilling Project (DSDP), managed by Scripps Institution of Oceanography. The drilling vessel was the *Glomar Challenger*. In 1974, five nations (France, West Germany, Japan, the United Kingdom, and the Soviet Union) joined the U.S.-led effort, which was completed by 1983.

After 1985 and until 2003 the international deep-sea drilling venture relied on the new drilling vessel *JOIDES Resolution*. Its efforts were administered by the Ocean Drilling Program (ODP). The program had its home at Texas A&M, under contract from Joint Oceanographic Institutions (JOI Inc.).[3] Funding came from the U.S. National Science Foundation and from equivalent agencies in other countries. Every two months, give or take a day or two, the drilling vessel (first the *Glomar Challenger* and then the *JOIDES Resolution*) set out for a new expedition, a "leg," to probe the ocean floor in a presurveyed area. Scientists and technicians assembled where the ship happened to berth after the previous leg—San Juan, Capetown, Perth, Guam, San Diego, Las Palmas, or some other port. They then spent the following 8 weeks on the floating drilling rig, in the shadow of the sky-high derrick, surrounded by the reassuring sounds of heavy machinery.[4]

The scientific party of an ODP leg worked in shifts of 12 hours (and often more), seven days a week. The same was true for their support teams (the ship's crew, the drilling crew, and the stewards and kitchen staff). About half of the scientists would be from the United States, the others from affiliated countries: the United Kingdom, France, Germany, Japan, Sweden, Norway, Denmark, the Netherlands, Switzerland, Italy, Spain, and several others. Strong bonds of long-term scientific collaboration were formed during these two months, and lifelong friendships emerged among former strangers. In the memory of the participants, such an expedition looms large forever, as a milestone in their career and their life.

Typically, each expedition of the drilling vessel focused on one of four general topics: 1) clues that allow reconstruction of the regional paleogeography, that is, the horizontal and vertical motions that keep changing the position and the topographic setting of the seafloor, including such complications as triple plate junctions, jumping ridge crests, and clogged subduction zones; 2) the nature of the basaltic basement below the sediment and the messages it bears regarding processes within the mantle and at the seafloor spreading centers, and regarding the history of volcanism; 3) the processes of exchange between seafloor and the ocean, including alteration of volcanic rocks and sediments, expulsion of solutions from sediments, and production of gas in the thick piles of deposits at the continental margins; 4) the record of change of climate and of evolution in the ocean, including changes in water temperature at the surface and at depth, circulation and productivity, extinction and diversification.

For a thorough understanding of the history of the ocean and of the life within it, all these aspects must be studied. The motions must be reconstructed to get the past geography right; the intensity of volcanism bears on the amount of carbon dioxide in the atmosphere and, hence, helps determine the overall climate state of the planet; the recycling of phosphorus and other nutrients is important for the productivity of the ocean. Thus, ocean historians (paleoceanographers) tend to have the broadest range of interests among the many specialists in a scientific shipboard party.

Drilling into basalt is slow, while drilling into sediments (usually) results in hundreds of cores recovered, from as many as a dozen sites during a single leg. Holes can be more than a mile deep; most end about half a mile or less below the seafloor. The material cored within any one hole comes up in plastic liners, within steel barrels. Soon after being brought on board, the core sections become the subject of intense studies by geophysicists, geochemists, and biostratigraphers. In the well-equipped labs of the drilling vessel, scientists and technicians measure density, sound velocity, magnetism, content of carbonate and organic carbon, and composition of the water in the pore space of the material, and paleontologists determine the fossil content of the sediments recovered.

The fossils are minute. To the uninitiated they look (through a microscope) like Christmas ornaments created by an imagination gone wild. To the experts, they look like familiar faces at the office. Each type has its own formal name, and a range in time. When certain species co-occur, the time period for that sediment is fixed. This information is passed on to the co–chief scientists, who decide whether to keep drilling for older material, or whether it is time to pull pipe and move to the next site.

Four great repositories (at Scripps, at Lamont, at Texas A&M, and in Bremen, Germany) received the cores recovered. To keep them safe for ongoing and future investigations, they need to be kept in giant walk-in refrigerators. Computers at the central facility and elsewhere handle the colossal flow of information generated each year by the many teams of scientists involved in deep ocean drilling.

It is difficult, some 40 years after deep drilling started in earnest, to reconstruct how little geologists actually knew about the history of the ocean before *Glomar Challenger* embarked on its first expedition. As far as the new global tectonics, with seafloor spreading and mantle convection and continental drift—all that was already known to the initiated, when the *Glomar Challenger* put to sea in August of 1968, from Galveston, Texas. To be sure, many geologists still were skeptical of the new ideas, and they looked forward to drilling results to remove or justify their doubts. In this sense, the early results of drilling, which strongly supported seafloor spreading, were welcome and were hailed as important achievements.

The true breakthroughs, however, did not come from hunting after confirmation of the new geophysical worldview. The breakthroughs in understanding the living planet came from discovering changes in the types of sediments accumulated by the seafloor through geologic time, and from studying the fossils themselves. From the sediments the history of the ocean emerged with all its wonders—the changing circulation, the changing chemistry and fertility, and the changing life-forms inhabiting it.

A DRILLING VESSEL JOINS THE ACADEMIC FLEET

In the 1960s a small number of geologists—those working with deep-sea cores—were aware of the great potential of the ocean's deep memory bank for the reconstruction of climate history, well beyond the ice ages. One of the big questions was just how the overall cooling in the Tertiary proceeded and how it brought the planet into the ice ages.[5] But the revolution in geologic thinking that started with seafloor spreading entirely dominated the 1960s. It brought very different questions into focus, having to do with the geophysical processes shaping oceans and continents.

Seismic studies at Lamont and at Scripps had shown that the seafloor is layered, and that the upper, sedimentary layer is rather thin. Why should this be so? What is the nature of the "basement" below? Is it basalt or other volcanic products? Or are there ancient hardened sedimentary rocks to be found below the soft cover? Answers would come from drilling. It was for this reason that the four major oceanographic institutions, Scripps, Woods Hole, Miami, and Lamont, combined forces as Joint Oceanographic Institutions for Deep Earth Sampling (JOIDES), in 1964. The agreement, signed by the directors of these institutes—Roger Revelle, Paul Fye, F. Walton Smith, and Maurice Ewing—specified the operation of a drilling vessel, for the purpose of sampling the sediments and rocks below the seafloor.

The JOIDES pact was made in the waning phase of an earlier drilling scheme, invented by several leading oceanographers who had founded the whimsically named American Miscellaneous Society (AMSOC). This was Project Mohole, which pursued the idea of drilling to the lower boundary of the crust to sample the mantle below the seafloor, several miles below the seafloor. The project failed for various reasons, partly political, partly managerial, partly technical.[6] The prospects for success were never really good: the task is exceedingly difficult. Even only drilling into the iron-rich gabbros of the lower crust[7] is a major effort and is possible only where such rocks have become accessible by uplift and thinning of the overburden. An important result of deep ocean drilling, at a very few sites, has been to check on the interpretation of uplifted rocks from the lower crust and upper mantle, in the ophiolite zones of collision belts exposed on land.[8] It is here that we get glimpses of what very deep rocks look like.[9]

One element of Project Mohole that proved very useful was the demonstration that a drilling vessel could be dynamically positioned in deep water and thus drill into the deep seafloor while on station without being at anchor.[10] In April 1961, a large barge (CUSS I) with a drilling rig and four large outboard motors, originally built for oil exploration, drilled the first deep-sea hole off Guadalupe Island, Mexico, in 3,800 meters of water.[11] The drill penetrated 183 meters below the seafloor, recovering pelagic sediments (the oldest being roughly 25 million years old) and several meters of basalt. The experiment was a spectacular success. Scripps paleontologists William Riedel, M. N. Bramlette, and Frances Parker, eager to study the history of life in the ocean, worked on the samples. They instantly grasped the significance of the new opportunities for reconstructing the history of the ocean.

In deep-sea cores, one would find the complete and continuous record of the evolution of marine plankton, free from the frustrations associated with the fossil sequences on land, where erosion tears out the pages of history.

After similarly successful drilling experiments in the Caribbean (Miami's LOCO program, run by Cesare Emiliani) and off Florida (by Lamont), the National Science Foundation encouraged the JOIDES consortium to develop a strategy for systematic exploration of the deep ocean floor. In 1965, marine geologists at the various institutions put together a proposal for "drilling of sediments and shallow basement rocks in the Pacific and Atlantic Oceans and adjacent seas," which was submitted as UCSD Proposal 1581.[12]

Running a drilling vessel is for experts. Scripps found an operator willing to provide both ship and services at an acceptable price.[13] From 1967, Global Marine, Inc., was the new partner in the venture; the name given the drill ship was *Glomar Challenger,* in reference to the legendary H.M.S. *Challenger,* which had established oceanography as a science a century earlier. With the ship made ready, everything was in place to begin exploration. JOIDES advised on the scientific objectives and the scientific staffing of each expedition of the drilling vessel, that is, each two-month leg. An important aspect of the program was a predrilling survey at the proposed site, to make sure the site was suitable for drilling.[14]

The initial JOIDES plan called for nine legs in the 18 months allotted by the National Science Foundation. The first leg was to drill in the Gulf of Mexico, to explore features associated with deep-ocean salt domes there. Next, one leg each across North Atlantic and South Atlantic were to test the new theory of seafloor spreading, by dating basement below the sediment cover. If seafloor spreading was correct, there should be a regular progression of ages away from the central rift. A leg in the Caribbean was to retrieve lots of fossil-rich sediment for biostratigraphy, for comparison with pelagic sediments exposed on Barbados Island, which had been studied in some detail. After transit through the Panama Canal, five legs were to be invested in the Pacific, including north-south transects across the equatorial Pacific. There a thick sediment cover was known to exist, ascribed to the elevated productivity at the equator, due to upwelling of nutrient-rich waters. A far-west leg was to search for old-age crust, at a maximum distance from the spreading center (the East-Pacific Rise).

The vessel sailed in August 1968, from Galveston, Texas. Maurice Ewing and his Lamont colleague J. Lamar Worzel were co–chief scientists. Ewing took the *Glomar Challenger* into the Gulf of Mexico, not far away from the oil fields of Texas.[15]

The ship drilled into the cap-rock of a salt dome in 3,570 meters of water. Salt domes are widespread on the margin off Texas, and they commonly mark the site of gas and oil accumulations. In fact, drilling had to be stopped when sediments rich in hydrocarbons were encountered. The ship was not equipped to explore for oil and gas; it would not have been able to deal with the pressures potentially encountered when tapping into hydrocarbon reservoirs.

The fact that hydrocarbons were encountered was certainly interesting, both from a scientific and an economic viewpoint. But it also raised a flag, right at the start, that deep drilling in the ocean may imply considerable danger. From now on, proposed drilling would have to be approved by a committee of experts familiar with drilling for oil, and familiar with the hazards associated with that activity.[16]

For the second leg, across the North Atlantic, Melvin Peterson,[17] the newly appointed top scientist in the DSDP project, and his associate Terence Edgar served as co–chief scientists. Peterson took the *Glomar Challenger* across the Mid-Atlantic Ridge in the central North Atlantic. The goal was to test the predictions from the hypothesis of seafloor spreading, and to explore the nature of a number of prominent acoustic reflectors in the basin off the East Coast. The reflectors—the layers that provided for strong echoes from below the seafloor—are widespread and show a characteristic pattern. Once their nature was known, it would be possible to make subsurface geologic maps.

To everyone's surprise (and some dismay), the drill encountered hard chert (flint) layers in early Cenozoic sediments, which proved very difficult to penetrate. Chert breaks like thick glass and porcelain, jamming the drill bit with jagged bits and pieces of brittle material. Trying to get through chert is like trying to drill through a thick stack of dinner plates.

The widespread occurrence of Eocene-age chert in the North Atlantic proved difficult to explain. Chert is almost pure quartz, that is, silicon dioxide, but it comes in layers and is clearly of sedimentary origin. The source of the silica can be either volcanic ash or the siliceous skeletons of organisms, including diatoms, radiolarians, and sponges. During the Eocene, was there unusual volcanism in the North Atlantic? Or great diatom blooms? Or deposition of vast amounts of radiolarians? Or an enormous forest of sponges on the deep-seafloor? The answer, on the whole, is yes, all of these factors contributed to making siliceous layers, which turned into chert with time. Eventually, it would emerge that the emphasis on source was probably misplaced when asking these questions. More likely, it was the sinks that changed the silica budget of the ocean: on a cooling planet, large-scale removal of silica by diatoms impoverished the deep sea with silicate, preventing the buildup of this type of deposit after the Eocene. It was the disappearance of chert after the Eocene that was in need of explanation, not its presence in the deposits of the ancient ocean.

For the third leg of DSDP, the *Glomar Challenger* sailed from Dakar, Senegal, to Rio de Janeiro, Brazil, across the South Atlantic, occupying 10 sites and drilling 17 holes, for a very respectable recovery of 761 meters of core. The basic idea was to test seafloor spreading; the geophysicists Arthur Maxwell and Richard von Herzen were co–chief scientists.[18] The leg was a spectacular success. Drilling discovered an excellent match between the age estimates based on the magnetic anomaly patterns on the seafloor (under the assumption of uniform spreading rates symmetric about the Mid-Atlantic Ridge) and the ages obtained by the paleontologists for the oldest sediment found in contact with basement.[19] The skeptics among the geologists (the many who were not willing to take the results of paleomagnetism at face value) now had their hard facts, complete with samples of fossils and basalt.

Leg 3 opened yet other vistas besides showing that the geophysicists had been right about seafloor spreading. Two geologists, Kenneth Hsü and James Andrews, were able to persuade their cochiefs to take lots of sediment cores on the way down to the all-important target, the contact of sediment and basalt. Their diligence resulted in the recovery of hundreds of meters of calcareous chalk, seemingly uniform and boring, but in fact containing a wealth of information about the history of the ocean.[20] The meticulous observations of these two geologists, regarding the degree to which carbonate dissolution on the seafloor had attacked the fossil assemblages, paved the way for reconstructing major changes in the carbonate chemistry of bottom waters through time. In turn, such information greatly aids in the reconstruction of deep-ocean circulation and also offers clues regarding the carbon dioxide content of the atmosphere, for millions of years into the past.

From the record of preservation of calcareous fossils, it emerged that toward the end of the Eocene, during the onset of major mountain building and accompanying fall of sea level, the deep waters became much less aggressive toward calcareous shells accumulating on the bottom. The ocean was moving the main sites of deposition of carbonate on the planet from the shelves to the deep sea, as the shelves were falling dry. Carbonate, in essence, was being dissolved from the limestone-covered regions on land, brought to the sea by rivers, and dumped on the deep seafloor, after conversion to plankton shells.[21]

The geologists on leg 3 showed the way for future drilling strategies. It was possible to satisfy the need for ground truth for the new geophysical knowledge, which called for dating basaltic basement below the sediment layer. At the same time, and without prejudice to this goal, it was possible to recover the fossil record of the history of the ocean contained in the sediments overlying the basement.

Marine geologists and geophysicists were hugely excited about the new results from drilling, and their support quickly translated into a number of extensions granted by the National Science Foundation. Thus, *Glomar Challenger* and DSDP kept working for another 87 legs after the initial 9, till 1983. Almost 100 kilometers of core was amassed. These priceless archives have been kept accessible in the refrigerated core lockers of Lamont and Scripps, which became visitor centers, as geologists from many countries sampled the cores for detailed studies.[22] By 1983, *Glomar Challenger* had traveled 600,000 kilometers in the service of science, equivalent to 15 circumnavigations of the Earth. Along the way, new drill bits had been developed to penetrate difficult layers, hydraulic piston coring operated from within the drill string had been instituted for recovery of undisturbed sediments, and downhole measurements of various sediment and rock properties (logging) had become routine. Success for deep drilling was guaranteed for years into the future, and when the *Glomar Challenger* finally retired, a new drilling vessel, larger and technologically more advanced, soon followed.[23]

WHAT MEMORIES ARE MADE OF

As mentioned, the way the ocean stores its memories within the sediment layers is mainly as fossils, that is, nannofossils, foraminifers, diatoms, radiolarians, and a few others. The bulk of the fossils are remains of plankton organisms; the contribution from benthos is generally minor (but not unimportant). Benthic foraminifers are the main fossil clues to the history of the deep-sea environment. Ostracods (a class of arthropods) are comparatively rare.

The ocean's memory banks are layered deposits on the seafloor—the oldest on the bottom of the stack, the youngest on top, like pancakes on a platter. Overall thickness varies between a hundred meters for extremely slowly accumulating clay in the central Pacific, away from productive regions, to many kilometers in deltas and deep-sea fans building up off large rivers such as the Ganges, Niger, Amazon, and Mississippi. The typical thickness of sediments in the deep sea is several hundred meters, with the higher values in the Atlantic and the lower ones in the Pacific.

The two chief types of sediments rich in fossils are "calcareous ooze" and "siliceous ooze," while the barren "red clay" has neither calcareous nor siliceous fossils but may have some fish teeth. Around the continents, we have "terrigenous mud," which is rich in minerals brought in from land, such as quartz and mica. Some mud is rich in volcanic materials. The admixtures to mud vary, and the name reflects the particular consistency: "calcareous mud," "siliceous mud," "organic-rich mud."[24] As these soft sediments harden, we speak of "chalk" and "limestone," of "opal" and "chert," and of "claystone" and "mudstone."

Calcareous ooze typically builds up at around 1 to 3 centimeters per thousand years, red clay at a rate 10 times lower. Siliceous ooze is in between. A typical rate for the muds in the continental slope is 10 centimeters per thousand years; that is, 4 inches per millennium. When studying a sequence, geologists typically use sampling intervals of between an inch and several inches, which yields information on scales between a few hundred years and a few thousand years, depending on type of sediment. This type of spacing used to be called "high-resolution" sampling, although with time the label has moved out of millennia and into centuries. To a geologist, then, if some big change happens within 100 to 1,000 years, it is essentially instantaneous.[25]

There are three main factors controlling what kind of sediment will accumulate at any one place. Distance from land is one. "Ooze" consists of fossils and accumulates far from land, usually. "Mud" needs a supply of material from land (or volcanoes); it accumulates on continental slope and rise. Another factor is depth: calcareous ooze covers only the shallower half of the deep seafloor. Shells made of calcite (foraminifers, nannofossils) dissolve at great depth, and the distribution of calcareous ooze has a lower depth limit, therefore. So we can speak of a "carbonate line," in analogy to a snow line in the mountains. (The technical term is *carbonate compensation depth.*) Because of dissolution at depth, carbonate ooze is typically found on the flanks of the Mid-Ocean Ridge and on top of submarine plateaus and seamounts, but not in the deep parts of basins. The third factor is productivity: siliceous fossils are abundant below areas of high production, which send plenty of diatoms and radiolarians to the seafloor. Elsewhere, while siliceous fossils are supplied in some measure also, they are largely dissolved. Red clay is what remains after both calcareous and siliceous fossils have been dissolved. Thus, clay is typical for depths below the carbonate line and away from productive regions.

There is one more item to consider before we can go ahead and interpret the patterns of the different types of deposits found in a drilled sequence. The carbonate line is not just a function of depth. Its depth level differs quite a bit between North Pacific and North Atlantic, being shallower in the Pacific by a mile or so. The reason is that the deep water in the North Pacific is slightly more acidic than that of the North Atlantic, owing to the higher carbon dioxide content in the deep North Pacific that comes with age. The "young" water in the deep North Atlantic has much of its original oxygen content, recently acquired at the surface. But the "old" water of the deep North Pacific has lost much of its oxygen on the long trek north from the Antarctic, and this lost oxygen has gone toward making carbon dioxide. Old water rich in carbon dioxide dissolves carbonate more easily, and so the position of the carbonate line frozen within the sedimentary record can tell us something about the past age distribution patterns of the deep water, that is, the way it moved through the basins.

With this tool kit of concepts, which was largely stocked by John Murray of the *Challenger* expedition and was well in place in the 1960s,[26] we can readily interpret simple patterns of sediment succession. For example, let's say the drill penetrates red clay and then solidified carbonate ooze, called chalk, before it hits basalt. We have a history involving two regimes: the older one shallow and low production, and the younger one deep and low production. This is exactly the sequence expected for sediment accumulating on a subsiding flank of the spreading center, in a central (low-production) region.

Now let's take a more interesting sequence. For example, let's say the drill first penetrates through red clay, then through siliceous clay and siliceous ooze, then through siliceous calcareous ooze, and then through solid calcareous material called chalk. The history (starting at the bottom) would prescribe changing regimes starting with shallow depth and low production (chalk) and going to shallow depth and medium-high production (picking up silica), to greater depth and high production (losing carbonate, retaining silica), to less production (less silica), and finally to great depth and low production (neither carbonate nor silica). Again, we have the overall increase with depth that is the hallmark of the subsiding ridge flank. But we also have the crossing of a high-production zone, at the same time. The place where such a sequence was discovered is north of the equator, in the eastern Pacific.[27]

The deep-sea sequences that are recovered by drilling, then, can be properly interpreted only if we allow for plate motions, both vertical and horizontal.[28] We might wish to look for a place that had very little vertical motion through time, and that did not move across the equator but stayed in a low-production zone. Such a sequence would reflect the changing chemistry and general fertility of the sea, rather than the changing regional production patterns and the motions of the seafloor. Fortunately, there is a place bearing such a sequence. It is called the Ontong Java Plateau, and it is located east of New Guinea and just south of the equator. When drilling the thick stack of deposits on top of the plateau (more than 1 kilometer of sediment draped over flood basalt), we find pure carbonate sediment all the way down till we reach deposits 40 million years old. Below that, we find alternating layers of chert (which originally started out as siliceous ooze) and limestones (which started out as calcareous ooze). Interestingly, this major shift in sedimentation patterns, which happened within the late Eocene, seems to reflect a general pattern that is seen in sequences in many different regions of the deep ocean, where pelagic sediments go back that far.[29]

The simplest explanation for the late Eocene sediment shift is that during the period containing the shift (called the Auversian, after a region in France), processes started to come into play that extracted silicate from the sea, which made it difficult to produce or preserve siliceous sediments in the open ocean. In addition, more carbonate became available. From then on, after 40 million years ago, the deep sea accumulated

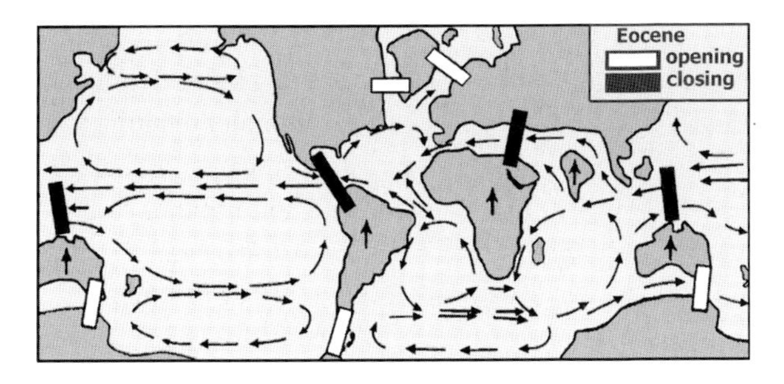

FIGURE 14.3. Seaways opening (white markers) and seaways closing (black markers) changed the circulation of the ocean in the Cenozoic. The straits at Tasmania and Drake Passage *(arrows)* are crucial in the history of the silica budget, which is dominated by extraction in the circum-polar current.

pure carbonates on its ridges and their flanks and on deep-sea platforms, because the silicate-starved deep water dissolved any siliceous fossils arriving on the seafloor. Siliceous sediments became much more restricted in their distribution and thus became markers for high-production areas (such as the equatorial zone in the Pacific and the great southern ring current around the Antarctic continent). Radiolarian skeletons started to get thinner, beginning at the end of the Eocene, as the extraction of silicate proceeded.[30] What had happened to change the chemistry of the ocean in this fashion?

What happened is that cooling set in—associated with the uplift of the Himalayas, accompanied by an overall regression of the sea, which exposed land areas that were covered by shallow seas up to that time. Thus the global transfer of carbonate from shelves to the deep sea was initiated, which accounts for the increased deposition of carbonate. The absence of the silica is a result of both the cooling of Antarctica and the opening of the seaways south of Australia and south of South America (Drake Passage) (fig. 14.3). The cooling produced strong winds, and the opening of seaways allowed development of a strong circumpolar current, where deep mixing supplied surface-living diatoms with dissolved silicate in summer. Thus, the Antarctic Ring became the production center for diatoms, and the associated sediment became the chief depository for silica, while the rest of the ocean was (and is) starved of silicate. Such are the types of memories contained in general sediment patterns.

The overall evolution from ancient to modern sediments, with the pivot point near 40 million years ago, reflects a shift from a warm world dominated by shelf seas to a world with continents that are largely emerged and have narrow shelves, that is, our world today. The shift to a new world without shelf seas initiated a general cooling of the planet.

Geography underwent other serious changes, as well. The ancient seaway that connected all ocean basins in the tropics—the Tethys—started to close (fig. 14.3). It had been running from the Pacific between today's India and China into the region of today's Alps and Apennine mountains and from there into the Atlantic and then back to the Pacific through the open gap between the Americas. Now, during the time of the great shift, India was slamming into Asia, closing the ancient seaway between these two very unequal continents, in a collision that would result in the uplift of the Himalayas, along with a general deepening of the ocean.

The closure of the Tethys was long completed when finally the Panama land bridge formed, denying to the Pacific the warm Caribbean waters that had previously bathed the shores of Baja California. With this closure of Middle America, starting some 10 million years ago and completed by 3.5 million years ago, the modern ocean acquired one of its most prominent characteristics—a feature well reflected in deep-sea sediments. It is the striking asymmetry in the deep-water properties of North Atlantic and North Pacific, expressed in a large depth difference of the carbonate line, with the Atlantic gaining the deeper line and thus acting as a carbonate trap. In addition, the deep Atlantic became poor in silica deposits, while the deep Pacific gained such deposits.

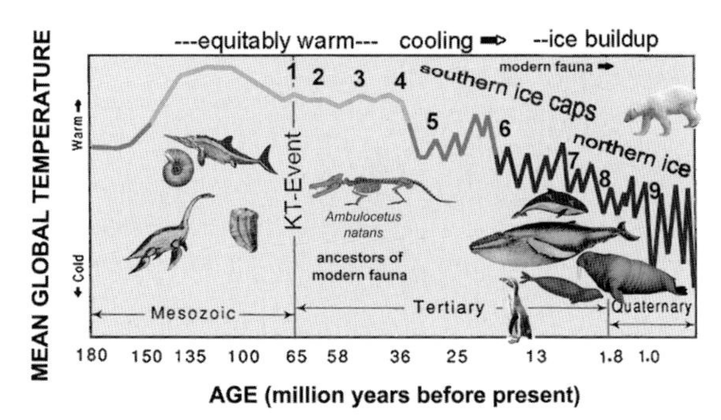

FIGURE 14.4. The modern fauna of marine mammals and birds evolved in a cooling ocean. The Mesozoic seas bore warm water creatures (ichthyosaurs, plesiosaurs, ammonites, and rudists as shown) that became extinct at the KT-Event (1). Others persisted and entered the earliest Tertiary (2). The onset of cooling in the Eocene (3), with a sudden temperature drop at its end (4) led to the buildup of ice and a corresponding drop in sea level (5). Twenty million years later (6) large ice sheets dominated Antarctica, and another 17 million years later (7) ice sheets expanded in Greenland, Arctic Canada, and Scandinavia. The evolution of marine mammals and birds is closely tied to this history of cooling steps, which resulted in the generation of highly productive regions in the ocean. The expansion of productive regions drove evolution from land-based mammals (note ancestral whale [3]) to fully marine forms (7, 8, 9).

TRENDS IN CLIMATE AND EVOLUTION

The overall cooling of surface temperatures on the planet, from the late Cretaceous to the present (the last 80 million years or so) has been part of textbook geology for quite some time (fig. 14.4). Likewise, it is conventional wisdom that this trend was interrupted by warm periods in the early Eocene and the early to middle Miocene, centered on 45 million and 15 million years ago, respectively. What is new, thanks to drilling, is the discovery that the climate cooled in distinct steps. Also, it turned out, the biota of the sea (as far as this can be read from the record) responded in rather surprising ways that could not have been predicted from the general trends alone.

The cooling largely proceeded in high latitudes.[31] In the 1970s, two teams, each with a deep-sea paleontologist and an isotope expert, made sufficient observations for a first comprehensive picture of climate trends in the Cenozoic, based on samples from deep-sea drilling.[32] One of these teams (Robert Douglas and Sam Savin) concentrated on low latitudes, the other on material from high southern latitudes (James Kennett and Nicholas Shackleton).

When comparing the records resulting from their work, a surprising fact emerges: for both the planktonic and benthic fossils from the earliest Tertiary, including the early Eocene, the low-latitude values from the central Pacific found by Douglas and Savin match very well the high-latitude values reported by Kennett and Shackleton[33] (fig. 14.5). In fact, there is no difference in the oxygen-18 values, and some of the plankton oxygen isotope values are actually "lighter" (that is, are more enriched in oxygen-16) in the high-latitude samples than in those from low latitudes. How could this make sense? Surely, in high latitudes the water was colder than in the central parts of the ocean, and we should see "heavier" values in high latitudes, since in cold water shells generally have a higher proportion of oxygen-18, compared with shells from warm water!

It was undoubtedly colder in the sub-Antarctic than in the central Pacific, always. Thus, to explain this discrepancy, we cannot call on temperature but must invoke precipitation: there was a rain belt associated with the rim of the

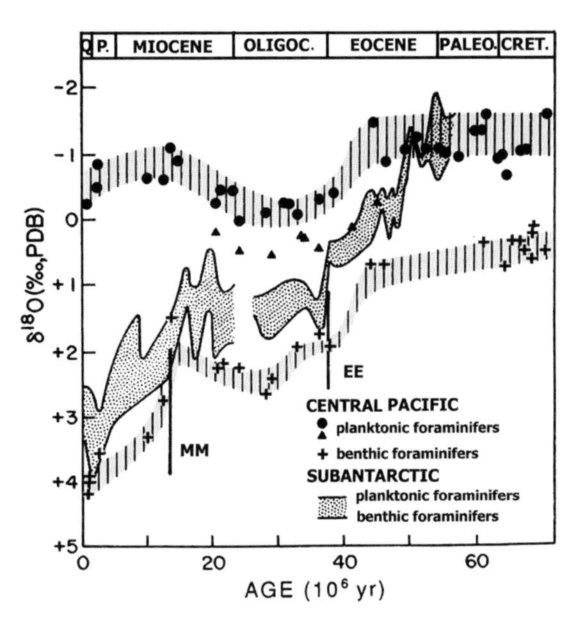

FIGURE 14.5. Inventory of climate change in the Tertiary ocean, in 1975, by R. Douglas and S. Savin (central Pacific) and by J. Kennett and N. Shackleton (sub-Antarctic), based on oxygen isotopes in foraminifers. Note overlap of plankton values in high and low latitudes in the early Eocene, and separation upon cooling in high latitudes.

Antarctic.[34] The necessary influx of freshwater in the rain (which is enriched with oxygen-16) ultimately came from the tropics. The associated flux of latent heat from the tropics toward the poles explains why high latitudes were warm in the early Tertiary. Fog-shrouded rainforests could develop under such conditions of high rainfall, along the coastal regions of Antarctica. We can see similar forests today, in southern New Zealand (along with pocket-size mountain glaciers and some boisterous parrots).[35]

Since about the mid-Eocene, based on these early data sets from the 1970s, high-latitude temperatures kept falling, on the whole (fig. 14.5). The oxygen isotope values of deep-sea benthic foraminifers (which are much the same for both series) show this trend especially clearly in subsequent compilations (fig. 14.2). The cooling trend is obvious in three series, then: in the high-latitude plankton record and in both the high-latitude and the low-latitude benthic records. Planktonic isotope records in low latitudes, on the other hand, did not change much, except for a tendency toward somewhat heavier values in the Oligocene. Thus, the tropics and subtropics stayed warm despite planetary cooling, which profoundly affected the deep ocean. The overall cooling trend is a high-latitude phenomenon and results in an increasing planetary temperature gradient during the entire Tertiary.

As mentioned, the striking thing about the overall cooling is that it occurs in marked steps. Fundamentally, the reason for a steplike response of the climate system to overall cooling indicates the crossing of thresholds. The main physical threshold in cooling is the transition from water to ice. The main geographic threshold is the point where large areas become available to receive and hold ice and snow. Together, these two factors produce sudden expansion of whitening at sea and on land, thereby greatly increasing the reflection of incoming sunlight, and thus accelerating the cooling process.

The first such step is in the latter part of the Eocene, and across the Eocene-Oligocene boundary. Kennett and Shackleton interpreted this shift as indicating the initiation of coldwater supply to the deep ocean.[36] To activate the cold-water sources around Antarctica, the rainfall in the region had to move way offshore: seawater with a high freshwater content cannot sink to fill the deep ocean. After the rain moved off, cooling could then proceed to make deep water, further aided by the deep mixing that set in around Antarctica after opening of the Drake seaway (mixing set the salinity toward average values). Also, sea level must have fallen at the time of this step, as the timing coincides with the aforementioned shift of carbonate from shelves to the deep-sea floor. Without the heat from the summer rain, snow and ice could now spread out down the flanks of the mountains, which provided an opportunity for the buildup of significant ice masses on Antarctica, forcing more cooling, more winds and mixing (from the stronger temperature gradient) and reenforcing the drop in global sea level. The shift in sedimentation patterns near 40 million years ago, seen all across the ocean, thus emerges as a signal not just of changes in the chemistry of the ocean, but of major shifts in the climate patterns of the planet.

The end of the Eocene, with the strong cooling of deep water, was a time for extinction of benthic organisms in the deep ocean; both foraminifers and ostracods (small shelled arthropods) were impacted.[37] We must assume that many or most of the other deep-sea organisms, albeit without shells and therefore without a record to study, were similarly impacted. This end-Eocene extinction, then, is the chief reason why the present-day abyssal fauna is not dominated by remnants from the Cretaceous but has fully modern aspects. The best places to look for such "living fossils," if we follow this logic, is not at abyssal depth, but in subsurface waters where low oxygen content of the water gives the edge to organisms that recall, in their genes, the low-oxygen conditions from ancient warm oceans. Candidates that come to mind are the lobe-finned fish *Latimeria* and the shelled cephalopod *Nautilus*, for example, and a host of eels breeding at midwater depths.

The deep ocean turned cold at the end of the Eocene, and glaciers grew on the Antarctic continent, but the time for large ice sheets on the lowlands and shelves of Antarctica had not yet arrived. It came more than 20 million years later, with the buildup of major ice caps in the middle Miocene, a climate step marked by a large drop in deep-water temperatures and by a general increase in oxygen isotope ratios in both benthic and planktonic fossils. Interestingly, in low latitudes there is no evidence for cooling at the time. To the contrary, temperatures held steady or increased somewhat. Perhaps the overall increase in zonal winds that accompanies high-latitude cooling enhances the contrast between tropics and polar regions. The fact that there is a step in ice buildup in the middle Miocene suggests that a large area suddenly became available as a base for such ice buildup; presumably, the lowlands and shelves of Antarctica provided that foundation. In the far north, with no land near the pole, the cooling could only produce sea ice and perhaps local mountain glaciers next to Arctic shores, as had happened already since the end of the Eocene, on a modest scale.

The high-latitude cooling, and the overall retreat of the sea from the low-lying portions of the continents had important consequences for the stratification and circulation in the ocean, for the intensity and distribution of upwelling, and for the chemistry and fertility of the sea. The increasing temperature gradient that resulted from the differential cooling (warm tropics, cold high latitudes) forced an overall strengthening of the zonal wind field. In turn, strong zonal winds (that is, winds blowing parallel to latitudes) forced strong zonal currents that thickened central gyres and powered narrow upwelling regions. These various changes affected the availability of nutrients to the sunlit zone to which production is confined, and the stability of the environment of growth of the plankton. The plankton responded with extinctions, and with the rise of new species able to take advantage of the new opportunities.

Evolution of the plankton is gradual at times and jumpy at others, reminiscent of the "punctuated equilibrium" concept urged by Niles Eldredge and Stephen Gould.[38] As far as the deep-sea record, it appears that the times of rapid change in both plankton and benthos are linked to rapid climatic change, except in special circumstances such as the catastrophe at the end of the Cretaceous. Thus, there are no mysterious internal forces of evolution at work here, as far as we can tell. Instead, the organisms adapt to external change, and such change proceeds in jumps for physical reasons, including thresholds in ice formation, thermocline generation, and primary productivity.

The rules of change of the deep-sea fossils are subtle but subject to observation because of the continuous records recovered by drilling. Radiolarians, a highly diverse group of planktonic organisms, when present, provide excellent clues to the age of sediments, since they evolve quite rapidly. Scripps paleontologist William Riedel found, for the Cenozoic, that the sense of evolution for some lineages follows some simple pattern (say, lengthening of spines) that is valid for many millions of years. This suggests that particular forces driving evolution are active for extended time spans, which would be the case if planetary cooling were at fault. We have mentioned one obvious example, that is, the evidence for silicate starvation of the sea. It results in an overall and progressive thinning of radiolarian skeletons within the Cenozoic. Quite obviously, Eocene sediments have abundantly thick and heavy radiolarians, while modern ones are thin and delicate. As pointed out by Ted Moore, then a graduate student at Scripps in the 1960s working with Riedel, this thinning may reflect a successively decreasing availability of silicate.[39] His suggestion remains eminently reasonable. Today, one-half or more of the silica lost from the ocean each year ends up in the sediments around Antarctica,[40] supporting the hypothesis that the evolution of the Circumpolar Current is the ultimate reason for the increasing shortage of silicate, which helped drive the evolution of radiolarians (and of pelagic diatoms, of course).

Planktonic foraminifers, being very abundant in the sediments but having relatively few species, are especially easy to categorize in terms of diversity. Within the Tertiary, their diversity changes greatly. In the earliest Tertiary, certainly, their diversity is low because of extinction during the end-Cretaceous catastrophe. But by the middle of the Eocene they had regained their former splendor, with a diverse fauna. This high diversity allows rather detailed subdivision of the middle Eocene into successive intervals with different fossils, on a scale of about 1 million years. At the end of the Eocene, diversity plummeted, and it stayed quite low during the entire Oligocene, both among planktonic foraminifers and among nannofossils. It is hard to tell time within deep-sea sediments of this epoch. At the transition from the Oligocene to the Miocene, there is again strong diversification. Also, within many species, there is an increase in morphological variation at that time. This suggests a proliferation of pelagic habitats.

It is not so difficult, perhaps, to rationalize the burst of speciation in the early Miocene that led to the rapid increase in diversity in the plankton. An answer was suggested early on, even before the relevant drilling results were in, since this time of speciation (radiation) was known from sediments taken from uplifted pelagic sediments. The paleontologist Jere Lipps at the University of California at Berkeley proposed that the proliferation of new species had something to do with establishing a thermocline, which stratified the habitat of plankton.[41] This proposition has become generally accepted and now provides the rationale for reconstructing thermocline history. Emiliani (in 1954) had shown earlier, based on isotopic analysis of the shells of different species of planktonic foraminifers, that these organisms do inhabit slightly different depths in the upper parts of the water column. Regarding this discovery, he wrote as follows: "I should say that the discovery that forams are stratified came as a big surprise to me, like the cyclic nature of the isotopic record downcore. But then again, surprises in science are usually heralding interesting discoveries. If one dismisses the surprising evidence, one surely misses out on the discovery."[42]

Besides stratification, there is yet another way to increase the diversity of pelagic habitats in order to provide for speciation. This is to increase the distinction between low- and high-production regions. Naturally, when a thermocline forms at the base of the sunlit layer, one finds both a stronger difference between surface and subsurface waters and greater horizontal contrast. The reason is that once a thermocline is not far below the ocean surface, it makes a great difference to the level of production just how thick the overlying mixed layer is, 100 meters or 10 meters. In contrast, if the thermocline is weak or far below the mixed layer, no such differences arise. Thus, thermocline development must have greatly influenced plankton evolution, both through stratification and through geographic differentiation.

Indeed, there is evidence that the thermocline, the transition between warm surface water and cold upper deep waters, became stronger beginning in the Miocene, and shallow enough to affect planktonic foraminifers, as seen in their isotopic compositions.[43] All sorts of life-styles open up when cold- and warm-water layers are close together, with a relatively narrow transition zone. Vertical migration as we know it today presumably started then. This greatly changed all interactions between species, while rearranging the food webs. Vertical migration, representing flexible behavior that is readily adjusted to new requirements and opportunities, opened up entirely new ways of making a living, thus encouraging rapid diversification. Similarly, in larger animals, horizontal migration between low-production areas and high-production areas would have provided for new opportunities and consequent speciation.

Habitat diversity implies species diversity. But what prevented the buildup of a biologically effective thermocline in the Oligocene, in low latitudes? Were there too many competing deep- and intermediate-water sources? Winds not strong enough to drive thermocline waters from high latitudes into the tropics? An excessively thick mixed layer? The problem is still waiting to be solved. Perhaps there was no fundamental change at the end of the Oligocene, in terms of physics, but only enough change to bring the top of the thermocline into the sunlit zone. This would satisfy Sverdrup's requirement that for plankton to grow, more time has to be spent within the sunlit zone than below it.[44] When this requirement was met, plankton productivity responded favorably.

One other trend that surely affected the evolution of plankton and other marine organisms was the increasing isolation of tropical regions from each other, as the ancient seaways of the Tethys began to close. In early Cenozoic time, the tropics were well connected. But in the Miocene these old connections broke down and new ones opened, in high latitudes. In addition, as we see in the benthic foraminifers, cold water at depth now connected the high-latitude regions of the two hemispheres. Thus, opportunities arose for tropical provinciality (most prominently expressed in the difference between Caribbean species and their cousins on the other side of the Isthmus of Panama) and for greater connectivity between cold-water faunas.

With the closing of the Tethys, the ocean's circulation system moved slowly from a warm-ring-dominated world toward the present cold-ring condition, where the waters of the ocean are well mixed at low temperatures, around the Antarctic.[45] The most common water is now cold, while thin warm-water lenses, precariously perched on their cold base, are all that remains of the once pervasive warm-water sphere. As a result, diversity has soared in low-latitude plankton, especially among the diatoms for which the newly wind-generated upwelling regions provided entirely new habitats.[46]

Without the detailed information from deep-ocean drilling, we would be at a loss to understand the history of baleen whales or of penguins (fig. 14.6), both of which depend on upwelling. The same is true for many other animals in the sea, including siliceous sponges making reefs around Antarctica, deep-water corals in the North Atlantic, giant amphipods in the deep sea, squid and lanternfish in the deep scattering layer, and the many herringlike fishes that use gill rakers to strain plankton. They all depend on a water structure, a circulation, and a geochemistry that is geologically quite young, as we now appreciate. Their stories begin in the late Eocene.

THE WARM WORLD OF ICHTHYOSAURS

When we think about the Mesozoic seas—especially those of the Jurassic and Cretaceous epochs—we are likely to come up with mental images of shelled cephalopods called ammonites, and of the large diving reptiles called ichthyosaurs. These images are deeply embedded in the textbook tradition of geologic history. There are good reasons that this should be so: the ammonites are the fossils most useful in telling time in the marine deposits of the Jurassic and the Cretaceous now exposed on land, and the

FIGURE 14.6. Penguins depend on highly productive waters in the Southern Hemisphere. Their evolution is linked to a cooling planet, with upwelling around an ice-covered Antarctic continent.

diving reptiles left spectacular fossils that caused a tremendous stir when they were first discovered, early in the nineteenth century. In addition, ammonites and ichthyosaurs well represent the fact that cephalopods were ubiquitous in Mesozoic seas, and that reptiles ruled the world.

However, what made the Cretaceous truly different from our own time is not so much the presence of cephalopods and marine reptiles (we have these today in squid and turtles), but the fact that this ancient world was warm and had an ocean that reached far into the interior of continents. For example, in the late Cretaceous a broad swath of west-central North America was a seaway from the Gulf of Mexico all the way north to the Arctic Sea. Before drilling, practically all the information we had on Cretaceous seas was from the deposits of this type of inland sea, on several continents.

These seas were truly extensive. Thus, the Cretaceous period left us a legacy of vast carbonate deposits—the chalks that are so closely associated with the epoch, even in name.[47] In addition, there are enormous sequences of black shale, source rocks for petroleum. Cretaceous oil and gas powers the world's economies, much of it extracted from the reservoirs in the Middle East. These treasures are bequests from a fertile Tethys Sea that piled up organic-rich deposits in certain oxygen-starved upwelling areas. On land, low-lying swamps bordering shelf seas produced organic-rich deposits that were to turn into coal-rich sediments (fig. 14.7). Vast areas in the Rocky Mountain region bear such coaly layers. The coal is low in sulfur, showing that seawater was largely kept out.[48]

The question that naturally arises from this impressionistic picture of the Cretaceous as an epoch of carbonate and carbon-rich deposits is this: where does all this carbon come from? The most likely answer is that the ultimate source of carbon—volcanic activity—was turned up quite a few notches compared with the Cenozoic. Volcanism was vigorous in the middle of the Cretaceous epoch. We find huge outpourings of basalt in the western Pacific: Ontong Java Plateau, Manihiki Plateau, Shatsky Rise, and Hess Rise are large volcanic edifices built at that time.[49] Dredging and drilling the basaltic edifices, and dating them, shows that they were built in a period centered around 100 million years ago. For tens of millions of years volcanism was intense. Volcanism delivers carbon dioxide to the ocean and atmosphere both directly and also indirectly through exhalation of acids that react with preexisting carbonates thus releasing carbon dioxide.

FIGURE 14.7. Typical marine deposits of Cretaceous age, exposed on land. The photo shows finely laminated black shale from oxygen-deprived shallow seas (southeastern France). Note the Jura Mountains in the background.

Because of the intense volcanism, we must assume that carbon dioxide values in the atmosphere were much higher than now. Such a scenario also readily explains the unusual warmth on the planet, from the greenhouse effect. As far as can be ascertained, there was little or no ice at the poles or anywhere else.

Obviously, given the fact that the planet is warming at present, and is projected by serious geophysicists to do so for the foreseeable future, we would like to know just how the climate system of the Cretaceous is different from that of today. What precisely was the role of volcanism in modifying conditions, including its role in the changing concentration of greenhouse gases? How was life different? We are interested not so much in the contrast between having ichthyosaurs rather than dolphins, or ammonites rather than squid, but in the chemistry of the ocean, concerning the rates of recycling of nutrients that govern the level of primary production on land and in the ocean, and in the mechanisms sustaining diversity in the marine realm.

There is no ready guide to the task of answering such questions, and the path through the thicket of information from the record is tortuous, to say the least. One useful approach to the complex of problems is to ask what is the single most important feature in the present ocean, concerning life in the open sea, and what is the equivalent dominant feature in the Cretaceous ocean. The answer for the present ocean is "the thermocline," and especially the fact that it is shallow enough to interact with the sunlit zone. This interaction informs the major patterns of productivity, as recognized around 1950 by Gordon Riley and Harald Sverdrup, among others.[50] Furthermore, its interaction with shelf seas controls living conditions on the shelves.

The answer for the Cretaceous ocean is "the extent and variability of the shelf seas" and "the extent and variability of the dead zone over the shelf seas," where the "dead zone" is a subsurface layer without oxygen. The extent of shelf seas is important because they can act as nutrient traps in estuarine settings, robbing from the deep ocean what is valuable and leaving much of the open sea to function as a desert, with reefs as oases. The extent of oxygen depletion is important because this controls whether benthic organisms can thrive and where, whether reefs can survive, and whether there are fish kills and harmful algal blooms.

The transitions, in the Mesozoic shallow seas, from oxygenated to sulfurated environments, back and forth, left a record in the episodes of black mud deposition, now found as black shale commonly loaded with sulfide minerals (mostly pyrite—fool's gold). The sulfurous aspect comes from the fact that when oxygen runs out, sulfur bacteria take over. They are able to use the abundant sulfate in seawater as an oxygen source. As the bacteria strip the oxygen from the sulfate, the water becomes poisonous with hydrogen sulfide (poisonous, that is, to animals that live on oxygen, not to the sulfur bacteria).

In the Cretaceous, deep waters were not especially cold. We know this from the isotopic chemistry of benthic foraminifers recovered by deep drilling.[51] In fact, for much of the Cretaceous, deep-water temperatures were around 15 °C (60 °F), with a cooling by a few degrees after the middle Cretaceous. Thus, the warm layer on top of the ocean was not an especially efficient lid preventing mixing with deeper waters. But zonal winds were much weaker than today, since the overall temperature gradient between high and low latitudes was weak. It is this gradient that drives zonal winds. Thus, rotation of central gyres (and associated eddy fields) was sluggish, and clearly defined upwelling regions were less well developed than today.

Storms stirred the ocean and brought up nutrients, generating plankton blooms. The storms, like today's tropical cyclones, were distributed over wide regions and could strike anywhere, at different times. The resulting disturbance patterns, like rainstorms in a desert, helped sustain diversity at a high level, through unpredictable changes in opportunity for growth, both in the plankton and on the seafloor below. Over large areas in the tropics and in high latitudes, the salinity of the surface layer was somewhat reduced by strong rainfall, and this low salinity maintained the stratification that is vital for growth. Along the Tethys and in shallow seas beyond the desert zone, high precipitation set up estuarine circulation within large regional seas, creating productive basins with organic-rich sediments and opaline deposits (that turned to chert with time). When the rainfall zones shifted out of such regions, long-term drought set in, and carbonate then accumulated. We see the result today in many places bearing Cretaceous shallow-water rock layers: alternating limestone and chert layers, and reefs contrasting with oil-bearing deposits.

The recycling of nutrients was greatly influenced by the limited availability of oxygen in deep waters. Warm waters hold less oxygen than cold. The renewal of deep waters in the Cretaceous owed much to sporadic outflow of saline waters from a multitude of shelf seas, with only a modest content of oxygen, in contrast to today's deep-water sources in high latitudes, where cold and oxygen-rich water sinks to depth at a high rate. One of the corollaries of oxygen depletion in intermediate waters, from oxidation of organic matter produced in coastal waters, is the reduction of nitrate, a vital nutrient of the production cycle.[52] Nitrate-depleted waters, when brought back to the surface into the light, cannot sustain growth of algae, except for that of cyanobacteria. These photosynthesize while binding nitrogen from dissolved nitrogen gas, of which there is plenty, supplied from the atmosphere.

Oxygen-depleted waters today are prime agents of fish kills under certain conditions. Perhaps we should visualize the large flying reptiles as vultures patrolling the edge of the sea in search of recent killings from oxygen depletion and related algal poisoning. Diving reptiles, such as ichthyosaurs, would have been little impacted from lack of oxygen in the water, since they breathed air. Thus, they had little competition when scavenging for dead ammonites and fishes on the seafloor, in regions of foul water, or when traversing foul water to find better hunting grounds.

What is the likelihood that the above life habitat scenarios are correct? Better than fair. Lack of oxygen as a major player in all marine ecosystems at that time is suggested by the presence of black shale layers and finely bedded marl layers in many parts of the Cretaceous (fig. 14.7). We can draw analogies with present-day fish kills in Walvis Bay[53] and in the Baltic Sea.[54] Additional clues come from the distribution of greensand layers and other iron-rich sediments, as well as pyritized fossils, in warm seas in general, from the middle Jurassic to the upper Cretaceous.[55] Also, in some places in the deep ocean, cyclic changes in sediment color point to high sensitivity of the deep-sea environment regarding the rate of deep-water replacement and, hence, oxygen content (fig. 14.8).

Thus, the supply of oxygen was likely crucial in determining the living conditions in the warm Cretaceous seas, over tens of millions of years. Fortunately, the results from deep ocean drilling provide many relevant details in the history of oxygen availability within the Cretaceous epoch. This allows us to appreciate, for the first time, the profound differences in marine life habitats on a warm planet compared with one (ours) that is cold at the poles. Evidence for low-oxygen conditions is especially abundant within certain well-defined periods, which gave rise to the concept of oceanic anoxic events.[56] Typically, evidence for such events was found in the Atlantic, but some evidence was found in the Pacific and in the Indian Ocean also. The term *event* may be somewhat misleading, since the anomalous conditions apparently persisted for considerable time spans. (The Aptian-Albian event lasted more than 10 million years, the Cenomanian-Turonian event 1 to 2 million years.) The reality of the oxygen stress in deep waters, on a global scale, is strongly supported by studies of the record of carbon isotopes, within foraminifers.[57]

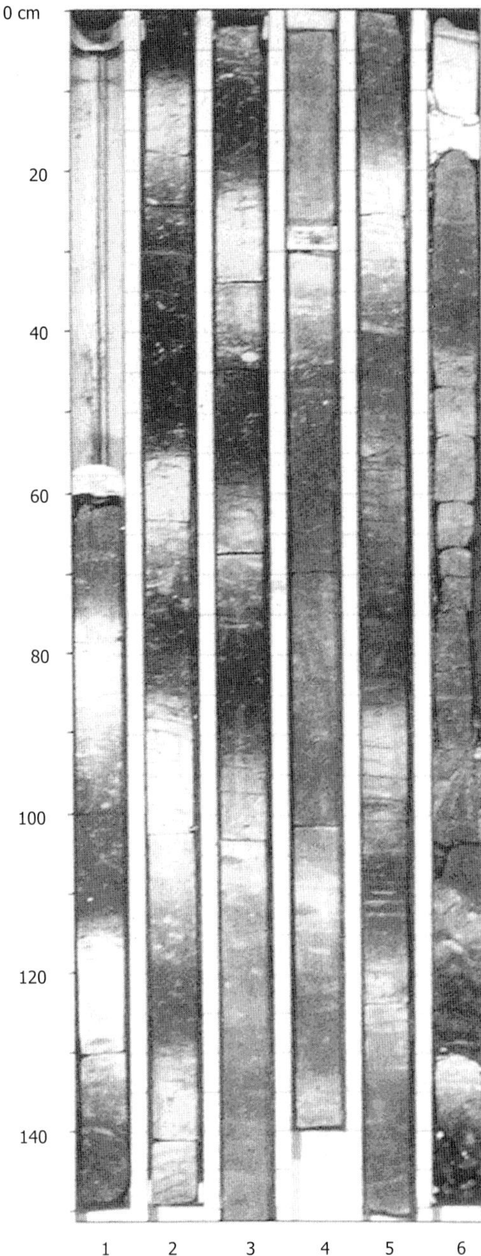

FIGURE 14.8. Cyclic changes in calcareous deep-sea sediments: evidence that subtle changes in seasonality (tied to the eccentricity of the orbit) produced cycles in deepwater production and ocean productivity in the Late Cretaceous.

One potential cause for the observed oxygen crises (as seen in black shale deposition in the deep sea) is the outpouring of flood basalts in the middle Cretaceous. A connection between Pacific volcanism and periods of anoxic intensity was proposed in the 1980s, by a number of geologists.[58] However, the idea apparently attracted little attention for the rest of the decade.[59] In 1991, the geophysicist Roger Larson took the topic up again and pointed out the coincidence of major volcanism, anaerobic deposition (and petroleum formation), and sea level change, in pregnant fashion.[60]

Without paying attention to oxygen stress in the Cretaceous, we might not understand some of the evolutionary history of today's marine fauna. For example, the reason why certain representatives of the eel family mastered the freshwater environment, and manage to travel between ponds on land, may go back to their deep-sea ancestors in the Cretaceous, which had to be able to close up gills and build up oxygen debts. Almost certainly, we will not understand the peculiar rudist reefs of the Tethys: giant structures of carbonate rock built by bivalves shaped like an urn with a cover on top. The locks on the cover are large: were they for keeping out predaceous crabs? Or did they close tightly whenever the water became ill tasting, from contamination with sulfur from below the sunlit zone? It would be interesting to know. An adaptation to "bad water" and oxygen stress would explain the advantage rudists had over corals, in places.

HAVOC FROM THE HEAVENS

In July 1994, the world was treated to a once-in-a-millennium celestial spectacle: On approaching Jupiter, a large comet (a dirty snowball) broke into pieces, which eventually resulted in a series of impacts hitting their huge gaseous target in a celestial bombing run lasting several days. This produced a row of incredibly powerful explosions with mushrooms as wide as the Earth.[61]

After this stunning display of collision, only a light-hour away, the notion that Earth is subject to bombardment from space seemed a lot less farfetched than before. In fact, it now seems unavoidable.[62]

No one would have been more satisfied to observe the Jupiter collision than the late Luis Alvarez (1911–1988), the tall and austere nuclear physicist who led the team that brought geology back into catastrophism. He was lead author of the crucial paper that revolutionized our understanding of mass extinction, specifically the mass extinction at the end of the Cretaceous.[63] The paper reported on a section of pelagic limestones in Italy, near the town of Gubbio, specifically on the transition between the Cretaceous and Tertiary epochs within that sequence. A layer of indurated clay marks the boundary. Above and below the clay layer, the limestone beds look much the same, but they contain entirely different microfossil assemblages. Alvarez and collaborators showed that the clay layer at the break in fossil content is highly enriched in iridium, a rare precious metal. This metal occurs in meteorites.

The conclusion put forward by the Alvarez team was that a very large meteorite (some 10 kilometers across) hit the Earth and caused enough particles to be thrown high into the atmosphere to block out the sunlight. The ensuing "impact winter" was sufficiently long and severe to bring about mass extinction.

After the Alvarez report, the nature of discussion about the history of life changed fundamentally. It became possible to talk about the "survival of the lucky"[64] as being equally important as the familiar "survival of the fittest."[65] Mass extinction, an ancient concept well ensconced in the geologic literature since the early nineteenth century, thanks to George Cuvier and Alcide d'Orbigny, became believable because a mechanism had been found. With the bombardment paradigm, natural selection, Darwin's paradigm for creating species, now has a dark-side partner, that is, wholesale destruction. The history of life cannot be understood when leaving out mass extinction. To make the attempt would be as futile as trying to understand human history by focusing on trade, commerce, and art, and leaving out war, famine, and pestilence.

That the Cretaceous differs fundamentally from the subsequent geologic epoch, the Tertiary, and that the transition is well defined—this much is old news. An introductory textbook on Earth history published in 1978 reflects the pre-Alvarez approach to mass extinction as follows: "Primarily on land but also at sea, gradual extinction overtook many seemingly secure groups of vertebrates and invertebrates. In the seas, the ichthyosaurs, plesiosaurs, and mosasaurs perished."[66] These statements are followed by a list of other organisms that disappeared from the sea: the ammonoid and belemnid cephalopods, the rudist bivalves, families of echinoids, bryozoans, planktonic foraminifers, and calcareous nanoplankton. The textbook author then notes the loss of the great reptiles on land—"Gone forever were the magnificent dinosaurs and soaring pterosaurs"—and asserts that extinctions eliminated about one-fourth of all known families of animals, "in the Late Cretaceous."[67] Note the phrase "gradual extinction" in the first sentence, which reminds us of the legacy of Charles Lyell and Charles Darwin. Of course, ascribing the extinctions to the "Late Cretaceous," rather than to the end of the Cretaceous, allows for plenty of time: millions of years.

The textbook then offers two fundamentally different classes of explanations for the demise of ichthyosaurs, ammonites and dinosaurs: "extraterrestrial interference, such as an influx of abnormally high amounts of cosmic radiation," and "phenomena that are intrinsic to the earth itself." The second class is favored, suggesting that an overall drop in sea level and the advent of harsher climates were to blame, and invoking collapse of the food chain.[68]

The concept that ocean plankton was catastrophically affected at the end of the Cretaceous had emerged in a number of studies, perhaps none more important than that of Hans Luterbacher and Isabella Premoli-Silva, in the early 1960s, whose work in the Gubbio section prepared the scene for the iridium discovery.[69] This team of paleontologists (one Swiss, the other Italian) showed that the transition between Cretaceous and Tertiary plankton is extremely abrupt, and that none of the abundant planktonic foraminifer species made it across the boundary. The section seemed undisturbed, consisting of uplifted pelagic sediments deposited in the Tethys Sea (which once connected Atlantic and Pacific along a belt defined by the Alps and the Himalayas). Assuming that deposition was continuous across the boundary (which seemed reasonable from the field evidence), here was strong evidence for abrupt extinction.

The deep-sea record confirmed the Gubbio story put forward by Luterbacher and Premoli-Silva. Wherever deep-sea drilling penetrated into the Cretaceous, a rather abrupt change in the sequence of planktonic nannofossils and microfossils was obvious (fig. 14.9). A large-scale, worldwide extinction event seemed the unavoidable conclusion. At Scripps, Hans Thierstein, an energetic young geologist with a penchant for Swiss precision and punctilio, made a detailed study of the transition as seen in nannofossils at a great number of DSDP sites and surveyed the available information for a 10-year drilling review held a combined conference of the Society of Economic Paleontologists and Mineralogists and the American Association of Petroleum Geologists in Houston in 1979.[70]

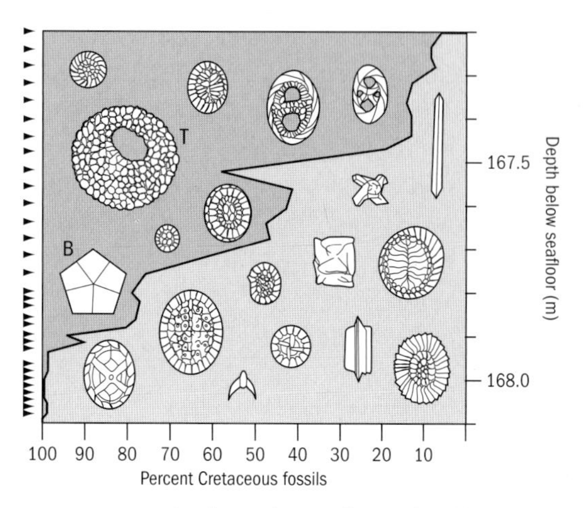

FIGURE 14.9. The abrupt change of nannofossil flora at the end of the Cretaceous, as documented by H. Thierstein and H. Okada, in site 384 of the Deep Sea Drilling Project, in 1979. Sample spacing to the left. B, *Braarudosphaera*, a stress-tolerant nannofossil; T, *Thoracosphaera*, a dinoflagellate cyst. Markers at left denote sample spacing.

Thierstein was interested in establishing the facts that need to be explained as far as the change in the nature of open-ocean phytoplankton. He considered that the record of nannofossils was sufficiently well known, after 10 years of drilling and intensive studies (including his own), to endorse the following propositions:

- We are dealing with the most severe plankton extinction event known in the geologic record;
- there is no hint of a coming catastrophe as the end of the Cretaceous is approached;

- the transition to the "new" earliest Tertiary nannofossil assemblages is not in any way gradual, but entirely abrupt (after taking into account the mixing of sediments on the seafloor);
- all Cretaceous species found above the boundary appear derived from older sediments (petering out from mixing processes);
- productivity apparently decreased drastically at the Cretaceous-Tertiary boundary; after the extinction there was a long-lasting interval of ecological instability;
- the new phytoplankton taxa look like immigrants from shelf seas in high latitudes where they were rare before the boundary event.[71]

Thierstein saw no evidence for a rapid increase in speciation near the boundary. He did see (as also reported by many of his colleagues) evidence for strange plankton blooms, after the catastrophe, from cyst-forming dinoflagellates. These various observations, and the fact that extinction was highly selective (warm-water planktonic species were greatly reduced, but not deep-sea benthic species, for example) made a buildup of gradual stress unlikely as a cause.

Once sudden extinction is accepted, there is no shortage of mechanisms ready to deliver catastrophe. Among the suggested processes is radiation from a nearby supernova explosion, poison gas from a comet entering the atmosphere, climate catastrophe from large-scale volcanism, and system collapse provoked by collision with enormous meteorites. The choice became clear with the evidence from high iridium values at the boundary in Gubbio, backed up by the presence of shocked quartz at other boundary locations.[72]

All through the 1980s, these findings led to lively discussions regarding the nature of the end-Cretaceous extinctions. In the meantime, a candidate for the end-Cretaceous impact crater (or at least one of the craters, assuming there might have been several) was found on the Yucatan peninsula in Mexico. It is the Chixculub crater, 200 kilometers in diameter. The size of the impactor could be estimated from the size of the hole it made: a diameter of between 10 and 20 kilometers has been suggested.[73]

The exact manner in which the global environment was changed as a consequence of the impact will always be a matter of speculation. A prolonged darkening of the Sun (from particles thrown into the stratosphere by the impact), acid rain (from the burning of molecular nitrogen when the air was heated), and drastic changes in temperature (from changes in greenhouse gases), are some of the possibilities that have been suggested. The acid rain scenario is supported by changes in weathering rates of granitic rock in the earliest Tertiary.[74] If rain rich in acids fell on the surface waters of the sea, we should see evidence of widespread carbonate dissolution for a short interval. This is indeed seen: a thin clay layer marks the boundary between late Cretaceous and early Tertiary limestone beds, both in Gubbio and in the corresponding sequence in Denmark (in a sea cliff called Stevns Klint). Carbon dioxide would have been released to the atmosphere, in the process of neutralizing the acid. For this reason (and some others as well, including flood basalts in India), we must consider the possibility that the short impact winter, from the darkening of the skies (lasting perhaps a year or so) would have been followed by a long period of strong greenhouse warming.

Many details of the impact event emerged from drilling into deep-sea sediments. In 1997, for example, the *JOIDES Resolution* drilled in the Atlantic Ocean, 300 miles off northeastern Florida (fig. 14.10). Three cores were taken across the boundary, revealing the sequence in fascinating detail. Impact ejecta, consisting of green, glassy globules of molten rock generated at the point of collision, together with other rock debris, overlie end-of-Cretaceous soft chalk containing a normal plankton assemblage. The chalk is disturbed—broken up—presumably from the strong earthquake felt at this site, not so far from the impact (about 1,000 miles). On top of the impact ejecta we find the accumulation of fallout, with hardly any fossils at all. In this fallout layer the rare metal iridium is

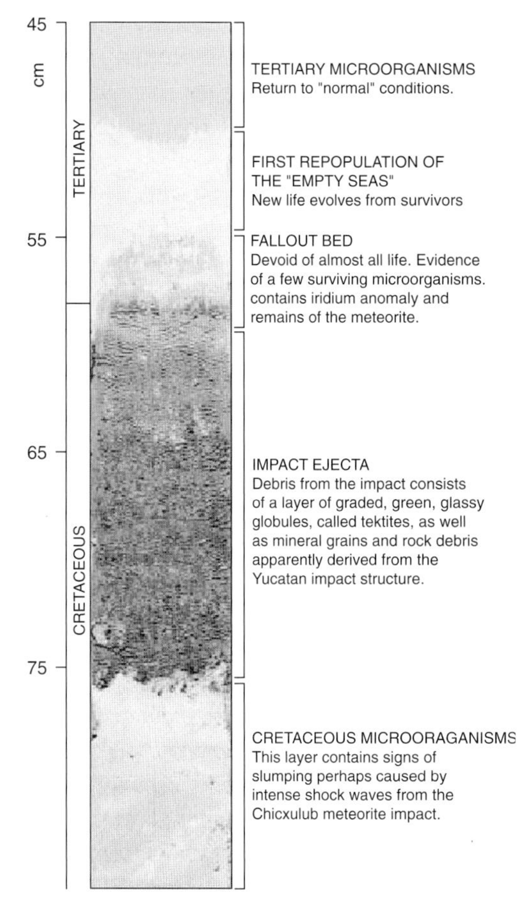

FIGURE 14.10. Details of the Cretaceous-Tertiary transition in a core taken by the Ocean Drilling Program, off northeast Florida.

greatly enriched, showing that material from the exploded meteorite is present. On top of the fallout, we find a return of plankton nannofossils and microfossils—an assemblage looking entirely different from the Cretaceous one that dominates only a foot deeper in the sediment, but yet made of a few selected survivors from that time. As one goes up in the section, new forms appear that are typical for the Tertiary, as evolution takes its course.

The impact destroyed more than half of all species extant on Earth, perhaps as many as 70 percent. It took several million years for the biosphere to return to something approaching its former complexity.

MORE THOUGHTS ON THE GREAT COOLING

The time since the Great Catastrophe, the Cenozoic, is referred to as the Age of Mammals. We could, with equal justification, refer to this era as the age of beetles or of grasses, or of pine forests, or of diatoms. Beetles show the greater diversity, grasses drove the evolution of many mammals (including horses), pine forests have the greater bulk, and diatoms characterize the productivity of the new ocean, flourishing and diversifying in the emerging upwelling regions that favored the evolution of baleen whales and penguins. The great leitmotifs of the Cenozoic are the recovery after the catastrophe and a return to biosphere opulence on a warm planet in the middle Eocene, but especially the metamorphosis into a planet with ice caps since the late Eocene. At the end of this transformation stood the Quaternary, with its orbitally driven ice age cycles, the time of human evolution (fig. 14.11). It is intriguing to reflect on how we arrived at that condition.

We have already, with a very broad brush, sketched out some trends for the great cooling that began in the late Eocene and led, via the Miocene into the late Pliocene, to the buildup first of Antarctic ice masses and then of large northern ice sheets.[75] This cooling is the central theme of the history of the ocean for the last 40 million years, and it is what set us on the path toward the modern ocean. In the beginning of the process, there were rainforests on Antarctica and around the Arctic, and the deep waters were relatively warm. At the end, Antarctica is covered with ice sheets and the Arctic Ocean with sea ice, and the deep ocean is frigid. And there is no longer strong evidence for switching back and forth between oxygenated and sulfurated waters in nearshore environments.[76]

Can we define the causes for the cooling? Perhaps not precisely, but there are some clues. First, there is the timing of the initial planetary cooling. Around 40 million years ago mountains started to grow around the closing Tethys, including the site of the future Himalayas, and

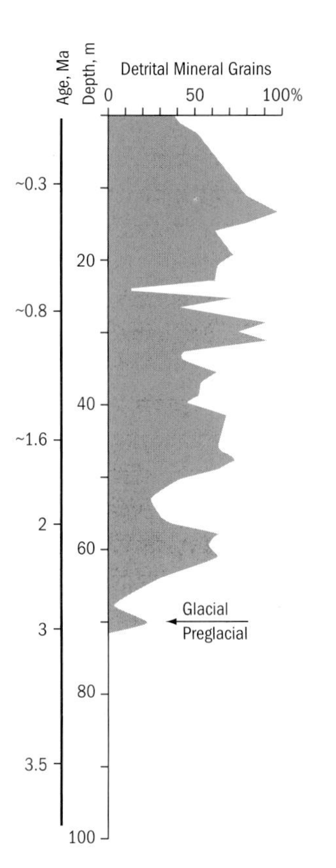

FIGURE 14.11. The onset of deposition of mineral grains in DSDP site 116, northwest Atlantic, marks the arrival of glacial conditions in North America, according to W. Berggren, 1972.

the reorganization of plate motions built a new Ring of Fire. When mountains rise, the sea retreats from the shelves. Turning sea into land changed the albedo of Earth: more sunlight was reflected, and less was absorbed. In addition, the newly exposed land surfaces provided sites for reactions with the carbon dioxide in the air. Slowly but steadily the carbon dioxide content of the ocean-atmosphere system was lowered by these increased opportunities for reaction. As the albedo increased and as carbon dioxide decreased, the planet cooled.

Quite early in the sequence, ice started to build up in the highlands of the Antarctic continent that is so conveniently centered on the South Pole. Cooling continued as the Tethys kept closing and the Antarctic circumpolar ocean kept opening, and by 10 million years ago the southern continent had its enormous ice cap. Mountains had replaced much of the former Tethys seaway, most prominent the Himalayas and the Alps. Mountains kept growing in the Tethys chain and elsewhere, around the Pacific, and cooling continued. About 7 to 6 million years ago, there was a pulse of ice buildup in regions around the Arctic Ocean. Sea level dropped, again. The Mediterranean, a paltry remnant of the Tethys, sporadically lost connection to the World Ocean. It dried out from time to time, accumulating large amounts of salt.[77]

Shortly after, the seaway between North and South America, which still allowed exchange between Atlantic and Pacific, began to close, starting on the path toward making the Panama Isthmus. Warm water, pushed into the Caribbean by the trade winds, lost its westward outlet and moved northward into an ancient Gulf Stream. The glaciers on Greenland melted. Sea level rose, and the rising waters filled the Mediterranean basin, reconnecting it to the ocean. For the next 2 million years, the heat brought with the Gulf Stream prevented the buildup of ice in Greenland, Canada, and Scandinavia. But the mountains kept rising, at the western coast of the Americas, in the Himalayas, in the Alps. Overall background cooling continued. Finally, the threshold for ice on Greenland was reached once more. Fed by the moist winds coming from the Gulf Stream, the winter storms brought plenty of snow. The northern ice age had begun, and it soon introduced large climate oscillations into the global ecosystems.[78]

The subtropics responded by expanding deserts and savannahs. From upright walking apes in Africa arose a new genus, *Homo*, with a propensity to use tools and language and keep evolving a larger brain. After 3 million years, its progeny would become a major geologic agent causing extinctions and climate change.

So much for a big picture narrative of how the Cenozoic developed since the great cooling started. We should now have a closer look at the first great cooling step, a period of dramatic change moving the planet into its present condition. To do this, we need tools from paleontology and isotope chemistry, among others.

The first striking observation related to cooling in the Cenozoic is that changes were rather large and some were abrupt, within the late Eocene. A marked drop in diversity is evident within the late Eocene.[79] Plankton diversity had been increasing before that turning point and had reached an acme in the late middle Eocene. The Eocene-Oligocene cooling step that marks the initial large step in the great planetary cooling trend shows well the new connection between abyssal water and one or more polar source regions for cold water. The ocean started to change from a warm ocean whose density structure was dominated by salinity differences, to a cold ocean dominated by temperature gradients. Many remarkable consequences followed from this change.

None was more important than the fact that conditions in the ocean became less predictable. During the time of the transition at the end of the Eocene and all through the early Oligocene, stratification collapsed sporadically, being maintained neither by the traditional salinity mechanism, nor yet by sufficiently strong temperature gradients. As a result, the diversity of plankton organisms greatly decreased, while blooms of opportunistic species became abundant. Both in the plankton and on the deep seafloor, opportunistic generalists won out over specialists adapted to well-defined conditions.

Why was the first cooling step rather pronounced and sudden? There are fundamentally two kinds of answers to the question of why there is an abrupt change along an overall trend: one invokes outside interference, from volcanisms, tectonics, or impact, and the other invokes internal positive feedback mechanisms. The first type of explanation, the Big Hammer, was much in vogue in the 1980s, following the spectacular success of the Alvarez hypothesis.[80] However, some paleontologists were very reluctant to apply this approach to their fossils. For example, Donald Prothero, a specialist in the Cenozoic mammalian fauna of North America, wrote in some frustration: "In the case of the Eocene-Oligocene transition, the stampede for impacts and periodicity threatened to overwhelm any sensible attempts to examine the data in detail. Once the iridium was found in upper Eocene rocks and the nearest extinction was attributed to the Raup and Sepkoski periodicity, some scientists treated the case as proven without further discussion."[81]

If the Big Hammer had indeed been the basic cause for the climate step, why did the system not revert to its previous condition? It did after the end-of-Cretaceous event—but not after the Eocene. Stepwise change along a trend suggests the crossing of a threshold. In the Eocene, there were rainforests along the shores of Antarctica, as mentioned. As long as heavy rains brought heat to the coast and freshened the water around Antarctica, no deep waters could be made there. But the situation changed when it became cold enough in the middle Eocene to make glaciers whose tongues reached the sea.[82] When glaciers started to grow, the effect was to push the rain belt well offshore, at least during winter. Strong winds associated with a strong temperature gradient provided for deep mixing offshore, aiding in making surface waters salty. Sea ice formation on salty water, in winter, produced heavy brines. Such brines, when mixed with cold surrounding seawater, can make heavy deep and bottom water. Thus, in this explanation, the fundamental threshold is the making of sea ice. There is nothing gradual about it. It either happens or it does not.

Once this step had been taken, and once the cold-water sources were powerful enough to compete for high density with the saline-water sources at the edge of the subtropics, the deep ocean started to fill with cold water. Closing down Tethyan deep-water sources most likely helped in bringing about the shift to high-latitude deep waters. Even today the densest deep water is made in the Mediterranean, but there is too little of it to compete as a deep-water source with the cold waters coming from polar regions.

The cooling of the tropics that resulted from the upwelling cold water, the decrease in water vapor in the atmosphere, the retreat of the sea and of swamps and wetlands, the expansion of arid and semiarid areas, the harsher seasonal contrasts—all of these factors combined to put stress on the biosphere, especially on land. Extinctions promptly followed. Extinction, on the whole, prevailed over speciation in an interval spanning several million years—suggesting that impact of an extraterrestrial body is not the issue here.[83]

The modern world started with the buildup of an ice cap on Antarctica. The major step in this process occurred around 14 million years ago, within the middle Miocene; it is marked by a large shift of oxygen isotopes in the sea toward enrichment with oxygen-18. From then on, the planet had a southern ice cap. Before that time, all through the Oligocene, the climate system vacillated between warm and cold—a situation unfavorable for stability of the water column and stratification, and one that hindered diversification of the plankton. But with the beginning of the Miocene, the high latitudes cooled further, while the low latitudes warmed in response to a modest flooding of continental lowlands. The increased planetary temperature gradient turned up the zonal winds (trades and westerlies), which in turn generated the familiar gyrating warm-water lenses rimmed by high-production zones. In this manner, the ocean started to look "modern," that is, familiar.[84]

The major first step toward a thoroughly modern world occurred only well within the Miocene, when summers on Antarctica could no longer melt all the winter snow. Rapidly, but with much vacillation, ice spread on the southern continent, until it reached almost the present extent. Once the ice cap was fully formed, as seen in the oxygen-isotope shift, the condition remained. There was no going back to a warmer time (as far as this can be ascertained). The shift in the isotope signal[85] contains two complementary types of information: a decrease in the temperature in high-latitude surface waters and bottom waters everywhere, and an increase in the amount of ice on the planet. Presumably, the cooling came first, and the ice buildup followed.[86] Sea level dropped by about 50 meters in consequence of the making of the ice cap. A sea level drop is seen in other types of evidence, as well, having to do with the type of layering of sediments in continental margins.[87]

What, precisely, triggered the relatively rapid buildup of ice within the middle Miocene is not known. One possibility, urged by the marine geologist James P. Kennett,[88] is the increasing isolation of Antarctica that resulted from the spin-up of an early circumpolar current in a widening southern ocean. Another possible triggering mechanism that must be considered is an overall decrease in carbon dioxide. The proposition that this was important arose from the observation that the oxygen isotope shift (signaling cooling and ice buildup) was preceded by a marked excursion in oceanic carbon isotope values, as seen in foraminifer shells.[89] The excursion is toward higher ratios of carbon-13 to carbon-12. The argument is that organic matter has high carbon-12 values, and that organic matter was vigorously sequestered during a time of strong thermocline buildup and upwelling preceding the cooling step.[90] The increased burial of organic carbon below upwelling areas would have lowered the atmospheric carbon-dioxide content, and kept it low. This would have led to additional cooling, parallel to the effects from uplift of mountains.[91]

The icing over of the Antarctic was soon followed by a major shift of opal sedimentation, out of the Atlantic into the Pacific. This is the Miocene "silica switch" discovered (and named) by the paleontologists Gerta Keller and John Barron.[92] The shift established the now reigning asymmetry in nutrient contents of Pacific and Atlantic waters, with a silicate-poor North Atlantic and a silicate-rich North Pacific. The asymmetry is owing to North Atlantic deep-water production. Thus, the silica switch marks the turning on of deep-water production in the North Atlantic. The associated changes in the carbon isotopes show that the shift was completed by about 10 million years ago, and that the asymmetry persisted since, getting stronger with additional cooling.[93]

The turning up of North Atlantic deep-water production 10 million years ago was accompanied by an increase in silica deposition around Antarctica. Interestingly, at the various drilling sites in the Southern Ocean, the increased accumulation of diatom debris precedes an increase in ice-rafted debris. This suggests (but does not prove, of course) that silica deposition is linked to further lowering of carbon dioxide values, through the carbon deposition that accompanies diatom sedimentation—with a corresponding drop in the greenhouse effect.[94] By the late Miocene, ice-rafted debris appeared even north of the polar front, in the South Atlantic. Ice on land and sea characterized the Antarctic.

As yet there were no northern ice sheets, as far as this can be ascertained. The ice had captured only one of the poles of the planet. For many millions of years the asymmetry persisted, and it was only about 3 million years ago that cooling had progressed to the point that the Northern Hemisphere could bear large ice sheets, at least temporarily.[95] Because the North Pole is centered on an ocean basin, the northern land surfaces initially were not available as repositories for ice. Also, once the northern ice sheets did build up, they were sufficiently far away from the pole to remain vulnerable to melting during periods when the summer Sun was unusually strong. During each glaciation cycle, the tendency for instability of the ice was increased as the land subsided under the weight of the ice. The result is cyclic waxing and waning of ice masses on the northernmost land areas, in an oscillation in tune with summer insolation, as elaborated in the previous chapter.

The northern climate oscillations soon took over much of the globe, steering conditions in the Southern Hemisphere through changes in sea level and carbon dioxide, as well as by fluctuations in heat exchange across the equator.

NOTES AND REFERENCES

1. L.W. Alvarez, W. Alvarez, F. Asaro, H.V. Michel, 1980, *Extraterrestrial cause for the Cretaceous-Tertiary extinction.* Science 208, 1095–1108. There may have been more than one impact, in rapid succession.

2. The dictum that "the present is the key to the past" defines the essence of *uniformitarianism,* the belief structure introduced by Charles Lyell in founding modern geology, in the early nineteenth century. He did so in opposition to the arbitrary approach to history associated with *catastrophism* and a belief in divine intervention.

3. For details, see essays by B. Malfait and by A.E. Maxwell and other articles in Oceanus 36 (4), Special Issue: *25 Years of Ocean Drilling* (1994).

4. On a drilling vessel, lack of noise spells trouble.

5. C. Emiliani, 1961, *Cenozoic climatic changes as indicated by the stratigraphy and chronology of deep-sea cores of Globigerina-ooze facies.* Annals of the New York Academy of Sciences 95 (1), 521–536.

6. For a history of these developments see E.N. Shor, 1985, *A chronology from Mohole to JOIDES.* Geological Society of America, Centennial Special Volume 1, 391–399.

7. Layer 3 in Raitt's scheme of the sound velocity structure of the crust; see chapter 11.

8. See J. Malpas, 1995, *Deep drilling of the ocean crust and upper mantle: A continuing scientific and technological challenge,* in M. Kastner, J.H. Natland, E.L. Winterer (eds.), *Proceedings of the Symposium on Future Challenges in Scientific Drilling.* SIO Ref. No. 95-15. SIO, La Jolla, Calif.

9. Y. Dilek, E.M. Moores, D. Elthon, A. Nicolas (eds.), 2000, *Ophiolites and oceanic crust: New insights from field studies and the Ocean Drilling Program.* Special Paper 349. Geological Society of America, Boulder, Colo., 552 pp.

10. Thanks to the efforts of the ocean engineer Willard Bascom (1916–2000). Bascom's design employed four 200-horsepower outboard engines on the ship's hull, operated by a central joystick on deck. The ship held position very well.

11. A month earlier, CUSS I had drilled several holes in San Diego Trough, as a test.

12. The new director at Scripps, William A. Nierenberg, signed on as the principal investigator. One year later, Scripps was chosen as the operating member of JOIDES and was charged with establishing the DSDP, under an initial 18-month contract worth $12.6 million. The task was to find a suitable ship and to recruit managers and technicians to run the project.

13. The ship had to be able to drill in deep water (more than 5,000 meters) and to penetrate deeply into the seafloor (more than 1,000 meters). Also, the ship was to keep station faithfully without being anchored to the bottom. The new positioning system eventually included an acoustic transponder that was sent to the seafloor when on station, from where it beamed sound pulses. Four underwater microphones were the ears of the ship. A computer used the differences in the arrival times of the sound pulses at the four microphones to calculate what the propulsion system needed to do to keep the ship on location.

14. A suitable site had to have a sediment cover appropriate for the purpose of the leg. Also, one

needed to make sure that there was no chance of drilling into trapped oil or gas that might blow out.

15. The idea, presumably, was to establish that scientific drilling could contribute to the prosperity of the nation by exploring the petroleum potential of the deep seafloor. It was reasonable to assume that such a contribution would be welcomed by funding agencies and industry alike and thus would help establish a long-term commitment to deep drilling as a useful (if expensive) academic activity.

16. The committee was duly constituted as the "Safety and Pollution Panel." Management and cochiefs could cut back on their anxiety level and lean on the experts to share responsibility. In fact, thanks to the diligence of the geologists on the safety panel, DSDP eventually concluded its 15-year operation with a flawless safety record. The same high standard was achieved by its successor, the Ocean Drilling Program.

17. Melvin N. A. Peterson (1929–1995), then a rising star in marine geochemistry recently hired to the faculty at Scripps, had a strong interest in the carbon cycle and in low-temperature transformation of sedimentary rocks, including chemical reactions near the spreading centers. He was asked to take the top scientist position at DSDP, in 1967. He accepted, abandoning the prospect of a brilliant career running his own laboratory, in order to serve as administrator in what promised to become the geology project of the century. His ideas figured prominently in the various proposals that carried DSDP forward.

18. A. E. Maxwell (Ph.D. 1959) and R. P. von Herzen (Ph.D. 1960)—both then at Woods Hole—are Scripps alumni. Maxwell had earlier participated in the famed MidPac Expedition and discovered, with Revelle, high heat flow from the seafloor. Von Herzen had studied heat flow on the East Pacific Rise. Both had made important contributions to the new global tectonics.

19. A. E. Maxwell, R. P. von Herzen, K. J. Hsü, J. E. Andrews, T. Saito, S. E. Percival, E. D. Milow, R. E. Boyce, 1970, *Deep sea drilling in the South Atlantic*. Science 168, 1047–1059.

20. K. J. Hsü, J. E. Andrews, 1970, *History of South Atlantic Basin*. Initial Reports of the Deep Sea Drilling Project 3, 464–467.

21. The explanation of the carbonate record invoking sea-level change is in W. H. Berger, E. L. Winterer, 1974, *Plate stratigraphy and the fluctuating carbonate line*. Special Publications of the International Association of Sedimentologists 1, 11–48. Hsü and Andrews had proposed vertical motions of the Mid-Atlantic Ridge to explain the changes in carbonate deposition observed. Mobility of the seafloor was the reigning paradigm, and anything was considered possible, even rapid uplift of enormous submarine mountain ranges.

22. With the advent of the Ocean Drilling Program (see next note), interest in the older materials waned and shifted to the newer ODP cores, first in Texas and then also in Bremen, especially since advanced technology produced cores with minimum disturbance from the drilling process.

23. In 1985, deep drilling resumed under the new management and with a bigger vessel, the *SEDCO/BP 471*, christened *JOIDES Resolution* by the scientific community. The project became the Ocean Drilling Program. Texas A&M, the new operator, was responsible to JOI (Joint Oceanographic Institutions), the contractor with the National Science Foundation. The scientific goals were defined in two international conferences, one at the University of Texas, in 1981, the other in Strasbourg, France, in 1987. Proposals from individual scientists from all participating countries then drove the ship, within the framework set by those conferences, and with the guidance of the Planning Committee, which was responsible to the Executive Committee (representing the United States and international members). Panels of experts (for lithology, tectonics, geochemistry, and paleoceanography) provided detailed scientific advice and ranked the proposals in the order of scientific appeal and promise. Thus, the program was run by scientists, from the bottom up. Operations, on the other hand, were run strictly by technical experts. This proved to be a highly successful organizational structure. The ODP terminated in September 2003; it has been replaced by the Integrated Ocean Drilling Program, which is driven by a need for several different drilling platforms that are capable of drilling in regions previously inaccessible.

24. The nomenclature is simple; it goes back to the original *Challenger* expedition. More technical definitions are available also.

25. Details on rates and distributions of sediment types are in E. Seibold, W. H. Berger, 1996, *The Sea Floor, An Introduction to Marine Geology*, 3rd ed. Springer, Berlin, 356 pp.

26. For example, see the various entries on deep-sea sediments in R. W. Fairbridge (ed.), 1966, *The Encyclopedia of Oceanography*. Reinhold Publishing, New York, 1021 pp.

27. W. H. Berger, E. L. Winterer, *Plate stratigraphy*; Tj. H. van Andel, G. R. Heath, T. C. Moore, 1975, *Cenozoic History and Paleoceanography of the Central Equatorial Pacific Ocean*. Geological Society of America Memoirs 143. GSA, Boulder, Colo., 134 pp.

28. The principles applicable to this task define the stratigraphic aspects of plate tectonics, referred to as *plate stratigraphy*.

29. The shift in sediment patterns ("facies shift") occurred in the time interval within the late Eocene

referred to as Auversian, hence the label *Auversian facies shift*. W. H. Berger, G. Wefer, 1996, *Expeditions into the past: Paleoceanographic studies in the South Atlantic*, in G. Wefer, W. H. Berger, G. Siedler, D. J. Webb (eds.), *The South Atlantic: Present and Past Circulation*. Springer-Verlag, Berlin, pp. 363–410.

30. T. C. Moore, 1969, *Radiolaria: Change in skeletal weight and resistance to solution*. Bulletin of the Geological Society of America 80, 2103–2108.

31. The response of the deep ocean to cooling since the Eocene was first outlined by Cesare Emiliani (1961, *Cenozoic climatic changes*). In 1954, he proposed a general cooling trend based on an increase of oxygen-18 in a few chance samples of benthic foraminifers from Tertiary sediments exposed on the sea floor and recovered by traditional methods. He postulated a temperature of greater than 10 °C for bottom waters in the Pacific, in the Eocene. His guess proved correct. A concept of relatively warm high-latitude bottom-water sources at the time agrees with evidence from fossil trees and reptiles, as offered in textbooks of geologic history.

32. The teams were Robert G. Douglas (University of Southern California) and Samuel M. Savin (Case Western Reserve), and James P. Kennett (then at the University of Rhode Island) and the late Nicholas J. Shackleton (Cambridge, U.K.).

33. R. G. Douglas, S. M. Savin, 1975, *Oxygen and carbon isotope analyses of Tertiary and Cretaceous microfossils from Shatsky Rise and other sites in the North Pacific Ocean*. Initial Reports of the Deep Sea Drilling Project 32, 509–520; N. J. Shackleton, J. P. Kennett, 1975, *Paleotemperature history of the Cenozoic and the initiation of Antarctic glaciation: Oxygen and carbon isotope analyses in DSDP Sites 277, 279, and 281*. Initial Reports of the Deep Sea Drilling Project 29, 743–755. The comparison is in W. H. Berger, 1979, *Impact of deep-sea drilling on paleoceanography*, in M. Talwani, W. Hay, W. B. F. Ryan (eds.), *Deep Drilling Results in the Atlantic Ocean: Continental Margins and Paleoenvironment*. Maurice Ewing Series vol. 3. American Geophysical Union, Washington D.C. pp. 297–314.

34. W. H. Berger, 1979, *Stable isotopes in foraminifera*. Society of Economic Paleontologists and Mineralogists, Short Course 6, 156–198.

35. The dominant type of tree in this type of rainforest is the southern beech, *Nothofagus*.

36. J. P. Kennett, N. J. Shackleton, 1976, *Oxygen isotope evidence for the development of the psychrosphere 38 m.y. ago*. Nature 260, 513–515.

37. R. H. Benson, 1975, *The origin of the psychrosphere as recorded in changes of deep-sea ostracod assemblages*. Lethaia 8, 69–83; R. G. Douglas, F. Woodruff, 1981, *Deep-sea benthic foraminifera*, in C. Emiliani (ed.), *The Oceanic Lithosphere. The Sea*, vol. 7. Wiley-Interscience, New York, pp. 1233–1327.

38. N. Eldredge, S. J. Gould, 1972, *Punctuated equilibria: An alternative to phyletic gradualism*, in T. J. M. Schopf (ed.), *Models in Paleobiology*. Freeman, Cooper and Co., San Francisco, pp. 82–115.

39. T. C. Moore Jr., Ph.D. 1968. T. C. Moore, *Radiolaria*. Moore identified the evolution of diatoms as an important factor in producing the trend.

40. S. E. Calvert, 1968, *Silica balance in the ocean and diagenesis*. Nature 219, 919–920.

41. J. H. Lipps, 1970, *Plankton evolution*. Evolution 24, 1–22.

42. Letter to W. H. B. cited in W. H. Berger, 2002, *Cesare Emiliani (1922–1995), pioneer of Ice Age studies and oxygen isotope stratigraphy*. Comptes Rendus Palevol 1, 479–487. The point about dismissal of surprising evidence is a reference to the skeptical reception of his results by Scripps's foraminifer experts Frances Parker and Fred Phleger.

43. For depth stratification of planktonic foraminifers in the Miocene, see J. T. Gaspari, J. P. Kennett, 1992, *Isotopic evidence for depth stratification and paleoecology of Miocene planktonic foraminifera: Western equatorial Pacific DSDP site 289*, in R. Tsuchi, J. C. Ingle (eds.), *Pacific Neogene, Environment, Evolution, and Events*. University of Tokyo Press, Tokyo, pp. 117–147; and articles in J. P. Kennett (ed.), 1985, *The Miocene Ocean: Paleoceanography and Biogeography*. Geological Society of America Memoirs 163. GSA, Boulder, Colo., 137 pp.

44. For pre-Oligocene time, presumably, salinity stratification provided habitat diversification. The Oligocene, in this view, was a bad period for plankton, the salinity stratification having been destroyed, and the thermocline being too deep to interact profitably with the sunlit zone.

45. W. A. Berggren, C. D. Hollister, 1974, *Paleogeography, paleobiogeography and the history of circulation in the Atlantic Ocean*, in W. W. Hay (ed.), *Studies in Paleo-Oceanography*. Special Publication 20. Society of Economic Paleontologists and Mineralogists, Tulsa, Oklahoma, pp. 126–186; B. U. Haq, 1984, *Paleoceanography: A synoptic overview of 200 million years of ocean history*, in B. U. Haq, J. D. Milliman (eds.), *Marine Geology and Oceanography of Arabian Sea and Coastal Pakistan*. Van Nostrand Reinhold Co., New York, pp. 201–231.

46. J. A. Barron, J. G. Baldauf, 1989, *Tertiary cooling steps and paleoproductivity as reflected by diatoms and biosiliceous sediments*, in W. H. Berger, V. S. Smetacek, G. Wefer (eds.), *Productivity of the Ocean: Present and Past*. John Wiley, Chichester, pp. 341–354.

47. Latin *creta*, meaning "chalk." The English word *crayon* is related to *creta*. In German, the word "Kreide" denotes both the Cretaceous period, and a piece of chalk.

48. Seawater brings sulfate, which provides the source for iron sulfides in coal, after the sulfate is stripped of oxygen by bacteria. When burning sulfide-rich coal, the effluent contains sulfuric acid. Thus, low-sulfur coal is a better fuel in this respect.

49. J. J. Mahoney, 1987, *An isotopic survey of Pacific oceanic plateaus: Implications for their nature and origin*, in B. Keating, P. Fryer, R. Batiza, G. Boethlert (eds.), *Seamounts, Islands, and Atolls*. Geophysical Monograph 43. American Geophysical Union, Washington, D.C., pp. 207–220; J. J. Mahoney, M. Storey, R. A. Duncan, K. J. Spencer, M. Pringle, 1993, *Geochemistry and geochronology of leg 130 basement lavas: Nature and origin of the Ontong Java Plateau*. Proceedings of the Ocean Drilling Program, Scientific Results 130, 3–22.

50. G. A. Riley, H. Stommel, D. F. Bumpus, 1949, *Quantitative ecology of the plankton of the western North Atlantic*. Bulletin of the Bingham Oceanographic Collection 12 (3), 1–169; H. U. Sverdrup, 1953, *On conditions for the vernal blooming of phytoplankton*. Journal of Conseil International pour l'Exploration de la Mer 18, 287–295.

51. R. G. Douglas, S. M. Savin, *Oxygen and carbon isotope analyses*.

52. Scripps chemist Norris Watson Rakestraw (1895–1982) worked on this topic to find out how fast nitrate is being destroyed in this situation. Others as well, including A. F. Carlucci and J. D. H. Strickland at Scripps, studied denitrification in the 1960s. In a low-oxygen ocean the process takes on huge significance.

53. M. Brongersma-Sanders, 1948, *The importance of upwelling water to vertebrate paleontology and oil geology*. Verhandelingen van de Koninklijke Nederlandse. Akademie van Wetenschapplu Afdeeling. Natuurkunde Sect 2, Pt. 45 (4), 1–112.

54. As reported in newspapers of the city of Kiel from time to time.

55. The iron, delivered to the sea by basalt-seawater interaction at depth, or by rivers, is kept mobile by low oxygen supply in the deep sea, and precipitates in shallow water in the presence of abundant oxygen.

56. S. O. Schlanger, H. C. Jenkyns, 1976, *Cretaceous oceanic anoxic events: Causes and consequences*. Geologie en Mijnbouw 55, 179–184.

57. P. A. Scholle, M. A. Arthur, 1980, *Carbon isotope fluctuations in Cretaceous pelagic limestones: Potential stratigraphic and petroleum exploration tool*. Bulletin of the American Association of Petroleum Geologists 64, 67–87; M. A. Arthur, W. E. Dean, S. O. Schlanger, 1985, *Variations in the global carbon cycle during the Cretaceous related to climate, volcanism, and changes in atmospheric CO_2*, in E. T. Sundquist, W. S. Broecker (eds.), *The Carbon Cycle and Atmospheric CO_2: Natural Variations Archean to Present*. Geophysical Monograph 32. American Geophysical Union, Washington, D.C., pp. 504–529.

58. For example, in the proposal to drill Shatsky Rise by S. O. Schlanger and W. V. Sliter, in the JOI-USSAC meeting 1985; and also in the paper by Arthur, Dean, and Schlanger, cited in the previous note.

59. Judging from the discussions in the Cosod II volume, Strasbourg 1987. X. Le Pichon (convener), 1987, *Report of the Second Conference on Scientific Ocean Drilling (Cosod II), Strasbourg, 6–8 July 1987*. European Science Foundation, Strasbourg, 142 pp.

60. R. L. Larson, 1991, *Geological consequences of superplumes*. Geology 19, 963–966. Larson obtained his Ph.D. from Scripps in 1970.

61. The object fated to hit Jupiter was named Shoemaker-Levy 9, after Eugene Merle Shoemaker, his wife Carolyn, and David H. Levy. They reported its approach long before it was near its target. It broke up in a close encounter with the great planet before its final rendezvous. D. H. Levy, 2000, *Shoemaker by Levy*. Princeton University Press, Princeton, N.J., 303 pp. Eugene M. Shoemaker (1928–1997), a geologist studying impacts, was a leading figure in NASA's program of planetary exploration.

62. The idea that Earth is bombarded from space has a long pedigree; the British astronomer William Herschel (1738–1822) suggested collisions of celestial objects with Earth two centuries ago. More recently, the Canadian geologist Digby McLaren and the marine geologist Robert Dietz of the Navy Electronics Lab in San Diego proposed that large impact structures left a mark on our planet. Dietz, in the 1970s, started to make a survey of impact structures as seen from the air and from space. Few took notice of his quest. Likewise, Eugene Shoemaker, who had studied the Great Meteor Crater in Arizona, had long claimed that collisions through geologic time are unavoidable, on the basis of calculations concerning the abundance of suitable celestial objects in the solar system.

63. L. W. Alvarez et al., *Extraterrestrial cause*. For background on the story of discovery see W. Alvarez, 1997, *T. rex and the Crater of Doom*. Princeton University Press, Princeton, N.J., 185 pp. The geologist Walter Alvarez, University of California, Berkeley, is the son of Luis Alvarez.

64. The phrase "survival of the lucky" was introduced by the Chinese-American-Swiss geologist Kenneth J. Hsü, codiscoverer of the drying-out of the Mediterranean. K. J. Hsu, 1986, *The Great Dying*. Harcourt Brace Jovanovich, San Diego, 292 pp. The importance of "bad luck" in evolution is discussed in D. M. Raup, 1991, *Extinction: Bad Genes or Bad Luck?* W. W. Norton, New York, 210 pp.

65. The phrase "survival of the fittest" was introduced by the English sociologist and philosopher Herbert Spencer (1820–1903). Encyclopaedia Britannica, 1974, vol. 14, p. 936. Darwin's original idea

might better be described by "survival through successful reproduction."

66. H. L. Levin, 1978, *The Earth through Time.* Saunders, Philadelphia, 530 pp. p. 433 ff.

67. The text is well written; no criticism of its contents is intended here.

68. The concept of a collapsing food chain in the sea was the brainchild of M. N. Bramlette; he introduced it in the 1960s. Bramlette was impressed with the evidence for mass extinction in nannofossils at the end of the Cretaceous. Obviously, whatever happened, it involved the entire ocean and had global consequences. Bramlette suggested that the productivity of the ocean suffered from a lack of supply of phosphate from the widely flooded continents. An oceanwide famine resulted, with dire consequences for larger organisms up the food chain. M. N. Bramlette, 1965, *Massive extinctions in the biota at the end of Mesozoic time.* Science 148, 1696–1699.

69. H. P. Luterbacher, I. Premoli-Silva, 1964, *Biostratigrafia del limite Cretaceo-Terziario.* Rivista Italiana di Paleontologia e Stratigrafia 70, 67–128. W. A. Berggren had earlier identified an abrupt change in the planktonic fauna, in Denmark. W. A. Berggren, 1962, *Some planktonic foraminifera from the Maestrichtian and the Danian Stages of southern Scandinavia.* Stockholm University Contributions in Geology 9, 1–106.

70. J. E. Warme, R. G. Douglas, E. L. Winterer, 1981, *The Deep Sea Drilling Project: A Decade of Progress.* Special Publication 32. Society of Economic Paleontologists and Mineralogists, Tulsa, Oklahoma, 564 pp. The iridium story is not mentioned in Thierstein's summary, but the work of Walter Alvarez and collaborators that prepared for the discovery is; this work was concerned with establishing, from paleomagnetic studies, that the Gubbio section is indeed continuous. Thanks to these efforts, and to measurements at Lamont, by the geophysicist Dennis V. Kent, the time period of the transition could be restricted sufficiently to say that extinction was geologically abrupt (10,000 years or less).

71. H. R. Thierstein, 1981, *Late Cretaceous nannoplankton and the change at the Cretaceous-Tertiary boundary,* in J. E. Warme, *The Deep Sea Drilling Project: A Decade of Progress.* Special Publication 32. Society of Economic Paleontologists and Mineralogists, Tulsa, Oklahoma, p. 371.

72. J. Smit, J. Hertogen, 1980, *An extraterrestrial event at the Cretaceous-Tertiary boundary.* Nature 285, 198–200.

73. See W. Alvarez, *T. rex and the Crater of Doom.*

74. Evidence for acid rain was presented by Scripps geochemist J. Douglas Macdougall, based on changes in the strontium isotope record. In his book, *A Short History of Planet Earth,* Macdougall emphasizes the potential importance of the meteor impacting deposits of gypsum (calcium sulfate) in making acid, and of flood basalts in India (the Deccan Traps, which released carbonic acid) in adding to the inclement conditions at end-Cretaceous time. J. D. Macdougall, 1996, *A Short History of Planet Earth.* John Wiley and Sons, New York), 266 pp.

75. The terms *Eocene, Miocene,* and *Pliocene* were originally introduced by Charles Lyell and refer to the proportion of mollusks from these periods that are readily recognized as modern. *Paleocene* and *Oligocene* were subsequently added, as well as *Pleistocene* and *Holocene.* The sequence is easily remembered: *pal-eos* means "early dawn"; *eo-,* "emerging"; *oligo-,* "few"; *mio-,* "medium"; *plio-,* "plenty"; *pleisto-,* "extra plenty"; *holo-,* "all." *Tertiary* is a term left over from the beginnings of geology, when we still had "primary," "secondary," "tertiary," and "quaternary" rocks and deposits. The sequence referred to induration. In this early classification scheme, tertiary rocks are the ones that are poorly cemented (friable or crumbly), and quaternary deposits are typically loose sand, mud, and gravel (which were then ascribed to the great Noachian Flood, by many natural philosophers).

76. Except in a few offshore basins with high production and a sill within the oxygen minimum, such as the Santa Barbara basin.

77. That the Mediterranean dried out repeatedly in the uppermost Miocene was discovered during leg 13 of the Deep Sea Drilling Project. William B. F. Ryan and Kenneth J. Hsü were the co–chief scientists; paleontologist Maria Bianca Cita determined the age of the foraminifers, which dated the deep-basin desiccation as Messinian in age, and hence correlative with well-known salt and gypsum deposits around the Mediterranean. K. J. Hsü, 1992, *Challenger at Sea: A Ship That Revolutionized Earth Science.* Princeton University Press, Princeton, N.J., 417 pp.

78. Some geologists see it differently; they think the northward intensification of the Gulf Stream *caused* the northern ice buildup. However, there is no evidence that an increase in precipitation is necessary to cause an ice age. Mainly, the need is for cooling below the freezing point over large areas. The narrative here presented is put forth in W. H. Berger, G. Wefer, 1996, *Expeditions into the past: Paleoceanographic studies in the South Atlantic,* in G. Wefer, W. H. Berger, G. Siedler, D. J. Webb (eds.), *The South Atlantic: Present and Past Circulation.* Springer-Verlag, Berlin, pp. 363–410.

79. B. U. Haq, 1978, *Calcareous nannoplankton,* in B. U. Haq, A. Boersma (eds.), *Introduction to Marine Micropaleontology.* Elsevier, New York, pp. 79–107; E. Vincent, W. H. Berger, 1981, *Planktonic foraminifera and their use in paleoceanography,* in C. Emiliani (ed.), *The Sea,* vol. 7. *The Oceanic Lithosphere.* Wiley-Interscience, New York, pp. 1025–1119. For extinctions on

land in the late Eocene see D. R. Prothero, 1994, *The Eocene-Oligocene Transition: Paradise Lost.* Columbia University Press, New York, 291 pp. Prothero discusses the various possible reasons other than cooling that have been suggested.

80. In the case at hand, extraterrestrial impact is attractive because there is evidence for increased influx of meteorites near the end of the Eocene. R. Ganapathy, 1982, *Evidence for a major meteorite impact on the earth 34 million years ago: Implications for Eocene extinctions.* Science 216, 885–886.

81. D.R. Prothero, 1994, *The Eocene-Oligocene Transition,* p. 135. The reference to Raup and Sepkoski is to the hypothesis of extinction cycles explored by the paleontologists David Raup and Jack Sepkoski, at the University of Chicago. D. M. Raup, J. J. Sepkoski, 1984, *Periodicity of extinctions in the geologic past.* Proceedings of the National Academy of Sciences 81, 801–805. They discussed a possible cycle of 26 million years, related to bombardment. The hypothesis implies external forcing.

82. W. Wei, 1989, *Reevaluation of the Eocene ice-rafting record from subantarctic cores.* Antarctic Journal of the U.S. 1989, 108–109.

83. The best evidence that extinctions are the result of impact comes not from iridium (iridium does not kill) but from the contrast of sudden versus gradual turnover in the fossil record. The abruptness of extinction is the issue when invoking impact as the mechanism.

84. The rest of the discussion follows the exposition in W. H. Berger, G. Wefer, 1996, *Expeditions into the past: Paleoceanographic studies in the South Atlantic,* in G. Wefer, W. H. Berger, G. Siedler, D. J. Webb (eds.), *The South Atlantic: Present and Past Circulation.* Springer-Verlag, Berlin, pp. 363–410.

85. N. J. Shackleton, J. P. Kennett, *Paleotemperature history of the Cenozoic.*

86. E. Vincent, J. S. Killingley, W. H. Berger, 1985, *Miocene oxygen and carbon isotope stratigraphy of the tropical Indian Ocean,* in J. P. Kennett (ed.), *The Miocene Ocean: Paleoceanography and Biogeography.* Geological Society of America Memoir 163, 103–130.

87. B.U. Haq, J. Hardenbol, P.R. Vail, 1987, *Chronology of fluctuating sea levels since the Triassic.* Science 235, 1156–1167.

88. J. P. Kennett, 1982, *Marine Geology.* Prentice-Hall, Englewood Cliffs, N.J., 813 pp., p. 724. Kennett is at University of California, Santa Barbara.

89. E. Vincent, W. H. Berger, 1985, *Carbon dioxide and polar cooling in the Miocene: The Monterey hypothesis,* in E. T. Sundquist, W. S. Broecker (eds.), *The Carbon Cycle and Atmospheric CO_2: Natural Variations Archean to Present.* Geophysical Monograph 32. American Geophysical Union, Washington, D.C., pp. 455–468.

90. The hypothesis of organic carbon sequestration is supported by the findings of Fay Woodruff and Samuel Savin (at the University of Southern California) who charted the development of the Miocene carbon isotopes in the sea as a function of water depth. They discovered a dramatic increase in carbon-13 at all depths, during the period in question, centered at 16 million years ago. F. Woodruff, S. M. Savin, 1989, *Miocene deepwater oceanography.* Paleoceanography 4, 87–140. The proposition linking the carbon isotope excursion with increased sequestration of organic matter is called the Monterey hypothesis, because it marks the period when source rock for hydrocarbons were formed that are now part of the Monterey formation of California.

91. Mountain uplift is recorded in the strontium isotope record, which can be extracted from calcareous fossils, all of which have strontium. H. Elderfield, 1986, *Strontium isotope stratigraphy.* Palaeogeography, Palaeoclimatology, Palaeoecology 57, 71–90.

92. G. Keller, J. A. Barron, 1983, *Paleoceanographic implications of Miocene deep-sea hiatuses.* Geological Society of America Bulletin 94, 590–613. Gerta Keller is at Princeton University; John Barron is with the U.S. Geological Survey in Menlo Park, California. See also J.G. Baldauf, J.A. Barron, 1990, *Evolution of biosiliceous sedimentation patterns Eocene through Quaternary: Paleoceanographic response to polar cooling,* in U. Bleil, J. Thiede (eds.), *The Geological History of Cenozoic Polar Oceans: Arctic versus Antarctic,* Kluwer Academic, Dordrecht, pp. 575–607.

93. The carbon isotope record is in F. Woodruff, S. M. Savin, *Miocene deepwater oceanography.* It tracks the apparent "age" of deep waters, that is, the time since sinking. Other studies support the rise of this asymmetry. M. L. Delaney, 1990, *Miocene benthic foraminiferal Cd/Ca records: South Atlantic and western equatorial Pacific.* Paleoceanography 5, 743–760.

94. The link between diatom deposition and the greenhouse effect is elaborated in W. H. Berger, 1991, *Produktivität des Ozeans aus geologischer Sicht: Denkmodelle und Beispiele.* Zeitschrift der Deutschen Gesellschaft für Geowissenschaften 142, 149–178.

95. The date of 3 million years is based on ice-derived material drilled during DSDP leg 12. W. A. Berggren, 1972, *Late Pliocene–Pleistocene glaciation.* Initial Reports of the Deep Sea Drilling Project 12, 953–963. As mentioned, earlier sporadic ice buildup during the late Miocene was stalled during most of the Pliocene. E. Jansen, J. Sjöholm, 1991, *Reconstruction of glaciation over the past 6 Myr from ice-borne deposits in the Norwegian Sea.* Nature 349, 600–603; H. C. Larsen, A. D. Saunders, P. D. Clift, J. Beget, W. Wei, S. Spezzaferi, ODP Leg 152 Scientific Party, 1994, *Seven million years of glaciation in Greenland.* Science 264, 952–955.

FIFTEEN

Global Warming and the Ocean

HUMAN IMPACT ON A GREENHOUSE PLANET

Our planet has life because of greenhouse gases in the air. They keep Earth warm, so that water can flow and clouds can form to bring rain. Without these gases, most of the planet would be covered with thick masses of ice, and the air would be dry. The two most important greenhouse gases are water vapor and carbon dioxide. Molecules of these gases intercept heat radiation that attempts to leave the planet. As a result, the planet's radiation balance is achieved several kilometers up in the atmosphere, and the ground and lower atmosphere, where we live, are pleasantly warm and suitable for growing things.

Within the last several decades, where we live has been getting warmer. Greenhouse gases have considerably increased since the industrial revolution. Each year, the burning of fossil fuels produces additional carbon dioxide—roughly 1 percent of what is in the air already. The ocean takes up a large portion of it, but nevertheless, the atmospheric content of this gas has been increasing by just under 0.5 percent annually. Methane, another greenhouse gas, also has been increasing substantially. In response to these changes, Earth is warming. Countervailing effects—heat uptake by the ocean, shading by effects from air pollution—are slowing the process somewhat, but warming continues.[1]

The physical interactions between the various elements of the climate system are very complex, and there are natural variations producing warming and cooling in the climate on the scale of decades and centuries. Thus, it is not possible to state precisely just how much of the observed warming owes to the man-made increase in greenhouse gases. Also, as the climate warms and the atmosphere changes composition (including an increase of particulate pollution), the rules are changing, especially concerning the formation of clouds and their role in the heat budget as heat traps and reflective umbrellas. Thus, projections of future warming effects are subject to considerable uncertainty—things could change more, or they could change less than the best guesses offered by experts indicate. The uncertainties (but not the fact that man-made greenhouse gases produce warming) are the subject of much research and discussion among active geophysicists. Others, with little or no research background in the relevant sciences, also participate in the discussion, motivated by various political or economic concerns. (Since global warming has political and economic impacts this is a perfectly reasonable development.)

What is of interest to ocean scientists is how the ocean will respond to warming. Since the ongoing warming is greater in high northern latitudes than in low latitudes, temperature gradients in the Northern Hemisphere will decrease, zonal winds will decrease correspondingly, and mixing and upwelling will decrease as a result. Productivity will drop. What will happen in the south is less clear: enormous ice masses on Antarctica are resistant to removal, which stabilizes the existing situation. Anticipated changes will affect the uptake of heat and carbon dioxide by the global ocean, and its patterns of productivity. In any case, sea level will continue to rise globally, and the rise might well accelerate to rates considerably higher than those of the twentieth century. On the whole, the weather will become less predictable. In addition, with continued warming from burning coal and petroleum, the risk from unanticipated and troublesome events will keep increasing.[2]

THE SCOPE OF THE GREENHOUSE PROBLEM

Our planet is habitable because certain trace gases in the atmosphere of Earth keep it from freezing over. This benign and welcome consequence of having the atmosphere we have is known as the *greenhouse effect*. The trace gases keeping the planet warm—the greenhouse gases—are water vapor and carbon dioxide, mainly. At the ground, the difference in temperature with and without greenhouse gases is around 32 °C.[3]

To satisfy our energy needs in a modern industrial society we burn fossil fuels: coal, oil, and natural gas. This produces prodigious amounts of carbon dioxide, a colorless and odorless gas that enters the atmosphere. As a greenhouse gas, carbon dioxide absorbs a portion of the heat radiation from the ground and lower atmosphere, radiation that would normally escape into space (fig. 15.1). Thus, the large-scale release of carbon dioxide into the air impacts the radiation balance of the planet in such a fashion as to warm the ground and lower atmosphere beyond the normal background greenhouse effect. Holding more water vapor, the warmed air traps yet more heat. The combined warming effect, which ultimately results from the burning of carbon-based fuels, has given rise to much concern.

Many questions arise in connection with the excess greenhouse effect. Potentially there are unwelcome side effects of global warming, with impacts on the use of energy resources, on agriculture and public health, on coastal habitats (from the rise of sea level), on water resources and forest survival, and even on economic stability and international affairs. These issues engender lively public debate, which started largely as a result of odd weather patterns in the 1980s, and has continued since that time.[4] Simultaneously, the scientific consensus has shifted over the decades, from a skepticism that an excess greenhouse effect can be detected (up to the mid-1980s), to determining that it exists and produces significant warming (from the mid-1990s), to a realization that further warming is inevitable and will call for adaptations, in the most recent discussions.[5]

The chief consequence of the massive release of carbon dioxide to the atmosphere, global warming of the planet, was pointed out as early as the end of the nineteenth century, by the Swedish chemist Svante Arrhenius.[6] In the 1930s, the British physicist G. S. Callendar presented evidence for ongoing warming and ascribed it to a rapid increase in carbon

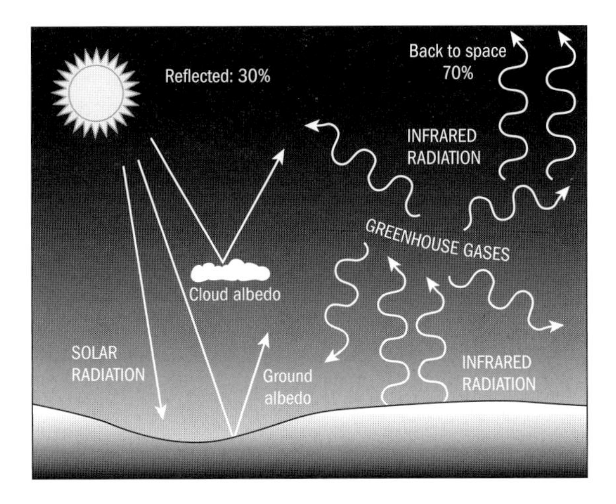

FIGURE 15.1. Radiation balance, light and heat radiation. Schematic.

dioxide in the atmosphere. Also, the American engineer and physicist G. Plass studied the physical details of the radiation balance involving greenhouse gases.[7] Plass stated, in 1956, that an uncontrolled experiment was being performed on the planet, which is bound to result in considerable warming that should be recognizable by the end of the century (that is, in the 1990s).

One widely quoted and paraphrased statement regarding the excess greenhouse effect is that of Roger Revelle and Hans Suess, published in 1957: Human beings are now carrying out a large-scale geophysical experiment of a kind that could not have happened in the past or be reproduced in the future. Within a few centuries we are returning to the atmosphere and oceans the concentrated organic carbon stored in sedimentary rocks of millions of years.[8] The conclusion was that detailed studies should be made to follow the course and outcome of the experiment. The first step in this process was to measure the carbon dioxide content of the air as precisely as possible. It was for this purpose that Revelle recruited Charles David Keeling to Scripps, who promptly started his systematic monitoring of carbon dioxide, producing the famous Keeling curve that now forms the basis of all discussions on the issue of global warming.

The statement about the global experiment is readily understood. We, the people living on this planet, are introducing more than 1 ton of carbon per person into the Earth's atmosphere, each year, mostly in the form of carbon dioxide.[9] The gas, as such, is natural and poses no particular problems in the environment at low concentrations. It is the same gas that we exhale from our lungs and is the result of combustion of carbon compounds with the air's oxygen. The burning within our bodies delivers the energy that keeps us warm and functioning, and the same is true for all other animals, from mammals to fishes to ants and snails. Green organisms use the carbon dioxide to grow and are in turn eaten again by animals; that is, the carbon moves in cycles and normally does not accumulate in the atmosphere. The problem is that we are burning fossil fuels (coal and petroleum) at a rate hugely greater than the natural cycles can accommodate. Much of this carbon is no longer readily recycled, therefore. It stays in the air, where it adds to the greenhouse effect.[10]

By 1960, scientists concerned with the relationship between carbon cycle and climate were aware of a number of crucial elements within the grand geophysical experiment: 1) warming by the added carbon dioxide would increase the amount of water vapor in the air, which would lead to further warming; 2) removal of snow and ice would darken the ground and the sea surface, which would lead to further warming; 3) reactions in seawater with the invading carbon dioxide would go in a direction such that additional invasion will be more difficult, leaving an increasing proportion of the carbon dioxide to accumulate in the atmosphere. These feedbacks in a warming climate are now very familiar and are considered basic to understanding the system.

Of the main two greenhouse gases, carbon dioxide and water vapor, the latter is actually the more powerful. The water vapor is delivered from the ocean, and also from plants and the soil. It returns to the ocean or to the ground as rain or snow, as part of the grand hydrological cycle. The warmer the air, the more water vapor it can hold, and the stronger, therefore, the greenhouse effect. Neither of the two major greenhouse gases interferes with the sunlight on its way to Earth's surface. However, when that same surface sends heat radiation toward space, the molecules of water and of carbon dioxide in the air absorb a certain portion of this radiation, and reradiate the energy back in all directions (fig. 15.1). As a consequence, the ground and the lower atmosphere heat up. The feedbacks that kick in as a result of this warming are called *positive* whenever they tend to favor additional warming, as is the case for albedo change involving sea ice, for example. If they oppose the warming, they are called *negative feedback*. An increase in reflective clouds in low latitudes might provide such negative feedback under certain conditions.

The addition of greenhouse gases to the atmosphere *must* produce a warming of the surface of the planet, on the ground and in the sea. This consequence cannot be argued; it is plain physics and not subject to ifs and buts. As it turns out, a rise in temperature near the ground also increases the methane content of the air from various biological processes in the soil. Each methane molecule (a carbon atom surrounded by four hydrogen atoms) is many times more effective than each carbon dioxide molecule in providing for additional interception of heat radiation trying to get out to space. None of these processes is in dispute (not among experts, anyway).

The open questions are how much warming will occur, precisely, upon increasing the carbon dioxide content of the atmosphere; where it will occur and in what season; and what kinds of changes we should expect as a consequence in the climatic conditions of the various regions where people live and grow food.

Among scientists, the basic question about the sensitivity of climate change to an increase in the greenhouse effect is commonly phrased in terms of doubling the amount of carbon dioxide in the atmosphere. Estimates for global warming for a doubling of carbon dioxide typically range from 1.5 to 4.5 °C.[11] If we (arbitrarily) ascribe all the observed warming since the industrial revolution to the introduction of carbon dioxide to the atmosphere, the resulting sensitivity emerges as 2 °C for a doubling of the carbon dioxide content.[12] Of course, given the short time scale for increasingly massive input (the last 50 to 60 years), we are dealing with a transient, and not with equilibrium. Thus, this value of 2 °C simply says that the observed warming is within the range expected and holds no surprise. But it does not really give us the sensitivity of the system, which by definition involves equilibrium.

As far as changes in regional conditions, they have to do with the patterns of warming (greatest in high northern latitudes, least in the tropics) and with changes in available moisture (expansion of the tropical rain belts, and poleward movement of the northern desert belt). Each individual region will react differently to such forcing, depending on the topography, the distance from the sea, the direction of winds, and other factors linked to regional conditions.

At the time Revelle started thinking about the implications of the grand experiment, there was little or no public or professional concern with the issue. In fact, little was known in the 1950s about the details of the composition of the atmosphere, the exchange of carbon between ocean and atmosphere, the way the ocean redistributes heat, and the role of water vapor and clouds in the climate machine.[13]

Ocean and atmosphere are in constant exchange, regarding carbon dioxide. The balance between the two reservoirs is such that the ocean has more than 50 times the carbon content of the atmosphere. Thus, an obvious expectation is that any carbon dioxide released to the atmosphere by human activities would be shared in these same proportions: 98 percent for the ocean, 2 percent for the air. If this expectation were realistic, on a human time scale, we would not have much of a problem. However, as Revelle and Suess realized, the ability of the ocean to take up excess carbon dioxide from the air is limited by the ocean's content of carbonate ion, which reacts with carbon dioxide to make bicarbonate.[14] The carbonate ion makes up only some 10 percent of the ocean's carbon content. If we accept the abundance of the carbonate ion as the limiting factor on relevant time scales, the sharing between ocean and atmosphere of newly introduced carbon dioxide is closer to 5 to 1, rather than to the overall 50 to 1. In addition, only the upper part of the ocean is well mixed and can respond readily to changes in the chemistry of the atmosphere. This additional constraint makes it likely that more than half of the carbon dioxide introduced to the atmosphere will stay within it. The evidence agrees; that is, observation indeed suggests that uptake by the ocean is comparatively modest, around one-third of the human input to the system.[15]

As time goes on, the proportion of the man-made carbon dioxide that stays in the air, the airborne fraction, is expected to increase. The atmospheric carbon content is now greater than the preindustrial value by a factor of 1.3, and this factor is being rapidly augmented by a growing input of carbon dioxide, which is approaching 8 billion tons (or gigatons) of carbon per year at present, and growing. This is slightly more than 1 percent of the original atmospheric reservoir of carbon, each year. As the upper ocean warms, the fraction taken up by the sea should decrease from a loss of carbonate ions in upper waters.

The fact that climate has been more or less in balance for the last 10,000 years suggests that there are negative feedbacks that keep the system from moving away from an average condition. The relative importance of negative and positive feedbacks determines the variability of the climate system. These factors are very difficult to measure, and there is consequently some considerable uncertainty about the sources of variability, which affects the construction of programs for simulating climate on large computers. Discussions among experts turn largely on the question of which feedbacks can be neglected, and which must be included in a model, and where the level of power is to be set for each one that is included.[16]

One way to obtain clues to the workings of the climate system from sources other than building models is to study the experiments that nature has performed in the past. Of special interest are past events that represent climate excursions that are out of the ordinary, since from such events one stands to learn the most. By elementary statistics, short time series—such as are available to meteorologists—have but a slim chance of containing large disturbances. Thus, models based on short experience (decades) tend to be well behaved, reflecting the humdrum of a modest range of variation. The longer time series, of course, offer a greater chance for finding surprises in the patterns of climate change. However, the price for using the longer repository of experience (looking back into centuries and millennia) is that measurements are by necessity indirect. No one measured temperature before the thermometer was invented. And even after there were thermometers, they were rarely systematically applied before the twentieth century. Tree rings and pollen, fossil beetles and fossil plankton, coral growth bands and isotopes—these must stand as proxies for temperature, precipitation, and other items related to weather (fig. 15.2).

Some of the results of proxy studies are truly remarkable. The most exciting discovery of the last three decades, in this context, is the finding that trace-gas content in the atmosphere can be faithfully reconstructed for tens and even hundreds of millennia, from the gas content trapped in polar ice.[17] It turns out that carbon dioxide values fluctuated greatly on a scale of 100,000 years, falling to levels below 200 parts per million during glacial periods, and reaching 300 parts per million during maximum warmth. The sense of this variation is as expected if carbon dioxide plays a strong role in climate

FIGURE 15.2. Trees, where growing seasonally, retain information in their growth rings about spring and summer conditions (precipitation and temperature) for many years. Such information is useful in reconstructing climate history well beyond instrumental measurements.

change. However, it is sobering to contemplate that the range of variation observed is as yet unexplained.[18]

The fact that there is no generally accepted explanation for the ice age variations discovered in fossil atmospheric carbon dioxide correctly reflects the state of the art regarding our understanding of the interplay between carbon dioxide and climate, when large changes are involved. Additional work will keep improving our knowledge, but large uncertainties will likely remain.

GLOBAL WARMING: FACTS AND GUESSES

No scientist who has anything to say on the matter now doubts any longer that global warming is real—we are in it. And few doubt that there is a strong human component to this warming.[19] The times when serious scientists could claim that global warming is mainly a political issue is definitely past.[20] To be sure, it *is* a political issue, because responsibilities and potential costs and benefits of ongoing climate change are distributed unequally, and there is the nagging question of what should be done about the warming, if anything. But to scientists it is first and foremost an issue in geophysics, as realized by Revelle half a century ago and by many active workers since. Announcing otherwise serves no useful purpose.

There is no surprise regarding the warming itself. The content of greenhouse gases in the air rose substantially in the past century, roughly by the equivalent of a factor of 1.5 over background, when accounting for the rise of methane.[21] A rise in temperature resulted, based on physical principles. These principles are so reliable that, even if a cooling is observed for a decade (as happened in the 1960s) we know that the warming is in fact proceeding in the background, albeit masked by some other factor, whose nature may be obscure. Furthermore, the physics involved is such that the observed warming is on the low side of what might be expected, because equilibrium is not achieved. Mainly, heat uptake by the ocean slows the process. In addition, human impacts are at work, increasing the amount of reflective particles in the air and thereby affecting the radiation balance.[22] The expression "global dimming" has been used to tag the effect. To derive comfort from seeing less of a change than expected may be premature: the winds preceding a storm may be tolerable, but it is the storm that matters.

How can we be sure that global warming has arrived, and how can we be reasonably certain that most of it is a result of human activities? The central piece of evidence comes from climate history. The crucial observation is that the present average temperature of the Northern Hemisphere has risen well beyond the usual fluctuations of the last thousand years.[23] It is important to realize that "unusual" is a statistical concept. Whether something is or is not unusual is not a matter of everyday geophysics and weather observation. The question can be assessed only against the background of history.[24] The likelihood of getting such unusually warm years as we have had in the last three decades by chance, *when such warming is actually expected*, is very small indeed. We are quite safe in postulating the connection between expectation and observation.[25]

Important evidence comes from monitoring the main driving factor, carbon dioxide. As mentioned, continuous monitoring of this gas started

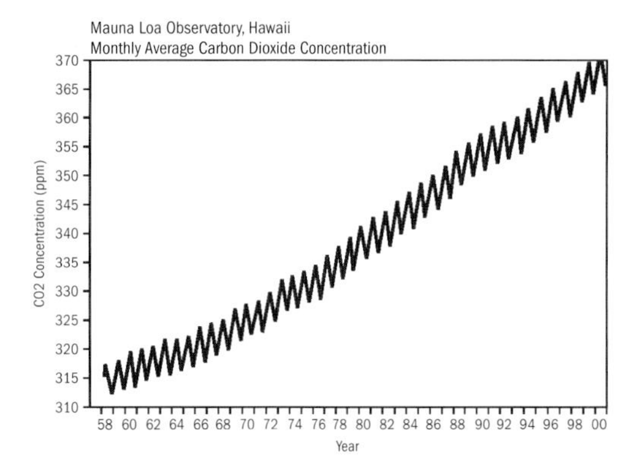

FIGURE 15.3. The Keeling curve, showing the inexorable rise of carbon dioxide, and the shack on Mauna Loa where the measurements were made—arguably the most important geophysical observatory of the twentieth century.

in 1957, with the pioneering efforts of C. D. Keeling, on Mauna Loa (fig. 15.3).[26] The fact that the data for successive measurements did not agree precisely worried him—he checked and rechecked his instruments to make sure that they were not the source of the variation.[27] What emerged is that the carbon dioxide in the atmosphere varies substantially with the seasons. It was a major discovery. Keeling realized that the variation reflects mainly the growth and decay of plant matter in the Northern Hemisphere. The continuous measurements he started mark the beginning of a systematic study of the chemistry of the atmosphere in the context of global change. This type of survey was then unprecedented. It was also unplanned. It was the discovery that carbon dioxide rose visibly from year to year, in addition to varying seasonally, that suggested monitoring at high altitudes as the only valid approach to the problem of atmospheric change.

No one doubts the reality of the carbon dioxide increase as described in the Keeling curve. And there is no doubt whatever that the rise produces warming, based on physics. Also, there is hardly any doubt that this carbon-driven warming is strong enough to be measurable. Contrary views do exist (as was the case with seafloor spreading, for about a decade after discovery), as should be expected for important issues that affect the ways we look at the planet. These views will fade away as reality takes over.

As mentioned, in 1996 the Intergovernmental Panel on Climatic Change (IPCC), an organization that advises governments on the progress of global warming and its implications, stated,[28] "The balance of evidence suggests that there is a discernible human influence on global climate." This cautiously phrased assessment brought forth a surprisingly vigorous attack from a number of dissenting critics. Knowledgeable scientists, however, simply saw the statement as reflecting fact; it seemed rational to assume that the distinct rise in greenhouse gases from human activities has something to do with the observation that temperatures on the Northern Hemisphere rose beyond values experienced since the invention of the thermometer.[29]

One issue that is commonly raised in the course of discussion, very reasonably, is the trustworthiness of models. It is quite obvious that computer models testing the system's response to the introduction of greenhouse gases do not agree in their output. It is clear that only one model can be right, which makes the others wrong. From there it takes no great logical leap to conclude that probably all of the models are wrong, on some level.

What to scientists are simply numerical experiments designed to deepen the understanding of climate mechanisms may be seen by intelligent lay observers as (failed) attempts to predict the future. However, there are points that the models do agree on: a strong warming of high northern latitudes, a moderate warming

FIGURE 15.4. In the Rocky Mountains, glaciers are retreating. This tongue of the Columbia Ice Field in the Canadian Rockies filled the entire valley below the ice a few decades ago. The moraine ridges mark the boundaries of the "ghost glacier."

of temperate latitudes, and relatively little warming in low latitudes. The discrepancies, mainly in the details and in the regional consequences, are to be expected. Discrepancies are normal: the models use different resolution, and they make different assumptions about the interactions of atmosphere and ocean, and about the role of clouds.

In summary, the most convincing evidence for the reality of global warming comes from the fact that the last quarter century has seen highly unusual warming, compared with the last thousand years, precisely as expected from the physics of radiation balance. To invoke causes other than human would be to substitute the contrived for the plausible.

The consequences of warming are not exactly obscure: they are obvious to anyone who takes the trouble to visit the Rocky Mountains, for example. Mountain glaciers, on the whole, are retreating rapidly (fig. 15.4).[30] In a few decades, the Glacier National Park in Montana will have little but the ghosts of former glaciers, vaguely outlined by moraines. The retreat, it is true, has been going on since the second half of the nineteenth century, which invites invoking natural causes such as a brightening of the Sun. But in the last few decades, the rates of retreat have greatly accelerated. Incidentally, when calling on the Sun to produce warming, it must be kept in mind that the Sun's brightness varies but modestly over the solar cycles. Changes in solar forcing, measured in terms of the energy received on the ground, are thought to be less than the additional forcing by the excess greenhouse effect, similarly measured. Thus, if the Sun is important, so is the greenhouse effect. This is not to say that the Sun's variation in energy output can be neglected. There is evidence that such variation matters.[31]

There has been concern, in some quarters, that some of the statements in the IPCC reports have been alarmist. There is in fact nothing alarming at all about the IPCC statement about a human influence on the recent warming. The warming could be either beneficial or detrimental or irrelevant to human affairs. The late Mikhail Budyko, a distinguished Russian hydrologist and climatologist, asserted that emitting carbon dioxide should allow the Siberian steppe to become a breadbasket for a hungry world. He was much in favor of burning more coal.[32] The German meteorologist Hans von Storch and the sociologist Nicholas Stehr state that they "strongly believe that near-surface temperatures are rising" [in response to the accumulation of greenhouse gases] but "are not convinced that present and future climate change will have a significant impact on society and global ecosystems."[33] Thus, basically, they see no problem.

While many thoughtful scientists do not agree with the sanguine stance of Stehr and von Storch,[34] their assessment correctly makes the point that the problem that needs addressing, in the context of sociology and ecology, is with the effects from warming, not so much with the precise cause of the warming. It is no longer necessary to argue about the physics. We can see trees topple and houses sink into the melting permafrost in Alaska; we can observe Arctic sea ice shrinking as never recorded before, and we are witness to the fact that tropical glaciers in the Andes are melting for the first time in 10,000 years. Clearly, changes are afoot, and no amount of wishful thinking will make them go away. It is time to face the facts.

It is time, as the American climate scientist James Hansen said some 20 years ago, to stop waffling and admit that global warming has arrived. A number of recent historical-geophysical studies have now delivered incontrovertible evidence that this is indeed the case. In the reconstruction of P. D. Jones and colleagues, in 1998 (who used information from tree-rings, ice-cores, corals and historical documents to reconstruct the global temperature history of the last millennium), the warming in the last two decades of the twentieth century is unprecedented over the time interval considered. Similar studies were made by others, with similar results.[35]

It appears that much of the feedback in a warming climate is positive, meaning that the changes reenforce the warming trend. Water vapor is the most important factor: it increases with warming and itself adds to the greenhouse effect. As the ocean warms, it can hold less carbon dioxide, making it more difficult to get rid of much of the man-made carbon dioxide through uptake in the ocean. As the ground warms in high latitudes, reflective snow is replaced with dark soil, which absorbs more sunlight. Sea ice forms later in the year and melts earlier—open waters are dark. Even increased winter snow (from increased moisture in the atmosphere) results in trapping heat in the ground from the previous summer. Such snow is a negative feedback only if it stays for spring and summer, to reflect sunlight. In addition, methane production appears to be increasing, on the whole, as the planet warms.

In any case, the Grand Experiment itself, and not our guesses, will decide the questions concerning the future developments associated with global warming. We shall have more accurate answers in the next several decades. The hope must be for a lack of unpleasant surprises.

ON THE OCEAN'S RESPONSE TO GLOBAL WARMING

We should expect the ocean to react to global warming, and we should be able to observe the reaction. The problem is how to sort out responses to the ongoing greenhouse-driven warming from the natural background fluctuations. This is difficult to solve, especially where natural fluctuations are large or reliable records of past ocean conditions are short. For physical conditions, we can now rely on satellite surveys yielding data on wind intensity, wave fields, strength of currents, and changing temperature and precipitation patterns on a scale of two or three decades. We can see the sea ice shrink in the Arctic, and we can observe the timing of plankton blooms in various regions of the sea.

But what about long-term variability? And what about biological responses, in some detail? So far, such data are available only for regions where systematic monitoring went on for many decades. The California Current and the North Atlantic are such regions.

With regard to plankton in the North Atlantic, there is in fact an interesting time series available for study. Alister C. Hardy, the famous British marine biologist, was very interested in the changing abundance ratios between phytoplankton and zooplankton, and in the patchy distribution of plankton in general, both geographically and through time.[36] Hardy was very aware that the success of spawning and recruitment of herring show large changes through the years. To find out why, one needed to study the year-to-year changes in the abundance of the plankton, that is, the food for herring. To this end, in 1931, he founded the Continuous Plankton Recorder Survey, based on the use of a plankton-filtering device he had invented in the 1920s.[37] The first surveys were between Hull at the Humber Estuary, in England, and Bremen on the Weser, in Germany. In the 1940s, the survey expanded to include, first, much of the northeastern North Atlantic and later the ship routes to New England and areas off the East Coast. An atlas of North Atlantic plankton was published in 1973.[38]

The impact of the plankton recorder on the world of biological oceanography appears to have been rather modest in its early decades.[39] However, recently the project has proved its value as a source of historic trends in plankton abundance: data collected since the late 1940s allow assessment of some long-term changes. It appears that there was a distinct decrease in phytoplankton abundance between 1960 and 1970 and that abundances stayed low in the following two decades. Zooplankton abundances follow this trend rather closely.[40] The question to what degree this change reflects some sort of response of the ocean to global change is open, of course. It is certainly possible, but it would be good to have other series for comparison. The point is, how are we to measure the response without the type of data that Hardy's multidecadal survey provides? Monitoring is not exciting science, and funding for this sort of thing is difficult to obtain.[41]

Observations should be set against expectations. Can we construct some reasonable expectations about the ocean's response by arguing from first principles? We know that the strength of currents depends on the strength of the winds driving them. One expectation is a weakening of zonal winds on the Northern Hemisphere, from a decrease in the temperature gradient between the tropics (where temperature change is small) and high latitudes (which are warming rapidly).[42] Zonal winds drive the central gyres, and so we should observe a broadening of these gyres, allowing warm water to spread into the coastal upwelling regions of the eastern boundary currents—such as off California. It so happens that we can test this idea, because here we have the longest-running record of temperature and productivity of any coastal region in the world.[43]

First of all, as described in chapter 7, the California Current has indeed experienced unusually warm temperatures in the last quarter of the twentieth century. A drastic drop in productivity accompanied the warming since the mid-1970s. Working with a data set spanning more than 40 years, oceanographer Dean Roemmich established a general warming across the current, by almost 1 °C, for the upper 100 meters or so.[44] A dramatic decline of zooplankton abundance, after 1975, has accompanied the warming, with effects reverberating up the food chain, as discussed by Scripps biological oceanographer John McGowan and associates.[45] Furthermore, the regime shift apparently concerns the entire West Coast, and indeed the entire North Pacific and the tropical Pacific involved in El Niño oscillations. Indeed, the first indication that a regime shift was underway emerged in a study of chlorophyll distributions in the central North Pacific.[46]

The discoveries of plankton ecologist Elizabeth L. Venrick and associates regarding chlorophyll distribution in the North Pacific suggested a strengthening of the Aleutian Low during winter (presumably from increased vapor supply from the tropics, via the Kuroshio pathway). Additional evidence for the vast scale of change came from warming of the surface waters off Peru and Chile, after the mid-1970s, with a concomitant decline in zooplankton abundance.

We might rightly conclude, from these various observations, that warming of coastal oceans is detrimental to zooplankton production there. The question that remains, of course, is whether the observed warming is part of the global trend that started a century ago, or whether it is mainly part of a natural background oscillation. In any case, it may be safely assumed that the overall warming of the ocean, from an increased concentration of greenhouse gases in the atmosphere, has consequences for the biological production in the surface waters. Also, the observed changes in productivity (a general decrease in regions depending on upwelling and upmixing of deep nutrient-rich waters) are such that they can be plausibly related to expectations.

At this juncture, the mechanisms by which biological changes are affected by climate change remain obscure. In their search for such mechanisms, McGowan and his colleagues have recently commented as follows:

> The biological consequences of climatic variability of the atmosphere and ocean are largely unknown. This is probably because of the mismatch between the scales of important atmospheric and oceanographic processes and the spatial and temporal dimensions of biological research programs. However, there is widespread consensus that marine populations respond to climatic events and that major changes have taken place in the past 20 years in the marine ecosystems of the Pacific.[47]

Assessing the ocean's response to the general warming seen in temperate to high latitudes in the Northern Hemisphere in the last several decades is difficult because of long-term climate fluctuations on a scale of decades to centuries. Such fluctuations are now well known from ice-core records.[48] There is yet another problem in correctly interpreting historical records bearing on marine biology: effects from climate change (whatever the origin) are not necessarily the main factors changing marine ecosystems at the present time. The factors depend on location and circumstances. In many coastal regions there is no question that overfishing has had major consequences for ecosystem structure.[49] Restructuring of the ecosystems (and, in some cases, collapse to primitive trophic levels) results from the removal of large predators and large grazers.

There is no reason to expect pelagic ecosystems to be insensitive to the removal of enormous masses of herring and cod and their relatives. For example, herring feed selectively on large calanoid copepods, given a choice.[50] Thus, after removal of herring from the North Atlantic, large calanoid copepods have improved survival chances, compared with their competitors. Consequently, any increase in *Calanus* would not necessarily signal a change in the conditions of production of the sea. In turn, *Calanus* feeds preferentially on certain species of diatoms, and so on. The point is this: we may notice the change (assuming that proper monitoring is in place), but we shall be unable to explain any of the more subtle changes convincingly.

How will global warming affect the exchange of carbon dioxide between the ocean and the atmosphere? What should be our expectation? We have already touched on this problem, when discussing the role of carbonate ions in the uptake of excess carbon dioxide by the ocean, and we have mentioned that it is primarily the upper layer of the sea that is available as a receptive reservoir. The first factor, the concentration of carbonate ions, describes the chemical state of the water; that is, its buffer capacity. When carbon dioxide reacts with water (which is why it invades water readily), the reaction produces carbonic acid. The carbonate neutralizes the acid, allowing additional reaction to proceed. It is used up in the process. Thus, acid builds up and this slows the invasion of the sea by carbon dioxide. With time, therefore, more of the carbon dioxide stays in the air. The second factor, the thickness of the water layer that actively exchanges gases with the atmosphere, is especially important, because it is sensitive to climate change. Of course, on long time scales (2,000 years) the entire ocean exchanges with the atmosphere. But on a scale of a decade to a century, only the uppermost few hundred meters are involved.[51] Warming the upper layer decreases its ability to exchange with deeper layers. Also, warming decreases the water's ability to hold carbon dioxide in solution. As the surface layer warms and acidifies, it becomes less able to act as a sink for excess carbon dioxide.

The acidification of the surface layer has consequences for the ecology of the ocean. A decrease in carbonate ion will require more effort from carbonate-precipitating organisms (coral, calcareous algae, mollusks, foraminifers, bryozoans, echinoderms, serpulid worms, and others) to build their shells and skeletons. This will give a competitive edge to those organisms that do not build calcareous hard parts, which is bound to change the overall functioning of shallow-water ecosystems, including coral reefs.

Much of the change seen in the Pacific, concerning the overall decrease of productivity, can be interpreted as a new prevalence of El Niño conditions, with such conditions having been favored since the early 1980s. As discussed in chapter 7, the El Niño phenomenon is tied to an oscillation in large-scale atmospheric pressure patterns in the tropical Pacific, which reflect changes in wind patterns and migration of the centers of drought and of precipitation. The two largest known El Niño events are those of 1982/83 and 1997. The fact that they occurred toward the end of the twentieth century fits the assumption that warming favors the development of extrastrong El Niño conditions. According to numerical climate experiments, the proposition that it is the global warming that produces extraordinary El Niño events cannot be dismissed. If it holds up, the widespread coral bleaching that accompanies such events must be assigned to global warming, albeit indirectly.[52] In any case, it is the compounding of stresses from different sources, both natural and human-induced, that will affect the ability of impacted ecosystems (such as reefs) to recover toward normal condition. When this rate slows sufficiently to prevent restoration before the following stressful event sets in, the affected ecosystems are bound to change markedly, usually in favor of opportunistic species.

SOME LESSONS FROM THE PAST

Two time scales are of special interest when attempting to assess the recent warming in context of history: the last several hundred years back to a thousand years ago, and the ice ages back through the last million years, and especially the transitions from glacial periods to warm periods (with emphasis on the last such transition).

Each of these time scales has lessons that are useful. The historical narratives expand our horizons well beyond the realm of known physical principles, especially when combined with rules derived from several decades of measurement. History adds a wealth of experience gleaned from experiments run by nature in the past. Such experiments tell us what is possible. If something has happened, it evidently can happen. And if it can happen, it will happen again, given the right conditions. Of course, since analogies are never perfect, we are still guessing. But the guesses gain in credibility if they can point to precedent.

The scale of history of the last thousand years is attractive because it includes the time when writing first became widely adopted in Europe, north of the Alps.[53] Written records can be found that are concerned with harvests and other matters related to climate; thus, they are potentially a rich source of information about long-period large-amplitude variations in the North Atlantic realm. The British meteorologist Hubert H. Lamb did the outstanding pioneer work in this field, and his books on the subject are classic references.[54] Perhaps the central discovery of this work was the realization that in northwestern Europe, the last millennium contains a highly unusual sequence of climatic conditions marked by many extreme winters and poor harvests, called the Little Ice Age. This period of harsh weather lasted roughly from the fourteenth century to 1850. During its reign, mountain glaciers expanded in the Northern Hemisphere, large permanent snowfields developed in Canada, and sea ice lingered unusually long around Iceland and in the Baltic Sea. Fierce winter storms battered the shores of northwestern Europe—to the extent that Denmark lost much coastal real estate.[55]

In assessing global warming and the human role in it, we have to deal with the fact that the early nineteenth century was especially cold. Naturally, this raises the question of whether much of the warming that Earth has recently experienced was to be expected anyway—the system simply started moving back toward some more "normal" condition. The answer depends on the reasons given as a cause for the Little Ice Age. Among the possible reasons are low solar activity and increased volcanism. If this is so, increased solar activity and reduced volcanism are mainly responsible for the warming in the nineteenth century. In this view, the Little Ice Age is an anomaly, and the warming after 1850 simply gets us back on track.[56] The argument is entirely reasonable, at least for a considerable portion of the observed warming into the middle of the twentieth century. However, we cannot exclude the possibility that the system was largely responding to the addition of carbon dioxide and other trace gases to the atmosphere, from human activities.[57] Whenever we aim to reconstruct conditions for the last millennium to learn about the ocean's response to warming, we come up against this Little Ice Age conundrum—the one clear example of warming in the last century also is the one example that has no good earlier analogs.

An intensely studied marine high-resolution record is the one in Santa Barbara basin, southern California. In the almost 600-meter (2,000-feet) deep center of the Santa Barbara basin, low-oxygen concentrations in near-bottom water exclude large burrowing animals, and this prevents destruction of detail within the sedimentary sequence. Millimeter-scale laminae of alternating light and dark sediment (reflecting the seasons) preserve a detailed record of productivity, on a year-by-year scale. The record is read from the content of plankton remains, such as diatoms and planktonic foraminifers. The recent decrease in productivity of the California Current, in the 1970s, is well represented in these sediments, as a drastic drop in diatom abundance. Concomitant increases in sea level height, coastal sea-surface temperature, frequency of occurrence of El Niño events, and a decrease in the upwelling index after the 1970s suggested to Scripps biologist Carina Lange and associates (in 1989) that they were observing the response of the California coastal ecosystem to a large-scale, long-term change in atmospheric and oceanic circulation of the northeastern Pacific. More recently, a careful study of changes in the abundance patterns of planktonic foraminifers by Scripps graduate student David Field indicated that conditions in the basin have favored an increase in warm-water species within the second half of the twentieth century, and that the change since the mid-1970s is unprecedented in the record of the last 1,400 years.[58]

Since the El Niño phenomenon is driven by atmosphere-ocean interactions in the tropical Pacific, coral skeletons in the tropical Pacific should be a prime source for the history of such phenomena. The systematic collection of coral records for the study of climate fluctuations in the tropical Pacific started relatively recently, in the 1980s.[59] The coral records are indeed useful in reconstructing periods of unusual conditions. Records are being collected in many different places—in Bermuda, in the Caribbean, all along the tropical Pacific, in the Great Barrier Reef, and in the Red Sea. Such records are analyzed for the history of the rate of growth, as well as for chemical changes in the carbonate skeleton, such as the oxygen and carbon isotope ratios.[60]

The largest ranges in past climate fluctuations that we might study to assess the response of the ocean to climate change are those associated with the ice ages, as discussed in chapter 13. For the last half million years or so, we have much fascinating information both on the variation of carbon dioxide and on the response of the ocean to warming (on a thousand-year scale, that is).

First of all, it is of interest to note that our own climate conditions are way off the average with respect to the conditions of the last 800,000 years or so, that is, the period of the grand ice age cycles. Within the range of oxygen isotope fluctuations as seen in the benthic foraminifers on the deep-sea floor, present values are high on the "light" (that is, warm) side. The sea level was as high as or higher than today only a few percent of the time over the period considered. What this means is that the future will be largely outside of experience with regard to climate fluctuations of the recent geologic past. The same is true for greenhouse forcing: values of atmospheric carbon dioxide as high as today simply did not exist over the last half million years, as we know from Antarctic ice core data.[61] Thus, again, whatever interesting patterns of response of the ocean to warming that we might extract from the ice age record is of limited relevance to the present and the future.

Nevertheless, of course, the information provided by the record is of great interest, because it has clues to the sense of changes that occur upon warming, and to the rates of change that might be expected. We may assume that studying the past can help us to get the sign of the change right, if little else.[62] The record, in principle, can offer clues on how to answer vitally important questions such as these: Will central gyres expand? Will the deep circulation become less vigorous? Will there be more fish or fewer? But we should realize that the answers are bound to be tentative.

In response to the discovery (published in 1980) that the carbon dioxide content of the atmosphere rose greatly at the end of the last glacial,[63] an enormous literature developed attempting to explain how the ocean shares carbon dioxide with the air.[64] Since the atmosphere is a relatively small reservoir for carbon and is intimately linked to the large reservoir that is the ocean, the uncertainties concerning the expected atmospheric level for different ocean conditions are quite large. We now know, thanks to the ice-core results in the last two decades, that the range of carbon dioxide values is rather well defined and steady through time.[65] There is a lower boundary near 180 parts per million (ppm)[66] and an upper one near 300 ppm. The rise from lower values in glacial periods (around 190 ppm) to higher values in interglacials (around 290 ppm) is remarkably rapid and is closely tied to the temperature changes seen in the ice. In turn, the temperature changes are highly correlated with the oxygen isotope record in deep-sea sediments. This allows a comparison of the climate record as seen in the ice cores with the changing conditions in the ocean, as documented in the deep-sea sediments.

Judging from the plethora of explanations put forward for the observed change in atmospheric carbon dioxide during transitions from glacial to postglacial conditions, it seems unlikely that the mix of mechanisms responsible for the rapid increase in this trace gas will emerge soon.[67] However, even without the benefit of a precise explanation, there is a message that can be read from the record: the system reacts to warming and sea level rise with an increase of atmospheric carbon dioxide. The transient nature of this response is not always sufficiently appreciated. It is tempting to assume some sort of equilibrium between the ocean and the atmosphere, in regard to carbon dioxide. However, equilibrium does not prevail during the great periods of melting, when the sea receives enormous amounts of freshwater that are difficult to mix downward.[68]

One corollary of the carbon dioxide story from the ice cores is that our inability to explain the past with respect to the variations observed must affect the confidence we can bring to guesses about the future course of ocean-atmosphere exchange. When working with known factors based on a few decades of precise measurements (and leaving out the various hidden factors that could help explain what we see in the record), we are bound to come up with predictions that are quite conservative.[69]

As concerns ocean productivity, especially in tropical and subtropical latitudes, it is closely tied to the response of the trade-wind system to warming and cooling, as mentioned. As the trade winds weaken in response to warming at the end of a glacial period, we see in the sedimentary record corresponding decreases in upwelling. This implies that apex consumers depending on high food density (such as marine mammals and seabirds) suffered decreases in populations in sensitive regions.

One concern, expressed by a number of ocean scientists, is that the ongoing warming might hold unpleasant surprises, in the shape of abrupt climate change from a reorganization of the ocean's deep circulation.[70] The last time the planet saw large-scale abrupt climate change—during the transition from the last glacial period to the present warm period—such change was intimately linked to a major extinction of large terrestrial animals (mammoth, wooly rhino, cave bear, sabertooth, and many others). It is fair to say that none of the sudden changes observed were predictable from the information available before they occurred.

As far as one can tell, abrupt climate changes during the period of melting had at least two causes: instability of ice sheets, and response of North Atlantic deepwater production to input of large amounts of meltwater. For both causes, one must assume that the abrupt change was forced when certain (unknown) threshold values were exceeded. Once initiated, the change was accelerated by positive feedback.

The concern arising, in particular, is that rapid heating in the far northern Atlantic, and the input of large amounts of meltwater, might prevent the sinking of surface waters in the Nordic Seas between Greenland, Iceland, and Norway. Such sinking, at present, is part of a global conveyor system involving the exchange of large amounts of water between Atlantic and Pacific and the transfer of heat into the North Atlantic, across the equator, and into far northern latitudes. When contemplating the much-reproduced graph describing the "global" conveyor (fig. 6.17), one has to keep in mind that it captures but part of what is important. Most importantly, it leaves out much of the dominant portion of the deep-sea circulation: the ring around Antarctica. In addition, it must be kept in mind that the transfer of heat across the Atlantic equator, from south to north, owes much to the north-south asymmetry in the radiation balance of the planet, which stems from the fact that Antarctica is fully glaciated while the Arctic is not. It is likely that this asymmetry will increase through global warming.[71] Thus, overall the heat transfer into the North Atlantic might well increase, rather than decrease as suggested on the basis of the conveyor-shutdown hypothesis.

Nevertheless, a shutdown event conceivably could occur, as a result of reaching a threshold in density contrast of surface and subsurface waters, and such an event undoubtedly would have untoward climatic consequences, especially in Scandinavia. Unfortunately, the outlook for predicting such an event is not encouraging. Threshold-driven events are familiar from earthquakes, where accumulated deformation leads to sudden rupture. The timing of threshold events is impossible to foresee except when the process has already started. Once it has started, it will run its course, urged on by positive feedback.

For relevant insights into the nature of threshold events we might again turn to the period of deglaciation, between 15,000 and 10,000 years ago. Threshold events apparently were important with respect to the stability of large ice sheets. A large ice sheet is self-stabilizing while undisturbed. It reflects sunlight, and the high-pressure zone it generates keeps any summer rain out. But once its glaciers speed up from warming, internal friction will further warm them. As they move, they draw down the ice of the mother field, which is similarly sensitive to frictional warming. In addition, cracks will appear that allow summer meltwater to enter the interior, accelerating the process of warming and decay. Small effects accumulated through time can have large consequences, and large amounts of ice can disappear quite rapidly: these are the chief lessons from the study of the end of the last ice age.[72]

THE BIOLOGICAL PUMP

Over most of the sea, the uppermost water layer is entirely depleted in nutrients through extraction by photosynthesizing organisms and by downward transport in settling organic-rich particles. The extraction of whatever nutrients were present originally (when the surface water first rose from below) is part of the process that fixes and removes organic carbon from this layer. Thus, nutrient-depleted waters are carbon-depleted, as well. The result is that the atmosphere sees a carbon-impoverished sea surface; consequently its carbon content is considerably lower than it would be for equilibrium with an average ocean (which would be mainly the deep ocean).[73]

This simple pattern has two corollaries. First, if the total amount of nutrients is increased in the ocean, this will tilt conditions further out of equilibrium, and the atmosphere will get even less carbon. Second, if the nutrient-depleted surface layer expands so that nutrient-rich waters take up a smaller area at the sea surface, less carbon dioxide becomes available to the atmosphere. Since these principles were elucidated by Lamont oceanographer Wallace S. Broecker,[74] they have become a standard tool in assessing changes in the strength of the biological pump through time; that is, the degree to which nutrient depletion in surface waters fosters lowering the share of carbon dioxide for the atmosphere.[75]

A third corollary is that if the areas now covered with nutrient-rich waters could be made to become nutrient-poor (like the other areas of the surface waters) then the carbon dioxide content of the atmosphere would have to drop. This is the concept that gave rise to the idea that adding iron to the high-nutrient regions of the sea surface would reduce the carbon dioxide in the atmosphere, as iron apparently is limiting to production in regions with a high nutrient content in surface waters.[76] The result of removing the limitation would be to strip carbon from these nutrient-rich areas and export it from surface waters to greater depth. In effect, so the iron hypothesis says, we would have strengthened the biological pump that removes carbon from contact with the atmosphere and sequesters it at depth, where the atmosphere cannot reach it.

The argument has merit. However, two types of concerns have been expressed in regard to this interesting idea. One is that the large-scale addition of iron to vast regions of the sea would constitute a grave interference in the natural ecology of these regions, with unforeseeable consequences.[77] Another is that the scheme envisaged would require continued application of iron because the original situation would establish itself rather readily whenever people ceased adding iron, with high-nutrient, high-carbon waters at the surface coming back from below.[78]

John H. Martin proposed a variant of his iron-fertilization hypothesis as a mechanism to decrease carbon dioxide during glacial periods: having wind-blown iron-rich dust fertilize the nutrient-rich surface waters around Antarctica.[79] A number of problems arise when taking a closer look at this quite attractive proposition.[80] For one, the evidence from sediments around Antarctica does not seem to bear out the main consequence of increased production that is expected, namely, the increased accumulation of diatoms. For another, an extensive glacial-age sea-ice cover should have blocked sunlight sufficiently to prevent a strong increase in production in the region.

Martin's iron hypothesis may bear importantly on increased glacial productivity in various other settings, including the northern and the equatorial Pacific. Iron fertilization by dust storms may be involved in a widespread reduction of nutrients in the upper 2 kilometers of the water column, in the glacial ocean. Also, increased dust in storms from emerging Asian highlands conceivably may have influenced the productivity of the North Pacific, beginning some 10 million years ago. On the whole, the effect would have been a general decrease of available phosphate in the ocean (and hence a decrease in the formation of marine phosphatic rocks, as indeed observed).

In summary, the efficiency of the biological pump in sequestering carbon at depth depends on the presence of nutrients in the ocean. Thus, not only iron, but also nitrate, phosphate, and silicate are important in fueling the pump and in helping to determine the level of atmospheric carbon dioxide (and hence the greenhouse effect).

To assess how the biological pump responds to warming, one has to bear in mind some of its quirks. It works as described as long as it runs on diatoms and dinoflagellates. But when it involves calcareous nannoplankton, carbonate is extracted from the surface waters and sent toward the seafloor. The removal of carbonate changes the chemistry of surface waters in such a fashion as to encourage escape of carbon dioxide to the atmosphere. This part of the pumping process, then, works counter to the general action. To obtain the net efficiency of the pump for given circumstances, the counterproductive carbonate precipitation has to be corrected for.

A shift from diatoms to nannoplankton is greatly favored by a shortage of silicate. In the water column, high silicate concentrations are found somewhat deeper than the maximum phosphate and nitrate concentrations. Thus, when cold waters with nutrients come to the surface in the upwelling process, the probability of getting higher silicate values increases with the depth range of the upwelling. Why is this important in the context of carbon dioxide and global warming? Because, as mentioned, diatoms are a mainstay of the biological pump, while calcareous nannoplankton are not. Now, when upwelling shallows, or draws from somewhat impoverished waters (as a result of the expansion of central gyres), the production of nannoplankton is favored over that of diatoms. The biological pump then weakens; in fact, the precipitation of carbonate engenders *release* of carbon dioxide, rather than uptake.

From the foregoing discussion it is clear that a change in the overall nutrient content of the ocean will affect the way ocean and atmosphere share carbon dioxide. Will warming affect nutrient content? Probably. The most vulnerable of the main nutrients is nitrate, which is destroyed wherever oxygen is in short supply. But just how nutrient availability would change in a warm ocean is an open question.

It may come as a surprise that nitrogen should be limiting at all. There is plenty of it in the atmosphere. Some of it gets "fixed" to oxygen in lightning and then rains out over the ocean. Once fixed to oxygen (nitrate) or hydrogen (ammonia), nitrogen is an important, even crucially important, nutrient. Some organisms have specialized in fixing nitrogen into organic compounds during photosynthesis. Well known among these are the unusually large colonial cyanobacterium *Trichodesmium* and the minute cyanobacterium *Richelia* that lives within some diatoms (e.g., *Rhizosolenia* and *Hemiaulus*), typically in warm and nutrient-poor waters of the open ocean.[81] *Trichodesmium* has recently been identified as a major contributor to primary production in the tropical North Atlantic, and the amount of nitrogen fixed was estimated as exceeding that delivered from below, through the thermocline.[82]

The main product of the biological pump, the export from the surface layer, has been studied since the early 1970s, by catching the material in traps moored at depth. The topic is discussed in chapter 9. Of special interest is the great contrast in the output from the coastal ocean versus that from the open sea, and the rapid decline of the amount of falling organic matter with depth, owing to the activity of bacteria. Another important theme is the pulsed nature of the output, that is, the strong seasonality and spikiness of the export.

Through the uppermost kilometer or so, the organic flux roughly decreases as a function of the inverse of the depth (that is, a doubling of the depth yields half as much flux), as a result of bacterial decay.[83] By the time the seafloor is reached (weeks later), only a small fraction remains of the original export. The flux of carbonate from sinking coccoliths and foraminifers, roughly 1 gram of carbon per square meter per year, is not attenuated to the same degree while sinking. The initial export out of the sunlit zone is predominantly organic carbon, rather than carbonate. But after sinking for a thousand meters or more, the ratio has reversed. For the carbonate (unlike for organic carbon) the reactions returning it to the water largely take place on the seafloor.

Regarding the proportion of organic matter exported from the productive zone, in regions with low silicate supply, about 10 percent of the primary production is exported. In regions with high supply, the value is closer to 30 percent, again emphasizing the importance of silicate to the workings of the biological pump.

Since silica is found in sediments, it is an important tracer for the geologic history of the biologic pump, and indirectly for the history of carbon dioxide, as well. A case in point is the surprising discovery that in many parts of the world the thermocline seems to have been depleted in silicate during glacial periods (times of vigorous upwelling).[84] There is a long-standing argument that changes in upwelling intensity should influence the level of carbon dioxide in the atmosphere, on long time scales.[85] Depletion of upwelling waters with silicate puts a special wrinkle into this argument. It suggests that carbon is sequestered within sediments together with silicate, during onset of a glacial period, and that this process can be recognized by the developing scarcity of silicate. If so, this would explain why the onset of a glacial period shows especially high productivity in the deep-sea record.

THE CARBONATE PUMP

As mentioned above, the discovery of substantial fluctuations in atmospheric carbon dioxide content, from analysis of ice cores, was one of the most exciting events in the history of the study of atmospheric chemistry.[86] Because the enormous carbon reservoir in the ocean is in exchange with the comparatively small reservoir in the atmosphere, the search was on immediately for mechanisms that could cause the ocean to give off or take up carbon dioxide on short notice. We have discussed how the addition of nutrients to the ocean would enhance the pumping of carbon into deeper waters, as organic debris. In addition, a focusing of organic carbon deposition (by increasing the intensity of upwelling) would enhance the dumping of organic carbon by overwhelming the local oxygen supply. Both processes would decrease the carbon dioxide available for the atmosphere.

Another mechanism has to do with the precipitation and redissolution of calcium carbonate in the sea, involving processes that may be subsumed under the heading of "carbonate pumping." When in chemical balance, the ocean has to deposit precisely the amount of dissolved calcium carbonate that enters it with the river influx and from volcanic sources. Deposition can take place both on the shelves and on the deep-sea floor. Normally, much more carbonate arrives at these sites of deposition than can stay there, if the balance is to be kept. Thus, the excess over the input has to be redissolved. In the grand cycle, the ocean adjusts its chemistry such that it is able to dissolve its own deposits, keeping the machine close to steady state. Because pressurized seawater dissolves carbonate more readily, most of the dissolution takes place at great depths in the ocean, while deposition takes place in the shallower regions, including the shelves.[87]

For a familiar analogy of the carbonate precipitation and dissolution patterns in the sea, one might think of the vertically mixing atmosphere, which is warmer in its lower portion than its upper one, because of the effects of compression and expansion on temperature. Snow precipitates in the upper regions of the mixed air layer. It falls everywhere but stays on the ground in the high mountains and dissolves on the lower flanks and in the valleys, returning the moisture to the air. The result is a snowline. In the deep sea, we have a carbonate line.[88]

In the carbonate budget of the ocean, a problem arises when the sea level drops during major buildup of ice, removing the shelves as sites of deposition. The balance now has to be kept by moving the sites of sedimentation into deeper waters. The carbonate line has to descend to greater depths, to make new seafloor available for depositing carbonate. To this end, the ocean needs to change its chemistry such that the level of saturation increases; that is, the seawater needs to become less aggressive toward carbonate deposits, at depth. The chemical adjustment that brings this about is an increase in alkalinity, that is, a higher content of calcium ions. This is as expected: when removing sites of easy precipitation, in shallow waters, it becomes more difficult to get rid of calcium ions, and as a result their abundance increases.

Increased alkalinity (that is, higher concentration of calcium ions) allows the ocean to hold more carbon dioxide in solution. Thus, the sea now claims a larger share of the total carbon dioxide in the ocean-atmosphere system. With the drying up of shelf seas from a drop in sea level, the content of carbon dioxide in the atmosphere must drop. By symmetry, with a rise of sea level, the carbon dioxide in the atmosphere is likely to rise.

Does the carbonate pump play some kind of role in the observed fluctuations of the carbon dioxide during the ice age cycles? If so, we should see evidence for this in a fluctuating carbonate line. We do indeed. In fact, the very first indication of ice age cycles came from carbonate cycles in the Pacific, discovered during the Swedish Deep Sea Expedition.[89] The sense of these cycles is such that, as postulated, the deep ocean accumulates more carbonate during glacial periods than during interglacials.[90] In principle, therefore, the shelf-poor glacial ocean is in equilibrium with a lower carbon dioxide content in the atmosphere, and the mechanism outlined is responsible for a portion of the ice age carbon dioxide fluctuations.

The biological pump and the carbonate pump are linked. When winds come up at the onset of a glacial period, the biological pump intensifies and additional organic carbon is delivered to deep waters in a major production pulse, enriching them with carbon dioxide. In turn, this enrichment results in a pulse of dissolution of deep-sea carbonate lasting several thousand years, as the additional carbon dioxide acquired at depth makes carbonic acid. This dissolution neutralizes much of the deep carbon dioxide generated by the production pulse, so it will not return to the atmosphere. Conversely, during periods of melting, the biological pump falters, because of warming of surface waters and because the rapid input of freshwater interferes with upwelling and ventilation of deep waters and, hence, with recycling of nutrients. Thus, we find a pulse of preservation of carbonate, centered on the period of warming and ice melting.

All the large ice age cycles, going back in time in the Pacific and Indian Ocean, show these pulses of dissolution and preservation at the onset and end of glacial periods, respectively. Their origin probably has additional elements besides that of the biological pump, such as the turning up and turning down of the production of North Atlantic Deep Water (NADW). In principle, turning up NADW production increases carbonate dissolution in the deep Pacific, and vice versa. The reason is that when the NADW runs strong, the deep-water asymmetry of Atlantic and Pacific is enhanced; the Pacific then collects the "old" waters rich in carbon dioxide, which are far from saturation.

In addition to the "regular" dissolution cycles associated with the ice age climate fluctuations, there is an overall long-term depression in carbonate preservation in the deep Pacific, centered around half a million years ago.[91] The reason for this anomaly is poorly understood. It may have to do with the buildup of the Great Barrier Reef at the time, whose carbonate had to come from somewhere. By making the entire ocean slightly less saturated with carbonate, the growth of the giant reef complex presumably initiated a large-scale transfer of deep-sea carbonates into the eastern margin of Australia. A slight increase in atmospheric carbon dioxide (on the order of 10 ppm) would have accompanied the process.

We see from these various illustrations of the carbon budget of the ocean that carbonate preservation patterns in the Quaternary have many interesting clues concerning the history of marine carbon chemistry. In a general way, we can say that just as the snowline on land reflects the overall temperature profile in the atmosphere, and therefore the climatic conditions on the planet, the snowline on the bottom of the sea—made of carbonate—reflects the chemical state of the ocean with respect to carbonate saturation. In turn, this state is relevant for the partitioning of carbon dioxide between the ocean and the atmosphere.[92]

The ongoing introduction of man-made carbon dioxide into the ocean will increase the rate of dissolution of carbonate, especially in those parts of the upper water column that are already close to or below saturation. This would be the case in high latitudes, for example in the fiords of Norway, where there are abundant shells of mollusks and even coral skeletons *(Lophelia)*. In low latitudes, the effect of adding carbon dioxide is a general decrease in excess saturation. Some organisms may depend on high excess saturation when making their shells and skeletons. But little is known for certain about the reaction of corals, algae, and mollusks to varying levels of saturation.[93]

DEBATING THE FUTURE

The topic of global warming has become the subject of numerous international conferences, of complicated discussions among scientists, and of heated debates in the public forum.[94] Carbon-based energy fuels the economies of most of the nations of the world and has done so for some considerable time. As the planet acquires additional greenhouse gases and warms up, there is growing concern about sea level rise and about changes in weather patterns, including patterns of floods, droughts, and hurricanes. However, the physical mechanisms linking warming and climate change are complex, so that there are large uncertainties in making the links for the purpose of assessing risk. The risks, then, are poorly defined. In addition, they are likely to be highly unequal, depending on the region considered, and the economy within that region. For example, agriculture in coastal plains is clearly more at risk than, say, making gadgets for export, in high country.

In the international forum, proposals have been made to cut back on carbon emissions, and a large number of nations have agreed to make efforts in that direction.[95] The reception accorded such proposals varies within and between nations, presumably depending on perceived risk of exposure to warming and on perceived cost and benefit of mitigation efforts to the economy. A common attitude is that others should lead the way. Naturally, those countries that have contributed little to the anthropogenic greenhouse load of the atmosphere and still have relatively low per-capita emissions tend to insist that those countries that have contributed most to the problem, and still do, should begin with the cutting of emissions. The suggestion is not necessarily welcomed by those thus exhorted. Many policymakers in the United States, for example, see no point in reducing emissions (which may be detrimental to the economy) when others can cancel any beneficial effect by increasing their emissions in turn. It is difficult to argue within this framework of discussion without calling on nonscientific concepts such as foresight, fairness, leadership, stewardship, justice, concern for future generations, and ethics. Thus, while the discussion is certainly appropriate given the seriousness of problems arising, and while it should be informed by science, it goes well beyond the scope of doing science.

From the point of view of climate science, the questions before us are these: Will the climate changes be drastic? Will they come quickly? How serious are the likely consequences? Such questions are no longer confined to the realm of academic discussion. Competent scientists across the world collaborate within the Intergovernmental Panel for Climate Change (IPCC) pondering them.[96] To provide answers with confidence presupposes a thorough understanding of the various feedback mechanisms governing climatic change. In fact, our understanding is inadequate.

Some things, as mentioned, are known for certain. The central fact that cannot be dislodged by any debate is this: the present carbon dioxide content of the atmosphere is higher by one-third over background, and its effect is exacerbated by the rise of methane, yielding a total effect corresponding to an increase of carbon dioxide by one-half over background. In addition, greenhouse gases are increasing rapidly. The symbol and signal of the grand experiment is the famous Keeling curve, showing the rapid overall rise of carbon dioxide, against the seasonal fluctuations. The annual addition to the atmosphere (which approaches 0.5 percent) is distinctly smaller than the large seasonal background fluctuation, which is about 3 percent of the total.[97] As a result of the year-to-year increase in carbon dioxide, the world has been getting warmer. As a result of further increase, it will warm some more.[98]

In contrast to the facts listed, future consequences are entirely in the realm of opinion. Not to disparage expertise, the future is *always* in the realm of opinion. By definition, there are no observations. However, some opinions are well founded, being supported by known physical principles, by observations of ongoing trends, by state-of-the-art forecasting, and by experience derived from climate history. Other opinions exist also, based on the questionable idea that what we do not know very well cannot hurt us, or on wishful thinking, or on a desire to take center stage. Among the various nonscientific opinions, those claiming uncertainty as an ally in diminishing the scope of dangers ahead are perhaps the most popular and may be found in many op-ed offerings.

So far, uncertainty is carrying the day. Without a clear idea of the dangers ahead, efforts at stemming the tide of greenhouse gases have been modest. The fundamental problem is that no one knows how much society should invest in switching out of a carbon-based economy, because no one knows what the risks and benefits may be that would accompany this type of decision. It is a matter of risk analysis, rather than just of geophysics. Geophysics tells us that sea level will continue to rise, and geologic experience indicates that it may well rise by several meters in the next few hundred years.[99] But geophysics and geology cannot tell us how many dollars should be invested now in mitigating the possible effects from sea level rise in the future.

Risk analysis is the staple of insurance companies, which bet against us, asserting that our house will not burn down. They have statistics showing the likelihood that they will be wrong. Accordingly, they set the premium to make a profit and stay in business. If your house is in a flood plain, they will insure it against fire but deny insurance for flooding, naturally. Bad bet to say there will be no flood. Since atmospheric carbon dioxide was never higher than now, for hundreds of thousands of years, we do not have similar statistics for climate change. It is not possible to predict the various effects of global warming at this time, or how fast it will proceed. We are betting in the dark.

In the present context we are especially interested in the response of the ocean to warming. We have considered various opinions on this issue, when discussing circulation, fertility and productivity, and various types of marine ecosystems. And we have called on principles (how the system should behave according to our understanding) and on history (how it did behave in the past, according to our reconstructions) to inform the opinions. One oceanographer who has prominently expressed opinions on the subject of the ocean's response to warming is Lamont's Wallace S. Broecker, who cautions that unpleasant surprises may be in store, if the climate switches to a different mode of operation.[100]

Broecker's warning concerns changes in deep circulation. As mentioned, his inspiration stems from sudden changes in climate that mark the transition from the last glacial period to the postglacial. There is evidence that meltwater input interfered with the production of NADW.[101] Meltwater has a low density and refuses to sink. In this it behaves like a warm-water layer. Rapid warming, then, could interfere with NADW production. Broecker has suggested that shutting down NADW production would result in cutting off the heat supply to high northern latitudes—warm water being no longer attracted from the south, to the site of deep-water formation. This type of shutdown mechanism, he proposes, is responsible for the origin of the Younger Dryas cold spell, some 13,000 years ago. In analogy, he proposed that rapid warming in high latitudes could shut down NADW production and thus cut back drastically on the supply of heat to Scandinavia. Glaciers might readvance in Norway.[102] While Broecker's specific concerns are open to discussion, no one can take issue with his assessment that the greenhouse effect may bring unpleasant surprises such as abrupt changes that nobody expected.

Likewise, there is no doubt that a focus on the consequences of the rapid melting of ice some 13,000 years ago is justified. We should carefully study the sequence of climatic events at that time. The input of meltwater is dynamically equivalent to warming, so these periods have much to tell us about possible consequences of rapid warming. Besides, input of meltwater results in a rise of sea level. The evidence for abrupt climatic change during the period of major melting is striking.[103]

We have already discussed possible consequences of high-latitude warming on a weakening of zonal winds, on gyre expansion, and on the reduced intensity of upwelling. The expectation is that focused production will drop, at least in the Northern Hemisphere, and that animals that depend on high food density (herringlike fish and their predators, as well as baleen whales and seabirds) will suffer as a consequence. Also, there is the likelihood that warming will tend to decouple upper waters from the deep ocean, so that the available reservoir for taking up carbon dioxide from the atmosphere will develop a tendency to shrink with time, by volume, in addition to building up resistance from the loss of carbonate ion.[104]

Such various expectations, based on simple principles, naturally need testing with computer models, which incorporate a host of feedback mechanisms that are impossible to keep track of in verbal reasoning. The building of complex models started in the 1970s, as computing power increased.[105] These were geochemical box models whose contents represented the efforts and results from various disciplines in oceanography and the Earth sciences; they were the first step toward a new type of scientific enterprise: Earth system science. The chief problem, then as now, in the attempt to make the models predict how atmospheric carbon will build up in response to carbon release is that the atmospheric reservoir is small, compared with the ocean's. So, a modest error in assessing the feedbacks within the ocean can make a large difference in the estimates of future atmospheric carbon dioxide content. The main point of interest is whether the ocean is slowing in its uptake of a large part of the excess carbon dioxide. It appears to be so.

Which are potentially the most serious threats? What are the most certain threats? Are there any silver linings in a dark sky? Is there remedial action that can be taken? Perhaps the main one of various conceivable major calamities is the acceleration of the greenhouse effect by massive release of methane. The silver linings consist mainly in the observation that nothing drastic seems to have happened so far, and the hope that this will hold true for the immediate future (defined, perhaps, as the lifetime of people now on the planet). Potential remedial actions include "fixes" (albeit not exactly "quick fixes") and the obvious device of cutting back on emissions.

Regarding the methane threat, what is of relevance is the widespread availability of methane ice (called clathrate) on many parts of the seafloor, especially around continents, below regions of upwelling.[106] Methane ice is a type of water ice that has an open structure, with space for enormous amounts of methane gas. When brought to the surface it quickly melts away, thus releasing the gas (fig. 15.5). (If lit with a match, the ice is seen to burn while melting.) The concern is that as the waters become warmer, below the shelf edge, methane ice will begin to melt, releasing large amounts of gas. At present, methane escapes from the seafloor in many places, but it comes out sufficiently slowly so that bacteria can consume it. In the process, the bacteria oxidize the methane to carbon dioxide and water.

Even without this particular threat from marine sources, it is noteworthy that methane has considerably increased in the atmosphere over the last several decades. The excess methane in the air (which is measured in parts per billion) is equivalent to a one-fourth higher amount of excess carbon dioxide, in terms of the greenhouse effect experienced from this rise.

The most reliable expectation, concerning future threats, is a rise in sea level. Mainly, such a rise is going on right now, the rate for the last century being put near 1.8 millimeters per year, and presumably increasing toward the end of the last century.[107] This higher rate, when extrapolated, yields a rise of about a foot per century. It may not seem much, but the impact depends on location. When living at sea level, as is the case for a surprisingly large number of

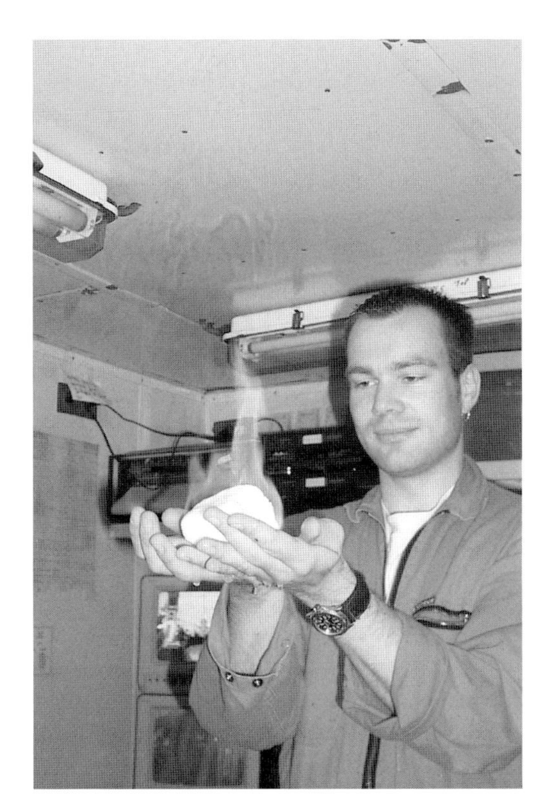

FIGURE 15.5. Massive methane hydrate recovered during DSDP leg 84, Middle America Trench. The methane is contained within a structure of water ice that is stable at high pressure and low temperature. Methane can burn.

people on the planet, a rise of the sea by one foot can mean the loss of homestead and real estate. The prime example for damages from this source to an urban environment is the ancient city of Venice, with its invaluable treasures of architecture and art.

An acceleration of sea level rise seems likely as the planet warms, both from expansion of the water, and from the melting of ice in high latitudes. During the course of the ice ages, a rate of rise of several feet per century lasting for many centuries has not been unusual, even at times when much of the vulnerable ice had already melted away and sea level stood as high as now. Thus, a rise of about 1 meter within this century seems well within the range of possibilities—although it might be deferred to the next century.[108]

As far as technological "fixes" of the global warming problem, schemes that have been proposed generally involve one of two concepts: sequestering carbon dioxide deep in the ground or deep in the ocean, or shading the surface of the planet by placing reflective particles in the upper atmosphere or into the space beyond, thus intercepting the sunlight. Marine sequestration using the power of the biologic pump to take carbon dioxide into the deep ocean is one possibility that has been discussed. Iron appears to be limiting to photosynthesis over wide areas of the ocean. By enriching surface waters rich in nutrients with trace amounts of iron, photosynthesis could be stimulated and the export of carbon from phytoplankton growth could be increased, over such areas. As discussed in chapter 8, there is much skepticism about the scheme.

The debate will continue. And so will the rise of greenhouse gases. Chances are, within less than two decades the climate system will be sufficiently out of the ordinary so that our meteorological experience—almost all based on observations in the past century—will be ever more difficult to apply to the changing situation. We will increasingly deal with a strange planet.[109]

For climate scientists, the object of their studies is becoming a moving target. As we learn more, prediction will become more difficult, not less so. As the uncertainties increase, it will become clear that motivations for stemming the tide of greenhouse gases will have to get help from broad political and economic considerations, well beyond the risk analysis associated with the physics and biology of climate change. Above all, risk analysis needs to focus on the use of carbon-based energy. And the issue is not just climate change and the likely response of the ocean. The issue includes public perception of the causes of natural hazards such as unusually large storms, floods, and drought, and the political ramifications of changing perceptions, both within nations and in the dealings between them.

The ocean has enormous energy resources: tides, winds, waves, currents, vertical temperature gradients, and salinity gradients where rivers enter the sea. Tides and sea winds are used to deliver power in some places. The other potential sources are being studied by research engineers here and there but have not been tapped in a serious manner. The reason is simple: as long as the use of carbon energy benefits from free disposal of its main waste product into the common atmosphere, other energy sources remain at a disadvantage. Given a choice, we prefer to use the energy that costs less. It is the rational thing to do. Unfortunately, it is rational behavior that drives the Tragedy of the Commons.[110]

NOTES AND REFERENCES

1. None of the statements in this paragraph are in doubt, and they were recognized as true 25 years ago. For example, see W. A. Nierenberg, members of the Carbon Dioxide Assessment Committee, 1983, *Changing Climate*. National Academy of Sciences, Washington, D.C., 496 pp.

2. None of the statements in this paragraph are in any manner controversial, although the fact that zonal winds must decrease when the temperature gradient weakens in northern latitudes has received surprisingly little attention. Controversy centers mainly on the last sentence, with regard to risk assessment, as it has economic implications.

3. W. W. Kellogg, 1996, *Greenhouse effect*, in S. H. Schneider (ed.), *Encyclopedia of Climate and Weather*, vol. 1. Oxford University Press, New York, pp. 368–371.

4. For scientific discussions of the carbon dioxide problem in the 1980s, see J. E. Hansen, T. Takahashi (eds.), 1984, *Climate Processes and Climate Sensitivity*. Geophysical Monograph 29. American Geophysical Union, Washington, D.C., 368 pp.; E. T. Sundquist, W. S. Broecker (eds.), 1985, *The Carbon Cycle and Atmospheric CO_2: Natural Variations Archean to Present*. Geophysical Monograph 32. American Geophysical Union, Washington, D.C., 627 pp. An international consensus of working scientists on the issues raised can be found in the reports of the IPCC, for example, J. T. Houghton, Y. Ding, D. J. Griggs, M. Noguer, P. J. van der Linden, X. Dai, K. Maskell, C. A. Johnson (eds.), 2001, *Climate Change 2001: The Scientific Basis. Contribution of Working Group I to the Third Assessment Report of the Intergovernmental Panel of Climate Change*. Cambridge University Press, Cambridge, U.K., 881 pp. There is now a more recent version, which basically confirms the findings of the earlier one. Naturally, the various statements from people without credentials in climate research, including best-selling authors of fiction, carry no weight in the realm of geophysics. Nevertheless, public reaction to their assertions is of some interest in the context of the sociology of climate change.

5. Leading climate scientists whose papers I have read or whose presentations I have listened to apparently do not hold strong opinions about what the future might bring, beyond a general warming, especially in high northern latitudes, and a rise in sea level. The reason is the substantial uncertainty associated with forecasting consequences in any detail. A lack of certainty is not the same as an optimistic outlook: many climate scientists are seriously concerned about the risks involved in global warming.

6. Svante Arrhenius (1859–1927) discovered the chemical nature of seawater as an ionic solution. Carbon dioxide was well known as a greenhouse gas in this time.

7. Guy Stewart Callendar (1897–1964), Gilbert Plass (1920–2004). For a summary of Callendar's and Plass's studies on the subject of carbon dioxide and global warming, see J. R. Fleming, 1998, *Historical Perspectives on Climate Change*. Oxford University Press, New York, 194 pp.

8. R. R. Revelle, H. E. Suess, 1957, *Carbon dioxide exchange between atmosphere and ocean and the question of an increase in CO_2 during the past decades*. Tellus 9, 18–27. Roger Randall Dougall Revelle (1909–1991) was director of Scripps Institution of Oceanography from 1951 to 1964, a period of major expansion in the institution. He did his Ph.D. work at Scripps and received the degree in 1936 from University of California, Berkeley (as was then the arrangement). The study was on deep-sea sediments. He published several articles during his time as graduate student, all related to the carbon dioxide problem that was to emerge later. Thus, he was well positioned to tackle the important problem of how the ocean would mitigate the rise of carbon dioxide in the atmosphere, when the question arose 20 years later. J. A. Knauss, 2003, *Roger Revelle, 1909–1991*, in R. L. Fisher, E. D. Goldberg, C. S. Cox (eds.), *Coming of Age, Scripps Institution of Oceanography, A Centennial Volume 1903–2003*. Scripps Institution of Oceanography, San Diego, pp. 119–137. The Austrian-American physical chemist Hans Eduard Suess (1909–1993) was a pioneer in the use of radiocarbon for dating archeological and geological objects. He came to La Jolla in 1955 from the U.S. Geological Survey, at the invitation of Revelle.

9. The overall average is close to 1.3 tons of carbon per person per year. The number is several times greater for the industrial nations, including the United States, Canada, Australia, and the members of the European Union. China's per capita emission is close to the overall average (and growing). India and many other developing nations release much smaller amounts per person, at this time.

10. That the carbon cycle, in consequence, is now dominated by human activities was pointed out by the geochemists James R. Arnold and Ernest C. Anderson, also in 1957, in the same volume that has the article by

Revelle and Suess (Tellus 9, 28–32). In a third essay in the same volume Harmon Craig (1926–2003) (who came to Scripps in 1955 at Revelle's invitation) further examined the issues brought up by Arnold and Anderson and by Revelle and Suess. Together, this set of articles opened a new field of carbon-cycle studies, including human impacts. That other biogeochemical cycles are affected by human impact also was recognized widely from the 1960s.

11. J. T. Houghton et al. *Climate Change 2001,* p. 525 ff.

12. This is an elementary exercise in curve-fitting in log-linear space, readily done in a spreadsheet. We are now at 1.3 times background; that is, halfway toward a doubling in log space. The warming observed is near 1 °C; a doubling produces the 2 °C. (In this exercise, all other effects are taken as part of internal feedback.) A value of 2.3 °C for a doubling of carbon dioxide was found in an early computer model of climate change, by Syukuro Manabe and Richard T. Wetherald, in 1967. S. Manabe, R. T. Wetherald, 1967, *Thermal equilibrium of the atmosphere with a given distribution of relative humidity.* Journal of the Atmospheric Science 24, 241–259.

13. As expressly stated by Revelle and Suess, in their 1957 paper.

14. R. R. Revelle, H. E. Suess, *Carbon dioxide exchange.*

15. J. T. Houghton et al., *Climate Change 2001,* p. 216 ff. The estimates are based on observed penetration of radiocarbon and other tracers, and estimates on the mixing behavior of water layers down to about 400 meters depth (which is the layer exchanging with the atmosphere, on time scales of decades).

16. The scientists building climate models know very well their limitations. Thus, the prominent climate researcher Stephen Schneider at Stanford University speaks of "educational machines" and warns of taking the output of models as gospel. Where they differ with critics of models ("climate skeptics") is in the assessment of the value of back-of-the-envelope estimates proposed as alternatives to mathematical modeling.

17. The French physicist Claude Lorius in Grenoble and the Swiss physicist Hans Oeschger in Bern were the leaders in this revolution.

18. The implication being that we do not understand the carbon cycle on a level necessary to go beyond the humdrum aspects familiar from present conditions.

19. The Summary for Policymakers in the 2001 IPCC report states (*ibid.* p. 10) that "there is new and strong evidence that most of the warming observed over the last 50 years is attributable to human activities." The previous report, 5 years earlier, merely concluded that the "balance of evidence suggests a discernible human influence on global climate." The reason for the change is a perception that the sources of natural variability (solar radiation, volcanism) are now better understood. According to this new understanding, man-made warming began in earnest in 1980 or thereabout (*ibid.* p. 11). In 1988 a strong statement to the effect that global warming had arrived was made by the prominent meteorologist James Hansen of the National Aeronautics and Space Administration (NASA). J. R. Fleming, *Historical Perspectives,* p. 134.

20. A statement of the 1990s attributed to a distinguished meteorologist working in Boston. A lack of interest in the matter of human climate modification is quite normal into the 1970s: an advanced textbook for climatology by R. G. Barry and A. H. Perry, first published in 1973 (*Synoptic Climatology,* Methuen, London), has no entry for "warming," "greenhouse," or "carbon dioxide," and no reference to Callendar, Plass, Revelle, or Keeling. Where it references Manabe, it is in the context of ocean-atmosphere interaction and glacial-age climate, but not for global warming. In contrast, the textbook by R. G. Barry and R. J. Chorley published a decade later (*Atmosphere, Weather, and Climate,* 4th ed., Methuen, London, 1982) does pay considerable attention to human impact. The obvious implication is that human impact was not part of the education of atmospheric geophysicists doing their graduate work in the 1970s. As is evident from contemplating initial opposition to the new concepts surrounding seafloor spreading and plate tectonics in the 1960s, educational background matters.

21. This is higher than the factor of increase of 1.3 for carbon dioxide, because of the addition of methane and other greenhouse gases.

22. J. F.B. Mitchell, T. C. Johns, 1997, *On the modification of global warming by sulphate aerosols.* Journal of Climatology 10, 245–267.

23. P. D. Jones, K. R. Briffa, T. P. Barnett, S. F. B. Tett, 1998, *High-resolution palaeoclimatic records for the last millennium: Interpretation, integration and comparison with general circulation model control-run temperatures.* The Holocene 8, 455–471.

24. Thus, opinions about the reality of man-made global warming that are not based on climate history lack authority; they are mainly of interest in a sociological context, as a measure of the intensity of ongoing political discussions.

25. In the 1970s, the climatologist J. M. Mitchell predicted that warming would move the measurements outside the natural background variation in the early years of the twenty-first century. Thus, to him the various record-breaking warm spells would hold not much surprise.

26. Charles David Keeling (1928–2005), pioneer of atmospheric chemistry and the carbon cycle,

joined Scripps in 1956. He came from the California Institute of Technology, where he had studied processes important in the terrestrial carbon cycle, as a postdoctoral researcher. He introduced a new level of precision to the measurement of carbon dioxide, using instruments of his own design. Keeling's time series of atmospheric carbon dioxide content spans half a century; it represents a resource of immense value.

27. C. D. Keeling, pers. comm., 2003.

28. J. T. Houghton, L. G. Meira Filho, B. A. Callander, N. Harriss, A. Kattenberg, K. Maskell (eds.), 1996, *Climate Change: The 1995 IPCC Assessment.* Cambridge University Press, Cambridge, U.K., 572 pp.

29. Proof that most of the warming is man-made is not possible, because the argument concerns probability. Thus, the statement by a former president of the U.S. National Academy of Sciences, made in 1997, that "we do not at present have convincing evidence of any significant climate change from other than natural causes," was and is difficult to refute, because "convincing" and "significant" are matters of judgment. F. Seitz, 1997, *Foreword,* in S. F. Singer, *Hot Talk, Cold Science.* The Independent Institute, Oakland, Calif., 120 pp. The problem is, of course, that when everyone agrees that the evidence is "convincing" it will then be very late in the game with regard to effective mitigation. Thus, "convincing evidence" is a red herring in the context of policy discussions. The real issue is anticipatory risk analysis in the context of economics.

30. This is especially true for glaciers in tropical regions. Ice that has persisted for thousands of years is now disappearing, since the 1980s. (As discovered by the glaciologist Lonnie Thompson of the Byrd Polar Research Center, Ohio State University.)

31. The occurrence of a spell of unusually cold winters and bad harvests in northern Europe, between 1350 and 1850, a period known as the Little Ice Age, has been linked to low solar output by many scientists, following a suggestion by astronomer J. A. Eddy. J. A. Eddy, 1976, *The Maunder Minimum.* Science 192, 1189–1202.

32. Dinner speech at the annual meeting of the American Geophysical Union, San Francisco, 1995.

33. H. von Storch, N. Stehr, 2000, *Climate change in perspective.* Nature 405, 615.

34. Obviously, a rise in sea level will affect societies living near sea level. Whether such impact will be "significant" is a matter of risk assessment, with results differing from one region to another, and strongly dependent on the ability of the people affected to deal with the problems arising.

35. P. D. Jones et al., *High-resolution palaeoclimatic records;* M. E. Mann, R. S. Bradley, M. K. Hughes, 1998, *Global-scale temperature patterns and climate forcing over the past six centuries.* Nature 392, 779–787; K. R. Briffa, P. D. Jones, F. H. Schweingruber, T. J. Osborn, 1998, *Influence of volcanic eruptions on Northern Hemisphere summer temperature over the past 600 years.* Nature 393, 450–455. Briffa et al. found that trees change their response through time, presumably in reaction to stress such as acid rain. This makes the task of reconstruction more difficult.

36. A. C. Hardy (1896–1985), professor of zoology at the University of Oxford from 1946 to 1985 (emeritus from 1961). Hardy is best known for his beautifully illustrated two-volume opus on the natural history of the open sea: 1956 and 1959, *The Open Sea: The World of Plankton* (vol. 1) and *Fish and Fisheries* (vol. 2), Collins, London, 335 pp. and 322 pp. The books are "written as much for the general reader as the zoologist."

37. The recorder is a filtering device that admits water at high speed and retains coarse plankton (greater than 280 micrometer) sandwiched between two layers of meshed fabric coming off two spools and rolled up on a third one running in preservative. The plankton recorder can be towed behind both research vessels and commercial ships and has provided the opportunity to greatly expand knowledge about the distribution of plankton in the North Atlantic.

38. The headquarters of the project are at the Plymouth Marine Laboratory. Project operator is the Sir Alister Hardy Foundation for Ocean Science.

39. The effort is not mentioned by Eric Mills in his important text on the history of biological oceanography, among various contemporary studies that are cited. E. L. Mills, 1989, *Biological Oceanography: An Early History, 1870–1960.* Cornell University Press, Ithaca, N.Y., 378 pp.

40. B. Planque, B. C. Reid, 2002, *What we have learned about plankton variability and its physical controls from 70 years of CPR records.* ICES Marine Science Symposia 215, 237–246; G. Beaugrand, P. C. Reid, 2003, *Long-term changes in phytoplankton, zooplankton and salmon related to climate.* Global Change Biology 9, 801–817.

41. In California, the CalCOFI program (chapter 7) has regularly encountered difficulties in finding funding for efforts involving long-term monitoring of the California Current.

42. Zonal winds run along latitude. They are strongest in the forties and fifties where temperature gradients are strong. Trade winds are the dominant zonal winds on the planet, relative to area. The strongest zonal winds blow around the Antarctic, where the temperature gradient is greatest, and where there are no obstacles in the path.

43. The records of the California Cooperative Oceanic Fisheries Investigation began in 1948 (see chapter 7).

44. D. Roemmich, 1992, *Ocean warming and sealevel rise along the southwest U.S. coast.* Science 257, 373–375.

45. D. Roemmich, J.A. McGowan, 1995, *Climate warming and the decline of zooplankton in the California Current.* Science 267, 1324–1326; J.A. McGowan, D.R. Cayan, L.M. Dorman, 1998, *Climate-ocean variability and ecosystem response in the northeast Pacific.* Science 281, 210–217.

46. The origin of the deep chlorophyll maximum at the bottom of the mixed layer is not well understood, and thus the change in its nature cannot be readily explained. From first principles, there is a level of favorable conditions for photosynthesis where the product of light (decreasing downward) and nutrient supply (increasing downward) goes through an optimum as far as appropriately adapted phytoplankton. However, other factors such as grazing and the maintenance of high chlorophyll levels by slowly growing forms interfere with a simple assessment. A shift in conditions in the mid-1970s was noted in E.L. Venrick, J.A. McGowan, D.R. Cayan, T.L. Hayward, 1987, *Climate and chlorophyll a: Long-term trends in the central North Pacific Ocean.* Science 238, 70–72.

47. J.A. McGowan et al., *Climate-ocean variability and ecosystem response,* p. 210.

48. W. Dansgaard and 10 others, 1993, *Evidence for general instability of past climate from a 250-kyr ice-core record.* Nature 364, 218–220; M. Stuiver, P.M. Grootes, T.F. Braziunas, 1995, *The GISP2 $\delta^{18}O$ climate record of the past 16,500 years and the role of the Sun, ocean and volcanoes.* Quaternary Research 44, 341–354; C. Appenzeller, T.F. Stocker, M. Anklin, 1998, *North Atlantic oscillation dynamics recorded in Greenland ice core.* Science 282, 446–449; M. Schulz, W.H. Berger, M. Sarnthein, P.M. Grootes, 1999, *Amplitude variations of 1470-yr climate oscillations during the last 100,000 years linked to fluctuations of continental ice mass.* Geophysical Research Letters 26, 3385–3388.

49. J.B.C. Jackson and 18 others, 2001, *Historical overfishing and the recent collapse of coastal ecosystems.* Science 293, 629–643.

50. A.C. Hardy, *Fish and Fisheries,* p. 61.

51. The rate of exchange of carbon between the atmosphere and the surface layer of the ocean was determined by Hans Suess, in the late 1950s, on the basis of radiocarbon measurements.

52. Coral bleaching accompanies pronounced warming of tropical surface waters. Off Tahiti (and elsewhere over large regions of French Polynesia), during El Niño conditions, corals are seen to become white during such periods; that is, the corals expel their dinoflagellate symbionts. Similar bleachings are observed around the Galapagos Islands, in Indonesia, in the Arabian Gulf, and in the Caribbean, whenever the sea is unusually warm. O. Hoegh-Guldberg, 1999, *Coral bleaching, climate change and the future of the world's coral reefs.* Marine and Freshwater Research 50, 839–866; O. Hoegh-Guldberg, R.J. Jones, S. Ward, W.K. Loh, 2002, *Is coral bleaching really adaptive?* Nature, 415, 601–602; G. Liu, A.E. Strong, W.J. Skirving, 2003, *Remote sensing of sea surface temperatures during 2002 Great Barrier Reef coral bleaching.* EOS 84, 137–144.

53. Relevant chronicles in the Middle Ages were largely written in Latin by clerics.

54. H.H. Lamb, 1972, *Climate: Present, Past and Future.* Vol. 1, *Fundamentals and Climate Now.* Methuen, London, 613 pp.; H.H. Lamb, 1977, *Climate: Present, Past and Future.* Vol. 2, *Climatic History and the Future.* Methuen, London, 835 pp.; H.H. Lamb, 1982, *Climate, History and the Modern World.* Methuen, London, 387 pp. In France, the historian Emmanuel Le Roy Ladurie collected information on history and harvests. E. Le Roy Ladurie, 1971, *Times of Feast, Times of Famine, A History of Climate Since the Year 1000.* Doubleday, Garden City, N.Y., 426 pp., first publ. in French, 1967. A broadly based look at climate change is by the British geographer C.E.P. Brooks, 1949, *Climate through the Ages,* 2nd ed. Republished by Dover Publishing, New York, in 1970, 395 pp.

55. J.M. Grove, 1988, *The Little Ice Age.* Methuen, New York, 498 pp.

56. About one-half of the warming could be assigned to the "back-to-normal" argument, at least until the 1960s. If one does this, the apparent sensitivity of the climate system (on the time scale of a century) drops to below 2 °C per doubling of carbon dioxide.

57. As mentioned, warming has proceeded, over the last 150 years, at a rate compatible with a 2 °C temperature increase for a doubling of carbon dioxide, compatible with expectations.

58. C.B. Lange, S.B. Burke, W.H. Berger, 1990, *Biological production off southern California is linked to climatic change.* Climatic Change 16, 319–329; D.B. Field, T.R. Baumgartner, C.D. Charles, V. Ferreira-Bartrina, M.D. Ohman, 2006, *Planktonic foraminifera of the California Current reflect twentieth century warming.* Science 311, 63–66.

59. J.E. Cole, R.G. Fairbanks, G.T. Shen, 1993, *The spectrum of recent variability in the Southern Oscillation: Results from a Tarawa Atoll coral.* Science 260, 1790–1793; E.R.M. Druffel, R.B. Dunbar, G.M. Wellington, S.A. Minnis, 1990, *Reef-building corals and identification of ENSO warming episodes,* in P.W. Glynn (ed.), *Global Ecological Consequences of the 1982–1983 El Niño-Southern Oscillation.* Elsevier, Holland, pp. 233–254; R.B. Dunbar, G.M. Wellington, M.W. Colgan, P.W. Glynn, 1994, *Eastern Pacific sea surface temperature since 1600 A.D.: The $\delta^{18}O$ record of*

climate variability in Galapagos corals. Paleoceanography 9, 291–315.

60. One such record is a *Montastrea* growth series from a colony that has many centuries of growth laminae. J. Pätzold, T. Bickert, B. Flemming, H. Grobe, G. Wefer, 1999, *Holozänes Klima des Nordatlantiks rekonstruiert aus massiven Korallen von Bermuda.* Natur und Museum 129, 165–177. It proved difficult to analyze the record of growth in terms of temperature, because the coral prefers warm water, but the cold water brings the nutrients and, hence, food. In addition, it appears that tidal action may be important in influencing growth, perhaps through nutrient supply. W. H. Berger, J. Pätzold, G. Wefer, 2002, *Times of quiet, times of agitation: Sverdrup's conjecture and the Bermuda coral record,* in G. Wefer, W. H. Berger, K.-E. Behre, E. Jansen (eds.), *Climate Development and History of the North Atlantic Realm.* Springer-Verlag, Berlin, pp. 89–99. Unfortunately, the growth record seems to have a gap in the seventeenth century (H. Kuhnert, Bremen, pers. comm. 2005), perhaps owing to unusually cool conditions.

61. J. R. Petit and 18 others, 1999, *Climate and atmospheric history of the past 420,000 years from the Vostok ice core, Antarctica.* Nature 399, 429–436.

62. This may not be a trivial question, considering that there are suggestions in the literature, beginning with M. Ewing and W. L. Donn, in 1956, that a warming of the Arctic Sea can trigger an ice age, by increasing snowfall around the Arctic. One would like to see definite evidence, from the record, that such a counterintuitive sequence is possible. M. Ewing, W. L. Donn, 1956, *A theory of ice ages.* Science 123, 1061–1066.

63. R. J. Delmas, J. M. Ascencio, M. Legrand, 1980, *Polar ice evidence that atmospheric CO_2 20,000 yr BP was 50% of present.* Nature 284, 155–157; W. Berner, H. Oeschger, B. Stauffer, 1980, *Information on the CO_2 cycle from ice core studies.* Radiocarbon 22, 227–235.

64. W. S. Broecker of the Lamont-Doherty Geological Observatory immediately pointed out that it has to be the chemistry of the ocean that controls large changes in carbon dioxide in the atmosphere on the time scales being considered. W. S. Broecker, 1982, *Glacial to interglacial changes in ocean chemistry.* Progress in Oceanography 11, 151–197.

65. J. M. Barnola, D. Raynaud, Y. Korotkevich, C. Lorius, 1987, *Vostok ice core provides 160,000-year record of atmospheric CO_2.* Nature 329, 408–414; J. R. Petit et al., *Climate and atmospheric history.*

66. The abbreviation *ppm* stands for "parts per million," hence the label "trace gas" for carbon dioxide.

67. Beyond the changes documented in the ice cores, there is a hidden change in the mass of carbon within forests and soils. Upon warming, the ocean has to deliver enough carbon dioxide to the atmosphere to increase its carbon dioxide content by 100 ppm, while at the same time delivering about half as much again for growing forests and increasing the carbon pool in soils.

68. On the contrary, transient phenomena, such as cessation of deep ventilation and loss of oxygen at depth, have to be considered. Also, a rise in sea level normally stimulates precipitation of carbonate on shelves. This decreases the ability of upper waters in the ocean to hold carbon dioxide, which results in increased supply to the atmosphere.

69. Statistically, the shorter the experience, the more restricted the choice of possible scenarios. This is one reason why some economists and political scientists (with a few years of personal experience with climate) insist on making conservative forecasts ("no problems ahead"), while many climate scientists working on the millennial scale, and climate geologists working with millions of years, are much less upbeat. Scientists trained in meteorology (time scale of decades up to a century) commonly choose a position between these end points, depending on background and inclination. I make this point to provide some structure in thinking about "pessimist" and "optimist" positions on the question of climate change and to forestall the mistaken idea that a lack of knowledge, and a consequent lack of confidence in predictions, lends support to an optimistic assessment of the future impact of global warming.

70. F. Bryan, 1986, *High latitude salinity effects and interhemispheric thermohaline circulations.* Nature 323, 301–304; S. Rahmstorf, 1994, *Rapid climate transitions in a coupled ocean-atmosphere model.* Nature 372, 82–85; S. Rahmstorf, 1995, *Bifurcations of the Atlantic thermohaline circulation in response to changes in the hydrological cycle.* Nature 378, 145–149; W. S. Broecker, 1997, *Thermohaline circulation, the Achilles heel of our climate system: Will man-made CO_2 upset the current balance?* Science 278, 1582–1588; T. F. Stocker, 2000, *Past and future reorganizations in the climate system.* Quaternary Science Review 19, 301–319.

71. This expectation is supported by reconstruction of conditions 4 million years ago, before buildup of northern ice masses. M.M. Robinson, H.J. Dowsett, M.A. Chandler, 2008, *Pliocene role in assessing future climate impacts.* EOS 89(49), 501–502.

72. G.H. Denton, T.J. Hughes (eds.), 1981, *The Last Great Ice Sheets.* Wiley-Interscience, New York, 484 pp.

73. In the process of stripping nutrients from surface layers, carbon is removed also, roughly in the Redfield proportion of 100 atoms of carbon for every 15 atoms of nitrogen. Another important process working in the same direction is the "physical

pump," whereby carbon gets concentrated in deep waters because of the higher solubility of carbon dioxide in cold water.

74. W. S. Broecker used the concept of the biological group to explain the variations in atmospheric carbon dioxide through the ice ages, postulating changes in nutrient content of the ocean. W. S. Broecker, *Glacial to interglacial changes.*

75. A survey of the marine carbon cycle is in E. T. Sundquist, W. S. Broecker (eds.), 1985, *The Carbon Cycle and Atmospheric CO_2: Natural Variations Archean to Present.* Geophysical Monograph 32. American Geophysical Union, Washington, D.C., 627 pp.

76. J. H. Martin, R. M. Gordon, S. E. Fitzwater, 1990, *Iron in Antarctic waters.* Nature 345, 156.

77. S.W. Chisholm, F. M. M. Morel (eds.), 1991, *What controls phytoplankton production in nutrient-rich areas of the open sea?* Limnology and Oceanography 36 (8), 1507–1970 (Special Issue). Concerns are summarized in the foreword by Chisholm and Morel. Also see J. A. Fuhrman, D. G. Capone, 1991, *Possible biogeochemical consequences of ocean fertilization.* Limnology and Oceanography 36 (8), 1951–1959.

78. A simple model of the Antarctic fertilization concept is offered in T.-H. Peng, W. S. Broecker, 1991, *Factors limiting the reduction of atmospheric CO_2 by iron fertilization.* Limnology and Oceanography 36 (8), 1919–1927.

79. J. H. Martin, 1990, *Glacial-interglacial CO_2 change: The iron hypothesis.* Paleoceanography 5, 1–13.

80. W. H. Berger, G. Wefer, 1991, *Productivity of the glacial ocean: Discussion of the iron hypothesis.* Limnology Oceanography 36, 1899–1918.

81. The cyanobacteria are ubiquitous and highly diverse. They used to be referred to as "blue-green algae," which is confusing, since they are procaryotes. N. G. Carr, M. Wyman, 1987, *Cyanobacteria: Their biology in relation to the oceanic picoplankton.* Canadian Bulletin of Fisheries and Aquatic Sciences 214, 159–204.

82. E. J. Carpenter, 1983, *Nitrogen fixation by marine* Oscillatoria (Trichodesmium) *in the world's ocean,* in E. J. Carpenter, D. G. Capone (eds.), *Nitrogen in the Marine Environment.* Academic Press, New York, pp. 65–103; D. G. Capone, J. P. Zehr, H. W. Paerl, B. Bergman, E. J. Carpenter, 1997, Trichodesmium: *A globally significant marine cyanobacterium.* Science 276, 1221–1229; D. G. Capone, E. J. Carpenter, 1999, *Nitrogen fixation by marine cyanobacteria: Historical and global perspectives.* Bulletin of the Institute of Oceanography (Monaco) 19, 235–256.

83. E. Suess, 1980, *Particulate organic carbon flux in the oceans: Surface productivity and oxygen utilization.* Nature 288, 260–263.

84. W. H. Berger, C. B. Lange, 1998, *Silica depletion in the thermocline of the glacial North Pacific: Corollaries and implications.* Deep-Sea Research II 45, 1885–1904.

85. E. Vincent, W. H. Berger, 1985, *Carbon dioxide and polar cooling in the Miocene: The Monterey hypothesis,* in E. T. Sundquist, W. S. Broecker (eds.), *The Carbon Cycle and Atmospheric CO_2: Natural Variations Archean to Present.* Geophysical Monograph 32. American Geophysical Union, Washington, D.C., pp. 455–468; M. Sarnthein, K. Winn, R. Zahn, 1987, *Paleoproductivity of oceanic upwelling and the effect on atmospheric CO_2 and climate change during deglaciation times,* in W. H. Berger, L. D. Labeyrie (eds.), *Abrupt Climatic Change: Evidence and Implications.* D. Reidel, Dordrecht, pp. 311–337.

86. The first data became available in 1980 and years following. Central figures in this discovery are the French physicist Claude Lorius and his associates, in Grenoble, and the late Swiss physicist Hans Oeschger and his coworkers in Bern. The results from Grenoble and from Bern supported an old idea that carbon dioxide was involved in ice age cycles. More than a century ago the British physicist John Tyndall (1820–1893) suggested that a decrease in carbon dioxide was involved in producing glacial periods. The Swedish chemist Svante Arrhenius (1859–1927) and the American geologist Thomas C. Chamberlin (1843–1928) made similar suggestions. The amplitude of carbon dioxide variation observed in the ice cores corresponds to a greenhouse effect of between 1 and 2 °C, according to recent climate models. Thus, roughly one-third of the variation in temperature between glacial and interglacial periods may be ascribed to changes in carbon dioxide.

87. The technical way of saying this is that the uppermost part of the water column is supersaturated with calcium carbonate. The chemical precipitation is inhibited by the presence of other ions in seawater. The deep part of the water column is undersaturated.

88. The level where carbonate first dissolves, which can be recognized by a deterioration of the quality of the preservation of foraminifer shells, is called the lysocline. It is like the boundary between fresh powder and old partially melted snow. The level where dissolution just balances the supply of carbonate is called the carbonate compensation depth, or CCD. It is like the snow line in the mountains.

89. G. O. S. Arrhenius, 1952, *Sediment cores from the east Pacific.* Reports of the Swedish Deep-Sea Expedition, 1947–1948 5 (1), 1–227.

90. The carbonate cycles agree with preservation cycles all across the Pacific and also in the Indian Ocean. Preservation is linked to saturation, which in turn is linked to the carbonate ion content in deep water. The changes are in the direction expected from the principles outlined.

91. This effect was discovered by Charles G. Adelseck, then a graduate student at Scripps (Ph.D. 1977).

92. The overall change in depth of the carbonate line during the ice ages roughly corresponds to 20 ppm of change in carbon dioxide in the atmosphere, other factors being equal. W. H. Berger, A. Spitzy, 1988, *History of atmospheric CO_2: Constraints from the deep-sea record.* Paleoceanography 3, 401–411. The vertical depth range of levels of equal preservation is between 500 and 1,000 meters.

93. The subject has attracted increasing attention over the last decade. Experiments suggest that a shortage of carbonate ions represents stress on carbonate-precipitating organisms.

94. The public debate is not always well informed. Routinely, the media give too much weight to views that disagree with the assessments of professional climate scientists. See, for example, N. Oreskes, 2004, *Beyond the ivory tower: The scientific consensus on climate change.* Science 306, 1886.

95. In Kyoto, Japan, in 1997. The Kyoto Protocol is the basis for the international agreement; it is an amendment to the U.N. Framework Convention on Climate Change. It requires signatory industrial nations to reduce carbon emissions by several percent below 1990 levels, by 2012. Recently (March 2007) the countries of the European Union have agreed to increase their use of renewable energy to 20percent of the total, by 2020. In the past, projections and pledges regarding decreases in carbon emissions proved unrealistic.

96. The IPCC has issued several major reports on the status of planetary warming and possible consequences. I have used the one from 2001. J. T. Houghton et al., *Climate Change 2001.* The most recent release (in 2007) largely confirms and extends earlier assessments. It also addresses the importance of the historical narrative. E. Jansen and 48 others, 2007, *Palaeoclimate,* in S. Solomon (cochair, Working Group I), *Climate Change 2007: The Physical Science Basis. Contribution of Working Group I to the Fourth Assessment Report of the Intergovernmental Panel on Climate Change.* Cambridge University Press, Cambridge, pp. 433–497. (The results reported could not be incorporated into the body of the present book, but nothing here said is in disagreement with the statements in the excecutive summary of this important review.)

97. Keeling's discovery of the large seasonal variation of carbon dioxide was a major event in the Earth sciences. It greatly stimulated the proper modeling of the carbon cycle.

98. This is totally predictable from elementary physics. Arguments arise about the exact value and pattern of warming across the globe.

99. W. H. Berger, *Sea level in the Late Quaternary: Patterns of variation and implications.* International Journal of Earth Sciences (in press; submitted 2007).

100. W. S. Broecker, *Thermohaline circulation.*

101. As first suggested by the Swedish geologist Eric Olausson and elaborated by the Woods Hole oceanographer Val Worthington, in the 1960s. At Scripps, evidence for a global meltwater effect was found in the deep-sea isotopic record of planktonic foraminifers, in the 1970s. E. Olausson, 1965, *Evidence of climatic changes in North Atlantic deep-sea cores, with remarks on isotopic paleotemperature analysis.* Progress in Oceanography 3, 221–252; L. V. Worthington, 1968, *Genesis and evolution of water masses.* Meteorological Monographs 8, 63–67; W. H. Berger, R. F. Johnson, J. S. Killingley, 1977, *"Unmixing" of the deep-sea record and the deglacial meltwater spike.* Nature 269, 661–663.

102. Or the glaciers might not advance, given the general warming in high latitudes from the greenhouse effect, or some of them might advance from an increase in snowfall, rather than from cooling. In any case, as outlined in chapter 12, there is no convincing evidence that a shutdown of NADW precipitated the Younger Dryas cold spell; alternative explanations are available.

103. Evidence for abrupt changes emerged in the 1970s. The central figure in this discovery was the Danish physicist Willi Dansgaard who assembled a team (in Copenhagen) to study the ice record on Greenland. In 1985, an assessment of the topic of abrupt change took place in a NATO conference near Grenoble in France, at the urging of the Belgian astronomer and climate scientist André Berger. The report was published in 1987. W. H. Berger, L. D. Labeyrie (eds.), 1987, *Abrupt climatic change: Evidence and implications.* Reidel, Dordrecht, 425 pp. Since then, the literature on abrupt change has greatly expanded. The evidence for sudden large changes is mounting and becoming more precise. J. P. Severinghaus, E. J. Brook, 1999, *Abrupt climate change at the end of the last glacial period inferred from trapped air in polar ice.* Science 286, 930–934.

104. Seasonal sea-ice formation, by release of brines that increase the density of near-surface polar waters, is an important process in the production of bottom water. Global warming presumably will decrease sea-ice formation, and this may prevent formation of sufficient brine to make bottom water. Such a development would foster decoupling of upper waters from the deep ocean. In any case, warming tends to interfere with deep convection. K. Bryan, F. G. Komro, C. Rooth, 1984, *The ocean's transient response to global surface temperature anomalies.* Geophysical Monograph 29, 29–38.

105. For examples of early modeling and references see E. T. Sundquist, W. S. Broecker, *The Carbon Cycle.*

106. K. A. Kvenvolden, 1988, *Methane hydrates and global climate.* Global Biogeochemical Cycles 2, 221–230; B. U. Haq, 1998, *Gas hydrates: Greenhouse nightmare, energy panacea or pipe dream?* GSA Today 8, 1–6.

107. See the discussion in *Nature,* Nov. 2005, 438, p. 35. The overall rise is largely a matter of expanding the water column in a warming ocean, plus additions of water from the melting of glaciers on land. Significant problems arise when attempting to determine changes in sea level rise at time periods of less than two decades, because of decadal-scale background fluctuations and disturbances from volcanism-caused climate effects. J. A. Church, N. J. White, J. M. Arblaster, 2005, *Significant decadal-scale impact of volcanic eruptions on sea level and ocean heat content.* Nature 438, 74–77.

108. There are indications that contributions to sea level rise from melting at the margins of the Greenland ice cap (which is much more vulnerable than the Antarctic ice) are largely compensated by increased snowfall in the center of the ice sheet. O. M. Johannessen, K. Khovorostovsky, M. W. Miles, L. P. Bobyev, 2005, *Recent ice-sheet growth in the interior of Greenland.* Science 310, 1013–1016. Clearly, this effect is of a transient nature: as warming proceeds, the area of increased snow buildup will shrink. Also, accelerated flow at the edges (H. J. Zwally, W. Abdalati, T. Herring, K. Larson, J. Saba, K. Steffen, 2002, *Surface melt-induced acceleration of Greenland ice-sheet flow.* Science 297, 218–222) will remove barriers to flow in the interior, favoring a lowering of the elevation and, hence, regional warming. But the time scale of these anticipated processes is highly uncertain; they might not materialize within this century, or they might already be operating. X. Fettweis, J.-P. van Ypersele, H. Gallée, F. Lefebre, W. Lefebvre, 2007, *The 1979–2005 Greenland ice sheet melt extent from passive microwave data using an improved version of the melt retrieval XPGR algorithm.* Geophysical Research Letters 34, L05502, 1–5; L. A. Stearns, G. S. Hamilton, 2007, *Rapid volume loss from two East Greenland outlet glaciers quantified using repeat stereo satellite imagery.* Geophysical Research Letters 34, L05503, 1–5.

109. According to Mojif Latif, a distinguished climate scientist at the Max Planck Institute in Hamburg, the system will move outside of the range of natural fluctuation in the 2020s (lecture at a meeting sponsored by the Leopoldina Academy of Sciences, in Halle, 2002).

110. The economic model for the tragedy of the commons is in G. Hardin, 1968, *The tragedy of the commons.* Science 162 (1), 243–248. See discussion in chapter 8. The observation is that the tragedy has led to an emptying of the ocean. R. Ellis, 2003, *The Empty Ocean—Plundering the World's Marine Life.* Island Press/Shearwater Books, Washington, D.C., 367 pp. Similar dynamics involving competition for resource benefits and externalization of environmental costs drive carbon-based climate modification.

Epilogue

THE GREAT TRENDS IN EXPLORATION AND THE CHALLENGES AHEAD

The difficulty in attempting to present a coherent picture of ocean sciences is that there is no such coherent picture—oceanography is whatever scientists interested in the ocean happen to do. Depending on the changing interests of the scientists practicing the art, ocean science can readily change focus and has done so many times over the last century. An early task was to make an inventory of what lives near the shore, and how. It was followed by exploration of coastal waters, and by systematic investigation of the global ocean—its physics, its chemistry, its biology, and its history and geologic framework. It is now the changing ocean that is at the center of attention. Motivations have varied as well. In the early years, questions concerning the rules and history of evolution were important, as well as the life histories of marine organisms, which documented in detail their exquisite adaptations to the environment and to each other. Also, it was the desire to understand, from first principles, why currents move the way they do, and how the physical environment controls the distribution of organisms.

The needs of fisheries and of navies provided the main impetus for expansion within the first half of the twentieth century, and after World War II. These needs supported growing efforts in physical, chemical, and biological expeditions and laboratory studies for more than half a century. Lately, with the increase in human impacts, first along the coast and then on a global scale, motivation has shifted again. Much of exploration now is linked to assessing the scope of the impacts of polluting coastal water bodies, of overfishing, and of changes in the physics and biology of the ocean that may be attributable to global warming.

When attempting such assessments, motivated by the desire to understand the response of the ocean to disturbance, and perhaps to guide remedial action where this is called for, a serious and pervasive problem soon emerges. It is that the natural condition, undisturbed by human influence, is no longer available for study and has not been available for several decades. Thus, it is not possible to clearly separate natural background from response to human disturbance, except in theory.

As the nature of the ocean has changed, so has the science that deals with it. The great trends are well summarized in a review by

Margaret Deacon, doyenne of the history of oceanography.[1] She takes the period between 1880 and 1930 as the decades that laid the foundations for modern oceanography. It is a time of great expeditions, from the British *Challenger* Expedition at the beginning, to the British *Discovery* Expeditions and the German *Meteor* Expedition at its end. Perhaps the most famous of all the expeditions within this period was the *Fram's*, venturing into the Arctic Sea, under the leadership of Fridtjof Nansen, explorer, all-around naturalist, and discoverer of important oceanographic processes, such as the drift-at-an-angle mathematically explained by Ekman. The needs of commercial fisheries and whaling, and a general desire to understand the physical and biological workings of the ocean, were important driving forces in this phase of research.

In her article, Deacon illustrates two discoveries from the founding period. One is the bottom topography of the Atlantic Ocean, as mapped by the *Challenger* scientists, which showed the existence of the Mid-Atlantic Rise, and different bottom-water temperatures on either side of the rise. The morphology of this mountain chain, elaborated in the early part of the twentieth century, posed a puzzle that would be solved, eventually, by the concept of seafloor spreading, in the 1960s. Deacon's other illustration shows longitudinal oceanographic sections through the western and eastern basins, for salinity, which demonstrate the complicated stratification of the Atlantic Ocean. The sections are from Wüst's synthesis of the *Meteor* survey. They are crucial for the understanding of the thermohaline circulation of the ocean. An Atlantic bias is evident in Deacon's review, and quite understandably so: important contributions to this founding phase of research came from British, Scandinavian, and German scientists.[2]

The 1930s saw great expansion into sophisticated instrumentation, including seismic surveying and gravity measurements, besides routine temperature measurements (by bathythermograph). Important insights came from experimental biological work on production in its dependency on light and nutrients, and on the vertical migration of zooplankton. In that decade, also, oceanographic work at North American institutions began to play an important role in the international scene, most prominently Scripps on the West Coast and Woods Hole on the East Coast.

In Woods Hole in Massachusetts, which had hosted the Marine Biological Laboratory since 1888, a new oceanographic research facility was created with the help of the Rockefeller Foundation: the Woods Hole Oceanographic Institution, in 1930. Henry Bryant Bigelow became its first director (from 1930 to 1940). His interests were focused on fisheries sciences. In a report to the National Academy of Sciences (which laid out a program for research endorsed by the leading oceanographers in the United States) Bigelow emphasized studies of economic value (fisheries and also navigation), basic physical oceanography (with applications to biology and navigation), and life in the sea (zoology and botany, marine physiology, and bacteriology). The report was ahead of its time in pointing to upwelling as an important process worthy of study, and in identifying bacteria as a vital part of the marine environment.[3] The insights regarding the importance of bacteria, one may assume, reflect the influence of two prominent founding members of Woods Hole, the chemist Norris Rakestraw, who pioneered comprehensive study of the marine nitrogen cycle, and Selman Waksman, a soil microbiologist. Other rising stars at the new institution were the Scandinavian meteorologist Carl Rossby, working at MIT, and the physiologist and all-around oceanographer Alfred Redfield.

In 1936, Harald Ulrik Sverdrup came to La Jolla to lead Scripps Institution of Oceanography into a new phase of exploration, with a new seagoing vessel, the *E. W. Scripps*.[4] He initiated expeditions into the Gulf of California and systematic surveys of currents and of plankton in the waters off California, Oregon, and Baja California. The peculiar whorls that characterize the California Current emerged, as well as the significance of upwelling. At the same time, Woods Hole scientists were discovering the

Gulf Stream as a boundary phenomenon along the great wall that separates the warm waters of the Sargasso Sea from the cold coastal waters.

The "Ocean Bible," the treatise by Harald U. Sverdrup, Martin W. Johnson, and Richard H. Fleming (1942), represents the most complete organization and summary of these founding years, in fact defining the field of oceanography for the next several decades. Its focus is on physical and biological oceanography and on that part of the chemistry of the ocean that is relevant to biological production.[5] However, the treatise did not just stand at the end of a pioneering development: it set a new tone compared with much earlier work. It put the emphasis on dynamics, rather than the traditional description of the physical environment, on relationships of organisms to their environment, rather than on taxonomy and individual life histories, and on basic research, rather than on applications to fisheries and navigation. A shift from an Atlantic perspective to a World Ocean perspective (that is, one centered on the Pacific) began to emerge in this new exposition of ocean sciences.

In the midst of these activities burst World War II, and a need to concentrate research on matters of concern to the navy, especially submarine and antisubmarine warfare. Two long-term modifications of the overall research paradigm emerged: a shift to acoustics, for sensing the environment in a dark sea, and a move from fisheries biology as a rationale for doing ocean science to marine physics as a means to enhance the effectiveness of naval operations.[6] Along with this expansion of the dominance of physics (including geophysics) came great advances in ocean engineering, which in turn benefited studies in marine biology including nutrient chemistry. For two to three decades after the war, this new pattern persisted. Gradually, the general ocean sciences regained ground as research support from the National Science Foundation grew, and as NSF split out separate programs for the various branches of oceanography.

In the 1960s and 1970s, after a time of unprecedented expansion of oceanographic research facilities, the landscape of doing ocean science changed again, and large-scale cooperative investigations became dominant features. At Scripps, these developments had been presaged by the CalCOFI program, aimed at studying the behavior of the California Current from a perspective of commercial fisheries, in collaboration with the state and federal agencies involved. The most important result of this program was the realization that climate change underlies the fluctuations of fish populations. With this realization, and with El Niño–Southern Oscillation events emerging as a concern in ocean and climate dynamics, the role of the ocean in determining weather patterns and all questions regarding air-sea interaction gained prominence again—closing the loop to the origins of dynamic oceanography in the Bergen group around the meteorologist Vilhelm Bjerknes, early in the twentieth century. In 1971, the meteorologist Jerome Namias, once mentored by Carl Rossby (himself an offspring of the Bergen group), made the transition from the U.S. Weather Bureau to Scripps, to begin building a climate research group at SIO. It became a fast growing part of Scripps and was quickly integrated into national and international efforts.

One highly successful international venture centered at Scripps was the Deep Sea Drilling Project, initiated in 1964. From 1968 to 1983, the *Glomar Challenger* crisscrossed the seas and drilled the seafloor in a great effort to map the geology of the crust of the submerged part of the Earth. While the geophysics of the crust became known in the 1950s, the geologic make-up, in the early 1960s, still was terra incognita, an unknown area as large as half of Earth's surface. Together, the geophysical discoveries of the 1950s and the drilling results of the *Glomar Challenger* gave us a radically new world view about the way the Earth works on long time scales.

In his brief review of the history of Scripps Institution of Oceanography, written at the end of his long and distinguished career, Roger Revelle singled out the following scientific topics for illustration: heat flow measurements by Arthur E. Maxwell (one of Scripps's contribution to the new global tectonics, along with Russell

Raitt's work on crustal thickness, and Ronald Mason's discovery of magnetic anomalies, among others), satellite-based productivity measurements of the California Current, the composition of the fauna near hydrothermal vents in the Galapagos Spreading Center, the rise of atmospheric carbon dioxide (Keeling curve), acoustic tomography of the ocean, satellite-based tracking of drifting buoys in the California Current (showing mesoscale eddies), the coverage of the World Ocean by the Deep Sea Drilling Project, and nearshore ecology.[7]

In his "final word" to this review, Revelle argued that while much of the work at Scripps centers on problems of great concern to human beings, "the human dimension is pretty much left out of the picture." He urged greater involvement in the study of human impacts on the oceans and the life within it, and a buildup of competence in questions of ocean policy.[8]

Revelle's concern with human impact has turned out to be most perceptive; such concern has received increasing attention within the last decade. We humans are now a geologic agent on a par with volcanoes and impacts from space; that is, we live in a new era that has no precedent. There now is an urgent need to understand the Earth's life-support systems, so we can mitigate our modifications of these systems and adapt to their changing conditions.

For a hundred thousand years, we humans have found ways to control natural forces and bend them to our advantage—using fire and fur against the cold, building shelters against bad weather, extracting sustenance from the biosphere on land and in the sea. Now we have to learn to cope with a new type of force of planetary proportions: our own activities. We have met the tragedy of the commons on a planetary scale.

Prime examples of the impact of this new force are in climate modification (that is, global warming) and in the overexploitation of biosphere resources, such as fishes, forests, and soil. Much experience has accumulated regarding the management of fisheries, efforts that have had mixed results, at best. The experiences gained can offer valuable insights for the managing of other types of resources in the context of sustainable development.[9] Knowing and avoiding past mistakes can improve chances for success.

Modern cultures—and thus our ideas concerning proper behavior—go back several thousand years. However, almost all of our knowledge about the conditions and processes supporting life on the planet was gathered in the past century. Only in the last few decades have we realized the extent of our own impact on these conditions and processes. Thus, it comes as no surprise that this new knowledge has not yet been integrated into our traditions. We are faced with unprecedented challenges, but our toolkit for coping, built over millennia, is sparsely endowed.[10]

Confronted with such problems, what is our task? The obvious answer—readily supported by scientists, businessmen, and politicians alike—is to increase knowledge about the workings of life-supporting systems. That the results of new scientific knowledge can improve decision making is well illustrated by the Montreal Protocol on the protection of the ozone layer. The lessons from the ozone episode are straightforward. Scientists exploring the chemistry of the atmosphere, for the purpose of broadening our understanding of what processes maintain its composition, found unexpected behavior, with great potential for damage to human health, as well as to livestock and to crops. Among the chief suspects were newly developed chemicals thought to be entirely harmless, because of their great stability (chloro-fluoro-carbon compounds). Their release to the atmosphere, in comparatively minute amounts, was of no interest whatever, until they were linked to observed hazards. At that point of crisis and high risk, action was taken quite promptly. The problem persists, but mitigation is in sight. The healing of the damage to the atmosphere will take place on the time scale of a century or so. In the meantime, we and later generations, and all living things, will bear the unintended consequences of one of the ongoing experiments, in this case releasing seemingly harmless substances to the environment.

Some problems, such as the loss of biodiversity, are amenable to mitigating action, because they can be attacked on a regional basis, for example, by setting aside refuges large enough to protect a given assemblage of organisms. One great benefit of such set-asides is that they can serve as a source of seeds for restocking surrounding areas. Other problems are much less tractable, mainly because they are intricately intertwined with economics, and include many different stakeholders from different political groupings. Perhaps the most serious of these difficult problems is the excess greenhouse effect. The great geophysical experiment is proceeding, and we do not know where it will take us. We do know that polar regions are changing in significant ways.[11]

As emphasized throughout in this book, the ability of the scientific community to predict the consequences of overuse of resources and of global warming is quite limited. This does not mean that there are no serious problems ahead. It just means that we cannot identify them to the satisfaction of all who need to get involved in mitigation. The political problem that arises in consequence of uncertainty is obviously this: we cannot tell which is potentially more painful, the change of environmental conditions, or the cost of the actions proposed.[12] In any case, the cost of mitigation would presumably mainly fall to those who use much energy and live well (and are able to make decisions), but the risks are distributed to all, without regard to use of resources, and including future generations. The concept of incurring serious costs for the benefit of others is not commonly employed in generating policy.[13] It is unlikely to prevail where benefits cannot be demonstrated and the potential beneficiaries are not yet born.

A situation where decisions have to be made in the fog, without a clear view of what is ahead, is not without precedent to officers at sea, on the bridge. We might learn much from their behavior. They do not assume that the fog holds no dangers. On the contrary, they assume it is not safe to proceed without relevant information on what lies ahead. Were we to adopt their precautionary stance in respect to the future of conditions on the planet, we should change our economic behavior to minimize all adverse human impact on ocean, atmosphere, and biosphere.

Our future, obviously, is rarely in the hands of well-trained and cautious navigators. Whatever will be done will have to be based on a broad consensus of an informed public actively pursuing the goals of sustainability within a general setting of economic constraints. Hence, public education in the pertinent Earth sciences, including oceanography, ecology, and atmospheric sciences, is one crucial ingredient of coping with future challenges. But equally important is an effective education toward environmental stewardship, that is, toward a sense of responsibility, as providing a balancing perspective on economic priorities.

The needs that will inform future research will focus on sustainability and on life-support systems. However, more than scientific understanding will be called for to meet the challenges, that is, a general desire to manage human activities for the benefit of a habitable planet. Without participation of a committed public, scientific knowledge will not translate into political action.[14]

What type of action might be useful in furthering stewardship and sustainability? Many different types of approaches can be envisaged and are being discussed. With regard to climate change, the focus is on replacing carbon-based energy with alternative energy sources including sun, wind, and nuclear power. In addition, the possibility of sequestering carbon dioxide underground, next to the power plants generating the gas, is being studied. As the climate warms, energy sources for air conditioning will be of increasing interest, including the cold water below the thermocline in the sea. Also, looming water shortages are directing discussion toward the need for desalinization of seawater, likewise an energy-intensive proposition. With regard to conservation, the shining example of the National Parks in the United States and elsewhere point the way: the need for setting aside sufficiently large regions to preserve a portion of the natural ecologic endowment of the planet. In the marine realm, the path toward set-asides for conservation includes generating the scientific underpinnings for judging the necessary size of restricted areas, and the nature of the proposed restrictions.[15]

What the various problems of sustainability have in common is that in each case we need to avoid Hardin's Tragedy of the Commons, that is, rational exploitation leading to collapse of the resource (16).

NOTES AND REFERENCES

1. M. B. Deacon, 1996, *How the science of oceanography developed,* in C. P. Summerhayes, S. A. Thorpe, *Oceanography, an Illustrated Guide.* Manson Publishing, London, pp. 9–26.

2. *Ibid.,* figures 1.12 and 1.20.

3. H. B. Bigelow, 1931, *Oceanography, Its Scope, Problems, and Economic Importance.* Houghton Mifflin, Boston, 263 pp.

4. Robert P. Scripps donated the ship, a two-masted schooner, christened the *E. W. Scripps* after his father and the man who helped get the institution started. Helen Raitt, B. Moulton, 1967, *Scripps Institution of Oceanography: First Fifty Years.* Ward Ritchie Press, Los Angeles, 217 pp.; R. Revelle, 1992, *The Scripps Institution of Oceanography, California,* in E. M. Borgese (ed.), *Ocean Frontiers, Explorations by Oceanographers on Five Continents.* Abrams, New York, pp. 14–53.

5. Geology appears only in the last chapter of 20, with a review of sedimentation in the sea. H. U. Sverdrup, M. W. Johnson, R. H. Fleming, 1942, *The Oceans, Their Physics, Chemistry, and General Biology.* Prentice-Hall, Englewood Cliffs, N.J., 1087 pp.

6. For many years, the vision and resources of the Office of Naval Research, established after the war, were crucial in the growth of both Scripps and Woods Hole, and in the growth of emerging oceanographic research institutes on all coasts of the United States.

7. R. Revelle, *The Scripps Institution of Oceanography.* Other items are photos of the institution and its ships, or similar images of a general nature.

8. *Ibid.,* p. 53.

9. Regarding attempts at management, the Northeast Atlantic Fisheries Commission stands out. It was established after World War II (with discussions going back into the 1930s), with the idea to facilitate international agreements to regulate such items as minimum fish size and use of fishing gear, and setting values for total allowable catch. Innumerable agreements were signed between the interested parties, regarding fishing in international waters. Many of these agreements did not take into account the great susceptibility of fish stocks to climate change. See K. Brander, 2003, *Fisheries and climate,* in G. Wefer, F. Lamy, F. Mantoura (eds.), *Marine Science Frontiers for Europe.* Springer-Verlag, Berlin, pp. 29–38. In any case, the familiar pattern of regulation is one of too little, too late: some of the most important fisheries in the Northeast Atlantic have since collapsed.

10. The lack of an appreciation of ecologic principles—the economy of nature—has serious consequences in human affairs. It is fundamentally the reason that intelligent people trained in sociology and political science, but with no background in biology or Earth sciences, can come to the type of conclusions proffered in the upbeat works of the late economist Julian Simon, with statements that emphasize recent progress in human welfare and belittle the thought that natural resources are used at an unsustainable pace. J. Simon (ed.), 1995, *The State of Humanity.* Blackwell, Oxford, 694 pp. More recently, the Danish political scientist Bjoern Lomborg has made statements reminiscent of Simon's style. B. Lomborg, 2001, *The Skeptical Environmentalist, Measuring the Real State of the World.* Cambridge University, Cambridge, U.K., 515 pp. When applied to anticipating future developments, Simon's approach resembles straight-line extrapolation in an environment without limits.

11. Changes include thinning of Arctic sea ice and widespread melting of permafrost in Siberia and Alaska, as well as the breakup of large ice-shelves off Antarctica. A summary of recent results of studies of changes in polar regions is in a special issue of the magazine *Science.* J. Smith, R. Stone, J. Fahrenkamp-Uppenbrink, 2002, *Trouble in polar paradise.* Science 297, 1489, and articles following.

12. The dilemma has been well recognized for some time. As stated in 1990 by the geophysicist Frank Press, then president of the National Academy of Sciences: "Thus difficult policy decisions must be made on the basis of judgments between dimly perceived future risks and possible economic or other consequences that may be more immediate." C. S. Silver, R. S. DeFries, 1990, *Preface,* in *One Earth, One Future.* National Academy Press, Washington, D.C., pp. iii–v, p iv.

13. These remarks, which touch on sociologic and economic issues, illustrate the new paradigm arising from human impact on the global environment. In fact, human behavior has become a branch of geophysics and ecology, and physical and biological scientists are thus forced to pay attention to sociology and economics when studying Earth. Without

drifting off into the jungle of political sciences, we can say this: At least in ecology, self-interest commonly dominates behavior, even where altruism is suspected from appearances. Likewise, economic systems appear to thrive on self-interest, as so forcefully claimed by Adam Smith (1723–1790) two and a half centuries ago. If this is so, when setting up collaborations for attacking environmental problems, the payoff to all players must be clearly recognized and tailored to the framework of assigned contributions to the common task.

14. The trends are not favorable. Emissions in 2005 were higher than those in 1990 by 27 percent, despite much discussion of the need for reducing the release of carbon dioxide to the atmosphere. The total input to the atmosphere each year, as a consequence, is now near 8 billion tons of carbon. Increased globalization of the economy is not helpful in stemming the trend: with increased competition in producing goods for the world market comes a growing need for cheap energy, for each participant. Thus, moving out of carbon energy comes with a price tag in terms of position in the competition. Payments for "emission rights" from prolific emitters to lesser emitters exacerbate this problem.

15. At Scripps, these questions are pursued at the Center for Marine Biology and Conservation, initiated and led by Nancy Knowlton, Jeremy Jackson, and Enric Sala. A substantial training program equips students with the background to tackle issues of conservation in the real world of public discussion and policy.

16. The remedy is regulation based on scientific understanding and sociological reality. If urged to give an opinion in this context, as an oceanographer, I would focus on the Antarctic Current as the region for which increased regulation of access would make much sense. To me, it seems likely that an enduring presence of the enormous ice cap on the southern continent will ensure strong winds and deep mixing into the foreseeable future. Thus, high productivity should persist during summer months, regardless of what happens elsewhere in the ocean. If so, the region could serve as a kind of Noah's Ark, in continuing to support high-energy apex consumers (that is, seabirds and marine mammals). As the fish removal operations of the industrial fleets complete their work across the seas, fish farming will continue to grow, and so will the demand for feed for such farming. It stands to reason, then, that the remaining high-production regions in the Antarctic will increasingly attract the attention of industrial feed producers, and that large-scale krill removal will soon be the result. If this is to be avoided, the time to restrict removal operations is now, before the vested interest of powerful stakeholders makes certain that the tragedy of the commons prevails, again.

Appendix One

Units Used in the Ocean Sciences

DISTANCE

m, meter; picometer (1/trillion m); nanometer (1/billion m); micrometer (1/million m); millimeter (1/1,000 m); centimeter (1/100 m); decimeter (1/10 m); kilometer (1,000 m); inch (2.54 cm); foot (30.48 cm); fathom (6 foot, 1.8288 m); yard (1/2 fathom); mile (1.61 km); nautical mile (1.852 km; 1 min. of latitude at 45° of lat.)

VOLUME

liter (33.814 U.S. fluid ounces; volume of 1 kg of water); cubic meter (1,000 liter); milliliter (1 ml; 1 cubic cm; 0.001 liter); cubic inch (16.387 cm^3); gallon (U.S.) (3.7854 liter); quart (0.95 l)

TIME

s, second; pico- to millisecond, see distance for meaning of prefix terms; minute (60 s); hour (60 min); solar day (24 hours); astronomical day (0.9973 solar days); lunar month (29.53 days); year (mean solar year, 365.2564 solar days; 31,558,153 seconds)

VELOCITY (DISTANCE PER TIME)

cm/s (1 knot = 51.444 cm/s), kph (27.78 cm/s; 1 kn = 1.852 kph), mph (1 kn = 1 naut. mile/h), m/s (3.6 kph; 1.944 knot)

TEMPERATURE

degrees centigrade (or celsius or kelvin); fahrenheit (1.8 centigrade; zero celsius at 32 °F)

MASS

gram (mass of 1 ml of water; 1/1,000 kg; 1/million ton); mg (1/1,000 g); μg (1/million g); metric ton (1 t = 1,000 kg = 2,205 lb).

TRANSPORT (VOLUME PER TIME)

sverdrup (1 sv = 1 million m^3/s)

ENERGY AND POWER

calorie (gram) (energy to raise 1 ml of water by 1 °C); watt (1 joule/s; 860 g-cal/h); horsepower (0.7457 watt)

IRRADIANCE: watts per square meter

PRESSURE (FORCE PER AREA)

atmosphere (1.0332 kg/cm^2); bar (0.9869233 atm); $lb/inch^2$ (14.22 $lb/inch^2$ = 1 kg/cm^2)

CONCENTRATION

grams per liter; mg/liter (milligrams per liter, parts per million); moles per liter (or gram-atom per liter, number of molecules or atoms per liter in terms of the ratio of mass over molecular weight; unity denotes Avogadro's number)

SOUND INTENSITY AND FREQUENCY

decibel (db), a measure of the ratio of power densities, on a logarithmic scale, whereby the intensity of interest is compared to a standard. In air, faint whispers are near 40 decibels, the average human voice is 60, a leaf blower is 100 (for the operator), and exposure above 115 is considered hazardous. In water, numbers for equivalent intensity levels are higher by 62 db, because of differences in reference level and in transmission efficiency.

hertz (Hz, cycles per second); pitch of sound; important in communication and orientation and echo hunting (acoustic biology), and in acoustic exploration (including seismic exploration); infrasound (low frequency, <20 Hz) carries over long distances; ultrasound (high frequency, >20 kHz) is useful in resolving small objects (such as swim-bladders); typical range of hearing in fishes (response to noise) ca. 100 to 1,000 Hz, but varies greatly depending on species (some respond to ultrasound); in baleen whales ca. 10 to 1,000 Hz; in dolphins ca. 100 to 100,000 Hz. (Maximum sensitivity in humans ca. 2,000 Hz.) Typical frequency in seismic exploration, 100 to 1,000 Hz.

SOURCES

Other than acoustics: C. Emiliani, 1992, *Planet Earth—Cosmology, Geology, and the Evolution of Life and Environment*. Cambridge University Press, Cambridge New York, 719 pp.; acoustics: various sources, scientific literature, also see articles on ocean acoustics in Steele, J. H., S. A. Thorpe, K. K. Turekian (eds.), 2001, *Encyclopedia of Ocean Sciences*, 6 vols., Academic Press, San Diego, 3399 pp.

Appendix Two

Aspects of Ocean Chemistry

COMPOSITION OF SEAWATER

Seawater is a solution of various SALTS, with about 96.5 percent water. The content of salt in seawater is called the *salinity;* it is normally near 3.5 percent (by weight, given as 35 per mil). The salinity is commonly measured by determining the conductivity of seawater (while correcting for the temperature effect). The salt is present as ions, that is, charged molecules. The ions are surrounded by water molecules (which have a dipole nature; that is, a negative and a positive pole). Thus, they do not readily combine to make salt, which explains the fact that water is of all natural liquids the best solvent.

The MAJOR IONS in seawater are the following (numbers are percent of total weight of the salt in seawater): chloride (Cl^-, 55.04), sodium (Na^+, 30.61), sulfate (SO_4^{2-}, 7.68), magnesium (Mg^{2+}, 3.69), calcium (Ca^{2+}, 1.16), potassium (K^+, 1.10), bicarbonate (HCO_3^-, 0.41) (sum, 99.69 percent). The next four components are bromide, boric acid, strontium, and fluoride. The first four are entirely dominant (97 percent); the main constituent of the salt formed when evaporating seawater is the familiar table salt. On the whole, the composition of seawater reflects the solubility of the substances involved, which are abundant in crustal rocks (sodium, magnesium, calcium, potassium) and in volcanic emissions (chlorine, sulfur, carbon). Long-term control of the composition of seawater involves reaction of volcanic gases with the continental crust (chemical weathering), deposition of enormous salt bodies through geologic time, and reactions of seawater with hot basalt in volcanically active regions.

Seawater contains dissolved GASES such as nitrogen (N_2), oxygen (O_2), argon (Ar), and carbon dioxide (CO_2). In the case of CARBON DIOXIDE, there is a reaction with the water, which generates a number of molecules, including ions (H_2CO_3, HCO_3^-, CO_3^{2-}). This greatly enhances the ability of the water to hold carbon dioxide in solution, which results in the fact that the ocean holds the bulk of the combined ocean-and-atmosphere mass of carbon. Of the three species, the BICARBONATE ion (HCO_3^-) is greatly dominant at the pH values (concentration of hydrogen ions, acidity) found in the ocean (slightly basic, PH NEAR 8). High alkalinity (high carbonate-ion content) enhances carbon dioxide uptake by seawater. Precipitation of solid carbonate lowers the alkalinity, while dissolution of solid carbonate increases it. When seawater takes up additional carbon dioxide, carbonate is converted to bicarbonate, and the pH drops somewhat. This process is currently driven by the ongoing increase in carbon dioxide in the atmosphere and referred to as *acidification.* As the concentration of carbonate ion drops, the ability of seawater to take up carbon dioxide decreases.

NUTRIENTS LIMITING OCEAN PRODUCTIVITY

In most of the surface waters of the ocean, wherever there is sufficient light for photosynthesis, the concentrations of NITRATE (NO_3^-), PHOSPHATE (PO_4^{3-}), and SILICATE (H_4SiO_4) are extremely low, these species having been removed by precipitation within photosynthesizing organisms and an associated "rain" of particles toward deeper waters. The ratio of the elements nitrogen and phosphorus, during such removal and regardless of concentrations, stays approximately constant (reflecting the mixture of phytoplankton using them); this is the so-called REDFIELD RATIO and is near a value of 16. The ratio of silicon to the other two nutrients varies considerably, which is of great interest. Silicate is crucially important in the growth of diatoms (the ratio of silicon to carbon in diatoms approaching unity); it is used in making the glass shells of the siliceous phytoplankton. Since diatoms are at the base of short food chains, silicate supply governs the distribution and abundance of high-energy apex consumers (such as marine mammals: no silicate means long food chains, little food for apex consumers at the top level).

Nutrients are remineralized below the sunlit zone by bacteria-facilitated processes involving oxygen; thus, the amount of oxygen in deep waters is important in the recycling of nutrients, as is the rate of replacement of deep waters (thermohaline circulation). Typical concentrations of PHOSPHATE in the ocean vary between 0.1 and 3.0 micromoles per liter (μg-atoms/l) with values below 2.0 common in the deep Atlantic, and values greater than 2.0 common in the deep Pacific. NITRATE concentrations are 16 times greater by number of molecules (Redfield ratio). There has been much discussion about which is the "true" limiting factor, nitrate or phosphate. Most oceanographers concerned with this problem agree that nitrate (along with other bioavailable nitrogen species such as nitrite and ammonia) are limiting on the short time scales considered by biologists, while phosphate abundance controls the overall productivity of the sea on geologic time scales (millennia).

IRON is one of the more abundant elements in the crust (4.3 percent in the continental crust, making the Earth look yellowish brown from space) and indeed in the solar system (it makes Mars look reddish). It is of very low abundance in seawater, but abundant in the sediment. It is intimately involved in biological processes, mainly because of its ability to readily change its oxidation state (FERRIC AND FERROUS IRON), but also in other contexts of biochemical reactions. Being in short supply in the oxidized uppermost waters of the sea, IRON IS A LIMITING NUTRIENT, a fact that has been verified by experiment. Iron is also involved in the productivity of the ocean through the adsorption of PHOSPHATE on freshly precipitated iron species within the uppermost sediment, especially in continental margins. The intimate connection of iron to the sulfur cycle (precipitation of iron sulfide) links the SULFUR cycle to ocean production, as well. Besides iron there are other trace elements that are necessary to life processes, including manganese, zinc, cobalt, molybdenum, and others. Controls on their cycling differ, but commonly involve microbe-dominated reactions with OXYGEN and SULFUR at the seafloor.

COMPOSITION OF VENT WATER

Reactions of seawater with hot basalt within hydrothermal cells provide important controls on the chemistry of the sea. The REMOVAL OF SULFATE AND MAGNESIUM is central, as well as the addition of HYDROGEN IONS (acidity) and of CALCIUM and SILICATE. IRON and MANGANESE are delivered in large amounts, but in the present OXYGEN-rich environment in the deep sea, these are rapidly precipitated, making ferromanganese concretions and crusts. Thus, the chemistry of seawater is not much affected.

SOURCES

P. K. Weyl, 1970, *Oceanography: An Introduction to the Marine Environment.* John Wiley & Sons, New York, 535 pp.; National Oceanic and Atmospheric Administration (U.S. Department of Commerce); H. U. Sverdrup, M. W. Johnson, R. H. Fleming, 1942, *The Oceans: Their Physics, Chemistry, and General Biology.* Prentice Hall, Englewood Cliffs, N.J., 1087 pp.; P. G. Falkowski (ed.), 1980, *Primary Productivity in the Sea.* Plenum Press, New York, 531 pp.; L. A. Codispoti, 1989, *Phosphorus vs. nitrogen limitation of new and export production,* in W. H. Berger, V. S. Smetacek, G. Wefer (eds.), *Productivity of the Ocean: Present and Past.* John Wiley & Sons, Chichester, pp. 377–394; R. R. Haese, 2000, *The reactivity of iron,* in H. D. Schulz, M. Zabel (eds.), *Marine Geochemistry.* Springer-Verlag, Berlin, pp. 233–261; J. H. Steele, S. A. Thorpe, K. K. Turekian (eds.), 2001, *Encyclopedia of Ocean Sciences,* 6 vols. Academic Press, San Diego, 3399 pp.; C. L. van Dover, 2000, *The Ecology of Deep-Sea Hydrothermal Vents.* Princeton University Press, Princeton, N.J., 424 pp.

Appendix Three

Overview of Major Groups of Important Marine Organisms

SINGLE-CELLED ORGANISMS ("MICROBES")

AUTOTROPHS (SYNTHESIZING)

PROKARYOTES: blue-green algae (*Prochlorococcus*, cyanobacteria); methanogens (archaea) (bacteria and archaea are kingdoms).

EUKARYOTES: coccolithophorids, "nannofossils" (Haptophyta); dinoflagellates (Pyrrhophyta); diatoms, silicoflagellates (Chrysophyta) (Haptophyta, Pyrrhophyta, Chrysophyta are "divisions" of the "plant" kingdom).

HETEROTROPHS (USING PREEXISTING ORGANIC MATTER)

PROKARYOTES: sulfate-reducing bacteria, nitrate-reducing bacteria, methane-oxidizing bacteria, fermenting bacteria, organic-matter-oxidizing bacteria.

EUKARYOTES: foraminifera (rhizopods); radiolarians (actinopods); tintinnids (ciliophora) (single-celled heterotrophic eukaryotes are "protozoans" if excluding forms with chlorophyll, otherwise "protists" on the level of kingdom).

MULTICELLULAR ORGANISMS (PLANTS, ANIMALS, FUNGI)

AUTOTROPHS (USING LIGHT AND CARBON DIOXIDE)

Giant kelp *(Macrocystis)*, elkhorn kelp *(Pelagophycus)*, *Fucus, Alaria, Sargassum* (brown algae, Phaeophyta); calcareous green algae (*Halimeda, Penicillus*, Chlorophyta); coralline and encrusting algae (*Lithothamnium, Lithophyllum*, Rhodophyta); "sea grasses" (*Zostera, Phyllospadix, Thalassia*, marine angiosperms); mangrove *(Rhizophora, Laguncularia, Avicennia)*, salt-marsh halophytes: cordgrass and pickleweed *(Spartina, Salicornia)* (terrestrial angiosperms, Spermatophyta).

HETEROTROPHS (CONSUMING PREEXISTING ORGANIC MATTER)

FUNGI: (decomposers of organic matter, or living in association with marine macroalgae or with animals, commonly as infectious parasites).

ANIMALS: invertebrates, numerous phyla, major phyla: sponges, cnidarians, mollusks, annelids, arthropods, echinoderms; vertebrates, a subdivision of the phylum chordates (see below).

MAJOR MARINE ANIMAL PHYLA

PORIFERA: sponges, organization of flagellated cells, structure of support for the cells with canals and chambers, no mouth or digestive cavity, skeletal elements: "spicules."

CNIDARIA: animals with stinging cells: jellyfish, stony corals, gorgonians, sea pens, sea anemones, hydroids.

CTENOPHORES: comb jellies (a small group but with common representatives).

PLATYHELMINTES: flatworms and flukes, thousands of species.

NEMATODA: nematode worms, thousands of species.

MOLLUSCA: mollusks (thousands of species): gastropods (prosobranch and opisthobranch forms, thousands of species each, with and without shell), bivalves (mussels, scallops, cockles, clams, geoducks; thousands of species), cephalopods (squids, octopods, cuttlefish, ammonites [X]), chitons, scaphopods, etc.

ANNELIDA: segmented worms (thousands of species): polychaetes (scale worms, paddleworms, fireworms, ragworms, lugworms, sabellarid worms, terebellid worms, sabellid fan worms, serpulid fan worms, etc.).

POGONOPHORA: giant tube worms (few species but spectacular).

ARTHROPODA: arthropods (thousands of species, jointed-limb animals): copepods (thousands, calanoid, harpacticoid, etc., also parasitic forms), isopods (thousands), amphipods (thousands), decapods (thousands, crabs, shrimps, lobsters), cumacea (hundreds), euphausids (few, but important as krill), pycnogonids (hundreds, sea spiders), ostracods (thousands, mussel shrimps), cirripeds (hundreds, barnacles), etc.; some break this superphylum into several phyla (crustaceans, chelicerates, uniramians, tardigrades and pentastomids); the largest animal class (insects, uniramians) only has a few water striders in the sea; trilobites and eurypterid sea scorpions are extinct, the horseshoe crab (related to eurypterids) is still extant.

BRYOZOA: moss animals, sea mats (thousands of species), colonies of small individuals (zooids) with a tentacled crown, mouth and anus, branching and encrusting forms, common in the intertidal zone.

BRACHIOPODA: lampshells (hundreds of species, common benthic organism, extensive fossil record) organism with a bivalve shell on a stalk (pedicle) and an internal bilateral tentacle crown (lophophore).

CHAETOGNATHA: arrow worms (few species, but abundant in the plankton).

ECHINODERMATA: echinoderms (thousands of species, spiny-skinned animals, benthic when adult): starfishes, brittle stars and basket stars (asterozoans), sea lilies and feather stars (crinozoans), sea urchins and sea cucumbers (echinozoans), and many extinct forms.

HEMICHORDATA: acorn worms and pterobranchs (few species, but important in an evolutionary context, connecting chordates to their distant relatives, the echinoderms and other groups).

CHORDATA: animals with a dorsal nerve chord (thousands of species): urochordates (sea squirts, salps, planktonic tunicates), cephalochordates (lancelets), vertebrates (mainly fishes, also reptiles, mammals, birds) (see below); the relationship between sea squirts and vertebrates is best seen in the "tadpole" larvae of the sea squirts (the saclike sessile adults do not resemble vertebrates), the fishlike lancelet is built much like a squirt larva and stays mobile throughout its life.

VERTEBRATES

Jawless fishes (hagfish and lampreys), cartilaginous fishes, bony fishes, and tetrapods (amphibians [freshwater], reptiles and birds, mammals); fishes have thousands of species, tetrapods have a few hundred within the marine realm (being mostly terrestrial). Some biologists include the tetrapods in a subclass with the lobe-finned fishes, that is, with the bony fishes. About half of all vertebrates are fishes, and there are about 24,000 species, of which 60 percent are marine. The largest fish is the whale shark *Rhincodon typus*. Giant sturgeons can reach an age of more than 100 years. Fishes have a remarkable range of adaptations to different environments. Some fishes, adapted to freezing waters in polar regions, have antifreeze in their blood. Others have adapted to hot springs and soda lakes, with rather high temperatures. Some have life cycles involving both the sea and freshwater (e.g., salmon and some eels). Some marine fishes can readily move out of water (e.g., mudskippers). Some can fly, while others (on land) can survive long periods of drought, burrowing into the ground. The following list mentions only major groups and is not comprehensive.

AGNATHA

Jawless fishes (few species, hagfishes and lampreys, sea lamprey: *Petromyzon marinus*); the two groups placed here are not closely related.

HAGFISHES: predators and scavengers (can twist the body into a knot to gain leverage in penetrating live or dead prey), eel-like, several blood-pumping organs are present, no paired fins, no eyes (but light-sensitive skin patches in places), highly developed sense of smell, with tentacles around the mouth, rasping teeth on the tongue only, defend themselves by extruding large amounts of slime that clogs the gills of potential predators.

LAMPREYS: predators (some attack larger fishes attaching to their sides or bellies, some feed only as larvae on smaller invertebrates), circular sucking mouth ringed with teeth that open access to the body of the prey, eel-like, no paired fins, with a pair of large eyes (and a light-sensitive pineal organ on top of the head), seven gill slits; lampreys breed in freshwater, but some spend their life at sea, as adults.

CHONDRICHTYES

Fishes with a cartilaginous skeleton, lacking a swimbladder, and with multiple gill slits. The skin is covered by denticles (rather than scales). Fertilization is internal, and eggs are few and large, while live births prevail in pelagic forms. The young look like small adults (no larvae as in bony fishes).

SHARKS AND RAYS: hundreds of species in 12 orders for sharks and 1 order for skates and rays; SHARKS have five gill slits (some have six or seven), a strong sense of smell; some sharks can sense very weak electric fields; teeth are replaced in rows (commonly there are hundreds of them in the mouth at any one time); they have a highly developed lateral line system for picking up vibrations and pressure fluctuations; sharks are found throughout the water column in all parts of the ocean where there is sufficient food; the largest fishes are plankton-eating sharks (whale shark, basking shark), while fish- (and meat-) eating predators (great white shark, tiger shark) overlap in size with the largest bony fishes (marlin, sturgeon); the mako shark is among the fastest fish extant (clocked at 60 mph; perhaps matched by swordfish); SKATES AND RAYS basically are flattened sharks, adapted for life in contact with the seafloor, where they feed on mollusks and crustaceans and other benthic organisms using buttonlike teeth for crushing, except for a few (like the plankton-eating Manta ray, some of which reach giant proportions) that opted for the open sea; their pectoral fins are modified to make wings; some (sawfish) have a modified head, carrying a "saw" with sharp teeth, which is used to slash prey and to defend against larger attackers; some rays are capable of producing electric shock. There are hundreds of species of rays.

CHIMAERAS: ratfish, few species, all marine; somewhat similar to sharks and rays, living in cold water and at depth, slow swimming, bottom-hugging, teeth fused into a solid beak, modified breathing apparatus (relative to sharks).

OSTEICHTHYES

Fishes with a bony skeleton, with an immense range of adaptations to various environments in the sea and on land; containing two major groups: the lobe-finned fishes (with lungfishes and coelacanths) and the ray-finned fishes (the great majority of fishes populating the sea and freshwater bodies); characteristic properties (with some exceptions): true scales on the skin, a protective structure for the gills, and movable rays in the fins and the tail. Swim-bladders are common. Reproduction commonly involves a large number of eggs, and fertilization is external. (Exceptions exist.) In many species there is a larval stage, linked to dispersion.

LUNGFISHES: few species, habitat is freshwater in drought-prone areas. Closest living relatives of tetrapods, among the fishes.

COELACANTHS: one genus only, *Latimeria*, a marine fish, rarely caught. Large eggs, few in number; swim-bladder filled with fat.

RAY-FINNED FISHES: an enormously large group, including the modern teleosts that dominate the scene throughout the Tertiary.

STURGEONS: bony fishes with a sharklike tail, less than 50 species, largest freshwater fishes, some live at sea when adult; millions of eggs ("caviar"); stocks greatly reduced from overfishing.

TELEOSTS: ray-finned fishes, thousands of species, two-thirds marine. Some important types of teleosts are eels and tarpons, herringlike fishes (herring, sardine, alewife, shad, anchovy, pilchard, sprat), salmons and argentines ("herring smelt"), codfishes and hakes and grenadiers and anglerfishes (including frogfishes and batfishes, and many deep-sea species), clingfishes, bristle mouth and dragonfishes and viperfishes (deep-sea fishes with luminous organs), lizard fishes and lanternfishes (the latter with hundreds of species in the deeper waters of all oceans), catfishes and carplike fishes (thousands of species with some marine representation), needlefishes and flying fishes and silversides (including the familiar grunion), flashlight fishes and squirrel fishes, pipefishes and shrimp fishes and seahorses (male brood pouch), zebrafishes and scorpionfishes (poison spines), gurnards and sculpins, sea basses, remoras (modified dorsal fin for attaching to free ride), jacks, dolphin fishes, snappers, drums, butterfly fishes, ribbonfishes (including the oarfish) and moonfishes, angelfishes, damselfishes, parrotfishes (fused beaklike teeth for biting coral, grinding teeth in the throat), clownfishes, wrasses, eelpouts, icefishes (with antifreeze in the blood), stargazers, blennies, dragonets, gobies, surgeonfishes, tunas and mackerels (built for speed), billfishes (including the fast-swimming marlins, sailfishes and swordfish), barracudas, flatfishes (hundreds of species, including soles and flounders; some change their colors to match the seafloor, on which they rest), triggerfishes, pufferfishes, boxfishes, sunfishes, and porcupine fishes.

TETRAPODS

Four-legged vertebrates; in essence, lobe-finned fishes adapted to a life without water (that is, breathing air), as most clearly seen in reptiles and the closely related birds and mammals. In most amphibians, the ancestral water-link is maintained through breeding and larval development within freshwater. The crab-eating frog *Rana cancrivora* inhabits mangrove swamps in southeastern Asia. In marine reptiles, birds, and mammals, there is a return to the water in cases. This shift is completed in whales and is partially completed in seals and penguins. It is most clearly seen happening at present in sea otters (which have closely related weasel cousins on land) and polar bears (which have terrestrial congeners). Marine turtles, crocodiles, and penguins continue to lay their eggs on land; a behavioral reminder of their terrestrial ancestry.

REPTILES: close to 100 marine species, with sea snakes dominant, followed by sea turtles. SEA SNAKES, leg-less reptiles with a flattened tail for swimming, for example, the yellow-bellied sea snake

(*Pelamis platurus,* the most abundant of reptiles, a relative of the cobra); some sea snakes lay eggs, others give live birth. SEA TURTLES: reptiles that make a shell from horny shields (except the Leatherback turtle, *Dermochelys coriacea,* which is covered by leathery skin). Representatives are: Loggerhead turtle, Ridley turtle, Flatback turtle, Green turtle, and Hawksbill turtle. Some turtles apparently have an internal magnetic compass useful in guiding migration. There are two species of (optionally) marine crocodiles (American crocodile: *Crocodylus acutus*; Indo-Pacific crocodile: *Crocodylus porosus,* the largest living reptile) and one of marine lizard (marine iguana, *Amblyrhynchus cristatus*) that feeds on algae in the Galapagos Islands.

BIRDS: warm-blooded egg-laying tetrapods resembling reptiles but bearing feathers; several hundred marine species. Most seabirds are capable of flight, except for penguins (cold southern waters) and some representatives of cormorants. Prominent among strictly marine birds are albatrosses, shearwaters and petrels, storm-petrels and diving-petrels, tropic-birds, gannets and boobies, frigatebirds, and auks (habitat: cold northern waters, with the familiar guillemots, murres, and puffins). Among cormorants and pelicans there are some species that are found with freshwater bodies. Some ducks and geese find food in saltwater lagoons. Skuas (jaegers), gulls, and terns are mainly marine, although some species prefer freshwater. Many freshwater birds can be observed along the seashore feeding, depending on the season. In terms of number of species, the most diverse are petrels and shearwaters, then gulls, then terns, then cormorants and shags, then auks. There are 17 species of penguins; the most abundant are the species of the genus *Pygoscelis* (adélie, chinstrap, gentoo), whose members are counted in the millions. Shearwaters, albatrosses, and terns, among other birds, undertake extensive migrations involving thousands of miles.

MAMMALS: warm-blooded tetrapods bearing live offspring that developed in utero, and which feed on their mother's milk. Marine forms include whales, pinnipeds, seacows, and otters, all of which are able to hold their breath to procure food below the sea surface.

WHALES: a diverse group of marine mammals that spend their entire lives at sea, from birth to mating to death, and whose members have a fishlike appearance. There are fewer than 100 species. The common ancestors (archaeocetes, e.g., *Pakicetus*) were terrestrial and lived in the earliest Tertiary, at the edge of land and water (Paleocene), in a life-style reminiscent of the modern hippo (which is in fact an extant relative of the cetacean ancestor). There are two major groupings: whales bearing teeth (ODONTOCETI) and whales bearing baleen (MYSTICETI). Baleen whales evolved from teeth-bearing ancestors in the middle of the Tertiary (Oligocene), as a result of the appearance of high-production upwelling areas with high plankton density, on a planet with cooling polar regions. The baleen is used to filter the water for food. The odontocetes contain the following major groups: delphinids (e.g., *Delphinus,* common dolphin, *Orcinus,* killer whale; *Tursiops,* bottle-nosed dolphin); monodontids (beluga and narwhal); phocoenids (porpoises); physeterids (sperm whales); "platanistids" (various types of river dolphins); ziphiids (beaked whales and bottlenose whales). The largest predatory diving mammals are in this group (sperm whale, *Physeter*). Toothed whales use echolocation to "see" using sound when submerged; they have an unusually large brain (compared with other tetrapods) for processing acoustic signals; they also communicate using such signals. The mysticetes contain the following major groups: balaenids (bowhead and right whales, *Balaena*), rorquals (*Balaenoptera,* blue, fin, minke, sei, bryde's whales; *Megaptera,* humpback whale), gray whale *(Esrichtius),* and pygmy right whale (*Caperea* in the family Neobalaenidae). The largest animals on Earth (blue whale, fin whale) are mysticetes. Large size is an asset when storing energy for long migrations and when nursing without feeding. In cold waters, large size is good for preserving heat. PINNIPEDS: four-legged marine carnivore mammals that feed at sea and breed on shore (or on ice); comprising the "true" seals. (PHOCIDAE: about 20 species of northern and southern seals, with the greatest abundance in the Antarctic lobodonts, which group contains the krill-eating seal, *Lobodon carcinophagus,* with a population estimated at more than 10 million), the "eared" seals (OTARIIDAE: fewer than 20 species of sea lions and fur seals), and the walrus (ODOBENIDAE: one species). Phocids rely on blubber for insulation; otarids have a thick fur. Phocids are the better divers, some can go without air for more than an hour, and Weddell seals and elephant seals have been observed to reach great depths (hundreds of meters, and even exceeding a mile in *Mirounga*). Terrestrial ancestors of pinnipeds existed in the late Oligocene; they were related to ancient types of dogs, bears, and weasels. It is not clear whether the last common ancestor was aquatic or terrestrial. SEA COWS: (sirenians) fully aquatic mammals feeding on plants; there are three species, one of which lives in the Amazon river system; dugongs (tropical Indo-Pacific) are marine; manatees (tropical Atlantic) are at home in coastal waters including associated estuaries. The large Steller's sea cow (a cold-water species in the North Pacific) was hunted to extinction in the eighteenth century. As in whales, there are no hind limbs, and there is a fluke for propulsion. Movement is comparatively sluggish. The terrestrial ancestor of sirenians lived in the earliest Tertiary (Paleocene). It was related to elephants. SEA OTTER: *Enhydra lutris*, a fully marine mammal, living in coastal waters of the North Pacific and feeding on sea urchins, mollusks, and crabs; closely related to river otters, carnivores of the weasel group. Pups are born at sea. Little is known about the habits of a coastal otter hunting in the waters off Peru and Chile, the "marine otter."

SOURCES

B. B. Joergensen, 2000, *Bacteria and marine biogeochemistry*, in H. D. Schulz, M. Zabel (eds.), *Marine Geochemistry*. Springer-Verlag, Berlin, pp. 173–207; E. Y. Dawson, 1966, *Marine Botany*. Holt, Rinehart, Winston, New York, 371 pp.; G. Waller (ed.), 1996, *SeaLife: A Complete Guide to the Marine Environment*. Smithsonian Institution Press, Washington, D.C., 504 pp.; K. Banister, A. Campbell (eds.), 1985, *The Encyclopedia of Aquatic Life*. Facts on File, Inc., New York, 349 pp.

Appendix Four

Geologic Time Scale

MAJOR SUBDIVISIONS OF THE GEOLOGIC TIME SCALE

PERIOD	START (10^6 years ago)	DURATION (10^6 years)	REMARKS
Quaternary	1.8	1.8	Ice age cycles; *Homo* spp.
Neogene	23.3	21.5	Antarctic ice shield expansion, thermocline
Paleogene	65	41.7	Circumpolar Current established, great cooling step
Cretaceous	146	81	Giant marine lizards, diving birds, warm ocean
Jurassic	208	64	Ichthyosaurs, ammonites abundant, warm ocean
Triassic	250	42	North America separating from Africa
Permian	295	45	Ice ages; great extinction at end
Carboniferous	360	65	Extensive coal formations
Devonian	412	42	Fishes thrive; first amphibians
Silurian	445	33	First bony fishes
Ordovician	490	45	First corals; first jawless fishes
Cambrian	550	60	Sudden expansion of shelled fossils
Precambrian	>550	~3,000	Poor fossil record; low oxygen

Note: The age of the Earth is currently estimated at 4.6 billion years.

MAJOR SUBDIVISIONS OF THE CRETACEOUS PERIOD

PERIOD	START (10^6 years ago)	DURATION (10^6 years)	REMARKS
Maastrichtian	71	6	Major extinctions at end of Maastrichtian
Campanian	83	12	Cooling trend
Santonian	86	3	Magnetically quiet zone ends
Coniacian	89	3	Mid-Cretaceous warm time ends
Turonian	94	5	Changes in deep circulation
Cenomanian	100	6	Radiation of plankton (starts in Albian)
Albian	112	12	Volcanism, high CO_2, maximum warmth
Aptian	125	13	Anoxia spreads, high extinction rates
Barremian	130	5	Sea level rises; magnetic quiet zone starts
Hauterivian	136	6	South Atlantic opens
Valanginian	140	4	Circum-tropical sea: Tethys dominant
Berriasian	146	6	North Atlantic keeps expanding

MAJOR SUBDIVISIONS OF THE CENOZOIC (PALEOGENE AND NEOGENE PERIODS)

PERIOD	START (10^6 years ago)	DURATION (10^6 years)	REMARKS
Paleogene			
Holocene	0.012	0.012	Agriculture, iron age
Pleistocene	1.8	1.8	Ice age cycles
Late Pliocene	3.0	1.2	Onset of northern ice ages
Middle Pliocene	3.4	0.4	Uplift of Himalayas, accelerates
Early Pliocene	5.2	1.8	Closing of Panama Isthmus
Late Miocene	12	6.8	N. Atl. Deep Water prod.; dry Mediterranean
Middle Miocene	16	4	Expansion of southern ice sheets
Early Miocene	23	7	Radiation of plankton; Tethys closes
Neogene			
Late Oligocene	29	6	Cold deep water; ancestral baleen whales
Early Oligocene	34	5	Major cooling; low-production ocean
Late Eocene	40	6	Major shift in climate; Great Cooling Step
Middle Eocene	49	9	High marine diversity; Tethyan seas
Early Eocene	56	7	Plankton radiation
Late Paleocene	61	5	End-of-Paleocene warm peak
Early Paleocene	65	4	Recovery from end-of cretaceans extinctions

Sources: C. Emiliani, 1992, *Planet Earth—Cosmology, Geology, and the Evolution of Life and Environment.* Cambridge University Press, Cambridge, 719 pp.; H. M. Bolli, J. B. Saunders, K. Perch-Nielsen, 1985, *Plankton Stratigraphy.* Cambridge University Press, Cambridge, 1032 pp.; B. McGowran, 2005, *Biostratigraphy—Microfossils and Geologic Time.* Cambridge University Press, Cambridge, 459 pp.; F. M. Gradstein, J. G. Ogg, 2004, *Geologic time scale 2004—why, how, and where next!* Lethaia 37, 175–181; W. H. Berger, J. C. Crowell (eds.), 1982, *Climate in Earth History.* National Academy Press, Washington, D.C., 198 pp.; X. LePichon (Chairman, Steering Committee), 1988, *Report of the Second Conference on Scientific Ocean Drilling (Cosod II).* European Science Foundation, Strasbourg, 142 pp.; K. Becker (Chair, Editorial Review Board), 2002, *Achievements and opportunities of scientific ocean drilling.* Joides Journal 28 (1, Special Issue), Joint Oceanographic Institutions for Deep Earth Sampling.

Appendix Five

Topographic Statistics

EARTH'S SIZE AND SURFACE

	NORTHERN HEMISPHERE		SOUTHERN HEMISPHERE	
LATITUDE (°)	(% of total area)	(% water)	(% of total area)	(% water)
0–5	4.34	78.6	4.34	75.9
5–10	4.31	75.7	4.31	76.9
10–15	4.24	76.5	4.24	79.6
15–20	4.15	70.8	4.15	76.4
20–25	4.02	65.2	4.02	75.4
25–30	3.86	59.6	3.86	78.4
30–35	3.68	57.7	3.68	84.2
35–40	3.46	56.8	3.46	93.4
40–45	3.22	51.2	3.22	96.4
45–50	2.96	43.8	2.96	97.5
50–55	2.665	40.7	2.665	98.5
55–60	2.355	45.0	2.355	99.9
60–65	2.025	31.2	2.025	99.7
65–70	1.68	28.7	1.68	79.5
70–75	1.32	65.5	1.32	38.6
75–80	0.95	77.1	0.95	10.7
80–85	0.575	85.2	0.575	0.0
85–90	0.19	100.0	0.19	0.0
Total	50	60.7	50	80.9

Note: The Earth's surface comprises 361.059 million square kilometers of ocean and sea (70.8 percent of the total area) and 148.892 million square kilometers of land (29.2 percent).

AREA, VOLUME, AND MEAN DEPTH OF OCEAN BASINS AND SOME SEAS

GEOGRAPHIC UNIT	AREA (million km²)	VOLUME (million km³)	MEAN DEPTH (m)
Atlantic Ocean	82.44	323.6	3,926
With adjacent seas	106.46	354.7	3,332
Pacific Ocean	165.25	707.6	4,282
With adjacent seas	179.68	723.7	4,028
Indian Ocean	73.44	291.0	3,963
With adjacent seas	74.92	291.9	3,897
Arctic Ocean	14.09	17.0	1,205
Caribbean and Gulf of Mexico	4.32	9.57	2,216
Mediterranean and Black seas	2.97	4.24	1,429
Hudson Bay	1.23	0.16	128
Bering Sea	2.27	0.69	1,437
Okhotsk Sea	1.53	3.26	838
Japan Sea	1.01	1.36	1,350
East China Sea	1.25	0.24	188
All oceans and seas	361.06	1,370.32	3,795

AREA OF DEPTH ZONES IN THE OCEANS (IN PERCENT, REFLECTING BENTHIC HABITAT)

DEPTH ZONE (m).	ATLANTIC OCEAN	PACIFIC OCEAN	INDIAN OCEAN	ALL OCEANS
0–200	13.3	5.7	4.2	7.6
200–1,000	7.1	3.1	3.1	4.3
1,000–2,000	5.3	3.9	3.4	4.2
2,000–3,000	8.8	5.2	7.4	6.8
3,000–4,000	18.5	18.5	24.0	19.6
4,000–5,000	25.8	35.2	38.1	33.0
5,000–6,000	20.6	26.6	19.4	23.3
6,000–7,000	0.6	1.6	0.4	1.1
>7,000	<.1	0.2	<0.1	0.1

GREATEST DEPTHS IN THE OCEAN BASINS AND SOME SEAS

Note: All values are approximate because of uncertainties in estimating sound velocity; sources commonly disagree. Also note that rough estimates measured in steps of hundreds of meters can acquire spurious precision when converted to feet.

PACIFIC OCEAN: Marianas Trench (Challenger Deep), 10,915 m (35,810 ft); Tonga Trench, 10,850 m (35,600 ft); Kuril-Kamchatka Trench, 10,500 m (34,400 ft); Philippine Trench (near Cape Johnson Deep), 10,200 m (33,500 ft); Kermadec Trench, 10,050 (33,000 ft); Japan Trench (Ramapo Deep), 9,700 m (32,000 ft); New Hebrides Trench, 9,165 m (30,070 ft); Peru-Chile Trench, 8,055 m (26,430 ft); Aleutian Trench, ca. 7,700 m (25,000 ft); Middle America Trench, 6,660 m (21,900 ft)

ATLANTIC OCEAN: Puerto Rico Trench, 8,400 m (27,600 ft); South Sandwich Trench, 9,300 m (30,500 ft)

INDIAN OCEAN: Java Trench, 7,450 m (24,440 ft)

ARCTIC OCEAN: Pole Abyssal Plain, 4,660 m (15,300 ft)

CARIBBEAN SEA: Cayman Trench, 7,680 m (25,200 ft)

MEDITERRANEAN: Hellenic Trough, 5,090 m (16,700 ft)

SOURCES

H. U. Sverdrup, M. W. Johnson, R. H. Fleming, 1942, *The Oceans: Their Physics, Chemistry, and General Biology.* Prentice Hall, Englewood Cliffs, N.J., 1087 pp.; J. F. Luhr, 2003, *Earth.* Smithsonian Institution, DK Publishing, New York, 520 pp.; M. Leier (ed.), 2001, *World Atlas of the Oceans.* Firefly Books, Buffalo, N.Y., 264 pp.; M. Bramwell (ed.), 1977, *The Rand McNally Atlas of the Oceans.* Rand McNally, New York, 208 pp.; R. L. Fisher, H. H. Hess, 1963, *Trenches,* in M. N. Hill (ed.), *The Sea,* vol. 3. John Wiley & Sons, New York, pp. 411–436; Puerto Rico Trench *fide* U. ten Brink.

FIGURE SOURCES AND REFERENCES

Below are sources by figure number; full citations of references follow. Most figures drawn from the literature, the Web, or museum exhibits have been modified considerably for present purposes. Drawings and photographs by the author are marked "orig."

FIGURE SOURCES

P.1 Ocean Drilling Program.
P.2 Orig.

1.1 Monterey Bay Aquarium (orig).
1.2 H. Murayama 1939 and Denmark's Aquarium (Copenhagen), orig.
1.3 R. Hessler, SIO.
1.4 Courtesy of S. W. Chisholm, MIT.
1.5 After NOAA.
1.6 M. Yasuda, SIO.
1.7 Challenger Reports, as adapted by Bailey 1953.
2.1 Globe Model, Danish Geological Survey, Copenhagen.
2.2 Orig.
2.3 Orig.
2.4 Orig.
2.5 Orig.
2.6 Orig.
2.7 Orig.
2.8 After JPL NASA.
2.9 Orig.
2.10 After NASA, with additions.
2.11 After Kuenzi 2002 and Wittheit Club, Bremen, arrows added.
2.12 Orig.
2.13 After NOAA, modified.
2.14 *Brehm's Tierleben*, SIO Explorations, H. Murayama (see chapter 1 for reference), Expedition Reports of Albert I Prince of Monaco, Los Angeles Natural History Museum, Bergen Aquarium.
2.15 From Sverdrup et al. 1942, Winther 1953, E. Haeckel, NOAA, *Brehm's Tierleben*, Hardy, and other sources.
2.16 E. Haeckel 1904; *Brehm's Tierleben*; Los Angeles County Natural History Museum (orig.); Expedition Reports of Albert I Prince of Monaco; Aquarium of the Pacific, Long Beach, Calif. (orig.); SIO Explorations; and other sources.
2.17 Orig.
2.18 Orig., with icons adapted from Sverdrup et al. 1942, CalCOFI reports, H. Murayama 1939.
2.19 Berger 1989, simplified.
2.20 Orig.
2.21 After J. and K. Imbrie 1979, with additions.
3.1 Orig.
3.2 Orig.
3.3 After MODIS Ocean Team NASA.
3.4 After SeaWIFS NASA; R. H. Stewart, NASA.
3.5 Orig.
3.6 Courtesy of Brooke Marx (Oceanside, Ca.).
3.7 Orig.
3.8 Orig.
3.9 Orig.
3.10 Orig.
3.11 Orig.
3.12 Berger 1976.
3.13 E. Haeckel 1904.
3.14 Calvin 1939.
3.15 SIO archives.

3.16 Sverdrup et al. 1942.
3.17 Natural History Museum, Hannover (Lower Saxon), by permission, with modifications.
3.18 Sverdrup et al. 1942 and Berger 1976.
4.1 Honolulu Aquarium (orig.).
4.2 Parker and Haswell 1921 and E. Haeckel, 1904 (center).
4.3 Daly 1936.
4.4 SIO Explorations.
4.5 After Chadwick-Furman 1996 and NOAA 1993.
4.6 After Kinzie and Buddemeier 1996.
4.7 SIO Explorations.
4.8 Orig.
4.9 After Kinzie and Buddemeier 1996, with additions.
4.10 Orig.
4.11 Orig.
4.12 Dana 1872, Daly 1936, and Darwin 1842.
4.13 Harriott and Banks 2002.
4.14 After Daly 1936.
4.15 Orig.
5.1 Orig.
5.2 Orig.
5.3 After Berger 1976.
5.4 Orig. (North County of San Diego).
5.5 Orig.
5.6 After Berger 1976.
5.7 After Shepard 1963, with modifications.
5.8 After Berger 1976.
5.9 Orig.
5.10 Orig.
5.11 Shepard and Curray 1967 and Fairbanks 1989.
5.12 Moore 1969.
6.1 Orig. sketch based, in a general way, on Iselin 1936, Iselin and Fuglister 1948, Sverdrup et al. 1942, and various satellite images (infrared, NOAA).
6.2 Orig.
6.3 After Dietrich 1957.
6.4 Krümmel 1907.
6.5 Sverdrup et al. 1942.
6.6 After NOAA, arrows added.
6.7 Sverdrup et al. 1942.
6.8 Universum Museum Bremen (upper) and orig. (lower).
6.9 Russell and Yonge 1936 (upper), with modifications, and Fisheries Museum, Bergen, Norway (lower).
6.10 Munich Re Group 1998, with modifications.
6.11 Stommel 1980.
6.12 After Legates 1996, simplified.
6.13 After Fleming 1957, by permission of the Geological Society of America.
6.14 TOPEX/Poseidon, ERS-1, ERS-2, Space Oceanography Division CLS Toulouse, France, (simplified).
6.15 G. Wüst 1936 (upper) and orig. (lower).
6.16 Stommel 1958, with modifications.
6.17 NOAA (upper) and Sverdrup et al. 1942 (lower).
7.1 After H. Thorade in Krümmel 1907.
7.2 After California Department of Fish and Game, McEvoy 1986, and CalCOFI.
7.3 After Dietrich et al. 1975.
7.4 Orig.
7.5 After Lluch-Belda et al. 2003.
7.6 After CalCOFI.
7.7 NOAA.
7.8 After J. Isaacs and A. Soutar.
7.9 After CalCOFI.
8.1 Reid 1962, by permission (author and the American Society of Limnology and Oceanography).
8.2 Fleming 1957, by permission of the Geological Society of America.
8.3 After Sumich 1976, from Hardy 1924.
8.4 Sverdrup et al. 1942.
8.5 Orig.
8.6 After E. Suess 1980, with additions.
8.7 Sverdrup et al. 1942, with additions.
8.8 Orig.
8.9 Orig., with drawings in Sverdrup et al. 1942, SIO Explorations, CalCOFI Atlas, H. Murayama 1939, and an exhibit in Los Angeles Museum of Natural History.
8.10 Orr and Marshall 1969, with modifications.
9.1 Courtesy of Karl Berger, Santa Clarita, Ca.
9.2 Sea Life Park, Oahu (orig.).
9.3 Skeleton of Minke: Zoological Institute and Museum, Hamburg (orig.); Zoo am Meer, Bremerhaven. Blue whale drawing: NOAA. Orca model: Museum of Natural History, University of Bergen (orig.).
9.4 NOAA.
9.5 After Bonner 1989, redrawn.
9.6 After Ciesielski and Weaver 1983, modified.
9.7 After Bonner 1989, redrawn.
9.8 Orig.
9.9 Los Angeles County Museum of Natural History (orig.), by permission.
9.10 Orig.
9.11 SIO Explorations.
9.12 Deutsches Meeresmuseum Stralsund (orig.), by permission.
9.13 Los Angeles County Museum of Natural History, by permission.
9.14 Bergen Museum of Natural History (upper), and Colbert 1955 (lower); with modifications.
9.15 After Colbert 1955.
10.1 Orig.
10.2 SIO Explorations.
10.3 Russell and Yonge 1936 (left) and Richard Ellis 1997 (umbrella squid), by permission.

10.4 SIO Explorations.
10.5 Albert I of Monaco.
10.6 Modified after drawings by C. P. Idyll 1964 and A. C. Hardy 1956.
10.7 E. Haeckel.
10.8 Orig.
10.9 After Russell and Yonge 1936.
10.10 Albert I of Monaco.
10.11 E. Haeckel 1904.
11.1 After Seibold and Berger 1993.
11.2 NOAA Historical Collections.
11.3 After Aquarium of the Pacific, Long Beach, Calif., by permission.
11.4 San Diego Natural History Museum (orig.), by permission.
11.5 After Dietz 1962, based on record of U.S. Navy Electronics Laboratory, San Diego.
11.6 SIO archives.
11.7 After Menard 1964.
11.8 V. Spiess 1996.
11.9 SIO archives.
12.1 Orig.
12.2 After SIO Explorations.
12.3 After SIO Explorations.
12.4 After SIO Explorations.
12.5 After E. Hamilton, from Hess 1945.
12.6 After Meyer and Gillis 1994.
12.7 Heirtzler et al. 1966.
12.8 Courtesy of David Sandwell, SIO.
12.9 NOAA.
12.10 Orig. after J. T. Wilson (1963) from Seibold and Berger 1993 (see chapter 4 for reference).
12.11 Orig.
12.12 After Berger et al. 1994.
12.13 After Herzig and Hannington 2000.
13.1 Orig.
13.2 Orig.
13.3 Orig.
13.4 Orig.
13.5 Orig.
13.6 Orig.
13.7 Upper left (four insets clockwise: GEO Bremen University; Thomson in Krümmel 1907; Seibold and Berger 1993); lower left: M. Yasuda, Scripps Institution of Oceanography; drawings on the right: F. L. Parker 1962, SIO.
13.8 Orig., data from J. Zachos et al. 2001.
13.9 Orig., data from A. Berger and M. Loutre 1991.
13.10 After Berger and Jansen 1995.
14.1 Ocean Drilling Program.
14.2 After Miller et al. 1987.
14.3 After Haq 1981, modified from Seibold and Berger 1993.
14.4 Orig., background graph from J. Thiede et al., AWI. Icons from various sources, including Neumayr 1895, Hauff Museum (Holzmaden), Mueller Museum Solnhofen, NOAA, SIO Explorations, and Nuremberg Zoo (penguin, polar bear). *Ambulocetus* courtesy J. G. M. Thewissen.
14.5 After Berger 1979, data from Douglas and Savin 1975 and Shackleton and Kennett 1975.
14.6 Orig.
14.7 Orig.
14.8 Deep Sea Drilling Project (T. Herbert).
14.9 After Thierstein and Okada 1979.
14.10 Ocean Drilling Program (R. Norris and Leg 171B Scientific Party).
14.11 After Berggren 1972.
15.1 After NOAA.
15.2 Orig.
15.3 After SIO Carbon Dioxide Laboratory. Shack: orig.
15.4 Orig.
15.5 GeoMar archives, Kiel; courtesy of G. Bormann, Bremen.

FIGURE REFERENCES

ARCHIVES

Deep Sea Drilling Project, SIO / University of California, San Diego (DSDP)
Jet Propulsion Laboratory (JPL NASA)
National Oceanic and Atmospheric Administration, U.S. Department of Commerce (NOAA)
Ocean Drilling Program,Texas A&M, College Station (ODP)
Scripps Institution of Oceanography (SIO)
U.S. National Aeronautics and Space Administration (NASA)

PRINTED WORKS

Albert I, Prince of Monaco, ca. 1900, *Résultats des Campagnes Scientifiques accomplies sur son yacht* (numerous volumes). Imprimerie de Monaco.
C. Augé, 1926, *Nouveau Petit Larousse Illustré, Dictionnaire Encyclopédique.* Paris, Librairie Larousse, 1760 pp.
H. S. Bailey 1953, The Voyage of the Challenger, as in *Ocean Science, Readings from Scientific American.* W. H. Freeman, San Francisco. Originally figured in the Challenger Report.
H. C. Berann (Nat. Geogr. Soc.), painting based on studies of B. C. Heezen and M. Tharp, cited in E. Seibold, W. H. Berger, 1996, *The Sea Floor, An Introduction to Marine Geology*, 3rd ed. Springer Verlag, Heidelberg, 356 pp.
A. Berger, M. F. Loutre, 1991, *Insolation values for the climate of the last 10 million years.* Quaternary Science Review 10, 297–317.
W. H. Berger, 1976, *Walk along the Ocean.* Solana Beach, Calif., 69 pp.
W. H. Berger, 1989, *Global maps of ocean productivity*, in W. H. Berger, V. S. Smetacek, G. Wefer (eds.), *Pro-*

ductivity of the Ocean: Present and Past. Dahlem Konferenzen, John Wiley, Chichester, pp. 429–455.

W. H. Berger, E. Jansen, 1995, *Younger Dryas episode: Ice collapse and super-fjord heat pump,* in S. R. Troelstra, J. E. van Hinte, G. M. Ganssen (eds.), The Younger Dryas. North-Holland, Amsterdam, pp. 61–105.

W. H. Berger et al. 1981 (data R. G. Douglas, S. M. Savin, 1975, and N. J. Shackleton, J. Kennett, 1975) cited in E. Seibold, W. H. Berger, 1993, *The Sea Floor, An Introduction to Marine Geology,* 2nd revised and updated ed. Springer-Verlag, Heidelberg, 356 pp.

W. Berger, T. Bickert, E. Jansen, G. Wefer, M. Yasuda, 1994, *The central mystery of the Quaternary ice age.* Oceanus 36 (4), 53–56.

W. A. Berggren, 1972, cited in E. Seibold, W. H. Berger, 1993, *The Sea Floor, An Introduction to Marine Geology,* 2nd revised and updated ed. Springer-Verlag, Heidelberg, 356 pp.

N. Bonner, 1989, *Whales of the World.* Facts on File, New York, 191 pp.

A. E. Brehm, *Brehm's Tierleben,* 2nd ed., as extracted by P. Rietschel et al. (eds.), *Das Tierreich nach Brehm.* Buch und Bild (Bertelsmann), Hamburg, no date (ca. 1955); and 4th ed. vol. 13, (ed. by Otto zur Strassen, 1911–1920), as extracted by R. Barth (ed.), 1953, *Brehm's Tierleben,* Verlag Zimmer and Herzog, Berchtesgaden.

W. S. Broecker, 1991, Oceanography 4, The Great Ocean Conveyor, 79–89; also see W. S. Broecker, G. H. Denton, 1989, *The role of ocean-atmosphere reorganizations in glacial cycles.* Geochimica Cosmochimica Acta 53, 2465–2501.

D. Bukry, M. N. Bramlette, 1969, *Coccolith age determinations: Leg 1, Deep Sea Drilling Project.* Initial Reports of the Deep Sea Drilling Project 1, 369–387.

California Cooperative Fisheries Investigations (CalCOFI), atlas series published by SIO report series published by Southern Fisheries Science Center, NOAA, LaJolla.

California Department of Fish and Game, 2003, "Review of Fisheries," CalCOFI Reports 44, 10; ISO 45, 13.

J. Calvin in E. F. Ricketts, J. Calvin, 1939, *Between Pacific Tides.* Stanford University Press, Stanford, Calif, 365 pp.

N. E. Chadwick-Furman, 1996, *Reef coral diversity and global change.* Global Change Biology 2, 559–568; and National Oceanic and Atmospheric Administration, U. S. Department of Commerce (sea-surface temperatures).

P. F. Ciesielski, F. M. Weaver 1983, cited in W. H. Berger, G. Wefer, 1996, *Expeditions into the past: Paleoceanographic studies in the South Atlantic,* in G. Wefer, W. H. Berger, G. Siedler, D. J. Webb (eds.), *The South Atlantic: Present and Past Circulation.* Springer-Verlag, Berlin, pp. 363–410.

E. H. Colbert, 1955, *Evolution of the Vertebrates.* John Wiley and Sons, New York, 479 pp.

R. A. Daly, 1936, *The Changing World of the Ice Age.* Yale University Press, New Haven, Conn., 271 pp.

J. D. Dana, 1872, *Corals and Coral Islands.* Dodd and Mead, New York, 398 pp.

C. Darwin, 1842, *The Structure and Distribution of Coral Reefs.* Smith Elder, London (3rd ed. 1889, 344 pp.), as interpreted in E. Seibold, W. H. Berger, 1993, *The Sea Floor, An Introduction to Marine Geology,* 2nd revised and updated ed. Springer-Verlag, Heidelberg, 356 pp.

G. Dietrich, 1957, *Allgemeine Meereskunde.* Bornträger, Berlin, 492 pp., cited in G. Dietrich, K. Kalle, W. Krauss, G. Siedler, 1975.

G. Dietrich, K. Kalle, W. Krauss, G. Siedler, 1975, *Allgemeine Meereskunde, Eine Einführung in die Ozeanographie.* Gebrüder Bornträger, Stuttgart, 593 pp.

R. S. Dietz, 1962, *The sea's deep scattering layers.* Scientific American 207 (2), 44–50.

S. Ekman, 1953, *Zoogeography of the Sea.* Sidgwick and Jackson, London, 417 pp.

R. Ellis, 1997, *Deep Atlantic: Life, Death, and Exploration in the Abyss,* A. Knopf, New York, 395 pp.

R. G. Fairbanks, 1989, *A 17,000-year long glacio-eustatic sea level record: Influence of glacial melting rates on the Younger Dryas event and deep-ocean circulation.* Nature 342, 637–643.

R. H. Fleming, 1957, *General features of the oceans,* in J. W. Hedgpeth (ed.), *Treatise on Marine Ecology and Paleoecology,* vol. 1. Geological Society of America Memoir 67. Geological Society of America, New York, pp. 87–107.

E. Haeckel, 1904, *Kunstformen der Natur.* Leipzig. Reprinted 2006, Prestel Verlag, München.

E. L. Hamilton, 1956. *Sunken islands of the Mid-Pacific Mountains.* Geol. Soc. America Mem. 64, 1–97.

B. U. Haq, 1981, cited in E. Seibold, W. H. Berger, 1993, *The Sea Floor, An Introduction to Marine Geology,* 2nd revised and updated ed. Springer-Verlag, Heidelberg, 356 pp.

A. C. Hardy, 1956, *The Open Sea: The World of Plankton,* vol. 1. Collins, London, 335 pp.

A. Hardy, 1959, *The Open Sea: Its Natural History.* Collins Clear-Type Press, London, 322 pp.

A. C. Hardy, 1924, The herring in relation to its animate environment, Part I, The food and feeding habits of the herring. Fish. Invest. London, Ser. II, 7 no. 3, 1–53. Cited in J. L. Sumich, 1976, *An Introduction to the Biology of Marine Life.* Wm. C. Brown, Dubuque (Iowa), 348 pp.

V. J. Harriott, S. A. Banks, 2002, Latitudinal variation in coral communities in eastern Australia: a qualitative bio-physical model of factors regulating coral reefs. Coral Reefs 21, 83–94, p. 85.

J. R. Heirtzler, X. LePichon, J. G. Baron, 1966, cited in E. Seibold, W. H. Berger, 1993, *The Sea Floor, An Introduction to Marine Geology*, 2nd revised and updated ed. Springer-Verlag, Heidelberg, 356 pp.

T. D. Herbert, S. L. D'Hondt, 1990, *Precessional climate cyclicity in late Cretaceous-early Tertiary marine sediments: A high resolution chronometer of Cretaceous-Tertiary boundary events.* Earth and Planetary Science Letters 99, 263–275.

P. M. Herzig, M. D. Hannington, 2000, in H. D. Schulz, M. Zabel (eds.), *Marine Geochemistry.* Springer Verlag, Berlin, 455 pp., p. 399.

H. Hess, cited in E. L. Hamilton, 1956, *Sunken islands of the Mid-Pacific Mountains.* Geological Society of America Memoir 64, 1–97.

C. P. Idyll, 1964, *Abyss: The Deep Sea and the Creatures That Live in It.* T. Y. Crowell Co., New York, 396 pp.

J. Imbrie, K. P. Imbrie, 1979, *Ice Ages, Solving the Mystery.* Enslow, Hillside, N.J., 224 pp.

J. D. Isaacs, A. Soutar, cited in T. R. Baumgartner, A. Soutar, V. Ferreira-Bartrina, 1992, *Reconstruction of the history of Pacific Sardine and Northern Anchovy populations over the past two millennia from sediments of the Santa Barbara Basin, California.* CalCOFI Report 30, 24–40, data here replotted.

C. O'D. Iselin, 1936, *A study of the circulation of the western North Atlantic.* Papers in Physical Oceanography and Meteorology 4 (4), 1–101.

C. O'D. Iselin, F. C. Fuglister, 1948, *Some recent developments in the study of the Gulf Stream.* Journal of Marine Research 7, 317–329.

R. A. Kinzie III, R. W. Buddemeier, 1996, *Reefs happen.* Global Change Biology 2, 479–494.

O. Krümmel, 1907, *Handbuch der Ozeanographie,* Band I. J. Engelhorn, Stuttgart, 526 pp.

K. Kuenzi, JPL NASA image in G. Hempel, F. Hinrichsen (eds.), 2002, *Der Ozean: Lebensraum und Klimasteuerung. Jahrbuch 2001/2002 der Wittheit zu Bremen.* Verlag Hausschild GmbH, Bremen, 147 pp.

D. R. Legates, 1996, *Precipitation,* in S. H. Schneider (ed.), *Encyclopedia of Climate and Weather.* Oxford University Press, New York, pp. 608–612.

D. Lluch-Belda, D. B. Lluch-Cota, S. E. Lluch-Cota, 2003, Scales of interannual variability in the California Current system: associated physical mechanisms and likely ecological impacts, 76–85. CalCOFI Reports 44, 76.

A. F. McEvoy, 1986, *The Fisherman's Problem: Ecology and Law in the California Fisheries 1850–1980.* Cambridge University Press, Cambridge, 368 pp.

H. W. Menard, 1964, *Marine Geology of the Pacific.* McGraw-Hill, New York, 271 pp.

P. S. Meyer, K. M. Gillis, 1994, *Oceanic crust composition and structure.* Oceanus 36 (4), 70–74 (Special Issue: 25 Years of Ocean Drilling).

K. G. Miller, R. G., Fairbanks, G. S., Mountain, 1987, *Tertiary oxygen isotope synthesis, sea level history, and continental margin erosion.* Paleoceanography 2, 1–19.

D. G. Moore 1969, cited in E. Seibold, W. H. Berger, 1993, *The Sea Floor, An Introduction to Marine Geology,* 2nd revised and updated ed. Springer-Verlag, Heidelberg, 356 pp.

H. Murayama in J. O. La Gorce (ed.), 1939, *The Book of Fishes.* National Geographic Society, Washington, D.C., 367 pp.

J. Murray, J. Hjort, 1912, *The Depths of the Ocean.* Macmillan, London, 821 pp.

M. Neumayr, 1895, Erdgeschichte, Bd. 2, *Beschreibende Geologie,* 2nd ed. Bibliographisches Institut, Leipzig, 700 pp.

A. P. Orr, S. M. Marshall, 1969, *The Fertile Sea.* Fishing News (Books) Ltd. London, 131 pp.

F. L. Parker, 1962, *Planktonic foraminiferal species in Pacific sediments.* Micropaleontology 8 (2), 219–254.

J. T. Parker, W. A. Haswell, 1921, *A Text-Book of Zoology,* vol. 1. McMillan and Co., London, 816 pp.

H. Pettersson, 1953, *Westward Ho with the Albatross.* E. P. Dutton, New York, 218 pp.

J. L. Reid, 1962, *On the circulation, phosphate-phosphorus content and zooplankton volumes in the upper part of the Pacific Ocean.* Limnology and Oceanography 7, 287–306.

F. S. Russell, C. M. Yonge, 1936, *The Seas, Our Knowledge of Life in the Sea and How It Is Gained.* Frederick Warne, London, 379 pp.

Scripps Institution of Oceanography (SIO).

E. Seibold and W. H. Berger, 1993, *The Sea Floor.* Springer-Verlag, Heidelberg, 356 pp.

F. P. Shepard, 1963, *Submarine Geology,* 2nd ed. Harper & Row, New York, 557 pp.

F. P. Shepard, J. R. Curray, 1967, *Carbon-14 determination of sea level changes in stable areas.* Progress in Oceanography 4, 283–291.

V. Spiess, *Meteor* Expedition in preparation of ODP leg 175 (G. Wefer, W. H. Berger, C. Richter, Scientific Shipboard Party, 1998, *Proceedings of the Ocean Drilling Program, Initial Reports* vol. 175. College Station, Texas).

R. H. Stewart, 1985, *Methods of Satellite Oceanography.* University of California Press, Berkeley, 360 pp.

SIO Explorations, *Explorations* (1995–2007), SIO office for public relations.

Space Oceanography Division, CLS, in R. E. Cheney, 2001, *Satellite altimetry,* in J. H. Steele, S. A. Thorpe,

K. K. Turekian (eds.), *Encyclopedia of Ocean Sciences*, vol. 5. Academic Press, San Diego, pp. 2504–2510.

R. H. Stewart, 1985, *Methods of satellite oceanography*. Scripps Studies in Earth and Ocean Sciences. University of California Press, Berkeley, 360 pp.

H. Stommel, 1958, *The abyssal circulation*. Deep-Sea Research 5, 80–82.

H. Stommel, 1980, *Asymmetry of interoceanic freshwater and heat fluxes*. Proceedings of the National Academy of Sciences USA, Geophysics 77(5), 2377–2381, cited in J. Woods, 1981, *The memory of the ocean*, in A. Berger (ed.), *Climatic Variations and Variability: Facts and Theories*. D. Reidel, Dordrecht, pp. 63–83.

E. Suess, 1980, *Particulate organic carbon flux in the oceans—surface productivity and oxygen utilization*. Nature 288, 260–263.

J. L. Sumich, 1976. *An Introduction to the Biology of Marine Life*. Wm. C. Brown, Dubuque, Iowa, 348 pp.

H. U. Sverdrup, M. W. Johnson, R. H. Fleming, 1942, *The Oceans, Their Physics, Chemistry, and General Biology*. Prentice-Hall, Englewood Cliffs, N.J., 1087 pp.

J. Thiede et al. 1992, cited in E. Seibold, W. H. Berger, 1993, *The Sea Floor, An Introduction to Marine Geology*, 2nd revised and updated ed. Springer-Verlag, Heidelberg, 356 pp.

H. R. Thierstein, H. Okada, 1979, cited in E. Seibold, W. H. Berger, 1993, *The Sea Floor, An Introduction to Marine Geology*, 2nd revised and updated ed. Springer-Verlag, Heidelberg, 356 pp.

W. Thomson in O. Krümmel, 1907, *Handbuch der Ozeanographie*, Bd. 1. J. Engelhorn, Stuttgart, 526 pp.

Poul Winther (1953) in A. F. Bruun, S. Greve, H. Mielche, R. Spärck, 1956, *The Galathea Deep Sea Expedition 1950–1952, Described by the Members of the Expedition*. Macmillan, New York, 296 pp. (First published in Danish, in 1953.)

G. Wüst, A. Defant, 1936, *Atlas zur Schichtung und Zirkulation des Atlantischen Ozeans*. Meteor Ergebn., 1925–1927, Wissenschaftl. Ergebn. Bd. 6, Atlas (103 pls.). Walter de Gruyter & Co., Berlin, cited in H. U. Sverdrup, M. W. Johnson, R. H. Fleming, 1942, *The Oceans, Their Physics, Chemistry, and General Biology*. Prentice-Hall, Englewood Cliffs, N.J., 1087 pp.

J. Zachos, M. Pagani, L. Sloan, E. Thomas, K. Billups, 2001, *Trends, rhythms, and aberrations in global climate 65 Ma to present*. Science 292, 686–693.

INDEX

Note: figures indicated by *f* following page number; notes indicated by *n*.